Regulatory T Cells and Clinical Application

Shuiping Jiang
Editor

Regulatory T Cells and Clinical Application

Springer

Editor
Shuiping Jiang
Department of Nephrology and Transplantation
Guy's Hospital
King's College London
London SE1 9RT, UK
sj4774@hotmail.com

ISBN: 978-0-387-77908-9   e-ISBN: 978-0-387-77909-6
DOI: 10.1007/978-0-387-77909-6

Library of Congress Control Number: 2008925544

© 2008 Springer Science+Business Media, LLC
All rights reserved. This work may not be translated or copied in whole or in part without the written permission of the publisher (Springer Science+Business Media, LLC, 233 Spring Street, New York, NY 10013, USA), except for brief excerpts in connection with reviews or scholarly analysis. Use in connection with any form of information storage and retrieval, electronic adaptation, computer software, or by similar or dissimilar methodology now known or hereafter developed is forbidden.
The use in this publication of trade names, trademarks, service marks, and similar terms, even if they are not identified as such, is not to be taken as an expression of opinion as to whether or not they are subject to proprietary rights.
While the advice and information in this book are believed to be true and accurate at the date of going to press, neither the authors nor the editors nor the publisher can accept any legal responsibility for any errors or omissions that may be made. The publisher makes no warranty, express or implied, with respect to the material contained herein.

Printed on acid-free paper

springer.com

Preface

A major process of rediscovery has taken place in the field of Cellular Immunology over the past 12 years—subsets of T lymphocytes exist that are specifically dedicated to regulation or as it should be more appropriately termed suppression of all aspects of immune responses. It is certainly appropriate at this time to recall some of the history that lead to the development of the concept of immune regulation/suppression. Shortly after the term helper T cells was coined to described lymphocytes that "helped" both humoral and cell-mediated responses, studies from the laboratory of the late Professor Richard Gershon demonstrated that under certain conditions, antigen recognition by T lymphocytes also resulted in the development of cells that are able to suppress immune responses. Unfortunately, research in this field rapidly shifted from studies of the function of the suppressor T cells to studies of their soluble products that were thought to be shed or secreted T cell receptors. A number of highly complex suppressor cell pathways and cell circuits were developed and were the subjects of more than 5,000 papers during this era. In 1983–1984, this field completely collapsed as studies called into question the existence of the I-J region of the mouse major histocompatibility complex that was thought to encode one of the major chains of the suppressor T cell factors. The cloning of the T cell receptor at that time firmly established that the T cell receptor genes were completely unrelated to the genes encoding immunoglobulin heavy chains calling into question the existence of soluble T cell factors that contained immunoglobulin VH gene products. The number of papers in the literature dealing with suppressor cells fell from a high of 1,300–1,500/year in 1981 to 150–200/year by the end of the 1980s. At this point in time, most immunologists felt it was even inappropriate to use the term suppressor cell!

Although a number of workers in the period of 1970–1995 continued to focus their studies on T suppressor cells rather than soluble factors, their work was largely ignored by the immunologic community. The detailed history of their pioneering work will be covered in Chapter 1 by Professor Sakaguchi. Immunologists are somewhat obsessed with dividing what initially appears to be homogeneous population of cells, e.g, $CD4^+$ T lymphocytes, into multiple subpopulations with distinct functional properties, e.g, Th1 and Th2 cells. Ideally, most immunologists desire that each subpopulation could easily be identified and separated by the expression of a cell surface antigen unique to that subpopulation. Although immunologic phenomena that appeared to be mediated by regulatory T cells were described in the literature in the early 1990s, what was really missing from this field was a cell surface marker that would allow immunologists to define a regulatory/suppressor cell. It was only after Prof. Sakaguchi identified the CD25 antigen in 1995 as a marker for a major population of T cells that had suppressor functions both in vitro and in vivo that the resurgence in the regulatory T cell area could begin.

The regulatory T cells field has grown dramatically over the past decade. It is now impossible to read a journal that does not contain numerous papers whose titles deal with regulatory T cells. More importantly, it is also difficult to submit a new research grant proposal in any area of immunologic research that does not include a section on analysis on the contribution of regulatory T cells to the subject matter under study. Regulatory/Suppressor T cells have come of age, again, hopefully this time to stay. Although it was initially thought that regulatory T cells functioned primarily in controlling autoreactive immune responses and several chapters in this volume are devoted to that topic, there is little doubt that the role of regulatory T cells in infection, cancer, and transplantation is just as important. Regulatory T cells even appear to play critical roles in cardiovascular disease in the pathogenesis of atherosclerosis. Many of the chapters in this volume with deal with the lineage of regulatory T cells that are defined by expression of CD25 and more importantly the transcription factor Foxp3. These cells were originally believed to be generated exclusively during T cell development in the thymus, but many recent studies indicate that they can be generated extrathymically. Cell types other than $CD4^+CD25^+Foxp3^+$ have also been shown to manifest regulatory properties and some of these unique cell types will be described in Chapters 23–30.

As in any rapidly moving field in science, many of the concepts and theories presented here will rapidly be modified or even discarded as new studies are performed and new questions are raised. For example, there are now at least a dozen proposed cellular mechanisms for the suppressive activity of the $CD4^+CD25^+Foxp3^+$ regulatory cells. Are all of these suppressive pathways actually used? Which ones are the most important? Which ones can be manipulated for therapeutic purposes? All of these questions should be answered in the next five years. Lastly, an important focus of this book is clinical application. Although numerous studies in animal models have strongly suggested that manipulation (augmentation or downregulation) of regulatory T cell function can be used for therapy of autoimmune, neoplastic, or infectious disease, we are now just on the threshold of translating some of the approaches from animals to man. Regulatory T cells can be best thought of today as "teenagers" ready to take on all the challenges of complex immune responses. In ten years, the field will certainly be more mature, and manipulation of regulatory T cell function by cellular biotherapy, antibodies and small molecules will be routine function of the clinical immunologist.

Bethesda, USA   Ethan M. Shevach

Contents

Part I Immunobiology of Regulatory T Cells

1 **Regulatory T Cells and the Control of Auto-Immunity: From day 3 Thymectomy to FoxP3+ Regulatory T Cells** 3
Makoto Miyara and Shimon Sakaguchi

2 **FoxP3 and Regulatory T Cells** 17
Karsten Kretschmer, Irina Apostolou, Panos Verginis and Harald von Boehmer

3 **Thymic and Peripheral Generation of $CD4^+$ $Foxp3^+$ Regulatory T Cells** 29
Paola Romagnoli, Julie Ribot, Julie Tellier, and Joost P.M. van Meerwijk

4 **The Role of IL-2 in the Development and Peripheral Homeostasis of Naturally Occurring $CD4^+CD25^+Foxp3^+$ Regulatory T Cells** 57
Allison L. Bayer and Thomas R. Malek

5 **IL-2 Signaling and CD4+ CD25+ Regulatory T Cells** 77
Louise M. D'Cruz and Ludger Klein

6 **TGF-Beta and Regulatory T Cells** 91
Yisong Y. Wan and Richard A. Flavell

7 **TGF-β Regulates Reciprocal Differentiation of $CD4^+CD25^+Foxp3^+$ Regulatory T Cells and IL-17-Producing Th17 Cells from Naïve $CD4^+CD25^-$ T Cells** 111
Wanjun Chen

8 **Molecular Signalling in T Regulatory Cells** 135
Natasha R. Locke, Natasha K. Crellin and Megan K. Levings

Part II Regulatory T Cells in Disease and Clinical Application

9   CD4+Foxp3+ Regulatory T Cells in Immune Tolerance............ 155
    Ciriaco A. Piccirillo

10 Regulatory T Cell Control of Autoimmune Diabetes and Their
    Potential Therapeutic Application................................ 199
    Qizhi Tang and Jeffrey A. Bluestone

11 CD4+CD25+ Regulatory T Cells as Adoptive
    Cell Therapy for Autoimmune Disease
    and for the Treatment of Graft-Versus-Host Disease................ 231
    Swati Acharya and C. Garrison Fathman

12 Natural CD4+CD25+ Regulatory T Cells in Regulation
    of Autoimmune Disease 253
    Adam P. Kohm and Stephen D. Miller

13 Multiple Sclerosis and Regulatory T Cells......................... 265
    Jonathon Hutton, Clare Baecher-Allan, and David A. Hafler

14 CD4+CD25+ Regulatory T Cells and TGF-Beta in Mucosal
    Inflammation.. 279
    M. Fantini and Markus F. Neurath

15 Induction of Adaptive CD4+CD25+Foxp3+ Regulatory T Cell
    Response in Autoimmune Disease 293
    Jian Hong, Sheri Skinner, and Jingwu Zhang

16 Regulatory T Cells in Transplantation 307
    Kathryn J Wood, Andrew Bushell, Manuela Carvalho-Gaspar,
    Gang Feng, Ross Francis, Nick Jones, Elaine Long,
    Shiqiao Luo, Ian Lyons, Satish Nadig, Birgit Sawitzki,
    Gregor Warnecke, Bin Wei, and Joanna Więckiewicz

17 Regulatory T-cells in Therapeutic Transplantation Tolerance 325
    Herman Waldmann, Elizabeth Adams, Paul Fairchild, and Stephen
    Cobbold

18 CD4+CD25+ Regulatory T Cell Therapy for the Induction of
    Clinical Transplantation Tolerance............................... 335
    David S. Game, Robert I. Lechler, and Shuiping Jiang

19 Regulatory T Cells in Allergic Disease 355
    Catherine Hawrylowicz

20	**Regulatory T Cells and Tumour Immunotherapy** .... 379 Ilona Kryczek and Weiping Zou	
21	**Regulatory T Cells in Hepatitis and Hepatocellular Carcinoma** .... 393 Fu-Sheng Wang and George F. Gao	
22	**CD4⁺CD25⁺ Regulatory T Cells in Viral Infections** .... 407 Wayne A. Tompkins, Mary B. Tompkins, Angela M. Mexas, and Jonathan E. Fogle	
23	**IL-10 and TGF-β-Producing Regulatory T Cells in Infection** .... 423 P.J. Dunne, A.G. Rowan, J.M. Fletcher, and Kingston H.G. Mills	
24	**Human Type 1 T Regulatory Cells** .... 455 Manuela Battaglia, Silvia Gregori, Rosa Bacchetta, and Maria Grazia Roncarolo	
25	**CD8⁺ T Regulatory Cells in Eye Derive Tolerance** .... 473 Joan Stein-Streilein and Hiroshi Keino	
26	**Immune Suppression by a Novel Population of CD8αα+TCRαβ+ Regulatory T cells** .... 489 Trevor R.F. Smith and Vipin Kumar	
27	**Innate Regulatory iNKT Cells** .... 501 Dalam Ly and Terry L. Delovitch	
28	**Natural Killer T Cells Regulate the Development of Asthma** .... 525 Muriel Pichavant, Rosemarie H. DeKruyff, and Dale T. Umetsu	
29	**The Development, Activation, Function and Mechanisms of Immunosuppressive Double Negative (DN) T Cells** .... 543 Megan S. Ford and Li Zhang	
30	**γδ T Cells in Immunoregulation** .... 563 Long Tang, Ning Kang, and Wei He	

Index 569

Contributors

Swati Acharya
School of Medicine, Stanford University, Palo Alto, CA 94305, USA

Elizabeth Adams
Sir William Dunn School of Pathology, University of Oxford, South Parks Road, Oxford OX1 3RE, UK

Irina Apostolou
Harvard Medical School, Dana-Farber Cancer Institute, Harvard University, 44 Binney Street, Boston, MA 02115, USA

Rosa Bacchetta
San Raffaele Telethon Institute for Gene Therapy (HSR-TIGET), Via Olgettina 58, Milan 20132, Italy

Clare Baecher-Allan
Division of Molecular Immunology, Center for Neurologic Diseases, Brigham and Women's Hospital, Harvard Medical School, Boston, MA 02115, USA

Manuela Battaglia
San Raffaele Telethon Institute for Gene Therapy (HSR-TIGET); San Raffaele Scientific Institute, Immunology of Diabetes Unit, Via Olgettina 58, Milan 20132, Italy

Allison L. Bayer
Department of Microbiology and Immunology, Diabetes Research Institute, Miller School of Medicine, University of Miami, Miami, FA 33136, USA

Jeffrey A. Bluestone
UCSF Diabetes Center, Department of Medicine and Department of Pathology, University of California, 513 Parnassus Avenue, HSW 1118, 513 Parnassus Avenue, San Francisco, CA, USA

Andrew Bushell
Transplantation Research Immunology Group, Nuffield Department of Surgery, John Radcliffe Hospital, University of Oxford, Oxford OX3 9DU, UK

Manuela Carvalho-Gaspar
Transplantation Research Immunology Group, Nuffield Department of Surgery, John Radcliffe Hospital, University of Oxford, Oxford OX3 9DU, UK

Wanjun Chen
Mucosal Immunology Unit, OIIB, National Institute of Dental and Craniofacial Research, National Institutes of Health, Bethesda, MD 20892, USA

Stephen Cobbold
Sir William Dunn School of Pathology, University of Oxford, Oxford OX1 3RE, UK

Natasha K. Crellin
Department of Surgery, University of British Columbia and Immunity and Infection Research Centre, Vancouver Coastal Health Research Institute, Vancouver, B.C., V6H 3Z6, Canada

Louise M. D'Cruz
Research Institute of Molecular Pathology, University of Vienna, Dr. Bohr-Gasse 7, Vienna 1030, Austria

Rosemarie H. DeKruyff
Karp Laboratories, Division of Immunology and Allergy, Children's Hospital, Harvard Medical School, Harvard University, Boston, MA 02115, USA

Terry L. Delovitch
Laboratory of Autoimmune Diabetes, Robarts Research Institute; Department of Microbiology and Immunology, University of Western Ontario; FOCIS Centre for Clinical Immunology and Immunotherapeutics, London, ON N6A5K8, Canada

P. J. Dunne
School of Biochemistry and Immunology, Trinity College Dublin 2, Dubin, Ireland

Paul Fairchild
Sir William Dunn School of Pathology, University of Oxford, Oxford OX1 3RE, UK

M. Fantini
Unit of Gasteoenterology, Department of Internal Medicine, Tor Vergata University of Rome, Rome, Italy; Laboratoty of Mucosal Immunology, Johannes Gitenberg University of Mainz, Mainz, Germany

C. Garrison Fathman
School of Medicine, Stanford University, Palo Alto, CA 94305, USA

Gang Feng
Transplantation Research Immunology Group, Nuffield Department of Surgery, John Radcliffe Hospital, University of Oxford, Oxford OX3 9DU, UK

Contributors

Richard A. Flavell
Section of Immunobiology, Yale University School of Medicine; Howard Hughes Medical Institute, 300 Cedar Street, New Haven, CT 06520, USA

J. M. Fletcher
School of Biochemistry and Immunology, Trinity College Dublin 2, Dubin, Ireland

Jonathan E. Fogle
Immunology Program, North Carolina State University, 4700 Hillsborough St, Raleigh, NC 27606, USA

Megan S. Ford
Departments of Laboratory Medicine and Pathobiology, Immunology, Toronto General Research Institute, University Health Network, University of Toronto, TMDT 2-807, 101 College Street, Toronto, ON, M5G, 1L7, Canada

Ross Francis
Transplantation Research Immunology Group, Nuffield Department of Surgery, John Radcliffe Hospital, University of Oxford, Oxford OX3 9DU, UK

David S. Game
Department of Renal Medicine, Imperial College London, Hammersmith Hospital, Du Cane Road, London W12 0NN, UK

George F. Gao
Institute of Microbiology, Chinese Academy of Sciences, Beijing 100101, PR China

Silvia Gregori
San Raffaele Telethon Institute for Gene Therapy (HSR-TIGET), Via Olgettina 58, Milan 20132, Italy

David A. Hafler
Division of Molecular Immunology, Center for Neurologic Diseases, Brigham and Women's Hospital, Harvard Medical School, 77 Ave Louis Pasteur, Boston, MA 02115, USA

Catherine Hawrylowicz
MRC and Asthma UK Centre in Allergic Mechanisms of Asthma, Department of Asthma, Allergy and Respiratory Science, Guy's Hospital, King's College London, 5th Floor Thomas Guy House, London SE1 9RT, UK

Wei He
Department of Immunology, Institute of Basic Medical Sciences, Chinese Academy of Medical Sciences (CAMS) and School of Basic Medicine, Peking Union Medical College (PUMC), 5 Dong Dan San Tiao, Beijing 100005, PR China

Jian Hong
Institute of Health Sciences and Shanghai Institute of Immunology, Shanghai JiaoTong University School of Medicine and Shanghai Institutes of Biological Sciences, Chinese Academy of Sciences, Shanghai, PR China

Jonathon Hutton
Division of Molecular Immunology, Center for Neurologic Diseases, Brigham and Women's Hospital, Harvard Medical School, Boston, MA 02115, USA

Shuiping Jiang
Department of Nephrology and Transplantation, Guy's Hospital, King's College London, London SE1 9RT, UK

Nick Jones
Transplantation Research Immunology Group, Nuffield Department of Surgery, John Radcliffe Hospital, University of Oxford, Oxford OX3 9DU, UK

Ning Kang
Department of Immunology, Institute of Basic Medical Sciences, Chinese Academy of Medical Sciences (CAMS) and School of Basic Medicine, Peking Union Medical College (PUMC), 5 Dong Dan San Tiao, Beijing 100005, PR China

Hiroshi Keino
Kyorin University School of Medicine, Tokyo, Japan

Ludger Klein
Institute for Immunology, Ludwig-Maximilians-University Munich, Goethestrasse 31, D-80336 Munich, Germany

Adam P. Kohm
Department of Microbiology-Immunology and Interdepartmental Immunobiology Center, Feinberg School of Medicine, Northwestern University, 303 E. Chicago Avenue, Chicago, IL 60611, USA

Karsten Kretschmer
Harvard Medical School, Dana-Farber Cancer Institute, Harvard University, 44 Binney Street, Boston, MA 02115, USA

Ilona Kryczek
Department of Surgery, University of Michigan, Ann Arbor, MI 48109, USA

Vipin Kumar
Laboratory of Autoimmunity, Torrey Pines Institute for Molecular Studies, 3550 General Atomics Court, San Diego, CA 92121, USA

Robert I. Lechler
Department of Nephrology and Transplantation, Guy's Hospital, King's College London, London SE1 9RT, UK

Contributors

Megan K. Levings
Department of Surgery, University of British Columbia and Immunity and Infection Research Centre, Vancouver Coastal Health Research Institute, 2660 Oak St., Vancouver, B.C., V6H 3Z6, Canada

Natasha R. Locke
Department of Surgery, University of British Columbia and Immunity and Infection Research Centre, Vancouver Coastal Health Research Institute, Vancouver, B.C., V6H 3Z6, Canada

Elaine Long
Transplantation Research Immunology Group, Nuffield Department of Surgery, John Radcliffe Hospital, University of Oxford, Oxford OX3 9DU, UK

Shiqiao Luo
Transplantation Research Immunology Group, Nuffield Department of Surgery, John Radcliffe Hospital, University of Oxford, Oxford OX3 9DU, UK

Dalam Ly
Laboratory of Autoimmune Diabetes, Robarts Research Institute; Department of Microbiology and Immunology, University of Western Ontario; FOCIS Centre for Clinical Immunology and Immunotherapeutics, London, ON N6A5K8, Canada

Ian Lyons
Transplantation Research Immunology Group, Nuffield Department of Surgery, John Radcliffe Hospital, University of Oxford, Oxford OX3 9DU, UK

Thomas R. Malek
Department of Microbiology and Immunology and Diabetes Research Institute, Miller School of Medicine, University of Miami, Miami, FL 33136, USA

Angela M. Mexas
Immunology Program, North Carolina State University, 4700 Hillsborough St, Raleigh, NC 27606, USA

Stephen D. Miller
Department of Microbiology-Immunology and Interdepartmental Immunobiology Center, Feinberg School of Medicine, Northwestern University, 303 E. Chicago Avenue, Chicago, IL 60611, USA

Kingston H. G. Mills
School of Biochemistry and Immunology, Trinity College Dublin 2, Dubin, Ireland

Makoto Miyara
Institute for Frontier Medical Sciences, Kyoto University, Kyoto, Japan

Satish Nadig
Transplantation Research Immunology Group, Nuffield Department of Surgery, John Radcliffe Hospital, University of Oxford, Oxford OX3 9DU, UK

Markus F. Neurath
Laboratory of Mucosal Immunology, Johannes Gutenberg University of Mainz, Mainz, Germany

Ciriaco A. Piccirillo
Departments of Microbiology and Immunology, McGill Centre for the Study of Host Resistance, McGill University Health Center, Montreal, QC, H3A 2B4, Canada

Muriel Pichavant
Karp Laboratories, Division of Immunology and Allergy, Children's Hospital, Harvard Medical School, Harvard University, Boston, MA 02115, USA

Julie Ribot
Tolerance and Autoimmunity Section, Centre de Physiopathologie de Toulouse Purpan, Institut National de la Santé et de la Recherche Médicale (Inserm) U563; University Paul Sabatier; IFR 30, Institut Claude de Preval, Toulouse, France

Paola Romagnoli
Tolerance and Autoimmunity Section, Centre de Physiopathologie de Toulouse Purpan, Institut National de la Santé et de la Recherche Médicale (Inserm) U563; University Paul Sabatier; IFR 30, Institut Claude de Preval, Toulouse, France

Maria Grazia Roncarolo
San Raffaele Telethon Institute for Gene Therapy (HSR-TIGET); Università Vita-Salute San Raffaele, Via Olgettina 58, Milan 20132, Italy

A. G. Rowan
School of Biochemistry and Immunology, Trinity College Dublin 2, Dubin, Ireland

Shimon Sakaguchi
Institute for Frontier Medical Sciences, Kyoto University, Kyoto, Japan

Birgit Sawitzki
Transplantation Research Immunology Group, Nuffield Department of Surgery, John Radcliffe Hospital, University of Oxford, Oxford OX3 9DU, UK

Ethan M. Shevach
Laboratory of Immunology, National Institutes of Allergy and Infectious Diseases, National Institutes of Health, Bethesda, MD 20892, USA

Sheri Skinner
Departments of Neurology and Immunology, Baylor College of Medicine, Houston, TX, USA

Trevor R.F Smith
Laboratory of Autoimmunity, Torrey Pines Institute for Molecular Studies, San Diego, CA 92121, USA

Contributors

Joan Stein-Streilein
Department of Ophthalmology, Schepens Eye Research Institute, Harvard Medical School, 20 Staniford Street, Boston, MA 02114, USA

Long Tang
Department of Immunology, Institute of Basic Medical Sciences, Chinese Academy of Medical Sciences (CAMS) and School of Basic Medicine, Peking Union Medical College (PUMC), 5 Dong Dan San Tiao, Beijing 100005, PR China

Qizhi Tang
UCSF Diabetes Center, Department of Medicine and Department of Pathology, University of California, HSW 1118, 513 Parnassus Avenue, San Francisco, CA 94143-0540, USA

Julie Tellier
Tolerance and Autoimmunity Section, Centre de Physiopathologie de Toulouse Purpan, Institut National de la Santé et de la Recherche Médicale (Inserm) U563; University Paul Sabatier, Toulouse, France; IFR 30, Institut Claude de Preval, Toulouse, France

Mary B. Tompkins
Immunology Program, North Carolina State University, 4700 Hillsborough St, Raleigh, NC 27606, USA

Wayne A. Tompkins
Immunology Program, North Carolina State University, 4700 Hillsborough St, Raleigh, NC 27606, USA

Dale T. Umetsu
Karp Laboratories, Division of Immunology and Allergy, Children's Hospital, Harvard Medical School, Harvard University, Boston, MA 02115, USA

Joost P.M. van Meerwijk
Tolerance and Autoimmunity Section, Centre de Physiopathologie de Toulouse Purpan, Institut National de la Santé et de la Recherche Médicale (Inserm) U563; University Paul Sabatier; IFR 30, Institut Claude de Preval; Institut Universitaire de France and Faculty of life-sciences (UFR-SVT), University Paul Sabatier, Toulouse, France

Panos Verginis
Harvard Medical School, Dana-Farber Cancer Institute, Harvard University, 44 Binney Street, Boston, MA 02115, USA

Harald von Boehmer
Harvard Medical School, Dana-Farber Cancer Institute, Harvard University, 44 Binney Street, Boston, MA 02115, USA

Herman Waldmann
Sir William Dunn School of Pathology, University of Oxford, South Parks Road, Oxford OX1 3RE, UK

Yisong Y. Wan
Section of Immunobiology, Yale University School of Medicine, 300 Cedar Street, New Haven, CT 06520, USA

Fu-Sheng Wang
Research Center for Biological Therapy, Beijing Institute of Infectious Diseases, Beijing 100039, PR China

Gregor Warnecke
Transplantation Research Immunology Group, Nuffield Department of Surgery, John Radcliffe Hospital, University of Oxford, Oxford OX3 9DU, UK

Bin Wei
Transplantation Research Immunology Group, Nuffield Department of Surgery, John Radcliffe Hospital, University of Oxford, Oxford OX3 9DU, UK

Joanna Wickiewicz
Transplantation Research Immunology Group, Nuffield Department of Surgery, John Radcliffe Hospital, University of Oxford, Oxford OX3 9DU, UK

Kathryn J. Wood
Transplantation Research Immunology Group, Nuffield Department of Surgery, John Radcliffe Hospital, University of Oxford, Oxford OX3 9DU, UK

Jingwu Zhang
GlaxoSmithKline R&D China, Shanghai 201203, China

Li Zhang
Departments of Laboratory Medicine and Pathobiology, Immunology Toronto General Research Institute, University Health Network, University of Toronto, TMDT 2-807, 101 College Street, Toronto, ON, M5G, 1L7, Canada

Weiping Zou
Department of Surgery, University of Michigan School of Medicine, Ann Arbor, MI 48109, USA

Part I
Immunobiology of Regulatory T Cells

Chapter 1
Regulatory T Cells and the Control of Auto-Immunity: From day 3 Thymectomy to FoxP3+ Regulatory T Cells

Makoto Miyara and Shimon Sakaguchi

Abstract Regulatory T-cell population is now widely accepted as an important component of the immune system as professional suppressors of immune responses. It was shown in the late sixties that some CD4+ T cells in normal mice were capable of suppressing autoimmunity. Efforts to characterize this autoimmune-suppressive CD4+ T cell population led to the identification of CD25 as a constitutional marker. Using this marker, it became possible to separate regulatory T cells from other CD4+ T cells, to further analyze their developmental pathways, especially in the thymus, and to better describe how they suppress immune responses in vivo and in vitro. The marker was also found to be useful to identify regulatory T cells with comparable suppressive function and phenotype in humans. It was recently shown that transcription factor Foxp3 was specifically expressed by CD25+ CD4+ regulatory T cells in rodents. Anomalies in FOXP3 gene are responsible for the development of an autoimmune and inflammatory disease in humans and rodents characterized by a deficiency in the development and function of CD25+CD4+ regulatory T cells. These recent findings provide clear evidence that Foxp3+CD25+CD4+ regulatory T cells are indispensable for the establishment and the maintenance of immunologic self-tolerance and immune homeostasis. Therefore, characterization of regulatory T cell mediated immune suppression should bring new clinical tools to control pathological immune responses.

The immune system is able to mount durable efficient destructive responses against exogenous pathogenic micro-organisms like viruses, bacteria, fungi and parasites and against endogenous pathogens like tumors. These responses are potent enough to destroy not only pathogens but also the host. This implies that immunity, like the other major systems in the body machinery, maintains essential regulatory mechanisms that prevent inappropriate harmful responses. These regulatory processes constitute the immunological tolerance. Several key mechanisms have been described in the last three decades to explain how the immune system is capable of preventing

---

S. Sakaguchi
Institute for Frontier Medical Sciences, Kyoto University, Kyoto, Japan
e-mail: shimon@frontier.kyoto_u.ac.jp

what Ehrlich denominated the *Horror autotoxicus* i.e. auto-immunity at the end of the nineteenth century [1]. One of these major mechanisms is the clonal deletion that occurs in the thymus. Clonal deletion leads to the destruction of most self reactive T cells [2–5] and is effective under the control of the gene AIRE [6–8]. Mutation of AIRE gene leads to the development of systemic auto-immune disease APECED that is characterized by the occurrence of multiple endocrine organ autoimmunity [9]. Another key mechanism in the maintenance of tolerance is the activation induced cell death (AICD) of T cells [10] through the interaction of Fas with its ligand Fas-L [11]. The auto-immune lymphoproliferative syndrome (ALPS) [12] which is characterized by the occurrence of auto-immune cytopenia including auto-immune hemolytic anemia and idiopathic thrombopenic purpura and sometimes of a systemic lupus erythematosus resembling disease is secondary to mutations in FAS or FAS-ligand gene.

In addition, it is now widely accepted that another mechanism involving T cells that actively suppress the activation and the proliferation of other immune cells is also key in the maintenance of tolerance to self constituents and in the prevention of auto-immune diseases. Circulating T cells that have specificity to self antigens can be detected in healthy individuals but usually do not trigger clinically patent auto-immunity [13]. This suggests that a dominant inhibitory phenomenon that controls these potentially harmful cells is operating permanently. It took a very long time for investigators to convince the scientific community that a distinct subset of CD4+ T cells, initially called suppressor T cells then rebaptized regulatory T cells, was the mediator of such permanent suppression of auto-immune responses in the periphery.

The Origins (1969–1982)

One major milestone discovery in immunology was made in 1966 by Claman et al. who showed that the cooperation of two distinct subsets of lymphocytes, one originating from the bone marrow and the other one derived from the thymus, was necessary for the production of antibodies [14]. Thymus derived lymphocytes named T cells were defined as the cells that highly responded to antigenic stimuli by mitosis and protein production [15]. Lymphocytes derived from the bone marrow were identified as the cells that produced antibodies [16]. As T cells were shown to be necessary for the induction of immune responses, it became critical to know whether the very same population could also dampen immune responses and how important T cells were in the induction or in the prevention of auto-immunity.

Two major types of experimental approaches were then developed to break or to induce tolerance in normal animals by manipulating T cells. Experiments conducted in the late 1960s by Nishizuka and Sakakura [17] consisted in the observation of mice that were thymectomized at day 3 of life whereas the other kind of experiments designed by Gershon and Kondo in 1970–1971 was meant to study the effects of cells derived from the thymus transferred into mice after they were immunized with foreign antigens [18, 19]. In these experiments, lymphocytes that were isolated from the thymus could inhibit the cooperation between T and B cells and inhibit the production of antibodies against foreign antigens. In their experiments, athymic mice were first irradiated and then reconstituted 10 days later with autologous hematopoietic stem cells and with autologous thymocytes. These mice were then immunized with sheep red blood cells (RBC). As expected, they produced allo-antibodies against sheep RBC. However, mice that received a second injection of autologous thymocytes before a second challenge with sheep RBC did not increase their allo-antibodies secretion [18]. Gershon and Kondo therefore concluded that induction of both immune responses and tolerance in bone marrow derived cell population required the co-operation of thymus derived T cells. One year later, Gershon and Kondo showed that the transfer of splenocytes in athymic mice that were isolated from tolerized mice could induce a similar tolerance in the recipient [19]. Hence, they termed this phenomenon "infectious tolerance" and the tolerizing thymocytes "suppressor T cells" [20].

In parallel, Nishizuka and Sakakura performed their experiments of day 3 neonatal thymectomy in mice at a period when it was not clearly demonstrated that thymus had an exclusive immunologic function. Thymus had "long been considered to be an endocrine organ somehow related to sexual physiology" because thymectomy at 3 days of age in certain strains was followed by an impaired development of mammary glands and to a reduction in the frequency of mammary cancers [17]. They also observed that female mice that were thymectomized at day 3 of life were sterile because of the destruction of their ovaries. Interestingly, this ovarian disease was not induced when the thymectomy was executed after the 7th day of life and it was also observed that the disease could be prevented by thymus grafting [17]. It came to light that the pathological mechanism that led to ovarian failure was not the lack of sexual hormone, supposedly secreted by the thymus, but rather due to auto-immune inflammation [21]. Moreover, depending on the mice strain, other auto-immune features were observed in day 3 thymectomized mice. For example, in BALB/c strain, about 25% of mice developed auto-immune oophoritis and one third developed gastritis. In A strain, oophoritis was more prevalent (90%) and gastritis was observed in 10% of mice. Thyroiditis was also found in 6% of mice and orchitis developed in 16% of male mice. However, in other strains such as C57BL/6, mice did not develop auto-immunity [22]. Altogether these findings raised the possibility that particular lymphocytes produced in the thymus from day 3 of life in mice were capable of preventing the emergence of auto-immune diseases.

Penhale et al. also confirmed in 1973 that thymectomy could induce organ specific auto-immunity. They showed that normal rats that were thymectomized at 6 weeks of life developed auto-immune thyroiditis with the production of anti-thyroglobulin antibodies, after they received four sublethal doses of X-irradiation every 2 weeks [23]. Following the same procedure in other strains, they confirmed several years later that thymectomy could also induce other autoimmune features such as diabetes mellitus [24].

Therefore, the bases of the concept of T cells that suppress immune responses against either foreign or self antigens were set.

Tracking Suppressor T Cells (1982–1995)

The fact that auto-immune diseases could only be induced when the thymectomy was performed between day 3 and day 7 suggests that a distinct population of thymus derived cells which can prevent auto-immunity arises starting from day 3 of life [17]. Importantly, auto-immunity did not occur when the mice were thymectomized after day 7. This suggests that tolerogenic T cells accumulate in the periphery between day 3 and 7, which would be sufficient for a long-term prevention of auto-immunity. Sakaguchi in Nishizuka's laboratory demonstrated in 1982 that auto-immunity provoked by neonatal thymectomy could indeed be prevented by the injection of normal adult mice T cells when performed within 2 weeks after thymectomy [21]. These findings suggested that a distinct population of T cells persisted in adult life and could prevent auto-immunity. Therefore, several groups attempted to better characterize the phenotype of the suppressor T cell subpopulation by identifying specific surface markers. They also attempted to prove the relevance of in vivo suppression by studying the consequences of the elimination of certain T-cell subpopulations in mice.

First, Sakaguchi et al. showed in 1985 that athymic BALB/c nude mice reconstituted with splenocytes without CD4+Lyt^{high} (CD5^{high}) cells developed multi-organ auto-immunity that included gastritis, thyroiditis, sialadenitis, diabetes, adrenalitis, oophoritis and testicular inflammation. They also observed that the co-transfer of all CD4+ T cells together with CD4+CD5^{low} T cells could prevent auto-immunity [22]. Five years later, Powrie and Mason demonstrated that athymic rats reconstituted with splenocytes without CD4+CD45RB^{low} T cells developed a graft versus host reaction like disease. They also observed a multiple auto-immune syndrome that included liver, lung, stomach, thyroid and endocrine pancreas autoimmune inflammation [25]. McKeever et al. also showed in 1990 that the transfer in PVG nude rats of RT6.1+ T cell depleted splenocytes led to the development of thyroiditis and diabetes mellitus [26]. In 1993, Powrie et al. on one hand and Morrissey et al. on the other hand showed that the adoptive transfer of CD4+CD45RB^{high} T cells in SCID BALB/c mice led to the development of exudative enteropathy [27]. The same year, Fowell and Mason demonstrated that the adoptive transfer of CD4+CD45RC^{low} RT6+ Thy-1- OX-40- T cells in thymectomized irradiated rats prevented the emergence of diabetes mellitus [28].

Meanwhile, in the late 1980s, two major discoveries and a scientific fiasco made the contribution of suppressor T cells meaningless in the explanation of tolerance mechanisms. First, the clonal deletion of auto-reactive T cells was shown to be effective in the thymus by Kappler in 1987 [3]. Therefore, the key role of the thymus in the prevention of autoimmunity was assumed to be exclusively related to the clonal deletion phenomenon. Second, Mosman and Coffman introduced the T helper 1 and 2 subsets dichotomy [29]. It has rapidly been shown that TH1 subset defined by TH1 cytokine secretion (interleukin 2, interferon-gamma, GM-CSF and IL-3) after stimulation by APC [30] could suppress TH2 cytokines secretion such as IL-4 [31] by TH2 cells through their cytokine secretion and vice-versa [32]. Thus, it was assumed that peripheral tolerance was rather due to the balance between TH1 and TH2 cells termed "immune deviation" rather than secondary to the action of a specific suppressor subset of T cells. At last, it had been first assumed that the mechanism of suppressor T cells mediated inhibition of other immune cells was secondary to the production of a soluble suppressor T-cell factor called "I-J protein" and I-J locus was thought to be located between I-A and I-E [33, 34]. However, when the DNA sequence of murine MHC was molecularly cloned, it appeared that such locus did not even exist [35, 36].

As a result, despite convincing results brought by independent laboratories, the concept of suppressor T cells had eventually become taboo [37].

The Rebirth of Suppressor-Regulatory T Cells (1995–2000)

Sakaguchi et al. showed in 1995 that a distinct population of CD4+ T cells which expresses the alpha chain of IL-2 receptor (CD25) that represents from 5 to 10% of CD4+ T cells could prevent the emergence of autoimmune diseases [38]. The injection of normal BALB/c mice splenocytes that were depleted of CD25+ CD4+ T cells in BALB/c nude mice was sufficient to induce a multiple organ auto-immune disease that included gastritis with anti-parietal cell auto-antibodies and oophoritis in almost all cases and in some cases thyroiditis, sialadenitis, glomerulonephritis, adrenalitis, insulitis and arthritis. Moreover, when injected within a limited period of time, autologous CD25+CD4+ T cells could prevent the development of diseases in nude mice reconstituted with CD25–CD4+T cells. The following year, Asano and Sakaguchi made the relationship between CD4+CD25+ suppressor T cells and neonatal day 3 thymectomy clearer [39]. They demonstrated that the inoculation of syngeneic CD4+ CD25+ T cells could prevent auto-immunity in thymectomized mice but more importantly, they brought the evidence that the emergence of CD25+CD4+ T cells in the periphery of normal mice started immediately after day 3, and then rapidly increased within 2 week to reach levels that were close to the ones observed in adult mice. They also showed that day 3 thymectomy was sufficient to eliminate CD25+ CD4+ T cells from the periphery for several days.

In 1998, Shevach's group and Sakaguchi's group made one step forward to better characterizing the function of suppressor CD4+ CD25+ T cells in vitro. They demonstrated that CD25+ CD4+ T cells, that were anergic upon stimulation, could suppress the proliferation and the production of IL-2 of activated CD4+ T cells in vitro in a contact-dependent manner [40, 41]. Thus, the characterization of CD25 as a reliable surface marker and the possibility to assess their function in vitro definitively pushed suppressor T cells thereafter called regulatory T cells (Tregs) out from oblivion.

The Hunt for CD25+ CD4+ Regulatory T Cells in Mice and Humans (2000–2003)

One key issue to address to better defining the Treg subset is the lack of a specific marker that allows the isolation of a pure homogeneous Treg population even upon activation. Although most CD25+ CD4+ T cells have regulatory properties in naïve mice, the isolation of a pure Treg population according to the expression of

CD25 is challenging in diseased conditions or after activation in vitro because activated CD4+ T cells upregulate their expression of CD25. Therefore, several groups attempted to find new additional molecules that would have allowed a more precise definition of Treg phenotype.

It was first reported that CD4+ T cells that expressed the adhesion molecule L-selectin (CD62-L) were capable of preventing the emergence of auto-immune diabetes in NOD mice [42, 43]. Then, it was reported that the transfer of CD25+ CD4+T cells in NOD mice prevented diabetes [44]. Among these cells, those that expressed CD25 and the highest levels of CD62-L were shown to be the most potent suppressor of the disease [45, 46]. Nevertheless, it was clear that CD62-L could not be used as a single marker to isolate Tregs because CD62-L was also expressed on other T cells subsets.

One relevant way to find Treg specific markers is to elucidate the molecular mechanisms of suppression by identifying molecules that are involved in cell-cell contact mediated suppression or in the regulation of suppression. Co-stimulatory molecule Cytotoxic T Lymphocyte Antigen 4 (CTLA-4) is a major molecule in the attenuation of immune responses. Although CTLA-4 was first described to be involved in the termination of costimulatory signals in T cells once their CD28 molecules had interacted with CD80-CD86 on APCs [47], it was questioned whether CTLA-4 could also be involved in Treg mediated suppression. CTLA-4 knock-out mice develop a fatal lymphoproliferative disease with systemic auto-immune features [48] that resemble the ones observed in nude mice reconstituted with splenocytes depleted of CD25+ T cell. Sakaguchi's group and Powrie's group showed that CTLA-4 was constitutively expressed by CD4+ CD25+ T cells [49, 50]. Moreover, the latter showed that the suppression mediated by CD4+CD25+CD45RB^{low} T cells, the depletion of which was responsible for the induction of the colitis described in the middle of the 1990s, required the signalling of CTLA-4 to be effective. Nevertheless, CTLA-4 is also expressed on activated cells, making it difficult to distinguish activated T cells from genuine Tregs.

The search for monoclonal antibodies that would break the suppression mediated by Tregs led to the identification of the Glucocorticoid-Induced Tumor necrosis factor Receptor family-related gene (GITR) in 2002. Shimizu et al. in Sakaguchi's lab demonstrated that the anti-GITR antibody they produced could deliver an active signal in Tregs through GITR that could attenuate suppression in vitro. Moreover, mice injected with anti-GITR antibodies developed gastritis with anti-parietal cells antibodies [51]. At the same time, another group led by Byrne and Shevach also identified GITR using DNA microarray as a gene that was specifically expressed by Tregs [52]. However, it rapidly appeared that GITR was also expressed on activated effector cells making it difficult to isolate Tregs from activated T cells using this marker alone [53].

In human research field, several groups (Baecher-Allan in Hafler's group, Jonuleit et al., Dieckmann et al., Levings et al. in Roncarolo's group, Taams et al. in Akbar's group and Ng et al. in Lechler's group) confirmed in 2001 that CD4+ CD25+ regulatory T cells were also prevalent in humans [54–59]. However, although it was demonstrated in healthy donors that the 1–2% of CD4+ T cells that expressed the highest levels of CD25 had the best suppressive capacity in vitro, it rapidly come to light that CD25 could not be used as a reliable marker for human Tregs, especially in inflammatory diseases in which circulating activated T cells expressing CD25 are also found. A specific marker for Tregs was indeed awaited until...

The Modern Age: FoxP3+ Regulatory T Cells (2003-now)

The Immune dysregulation Polyendocrinopathy Enteropathy X linked syndrome also known as IPEX is a severe multisystemic autoimmune and inflammatory disease which onset usually occurs in the neonatal period. The first description of the disease was made in 1982 by Powell et al. [60]. The murine equivalent of IPEX is the scurfy mouse that was described by Godfrey et al. [61]. Three teams simultaneously brought the evidence in 2001 that IPEX and scurfy were the consequence of a deficiency in the gene expression of the transcription factor Foxp3 (Forkhead box protein 3) [62–64]. The following year, Schubert et al. showed that Foxp3 was a transcription inhibitor that regulated T cell activation [65]. Three independent teams demonstrated in 2003 that the expression of the transcription factor FoxP3 was required for the development and the function of regulatory T cells [66–68]. Not only Foxp3 was found to be specifically expressed in CD4+CD25+ Tregs in mice but also the forced expression of FoxP3 in naive CD4+T cells was sufficient to make them suppressive in vitro and in vivo. Moreover, several genes preferentially expressed by CD4+CD25+ Treg cells such as CD25, CTLA-4, and GITR were shown to be directly controlled by FoxP3 [67]. Therefore, these findings supported the role of Foxp3 as a master control gene in the development and the function of the Treg population. In 2005, using a GFP reporter mouse expressing a fusion GFP-FoxP3 protein, Fontenot et al. in Rudensky's group showed that the expression of Foxp3 could be detected in the thymus starting from day 3 of life. They also showed that the peripheral colonization by FoxP3+ cells was seen starting from day 3 too [69]. These findings ultimately demonstrated that the induction of autoimmunity by day 3 thymectomy was indeed caused by the depletion of Foxp3 expressing natural regulatory T cells.

The most recent studies are mainly focused on the relationship between FoxP3 and the other transcription factors that are involved in the functions of T cells. Several groups showed that Foxp3 acts as a repressor of the expression of IL-2, IL-4 and IFN-γ through direct physical interactions with transcription factors NF-κB and NF-AT on one hand [70, 71] and AML1/Runx1 on the other hand. Furthermore, it has been demonstrated that the suppressive function of Tregs was dependent on the interaction of FoxP3 with NF-AT and/or AML1/Runx1 proteins. Other recent studies have dealt with the genes that are controlled by FoxP3 [72–75] in order to figure out the molecular mechanisms of suppression and to find other molecules that are specifically expressed by Tregs. Despite those efforts, it is still unclear how Tregs suppress immune responses and it is still not determined which surface proteins can be used as a mere marker for Tregs.

Tregs and Auto-Immunity

Regulatory T cells can downregulate a wide spectrum of physiological and pathological immune responses and many studies raised the possibility that Tregs could be involved in the evolution of diseases that have immune components such as infection, cancer, transplantation, allergy and inflammatory/autoimmune diseases. However, what has been shown in the last 30 years is that the primary key role of Tregs is to control auto-immune responses.

Day 3 thymectomy can be considered the first model of Treg depletion in vivo. Most mice that are thymectomized develop, depending on their strain, oophoritis and sometimes gastritis or thyroiditis. Of note, these auto-immune diseases are mainly concerning endocrine organs [17]. In the model of Sakaguchi in which nude mice were reconstituted with CD25-T cells, it was also observed that most mice developed auto-immune gastritis and that other endocrine glands were diseased such as ovaries, thyroid, adrenal glands, endocrine pancreas but also exocrine glands i.e. salivary glands [38]. Taken together, these auto-immune features is reminiscent of human diseases APECED and type II autoimmune polyglandular syndrome. The first syndrome is rather seen in young children whereas the second is prevalent in middle age adults. APECED (autoimmune polyendocrinopathy, candidiasis, ectodermal dystrophy) disease is usually characterized by the association of several endocrine autoimmune disorders such as type I diabetes, thyroiditis, gonadic deficiency and pernicious anaemia due to autoimmune gastritis. The cause of APECED syndrome has been recently identified as a mutation in the gene AIRE (autoimmune regulator) [6, 9] which is necessary for the presentation of self antigens to thymocytes in order to eliminate self reactive ones. It has been recently demonstrated that Aire was also involved in the selection of Tregs in the thymus [76]. Moreover, Treg function has been shown to be deficient in patients with APECED [77]. In type II APS, one major feature is the appearance of auto-immune gastritis which is associated with other glandular diseases such as adrenalitis, parathyroiditis, thyroiditis and type I diabetes. A deficiency in Treg suppressive function has also been described in patients with the disease [78]. Taken together, those experimental findings in mice and translational studies in human diseases strongly suggest that Tregs play a key role in the prevention of polyglandular auto-immune syndromes.

The role of Regulatory T cell subset has also been extensively studied in the pathophysiology of type I diabetes. CD25+CD4+ Treg can prevent the disease especially if NOD mice are transferred with CD25+CD4+ T cells that have high levels of CD62-L [43–45]. Studies in humans suggest that Tregs might play a role in type I diabetes. Several groups have raised the possibility that Treg suppressive function could be impaired [79, 80] or that the number of Tregs could be diminished [81]. Nevertheless, due to the lack of a reliable marker for human Treg, the conclusions on the function or the quantity of Tregs are likely to change in the next years. For example, according to the expression of FoxP3, a recent report concluded that the proportion of Treg was not modified in type I diabetes patients [82]. But again, these conclusions might be someday obsolete since FoxP3 has recently been shown to be also upregulated in activated CD4+ T cells, exactly like CD25 or GITR [83, 84].

Powrie's group showed in the 1990s that the adoptive transfer of CD4+ CD45RB^{high} T cells in SCID mice led to the development of a severe wasting disease that was secondary to colitis [25]. They could demonstrate that Tregs could prevent the disease but they also brought the evidence that Treg could cure the colitis [85]. In humans, Maul et al. showed that patients with inflammatory bowel disease including Crohn's disease and ulcerative colitis had less circulating CD4+CD25+ T cells when compared to healthy donors. However, they did not notice any impairment in the suppressive function of Treg in vitro [86].

In some nude mice reconstituted with CD25– T cells, in addition to polyglandular syndrome, one could observe systemic auto-immune features such as arthritis that is reminiscent of rheumatoid arthritis (RA), or glomerulonephritis and the presence of anti-double strand DNA antibodies that are characteristic of systemic lupus erythematosus(SLE) [38]. Several groups observed that the proportion of circulating Treg was decreased among CD4+ T cells in patients with SLE during flares whereas no modification was noted when the disease was inactive [87–89]. In RA, both quantitative deficiency [90] or functional impairment [91] in Treg cell population were reported. Whereas Tregs have been shown to be reduced in the periphery, high prevalence of Treg was noted in the inflamed joints [90]. In vitro, it has been shown that Treg that were capable of suppressing the proliferation of activated cells on one hand, were not capable of controlling the secretion of TNF alpha on the other hand [91]. These findings suggest that Tregs in RA might migrate and accumulate in the joints but would not be efficient enough to control inflammation. Interestingly, the suppression potency was recovered in patients that responded clinically to treatments with monoclonal anti-TNF alpha antibodies [91].

Similar functional deficiency was described in sarcoidosis which is an inflammatory disease characterized by the presence of granuloma in diseased organs. TNF alpha also plays a major role in the pathophysiology of the disease. An accumulation of Tregs has been observed in diseased organs but, in opposite to RA, an expansion of Treg was also observed in the periphery [92]. Treg function has been shown to be impaired in two other immune diseases which are multiple sclerosis [93] and psoriasis [94]. In both cases, Treg isolated from patients were shown to be incapable of completely suppressing the proliferation of effector cells in vitro.

These numerous examples of animal models and human diseases strongly argue for the essential role of Tregs in the prevention of aberrant immune responses. Therefore, it is becoming more and more evident that the manipulation of Treg cells would be a promising approach to treat auto-immune diseases.

Conclusion

Since the seminal articles by Nishizuka et al. and by Gershon et al., it has been for long debated whether suppressor/regulatory T cells did actually even exist. Today, this question is no longer accurate and the subject of regulatory T cells is the object of a constantly growing interest in immunologists and clinicians. However, even if significant progresses have been made since the definition of CD4+C25+ T cells in

1995 and of foxP3 seven years later, the most important questions about Tregs are still unanswered: we still do not know what are the best surface markers for Tregs in rodents and in humans and more importantly, we still do not know the exact mechanisms of suppression. There is no doubt that the history of Tregs will make a new big leap ahead when the answers of this crucial question come.

References

1. Ehrlich P, Morgenroth J. Über Hämolysine. Dritte Mittheilung. Berl Klin Wochenschr 1900; 37:453–8.
2. Hengartner H, Odermatt B, Schneider R, et al. Deletion of self-reactive T cells before entry into the thymus medulla. Nature 1988; 336(6197):388–90.
3. Kappler JW, Roehm N, Marrack P. T cell tolerance by clonal elimination in the thymus. Cell 1987; 49(2):273–80.
4. Kisielow P, Teh HS, Bluthmann H, von Boehmer H. Positive selection of antigen-specific T cells in thymus by restricting MHC molecules. Nature 1988; 335(6192):730–3.
5. Marusic-Galesic S, Stephany DA, Longo DL, Kruisbeek AM. Development of CD4–CD8+ cytotoxic T cells requires interactions with class I MHC determinants. Nature 1988; 333(6169):180–3.
6. Anderson MS, Venanzi ES, Klein L, et al. Projection of an immunological self shadow within the thymus by the aire protein. Science 2002; 298(5597):1395–401.
7. Liston A, Gray DH, Lesage S, et al. Gene dosage – limiting role of Aire in thymic expression, clonal deletion, and organ-specific autoimmunity. J Exp Med 2004; 200(8):1015–26.
8. Liston A, Lesage S, Wilson J, Peltonen L, Goodnow CC. Aire regulates negative selection of organ-specific T cells. Nat Immunol 2003; 4(4):350–4.
9. Nagamine K, Peterson P, Scott HS, et al. Positional cloning of the APECED gene. Nat Genet 1997; 17(4):393–8.
10. Liu Y, Janeway CA, Jr. Interferon gamma plays a critical role in induced cell death of effector T cell: a possible third mechanism of self-tolerance. J Exp Med 1990; 172(6):1735–9.
11. Alderson MR, Armitage RJ, Maraskovsky E, et al. Fas transduces activation signals in normal human T lymphocytes. J Exp Med 1993; 178(6):2231–5.
12. Fisher GH, Rosenberg FJ, Straus SE, et al. Dominant interfering Fas gene mutations impair apoptosis in a human autoimmune lymphoproliferative syndrome. Cell 1995; 81(6): 935–46.
13. Danke NA, Koelle DM, Yee C, Beheray S, Kwok WW. Autoreactive T cells in healthy individuals. J Immunol 2004; 172(10):5967–72.
14. Claman HN, Chaperon EA, Triplett RF. Thymus-marrow cell combinations. Synergism in antibody production. Proc Soc Exp Biol Med 1966; 122(4):1167–71.
15. Leuchars E, Morgan A, Davies AJ, Wallis VJ. Thymus grafts in thymectomized and normal mice. Nature 1967; 214(90):801–2.
16. Mitchell GF, Miller JF. Cell to cell interaction in the immune response. II. The source of hemolysin-forming cells in irradiated mice given bone marrow and thymus or thoracic duct lymphocytes. J Exp Med 1968; 128(4):821–37.
17. Nishizuka Y, Sakakura T. Thymus and reproduction: sex-linked dysgenesia of the gonad after neonatal thymectomy in mice. Science 1969; 166(906):753–5.
18. Gershon RK, Kondo K. Cell interactions in the induction of tolerance: the role of thymic lymphocytes. Immunology 1970; 18(5):723–37.
19. Gershon RK, Kondo K. Infectious immunological tolerance. Immunology 1971; 21(6): 903–14.
20. Gershon RK, Cohen P, Hencin R, Liebhaber SA. Suppressor T cells. J Immunol 1972; 108(3):586–90.

21. Sakaguchi S, Takahashi T, Nishizuka Y. Study on cellular events in postthymectomy autoimmune oophoritis in mice. I. Requirement of Lyt-1 effector cells for oocytes damage after adoptive transfer. J Exp Med 1982; 156(6):1565–76.
22. Sakaguchi S, Fukuma K, Kuribayashi K, Masuda T. Organ-specific autoimmune diseases induced in mice by elimination of T cell subset. I. Evidence for the active participation of T cells in natural self-tolerance; deficit of a T cell subset as a possible cause of autoimmune disease. J Exp Med 1985; 161(1):72–87.
23. Penhale WJ, Farmer A, McKenna RP, Irvine WJ. Spontaneous thyroiditis in thymectomized and irradiated Wistar rats. Clin Exp Immunol 1973; 15(2):225–36.
24. Penhale WJ, Stumbles PA, Huxtable CR, Sutherland RJ, Pethick DW. Induction of diabetes in PVG/c strain rats by manipulation of the immune system. Autoimmunity 1990; 7(2–3): 169–79.
25. Powrie F, Mason D. OX-22high CD4+ T cells induce wasting disease with multiple organ pathology: prevention by the OX-22low subset. J Exp Med 1990; 172(6):1701–8.
26. McKeever U, Mordes JP, Greiner DL, et al. Adoptive transfer of autoimmune diabetes and thyroiditis to athymic rats. Proc Natl Acad Sci USA 1990; 87(19):7618–22.
27. Powrie F, Leach MW, Mauze S, Caddle LB, Coffman RL. Phenotypically distinct subsets of CD4+ T cells induce or protect from chronic intestinal inflammation in C. B-17 scid mice. Int Immunol 1993; 5(11):1461–71.
28. Fowell D, Mason D. Evidence that the T cell repertoire of normal rats contains cells with the potential to cause diabetes. Characterization of the CD4+ T cell subset that inhibits this autoimmune potential. J Exp Med 1993; 177(3):627–36.
29. Mosmann TR, Coffman RL. TH1 and TH2 cells: different patterns of lymphokine secretion lead to different functional properties. Annu Rev Immunol 1989; 7:145–73.
30. Mosmann TR, Cherwinski H, Bond MW, Giedlin MA, Coffman RL. Two types of murine helper T cell clone. I. Definition according to profiles of lymphokine activities and secreted proteins. J Immunol 1986; 136(7):2348–57.
31. Coffman RL, Shrader B, Carty J, Mosmann TR, Bond MW. A mouse T cell product that preferentially enhances IgA production. I. Biologic characterization. J Immunol 1987; 139(11):3685–90.
32. Fernandez-Botran R, Sanders VM, Mosmann TR, Vitetta ES. Lymphokine-mediated regulation of the proliferative response of clones of T helper 1 and T helper 2 cells. J Exp Med 1988; 168(2):543–58.
33. Tada T, Taniguchi M, David CS. Properties of the antigen-specific suppressive T-cell factor in the regulation of antibody response of the mouse. IV. Special subregion assignment of the gene(s) that codes for the suppressive T-cell factor in the H-2 histocompatibility complex. J Exp Med 1976; 144(3):713–25.
34. Asherson GL, Zembala M. The role of the T acceptor cell in suppressor systems. Antigen-specific T suppressor factor acts via a T acceptor cell; this releases a nonspecific inhibitor of the transfer of contact sensitivity when exposed to antigen in the context of I-J. Ann NY Acad Sci 1982; 392:71–89.
35. Steinmetz M, Minard K, Horvath S, et al. A molecular map of the immune response region from the major histocompatibility complex of the mouse. Nature 1982; 300(5887):35–42.
36. Kronenberg M, Steinmetz M, Kobori J, et al. RNA transcripts for I-J polypeptides are apparently not encoded between the I-A and I-E subregions of the murine major histocompatibility complex. Proc Natl Acad Sci USA 1983; 80(18):5704–8.
37. Rocken M, Shevach EM. Immune deviation – the third dimension of nondeletional T cell tolerance. Immunol Rev 1996; 149:175–94.
38. Sakaguchi S, Sakaguchi N, Asano M, Itoh M, Toda M. Immunologic self-tolerance maintained by activated T cells expressing IL-2 receptor alpha-chains (CD25). Breakdown of a single mechanism of self-tolerance causes various autoimmune diseases. J Immunol 1995; 155(3):1151–64.
39. Asano M, Toda M, Sakaguchi N, Sakaguchi S. Autoimmune disease as a consequence of developmental abnormality of a T cell subpopulation. J Exp Med 1996; 184(2):387–96.

40. Thornton AM, Piccirillo CA, Shevach EM. Activation requirements for the induction of CD4+CD25+ T cell suppressor function. Eur J Immunol 2004; 34(2):366–76.
41. Takahashi T, Kuniyasu Y, Toda M, et al. Immunologic self-tolerance maintained by CD25+CD4+ naturally anergic and suppressive T cells: induction of autoimmune disease by breaking their anergic/suppressive state. Int Immunol 1998; 10(12):1969–80.
42. Herbelin A, Gombert JM, Lepault F, Bach JF, Chatenoud L. Mature mainstream TCR alpha beta+CD4+ thymocytes expressing L-selectin mediate "active tolerance" in the nonobese diabetic mouse. J Immunol 1998; 161(5):2620–8.
43. Lepault F, Gagnerault MC. Characterization of peripheral regulatory CD4+ T cells that prevent diabetes onset in nonobese diabetic mice. J Immunol 2000; 164(1):240–7.
44. Salomon B, Lenschow DJ, Rhee L, et al. B7/CD28 costimulation is essential for the homeostasis of the CD4+CD25+ immunoregulatory T cells that control autoimmune diabetes. Immunity 2000; 12(4):431–40.
45. Chatenoud L, Salomon B, Bluestone JA. Suppressor T cells – they're back and critical for regulation of autoimmunity! Immunol Rev 2001; 182:149–63.
46. Szanya V, Ermann J, Taylor C, Holness C, Fathman CG. The subpopulation of CD4+CD25+ splenocytes that delays adoptive transfer of diabetes expresses L-selectin and high levels of CCR7. J Immunol 2002; 169(5):2461–5.
47. Thompson CB, Allison JP. The emerging role of CTLA-4 as an immune attenuator. Immunity 1997; 7(4):445–50.
48. Waterhouse P, Penninger JM, Timms E, et al. Lymphoproliferative disorders with early lethality in mice deficient in Ctla-4. Science 1995; 270(5238):985–8.
49. Read S, Malmstrom V, Powrie F. Cytotoxic T lymphocyte-associated antigen 4 plays an essential role in the function of CD25(+)CD4(+) regulatory cells that control intestinal inflammation. J Exp Med 2000; 192(2):295–302.
50. Takahashi T, Tagami T, Yamazaki S, et al. Immunologic self-tolerance maintained by CD25(+)CD4(+) regulatory T cells constitutively expressing cytotoxic T lymphocyte-associated antigen 4. J Exp Med 2000; 192(2):303–10.
51. Shimizu J, Yamazaki S, Takahashi T, Ishida Y, Sakaguchi S. Stimulation of CD25(+)CD4(+) regulatory T cells through GITR breaks immunological self-tolerance. Nat Immunol 2002; 3(2):135–42.
52. McHugh RS, Whitters MJ, Piccirillo CA, et al. CD4(+)CD25(+) immunoregulatory T cells: gene expression analysis reveals a functional role for the glucocorticoid-induced TNF receptor. Immunity 2002; 16(2):311–23.
53. Stephens GL, McHugh RS, Whitters MJ, et al. Engagement of glucocorticoid-induced TNFR family-related receptor on effector T cells by its ligand mediates resistance to suppression by CD4+CD25+ T cells. J Immunol 2004; 173(8):5008–20.
54. Baecher-Allan C, Brown JA, Freeman GJ, Hafler DA. CD4+CD25high regulatory cells in human peripheral blood. J Immunol 2001; 167(3):1245–53.
55. Dieckmann D, Plottner H, Berchtold S, Berger T, Schuler G. Ex vivo isolation and characterization of CD4(+)CD25(+) T cells with regulatory properties from human blood. J Exp Med 2001; 193(11):1303–10.
56. Jonuleit H, Schmitt E, Stassen M, Tuettenberg A, Knop J, Enk AH. Identification and functional characterization of human CD4(+)CD25(+) T cells with regulatory properties isolated from peripheral blood. J Exp Med 2001; 193(11):1285–94.
57. Ng WF, Duggan PJ, Ponchel F, et al. Human CD4(+)CD25(+) cells: a naturally occurring population of regulatory T cells. Blood 2001; 98(9):2736–44.
58. Taams LS, Smith J, Rustin MH, Salmon M, Poulter LW, Akbar AN. Human anergic/suppressive CD4(+)CD25(+) T cells: a highly differentiated and apoptosis-prone population. Eur J Immunol 2001; 31(4):1122–31.
59. Levings MK, Sangregorio R, Roncarolo MG. Human cd25(+)cd4(+) t regulatory cells suppress naive and memory T cell proliferation and can be expanded in vitro without loss of function. J Exp Med 2001; 193(11):1295–302.

60. Powell BR, Buist NR, Stenzel P. An X-linked syndrome of diarrhea, polyendocrinopathy, and fatal infection in infancy. J Pediatr 1982; 100(5):731–7.
61. Godfrey VL, Wilkinson JE, Russell LB. X-linked lymphoreticular disease in the scurfy (sf) mutant mouse. Am J Pathol 1991; 138(6):1379–87.
62. Bennett CL, Christie J, Ramsdell F, et al. The immune dysregulation, polyendocrinopathy, enteropathy, X-linked syndrome (IPEX) is caused by mutations of FOXP3. Nat Genet 2001; 27(1):20–1.
63. Brunkow ME, Jeffery EW, Hjerrild KA, et al. Disruption of a new forkhead/winged-helix protein, scurfin, results in the fatal lymphoproliferative disorder of the scurfy mouse. Nat Genet 2001; 27(1):68–73.
64. Wildin RS, Ramsdell F, Peake J, et al. X-linked neonatal diabetes mellitus, enteropathy and endocrinopathy syndrome is the human equivalent of mouse scurfy. Nat Genet 2001; 27(1): 18–20.
65. Schubert LA, Jeffery E, Zhang Y, Ramsdell F, Ziegler SF. Scurfin (FOXP3) acts as a repressor of transcription and regulates T cell activation. J Biol Chem 2001; 276(40):37672–9.
66. Fontenot JD, Gavin MA, Rudensky AY. Foxp3 programs the development and function of CD4+CD25+ regulatory T cells. Nat Immunol 2003; 4(4):330–6.
67. Hori S, Nomura T, Sakaguchi S. Control of regulatory T cell development by the transcription factor Foxp3. Science 2003; 299(5609):1057–61.
68. Khattri R, Cox T, Yasayko SA, Ramsdell F. An essential role for Scurfin in CD4+CD25+ T regulatory cells. Nat Immunol 2003; 4(4):337–42.
69. Fontenot JD, Dooley JL, Farr AG, Rudensky AY. Developmental regulation of Foxp3 expression during ontogeny. J Exp Med 2005; 202(7):901–6.
70. Bettelli E, Dastrange M, Oukka M. Foxp3 interacts with nuclear factor of activated T cells and NF-kappa B to repress cytokine gene expression and effector functions of T helper cells. Proc Natl Acad Sci USA 2005; 102(14):5138–43.
71. Wu Y, Borde M, Heissmeyer V, et al. FOXP3 controls regulatory T cell function through cooperation with NFAT. Cell 2006; 126(2):375–87.
72. Sugimoto N, Oida T, Hirota K, et al. Foxp3-dependent and -independent molecules specific for CD25+CD4+ natural regulatory T cells revealed by DNA microarray analysis. Int Immunol 2006; 18(8):1197–209.
73. Marson A, Kretschmer K, Frampton GM, et al. Foxp3 occupancy and regulation of key target genes during T-cell stimulation. Nature 2007; 445(7130):931–5.
74. Gavin MA, Rasmussen JP, Fontenot JD, et al. Foxp3-dependent programme of regulatory T-cell differentiation. Nature 2007; 445(7129):771–5.
75. Zheng Y, Josefowicz SZ, Kas A, Chu TT, Gavin MA, Rudensky AY. Genome-wide analysis of Foxp3 target genes in developing and mature regulatory T cells. Nature 2007; 445(7130): 936–40.
76. Aschenbrenner K, D'Cruz LM, Vollmann EH, et al. Selection of Foxp3(+) regulatory T cells specific for self antigen expressed and presented by Aire(+) medullary thymic epithelial cells. Nat Immunol 2007 Apr; 8(4):351–8.
77. Kekalainen E, Tuovinen H, Joensuu J, et al. A defect of regulatory T cells in patients with autoimmune polyendocrinopathy-candidiasis-ectodermal dystrophy. J Immunol 2007; 178(2):1208–15.
78. Kriegel MA, Lohmann T, Gabler C, Blank N, Kalden JR, Lorenz HM. Defective suppressor function of human CD4+ CD25+ regulatory T cells in autoimmune polyglandular syndrome type II. J Exp Med 2004; 199(9):1285–91.
79. Brusko TM, Wasserfall CH, Clare-Salzler MJ, Schatz DA, Atkinson MA. Functional defects and the influence of age on the frequency of CD4+ CD25+ T-cells in type 1 diabetes. Diabetes 2005; 54(5):1407–14.
80. Lindley S, Dayan CM, Bishop A, Roep BO, Peakman M, Tree TI. Defective suppressor function in CD4(+)CD25(+) T-cells from patients with type 1 diabetes. Diabetes 2005; 54(1): 92–9.

81. Kukreja A, Cost G, Marker J, et al. Multiple immuno-regulatory defects in type-1 diabetes. J Clin Invest 2002; 109(1):131–40.
82. Brusko T, Wasserfall C, McGrail K, et al. No Alterations in the Frequency of FOXP3+ Regulatory T-Cells in Type 1 Diabetes. Diabetes 2007; 56(3):604–12.
83. Gavin MA, Torgerson TR, Houston E, et al. Single-cell analysis of normal and FOXP3-mutant human T cells: FOXP3 expression without regulatory T cell development. Proc Natl Acad Sci USA 2006; 103(17):6659–64.
84. Allan SE, Crome SQ, Crellin NK, et al. Activation-induced FOXP3 in human T effector cells does not suppress proliferation or cytokine production. Int Immunol 2007 Apr; 19(4):345–54.
85. Uhlig HH, Coombes J, Mottet C, et al. Characterization of Foxp3+CD4+CD25+ and IL-10-secreting CD4+CD25+ T cells during cure of colitis. J Immunol 2006; 177(9):5852–60.
86. Maul J, Loddenkemper C, Mundt P, et al. Peripheral and intestinal regulatory CD4+CD25(high) T cells in inflammatory bowel disease. Gastroenterology 2005; 128(7):1868–78.
87. Crispin JC, Martinez A, Alcocer-Varela J. Quantification of regulatory T cells in patients with systemic lupus erythematosus. J Autoimmun 2003; 21(3):273–6.
88. Miyara M, Amoura Z, Parizot C, et al. Global natural regulatory T cell depletion in active systemic lupus erythematosus. J Immunol 2005; 175(12):8392–400.
89. Liu MF, Wang CR, Fung LL, Wu CR. Decreased CD4+CD25+ T cells in peripheral blood of patients with systemic lupus erythematosus. Scand J Immunol 2004; 59(2):198–202.
90. Cao D, van Vollenhoven R, Klareskog L, Trollmo C, Malmstrom V. CD25brightCD4+ regulatory T cells are enriched in inflamed joints of patients with chronic rheumatic disease. Arthritis Res Ther 2004; 6(4):R335–46.
91. Ehrenstein MR, Evans JG, Singh A, et al. Compromised function of regulatory T cells in rheumatoid arthritis and reversal by anti-TNFalpha therapy. J Exp Med 2004; 200(3):277–85.
92. Miyara M, Amoura Z, Parizot C, et al. The immune paradox of sarcoidosis and regulatory T cells. J Exp Med 2006; 203(2):359–70.
93. Viglietta V, Baecher-Allan C, Weiner HL, Hafler DA. Loss of functional suppression by CD4+CD25+ regulatory T cells in patients with multiple sclerosis. J Exp Med 2004; 199(7):971–9.
94. Sugiyama H, Gyulai R, Toichi E, et al. Dysfunctional blood and target tissue CD4+CD25high regulatory T cells in psoriasis: mechanism underlying unrestrained pathogenic effector T cell proliferation. J Immunol 2005; 174(1):164–73.

Chapter 2
FoxP3 and Regulatory T Cells

Karsten Kretschmer, Irina Apostolou, Panos Verginis and Harald von Boehmer

Abstract Some regulatory T cells express the Foxp3 transcription factor and such Tregs have an essential function of preventing autoimmune disease in man and mouse. Foxp3 binds to Forkhead motifs of about 1100 genes and the strength of binding increases when Foxp3-expressing T cells are stimulated by PMA and ionomycin. In Foxp3-expressing T cell hybridomas, Foxp3 binding to DNA does not lead to the activation or suppression of genes which becomes only visible after T cell activation. These findings are in line with observations by others that Foxp3 exerts important functions through association with T cell receptor-dependent transcription factors in a DNA-binding complex.

Tregs can be generated when developing T cells encounter TCR agonist ligands in the thymus. This process does not require TGF-β signaling in the T cells but requires costimulatory signals. In contrast, the conversion of naïve T cells into Tregs in peripheral lymphoid tissue essentially depends on TGF-β and is inhibited by costimulation. In fact retinoic acid, produced by some dendritic cells, helps the conversion process by counteracting the negative impact of costimulation on the conversion process. Since AP-1 is produced after costimulation and appears to interfere with a Foxp3-NFAT transcription complex, it is of interest to note that retinoic acid interferes with AP-1-dependent transcription. Thus, retinoic acid may interfere with the negative impact of costimulation on Treg conversion by interfering with the generation and/or function of AP-1.

Peripherally converted Tregs have a stable Foxp3$^+$ phenotype and in mice can survive for several months in the absence of the antigen that induced their formation. In fact the prospective induction of Tregs can be used to generate antigen-specific tolerance that relies on immunosuppression of neighboring CD4 and CD8 T cells by Foxp3$^+$ Tregs in antigen-draining lymph nodes. The mechanisms of suppression may involve cytokines such as TGF-β and IL-10 but also

---

H. von Boehmer
Harvard Medical School, Dana-Farber Cancer Institute, Harvard University, Boston, MA 02115, USA
e-mail: Harald_von_Boehmer@dfci.harvard.edu

other mechanisms that involve suppressive purine-metabolites such as adenosine or adenosine-monophosphate.

Introduction

Cellular therapy employing Foxp3-expressing regulatory T cells (Tregs) holds the promise to replace and/or supplement indiscriminatory immunosuppression by drugs. In order to achieve this goal in the clinic we need to learn more about the generation, lifestyle and function of Tregs. One way to generate Tregs of any desired antigen specificity is the retroviral introduction of the Foxp3 gene into activated CD4 T cells. Foxp3 is a transcriptional repressor and activator that interferes with T cell receptor (TCR) –dependent activation of genes and may exert its effect, at least in part, by compromising NFAT-dependent gene activation. Another way of generating Tregs extrathymically in vivo is the introduction of low amounts of peptides under subimmunogenic conditions. Such artificially induced Tregs have a long lifespan in the absence of the inducing antigen and can thus mediate antigen-specific tolerance. Antigen specificity of Treg-mediated immunosuppression is due to effective co-recruitment and expression of Tregs and T effector cells to antigen-draining lymph nodes and sites of inflammation such that Tregs effectively suppress neighboring effector T cells at early or late stages of their differentiation. The latter allows for interference with already established unwanted immunity and may thus be employed to treat rather than prevent unwanted immune reactions.

The notion that the immune system employs different mechanisms to prevent autoimmune disease or maintain self-tolerance has been around for decades but definitive evidence emphasizing the essential role of negative selection as well as that of suppressor or regulatory T cells is of more recent origin. Today we distinguish negative selection in the form of deletion [1] of certain antigen-specific cells as well as in the form of "anergy" [2] by cell-autonomous mechanisms, also referred to as "recessive" tolerance, from tolerance that relies on the silencing of immune cells by regulatory or suppressor T cells by non-cell-autonomous mechanisms [3], also referred to as "dominant" tolerance. Both forms of tolerance can achieve antigen-specific non-responsiveness of the immune system in contrast to pharmacological interventions that usually result in undesirable general immunosuppression with potentially deadly side effects. In many clinical situations antigen-specific non-responsiveness represents the desired goal but present day treatment does generally not achieve that goal. For that reason it remains a great challenge for immunologists to design strategies and protocols that achieve antigen-specific non-responsiveness since there is little hope that the pharmaceutical industry will come up with suitable procedures to effectively and specifically interfere with unwanted immunity in the near future. Given this goal, it appears a reasonable strategy to exploit evolutionarily selected mechanisms effective in self-tolerance for clinical purposes. This requires a thorough understanding of how the immune system manages to avoid self-aggression. It is now appreciated that so-called negative selection of potentially self-reactive T cells by antigens in- and probably also out-side the thymus essentially contributes to self-tolerance [4]. Likewise it has become clear that the generation of Foxp3-expressing regulatory T cells is mandatory to achieve self-tolerance [5]. The progress in understanding the contribution of such reasonably well-defined mechanisms to tolerance has thus established the somewhat limited usefulness of models that solely consider the absence of "danger" signals as an essential feature of self-tolerance.

While we have some basic ideas about mechanisms that can be exploited to induce antigen-specific non-responsiveness much needs to be learned in detail before this will become clinically applicable. Experiments have shown that overexpression of certain crucial self-antigens (such as insulin) that results in more profound tolerance by negative selection [6], can be helpful in preventing autoimmune disease, perhaps because certain autoimmune diseases, such as in type 1 diabetes, begin with a rather limited autoimmune response to antigens such as insulin [6,7], while later on a variety of other antigens in pancreatic β cells are recognized. However, clinically, such maneuvers would be limited to introducing such antigens prior to disease outbreak or when the immune system is "reset" after elimination of mature lymphocytes by x-irradiation and/or cytotoxic drugs.

In contrast, the manipulation of regulatory T cells appears to represent a more widely applicable approach to not only prevent but potentially also interfere with already ongoing unwanted immunity. With such a clinical goal in mind it is clear that we need to have a much better understanding of how antigen-specific regulatory T cells are and can be generated and/or amplified and how they can achieve antigen-specific non-responsiveness. It is the purpose of this little chapter to review recent progress in the understanding of several aspects of regulatory T cells with the hope that some of this information may find its way into the clinic with the challenge that ensuing procedures will eventually replace or at least supplement the present day practice of indiscriminatory immunosuppression.

Characteristics of Regulatory T Cells

Recent years have seen rapid progress in the characterization of regulatory T cells (Tregs). There is not one particular cell surface marker that defines Tregs but the CD25 surface molecule is at least expressed on the vast majority of cells that express the Foxp3 transcription factor, which has become a signature gene expressed in Tregs. The recognition that $CD25^+$ cells are enriched in Tregs has thus contributed considerably to establishing their role in suppressing activation and function of other lymphocytes [8]. In the meantime other molecules such as neuropilin 1[9], CD103 [10], GPR83 [11], GITR [12] and CTL-A4 [13] have been shown to have a characteristic expression profile in Tregs and thus can be helpful in achieving optimal purification in combination with the CD25 marker. Recent evidence shows that $CD4^+25^+$ Tregs are IL-7R-negative in contrast to $CD4^+25^+$ cells that just represent activated T cells without obvious regulatory function [14]. Intracellular staining by Foxp3 antibodies represents a useful means to identify Tregs in various tissues [15] and in the meantime various Foxp3 reporter mice [16,17] have become available which allow functional purification of Foxp3-expressing cells. While Foxp3 expression represents a good signature for Tregs it can have its drawbacks because Foxp3 can be transiently expressed in activated T cells that, however, do not qualify as stable Tregs [15].

A variety of studies indicate that stable Foxp3 expression is sufficient to confer a regulatory T cell phenotype to CD4 T cells [18–20]. Thus retroviral Foxp3 transduction is a valuable means to endow antigen-specific T cells with a regulatory phenotype. This represents an important tool because unlike the in vitro expansion [21,22] of Tregs preformed in vivo it allows to produce Tregs of any desired specificity.

Recent data suggest that Foxp3 can interact with NFAT to regulate gene expression such as downregulation of the IL-2 gene and upregulation of CTL-A4 and CD25 molecules [23]. It is presently not clear whether all Foxp3-dependent gene regulation involves NFAT and whether NFAT plays a crucial role in the generation of Tregs. It has become clear from the combined analysis of Foxp3 binding and genome-wide gene expression, however, that Foxp3 is predominantly but not exclusively a repressor that silences genes that are normally activated after T cell stimulation, especially genes associated with T cell receptor (TCR) signaling [24]. This fact may contribute to the relatively poor response of Tregs in response to antigenic stimulation in vitro while exogenous growth factors may permit effective clonal expansion in vivo. The latter feature that is likely essential for effective in vivo suppression.

Among the genes that fail to be upregulated in Foxp3-expressing cells is the PTPN22 phosphatase that has a role in dephosphorylating $p56^{lck}$ and Zap-70 [24]. Interestingly a gain of function mutation of this gene has been postulated to affect several autoimmune diseases and it is presently not clear whether this mutant affects Tregs that control autoimmune disease or effector T cells that cause autoimmune disease [25].

Another important characteristic of Tregs is that they do express an $\alpha\beta$TCR that confers antigen specificity. This is worthwhile pointing out since many studies on Tregs ignore this fact. It is our belief that antigen specificity of Tregs is absolutely crucial for antigen-specific suppression of immune responses and hence considerable attention has to be paid to the role of TCR specificity in the generation, homing and effector function of Tregs [26]. As all T cells with $\alpha\beta$TCRs, Tregs also undergo stringent TCR-dependent selection in primary and secondary lymphoid organs [27] which eventually may be exploited to generate Tregs of any desired specificity and to interfere specifically with unwanted immune responses in the clinic.

Intra- and Extra-Thymic Generation of Tregs

Experiments in TCR transgenic mice in which the transgenic TCR was the only TCR expressed by developing T cells have clearly shown that ligation of the $\alpha\beta$TCR by strong agonist ligands plays an essential role in the intrathymic generation of

Tregs [28,29]. These results are compatible with analysis of the Treg TCR repertoire in normal mice suggesting a focus on self-antigens [30]. It became especially obvious that expression of TCR ligands by thymic epithelial cells represented a powerful means to commit developing $CD4^+$ T cells to the Treg lineage [29]. In this context it is of considerable interest to note that thymic epithelial cells and especially thymic medullary epithelial cells can express "ectopically" a variety of proteins that otherwise would be considered "organ-specific" such as preproinsulin 2 that is expressed in pancreatic β cells but also in thymic medullary epithelial cells [31,32]. Such ectopic expression can be regulated, at least in part, by the AIRE (for *a*utoimmune *i*mmune *r*egulation) transcription factor [33] and it is thus conceivable that the ectopic expression of "organ-specific" antigen by thymic epithelium plays a decisive role in the generation of Tregs specific for such antigens, even though experiments addressing that question have so far yielded negative results [34,35]. However, negative results by no means rule out that AIRE-regulated antigens contribute to the generation of Tregs under more favorable experimental conditions.

The intrathymic generation of Tregs by strong agonist ligands appears to require costimulation of developing cells by B7-1 (CD80) [36] ligands that are expressed on thymic epithelial cells as well as on antigen-presenting cells of hemopoietic origin at least under certain experimental conditions. This is a somewhat astonishing observation in the light of findings that Treg generation in peripheral lymphoid tissue is most effective under conditions that avoid costimulation (see below). Conceivably this could be due to the different stages of development of thymic and extrathymic T cells which may require different signaling inputs for Treg commitment. From thymus transplantation experiments it is clear that Treg-generated by ligands expressed on thymic epithelium only, can migrate into peripheral lymphoid tissue and patrol the body for long periods of time without being confronted with the same ligand that was involved in their generation [29,37]. This does not exclude that lower affinity ligands in peripheral lymphoid tissue may contribute to survival much like they can contribute to survival of CD4 and CD8 conventional T cells [38].

Considering the intrathymic generation of Tregs it is of interest to note that generation of Tregs from cells with one particular $\alpha\beta$TCR is not mutually exclusive to deletion of some of these cells [29]. Thus both processes depend on recognition of agonist ligands by developing $CD4^+$ T cells but under some conditions such recognition results in deletion and under other conditions in Treg generation even within the same thymus, perhaps because some of these cells encounter their TCR ligands on different cells i.e. either on cross-presenting dendritic cells or directly on thymic epithelial cells [39].

Whereas the intrathymic generation of Tregs would mostly depend on instruction of lineage commitment by self-antigens, the peripheral generation of Tregs may also include instruction by foreign antigens. It is therefore of considerable interest to define conditions permissible for extrathymic Treg generation. To this end we have exploited protocols of subimmunogenic antigen presentation because circumstantial and historic evidence suggested that one might be able to induce "dominant" tolerance in this way. Indeed it was found that either constant delivery of peptides by osmotic mini-pumps [40] or by targeting dendritic cells with peptide-containing fusion antibodies directed against the DEC205 endocytic receptor on dendritic cells [15] allowed the conversion of naïve T cells into Foxp3 regulatory T cells. The conversion process depended on an intact TGF-βRII receptor on naïve T cells and conditions that avoided activation of dendritic cells as well as IL-2 production by naïve T cells. It was clear that Tregs were generated by conversion rather than expansion of already committed Tregs since the experiments were performed in mice expressing only one particular transgenic TCR in the absence of coexpression of a TCR agonist ligand resulting in the unique constellation that none of the generated CD4$^+$ T cells exhibited initially a Treg phenotype and only a certain percentage ($\sim$30%) assumed it after the artificial introduction of the respective TCR agonist ligand [15].

The generation of Treg by subimmunogenic antigen delivery and the negative impact of strong costimulation on conversion of naïve T cells into Treg [15] correlates with the fact that costimulation results in the accumulation of Fos and Jun containing AP-1 that interacts with NFAT and thereby blocks the formation of a Foxp3-NFAT complex that is required for the generation of functional Treg [23]. The latter data are well in line with the observation that Foxp3 overexpression inc T cell hybridomas, that are not activated through their TCR results in binding of Foxp3 to target genes but very little regulation of target genes, while in contrast activation of such hybridomas results in activation as well as Foxp3 dependent downregulation and upregulation of genes [24]. Thus TCR signals and Foxp3 need to synergize in the generation of functional Tregs [23,24].

Importantly, the peripherally generated Tregs exhibited the same global gene expression pattern as intrathymically generated Tregs [39] and much like intrathymically generated Tregs exhibited a long lifespan that was independent on further supply of the TCR agonist ligand. Thus by these maneuvers a Treg "memory" to external TCR ligands could be induced, resulting in the subsequent suppression of immune responses elicited by the same agonist ligand i.e. this protocol succeeded in generating specific immunological tolerance to one particular antigen. Hopefully this protocol can be extended to many other antigens and thus help the prevention of unwanted immune responses. Recently some of us (I.A., P.V., H. v. B.) succeeded to induce transplantation tolerance in wt female mice by infusing them with male peptide, resulting in the generation of male-specific Foxp3$^+$ regulatory T cells. Of note this particular protocol only works with naïve T cells and not with T cells that have been already activated in vivo and thus can presumably not be used to suppress already established autoimmunity in which most antigen-specific T cells are already activated. In such cases the in vitro generation of Tregs by Foxp3 transduction would likely be more appropriate (see below) [39].

Recently it has been reported that retinoic acid generated in CD103 positive DC in the gut helps the conversion of naïve T cells into Foxp3$^+$ Tregs thereby giving credibility to the disputed concept of oral tolerance [41–43]. Interestingly, retinoic acid appears to interfere with the negative effect of costimulation on TGF-β dependent conversion of naïve T cells into Tregs [43] providing a possible (only) mechanism for its effect, since by itself in the absence of TGF-β retinoic acid does not affect conversion. Thus, there is an interesting difference between intra-thymic and extra-thymic Treg generation: While the former requires costimulation and takes place even in the absence of TGF-β, the latter is essentially dependent on TGF-β and takes place in the absence of costimulation. It is worthwhile pointing out that even in the presence of retinoic acid, the TGF-β-dependent conversion works effectively only with naïve T cells while preactivated T cells, perhaps because of their high AP-1 content, are relatively resistant to a TCR-induced conversion process.

At present one can only speculate why costimulation and AP-1 generation interferes with the upregulation of Foxp3 in naïve T cells: it may be that AP-1 interferes with regulation of the Foxp3 locus by TCR and TGF-β-dependent signals that eventually results in stable Foxp3 expression. Whether it does so by interfering with the action of TGF-β–induced Foxp3 regulation or autoregulation by Foxp3 is unknown. It is also possible that AP-1 interferes somehow with demethylation of the Foxp3 locus. In this context it is of interest to note that the slow in vivo conversion process generates long-lived and stable Foxp3-expressing Tregs whereas the conversion process in vitro utilizing costimulation and TGF-β often results in cells with unstable Foxp3 expression, at least when these cells are analyzed during antigenic stimulation.

Lifestyle of Tregs

As pointed out above, Tregs can survive for relatively long periods of time as resting cells at an intermitotic stage but as soon as they encounter their TCR agonist ligand they will express activation markers and begin to home to antigen-draining lymph nodes and undergo considerable expansion [21,22,37]. This is usually accompanied by loss of CD62L and acquisition of CD44 expression and followed by expression of the α_E integrin (CD103) (at least in the mouse). Such activated cells extravasate and accumulate together with other T effector cells in inflamed tissue [10]. It is in fact the co-recruitment of CD4 and/or CD8 effector cells with activated Tregs in draining lymph nodes and/or inflamed tissue which determines the specificity of immunosuppression [37]: since Tregs suppress neighboring T cells in a "bystander" fashion it can only be effective when most antigen-specific effector cells are co-recruited to the same anatomical location, which depends on presentation of TCR ligands in these places, such as antigen-draining lymph nodes [20]. Thus while Tregs may suppress "innocent" bystanders that happen to be in their vicinity this will not result in general immunosuppression because the majority of such "innocent" cells will be distributed throughout the body and not recruited by antigen such that they will not be subject to suppression. It is for this reason that injection of Tregs specific for a pancreas-derived antigen is far more effective in suppressing diabetes than polyclonal Tregs that will all not accumulate in pancreatic lymph nodes [20].

"Bystander suppression" is well documented by the fact that for instance $CD4^+$ Tregs recognizing a class II MHC-presented epitope from one particular protein can suppress CD8 T cells recognizing a different class I MHC-presented epitope from the same protein [44]. Thus the antigen specificity of Tregs and effector T cells does not need to match in order for effective immunosuppression to occur: it is sufficient that the two cell types are co-recruited to the same tissue. This of course is good news since this will permit a Treg of one particular specificity to suppress a variety of effector cells with different specificity as long as all these different epitopes are present within the same draining lymph node or anatomical site.

Since many intrathymically generated Tregs are specific for self-antigen it is perhaps not surprising that normally there are always "activated" Tregs present in the organism [45] and some of these Tregs may be engaged in locally preventing autoimmunity. In fact neonatal removal of Tregs will result in the "scurfy" phenotype associated with multi-organ-specific autoimmunity [46,47]. Other Tregs are apparently not "in action" and patrol the body by exhibiting a phenotype of naïvec T cells that do not divide [31,45].

Function of Tregs

One of the questions that has remained rather elusive concerns the molecular mechanisms by which Tregs control other T cells. There are probably several not mutually exclusive mechanisms, some of which may dominate in certain situations [26]. In vitro data have emphasized the role of close cell-to-cell contact and dispensable cytokines such as IL-10 or TGF-β. All in vivo data published so far have emphasized the crucial role of the TGF-βRII on suppressed cells since a dominant negative form of that receptor is usually associated with ineffective Treg suppression and with generalized autoimmunity. It is still not clear whether this results from the fact that Tregs produce TGF-β (which they do but only in moderate amounts) or whether in general TGF-β-induced signaling "conditions" effector cells for more stringent suppression by a mechanism that does not involve increased TGF-β production but depends on specific Treg activation [26]. A good example for such a scenario is the suppression of tumor-specific CD8 T cells by CD4 Tregs that crucially depends on an intact TGF-βRII receptor on the CD8 T cells: In this particular model the suppression affects the function of fully differentiated cytotoxic T lymphocytes (CTL), notably the secretion of cytolytic granules. However, in vitro experiments with fully differentiated CTL have shown that TGF-β does not have any negative impact on cytolysis when added during the effector phase. This is consistent with the hypothesis that TGF-β-dependent signaling "conditions" the CD8 T cells for Treg suppression rather than representing the sole suppressor mechanism [44].

These experiments also make another important point, namely that it is apparently never too late to interfere with an immune response by Treg suppression since the experiments show that suppression can affect fully differentiated effector cells. This is good news in the sense that the obviously effective suppression late during an immune response can revert rather than prevent unwanted immunity, a concept that may become extremely useful in the clinic.

Different experiments attempting to reverse rather than prevent diabetes are fully consistent with that view: CD4 T cells specific for an islet-derived antigen of unknown nature could be activated in vitro and retrovirally transduced with Foxp3 such that within 24 hours they assumed a phenotype of Tregs. When 10^5 of such converted cells were injected into NOD mice that had become just diabetic because of beginning destruction of their islet cells, these islet-specific Tregs cured the mice of diabetes and they remained diabetes-free for at least three months when the experiment was terminated. Again this experiment suggests that Tregs can silence already fully developed effector cells [20].

Additional controls make important points with regard to the role of Treg antigen receptors in this process and hence the specificity of immunosuppression: while the injection of 10^5 cells with islet-antigen specificity was sufficient to abolish disease, the injection of even 10^6 Tregs with specificity for a large variety of different antigens or the injection of Tregs with specificity for an antigen not present in the pancreatic lymph node did not have any effect and the animals died several days later from complete destruction of β cells and resulting diabetes that obviously at this point could be no longer reversed by Tregs [20]. These results and similar results by others employing in vitro expanded Tregs [21,22] are very encouraging since they suggest that by adoptive Treg therapy early-diagnosed diabetes may be cured, in spite of the fact that the generation of sufficient numbers of islet-antigen-specific Tregs still represents a staggering logistic problem.

Thus in spite of our ignorance concerning molecular mechanisms of Treg-mediated suppression (even though a variety has been proposed [26]) we have promising evidence from murine models of disease that Tregs have the capacity to interfere with unwanted immunity early and/or late during the immune response in an antigen-specific way since they interfere with such immunity in a local milieu only while leaving the rest of the immune system intact.

There is also no compelling reason why the findings made in the somewhat popular models of type 1 diabetes should not be extended to other autoimmune diseases such as rheumatic diseases provided that there are clues about relevant antigens that are presented in local lymphoid tissue.

Concluding Remarks

The described properties of Tregs i.e. the possibility to generate them extrathymically in vivo or in vitro with any desired antigen specificity, their ability to co-home with T effector cells into antigen-draining lymph nodes and/or sites of inflammation, their potential to suppress effector cells at early and late stages of differentiation and last but not least to suppress neighboring T effector cells of any antigenic specificity, make these cells an ideal tool to intervene with unwanted immunity in an antigen-specific way. Thus one would eventually hope that the exploitation of evolutionarily selected mechanisms to deal with unwanted immune responses against self will replace indiscriminatory immunosuppression by drugs with potentially deadly side effects. This is not to say that such drugs may be completely useless: their transient application may help to set the immune system to a stage where Tregs can be more effective in dealing specifically with unwanted immunity. What should be avoided, however, is the long-term indiscriminatory use of the drugs that eventually will ruin the protection against infections and malignant disease afforded by the immune system.

References

1. von Boehmer, H., and P. Kisielow. 2006. Negative selection of the T-cell repertoire: where and when does it occur? *Immunol Rev* 209:284–289.
2. Rocha, B., and H. von Boehmer. 1991. Peripheral selection of the T cell repertoire. *Science* 251:1225–1228.
3. Sakaguchi, S., M. Ono, R. Setoguchi, H. Yagi, S. Hori, Z. Fehervari, J. Shimizu, T. Takahashi, and T. Nomura. 2006. Foxp3CD25CD4 natural regulatory T cells in dominant self-tolerance and autoimmune disease. *Immunol Rev* 212:8–27.
4. von Boehmer, H., I. Aifantis, F. Gounari, O. Azogui, L. Haughn, I. Apostolou, E. Jaeckel, F. Grassi, and L. Klein. 2003. Thymic selection revisited: how essential is it? *Immunol Rev* 191:62–78.
5. Khattri, R., T. Cox, S.A. Yasayko, and F. Ramsdell. 2003. An essential role for Scurfin in CD4+CD25+ T regulatory cells. *Nat Immunol* 4:337–342.
6. Jaeckel, E., M.A. Lipes, and H. von Boehmer. 2004. Recessive tolerance to preproinsulin 2 reduces but does not abolish type 1 diabetes. *Nat Immunol* 5:1028–1035.
7. Nakayama, M., N. Abiru, H. Moriyama, N. Babaya, E. Liu, D. Miao, L. Yu, D.R. Wegmann, J.C. Hutton, J.F. Elliott, and G.S. Eisenbarth. 2005. Prime role for an insulin epitope in the development of type 1 diabetes in NOD mice. *Nature* 435:220–223.
8. Itoh, M., T. Takahashi, N. Sakaguchi, Y. Kuniyasu, J. Shimizu, F. Otsuka, and S. Sakaguchi. 1999. Thymus and autoimmunity: production of CD25+CD4+ naturally anergic and suppressive T cells as a key function of the thymus in maintaining immunologic self-tolerance. *J Immunol* 162:5317–5326.
9. Bruder, D., M. Probst-Kepper, A.M. Westendorf, R. Geffers, S. Beissert, K. Loser, H. von Boehmer, J. Buer, and W. Hansen. 2004. Neuropilin-1: a surface marker of regulatory T cells. *Eur J Immunol* 34:623–630.
10. Huehn, J., K. Siegmund, J.C. Lehmann, C. Siewert, U. Haubold, M. Feuerer, G.F. Debes, J. Lauber, O. Frey, G.K. Przybylski, U. Niesner, M. de la Rosa, C.A. Schmidt, R. Brauer, J. Buer, A. Scheffold, and A. Hamann. 2004. Developmental stage, phenotype, and migration distinguish naive- and effector/memory-like CD4+ regulatory T cells. *J Exp Med* 199: 303–313.
11. Hansen, W., K. Loser, A.M. Westendorf, D. Bruder, S. Pfoertner, C. Siewert, J. Huehn, S. Beissert, and J. Buer. 2006. G protein-coupled receptor 83 overexpression in naive CD4+CD25- T cells leads to the induction of Foxp3+ regulatory T cells in vivo. *J Immunol* 177:209–215.
12. Shimizu, J., S. Yamazaki, T. Takahashi, Y. Ishida, and S. Sakaguchi. 2002. Stimulation of CD25(+)CD4(+) regulatory T cells through GITR breaks immunological self-tolerance. *Nat Immunol* 3:135–142.
13. Bachmann, M.F., G. Kohler, B. Ecabert, T.W. Mak, and M. Kopf. 1999. Cutting edge: lymphoproliferative disease in the absence of CTLA-4 is not T cell autonomous. *J Immunol* 163:1128–1131.
14. Liu, W., A.L. Putnam, Z. Xu-Yu, G.L. Szot, M.R. Lee, S. Zhu, P.A. Gottlieb, P. Kapranov, T.R. Gingeras, B.F. de St Groth, C. Clayberger, D.M. Soper, S.F. Ziegler, and J.A. Bluestone. 2006. CD127 expression inversely correlates with FoxP3 and suppressive function of human CD4+ T reg cells. *J Exp Med* 203:1701–1711.
15. Kretschmer, K., I. Apostolou, D. Hawiger, K. Khazaie, M.C. Nussenzweig, and H. von Boehmer. 2005. Inducing and expanding regulatory T cell populations by foreign antigen. *Nat Immunol* 6:1219–1227.

16. Fontenot, J.D., J.P. Rasmussen, L.M. Williams, J.L. Dooley, A.G. Farr, and A.Y. Rudensky. 2005. Regulatory T cell lineage specification by the forkhead transcription factor foxp3. *Immunity* 22:329–341.
17. Wan, Y.Y., and R.A. Flavell. 2005. Identifying Foxp3-expressing suppressor T cells with a bicistronic reporter. *Proc Natl Acad Sci USA* 102:5126–5131.
18. Hori, S., T. Nomura, and S. Sakaguchi. 2003. Control of regulatory T cell development by the transcription factor Foxp3. *Science* 299:1057–1061.
19. Fontenot, J.D., M.A. Gavin, and A.Y. Rudensky. 2003. Foxp3 programs the development and function of CD4+CD25+ regulatory T cells. *Nat Immunol* 4:330–336.
20. Jaeckel, E., H. von Boehmer, and M.P. Manns. 2005. Antigen-specific FoxP3-transduced T-cells can control established type 1 diabetes. *Diabetes* 54:306–310.
21. Tang, Q., K.J. Henriksen, M. Bi, E.B. Finger, G. Szot, J. Ye, E.L. Masteller, H. McDevitt, M. Bonyhadi, and J.A. Bluestone. 2004. In vitro-expanded antigen-specific regulatory T cells suppress autoimmune diabetes. *J Exp Med* 199:1455–1465.
22. Tarbell, K.V., S. Yamazaki, K. Olson, P. Toy, and R.M. Steinman. 2004. CD25+ CD4+ T Cells, expanded with dendritic cells presenting a single autoantigenic peptide, suppress autoimmune diabetes. *J Exp Med* 199:1467–1477.
23. Wu, Y., M. Borde, V. Heissmeyer, M. Feuerer, A.D. Lapan, J.C. Stroud, D.L. Bates, L. Guo, A. Han, S.F. Ziegler, D. Mathis, C. Benoist, L. Chen, and A. Rao. 2006. FOXP3 controls regulatory T cell function through cooperation with NFAT. *Cell* 126:375–387.
24. Marson, A., K. Kretschmer, G.M. Frampton, E.S. Jacobsen, J.K. Polansky, K.D. MacIsaac, S.S. Levine, E. Fraenkel, H. von Boehmer, and R.A. Young. 2007. Foxp3 occupancy and regulation of key target genes during T-cell stimulation. *Nature* 445:931–935.
25. Bottini, N., T. Vang, F. Cucca, and T. Mustelin. 2006. Role of PTPN22 in type 1 diabetes and other autoimmune diseases. *Semin Immunol* 18:207–213.
26. von Boehmer, H. 2005. Mechanisms of suppression by suppressor T cells. *Nat Immunol* 6:338–344.
27. von Boehmer, H. 2004. Selection of the T-cell repertoire: receptor-controlled checkpoints in T-cell development. *Adv Immunol* 84:201–238.
28. Jordan, M.S., A. Boesteanu, A.J. Reed, A.L. Petrone, A.E. Holenbeck, M.A. Lerman, A. Naji, and A.J. Caton. 2001. Thymic selection of CD4+CD25+ regulatory T cells induced by an agonist self-peptide. *Nat Immunol* 2:301–306.
29. Apostolou, I., A. Sarukhan, L. Klein, and H. von Boehmer. 2002. Origin of regulatory T cells with known specificity for antigen. *Nat Immunol* 3:756–763.
30. Hsieh, C.S., Y. Liang, A.J. Tyznik, S.G. Self, D. Liggitt, and A.Y. Rudensky. 2004. Recognition of the peripheral self by naturally arising CD25+ CD4+ T cell receptors. *Immunity* 21:267–277.
31. Derbinski, J., A. Schulte, B. Kyewski, and L. Klein. 2001. Promiscuous gene expression in medullary thymic epithelial cells mirrors the peripheral self. *Nat Immunol* 2:1032–1039.
32. Vafiadis, P., S.T. Bennett, J.A. Todd, J. Nadeau, R. Grabs, C.G. Goodyer, S. Wickramasinghe, E. Colle, and C. Polychronakos. 1997. Insulin expression in human thymus is modulated by INS VNTR alleles at the IDDM2 locus. *Nat Genet* 15:289–292.
33. Anderson, M.S., E.S. Venanzi, L. Klein, Z. Chen, S.P. Berzins, S.J. Turley, H. von Boehmer, R. Bronson, A. Dierich, C. Benoist, and D. Mathis. 2002. Projection of an immunological self shadow within the thymus by the aire protein. *Science* 298:1395–1401.
34. Liston, A., D.H. Gray, S. Lesage, A.L. Fletcher, J. Wilson, K.E. Webster, H.S. Scott, R.L. Boyd, L. Peltonen, and C.C. Goodnow. 2004. Gene dosage – limiting role of aire in thymic expression, clonal deletion, and organ-specific autoimmunity. *J Exp Med* 200:1015–1026.
35. Anderson, M.S., E.S. Venanzi, Z. Chen, S.P. Berzins, C. Benoist, and D. Mathis. 2005. The cellular mechanism of aire control of T cell tolerance. *Immunity* 23:227–239.
36. Tai, X., M. Cowan, L. Feigenbaum, and A. Singer. 2005. CD28 costimulation of developing thymocytes induces Foxp3 expression and regulatory T cell differentiation independently of interleukin 2. *Nat Immunol* 6:152–162.

37. Klein, L., K. Khazaie, and H. von Boehmer. 2003. In vivo dynamics of antigen-specific regulatory T cells not predicted from behavior in vitro. *Proc Natl Acad Sci USA* 100:8886–8891.
38. Hao, Y., N. Legrand, and A.A. Freitas. 2006. The clone size of peripheral CD8 T cells is regulated by TCR promiscuity. *J Exp Med* 203:1643–1649.
39. Kretschmer, K., I. Apostolou, E. Jaeckel, K. Khazaie, and H. von Boehmer. 2006. Making regulatory T cells with defined antigen specificity: role in autoimmunity and cancer. *Immunol Rev* 212:163–169.
40. Apostolou, I., and H. Von Boehmer. 2004. In vivo instruction of suppressor commitment in naive T cells. *J Exp Med* 199:1401–1408.
41. Coombes, J. L., K. R. Siddiqui, C. V. Arancibia-Cárcamo, J. Hall, C. M. Sun, Y. Belkaid, and F. Powrie. 2007. A functionally specialized population of mucosal CD103+ DCs induces Foxp3+ regulatory T cells via a TGF-beta and retinoic acid-dependent mechanism. *J Exp Med* 204:1757–1764.
42. Sun, C. M., J. A. Hall, R. B. Blank, N. Bouladoux, M. Oukka, J. R. Mora, and Y. Belkaid. 2007. Small intestine lamina propria dendritic cells promote de novo generation of Foxp3 T reg cells via retinoic acid. *J Exp Med* 204:1775–1785.
43. von Boehmer, H. 2007. Oral tolerance: is it all retinoic acid? *J Exp Med* 204:1737–1739.
44. Mempel, T.R., M.J. Pittet, K. Khazaie, W. Weninger, R. Weissleder, H. von Boehmer, and U.H. von Andrian. 2006. Regulatory T cells reversibly suppress cytotoxic T cell function independent of effector differentiation. *Immunity* 25:129–141.
45. Fisson, S., G. Darrasse-Jeze, E. Litvinova, F. Septier, D. Klatzmann, R. Liblau, and B.L. Salomon. 2003. Continuous activation of autoreactive CD4+ CD25+ regulatory T cells in the steady state. *J Exp Med* 198:737–746.
46. Lahl, K., C. Loddenkemper, C. Drouin, J. Freyer, J. Arnason, G. Eberl, A. Hamann, H. Wagner, J. Huehn, and T. Sparwasser. 2007. Selective depletion of Foxp3+ regulatory T cells induces a scurfy-like disease. *J Exp Med* 204:57–63.
47. Kim, J.M., J.P. Rasmussen, and A.Y. Rudensky. 2007. Regulatory T cells prevent catastrophic autoimmunity throughout the lifespan of mice. *Nat Immunol* 8:191–197.

Chapter 3
Thymic and Peripheral Generation of CD4+ Foxp3+ Regulatory T Cells

Paola Romagnoli, Julie Ribot, Julie Tellier, and Joost P.M. van Meerwijk

Abstract The existence of regulatory T lymphocytes ("Treg") was suspected more than twenty years ago from seminal experiments on induction of transplantation tolerance in chick-quail chimeras. Much more recently, naturally occurring thymus-derived Treg were characterized phenotypically and functionally. It became clear that these cells are critically involved in prevention of autoimmune disease and, accordingly, it was found that the Treg repertoire is enriched in autospecific cells. The latter observation incited substantial work on selection and thymic lineage commitment of Treg. Some results supported an "instructive" model for Treg commitment in which precursors expressing a TCR with high affinity for self-ligands are directed to the Treg lineage. Other evidence supported a "stochastic" model in which commitment to the Treg lineage occurs independently of TCR specificity. The autospecific Treg-repertoire appears to be shaped in the thymus through a two-step positive selection process, one step occurring in the thymic cortex and the other in the medulla, combined with reduced sensitivity to negative selection. Also peripheral and therefore fully mature conventional T cells can, under certain experimental conditions, be converted to the Treg lineage and thus contribute to establishing tolerance to innocuous, e.g. intestinal, antigens. Understanding the thymic and peripheral mechanisms of Treg-induction will contribute to development of therapies involving modulation of these cells in pathologies ranging from cancer, via autoimmunity, to graft-rejection and Graft-versus-host disease.

Introduction

Adaptive immunity depends on the activity of B and T lymphocytes. These cells clonally express receptors for antigen whose specificity is determined by somatic rearrangements of their genes. Since these rearrangements are stochastic in nature, immature B and T cell precursors display a very wide repertoire of B and

---

J.P.M. van Meerwijk
Tolerance and Autoimmunity section, Centre de Physiopathologie de Toulouse Purpan, Institut National de la Santé et de la Recherche Médicale (Inserm) U563, Toulouse, France
e-mail: Joost.van-Meerwijk@toulouse.inserm.fr

T-cell receptor specificities. A very substantial proportion of these precursors is autospecific and, therefore, potentially dangerous for the organism [1,2]. To limit the dangerous nature of mature B and T lymphocyte repertoires, their immature precursors are submitted to negative selection processes in primary lymphoid organs. Recognition of cognate ligand by these immature cells leads to induction of apoptosis or of an anergic state [3–5]. However, despite the relative efficiency of central tolerance [1,2], the peripheral mature B and T cell-repertoires contain significant numbers of autospecific cells [6]. To avoid immunopathology, so-called peripheral tolerance mechanisms control activity of such potentially dangerous cells. Induction of apoptosis and anergy, leading to a state of so-called "recessive tolerance" [7], has been implicated in maintenance of peripheral tolerance. However, the activity of "regulatory T cells" (Treg), responsible for "dominant tolerance" [7], plays a major role in peripheral tolerance.

Several populations of T lymphocytes with regulatory capacities are currently known [8–16]. The most extensively studied of these populations is characterized by expression of CD4 and the forkhead/winged-helix transcription factor Foxp3 [13–15]. Mutations in the gene encoding Foxp3 lead to lethal autoimmune syndromes in Man and in the Mouse, unambiguously demonstrating the central role of Foxp3-expressing Treg in maintenance of peripheral tolerance [17,18]. Since the vast majority of Foxp3-expressing cells also express CD4 [19,20], these observations very strongly suggest a critical and central role of $CD4^{pos}Foxp3^{pos}$ cells. Whereas $CD4^{pos}Foxp3^{pos}$ cells are also characterized by a particular expression pattern of other cell-surface molecules (e.g. $CTLA-4^{pos}$, $GITR^{pos}$, $CD127^{neg}$ [21–23]), they are at present best identified by their expression of CD4, very high levels of CD25, and Foxp3.

Thymic Origin of Regulatory T Cells

The major source of Treg appears to be the thymus, the very same organ as that in which conventional T cells develop. Thymic origin of cells responsible for a state of dominant tolerance was suspected more than twenty years ago from seminal experiments on induction of transplantation tolerance in chick-quail chimeras [24]. In these chimeric chicks, quail thymic epithelium induced tolerance to quail donor tissue. T cells developed in quail as well as in chick thymi. It was therefore hypothesized that xenoreactive T cells from chick thymi were controlled by Treg that had developed in quail thymi [25]. Later experiments established that a state of dominant tolerance could also be induced in mice by transplantation of allogeneic donor thymic epithelium [7,26,27].

An independent line of evidence for thymic origin of Treg came from experiments on autoimmune pathology induced by neonatal thymectomy. Thymectomy at days 2–4 after birth leads to various auto-immune manifestations, the precise nature of which depends on the particular strain of mice analyzed [28–35]. Sakaguchi and colleagues had observed that nude mice reconstituted with CD25-depleted cells from lymph nodes or spleen developed autoimmune diseases (such as thyroiditis, gastritis, insulitis, sialoadenitis, adrenalitis, oophoritis, glomerulonephritis, and polyarthritis). They therefore hypothesized that CD25^{pos} T cells may have regulatory potential [36]. This hypothesis was tested by injecting neonatally thymectomized mice with CD4^{pos}CD25^{pos} T cells and it was observed that they efficiently prevented autoimmune pathology [37]. It was shown that CD4^{pos}CD25^{pos} T cells appeared in the spleen starting at day 4 after birth, several days after appearance of conventional T cells. Combined, these results suggested that neonatal thymectomy caused a quantitative lack of thymus-derived CD25^{pos} Treg leading to autoimmune-disease (but this view was later challenged, see ref. [38]). More direct evidence that CD4^{pos}CD25^{pos} Treg are of thymic origin came from experiments by the same group showing that thymic CD4^{pos}CD25^{pos} T cells inhibited autoimmune-manifestations induced by T cell transfer into athymic mice [39]. These data were consistent with earlier observations from the Mason laboratory showing that in the rat thymus-derived CD4^{pos}CD45RC^{low} cells prevented autoimmune disease (diabetes and thyroiditis) induced by rendering adult rats lymphopenic by a combined treatment of thymectomy and sublethal γ-irradiation [40,41]. Combined, these results strongly suggested that Treg can develop in the thymus and that quantitative and/or qualitative defects in the Treg-repertoire lead to autoimmune manifestations even in absence of immunization, emphasizing the crucial physiological role these cells play in maintaining immunological homeostasis.

In the before-mentioned reports it is formally impossible to exclude the possibility that (e.g. activated) peripheral T cells differentiate into CD4^{pos}CD25^{pos} Treg and then home back to the thymus. During ontogeny, CD25^{pos}Foxp3^{pos} CD4SP cells appear in the thymus a few days later than mature CD4^{pos}CD25^{neg} cells [42]. Also in adult animals, CD25^{pos}Foxp3^{pos} cells develop later than conventional T cells from dividing precursors (PR, unpublished data). Moreover, whereas immature conventional T cells can be identified using cell-surface markers as HSA, CD69, and MHC class I, CD25^{high}Foxp3^{pos} CD4SP cells uniformly have a relative mature HSA^{low}, CD69^{low}, and MHC class I^{high} phenotype [42,43]. These data therefore are consistent with the notion that Treg differentiate in the periphery and recirculate to the thymus. It has indeed been reported that in genetically lymphopenic mice increased proportions of Treg were found in the peripheral lymphoid organs and that they recirculate to the thymus [44]. However, peripheral activation and differentiation would require TCR-mediated interaction with MHC class II molecules. In radiation chimeras in which radioresistant cells normally express MHC (class I and class II) molecules but hematopoietic cells do not, normal percentages and even increased absolute numbers of CD4^{pos}CD25^{pos}Foxp3^{pos} T cells were found in the thymus [45]. Also transgenic mice in which MHC class II expression is limited to squamous (in the thymus cortical) epithelial cells largely normal numbers of thymic CD4^{pos}CD25^{pos} cells developed [43,46]. Moreover, despite the observation that thymic Treg have a relatively mature phenotype, it is substantially less mature than that of peripheral Treg. For example, in contrast to peripheral Treg, thymic cells display evidence of recent RAG-expression and express low levels of CD8 and HSA ([47] and our unpublished results). In mice in which the diphtheria toxin receptor is expressed under control of the endogenous *foxp3*-promoter, Treg can be eliminated by injection of the toxin. It was observed that upon depletion of Treg, thymic Foxp3$^+$ cell-levels were reconstituted well before peripheral levels, suggesting that thymic Treg do not recirculate from the periphery [48]. These observations therefore strongly argue against a model in which CD4^{pos}CD25^{pos}Foxp3^{pos} thymocytes are recirculating cells that in the periphery had differentiated into Treg. However, to formally evaluate if Treg can develop in the thymus, one would have to perform *in vitro* organ culture of thymi that only contain immature precursors. Such experiments have been performed and it was observed that CD4^{pos}CD25^{pos}Foxp3^{pos} T cells developed in fetal thymus organ cultures (FTOC) as well as in reaggregate thymus organ cultures (RTOC) ([39,49,50] and our unpublished observations). These results therefore formally demonstrate that CD4^{pos}CD25^{pos}Foxp3^{pos} Treg can develop in the thymus.

Peripheral Induction of Regulatory T Cells

The observation that Treg can develop in the thymus does not exclude the possibility that regulatory phenotype can also be induced in the periphery. Several immunomodulatory T cell populations induced in peripheral lymphoid organs and/or *in vitro* have been described (e.g. Th3, Tr1, refs. [51,52]), but here we will limit the discussion to Treg of CD4^{pos}CD25^{pos}Foxp3^{pos} phenotype.

In "the early days" of CD4^{pos}CD25^{pos}Foxp3^{pos} Treg, the most useful marker of Treg was CD25. Several groups studied stability of expression of this marker by injecting CD4^{pos}CD25^{pos} and CD4^{pos}CD25^{neg} cells into lymphopenic (RAG-deficient) mice. It was observed that CD25^{pos} cells lost expression of this cell-surface marker, and that CD25^{neg} cells could upregulate it. As a matter of fact, both populations assumed similar CD25-phenotypes upon homeostatic proliferation *in vivo* [53,54]. While in most, but not all [55], experimental systems suppressive activity is found in the CD25^{pos}, but not CD25^{neg}, fraction, it was shown that CD25^{pos} cells downmodulating expression of this marker upon *in vivo* proliferation did not lose their suppressive activity [56]. Moreover, analysis of Foxp3 mRNA-expression levels suggested that the CD25^{neg} cells upregulating CD25-expression upon transfer into lymphopenic mice were recruited from the Foxp3^{pos} CD25^{neg} fraction [57]. Using knock-in mice in which a red fluorescent protein is expressed under control of the Foxp3 promoter, it was shown that Foxp3 negative cells do not upregulate expression of this transcription factor upon transfer into lymphopenic mice [20]. In similar mice expressing green-fluorescent protein under control of the Foxp3 promoter, Leishmania major infection did not lead to generation of Foxp3^{pos} from Foxp3^{neg} T cells responding to the infection [19]. Combined, these data show that CD25-expression by Treg is subject to changes, but did not provide direct evidence that Foxp3-expression can be induced in the periphery from conventional T cells.

However, it has since been demonstrated that mature conventional Foxp3^{neg} T cells can, under certain experimental conditions, acquire Foxp3-expression and suppressive activity upon *in vitro* or *in vivo* stimulation. It was thus shown that $CD4^{pos}CD25^{neg}Foxp3^{neg}$ T cells acquire Foxp3-expression and suppressor activity when stimulated *via* their TCR in presence of TGF-β *in vitro* [20,58–60]. Among $CD4^{pos}$ T cells exclusively expressing a transgenic TCR specific for an influenza hemagglutinin (HA)-derived peptide presented by I-E^d (HA-TCR), no $CD25^{pos}$ cells can be detected. When transferred into mice expressing transgenic HA by cells of hematopoietic origin (B cells, macrophages, and DC), mature HA-TCR T cells underwent an expansion and retraction phase. Remaining cells were in majority $CD25^{neg}$ but some $CD25^{pos}$ cells had also developed. Both populations had *in vitro* suppressive activity [61]. Subcutaneous administration of minute doses of HA in mice expressing the HA-TCR led to strongly increased numbers of CD25-expressing cells with *in vitro* and *in vivo* suppressive activity [62]. By analyzing BrdU-incorporation, the authors excluded the possibility that expansion of pre-existing Treg was responsible. Moreover, the experiments were performed in thymectomized mice, and Treg induction had therefore occurred extrathymically. In the two cited reports, Foxp3-expression was not analyzed and it remained therefore uncertain if expression of this transcription factor could be induced in mature T cells. This issue was directly addressed in a report in which MHC class II-mediated presentation of HA was targeted to dendritic cells (DC) by injection of the HA-peptide conjugated to an antibody specific for the DC surface antigen DEC (αDEC) [59]. HA-TCR T cells, that were isolated from RAG-deficient mice and did not express Foxp3, were adoptively transferred into normal syngeneic mice. These mice were injected with αDEC-HA, and two weeks later substantial numbers of Foxp3-expressing cells were found among the HA-TCR^{pos} T cells. The authors showed that thus generated Treg suppressed *in vivo* proliferation and IL-2 production by HA-TCR T cells. In agreement with the above-cited data on *in vitro* induction of Foxp3-expression in presence of TGF-β, it was shown that this immunosuppressive cytokine was also involved in αDEC-HA mediated *in vivo* induction of Foxp3-expression.

Oral antigen-administration, known to induce tolerance, was shown to lead to an increase in the proportion of $CD4^+CD25^+$ T cells with suppressive activity [63]. In transgenic mice exclusively expressing a TCR specific for an ovalbumin-derived peptide presented by MHC class II molecules (and therefore lacking thymus-derived Treg), antigen feeding led to conversion of $CD25^-Foxp3^-$ conventional to $CD25^+Foxp3^+$ regulatory T cells [64–67]. The vitamin A metabolite retinoic acid and TGF-β, both produced by mucosal $CD103^+$ (but not $CD103^-$) DC [66], played important roles in this process [66–68].

Combined with other reports [20,69,70], these results showed that antigen-specific Foxp3-expressing Treg could be induced from mature conventional T cells under specific experimental conditions. TGF-β appears to play a critical role in induction of Foxp3-expression and suppressive activity. These observations may have substantial clinical relevance. It has indeed been shown that *in vivo* expression of TGF-β targeted to pancreatic β-cells protects from diabetes onset *via* accumulation of $Foxp3^{pos}$ Tregs [71].

Thymic *vs*. Peripheral Regulatory T Cell Differentiation

It appears therefore that CD4^{pos}Foxp3^{pos} Treg can differentiate in the thymus as well as in the periphery. Sequence analysis revealed considerable but incomplete overlap of TCR-repertoires of thymic and peripheral Treg in mice [47,72]. This observation may suggest that peripheral differentiation of Treg contributes to the Treg pool. However, despite the "anergic" (i.e. non-proliferative) phenotype of Treg *in vitro*, these cells proliferate upon adaptive transfer into lymphopenic mice [54,73]. More importantly, a sizeable proportion of Treg proliferates in the periphery of normal mice [74]. Using TCR-transgenic mice, it was shown that Treg proliferate *in vivo* if they encounter cognate ligand [59,73,75,76]. Therefore, antigen-driven peripheral expansion of Treg may substantially contribute to the difference between thymic and peripheral Treg repertoires, and the role of peripheral induction of Foxp3-expression remains, for the moment, unclear. Importantly, antigens capable of inducing peripheral Treg differentiation and/or proliferation are either (neo-)self-antigens [61,73,75,76] or antigens delivered in a non-inflammatory context [59,62]. In contrast, immunization in complete Freund's adjuvant, i.e. in a strong inflammatory context, led to only very limited proliferation of specific Treg [54]. Thus, peripheral expansion of Treg may be largely limited to self-antigen specific cells thereby contributing to tolerance to self while allowing useful immune-responses (i.e. those to infectious agents) to occur.

An issue that will need more investigation concerns the similarity between thymic *vs*. peripherally induced Treg. Phenotypically and functionally these two populations are very similar [62]. However, whereas Foxp3-expression is stable in thymus derived Treg, TGF-β induced Treg lose Foxp3-expression upon *in vitro* stimulation in absence of this cytokine [77] and upon adoptive transfer into mice [68]. The instability of Foxp3-expression in TGF-β induced Treg correlates with weak CpG-demethylation in the *foxp3*-locus [77]. Interestingly, retinoic acid appeared to enhance the stability of TGF-β-induced Foxp3-expression [68]. These data show that thymic and peripherally induced Treg are not identical and may have important implications for potential clinical applications of Treg.

The Regulatory T Cell-Repertoire is Enriched in Autospecific Cells

Given the major role of Treg, i.e. suppression of immune responses to self-antigens, it was thought that their peripheral repertoire is enriched in autospecific cells. This hypothesis was confirmed in an analysis of the Treg repertoire in unmanipulated mice. Using a limiting dilution analysis in which proliferation of T cells was assessed by flow-cytometry, it was found that the repertoire of (CD4^{pos}CD25^{pos}) Treg contains substantially more autospecific than allospecific cells [78]. In contrast (and as expected), the repertoire of conventional (i.e. CD4^{pos}CD25^{neg}) T cells contained more allospecific cells. The conclusion of these observations was that the Treg repertoire is relatively autospecific. This conclusion was later confirmed by a study in which the TCRα-repertoire of Treg from TCRβ transgenic mice was analyzed [79]. Sequence analysis showed that the Treg repertoire was very different from the repertoire of conventional T cells. To assess autospecificity of the Treg repertoire, Treg-derived TCRα-chain encoding sequences were retrovirally transduced into conventional, transgenic TCRβ-expressing T cells. When injected into lymphopenic mice, these cells strongly expanded and caused wasting disease. Moreover, in Foxp3-deficient mice, which develop a lethal autoimmune syndrome, pathogenic (self-reactive) T cells express a TCR-repertoire that has similarity with wildtype Treg [47]. These data therefore appeared to confirm the autospecific nature of the Treg repertoire. However, whereas in a recent study a small fraction of T-hybrids derived from Treg produced IL-2 upon stimulation with allogeneic APC, none were activated upon stimulation with syngeneic APC [80]. Combined, the distinct reports suggest that the Treg repertoire is autoreactive [78] and that TCR expressed by Treg have a higher afinity for self than TCR from conventional T cells [79], but that this affinity is not high enough to activate effector functions of T hybrids [80]. The autoreactivity of Treg may, therefore, in part be due to an intrinsic higher intrinsic sensitivity of regulatory than of conventional T cells, e.g. owing to differences in signal transduction. This intriguing possibility merits further investigation.

The Thymus Produces an Autospecific Regulatory T Cell-Repertoire

The data discussed thus far indicate that Treg differentiate in part in the thymus and in part in the periphery. As discussed above, sequence analysis revealed considerable overlap of TCR-repertoires of thymic and peripheral Treg [47,72]. Since the majority of thymic Treg differentiate (as opposed to re-circulate) in this organ, this result suggests that the thymus produces autospecific Treg.

The autospecific TCR-repertoire of thymic Treg is fundamentally different from that of conventional T cells, from which autospecific cells are pruned by the process of negative selection. It appears therefore that thymic negative selection of developing Treg must somehow be defective. On the other hand, positive selection of regulatory *vs.* conventional T cell-precursors may or may not be governed by different rules.

Thymic Regulatory T Cell Precursors are Less Sensitive to Negative Selection

Since the Treg repertoire is enriched in autospecific cells, thymic negative selection of Treg precursors must somehow be perturbed. The repertoire of conventional T cell-precursors is pruned of autospecific cells by at least two distinct mechanisms [5]. Recognition of autoantigens presented by antigen presenting cells of bone-marrow origin (i.e. thymic DC, [81]), induces apoptosis of autospecific

T cell precursors [1]. On the other hand, recognition of autoantigen presented by thymic epithelial cells mainly leads to induction of a reversible unresponsive state known as anergy [82–84]. However, it has been shown that thymic epithelial cells can, to a limited extent, also induce deletion of autospecific T cell precursors [85].

The sensitivity of developing Treg to deletion induced by thymic DC has been analyzed in different manners. Superantigen-specific conventional but also regulatory T cell-precursors are deleted when they encounter ligand at the surface of thymic antigen-presenting cells of bone-marrow origin [45,78,86,87]. Evidence was published that TCR-transgenic $CD4^{pos}CD25^{pos}$ thymocytes are deleted when they encounter cognate ligand expressed by antigen-presenting cells of hematopoietic origin [88]. However, the physiological relevance of data on thymic deletion of TCR-transgenic thymocytes remains unclear [89,90].

To study DC-mediated deletion of MHC/peptide specific thymocytes with a naturally diverse TCR-repertoire, bone marrow chimeras were generated in which cells of hematopoietic origin did not express MHC. In these chimeras, a two to three fold increased generation of mature T cells was observed, indicating that DC-mediated deletion eliminates half to two-thirds of developing conventional T cells [1]. In similar chimeras, also substantially increased numbers of Treg developed, indicating that Treg-precursors are sensitive to DC mediated deletion [78]. Deletion of MHC/peptide specific Treg precursors was also observed using mice expressing a single MHC/peptide ligand [91].

It remained important to assess negative selection of Treg-precursors in unmanipulated mice. Given the large diversity of peptide/MHC ligands and of TCR expressed in the thymus, this issue is intrinsically more complex to study. It was addressed in a study in which transfer of MHC class II molecules from thymic epithelium to developing thymocytes was assessed [92]. Mouse T cells and their precursors do not express endogenous MHC class II molecules [93] but they can acquire them, in an activation-dependent manner, from antigen-presenting cells [94,95]. By following the acquisition of MHC class II by developing conventional vs. regulatory T cell-precursors, it was shown that thymic DC induced negative selection of autospecific $CD25^{pos}$ precursors [92].

Therefore, despite some limited contradicting observations [91,96], it appears that autospecific Treg precursors are sensitive to deletion mediated by DC. However, it has to be taken into consideration that careful titrations of ligands expressed by DC have never been performed and it remains therefore possible that DC-mediated deletion of Treg-precursors is less efficient than that of conventional T cell-precursors.

Also thymic epithelial cells can, to some extent, induce deletion of autospecific thymocytes [85]. Since thymic DC and (medullary) epithelial cells express very different repertoires of self-antigens [97], it was important to understand the role of epithelial cells in negative selection of Treg-precursors. Jordan and colleagues were the first to address this question by analyzing I-E^d restricted TCR-transgenic mice also expressing cognate (influenza virus hemagglutinin, HA) ligand. In doubly transgenic mice, transgenic TCR-expressing $CD4^{pos}$ T cells were not deleted but had an anergic phenotype, expressed high levels of CD25, low levels of CD45RB, proliferated only in presence of IL-2, and suppressed proliferation of other T cells [98]. It was later shown that in the thymus only radioresistant (and therefore epithelial) cells expressed the HA-ligand [99]. Later studies on transgenic mice expressing MHC class II-restricted TCR specific for HA, chicken egg ovalbumin, or pigeon cytochrome C, as well as their ligands, confirmed that recognition of agonist ligand expressed by thymic epithelial cells does not induce deletion of $CD25^{high}Foxp3^{pos}$ Treg-precursors [61,100–103].

Deletion of autospecific conventional and regulatory T cell precursors by ligands expressed on epithelial cells was also studied by following the fate of superantigen-specific cells [45]. In bone-marrow chimeras in which superantigen was presented in the thymus exclusively by radioresistant (epithelial) cells, (limited) deletion of conventional (i.e. $CD25^{neg}$) T cell precursors was observed. In contrast, superantigen-specific Treg ($CD25^{high}Foxp3^{pos}$) precursors were not deleted.

In all these experimental systems, deletion of high frequency (TCR-transgenic or superantigen-specific) precursors was studied, and it was therefore important to evaluate negative selection of precursors with a normally diverse repertoire of TCR specific for peptide/MHC ligands. This issue was addressed in the above-cited study assessing transfer of MHC class II molecules from thymic epithelium to developing thymocytes [92]. By following the acquisition of MHC class II by developing conventional *vs.* regulatory T cell-precursors in bone marrow chimeras in which hematopoietic cells did not express MHC, it was shown that thymic epithelial cells did not negatively select autospecific $CD25^{pos}$ precursors.

It appears therefore that negative selection of conventional and regulatory T cell precursors is governed by distinct rules. Thymic cells of hematopoietic origin, i.e. mainly DC, can delete autospecific precursors for conventional and regulatory T cells. On the other hand, epithelial cells, which induce negative selection of conventional T cells, appear incapable of pruning the immature Treg repertoire of autospecific cells. Whereas thymic DC express ubiquitous and DC-specific antigens, thymic (medullary) epithelium ectopically expresses a large variety of tissue-specific antigens. Conventional T cell-precursors are therefore pruned of cells specific for all these self-antigens, leading to a self-tolerant mature T cell repertoire. On the other hand, Treg-precursors appear to be pruned of cells specific for ubiquitous, but not tissue-specific antigens. Thus, negative selection substantially contributes to the generation of mature Treg that inhibit tissue-specific autoimmune responses without totally paralyzing the immune system.

Positive Selection of Regulatory and Conventional T Cell Precursors Appears to be Governed by Distinct Rules

Thymic positive selection of Treg has mainly been analyzed using TCR-transgenic mice. Lafaille and colleagues generated mice expressing a transgenic TCR specific for a myelin basic protein-derived peptide presented by MHC class II [104].

When crossed to a RAG-deficient background, all transgenic mice spontaneously developed autoimmune encephalomyelitis. Very interestingly, RAG-sufficient transgenics were protected from this disease. These data indicated that lymphocytes expressing antigen-receptors other than the transgenic TCR controlled encephalitogenic $CD4^{pos}$ T cells. It was later shown that thymus-derived $CD4^{pos}CD25^{pos}$ Treg that developed in RAG-sufficient but not RAG-deficient mice protected from disease [74,105,106]. Also in mice exclusively expressing transgenic MHC class II-restricted TCR specific for HA and OVA very strongly reduced numbers of Treg were found [61,75,100]. Since it is not obvious how thymic negative selection could explain these data, it appeared that positive selection of regulatory *vs.* conventional T cells required distinct TCR-mediated signals.

An interesting observation was made in the TCR/ligand doubly transgenic systems discussed in the section on negative selection [61,98–103,107]. TCR-transgenic Treg precursors were not only spared from deletion by epithelial cells, but their proportions (among mature CD4SP thymocytes) were substantially increased in presence of agonist ligand (Table 3.1, columns 3, 4, 5). Since in TCR/ligand transgenic mice thymic deletion has dramatic effects on total cell-numbers, it was important to assess the absolute cell numbers of Treg in absence or presence of transgenic agonist ligand. It was observed that also the cell-numbers of Treg substantially increased in presence of transgenic cognate ligand (Table 3.1, columns 6, 7, 8). These data strongly suggested that positive selection of Treg-precursors could be mediated by agonist ligand. However, a caveat of this conclusion is that specificity of the observed effects on Treg development was, in most reports, not studied. Indeed, in the one study in which this issue was directly assessed, it was shown that the increase in Treg numbers was not limited to transgenic ligand specific precursors [101]. Whereas this important point will need further clarification, it suggests that homeostatic mechanisms may, at least in part, be responsible of the observed increased Treg development in mice expressing a transgenic cognate ligand.

One way to solve this caveat is to study Treg development in mice harboring thymic precursors with a naturally diverse TCR-repertoire. To this end, selection of superantigen-specific Treg precursors was studied in hematopoietic chimeras in which, in the thymus, exclusively epithelial cells expressed and presented superantigens [45]. Thus, negative selection of specific precursors by thymic DC was avoided and positive selection could be assessed. Substantially increased relative but also absolute numbers of superantigen-specific Tregs developed in chimeras (Table 3.1, lower part). Increased positive selection of Treg precursors was specific since it was limited to superantigen-specific cells.

Positive selection of Treg precursors was also studied in non-manipulated mice by assessment of MHC class II transfer to developing thymocytes [92]. Recently positively selected Treg precursors, i.e. $CD4^{pos}CD8^{low}CD25^{high}$ cells, acquired substantially (approximately four-fold) more MHC class II than their $CD25^{neg}$ counterparts. These results indicated that positive selection of Treg-precursors required a substantially higher avidity interaction with thymic stroma than selection of conventional T cell precursors.

3 Thymic and Peripheral Generation of CD4$^+$ Foxp3$^+$ Regulatory T Cells

Table 3.1 Summary of published data on involvement of agonist ligand in thymic development of Treg

TCR	Ligand	% Treg w/o Ag	% Treg w/Ag	Fold increase	Treg w/o Ag	Treg w/Ag	Fold increase	Treg definition	CD4SP w/o Ag	Ratio****	Ref.
		Among CD4SP thymocytes			Absolute numbers						
Part I: transgenic TCRαβ* specific for transgenic ligands											
6.5	pSV40-HA**	7	30	4.3×	0.5 × 10^6	1.2 × 10^6	2.4×	CD25	7 × 10^6	0.17	[98,99]
6.5	pIgκ-HA	0.1	55	550×				CD25			[61]
6.5	pIgκ-HA (in thymus-graft)	0.5	44	88×				CD25			[61]
6.5	pPgk1-HA	0.4	10	25×	4.2 × 10^4	6.4 × 10^5	15×	CD25			[102]
6.5	pGFAP-HA	0.01	19.7	19700×	0	18.8 × 10^3	>>×	Foxp3			[103]
6.5	pAIRE-HA	0.04	23.9	598×				Foxp3			[115]
DO11.10	pLd-OVA	1.2	4	3.3×	2.0 × 10^5	2.7 × 10^5	1.4×	CD25	1.7 × 10^7	0.02	[100]
DO11.10	pAIRE-OVA	0	0.84	>>				Foxp3			[115]
AND	pEα-TetR & TetO-pCMV-liMCC	0.6	24	40×	1.5 × 10^5	4 × 10^5	3×	CD25	3–6 × 10^7	0.01	[101]
AND	pH2Kb-PCC	0	6.7	>>				CD25			[107]
Part II: endogenous TCRβ*** specific for superantigens											
Vβ3	sAg on TEC	4.3	11.3	2.6×	9.3 × 10^3	32.5 × 10^3	3.5×	Foxp3	366 × 10^3	0.09	[45]
Vβ5	sAg on TEC	8.0	14.2	1.8×	18.4 × 10^3	47.1 × 10^3	2.6×	Foxp3	819 × 10^3	0.06	[45]
Vβ6	sAg on TEC	8.3	11.3	1.4×	17.2 × 10^3	40.3 × 10^3	2.3×	Foxp3	912 × 10^3	0.04	[45]

* Specificities of transgenic TCR used: 6.5, I-E^d/HA; DO11.10; I-A^d/OVA; AND, I-E^k/PCC
** "p" indicates promoter used
*** Specificities of endogenous TCR used: Vβ3, Mtv-1, 6, 13; Vβ5, Mtv-6, 8; Vβ6, Mtv-7
**** Ratio #Treg w/Ag/total #CD4SP w/o Ag

A next intriguing question concerns the precise and respective roles of cortical *vs.* medullary epithelial cells in positive selection of the Treg repertoire. On one hand, cortical but not medullary epithelial cells are capable of positive selection of T cell precursors [108–110]. On the other hand, medullary, but not cortical epithelial cells express tissue specific antigens, and would therefore ideally somehow play a positive role in shaping the Treg repertoire [97]. To study the role of cortical epithelium in positive selection of Treg precursors, we generated transgenic mice expressing a single MHC class II/peptide ligand exclusively on cortical thymic epithelial cells [43]. In the thymi of these mice, preferentially single MHC class II/peptide-specific Treg developed from precursors with a naturally diverse TCR-repertoire. These data therefore unambiguously demonstrate that cortical positive selection of Treg requires high avidity interactions and thus substantially contributes to the generation of an autoreactive Treg repertoire. The specific contribution of medullary epithelium to shaping the autoreactive Treg-repertoire was studied in two reports. The transcription factor AIRE (AutoImmune REgulator) is involved in ectopic expression of tissue-specific antigens by medullary epithelial cells [111]. Patients affected with autoimmune polyendocrinopathy-candidiasis-ectodermal dystrophy (APECED) carry mutations in the gene encoding this transcription factor [112,113], and AIRE deficient mice develop an autoimmune disorder [111,114]. AIRE-dependent medullary expression of tissue-specific antigen could therefore be involved in Treg development. This possibility was studied using athymic mice grafted with wildtype and AIRE-deficient thymi [111]. The authors observed that the chimeric mice developed a similar autoimmune disorder as mice grafted with an AIRE-deficient thymus only and concluded therefore that the disease was due to defective thymic negative selection and not to perturbed Treg-development. A role for medullary epithelial cells in Treg selection was more directly studied using TCR/ligand doubly transgenic mice [115] (see Table 3.1). In these mice, expression of the ligand was targeted to medullary epithelium using the AIRE promoter. Whereas in TCR singly transgenic mice no mature CD4SP Foxp3^{pos} thymocytes expressing transgenic TCR developed, in mice expressing ligand in the medulla substantial numbers of Treg were observed. Using an elegant chimeric approach, the authors also showed that in medullary islets not expressing MHC class II less Treg were found. These data therefore showed that medullary epithelium can somehow contribute to development of autoantigen specific Treg. Combined, the existing data indicates that cortical and medullary epithelial cells contribute to positive selection of high avidity Treg. Whereas cortical positive selection of high avidity Treg appears conceptually straightforward, it remains less clear how medullary cells favor Treg accumulation. Cortical MHC class II expression is required for positive selection of conventional and, until the contrary is proven, also of regulatory T cells [108]. High avidity Treg precursors could therefore be positively selected in the cortex and high avidity interactions in the medulla may favor their survival and/or proliferation. In the study on TCR/ligand doubly transgenic mice [115], endogenous TCR could have been involved in cortical selection and the transgenic TCR would be responsible for the medullary "positive selection". To assess this possibility, it will be critical to study Treg development in TCR/ligand transgenic mice that are RAG-deficient and therefore exclusively express the transgenic TCR. On the other hand, these conflicting data may also be explained in a model in which two distinct developmental pathways lead to differentiation of Treg, one requiring high avidity interactions in the cortex and the other one in the medulla.

Combined, the presently available data show that positive selection of Treg-precursors requires high avidity interactions with thymic cortical epithelial cells. Interaction with medullary epithelium appears to favor survival and/or proliferation of autospecific Treg. Thus, a two-step positive selection process appears to substantially contribute to the generation of an autospecific Treg repertoire. Subsequent negative selection prunes the developing Treg repertoire of cells specific for ubiquitous antigens but spares cells specific for tissue-specific antigens. The Treg repertoire is thus exquisitely shaped to prevent autoimmune attacks on tissues and to allow development of useful immune responses (Fig. 3.1).

Some details of this scheme will require further investigation. It will be important to assess the relation, if any, between cortical and medullary positive selection. The molecular mechanisms responsible for the observation that Treg precursors are normally sensitive to DC-mediated deletion but appear to resist negative selection induced by epithelial cells, will need to be clarified. Finally, in the thymus exclusively (medullary) epithelial cells may express tissue specific antigens, but DC are known to capture at least part of these antigens and induce deletion of specific precursors [4,85]. It will be important to assess to which extent this transfer prunes the developing Treg repertoire of cells specific for tissue-specific antigens.

Thymic Commitment of Precursors to the Regulatory T Cell Lineage

$CD4^{pos}$ thymus-derived regulatory T lymphocytes constitute a lineage clearly distinct from conventional T cells. Mature conventional T lymphocytes can differentiate into Foxp3-expressing cells under influence of antigen-specific stimulation and TGF-β (as described in section "Peripheral induction of regulatory T cells"). This observation may therefore suggest that regulatory activity reflects a differentiation state rather than a feature of a distinct T cell-lineage, much like Th1/Th2/Th17 differentiation. However, naturally occurring and peripherally induced Treg may share many characteristics, but they are not identical. It appears that expression of Foxp3 induced under influence of TGF-β *in vitro* is lost upon culture in absence of this cytokine [77]. In contrast, naturally occurring Treg do not lose Foxp3 expression upon proliferation *in vitro* [77] and retain their regulatory potential upon proliferation in lymphopenic mice [54]. This fundamental difference in phenotype correlates with and is most probably caused by differences in the methylation status of the *foxp3*-promoter [77]. Therefore, whereas Treg experimentally induced in the periphery appear to reflect a particular differentiation state, thymus-derived Treg clearly constitute a distinct T cell lineage.

An important question in Treg-biology is how and when during thymic differentiation of precursor cells the Treg lineage-choice is made. Thymic CD4/CD8 lineage

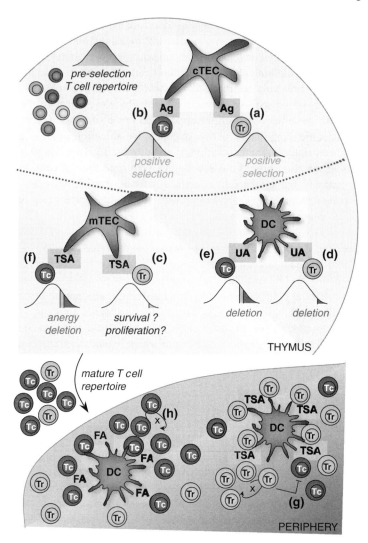

Fig. 3.1 Thymic Selection Positive selection of Treg-precursors (Tr) (**a**) requires higher avidity interactions with cortical thymic epithelial cells (cTEC) than positive selection of conventional T-cell precursors (Tc) (**b**). Medullary thymic epithelial cells (mTEC) somehow secondarily positively select Tr specific for autoantigens (**c**). Thus, a two-step positive selection process appears to substantially contribute to the generation of a Tr repertoire specific for autoantigens. Negative selection prunes the developing Tr and Tc repertoires of cells specific for ubiquitous antigens (UA) expressed by dendritic cells (DC) (**d, e**), neutralizes Tc specific for tissue-specific antigen (TSA) expressed by mTEC (**f**), but spares TSA-specific Tr (**c**). The Tr repertoire is thus exquisitely shaped to prevent autoimmune attacks on tissues (**g**) and to allow development of useful immune responses against foreign antigen (FA) (**h**). The bell-shaped curves indicate the assumed normal distribution of the avidity of pre-selection thymocytes' TCR for self MHC/peptide ligands *(See also Color Insert)*

commitment has been studied since over ten years using a variety of KO and transgenic mouse systems as well as extensive *in vitro* T cell differentiation approaches. A consensus model that has developed states that CD4-commitment is mediated by relatively strong and long-lasting TCR-derived signals whereas precursors that receive weaker and shorter-lasting signals choose the CD8-lineage [116–118]. The CD4 co-receptor binds more tyrosine kinase p56^{lck} than does CD8. Therefore, in MHC class II-restricted precursors (which would engage their CD4 coreceptor), strong TCR-derived signals are transmitted and these cells would commit to the CD4-lineage. On the other hand, in MHC class I-restricted precursors weaker signals are transmitted and these cells choose the CD8 lineage.

Could a variant of this model explain commitment to the Treg-lineage? This question was addressed using TCR/ligand doubly transgenic mice. Several laboratories showed that in these mice substantially more Treg developed than in TCR singly transgenic animals (Table 3.1). As discussed above, these data strongly suggested that positive selection of Treg requires high avidity interactions. However, this does not necessarily mean that a high avidity interaction redirects precursors to the Treg lineage (and is therefore involved in the lineage choice). If this were the case, virtually all precursors that in absence of ligand develop into conventional T cells, in presence of ligand would develop into Treg. However, the data on TCR/ligand doubly transgenic mice shows that in most cases only relatively limited numbers of Treg develop (Table 3.1, column 11). As earlier proposed [101], they therefore do not support an instructional model in which very strong TCR-derived signals redirect precursors to the Treg lineage. This conclusion was also supported by the observation that only a rather limited proportion of superantigen-specific precursors developed into Treg in mice in which superantigen was, in the thymus, exclusively expressed by thymic epithelial cells (Table 3.1, ref. [45]). Moreover, Hsieh and colleagues found that precursors expressing (retrovirally transduced) Treg-derived TCR develop efficiently, but incompletely, into the Treg lineage [79]. The data on TCR/ligand doubly transgenic mice, chimeras expressing superantigen ligands exclusively on thymic epithelial cells, and precursors expressing Treg-derived TCR suggest therefore that other factors than TCR-specificity must be involved in Treg commitment.

The question arises as to when during development these "other factors" would direct precursors to the Treg lineage. Foxp3 is a key protein involved in Treg function and forced expression of Foxp3 imparts a Treg phenotype on conventional T cells [119–121]. Foxp3 was therefore thought to constitute a "lineage-switch". Foxp3-expression in the thymus is a relatively late event and cells expressing this transcription factor appear in the thymus several days later than mature conventional T cells [42]. Very few Foxp3^{pos} cells are found in the cortex while they are abundant in the medulla [19]. Moreover, Foxp3-expressing cells have a relatively mature phenotype [42,43]. Induction of Foxp3-expression appeared therefore to take place in the medulla. However, in mice expressing MHC class II molecules exclusively in the cortex Treg develop normally [46], suggesting that Treg commitment takes place in the cortex. The delay between Treg-commitment (in the cortex) and Foxp3-expression (in the medulla) suggests that Foxp3 may not be the

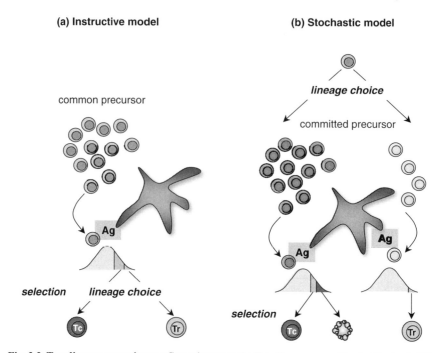

Fig. 3.2 Treg lineage commitment Commitment to the Treg lineage may either be "instructed" by high affinity interaction of common precursors with thymic stroma (**a**) or be caused by any other factor (**b**). In the instructive model, positive selection would therefore take place concomitant with lineage commitment. In the stochastic model for lineage commitment, positive selection (rather than lineage commitment) would depend on high avidity interaction of the thymocyte's TCR with self MHC class II/peptide ligand. It appears at present more likely that precursors choose the Treg lineage independently of the TCR's specificity *(See also Color Insert)*

earliest event in Treg commitment. This hypothesis is supported by the observation that in "knock-in" mice in which GFP was expressed instead of Foxp3, green cells with a Treg-like phenotype and gene-expression signature developed [122–124]. Treg-lineage commitment can therefore take place in absence of Foxp3. Whereas induction of Foxp3-expression requires TCR-mediated signaling [19], Treg commitment may therefore take place independently of TCR-signaling and may even precede positive selection. In this "stochastic" (i.e. non-instructional) model, high avidity TCR/ligand interactions would be involved in positive selection of predetermined Treg-precursors (Fig. 3.2). The stochastic model of Treg lineage commitment therefore separates the processes of selection and conventional *vs.* regulatory T cell-commitment.

What experimental evidence exists for a "stochastic" model for Treg-commitment? It was described that polymorphic genes control thymic Treg development [125,126]. The quantitative trait was due to thymocyte-intrinsic factors and modulated differentiation (rather than accumulation) of Treg. Whereas it cannot be excluded that differences in selection are responsible, the most attractive explanation is that they are due to differences in commitment. More definitive support for the stochastic model for Treg lineage choice came from studies on pTα-deficient mice [127]. In the thymi of these mice, precursors cannot express the pre-TCR required for efficient progression to the $CD4^{pos}CD8^{pos}$ "double positive" stage of development. However, some cells do progress through differentiation, and it was observed that substantially higher numbers of Treg developed in these mice. In contrast, in hemopoietic chimeras in which pTα-deficient and wildtype precursors developed side-by-side, and in which large numbers of (wt) double positive thymocytes developed, the numbers of mature wt and pTα-deficient Treg were similar. It was therefore concluded that double positive precursors somehow inhibit Treg development. It was shown that a particular immature thymocyte population (DN2-S) was enriched in Treg-precursors. Since these cells do not yet express the αβ TCR, Treg-lineage choice took place independently of TCR-specificity and prior to positive selection.

Molecules of the very conserved Notch family are involved in a large variety of binary cell-fate decisions during development and in the adult (for reviews see refs. [128,129]). In the thymus, Notch 1 is critically involved in B vs. T lymphocyte lineage-choice and may play a role in TCRαβ vs. TCRγδ commitment (for a review see ref. [130]). In the periphery, Notch appears to be involved in Th1/Th2 differentiation [131]. Notch may also be involved in Treg development in the thymus. In mice expressing a transgenic constitutively active form of Notch3, increased numbers of apparently functional Treg developed [132]. However, whereas constitutively active Notch1 was shown to modulate CD4 vs. CD8 lineage commitment [133], using inducible Notch1-deficient mice it was later demonstrated that Notch1 does not influence this binary cell-fate [134]. More work is therefore required before a definitive conclusion can be reached on involvement of Notch3 in Treg differentiation.

Are cytokines involved in thymic development of Treg? Most $CD4^{pos}Foxp3^{pos}$ Treg express high levels of CD25, the IL-2R α-chain [19,20], suggesting that IL-2 may be important in development and/or function of Treg. In mice deficient for IL-2 [96,107,135] or in which IL-2 was neutralized with antibody [136], reduced numbers of $CD4^{pos}CD25^{pos}Foxp3^{pos}$ Treg are found in the thymus. Also mice lacking expression of different components of the IL-2 receptor complex display different levels of Treg-deficiency [135,137]. However, Treg development was not totally blocked in these different experimental settings, suggesting that IL-2 is largely dispensable for Treg development. In hematopoietic chimeras in which TCR-transgenic IL-2Rα-deficient and IL-2Rα-sufficient cells developed side-by-side in a host expressing cognate ligand, no effect of IL-2Rα deficiency on Treg development was observed [102]. Combined, these data show that IL-2 is dispensable for thymic Treg development but that it somehow enhances it.

As discussed in section "Peripheral induction of regulatory T cells", Foxp3-expression can be induced in peripheral T cells in presence of TGF-β [20,58,59]. Moreover, mice deficient in TGF-β1 or expressing a dominant negative mutant of

TGF-βRII ("dnTβRII"), and therefore incapable of responding to this cytokine, develop a lethal auto-immune syndrome [138,139]. TGF-β may therefore be involved in development and/or function of Treg. In many experimental systems it has indeed been shown that suppression of immune-responses by Treg requires that pathogenic T cells are responsive to TGF-β (for review see ref. [140]). Moreover, TGF-β1 is required for peripheral maintenance of Treg [71,141]. However, in contrast to its crucial role in peripheral Treg maintenance and function, TGF-β1 is not required for development of normal numbers of Treg in the thymus [141]. Thymic Treg incapable of responding to TGF-β also efficiently prevent experimental inflammatory bowel disease in mice suggesting that Treg normally develop [142]. We have generated bone marrow chimeras in which wildtype and dnTβRII transgenic Treg develop simultaneously. Even in this competitive setting normal percentages and numbers of Treg developed from dnTβRII transgenic precursors (PR, unpublished data). Combined, these data clearly indicated that, in contrast to peripheral Treg-induction, TGF-β is not involved in thymic Treg differentiation.

Thymic stromal lymphopoietin (TSLP), a cytokine produced by Hassall's corpuscles in the thymus, is a strong activator of DC [143]. It was shown that thymic DC upregulated CD80 and CD86 upon stimulation with TSLP. Upon culture with TSLP-activated DC, thymic (but not peripheral) $CD4^{pos}CD25^{neg}$ cells upregulated CD25 and Foxp3 expression, and acquired suppressive potential. Moreover, medullary $CD4^{pos}CD25^{pos}CTLA-4^{pos}$ Treg are preferentially localized near DC and Hassall's corpuscles [144]. The authors therefore postulated that uncommitted high affinity $CD4^{pos}$ T cell precursors would differentiate into Treg upon interaction with TSLP-activated DC in the medulla. However, thymic development of Treg does not require MHC expression by DC and Treg lineage commitment can take place in the thymic cortex [45,46,78]. Moreover, whereas induction of Treg by TSLP activated DC required IL-2, thymic Treg differentiate in absence of IL-2 signaling [102,135,137]. Treg lineage commitment therefore rather unlikely depends on TSLP. It appears more likely that this cytokine indirectly causes amplification of already committed precursors. It would be important to analyze Treg development in TSLP-receptor deficient mice [145,146].

Co-stimulation is required for negative selection of conventional T cell-precursors [147–150]. Rather unexpectedly therefore, co-stimulation *via* CD28 engagement has a substantial enhancing effect on thymic Treg development [107,151,152]. Co-stimulation is required for IL-2 production by conventional T cells and it has been hypothesized that in the thymus it is required for production of the IL-2 involved in Treg development. Using elegant mixed bone marrow chimeras, Tai and colleagues showed that whereas CD28 is involved in production of the IL-2 that enhances Treg development, it also critically contributes in another, still unresolved, manner to differentiation of Foxp3-expressing and functional Treg in the thymus [107]. Intriguingly, both IL-2 production and Treg development required the $p56^{lck}$ (but not PI3-K or Itk) binding site on the cytoplasmic tail of CD28. The requirement for CD28 could not be overcome with stronger TCR-engagement in an *in vitro* Treg differentiation system and therefore appears to deliver a qualitatively distinct signal. At present it is difficult to distinguish between a role for CD28 in

Treg lineage choice and induction of Treg phenotype (e.g. Foxp3-expression) in pre-committed precursors. However, the observation that cortical epithelial cells express low but significant levels of CD80 and CD86 suggests that CD28-signaling may be involved in the commitment process itself [153]. Importantly, whereas CD28-deficient mice develop substantially lower numbers of Treg, some still develop. It will therefore be important to assess involvement of other cell-surface molecules known to be involved in thymic negative selection (e.g. CD5 and CD43 [147]), in differentiation of Treg.

Combined, these data do not allow for a definitive choice between the stochastic and instructional models for Treg lineage commitment. It appears rather unlikely that the mere TCR-specificity of precursors instructs lineage choice. However, none of the studied soluble or cell-surface molecules are strictly required for Treg development, they rather appear to enhance it. Ultimate comprehension of Treg lineage choice may depend on identification of the genes involved in the quantitative differences in Treg development observed in mice.

Conclusion

In the last several years it has become increasingly clear that thymus-derived Treg have an autospecific TCR-repertoire. This conclusion appears coherent with the main physiological function of Treg, i.e. prevention of autoimmune-disease. Autospecific Treg appear to be positively selected in the thymus upon high avidity TCR/ligand interactions in the cortex. Medullary negative selection by DC prunes the developing Treg repertoire of cells specific for antigens presented by these cells, i.e. mainly ubiquitous antigens. In contrast, presentation of self-antigen by medullary epithelial cells, which ectopically express a considerable variety of tissue-specific antigens, does not lead to deletion of specific Treg and even appears to somehow secondarily positively select these cells. Combined, these mechanisms shape a Treg repertoire specific for tissue-specific antigens exquisitely adapted to inhibit autoimmune disease.

Several issues concerning thymic Treg development remain incompletely understood. It will be important to better understand the precise roles of cortical and medullary epithelial cells in selection of Treg. Given the exquisite capacity of cortical epithelium to induce positive selection and of medullary epithelial cells to ectopically express tissue-specific antigen, this issues merits further work. Differences in positive and negative selection of conventional *vs.* regulatory T cells imply that these lineages diverge at or prior to positive selection in the thymus. Relatively little is known about Treg lineage-choice but an increasing body of data suggests that it occurs independently of TCR-specificity. However, not all available experimental data fit this conclusion, and more work will have to be performed before a definitive conclusion can be reached.

Acknowledgments This work was supported by institutional grants from the Inserm and by the EuroThymaide consortium (contract # LSHB-CT-2003-503410).

References

1. van Meerwijk, J. P. M., S. Marguerat, R. K. Lees, R. N. Germain, B. J. Fowlkes, and H. R. MacDonald. 1997. Quantitative impact of thymic clonal deletion on the T cell repertoire. *J. Exp. Med.* 185:377–383.
2. Wardemann, H., S. Yurasov, A. Schaefer, J. W. Young, E. Meffre, and M. C. Nussenzweig. 2003. Predominant autoantibody production by early human B cell precursors. *Science* 301:1374–1377.
3. Hardy, R. R., and K. K. Hayakawa. 2001. B cell development pathways. *Annu. Rev. Immunol.* 19:595–621.
4. Gallegos, A. M., and M. J. Bevan. 2006. Central tolerance: good but imperfect. *Immunol. Rev.* 209:290–296.
5. Hogquist, K. A., T. A. Baldwin, and S. C. Jameson. 2005. Central tolerance: learning self-control in the thymus. *Nat. Rev. Immunol.* 5:772–782.
6. Bouneaud, C., P. Kourilsky, and P. Bousso. 2000. Impact of negative selection on the T cell repertoire reactive to a self-peptide: A large fraction of T cell clones escapes clonal deletion. *Immunity* 13:829–840.
7. Le Douarin, N., C. Corbel, A. Bandeira, V. Thomas-Vaslin, Y. Modigliani, A. Coutinho, and J. Salaun. 1996. Evidence for a thymus-dependent form of tolerance that is not based on elimination or anergy of reactive T cells. *Immunol. Rev.* 149:35–53.
8. Ménager-Marcq, I., P. Pomié, P. Romagnoli, and J. P. M. van Meerwijk. 2006. CD8+CD28- regulatory T-lymphocytes prevent experimental inflammatory bowel disease in mice. *Gastroenterology* 131:1887–1885.
9. Xystrakis, E., A. S. Dejean, I. Bernard, P. Druet, R. Liblau, D. Gonzalez-Dunia, and A. Saoudi. 2004. Identification of a novel natural regulatory CD8 T-cell subset and analysis of its mechanism of regulation. *Blood* 104:3294–3301.
10. Endharti, A. T., I. M. S. Rifa, Z. Shi, Y. Fukuoka, Y. Nakahara, Y. Kawamoto, K. Takeda, K. Isobe, and H. Suzuki. 2005. Cutting edge: CD8+CD122+ regulatory T cells produce IL-10 to suppress IFN-gamma production and proliferation of CD8+ T cells. *J. Immunol.* 175:7093–7097.
11. Bendelac, A., P. B. Savage, and L. Teyton. 2006. The Biology of NKT Cells. *Annu. Rev. Immunol.* 25:297–336.
12. Vlad, G., R. Cortesini, and N. Suciu-Foca. 2005. License to heal: bidirectional interaction of antigen-specific regulatory T cells and tolerogenic APC. *J. Immunol.* 174:5907–5914.
13. Kim, J. M., and A. Rudensky. 2006. The role of the transcription factor Foxp3 in the development of regulatory T cells. *Immunol. Rev.* 212:86–98.
14. Sakaguchi, S., M. Ono, R. Setoguchi, H. Yagi, S. Hori, Z. Fehervari, J. Shimizu, T. Takahashi, and T. Nomura. 2006. Foxp3+ CD25+ CD4+ natural regulatory T cells in dominant self-tolerance and autoimmune disease. *Immunol. Rev.* 212:8–27.
15. Shevach, E. M., R. A. DiPaolo, J. Andersson, D. M. Zhao, G. L. Stephens, and A. M. Thornton. 2006. The lifestyle of naturally occurring CD4+ CD25+ Foxp3+ regulatory T cells. *Immunol. Rev.* 212:60–73.
16. Thomson, C. W., B. P. Lee, and L. Zhang. 2006. Double-negative regulatory T cells: non-conventional regulators. *Immunol. Res.* 35:163–178.
17. Wildin, R. S., F. Ramsdell, J. Peake, F. Faravelli, J. L. Casanova, N. Buist, E. Levy-Lahad, M. Mazzella, O. Goulet, L. Perroni, F. D. Bricarelli, G. Byrne, M. McEuen, S. Proll, M. Appleby, and M. E. Brunkow. 2001. X-linked neonatal diabetes mellitus, enteropathy and endocrinopathy syndrome is the human equivalent of mouse scurfy. *Nat. Genet.* 27:18–20.
18. Brunkow, M. E., E. W. Jeffery, K. A. Hjerrild, B. Paeper, L. B. Clark, S. A. Yasayko, J. E. Wilkinson, D. Galas, S. F. Ziegler, and F. Ramsdell. 2001. Disruption of a new forkhead/winged-helix protein, scurfin, results in the fatal lymphoproliferative disorder of the scurfy mouse. *Nat. Genet.* 27:68–73.

19. Fontenot, J. D., J. P. Rasmussen, L. M. Williams, J. L. Dooley, A. G. Farr, and A. Y. Rudensky. 2005. Regulatory T cell lineage specification by the forkhead transcription factor foxp3. *Immunity* 22:329–341.
20. Wan, Y. Y., and R. A. Flavell. 2005. Identifying Foxp3-expressing suppressor T cells with a bicistronic reporter. *Proc. Natl. Acad. Sci. U.S.A.* 102:5126–5131.
21. Takahashi, T., T. Tagami, S. Yamazaki, T. Uede, J. Shimizu, N. Sakaguchi, T. W. Mak, and S. Sakaguchi. 2000. Immunologic self-tolerance maintained by CD25(+)CD4(+) regulatory T cells constitutively expressing cytotoxic T lymphocyte-associated antigen 4. *J. Exp. Med.* 192:303–310.
22. McHugh, R. S., M. J. Whitters, C. A. Piccirillo, D. A. Young, E. M. Shevach, M. Collins, and M. C. Byrne. 2002. CD4(+)CD25(+) immunoregulatory T cells: gene expression analysis reveals a functional role for the glucocorticoid-induced TNF receptor. *Immunity* 16: 311–323.
23. Liu, W., A. L. Putnam, Z. Xu-Yu, G. L. Szot, M. R. Lee, S. Zhu, P.A. Gottlieb, P. Kapranov, T. R. Gingeras, B. F. de St Groth, C. Clayberger, D. M. Soper, S. F. Ziegler, and J. A. Bluestone. 2006. CD127 expression inversely correlates with FoxP3 and suppressive function of human CD4+ T reg cells. *J. Exp. Med.* 203:17-1-1711.
24. Ohki, H., C. Martin, C. Corbel, M. Coltey, and N. M. Le Douarin. 1987. Tolerance induced by thymic epithelial grafts in birds. *Science* 237:1032–1035.
25. Coutinho, A., J. Salaun, C. Corbel, A. Bandeira, and N. Le Douarin. 1993. The role of thymic epithelium in the establishment of transplantation tolerance. *Immunol. Rev.* 133:225–240.
26. Modigliani, Y., V. Thomas-Vaslin, A. Bandeira, M. Coltey, N. M. Le Douarin, A. Coutinho, and J. Salaun. 1995. Lymphocytes selected in allogeneic thymic epithelium mediate dominant tolerance toward tissue grafts of the thymic epithelium haplotype. *Proc. Natl. Acad. Sci. U.S.A.* 92:7555–7559.
27. Salaun, J., A. Bandeira, I. Khazaal, F. Calman, M. Coltey, A. Coutinho, and N. M. Le Douarin. 1990. Thymic epithelium tolerizes for histocompatibility antigens. *Science* 247:1471–1474.
28. Yunis, E. J., R. Hong, M. A. Grewe, C. Martinez, E. Cornelius, and R. A. Good. 1967. Postthymectomy wasting associated with autoimmune phenomena. I. Antiglobulin-positive anemia in A and C57BL-6 Ks mice. *J. Exp. Med.* 125:947–966.
29. Kojima, A., Y. Tanaka-Kojima, T. Sakakura, and Y. Nishizuka. 1976. Spontaneous development of autoimmune thyroiditis in neonatally thymectomized mice. *Lab Invest* 34:550–557.
30. Kojima, A., O. Taguchi, and Y. Nishizuka. 1980. Experimental production of possible autoimmune castritis followed by macrocytic anemia in athymic nude mice. *Lab Invest* 42:387–395.
31. Taguchi, O., Y. Nishizuka, T. Sakakura, and A. Kojima. 1980. Autoimmune oophoritis in thymectomized mice: detection of circulating antibodies against oocytes. *Clin. Exp. Immunol.* 40:540–553.
32. Taguchi, O., and Y. Nishizuka. 1981. Experimental autoimmune orchitis after neonatal thymectomy in the mouse. *Clin. Exp. Immunol.* 46:425–434.
33. Kojima, A., and R. T. Prehn. 1981. Genetic susceptibility to post-thymectomy autoimmune diseases in mice. *Immunogenetics* 14:15–27.
34. Tung, K. S., S. Smith, P. Matzner, K. Kasai, J. Oliver, F. Feuchter, and R. E. Anderson. 1987. Murine autoimmune oophoritis, epididymoorchitis, and gastritis induced by day 3 thymectomy. Autoantibodies. *Am. J. Pathol.* 126:303–314.
35. Tung, K. S., S. Smith, C. Teuscher, C. Cook, and R. E. Anderson. 1987. Murine autoimmune oophoritis, epididymoorchitis, and gastritis induced by day 3 thymectomy. Immunopathology. *Am. J. Pathol.* 126:293–302.
36. Sakaguchi, S., N. Sakaguchi, M. Asano, M. Itoh, and M. Toda. 1995. Immunologic self-tolerance maintained by activated T cells expressing IL-2 receptor alpha-chains (CD25). Breakdown of a single mechanism of self-tolerance causes various autoimmune diseases. *J. Immunol.* 155:1151–1164.

37. Asano, M., M. Toda, N. Sakaguchi, and S. Sakaguchi. 1996. Autoimmune disease as a consequence of developmental abnormality of a T cell subpopulation. *J. Exp. Med.* 184:387–396.
38. Dujardin, H. C., O. Burlen-Defranoux, L. Boucontet, P. Vieira, A. Cumano, and A. Bandeira. 2004. Regulatory potential and control of Foxp3 expression in newborn CD4+ T cells. *Proc. Natl. Acad. Sci. U.S.A.* 101:14473–14478.
39. Itoh, M., T. Takahashi, N. Sakaguchi, Y. Kuniyasu, J. Shimizu, F. Otsuka, and S. Sakaguchi. 1999. Thymus and autoimmunity: production of CD25+CD4+ naturally anergic and suppressive T cells as a key function of the thymus in maintaining immunologic self-tolerance. *J. Immunol.* 162:5317–5326.
40. Saoudi, A., B. Seddon, V. Heath, D. Fowell, and D. Mason. 1996. The physiological role of regulatory T cells in the prevention of autoimmunity: the function of the thymus in the generation of the regulatory T cell subset. *Immunol. Rev.* 149:195–216.
41. Fowell, D., and D. Mason. 1993. Evidence that the T cell repertoire of normal rats contains cells with the potential to cause diabetes. Characterization of the CD4+ T cell subset that inhibits this autoimmune potential. *J. Exp. Med.* 177:627–636.
42. Fontenot, J. D., J. L. Dooley, A. G. Farr, and A. Y. Rudensky. 2005. Developmental regulation of Foxp3 expression during ontogeny. *J. Exp. Med.* 202:901–906.
43. Ribot, J., G. Enault, S. Pilipenko, A. Huchenq, M. Calise, D. Hudrisier, P. Romagnoli, and J. P. M. van Meerwijk. 2007. Shaping of the autoreactive regulatory T cell repertoire by thymic cortical positive selection. *J. Immunol.* 179:6741–6748.
44. Bosco, N., F. Agenes, A. G. Rolink, and R. Ceredig. 2006. Peripheral T cell lymphopenia and concomitant enrichment in naturally arising regulatory T cells: the case of the pre-Talpha gene-deleted mouse. *J. Immunol.* 177:5014–5023.
45. Ribot, J., P. Romagnoli, and J. P. M. van Meerwijk. 2006. Agonist Ligands Expressed by Thymic Epithelium Enhance Positive Selection of Regulatory T Lymphocytes from Precursors with a Normally Diverse TCR Repertoire. *J. Immunol.* 177:1101–1107.
46. Bensinger, S. J., A. Bandeira, M. S. Jordan, A. J. Caton, and T. M. Laufer. 2001. Major Histocompatibility Complex Class II-positive Cortical Epithelium Mediates the Selection of CD4+25+ Immunoregulatory T Cells. *J. Exp. Med.* 194:427–438.
47. Hsieh, C. S., Y. Zheng, Y. Liang, J. D. Fontenot, and A. Y. Rudensky. 2006. An intersection between the self-reactive regulatory and nonregulatory T cell receptor repertoires. *Nat. Immunol.* 7:401–410.
48. Kim, J. M., J. P. Rasmussen, and A. Y. Rudensky. 2007. Regulatory T cells prevent catastrophic autoimmunity throughout the lifespan of mice. *Nat. Immunol.* 8:191–197.
49. Jiang, Q., H. Su, G. Knudsen, W. Helms, and L. Su. 2006. Delayed functional maturation of natural regulatory T cells in the medulla of postnatal thymus: role of TSLP. *BMC Immunol* 7:6.
50. Carter, J. D., G. M. Calabrese, M. Naganuma, and U. Lorenz. 2005. Deficiency of the Src homology region 2 domain-containing phosphatase 1 (SHP-1) causes enrichment of CD4+CD25+ regulatory T cells. *J. Immunol.* 174:6627–6638.
51. Faria, A. M., and H. L. Weiner. 2005. Oral tolerance. *Immunol. Rev.* 206:232–259.
52. Roncarolo, M. G., S. Gregori, M. Battaglia, R. Bacchetta, K. Fleischhauer, and M. K. Levings. 2006. Interleukin-10-secreting type 1 regulatory T cells in rodents and humans. *Immunol. Rev.* 212:28–50.
53. Annacker, O., R. Pimenta-Araujo, O. Burlen-Defranoux, T. C. Barbosa, A. Cumano, and A. Bandeira. 2001. CD25+ CD4+ T cells regulate the expansion of peripheral CD4 T cells through the production of IL-10. *J. Immunol.* 166:3008–3018.
54. Gavin, M. A., S. R. Clarke, E. Negrou, A. Gallegos, and A. Rudensky. 2002. Homeostasis and anergy of CD4+CD25+ suppressor T cells in vivo. *Nat. Immunol.* 3:33–41.
55. Graca, L., S. Thompson, C. Y. Lin, E. Adams, S. P. Cobbold, and H. Waldmann. 2002. Both CD4(+)CD25(+) and CD4(+)CD25(–) regulatory cells mediate dominant transplantation tolerance. *J. Immunol.* 168:5558–5565.
56. Nishimura, E., T. Sakihama, R. Setoguchi, K. Tanaka, and S. Sakaguchi. 2004. Induction of antigen-specific immunologic tolerance by in vivo and in vitro antigen-specific expansion of naturally arising Foxp3+CD25+CD4+ regulatory T cells. *Int. Immunol.* 16:1189–1201.

57. Zelenay, S., T. Lopes-Carvalho, I. Caramalho, M. F. Moraes-Fontes, M. Rebelo, and J. Demengeot. 2005. Foxp3+ CD25- CD4 T cells constitute a reservoir of committed regulatory cells that regain CD25 expression upon homeostatic expansion. *Proc. Natl. Acad. Sci. U.S.A.* 102:4091–4096.
58. Chen, W., W. Jin, N. Hardegen, K.-j. Lei, L. Li, N. Marinos, G. McGrady, and S. M. Wahl. 2003. Conversion of Peripheral CD4+CD25- Naive T Cells to CD4+CD25+ Regulatory T Cells by TGF-{beta} Induction of Transcription Factor Foxp3. *J. Exp. Med.* 198: 1875–1886.
59. Kretschmer, K., I. Apostolou, D. Hawiger, K. Khazaie, M. C. Nussenzweig, and H. von Boehmer. 2005. Inducing and expanding regulatory T cell populations by foreign antigen. *Nat. Immunol.* 6:1219–1227.
60. Fantini, M. C., C. Becker, G. Monteleone, F. Pallone, P. R. Galle, and M. F. Neurath. 2004. Cutting edge: TGF-beta induces a regulatory phenotype in CD4+CD25- T cells through Foxp3 induction and down-regulation of Smad7. *J. Immunol.* 172:5149–5153.
61. Apostolou, I., A. Sarukhan, L. Klein, and H. von Boehmer. 2002. Origin of regulatory T cells with known specificity for antigen. *Nat. Immunol.* 3:756–763.
62. Apostolou, I., and H. von Boehmer. 2004. In vivo instruction of suppressor commitment in naive T cells. *J. Exp. Med.* 199:1401–1408.
63. Zhang, X. 2001. Activation of CD25(+)CD4(+) regulatory T cells by oral antigen administration. *J. Immunol.* 167:4245–4253.
64. Thorstenson, K. M., and A. Khoruts. 2001. Generation of anergic and potentially immunoregulatory CD25+CD4 T cells in vivo after induction of peripheral tolerance with intravenous or oral antigen. *J. Immunol.* 167:188–195.
65. Sun, J. B., S. Raghavan, A. Sjoling, S. Lundin, and J. Holmgren. 2006. Oral tolerance induction with antigen conjugated to cholera toxin B subunit generates both Foxp3+CD25+ and Foxp3-CD25- CD4+ regulatory T cells. *J. Immunol.* 177:7634–7644.
66. Coombes, J. L., K. R. Siddiqui, C. V. Arancibia-Carcamo, J. Hall, C. M. Sun, Y. Belkaid, and F. Powrie. 2007. A functionally specialized population of mucosal CD103+ DCs induces Foxp3+ regulatory T cells via a TGF-{beta} and retinoic acid dependent mechanism. *J. Exp. Med.* 204:1757–1764.
67. Sun, C. M., J. A. Hall, R. B. Blank, N. Bouladoux, M. Oukka, J. R. Mora, and Y. Belkaid. 2007. Small intestine lamina propria dendritic cells promote de novo generation of Foxp3 T reg cells via retinoic acid. *J. Exp. Med.* 204:1775–1785.
68. Benson, M. J., K. Pino-Lagos, M. Rosemblatt, and R. J. Noelle. 2007. All-trans retinoic acid mediates enhanced T reg cell growth, differentiation, and gut homing in the face of high levels of co-stimulation. *J. Exp. Med.* 204:1765–1774.
69. Fu, S., N. Zhang, A. C. Yopp, D. Chen, M. Mao, D. Chen, H. Zhang, Y. Ding, and J. S. Bromberg. 2004. TGF-beta induces Foxp3 + T-regulatory cells from CD4 + CD25 - precursors. *Am J Transplant* 4:1614–1627.
70. Rao, P. E., A. L. Petrone, and P. D. Ponath. 2005. Differentiation and expansion of T cells with regulatory function from human peripheral lymphocytes by stimulation in the presence of TGF-{beta}. *J. Immunol.* 174:1446–1455.
71. Peng, Y., Y. Laouar, M. O. Li, E. A. Green, and R. A. Flavell. 2004. TGF-beta regulates in vivo expansion of Foxp3-expressing CD4+CD25+ regulatory T cells responsible for protection against diabetes. *Proc. Natl. Acad. Sci. U.S.A.* 101:4572–4577.
72. Pacholczyk, R., H. Ignatowicz, P. Kraj, and L. Ignatowicz. 2006. Origin and T cell receptor diversity of Foxp3+CD4+CD25+ T cells. *Immunity* 25:249–259.
73. Cozzo, C., J. I. Larkin, and A. J. Caton. 2003. Cutting edge: Self-peptides drive the peripheral expansion of CD4+CD25+ regulatory T cells. *J. Immunol.* 171:5678–5682.
74. Hori, S., M. Haury, J. J. Lafaille, J. Demengeot, and A. Coutinho. 2002. Peripheral expansion of thymus-derived regulatory cells in anti-myelin basic protein T cell receptor transgenic mice. *Eur. J. Immunol.* 32:3729–3735.
75. Walker, L. S. K., A. Chodos, M. Eggena, H. Dooms, and A. K. Abbas. 2003. Antigen-dependent Proliferation of CD4+ CD25+ Regulatory T Cells In Vivo. *J. Exp. Med.* 198: 249–258.

76. Fisson, S., G. Darrasse-Jeze, E. Litvinova, F. Septier, D. Klatzmann, R. Liblau, and B. L. Salomon. 2003. Continuous Activation of Autoreactive CD4+ CD25+ Regulatory T Cells in the Steady State. *J. Exp. Med.* 198:737–746.
77. Floess, S., J. Freyer, C. Siewert, U. Baron, S. Olek, J. Polansky, K. Schlawe, H.-D. Chang, T. Bopp, E. Schmitt, S. Klein-Hessling, E. Serfling, A. Hamann, and J. Huehn. 2007. Epigenetic control of the foxp3 locus in regulatory T cells. *PLoS Biol.* 5:e38.
78. Romagnoli, P., D. Hudrisier, and J. P. M. van Meerwijk. 2002. Preferential recognition of self-antigens despite normal thymic deletion of CD4+CD25+ regulatory T cells. *J. Immunol.* 168:1644–1648.
79. Hsieh, C. S., Y. Liang, A. J. Tyznik, S. G. Self, D. Liggitt, and A. Y. Rudensky. 2004. Recognition of the peripheral self by naturally arising CD25+ CD4+ T cell receptors. *Immunity* 21:267–277.
80. Pacholczyk, R., J. Kern, N. Singh, M. Iwashima, P. Kraj, and L. Ignatowicz. 2007. Nonself-antigens are the cognate specificities of Foxp3+ regulatory T cells. *Immunity* 27:493–504.
81. Matzinger, P., and S. Guerder. 1989. Does T-cell tolerance require a dedicated antigen-presenting cell? *Nature* 338:74–76.
82. Ramsdell, F., T. Lantz, and B. J. Fowlkes. 1989. A nondeletional mechanism of thymic self tolerance. *Science* 246:1038–1041.
83. Hudrisier, D., S. Feau, V. Bonnet, P. Romagnoli, and J. P. M. van Meerwijk. 2003. In vivo unresponsiveness of T lymphocyte induced by thymic medullary epithelium requires antigen presentation by radioresistant cells. *Immunol.* 108:24–31.
84. Ramsdell, F., and B. J. Fowlkes. 1992. Maintenance of in vivo tolerance by persistence of antigen. *Science* 257:1130–1134.
85. Gallegos, A. M., and M. J. Bevan. 2004. Central Tolerance to Tissue-specific Antigens Mediated by Direct and Indirect Antigen Presentation. *J. Exp. Med.* 200:1039–1049.
86. MacDonald, H. R., R. Schneider, R. K. Lees, R. C. Howe, H. Acha-Orbea, H. Festenstein, R. M. Zinkernagel, and H. Hengartner. 1988. T-cell receptor V beta use predicts reactivity and tolerance to Mlsa-encoded antigens. *Nature* 332:40–45.
87. Kappler, J. W., N. Roehm, and P. Marrack. 1987. T cell tolerance by clonal elimination in the thymus. *Cell* 49:273–280.
88. Shih, F. F., L. Mandik-Nayak, B. T. Wipke, and P. M. Allen. 2004. Massive Thymic Deletion Results in Systemic Autoimmunity through Elimination of CD4+ CD25+ T Regulatory Cells. *J. Exp. Med.* 199:323–335.
89. Baldwin, T. A., M. M. Sandau, S. C. Jameson, and K. A. Hogquist. 2005. The timing of TCR alpha expression critically influences T cell development and selection. *J. Exp. Med.* 202:111–121.
90. Takahama, Y., E. W. Shores, and A. Singer. 1992. Negative selection of precursor thymocytes before their differentiation into CD4+CD8+ cells. *Science* 258:653–656.
91. Pacholczyk, R., P. Kraj, and L. Ignatowicz. 2002. Peptide specificity of thymic selection of CD4+CD25+ T cells. *J. Immunol.* 168:613–620.
92. Romagnoli, P., D. Hudrisier, and J. P. M. van Meerwijk. 2005. Molecular signature of recent thymic selection events on effector and regulatory CD4+ T lymphocytes. *J. Immunol.* 175:5751–5758.
93. Benoist, C., and D. Mathis. 1990. Regulation of major histocompatibility complex class-II genes: X, Y and other letters of the alphabet. *Annu. Rev. Immunol.* 8:681–715.
94. Sharrow, S. O., B. J. Mathieson, and A. Singer. 1981. Cell surface appearance of unexpected host MHC determinants on thymocytes from radiation bone marrow chimeras. *J. Immunol.* 126:1327–1335.
95. Hudrisier, D., and P. Bongrand. 2002. Intercellular transfer of antigen-presenting cell determinants onto T cells: molecular mechanisms and biological significance. *Faseb J.* 16:477–486.
96. Papiernik, M., M. L. de Moraes, C. Pontoux, F. Vasseur, and C. Penit. 1998. Regulatory CD4 T cells: expression of IL-2R alpha chain, resistance to clonal deletion and IL-2 dependency. *Int. Immunol.* 10:371–378.

97. Kyewski, B., and J. Derbinski. 2004. Self-representation in the thymus: an extended view. *Nat. Rev. Immunol.* 4:688–698.
98. Jordan, M. S., M. P. Riley, H. von Boehmer, and A. J. Caton. 2000. Anergy and suppression regulate CD4(+) T cell responses to a self peptide. *Eur. J. Immunol.* 30:136–144.
99. Jordan, M. S., A. Boesteanu, A. J. Reed, A. L. Petrone, A. E. Holenbeck, M. A. Lerman, A. Naji, and A. J. Caton. 2001. Thymic selection of CD4+CD25+ regulatory T cells induced by an agonist self-peptide. *Nat. Immunol.* 2:301–306.
100. Kawahata, K., Y. Misaki, M. Yamauchi, S. Tsunekawa, K. Setoguchi, J.-I. Miyazaki, and K. Yamamoto. 2002. Generation of CD4+CD25+ Regulatory T Cells from Autoreactive T Cells Simultaneously with Their Negative Selection in the Thymus and from Nonautoreactive T Cells by Endogenous TCR Expression. *J. Immunol.* 168:4399–4405.
101. van Santen, H.-M., C. Benoist, and D. Mathis. 2004. Number of T Reg Cells That Differentiate Does Not Increase upon Encounter of Agonist Ligand on Thymic Epithelial Cells. *J. Exp. Med.* 200:1221–1230.
102. D'Cruz, L. M., and L. Klein. 2005. Development and function of agonist-induced CD25+Foxp3+ regulatory T cells in the absence of interleukin 2 signaling. *Nat. Immunol.* 6:1152–1159.
103. Cabarrocas, J., C. Cassan, F. Magnusson, E. Piaggio, L. Mars, J. Derbinski, B. Kyewski, D.-A. Gross, B. L. Salomon, K. Khazaie, A. Saoudi, and R. S. Liblau. 2006. Foxp3+ CD25+ regulatory T cells specific for a neo-self-antigen develop at the double-positive thymic stage. *Proc. Natl. Acad. Sci. U.S.A.* 103:8453–8458.
104. Lafaille, J. J., K. Nagashima, M. Katsuki, and S. Tonegawa. 1994. High incidence of spontaneous autoimmune encephalomyelitis in immunodeficient anti-myelin basic protein T cell receptor transgenic mice. *Cell* 78:399–408.
105. Olivares-Villagomez, D., Y. Wang, and J. J. Lafaille. 1998. Regulatory CD4(+) T cells expressing endogenous T cell receptor chains protect myelin basic protein-specific transgenic mice from spontaneous autoimmune encephalomyelitis. *J. Exp. Med.* 188:1883–1894.
106. Furtado, G. C., M. A. Curotto de Lafaille, N. Kutchukhidze, and J. J. Lafaille. 2002. Interleukin-2 signaling is required for CD4+ regulatory T cell function. *J. Exp. Med.* 196:851–857.
107. Tai, X., M. Cowan, L. Feigenbaum, and A. Singer. 2005. CD28 costimulation of developing thymocytes induces Foxp3 expression and regulatory T cell differentiation independently of interleukin 2. *Nat. Immunol.* 6:152–162.
108. Cosgrove, D., S. H. Chan, C. Waltzinger, C. Benoist, and D. Mathis. 1992. The thymic compartment responsible for positive selection of CD4+ T cells. *Int. Immunol.* 4:707–710.
109. Laufer, T. M., J. DeKoning, J. S. Markowitz, D. Lo, and L. H. Glimcher. 1996. Unopposed positive selection and autoreactivity in mice expressing class II MHC only on thymic cortex. *Nature* 383:81–85.
110. Capone, M., P. Romagnoli, F. Beermann, H. R. MacDonald, and J. P. M. van Meerwijk. 2001. Dissociation of thymic positive and negative selection in transgenic mice expressing major histocompatibility complex class I molecules exclusively on thymic cortical epithelial cells. *Blood* 97:1336–1342.
111. Anderson, M. S., E. S. Venanzi, L. Klein, Z. Chen, S. Berzins, S. J. Turley, H. von Boehmer, R. Bronson, A. Dierich, C. Benoist, and D. Mathis. 2002. Projection of an Immunological Self-Shadow Within the Thymus by the Aire Protein. *Science* 298:1395–1401.
112. Consortium, T. F.-G. A. 1997. An autoimmune disease, APECED, caused by mutations in a novel gene featuring two PHD-type zinc-finger domains. *Nat. Genet.* 17:399–403.
113. Nagamine, K., P. Peterson, H. S. Scott, J. Kudoh, S. Minoshima, M. Heino, K. J. Krohn, M. D. Lalioti, P. E. Mullis, S. E. Antonarakis, K. Kawasaki, S. Asakawa, F. Ito, and N. Shimizu. 1997. Positional cloning of the APECED gene. *Nat. Genet.* 17:393–398.
114. Ramsey, C., O. Winqvist, L. Puhakka, M. Halonen, A. Moro, O. Kampe, P. Eskelin, M. Pelto-Huikko, and L. Peltonen. 2002. Aire deficient mice develop multiple features of APECED phenotype and show altered immune response. *Hum Mol Genet* 11:397–409.

115. Aschenbrenner, K., L. M. D'Cruz, E. H. Vollmann, M. Hinterberger, J. Emmerich, L. K. Swee, A. Rolink, and L. Klein. 2007. Selection of Foxp3(+) regulatory T cells specific for self antigen expressed and presented by Aire(+) medullary thymic epithelial cells. *Nat. Immunol.* 8:351–358.
116. Singer, A., and R. Bosselut. 2004. CD4/CD8 coreceptors in thymocyte development, selection, and lineage commitment: analysis of the CD4/CD8 lineage decision. *Adv. Immunol.* 83:91–131.
117. Laky, K., C. Fleischacker, and B. J. Fowlkes. 2006. TCR and Notch signaling in CD4 and CD8 T-cell development. *Immunol. Rev.* 209:274–283.
118. Kappes, D. J., X. He, and X. He. 2005. CD4-CD8 lineage commitment: an inside view. *Nat. Immunol.* 6:761–766.
119. Khattri, R., T. Cox, S. A. Yasayko, and F. Ramsdell. 2003. An essential role for Scurfin in CD4(+)CD25(+) T regulatory cells. *Nat. Immunol.* 3:3.
120. Fontenot, J. D., M. A. Gavin, and A. Y. Rudensky. 2003. Foxp3 programs the development and function of CD4(+)CD25(+) regulatory T cells. *Nat. Immunol.* 3:3.
121. Hori, S., T. Nomura, and S. Sakaguchi. 2003. Control of regulatory T cell development by the transcription factor Foxp3. *Science* 299:1057–1061.
122. Lin, W., D. Haribhai, L. Relland, N. Truong, M. Carlson, C. Williams, and T. Chatila. 2007. Regulatory T cell development in the absence of functional Foxp3. *Nat. Immunol.*
123. Zheng, Y., S. Z. Josefowicz, A. Kas, T. T. Chu, M. A. Gavin, and A. Y. Rudensky. 2007. Genome-wide analysis of Foxp3 target genes in developing and mature regulatory T cells. *Nature* 445:936–940.
124. Gavin, M. A., J. P. Rasmussen, J. D. Fontenot, V. Vasta, V. C. Manganiello, J. A. Beavo, and A. Y. Rudensky. 2007. Foxp3-dependent programme of regulatory T-cell differentiation. *Nature* 445:771–775.
125. Romagnoli, P., J. Tellier, and J. P. M. van Meerwijk. 2005. Genetic control of thymic development of CD4+CD25+FoxP3+ regulatory T lymphocytes. *Eur. J. Immunol.* 35:3525–3532.
126. Tellier, J., J. P. van Meerwijk, and P. Romagnoli. 2006. An MHC-linked locus modulates thymic differentiation of CD4+CD25+Foxp3+ regulatory T lymphocytes. *Int. Immunol.* 18:1509–1519.
127. Pennington, D. J., B. Silva-Santos, T. Silberzahn, M. Escorcio-Correia, M. J. Woodward, S. J. Roberts, A. L. Smith, P. J. Dyson, and A. C. Hayday. 2006. Early events in the thymus affect the balance of effector and regulatory T cells. *Nature* 444:1073–1077.
128. Artavanis-Tsakonas, S., M. D. Rand, and R. J. Lake. 1999. Notch signaling: cell fate control and signal integration in development. *Science* 284:770–776.
129. Greenwald, I. 1998. LIN-12/Notch signaling: lessons from worms and flies. *Genes Dev.* 12:1751–1762.
130. Radtke, F., A. Wilson, S. J. Mancini, and H. R. MacDonald. 2004. Notch regulation of lymphocyte development and function. *Nat. Immunol.* 5:247–253.
131. Amsen, D., J. M. Blander, G. R. Lee, K. Tanigaki, T. Honjo, and R. A. Flavell. 2004. Instruction of distinct CD4 T helper cell fates by different notch ligands on antigen-presenting cells. *Cell* 117:515–526.
132. Anastasi, E., A. F. Campese, D. Bellavia, A. Bulotta, A. Balestri, M. Pascucci, S. Checquolo, R. Gradini, U. Lendahl, L. Frati, A. Gulino, U. Di Mario, and I. Screpanti. 2003. Expression of activated Notch3 in transgenic mice enhances generation of T regulatory cells and protects against experimental autoimmune diabetes. *J. Immunol.* 171:4504–4511.
133. Robey, E., D. Chang, A. Itano, D. Cado, H. Alexander, D. Lans, G. Weinmaster, and P. Salmon. 1996. An activated form of Notch influences the choice between CD4 and CD8 T cell lineages. *Cell* 87:483–492.
134. Wolfer, A., T. Bakker, A. Wilson, M. Nicolas, V. Ioannidis, D. R. Littman, C. B. Wilson, W. Held, H. R. MacDonald, and F. Radtke. 2001. Inactivation of Notch 1 in immature thymocytes does not perturb CD4 or CD8T cell development. *Nat. Immunol.* 2:235–241.

135. Fontenot, J. D., J. P. Rasmussen, M. A. Gavin, and A. Y. Rudensky. 2005. A function for interleukin 2 in Foxp3-expressing regulatory T cells. *Nat. Immunol.*
136. Bayer, A. L., A. Yu, D. Adeegbe, and T. R. Malek. 2005. Essential role for interleukin-2 for CD4(+)CD25(+) T regulatory cell development during the neonatal period. *J. Exp. Med.* 201:769–777.
137. Malek, T. R., A. Yu, V. Vincek, P. Scibelli, and L. Kong. 2002. CD4 regulatory T cells prevent lethal autoimmunity in IL-2Rbeta-deficient mice. Implications for the nonredundant function of IL-2. *Immunity* 17:167–178.
138. Shull, M. M., I. Ormsby, A. B. Kier, S. Pawlowski, R. J. Diebold, M. Yin, R. Allen, C. Sidman, G. Proetzel, D. Calvin, N. Annunziata, and T. Doetschman. 1992. Targeted disruption of the mouse transforming growth factor-b1 gene results in multifocal inflammatory disease. *Nature* 359:693–699.
139. Gorelik, L., and R. A. Flavell. 2000. Abrogation of TGFbeta signaling in T cells leads to spontaneous T cell differentiation and autoimmune disease. *Immunity* 12:171–181.
140. Li, M. O., Y. Y. Wan, S. Sanjabi, A. K. Robertson, and R. A. Flavell. 2006. Transforming growth factor-beta regulation of immune responses. *Annu. Rev. Immunol.* 24:401–448.
141. Marie, J. C., J. J. Letterio, M. Gavin, and A. Y. Rudensky. 2005. TGF-beta1 maintains suppressor function and Foxp3 expression in CD4+CD25+ regulatory T cells. *J. Exp. Med.* 201:1061–1067.
142. Fahlen, L., S. Read, L. Gorelik, S. D. Hurst, R. L. Coffman, R. A. Flavell, and F. Powrie. 2005. T cells that cannot respond to TGF-beta escape control by CD4(+)CD25(+) regulatory T cells. *J. Exp. Med.* 201:737–746.
143. Ziegler, S. F., and Y. J. Liu. 2006. Thymic stromal lymphopoietin in normal and pathogenic T cell development and function. *Nat. Immunol.* 7:709–714.
144. Watanabe, N., Y. H. Wang, H. K. Lee, T. Ito, Y. H. Wang, W. Cao, and Y. J. Liu. 2005. Hassall's corpuscles instruct dendritic cells to induce CD4+CD25+ regulatory T cells in human thymus. *Nature* 436:1181–1185.
145. Al-Shami, A., R. Spolski, J. Kelly, T. Fry, P. L. Schwartzberg, A. Pandey, C. L. Mackall, and W. J. Leonard. 2004. A role for thymic stromal lymphopoietin in CD4(+) T cell development. *J. Exp. Med.* 200:159–168.
146. Carpino, N., W. E. Thierfelder, M. S. Chang, C. Saris, S. J. Turner, S. F. Ziegler, and J. N. Ihle. 2004. Absence of an essential role for thymic stromal lymphopoietin receptor in murine B-cell development. *Mol. Cell. Biol.* 24:2584–2592.
147. Kishimoto, H., and J. Sprent. 1999. Several different cell surface molecules control negative selection of medullary thymocytes. *J. Exp. Med.* 190:65–73.
148. Punt, J. A., B. A. Osborne, Y. Takahama, S. O. Sharrow, and A. Singer. 1994. Negative selection of CD4+CD8+ thymocytes by T cell receptor-induced apoptosis requires a costimulatory signal that can be provided by CD28. *J. Exp. Med.* 179:709–713.
149. Degermann, S., C. D. Surh, L. H. Glimcher, J. Sprent, and D. Lo. 1994. B7 expression on thymic medullary epithelium correlates with epithelium-mediated deletion of V beta 5+ thymocytes. *J. Immunol.* 152:3254–3263.
150. Buhlmann, J. E., S. K. Elkin, and A. H. Sharpe. 2003. A Role for the B7-1/B7-2:CD28/CTLA-4 pathway during negative selection. *J. Immunol.* 170:5421–5428.
151. Salomon, B., D. J. Lenschow, L. Rhee, N. Ashourian, B. Singh, A. Sharpe, and J. A. Bluestone. 2000. B7/CD28 costimulation is essential for the homeostasis of the CD4+CD25+ immunoregulatory T cells that control autoimmune diabetes. *Immunity* 12:431–440.
152. Tang, Q., K. J. Henriksen, E. K. Boden, A. J. Tooley, J. Ye, S. K. Subudhi, X. X. Zheng, T. B. Strom, and J. A. Bluestone. 2003. Cutting edge: CD28 controls peripheral homeostasis of CD4+CD25+ regulatory T cells. *J. Immunol.* 171:3348–3352.
153. Gray, D. H., N. Seach, T. Ueno, M. K. Milton, A. Liston, A. M. Lew, C. C. Goodnow, and R. L. Boyd. 2006. Developmental kinetics, turnover, and stimulatory capacity of thymic epithelial cells. *Blood* 108:3777–3785.

Chapter 4
The Role of IL-2 in the Development and Peripheral Homeostasis of Naturally Occurring CD4$^+$CD25$^+$Foxp3$^+$ Regulatory T Cells

Allison L. Bayer and Thomas R. Malek

Abstract Naturally occurring CD4$^+$CD25$^+$Foxp3$^+$ regulatory T cells (T$_{reg}$) actively suppress autoreactive T cells that escape negative selection in the thymus, preventing a wide variety of autoimmune diseases. Along with Foxp3, the high affinity IL-2R represents one of the better characterized molecules of T$_{reg}$ cells. Current data support models where the IL-2/IL-2R interaction, primarily through STAT5 activation, controls T$_{reg}$ cell development in the thymus and their growth and maintenance in peripheral immune tissues. The recent link between IL-2R signaling and upregulation of Foxp3 also raises the possibility that IL-2 is important for T$_{reg}$ cell suppressive function. Other important issues concern the cellular source of IL-2 and novel aspects that potentially regulate and limit IL-2 to T$_{reg}$ cells. Although IL-2 is essential, other cytokines within and outside the γc family likely contribute to T$_{reg}$ cell production.

Introduction

A central issue of modern day medicine is how peripheral self-tolerance is achieved and maintained as increasing evidence has clearly demonstrated that potentially self-destructive T cell clones escape central tolerance and circulate in the periphery of healthy individuals. A pivotal finding by Sakaguchi et al. demonstrated that a subpopulation of CD4$^+$ T cells that expressed the α-chain (CD25) of the IL-2R plays an important function in peripheral tolerance by suppressing autoimmune disease in mice [1]. Since that time there has been extensive study of CD4$^+$CD25$^+$ T regulatory cells (T$_{reg}$) leading to the discovery that these cells are present in humans and their function extends to many aspects of immune regulation, including responses to autologous, allogeneic, pathogen-derived, and tumor antigens. Moreover, these CD4$^+$CD25$^+$ T$_{reg}$ cells naturally develop in the thymus as a fully functional distinct subset of CD4$^+$ T cells that migrate to the periphery to actively suppress

---

T.R. Malek
Department of Microbiology and Immunology, Diabetes Research Institute, Miller School of Medicine, University of Miami, Miami, FL 33136, USA
e-mail: tmalek@med.miami.edu

auto-reactive T cells. This population of naturally occurring T_{reg} cells bears the high affinity IL-2R, which is comprised of three subunits known as the α-chain (CD25), β-chain (CD122), and the common cytokine-receptor γ-chain (γc, CD132) [2,3]. Although all three chains contribute to IL-2 binding, only the β- and γ-chains are responsible for signal transduction following IL-2 ligand binding to the receptor. There is mounting evidence that points to signaling through the IL-2R as a critical event for naturally occurring $CD4^+CD25^+$ T_{reg} cell development, maintenance, and function.

IL-2/IL-2R Interaction Controls Peripheral Self-Tolerance

The notion that the non-redundant function of the IL-2/IL-2R interaction was for self-tolerance was supported by the unexpectedly finding that IL-2-deficient mice contained hyperproliferative rather than hyporesponsive T cells which led to a fatal autoimmune disease [4,5]. Mice deficient in IL-2Rα or IL-2Rβ creates a similar lymphoproliferative disorder and lethal autoimmunity [6–8]. At that time an unresolved issue was whether this lack of tolerance in IL-2- and IL-2R-deficient mice resided at the level of central or peripheral tolerance. An increased escape of autoreactive T cells from thymic deletion might have accounted for this aberrant accumulation of autoreactive T cells. However, thymic negative selection was shown to be largely normal in both IL-2- or IL-2Rβ-deficient mice [9,10]. Since cells that respond to IL-2 are very sensitive to activation-induced cell death (AICD) upon restimulated through their T cell receptors (TCR), it was also originally thought that the accumulation of activated T cells in IL-2- and IL-2R-deficient mice might be due to lack of sensitization to apoptosis [11–13]. However, we found that thymus-specific transgenic expression of IL-2Rβ in IL-2Rβ$^{-/-}$ mice prevented lethal autoimmunity while the peripheral T cells remained unresponsive to IL-2. This finding strongly favors a mechanism other than peripheral AICD for tolerance [10,14]. Importantly, studies with mixtures of bone marrow (BM) or T cells from either IL-2Rβ$^{-/-}$ or IL-2$^{-/-}$ cells with wild-type (WT) cells showed that the autoimmunity in these deficient mice was not due to intrinsic defects but rather lack of a regulatory population [15,16]. This point is highlighted in the mixture experiments where IL-2Rβ$^{-/-}$ T cells cannot bind to IL-2 produced by WT cells, yet these mice were protected from autoimmunity [15].

It would be some time before the underlying mechanism causing this autoimmunity in IL-2/IL-2R-deficient mice was attributed to lack of functional $CD4^+CD25^+$ T_{reg} cells. The first hint that IL-2 may function in generating regulatory cells came from adoptive transfer studies where large numbers of thymocytes or splenocytes from IL-2-treated IL-2$^{-/-}$ mice resulted in delayed onset of autoimmunity upon a subsequent transfer into IL-2$^{-/-}$ neonates [17]. Moreover, the regulatory population was likely of T cell origin because antigen-challenged IL-2$^{-/-}$ OVA-specific T cells continued to accumulate within athymic but efficiently contracted within euthymic hosts [16]. Furthermore, athymic mice developed autoimmunity when adoptively transferred with lymph node cells from IL-2$^{-/-}$ mice, which lack functional T$_{reg}$ cells, while cells from IL-2$^{+/+}$ mice prevented disease [9].

Our laboratory was the first to clearly demonstrate that the non-redundant role for the IL-2/IL-2R interaction in the regulation of self-reactivity was through the generation and maintenance of CD4$^+$CD25$^+$ T$_{reg}$ cells. We noted that thymic restricted expression of IL-2Rβ restored the production and number of functional CD4$^+$CD25$^+$ T$_{reg}$ cells in both the thymus and periphery of IL-2Rβ$^{-/-}$ mice, correlating T$_{reg}$ cell production with protection to autoimmunity [10,14,18]. The decisive experiment was to show that the adoptive transfer of a small number of highly purified WT T$_{reg}$ cells into neonatal IL-2Rβ$^{-/-}$ mice resulted in the long-term persistence of the donor T$_{reg}$ cells, with these mice having a normal disease-free lifespan. Importantly, in order for this adoptive transfer to prevent autoimmunity, the donor T$_{reg}$ cells must express a fully functional IL-2R and the host must produce IL-2, which caused substantial IL-2-dependent expansion by donor T$_{reg}$ cells [10].

At about the same time, several other studies also showed an essential role for IL-2R signaling in the development and survival of CD4$^+$CD25$^+$ T$_{reg}$ cells and in maintaining peripheral self-tolerance. Adoptive transfer of CD4$^+$CD25$^+$ T$_{reg}$ cells in mice recently reconstituted with IL-2Rα-deficient BM cells or co-transfer with IL-2-deficient T cells into athymic hosts controlled the accumulation of activated T cells and led to full protection from autoimmunity [16,19]. Interestingly, in chimeric mice derived from a mixture CD25$^{-/-}$ and IL-2$^{-/-}$ BM, a normal immune compartment developed and protected the chimeric mice from lethal autoimmunity associated with IL-2/IL-2R deficiency [19]. Thus, T$_{reg}$ cells were generated by restoring the IL-2/IL-2R interaction on IL-2$^{-/-}$ T cells through the production of IL-2 from CD25$^{-/-}$ T cells. In an analogous fashion, adoptive transfer of thymic or splenic IL-2$^{-/-}$ CD4$^+$ T cells protected IL-2-sufficient mice from spontaneous experimental autoimmune encephalomyelitis, while CD25$^{-/-}$ thymic or splenic CD4$^+$ T cells conferred little or no protection [20]. A mutation within the human CD25 gene has also been found that resulted in an immune disorder with similar symptoms as found in IL-2/IL-2R-deficient mice which was completely resolved after allogeneic BM transplantation, probably through production of T$_{reg}$ cells [21]. Taken together, these studies provide strong evidence for an essential non-redundant role of IL-2/IL-2R in the development and maintenance of naturally occurring T$_{reg}$ cells.

Although CD25 was the initial marker that identified a population of T$_{reg}$ cells, upregulation or induction of CD25 is also seen upon activation of naïve T cells, making it difficult to unambiguously identify regulatory T cells during ongoing immune responses. However, T$_{reg}$ cells were later discovered in mice to uniquely express the forkhead transcription factor Foxp3, [22–24] allowing more precise examination of their basic biology. Similar to mice that lack a functional IL-2R, mutations or absence of Foxp3 leads to an immune compartment devoid of CD4$^+$CD25$^+$ T$_{reg}$ cells and these mice die within one month of birth. The identification of mutations in the Foxp3 gene in scurfy mice and in Immune dysregulation, Polyendocrinopathy, Enteropathy, X-linked (IPEX) syndrome in humans, both of which succumb to lethal autoimmune disease early in life, was critical in establishing an essential role of CD4$^+$CD25$^+$ T$_{reg}$ cells in the maintenance of peripheral self-tolerance. Furthermore, mixed BM chimeric mice inoculated with both WT and Foxp3-deficient BM cells demonstrated that only WT BM was capable of generating $CD4^+CD25^+Foxp3^+$ T_{reg} cells [22]. Using Foxp3-GFP ($Foxp3^{gfp}$) reporter mice to investigate T_{reg} ontogeny, $Foxp3^{gfp}$ cells development is considerably delayed compared to thymic nonregultory T cells [25]. This finding may explain the development of multi-organ autoimmune disease following thymectomy of 3-day old mice in that potential autoreactive T cells likely migrate to the periphery of neonatal mice prior to the appearance of a significant number of T_{reg} cells. Taken together, these data indicate that along with an essential role of IL-2R signaling, Foxp3 is a key protein involved in the development and function of naturally occurring $CD4^+CD25^+$ T_{reg} cells.

IL-2R Signaling in T_{reg} Cells

For antigen-activated conventional and regulatory T cells, binding of IL-2 to IL-2Rα drives the oligomerization of all of the IL-2R subunits (Fig. 4.1), leading to signal transduction mediated by the cytoplasmic tails of the IL-2R β- and γ-chains, as all the signaling molecules associated with the IL-2R are connected to these two subunits. As extensively characterized for antigen-activated T cells [26], Janus activated kinase (JAK) 1 and 3 associate with the β-and γ-chain, respectively, and phosphorylate key residues in the cytoplasmic tails of the β and γc subunits, the JAKs themselves, and signal transducer and activator of transcription 5 (STAT5). Phosphorylation of STAT5 leads to its dimerization and translocation to the nucleas where it activates a number of genes, including those for cell cycle progression and survival. Phosphorylation of the IL-2R also permits recruitment of phosphatidylinositol 3-kinase (PI3K) and mitogen-activated protein kinase (MAPK) signaling pathways that also induce proliferation and promote cell survival.

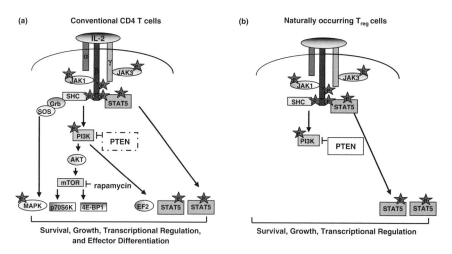

Fig. 4.1 IL-2R signaling in conventional CD4 T cells and naturally occurring T_{reg} cells *(See also Color Insert)*

Although multiple signaling pathways emanate after engaging the IL-2R of conventional activated T cells, the STAT5 pathway appears to be most important for CD4$^+$CD25$^+$ T$_{reg}$ cells. STAT5-deficient mice have markedly reduced numbers of T$_{reg}$ cells and exhibit a phenotype similar to IL-2/IL-2R-deficient mice, with uncontrolled lymphocyte proliferation and death early in life from autoimmune disease [27–30]. Conversely, when transgenic constitutively active STAT5b was expressed in normal and IL-2-deficient mice, CD4$^+$CD25$^+$ T$_{reg}$ cells selectively expanded [28,30]. Thus, active STAT5 bypassed the requirement for IL-2 and rescued the production of T$_{reg}$ cells without obvious activation of the IL-2R-dependent PI3$\notin$K or MAPK pathways. Interestingly, short-term treatment of WT or IL-2$^{-/-}$ mice with recombinant IL-2 resulted in upregulation CD25 and Foxp3 in CD4$^+$Foxp3$^+$ T cells suggesting that IL-2R signaling may increase Foxp3 expression [31]. Consistent with this notion, a patient was recently described with a mutation in STAT5 that displayed immune dysregulation and reduced numbers of CD4$^+$CD25^{bright} T cells with low Foxp3 levels and suppressor function [32]. Indeed, the Foxp3 gene contains consensus binding sequences for STAT5 that are required for transcription, further supporting an important role STAT5 in Foxp3 expression [29,33]. IL-2 was shown to regulate Foxp3 expression in both mouse and human CD4$^+$CD25$^+$ T$_{reg}$ cells through STAT5-dependent signals, revealing a critical link between IL-2 and STAT5 signaling in the maintenance of Foxp3-expressing T$_{reg}$ cells [29,33,34].

Although T$_{reg}$ cells express the high affinity IL-2R and IL-2 is required for their development, peripheral survival, and perhaps function, many studies have shown that T$_{reg}$ cells are hypoproliferative in response to IL-2 in vitro, further indicating that IL-2R signaling is distinct in T$_{reg}$ cells compared to conventional activated T lymphocytes that undergo robust IL-2-dependent proliferation. In this regard, T$_{reg}$ cells contain relative high levels of suppressor of cytokine signaling-2 (SOCS-2), which might temper IL-2-induced signaling [35]. Furthermore, engagement of the IL-2R on the surface of T$_{reg}$ cells activates JAK and STAT signaling molecules, but fails to fully stimulate the downstream targets of PI3K, Akt and p70^{s6kinase} [36]. Moreover, this inability to stimulate Akt and p70^{S6Kinase} was seen with a relatively high expression of the ubiquitously expressed negative regulator of PI3K, phosphatase and tensin homologue deleted chromosome 10 (PTEN). However, TCR stimulation resulted in the downregulation of PTEN in CD4$^+$CD25$^+$ T$_{reg}$ cells and may explain the observation that TCR and IL-2 signals can restore proliferative capacity of these cells in vitro. T$_{reg}$ cells from mice in which PTEN was specifically deleted in T cells also exhibited enhanced proliferation in vivo and readily expanded directly to IL-2 stimulation in vitro [37]. Importantly, PTEN-deficient T$_{reg}$ cells are potent suppressor cells that readily prevented colitis [37], consistent with the idea that an intact JAK and STAT cascade, which supports Foxp3 expression, is sufficient for fully functional T$_{reg}$ cells.

Incidentally, the immunosuppressant drug rapamycin, which inhibits IL-2-dependent proliferation by conventional activated T cells, does not affect proliferation by either mouse or human CD4$^+$CD25$^+$ T$_{reg}$ cells in vitro, thereby selectively promoting T$_{reg}$ cell growth in the presence of other T cells [38–40]. Unlike cyclosporin which blocks down-stream events of TCR signaling, rapamycin inhibits signaling in responses to cytokines and growth factors by suppressing protein synthesis and cell cycle progression through the binding to the mammalian target of rapamycin (mTOR) and thereby inhibiting PI3K-induced signaling. Moreover, rapamycin spares the JAK and STAT signaling pathways, maintaining Foxp3 expression in T_{reg} cells (Fig. 4.1). Since the engagement of the IL-2R on T_{reg} cells fails to activate the downstream targets of PI3K, it is thought that rapamycin may specifically target PI3K-sensitive T effector cells rather than $CD4^+CD25^+$ T_{reg} cells [38].

Requirements for IL-2 Within the Thymus and Peripheral Lymphoid Tissues

Studies showing that IL-2 production or IL-2R responsiveness restores a functional population of $CD4^+CD25^+$ T cells provide evidence for an essential role for the IL-2/IL-2R interaction in the generation and maintenance of naturally occurring T_{reg} cells. It was unclear from these studies, whether T_{reg} cell reconstitution by IL-2 was due to their production from precursor cells, their expansion from a small pool of surviving mature T_{reg} cells, or their enhanced function. Furthermore, whether IL-2 functions in T_{reg} cell thymic development, peripheral homeostasis, or both has been an unresolved issue. Two recent studies have demonstrated the presence of polyclonal or TCR-specific Foxp3-expressing thymic and peripheral $CD4^+$ T cells in the context of IL-2 or IL-2Rα deficiency [31,41]. These studies led the authors to conclude that IL-2 is not necessary for T_{reg} development within the thymus. However, $Foxp3^+$ thymocytes likely depend upon IL-2 because the polyclonal thymic IL-$2^{-/-}$ and IL-2Rβ$^{-/-}$ Foxp3$^+$ T cells are at a reduced number and expressed a lower level of both Foxp3 and CD25 when compared to T_{reg} cells from control mice [18,31,41]. Although Foxp3^{low} CD4 T cells in IL-2- or IL-2R-deficient mice suppressed responding T cells in vitro, these cells fail to effectively control autoimmunity resulting in death of these deficient mice. It may be noteworthy that the relative low levels of Foxp3 associated with T cells from IL-2- and IL-2R-deficient mice is analogous to human Foxp3^{low} CD4 T cells, which is associated with effector rather than regulatory function.

We have made two important observations indicating that the IL-2/IL-2R interaction is active within the thymus. First, we found that administration of anti-IL-2 to neonatal mice reduced T_{reg} cells within the thymus of WT or IL-2Rβ$^{-/-}$ mice, the latter which expressed thymic targeted transgenic IL-2Rβ [18]. Second, expression of this thymic transgenic IL-2Rβ reconstituted the production of functional $CD4^+CD25^+$ T_{reg} cells in IL-2Rβ$^{-/-}$ mice that prevented autoimmunity [10]. This activity is in line with the recent discovery of the link between IL-2R-dependent STAT5 signaling and its regulation of Foxp3 expression [32–34]. Thymic $CD4^+Foxp3^+$ cells are seen within subsets of double positive (DP) $CD4^+CD8^+$ and single positive (SP) $CD4^+CD8^-$ cells [18,25]. Somewhat surprisingly, for IL-2Rβ$^{-/-}$ mice that expressed the transgenic thymic IL-2Rβ, normal CD122 expression was only seen in the most immature $CD24^+$ (HSA) DP Foxp3$^+$ thymocytes, with nearly undetectable levels in CD4$^+$ SP T$_{reg}$ cells. This restricted IL-2Rβ expression, nevertheless, was sufficient to increase the number of Foxp3$^+$ cells and upregulate Foxp3 and CD25 to levels equivalent to that of T$_{reg}$ cells within the thymus of WT mice [18]. This finding raises the idea that IL-2R signaling may represent an important checkpoint that supports CD4$^+$CD25$^+$ Foxp3$^+$ T cell development in the thymus.

IL-2 also importantly functions for peripheral T$_{reg}$ cells. In a variety of experimental settings, adoptive transfer of purified WT CD4$^+$CD25$^+$ T$_{reg}$ cells suppressed symptoms associated with autoimmunity in IL-2 or IL-2R-deficient mice. For example, adoptive transfer of T$_{reg}$ cells into neonatal IL-2Rβ$^{-/-}$ mice results in substantial IL-2-dependent proliferation leading to stable engraftment of donor T$_{reg}$ cells that completely bypassed the involvement of thymus [10]. In this setting, the donor T$_{reg}$ cells must express a functional IL-2R and the host must produce IL-2 indicating that IL-2R signaling is required for T$_{reg}$ cell proliferation in peripheral immune compartment [10]. In support of this finding, IL-2 blockade also efficiently inhibited this extensive T$_{reg}$ cell proliferation [42]. Furthermore, anti-IL-2 treatment during the neonatal period resulted in autoimmune gastritis in BALB/c mice and early diabetes, peripheral neuritis, gastritis, and thyroiditis in NOD mice which was accompanied by decreased numbers of CD4$^+$CD25$^+$Foxp3$^+$ T$_{reg}$ cells [43]. Similarly, anti-IL-2 also reduced T$_{reg}$ cells in the periphery of normal and thymectomized adult mice [44].

Interestingly, in most of the above studies where IL-2 blockade inhibited T$_{reg}$ cell production, anti-IL-2 was administered to neonatal mice. It may be relevant that when we treated adult C57BL/6 mice with anti-IL-2 in a fashion that effectively blocked T$_{reg}$ cells within the thymus and lymph nodes of neonatal mice [42], CD4$^+$CD25$^+$ T cells were reduced in both the spleen and lymph nodes, with an obvious decrease in the CD4$^+$ CD25^{hi} GITR^{hi} population (Fig. 4.2a and b and data not shown). This result is in line with what was previously reported for short-term anti-IL-2 treatment in adult mice and showed that anti-IL-2 was inhibitory in these mice. However, when we enumerated Foxp3$^+$ T cells following anti-IL-2 treatment, no decrease was detected (Fig. 4.2c). This result suggests that mature T$_{reg}$ cells are IL-2-independent or require minimal IL-2 signaling which essentially rendered them resistant to anti-IL-2 blockade.

Our finding of a normal number of functional T$_{reg}$ cells in the periphery of IL-2Rβ$^{-/-}$ mice after thymic-targeted expression of IL-2Rβ is also consist with the notion that T$_{reg}$ cell homeostasis in adults does not required IL-2 or just minimal IL-2 signaling. To further address the requirements for IL-2 in the periphery, mixed chimeras were generated by transferring a 1:1 mixture of BM cells from WT and IL-2Rβ$^{-/-}$ mice, the latter expressing thymus targeted transgenic IL-2Rβ, into lethally irradiate recipients. When analyzed 7–8 weeks latter, there were equivalent number of T$_{reg}$ cells of both donor types within the thymus, corresponding to expression of IL-2R in this tissue. In contrast, the WT-derived T$_{reg}$ cells were substantially more effective in populating peripheral immune tissue than T$_{reg}$ cells with impaired IL-2 signaling [18]. Thus, this finding indicates that under normal circumstances IL-2R signaling remains the dominant mechanism for T$_{reg}$ expansion and peripheral homeostasis. This preference for WT T$_{reg}$ cells was also seen

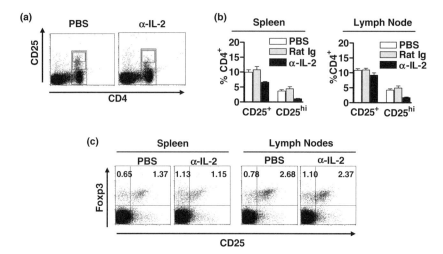

Fig. 4.2 Blockade of IL-2 in adult mice decreases CD25$^+$, but not Foxp3$^+$, CD4$^+$ T cells. C57BL/6 mice (8–12 weeks of age) were injected with PBS, rat Ig (3 mg), or anti∀-IL-2 (S4B6, 3 mg) i.p. on day 0 and day 3. Five days after the first injection, spleen and LN cells were subjected to FACS analysis. Shown is a representative FACS dot plots (**a**) and the % of total CD4$^+$ T cells (**b**) expressing CD25 as gated in (a). Data are mean ± SEM for 5–7 mice per treatment. (**c**) Representative (n=2) dot plots for CD25 and Foxp3 expression

when mixed BM chimeras were generated with WT and IL-2Rα$^{-/-}$ BM cells [31]. These data corroborate that IL-2/IL-2R interaction remains a major component of peripheral homeostasis of naturally occurring T$_{reg}$ cells. However, these data cannot distinguish whether this dominance is at the level of their early expansion in LN during the initial stages of their production or for their subsequent homeostasis in the steady state.

Most Foxp3$^+$ T$_{reg}$ cells in the lymph nodes of WT C57BL/6 mice co-express CD4 and CD25. There are far fewer CD4$^+$Foxp3$^+$ cells in IL-2Rβ$^{-/-}$ mice and these express a lower level of Foxp3 and largely do not express CD25 (Fig. 4.3a). It has been proposed that these CD4$^+$Foxp3^{low} T cells are functional regulatory cells but are less fit and cannot keep-up with the autoreactive T cells causing systemic autoimmunity [31]. Interestingly, when neonatal IL-2Rβ$^{-/-}$ mice were adoptively transferred with congenic CD45.1$^+$, WT CD4$^+$CD25$^+$ T$_{reg}$ cells, the major population of T$_{reg}$ cells in the recipients remained donor-derived WT cells, which persisted for the life of these autoimmune free mice (Fig. 4.3b). Thus, even after correcting autoimmune disease, the endogenous Foxp3^{low} cells of these IL-2Rβ$^{-/-}$ recipients remained a minor component of the total T$_{reg}$ pool even though one prediction is these recipient cells might have a competitive advantage over the donor T$_{reg}$ cells due to the constant thymic output of only the former. These findings suggest that the IL-2Rβ$^{-/-}$ CD4$^+$Foxp3^{low} T cells are severely functionally impaired or cannot enter the niche where T$_{reg}$ cell suppression occurs.

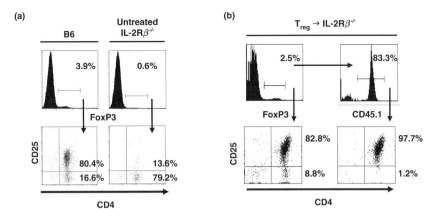

Fig. 4.3 IL-2/IL-2R interaction remains a major component of peripheral homeostasis of naturally occurring T_{reg} cells. (**a**) Representative histogram for Foxp3 staining with corresponding dot plots for CD4 and CD25 expression after gating on Foxp3+ LN cells from C57BL/6 and untreated IL-2Rβ−/− mice. (**b**) Representative histogram for Foxp3 staining with corresponding histogram for CD45.1 staining or dot plots for CD4 and CD25 expression after gating on Foxp3+ or donor CD45.1+Foxp3+ cells 12 weeks after neonatal IL-2Rβ−/− mice received purified WT CD45.1 congenic CD4+CD25+ T_{reg} cells (Treg → IL-2Rβ−/−)

Considering the cohort of data, we favor a model (Fig. 4.4) in which IL-2 critically acts in the thymus where it plays an essential role in the earliest stages of the production of T_{reg} cells to promote their expansion and upregulate Foxp3 and CD25. IL-2 is also essential in the neonatal lymph nodes where it acts as a potent

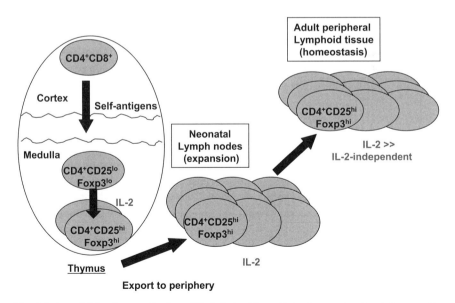

Fig. 4.4 A model for the contribution of IL-2 during T_{reg} cell development and maintenance *(See also Color Insert)*

growth factor to promote extensive T_{reg} cell expansion. Moreover, if IL-2-signals are delivered to immature thymic T_{reg} cells, peripheral expansion and homeostasis of mature T_{reg} cells can occur without a requirement for sustained IL-2R signaling and may essentially be IL-2-independent. Since many other interactions have been implicated in controlling aspects of T_{reg} cell growth or function, including TCR and co-stimulatory interactions through CD28, CD40L, 4-1BB, and CD7, or IL-4 [45–50], signaling through these molecules may compensate for poor IL-2R signaling to maintain peripheral T_{reg} cells.

IL-2 and T_{reg} Cell Function

Until recently, there has been little known about whether IL-2 is required for T_{reg} cell suppressor function. Two studies support of a role for IL-2 in the suppression by T_{reg} cells in vitro. Pre-culturing of T_{reg} cells with anti-CD3 in the presence of IL-2 enhanced their suppressor capacity [51]. Furthermore, T_{reg} cell suppressor function also relied upon transient IL-2 production by the responder T cells [52,53]. The addition of IL-4 to pre-cultures of T_{reg} cells similarly enhanced their function [51], indicating that there are redundant IL-2-independent means for such T_{reg} suppressor activity in vitro. Several other studies, however, are consistent with no requirements for IL-2 or other exogenous cytokines for T_{reg} cell suppression in vitro. For example, peripheral T_{reg} cells from our thymic IL-2Rβ transgenic model, which do not express a functional IL-2R, exhibited suppressor cells activity in vitro, albeit at a slightly lower level when compared to WT T_{reg} cells [10,18]. These peripheral T_{reg} cells express essentially undetectable IL-2Rβ, but upon IL-2 treatment generated very low and transient STAT5 phosphorylation in comparison to WT T_{reg} cells. It remains possible, therefore, that this minimal signaling accounts for their in vitro suppressor activity. However, this minimal signaling did not support down-stream IL-2-dependent functional responses by these T_{reg} cells in other assays in vitro and in vivo. Furthermore, Foxp3^{low} cells from IL-2Rα$^{-/-}$ or IL-2$^{-/-}$ mice were suppressive in vitro [31,41]. The suppressor activity by Foxp3^{low} IL-2$^{-/-}$ T cells might reflect IL-2 produced from the IL-2-sufficient responder population that can still act on the IL-2$^{-/-}$ Foxp3$^+$ T cells. Notably, CD25$^{-/-}$ Foxp3^{low} are not expected to bind IL-2 but were competent suppressors. Thus, it is unlikely that IL-2R signaling is required for T_{reg} cell suppressor activity in vitro.

These above data, especially the suppressive activity from CD4$^+$Foxp3^{low} T cells from IL-2Rα-deficient mice, argue autoimmunity associated with IL-2 and IL-2R-deficient mice is not a failure in T_{reg} cell function. Rather, impaired survival and/or inefficient expansion of CD4$^+$Foxp3^{low} T_{reg} within IL-2-and IL-2R-deficient mice accounts for autoimmunity. Another argument for some suppressive activity by IL-2- or IL-2R-deficient CD4$^+$Foxp3^{low} cells is that Foxp3-deficient mice die more quickly than IL-2- or IL-2R-deficient mice. Alternatively, the longer course of lethal autoimmunity in IL-2/IL-2R-deficient mice might not be due to some intrinsic regulatory activity by their Foxp3$^+$ T cells, but instead reflects less robust autoimmune effector activity by the IL-2/IL-2R-deficient T cells [14] than the IL-2/IL-2R-sufficient T cells in Foxp3-deficient mice.

Mice with restricted thymic IL-2Rβ expression in IL-2Rβ-deficient mice have a functional population of CD4$^+$CD25$^+$ Foxp3$^+$ T$_{reg}$ cells such that they live a normal life-span and are vigorous breeders consistent with the overall health of these mice, suggesting that IL-2R signaling in the periphery may not be essential for T$_{reg}$ cells suppressor function. However, when these mice were examined more carefully, there is some evidence for mild autoimmune symptoms, especially in older mice [14]. For example, their lymph nodes are somewhat larger due to increased number of CD4 T cells. An occasional elevated auto-antibody titer and minimal elevated serum IgG1 levels were also detected. Inflammatory infiltrates have also occasionally been noted in non-lymphoid tissues of aging animals. One interpretation of these data is that IL-2 non-responsive T$_{reg}$ cells are not fully functional. In support of this idea, their IL-10 mRNA is lower than found in WT T$_{reg}$ cells by DNA microarray analysis (T.R. Malek, unpublished data). Furthermore, the recently described link between IL-2-dependent STAT5 activation in the regulation of Foxp3 expression associates IL-2 in T$_{reg}$ cell function. Collectively, there is suggestive, but no clear cut evidence for a direct involvement of IL-2 in promoting T$_{reg}$ cell function.

Cellular Source of IL-2

Another important issue concerns the cellular source supplying IL-2 to naturally occurring T$_{reg}$ cells for their survival and growth in the thymus and periphery. The likely candidate is auto-reactive or antigen-stimulated T cells. In support of this notion, examination of different cell populations from the spleens of naïve normal mice revealed that predominately CD4$^+$CD25^{low} T cells, which are largely Foxp3^{neg} and are likely effector T cells, actively transcribe the IL-2 gene and produce IL-2 [43]. Moreover, the ontogeny of CD4$^+$Foxp3$^+$ T$_{reg}$ cells is delayed compared to thymic nonregulatory T cells [25]. IL-2 from autoreactive T cells may be especially important in neonatal mice where it has been demonstrated that thymic negative selection is less stringent [54,55] and hence there may be more T cells that are readily activated to produce cytokines, including IL-2. Therefore, one model is that autoreactive T cells that escape negative selection exit the thymus to the periphery where they encounter self-peptides and produce IL-2 allowing rapid expansion of T$_{reg}$ cells, which leads to sufficient numbers of T$_{reg}$ cells to then suppress the autoreactive T cells.

In addition to activated CD4 T cells, other lymphoid populations express IL-2 mRNA including CD8, NK, NKT, and DC cells, but at much lower levels than these CD4$^+$CD25^{low} T cells [43]. Moreover, DC have been shown to maintain and expand T$_{reg}$ cell in vitro and in vivo [49,56–59] and to produce IL-2 promoting T$_{reg}$ cell expansion in vitro, with IL-2 production mediated through a CD40/CD40L interaction [46,60,61]. DC-derived IL-2 was necessary as an anti-CD25 blocking antibody severely impaired this T$_{reg}$ cell expansion [46].

Analysis of T_{reg} cell development in IL-2$^{+/+}$ lethally irradiated Rag2$^{-/-}$ host after BM transfers was consistent with the possibility that non-T cells are an effective source of IL-2 for T_{reg} cells [19]. To more carefully examine this notion, T_{reg} cell production was then examined in lethally irradiated Rag2$^{-/-}$ IL-2$^{-/-}$ hosts that received mixed BM from IL-2$^{-/-}$ and TCRα$^{-/-}$ IL-2$^{+/+}$, so there are no T cells that produce IL-2 from any source in these recipient mice [62]. These mixed chimeras contained slightly more CD4$^+$CD25$^+$ T cells, that were CD25^{low}, than detected in hosts that received only IL-2$^{-/-}$ BM cells [62], but were not protected from death, indicating that in this setting functional T_{reg} cells were not produced. Although Foxp3 expression was not examined, the low level of CD25 expression is analogous to the CD4$^+$Foxp3^{low} T cells in IL-2- and IL-2R-deficient mice that do not prevent lethal autoimmunity. In contrast, Rag2$^{-/-}$ IL-2$^{-/-}$ chimeras were protected from death when they received IL-2$^{-/-}$ and CD25$^{-/-}$ BM cells, where T cells derived from CD25$^{-/-}$ BM produce IL-2. Collectively, these data indicate that IL-2 production from non-T cells is not sufficient to support fully functional T_{reg} cells and that the likely source of IL-2 is indeed from activated T cells.

This critical requirement for IL-2 may depend on IL-2 being produced in a distinct microenvironment that supports T_{reg} cells. To begin to test this notion, we evaluated the extent that the cytoplasmic domains of IL-2Rβ and IL-7Rα were interchangeable for T_{reg} cell production as both receptors activate STAT5. Transgenic cytokine receptors were produced such that the extracellular portion of IL-2Rβ was linked to the cytoplasmic portion of IL-7Rα (2β/7α) or the extracellular portion of IL-7Rα was linked to the cytoplasmic tail of IL-2Rβ (7α/2β) and these were expressed in T lineage cells, including CD4$^+$Foxp3$^+$ T cells, in IL-2Rβ$^{-/-}$ mice. The expression of 2β/7α receptor restored production of CD4$^+$CD25$^+$ Foxp3$^+$ T_{reg} cells in the thymus and periphery, leading to prevention of lethal autoimmunity [63]. However, transgenic expression of 7α/2β or the WT IL-7Rα failed to restore T_{reg} cell production and prevent autoimmunity. Thus, these data indicate that IL-7Rα-chain signaling is sufficient to produce T_{reg} cells and inhibit autoimmunity in IL-2Rβ-deficient mice, probably through the redundant activation of STAT5, but only when triggered by IL-2. Failure of 7α/2β and WT IL-7Rα is most likely due to lack of IL-7 in the inner thymus where T_{reg} cells preferentially develop.

A novel autoregulatory loop has recently been described where IL-2 inhibits its own production [64,65]. This existence of this autoregulatory loop raises the possibility that autocrine IL-2 production will be very short-lived. We found that the transcriptional repressor B lymphocyte-induced maturation protein (Blimp-1), that drives terminal differentiation of B cells into plasma cells [66,67], is involved in the inhibition of IL-2 [64]. Blimp-1 was also recently found to be involved in the regulation of T cell activation and homeostasis and highly expressed in CD4$^+$CD25$^+$ T_{reg} cells [68,69]. Interestingly, mice in which Blimp-1 was specifically deleted in T cells resulted in severe colitis, indicating a role for Blimp-1 in regulating self-tolerance.

There are two interesting issues with respect to Blimp-1 and T cell tolerance. One issue is that a property of T_{reg} cells is the inability to produce IL-2. As Blimp-1 is expressed in T_{reg} cells, one role for Blimp-1 might be to directly repress IL-2 and other target genes. Hence Blimp-1 deficiency might lead to an intrinsic defect in T_{reg} cells. The other issue is that IL-2 levels may be tightly regulated to control the balance of T_{reg} cells versus autoreactive pathogenic T cells. In the absence of Blimp-1, IL-2 levels might greatly increase, perhaps tipping the balance toward an auto-aggressive response. In any case, much more work is needed to define the role of Blimp-1 in T cells, including the biological relevance of the Blimp-1-dependent repressive IL-2 autoregulatory loop.

Other Cytokines Involved in the Development and Homeostasis of $CD4^+CD25^+$ T_{reg} Cells

One needs to consider that the β- and γ-chains of the high affinity IL-2R are also subunits to other cytokine receptors. In addition to the IL-2R, the common cytokine receptor γc is a signaling component to the IL-4, IL-7, IL-9, IL-15 and IL-21 receptors. Mutations in the γc or JAK3 gene results in X-linked severe combined immunodeficiency (XSCID) characterized by a lack of T and NK cells, with near normal numbers of B cells that primarily secrete IgM, but cannot undergo class switching [70–73]. Interestingly, a few T cells develop in γc-deficient mice with peripheral CD4 T cells that have an activated/memory phenotype [74]. Importantly, γc knockout mice have a more striking defect in T_{reg} cell production than IL-2- or IL-2R-deficient mice, with a near complete absence of $Foxp3^+$ T_{reg} cells in both the thymus and periphery [29,31]. Thus, besides the IL-2/IL-2R interaction, another γc-cytokine(s) must be critically important during thymic development of $CD4^+CD25^+Foxp3^+$ T_{reg} cells.

Two other γc cytokines that are key candidates for T_{reg} cells are IL-7 and IL-15 as they both activate STAT5. With respect to IL-7, two recent studies have shown that despite the dramatic reduction of T cells, IL-7- and IL-7Rα-knockout mice contain a normal fraction of $Foxp3^+$ T cells within the total CD4 T cell pool and do not exhibit autoimmunity [29,75]. IL-15 and IL-15Rα knockout mice also do not exhibit symptoms of autoimmunity but rather are characterized by impaired development of NK and NKT cells and abnormal homeostasis of memory CD8 T cells [76,77]. Moreover, $IL-15^{-/-}$ and $IL-15R\alpha^{-/-}$ mice have near normal numbers of $CD4^+CD25^+$ $Foxp3^+$ T_{reg} cells [[29] and A.L. Bayer, unpublished data]. Mice deficient in other γc cytokines including IL-4, IL-9, and IL-21 do not exhibit autoimmunity and live a normal lifespan. Thus, the lack of $Foxp3^+$ cells in γc-knockout mice results from the absence of IL-2/IL-2 interaction in combination with defect(s) in other γc cytokine receptor signaling pathways that remain to be defined.

As mentioned above, mice deficient in IL-7 and IL-7Rα have a substantial reduction in the number of thymocytes and peripheral lymphocytes with IL-7Rα-deficiency resulting in a somewhat greater block in the T and B cells compartments than seen in IL-7 knockout mice. This difference is probably because the IL-7Rα subunit is also a component of the thymic stromal lymphopoietin (TSLP) receptor complex [78,79]. Even though both the IL-7 and TSLP receptors activate the transcription factor STAT5 and induce the expression of common genes, there is little evidence that TSLP participates in T cell development. However, human Hassell's corpuscles, which is located in the thymic medulla and is composed of epithelial cells, was found to express TSLP and induced myeloid DC to express CD80 and CD86 [80]. Interestingly, these TSLP-stimulated DC induced proliferation and differentiation of CD4$^+$CD8$^-$CD25$^-$ thymic cells into CD4$^+$CD25$^+$Foxp3$^+$ T cells that depended upon MHC II, CD80/86, and IL-2. These DCs have been proposed to positively select medium- to high-affinity self-reactive thymocytes into T_{reg} cells [80]. These data, taken together with the fact that IL-7Rα-deficient mice have a normal functional population of T_{reg} cells and do not develop autoimmunity, indicate other molecules apart from TSLP are capable of supporting normal T_{reg} selection and development.

The role of transforming growth factor beta (TGFβ) in development, maintenance, and function of naturally occurring T_{reg} cells remains controversial. However, TGFβ has been shown to regulate the level of Foxp3 expression [81–84]. TGFβ signals through Smad proteins which upon their activation translocate to the nucleas and regulate the transcription of TGFβ-targeted genes [85,86]. However, aspects of TGFβ signaling is Smad-independent, in part through activation of the TGFβ-activating kinase (TAK1) [87,88]. TAK1 is a serine/threonine kinase and belongs to the family of MAPKKK family (mitogen activated protein kinase kinase kinase) and also mediates signaling initiated through IL-1β, TNFα, and toll-like receptors. Mice with a T cell-specific deletion of TAK1 exhibited colitis, splenomegaly and lymphadenopathy, with a high percentage of activated/memory T cells [89]. Importantly, these mice also exhibit a severe defect in the development of naturally occurring T_{reg} cells, with near complete absence of T_{reg} cells in thymus and periphery [89–91]. In mixed BM chimeras generated from WT and T cell-specific deletion of TAK1 donor mice, only WT BM was capable of generating CD4$^+$CD25$^+$Foxp3$^+$ T_{reg} cells [91]. Thus, TAK1 mediates an essential signaling pathway for Foxp3 expression and thymic development of T_{reg} cells. TAK1 may also indirectly control T_{reg} cell production and homeostasis though regulation of IL-2 and IL-2R expression. TAK1 was shown to be required for TCR-induced activation of NF-$\kappa\beta$ and JNK, upregulation of CD25, and IL-2 production [91,92]. Although TAK1 mediated signaling is critical for T_{reg} development, the precise pathways by which TAK1 promotes their generation is an open question and will require further investigation.

Concluding Remarks

It is now established that a critical non-redundant function of the IL-2/IL-2R interaction lies in controlling T_{reg} cells. Based on current data, IL-2 is necessary for thymic development and peripheral homeostasis of T_{reg} cells. We propose that there may be two important roles for IL-2 for T_{reg} cells. One is that it may act as a developmental checkpoint, essential for proper thymic development of T_{reg} cells. In the peripheral immune compartment, IL-2 then functions as a dominant growth and homeostatic regulator for these cells. This dual biological role for IL-2 is very analogous to IL-7 and IL-15 where they each control key early developmental steps within the thymus, only later to then importantly regulate homeostasis of mature conventional naïve and memory T cells. The link between the IL-2/IL-2R interaction, STAT5 activation, and upregulation of Foxp3 expression also raises the possibility that IL-2 may be also be critical for T_{reg} cell functional activity, as the Foxp3^{high} phenotype is associated with active suppressor T cells.

There is much interest to manipulate T_{reg} cells in vitro or in vivo for therapeutic purposes. With respect to IL-2, it is necessary for the growth of T_{reg} cell in vitro and may be an ideal molecule to support their proliferation in vivo. In this regard, there has been some progress in the last several years to expand naturally occurring T_{reg} cells with IL-2 and TCR stimulation in vitro, with antigen-specific rather than polyclonal T_{reg} cells being more efficacious for immunotherapy of autoimmune disease in several mouse models [93]. An issue of some interest is that signal transduction through the IL-2R for T_{reg} cells appears to be dominated by STAT5 activation whereas there is more complex signaling by conventional activated T lymphocytes. Thus, the selective blockade of these other pathways by use of rapamycin should facilitate IL-2-dependent T_{reg} cell growth while inhibiting typical activated T cells. Conversely, drugs that might target PTEN, which seems to be responsible for inhibiting IL-2-dependent PI3K activation in T_{reg} cells, might prove to be useful to promote T_{reg} cells growth in vitro or in vivo. Needless to say, future successful application of T_{reg} cells in the treatment of autoimmune diseases or tolerance induction to allogeneic transplants will likely parallel a much better understanding of contribution of IL-2 for T_{reg} cell growth and function, the relationship of T_{reg} cells to responding T cells in vivo, and the means by which T_{reg} cells suppress these activated cells.

Acknowledgments Our work is support by grants from the NIH and JDRF.

References

1. Sakaguchi, S., N. Sakaguchi, M. Asano, M. Itoh, and M. Toda. 1995. Immunologic self-tolerance maintained by activated T cells expressing IL-2 receptor α-chains (CD25). Breakdown of a single mechanism of self-tolerance causes various autoimmune diseases. *J Immunol* 155:1151–1164.
2. Malek, T.R., and A.L. Bayer. 2004. Tolerance, not immunity, crucially depends on IL-2. *Nat Rev Immunol* 4:665–674.
3. Shevach, E.M. 2000. Suppressor T cells: rebirth, function and homeostasis. *Curr Biol* 10:R572–575.
4. Sadlack, B., J. Lohler, H. Schorle, G. Klebb, H. Haber, E. Sickel, R.J. Noelle, and I. Horak. 1995. Generalized autoimmune disease in interleukin-2-deficient mice is triggered by an uncontrolled activation and proliferation of CD4$^+$ T cells. *Eur J Immunol* 25:3053–3059.
5. Sadlack, B., H. Merz, H. Schorle, A. Schimpl, A.C. Feller, and I. Horak. 1993. Ulcerative colitis-like disease in mice with a disrupted interleukin-2 gene. *Cell* 75:253–261.
6. Papiernik, M., M.L. de Moraes, C. Pontoux, F. Vasseur, and C. Penit. 1998. Regulatory CD4 T cells: expression of IL-2Rα chain, resistance to clonal deletion and IL-2 dependency. *Int Immunol* 10:371–378.

7. Suzuki, H., T.M. Kundig, C. Furlonger, A. Wakeham, E. Timms, T. Matsuyama, R. Schmits, J.J. Simard, P.S. Ohashi, H. Griesser et al. 1995. Deregulated T cell activation and autoimmunity in mice lacking interleulin 2 receptor β. *Science* 268:1472–1476.
8. Willerford, D.M., J. Chen, J.A. Ferry, L. Davidson, A. Ma, and F.W. Alt. 1995. Interleukin-2 receptor α-chain regulates the size and content of the peripheral lymphoid compartment. *Immunity* 3:521–530.
9. Kramer, S., A. Schimpl, and T. Hunig. 1995. Immunopathology of interleukin (IL) 2-deficient mice: thymus dependence and suppression by thymus-dependent cells with an intact IL-2 gene. *J Exp Med* 182:1769–1776.
10. Malek, T.R., A. Yu, V. Vincek, P. Scibelli, and L. Kong. 2002. CD4 regulatory T cells prevent lethal autoimmunity in IL-2Rβ-deficient mice. Implications for the nonredundant function of IL-2. *Immunity* 17:167–178.
11. Kneitz, B., T. Herrmann, S. Yonehara, and A. Schimpl. 1995. Normal clonal expansion but impaired Fas-mediated cell death and anergy induction in interleukin-2-deficient mice. *Eur J Immunol* 25:2572–2577.
12. Lenardo, M.J. 1991. Interleukin-2 programs mouse alpha beta T lymphocytes for apoptosis. *Nature* 353:858–861.
13. van Parijs, L., V.L. Perez, and A.K. Abbas. 1998. Mechanisms of peripheral T cell tolerance. *Novartis Found Symp* 215:5–14; discussion 14–20, 33–40.
14. Malek, T.R., B.O. Porter, E.K. Codias, P. Scibelli, and A. Yu. 2000. Normal lymphoid homeostasis and lack of lethal autoimmunity in mice containing mature T cells with severely impaired IL-2 receptors. *J Immunol* 164:2905–2914.
15. Suzuki, H., Y.W. Zhou, M. Kato, T.W. Mak, and I. Nakashima. 1999. Normal regulatory α/β T cells effectively eliminate abnormally activated T cells lacking the interleukin 2 receptor β in vivo. *J Exp Med* 190:1561–1572.
16. Wolf, M., A. Schimpl, and T. Hunig. 2001. Control of T cell hyperactivation in IL-2-deficient mice by $CD4^+CD25^-$ and $CD4^+CD25^+$ T cells: evidence for two distinct regulatory mechanisms. *Eur J Immunol* 31:1637–1645.
17. Klebb, G., I.B. Autenrieth, H. Haber, E. Gillert, B. Sadlack, K.A. Smith, and I. Horak. 1996. Interleukin-2 is indispensable for development of immunological self-tolerance. *Clin Immunol Immunopathol* 81:282–286.
18. Bayer, A., A. Yu, and T. Malek. 2007. Function of the IL-2R for thymic and peripheral $CD4^+CD25^+$ Foxp3$^+$ T regulatory cells. *J Immunol* 178:4062–4071.
19. Almeida, A.R., N. Legrand, M. Papiernik, and A.A. Freitas. 2002. Homeostasis of peripheral $CD4^+$ T cells: IL-2Rα and IL-2 shape a population of regulatory cells that controls $CD4^+$ T cell numbers. *J Immunol* 169:4850–4860.
20. Furtado, G.C., M.A. Curotto de Lafaille, N. Kutchukhidze, and J.J. Lafaille. 2002. Interleukin 2 signaling is required for $CD4^+$ regulatory T cell function. *J Exp Med* 196:851–857.
21. Sharfe, N., H.K. Dadi, M. Shahar, and C.M. Roifman. 1997. Human immune disorder arising from mutation of the alpha chain of the interleukin-2 receptor. *Proc Natl Acad Sci USA* 94:3168–3171.
22. Fontenot, J.D., M.A. Gavin, and A.Y. Rudensky. 2003. Foxp3 programs the development and function of $CD4^+CD25^+$ regulatory T cells. *Nat Immunol* 4:330–336.
23. Hori, S., T. Nomura, and S. Sakaguchi. 2003. Control of regulatory T cell development by the transcription factor Foxp3. *Science* 299:1057–1061.
24. Khattri, R., T. Cox, S.A. Yasayko, and F. Ramsdell. 2003. An essential role for Scurfin in $CD4^+CD25^+$ T regulatory cells. *Nat Immunol* 4:337–342.
25. Fontenot, J.D., J.L. Dooley, A.G. Farr, and A.Y. Rudensky. 2005. Developmental regulation of Foxp3 expression during ontogeny. *J Exp Med* 202:901–906.
26. Nelson, B.H. 2002. Interleukin-2 signaling and the maintenance of self-tolerance. *Curr Dir Autoimmun* 5:92–112.

27. Antov, A., L. Yang, M. Vig, D. Baltimore, and L. Van Parijs. 2003. Essential role for STAT5 signaling in CD25$^+$CD4$^+$ regulatory T cell homeostasis and the maintenance of self-tolerance. *J Immunol* 171:3435–3441.
28. Burchill, M.A., C.A. Goetz, M. Prlic, J.J. O'Neil, I.R. Harmon, S.J. Bensinger, L.A. Turka, P. Brennan, S.C. Jameson, and M.A. Farrar. 2003. Distinct effects of STAT5 activation on CD4$^+$ and CD8$^+$ T cell homeostasis: development of CD4$^+$CD25$^+$ regulatory T cells versus CD8$^+$ memory T cells. *J Immunol* 171:5853–5864.
29. Burchill, M.A., J. Yang, C. Vogtenhuber, B.R. Blazar, and M.A. Farrar. 2007. IL-2 receptor beta-dependent STAT5 activation is required for the development of Foxp3$^+$ regulatory T cells. *J Immunol* 178:280–290.
30. Snow, J.W., N. Abraham, M.C. Ma, B.G. Herndier, A.W. Pastuszak, and M.A. Goldsmith. 2003. Loss of tolerance and autoimmunity affecting multiple organs in STAT5A/5B-deficient mice. *J Immunol* 171:5042–5050.
31. Fontenot, J.D., J.P. Rasmussen, M.A. Gavin, and A.Y. Rudensky. 2005. A function for interleukin 2 in Foxp3-expressing regulatory T cells. *Nat Immunol* 6:1142–1151.
32. Cohen, A.C., K.C. Nadeau, W. Tu, V. Hwa, K. Dionis, L. Bezrodnik, A. Teper, M. Gaillard, J. Heinrich, A.M. Krensky, R.G. Rosenfeld, and D.B. Lewis. 2006. Cutting edge: decreased accumulation and regulatory function of CD4$^+$ CD25$^{(high)}$ T cells in human STAT5b deficiency. *J Immunol* 177:2770–2774.
33. Zorn, E., E.A. Nelson, M. Mohseni, F. Porcheray, H. Kim, D. Litsa, R. Bellucci, E. Raderschall, C. Canning, R.J. Soiffer, D.A. Frank, and J. Ritz. 2006. IL-2 regulates FOXP3 expression in human CD4$^+$CD25$^+$ regulatory T cells through a STAT-dependent mechanism and induces the expansion of these cells in vivo. *Blood* 108:1571–1579.
34. Murawski, M.R., S.A. Litherland, M.J. Clare-Salzler, and A. Davoodi-Semiromi. 2006. Upregulation of Foxp3 expression in mouse and human Treg is IL-2/STAT5 dependent: implications for the NOD STAT5B mutation in diabetes pathogenesis. *Ann N Y Acad Sci* 1079: 198–204.
35. Sugimoto, N., T. Oida, K. Hirota, K. Nakamura, T. Nomura, T. Uchiyama, and S. Sakaguchi. 2006. Foxp3-dependent and -independent molecules specific for CD25$^+$CD4$^+$ natural regulatory T cells revealed by DNA microarray analysis. *Int Immunol* 18:1197–1209.
36. Bensinger, S.J., P.T. Walsh, J. Zhang, M. Carroll, R. Parsons, J.C. Rathmell, C.B. Thompson, M.A. Burchill, M.A. Farrar, and L.A. Turka. 2004. Distinct IL-2 receptor signaling pattern in CD4$^+$CD25$^+$ regulatory T cells. *J Immunol* 172:5287–5296.
37. Walsh, P.T., J.L. Buckler, J. Zhang, A.E. Gelman, N.M. Dalton, D.K. Taylor, S.J. Bensinger, W.W. Hancock, and L.A. Turka. 2006. PTEN inhibits IL-2 receptor-mediated expansion of CD4$^+$ CD25$^+$ Tregs. *J Clin Invest* 116:2521–2531.
38. Battaglia, M., A. Stabilini, B. Migliavacca, J. Horejs-Hoeck, T. Kaupper, and M.G. Roncarolo. 2006. Rapamycin promotes expansion of functional CD4$^+$CD25$^+$FOXP3$^+$ regulatory T cells of both healthy subjects and type 1 diabetic patients. *J Immunol* 177:8338–8347.
39. Battaglia, M., A. Stabilini, and M.G. Roncarolo. 2005. Rapamycin selectively expands CD4$^+$CD25$^+$FoxP3$^+$ regulatory T cells. *Blood* 105:4743–4748.
40. Strauss, L., T.L. Whiteside, A. Knights, C. Bergmann, A. Knuth, and A. Zippelius. 2007. Selective survival of naturally occurring human CD4$^+$CD25$^+$Foxp3$^+$ regulatory T cells cultured with rapamycin. *J Immunol* 178:320–329.
41. D'Cruz, L.M., and L. Klein. 2005. Development and function of agonist-induced CD25$^+$Foxp3$^+$ regulatory T cells in the absence of interleukin 2 signaling. *Nat Immunol* 6:1152–1159.
42. Bayer, A.L., A. Yu, D. Adeegbe, and T.R. Malek. 2005. Essential role for interleukin-2 for CD4$^+$CD25$^+$ T regulatory cell development during the neonatal period. *J Exp Med* 201: 769–777.
43. Setoguchi, R., S. Hori, T. Takahashi, and S. Sakaguchi. 2005. Homeostatic maintenance of natural Foxp3$^+$ CD25$^+$ CD4$^+$ regulatory T cells by interleukin (IL)-2 and induction of autoimmune disease by IL-2 neutralization. *J Exp Med* 201:723–735.

44. Murakami, M., A. Sakamoto, J. Bender, J. Kappler, and P. Marrack. 2002. CD25$^+$CD4$^+$ T cells contribute to the control of memory CD8$^+$ T cells. *Proc Natl Acad Sci USA* 99: 8832–8837.
45. Cozzo, C., J. Larkin, 3rd, and A.J. Caton. 2003. Cutting edge: self-peptides drive the peripheral expansion of CD4$^+$CD25$^+$ regulatory T cells. *J Immunol* 171:5678–5682.
46. Guiducci, C., B. Valzasina, H. Dislich, and M.P. Colombo. 2005. CD40/CD40L interaction regulates CD4$^+$CD25$^+$ Treg homeostasis through dendritic cell-produced IL-2. *Eur J Immunol* 35:557–567.
47. Sempowski, G.D., S.J. Cross, C.S. Heinly, R.M. Scearce, and B.F. Haynes. 2004. CD7 and CD28 are required for murine CD4$^+$CD25$^+$ regulatory T cell homeostasis and prevention of thyroiditis. *J Immunol* 172:787–794.
48. Tang, Q., K.J. Henriksen, E.K. Boden, A.J. Tooley, J. Ye, S.K. Subudhi, X.X. Zheng, T.B. Strom, and J.A. Bluestone. 2003. Cutting edge: CD28 controls peripheral homeostasis of CD4$^+$CD25$^+$ regulatory T cells. *J Immunol* 171:3348–3352.
49. Yamazaki, S., T. Iyoda, K. Tarbell, K. Olson, K. Velinzon, K. Inaba, and R.M. Steinman. 2003. Direct expansion of functional CD25$^+$ CD4$^+$ regulatory T cells by antigen-processing dendritic cells. *J Exp Med* 198:235–247.
50. Zheng, G., B. Wang, and A. Chen. 2004. The 4-1BB costimulation augments the proliferation of CD4$^+$CD25$^+$ regulatory T cells. *J Immunol* 173:2428–2434.
51. Thornton, A.M., C.A. Piccirillo, and E.M. Shevach. 2004. Activation requirements for the induction of CD4$^+$CD25$^+$ T cell suppressor function. *Eur J Immunol* 34:366–376.
52. de la Rosa, M., S. Rutz, H. Dorninger, and A. Scheffold. 2004. Interleukin-2 is essential for CD4$^+$CD25$^+$ regulatory T cell function. *Eur J Immunol* 34:2480–2488.
53. Thornton, A.M., E.E. Donovan, C.A. Piccirillo, and E.M. Shevach. 2004. Cutting Edge: IL-2 Is Critically Required for the In Vitro Activation of CD4$^+$CD25$^+$ T Cell Suppressor Function. *J Immunol* 172:6519–6523.
54. Min, B., R. McHugh, G.D. Sempowski, C. Mackall, G. Foucras, and W.E. Paul. 2003. Neonates support lymphopenia-induced proliferation. *Immunity* 18:131–140.
55. Schneider, R., R.K. Lees, T. Pedrazzini, R.M. Zinkernagel, H. Hengartner, and H.R. MacDonald. 1989. Postnatal disappearance of self-reactive (Vβ6$^+$) cells from the thymus of Mls^a mice. Implications for T cell development and autoimmunity. *J Exp Med* 169: 2149–2158.
56. Tarbell, K.V., L. Petit, X. Zuo, P. Toy, X. Luo, A. Mqadmi, H. Yang, M. Suthanthiran, S. Mojsov, and R.M. Steinman. 2007. Dendritic cell-expanded, islet-specific CD4$^+$ CD25$^+$ CD62L$^+$ regulatory T cells restore normoglycemia in diabetic NOD mice. *J Exp Med* 204:191–201.
57. Tarbell, K.V., S. Yamazaki, K. Olson, P. Toy, and R. Steinman. 2004. CD25$^+$CD4$^+$ T cells, expanded with dendritic cells presenting a single autoantigenic peptide, suppress autoimmune diabetes. *J Exp Med* 199:1467–1477.
58. Yamazaki, S., K. Inaba, K.V. Tarbell, and R.M. Steinman. 2006. Dendritic cells expand antigen-specific Foxp3$^+$ CD25$^+$ CD4$^+$ regulatory T cells including suppressors of alloreactivity. *Immunol Rev* 212:314–329.
59. Yamazaki, S., M. Patel, A. Harper, A. Bonito, H. Fukuyama, M. Pack, K.V. Tarbell, M. Talmor, J.V. Ravetch, K. Inaba, and R.M. Steinman. 2006. Effective expansion of alloantigen-specific Foxp3$^+$ CD25$^+$ CD4$^+$ regulatory T cells by dendritic cells during the mixed leukocyte reaction. *Proc Natl Acad Sci U S A* 103:2758–2763.
60. Granucci, F., S. Feau, V. Angeli, F. Trottein, and P. Ricciardi-Castagnoli. 2003. Early IL-2 production by mouse dendritic cells is the result of microbial-induced priming. *J Immunol* 170:5075–5081.
61. Granucci, F., C. Vizzardelli, N. Pavelka, S. Feau, M. Persico, E. Virzi, M. Rescigno, G. Moro, and P. Ricciardi-Castagnoli. 2001. Inducible IL-2 production by dendritic cells revealed by global gene expression analysis. *Nat Immunol* 2:882–888.

62. Almeida, A.R., B. Zaragoza, and A.A. Freitas. 2006. Indexation as a novel mechanism of lymphocyte homeostasis: the number of CD4+CD25+ regulatory T cells is indexed to the number of IL-2-producing cells. *J Immunol* 177:192–200.
63. Yu, A., and T.R. Malek. 2006. Selective availability of IL-2 is a major determinant controlling the production of CD4+CD25+Foxp3+ T regulatory cells. *J Immunol* 177:5115–5121.
64. Gong, D., and T.R. Malek. 2007. Cytokine-dependent Blimp-1 expression in activated T cells inhibits IL-2 production. *J Immunol* 178:242–252.
65. Villarino, A.V., C.M. Tato, J.S. Stumhofer, Z. Yao, Y.K. Cui, L. Hennighausen, J. O'Shea J, and C.A. Hunter. 2007. Helper T cell IL-2 production is limited by negative feedback and STAT-dependent cytokine signals. *J Exp Med* 204:65–71.
66. Shapiro-Shelef, M., K.I. Lin, L.J. McHeyzer-Williams, J. Liao, M.G. McHeyzer-Williams, and K. Calame. 2003. Blimp-1 is required for the formation of immunoglobulin secreting plasma cells and pre-plasma memory B cells. *Immunity* 19:607–620.
67. Turner, C.A. Jr., D.H. Mack, and M.M. Davis. 1994. Blimp-1, a novel zinc finger-containing protein that can drive the maturation of B lymphocytes into immunoglobulin-secreting cells. *Cell* 77:297–306.
68. Kallies, A., E.D. Hawkins, G.T. Belz, D. Metcalf, M. Hommel, L.M. Corcoran, P.D. Hodgkin, and S.L. Nutt. 2006. Transcriptional repressor Blimp-1 is essential for T cell homeostasis and self-tolerance. *Nat Immunol* 7:466–474.
69. Martins, G.A., L. Cimmino, M. Shapiro-Shelef, M. Szabolcs, A. Herron, E. Magnusdottir, and K. Calame. 2006. Transcriptional repressor Blimp-1 regulates T cell homeostasis and function. *Nat Immunol* 7:457–465.
70. Noguchi, M., H. Yi, H.M. Rosenblatt, A.H. Filipovich, S. Adelstein, W.S. Modi, O.W. McBride, and W.J. Leonard. 1993. Interleukin-2 receptor gamma chain mutation results in X-linked severe combined immunodeficiency in humans. *Cell* 73:147–157.
71. Nosaka, T., J.M. van Deursen, R.A. Tripp, W.E. Thierfelder, B.A. Witthuhn, A.P. McMickle, P.C. Doherty, G.C. Grosveld, and J.N. Ihle. 1995. Defective lymphoid development in mice lacking Jak3. *Science* 270:800–802.
72. Russell, S.M., N. Tayebi, H. Nakajima, M.C. Riedy, J.L. Roberts, M.J. Aman, T.S. Migone, M. Noguchi, M.L. Markert, R.H. Buckley, J.J. O'Shea, and W.J. Leonard. 1995. Mutation of Jak3 in a patient with SCID: essential role of Jak3 in lymphoid development. *Science* 270: 797–800.
73. Thomis, D.C., C.B. Gurniak, E. Tivol, A.H. Sharpe, and L.J. Berg. 1995. Defects in B lymphocyte maturation and T lymphocyte activation in mice lacking Jak3. *Science* 270:794–797.
74. Cao, X., E.W. Shores, J. Hu-Li, M.R. Anver, B.L. Kelsall, S.M. Russell, J. Drago, M. Noguchi, A. Grinberg, E.T. Bloom et al. 1995. Defective lymphoid development in mice lacking expression of the common cytokine receptor gamma chain. *Immunity* 2:223–238.
75. Peffault de Latour, R., H.C. Dujardin, F. Mishellany, O. Burlen-Defranoux, J. Zuber, R. Marques, J. Di Santo, A. Cumano, P. Vieira, and A. Bandeira. 2006. Ontogeny, function, and peripheral homeostasis of regulatory T cells in the absence of interleukin-7. *Blood* 108:2300–2306.
76. Kennedy, M.K., M. Glaccum, S.N. Brown, E.A. Butz, J.L. Viney, M. Embers, N. Matsuki, K. Charrier, L. Sedger, C.R. Willis, K. Brasel, P.J. Morrissey, K. Stocking, J.C. Schuh, S. Joyce, and J.J. Peschon. 2000. Reversible defects in natural killer and memory CD8 T cell lineages in interleukin 15-deficient mice. *J Exp Med* 191:771–780.
77. Lodolce, J.P., D.L. Boone, S. Chai, R.E. Swain, T. Dassopoulos, S. Trettin, and A. Ma. 1998. IL-15 receptor maintains lymphoid homeostasis by supporting lymphocyte homing and proliferation. *Immunity* 9:669–676.
78. Liu, Y.J., V. Soumelis, N. Watanabe, T. Ito, Y.H. Wang, R.D. Malefyt, M. Omori, B. Zhou, and S.F. Ziegler. 2006. TSLP: an Epithelial Cell Cytokine that Regulates T Cell Differentiation by Conditioning Dendritic Cell Maturation. *Annu Rev Immunol* 25:193–219.
79. Ziegler, S.F., and Y.J. Liu. 2006. Thymic stromal lymphopoietin in normal and pathogenic T cell development and function. *Nat Immunol* 7:709–714.

80. Watanabe, N., Y.H. Wang, H.K. Lee, T. Ito, Y.H. Wang, W. Cao, and Y.J. Liu. 2005. Hassall's corpuscles instruct dendritic cells to induce CD4+CD25+ regulatory T cells in human thymus. *Nature* 436:1181–1185.
81. Chen, W., W. Jin, N. Hardegen, K.J. Lei, L. Li, N. Marinos, G. McGrady, and S.M. Wahl. 2003. Conversion of peripheral CD4+CD25- naive T cells to CD4+CD25+ regulatory T cells by TGF-beta induction of transcription factor Foxp3. *J Exp Med* 198:1875–1886.
82. Fu, S., N. Zhang, A.C. Yopp, D. Chen, M. Mao, D. Chen, H. Zhang, Y. Ding, and J.S. Bromberg. 2004. TGF-beta induces Foxp3 + T-regulatory cells from CD4 + CD25 – precursors. *Am J Transplant* 4:1614–1627.
83. Marie, J.C., J.J. Letterio, M. Gavin, and A.Y. Rudensky. 2005. TGF-beta1 maintains suppressor function and Foxp3 expression in CD4+CD25+ regulatory T cells. *J Exp Med* 201:1061–1067.
84. Schramm, C., S. Huber, M. Protschka, P. Czochra, J. Burg, E. Schmitt, A.W. Lohse, P.R. Galle, and M. Blessing. 2004. TGFbeta regulates the CD4+CD25+ T-cell pool and the expression of Foxp3 in vivo. *Int Immunol* 16:1241–1249.
85. Gorelik, L., and R.A. Flavell. 2002. Transforming growth factor-beta in T-cell biology. *Nat Rev Immunol* 2:46–53.
86. Letterio, J.J. 2005. TGF-beta signaling in T cells: roles in lymphoid and epithelial neoplasia. *Oncogene* 24:5701–5712.
87. Watkins, S.J., L. Jonker, and H.M. Arthur. 2006. A direct interaction between TGFbeta activated kinase 1 and the TGFbeta type II receptor: implications for TGFbeta signalling and cardiac hypertrophy. *Cardiovasc Res* 69:432–439.
88. Yu, L., M.C. Hebert, and Y.E. Zhang. 2002. TGF-beta receptor-activated p38 MAP kinase mediates Smad-independent TGF-beta responses. *Embo J* 21:3749–3759.
89. Sato, S., H. Sanjo, T. Tsujimura, J. Ninomiya-Tsuji, M. Yamamoto, T. Kawai, O. Takeuchi, and S. Akira. 2006. TAK1 is indispensable for development of T cells and prevention of colitis by the generation of regulatory T cells. *Int Immunol* 18:1405–1411.
90. Liu, H.H., M. Xie, M.D. Schneider, and Z.J. Chen. 2006. Essential role of TAK1 in thymocyte development and activation. *Proc Natl Acad Sci USA* 103:11677–11682.
91. Wan, Y.Y., H. Chi, M. Xie, M.D. Schneider, and R.A. Flavell. 2006. The kinase TAK1 integrates antigen and cytokine receptor signaling for T cell development, survival and function. *Nat Immunol* 7:851–858.
92. Sakurai, H., P. Singhirunnusorn, E. Shimotabira, A. Chino, S. Suzuki, K. Koizumi, and I. Saiki. 2005. TAK1-mediated transcriptional activation of CD28-responsive element and AP-1-binding site within the IL-2 promoter in Jurkat T cells. *FEBS Lett* 579:6641–6646.
93. Tang, Q., and J.A. Bluestone. 2006. Regulatory T-cell physiology and application to treat autoimmunity. *Immunol Rev* 212:217–237.

Chapter 5
IL-2 Signaling and CD4+ CD25+ Regulatory T Cells

Louise M. D'Cruz and Ludger Klein

Abstract Over the last two decades, our understanding of the function of Interleukin-2 (IL-2) has experienced several paradigmatic shifts. Although IL-2 was initially identified as a T cell growth factor, loss of function experiments clearly showed that it rather acts as a gatekeeper of immune homeostasis and tolerance. It is now widely accepted that the major non-redundant function of IL-2 is the maintenance of naturally occurring CD25+ Foxp3+ regulatory T cells (T_{reg}). Importantly, this role as an essential survival factor may blur the interpretation of loss-of-function studies that tried to address other, mutually not exclusive functions of IL-2 in T_{reg} biology, such as development and function. This chapter will summarize our current understanding of how IL-2 signaling may relate to these aspects of immune regulation by T_{reg}.

Introduction

Interleukin-2 (IL-2) was identified over twenty years ago [1]. At the time, it was common practice to use supernatant of activated T cells to enhance the growth of T cell lines and clones *in vitro*. One of the factors from this supernatant was cloned and identified as IL-2, giving rise to the notion that IL-2 acted as a growth factor for T cells, at least *in vitro* [1]. Its structure is that of a small four-helix bundle cytokine with a molecular weight of 15.5 kDa (133 amino acids). IL-2 production is generally believed to be largely confined to activated T cells, although some reports exist on IL-2 production by mesenchymal cells of the skin, intestinal epithelium and activated dendritic cells, whereby the biological significance of IL-2 sources other than T cells has remained unclear [2, 3]. The IL-2 receptor (IL-2R) is a heterotrimeric complex consisting of a unique IL-2Rα chain (CD25), the IL-2Rβ chain (CD122), which is also a component of the IL-15 receptor (therefore also

L. Klein
Institute for Immunology, Ludwig-Maximilians-University Munich, Goethestrasse 31, D-80336 Munich, Germany
e-mail: ludger.klein@med.uni-muenchen.de

known as IL-15Rβ), and the common gamma chain γc (CD132), that is shared by the receptors for IL-4, IL-7, IL-9, IL-15 and IL-21 [4, 5]. Consistent with a role of IL-2 signaling in an autocrine feedback loop to enhance T cell expansion, the high affinity IL-2R is expressed on the surface of T cells within hours after activation. This is mainly achieved through up-regulation of CD25, whereas the IL-2Rβ chain and to a lesser extent γc are expressed in a more constitutive way. The function of IL-2 in the promotion of T cell proliferation was the rationale for a number of experimental therapies aiming to boost IL-2 signaling in tumor-therapy [6] or immunodeficiency [7], as well as for pharmacological intervention with IL-2 signaling in autoimmune or transplant settings [8]. The way in which data from the past few years have changed our view of IL-2, evolving from that of a simple pacemaker of immune responses to a multifaceted regulator of tolerance and immunity, is a very lucid example of the complexities inherent to translating *in vitro* observations into therapeutical applications.

The simple view that IL-2 only acts as a T cell growth factor was challenged already in the early 1990s, when it was found that IL-2 knock-out mice did not show obvious immuno-deficiencies [9]. Instead, these mice as well as mice with targeted mutations in the genes coding for IL-2Rα or IL-2Rβ succumbed early in life to what was called "IL-2 deficiency syndrome" [10–12]. This immunopathological condition was characterized by autoimmune manifestations such as severe splenomegaly and lymphadenopathy, as well as colitis in the case of C57BL/6 mice or anemia and wasting in the case of BALB/c mice. It was clearly T cell dependent because athymic Foxn1$^{-/-}$ mice, also known as *nude*, or recombination activating gene 2 (RAG-2) deficient mice (i.e. lacking T and B cells) did not develop disease when bred onto an IL-2 deficient background [13, 14]. Together, these experiments underscored that IL-2 signaling critically impinges on the maintenance of immune homeostasis and self-tolerance. At the same time, the growth promoting function of IL-2 obviously operates in a redundant context, whereby IL-2 deficiency may *in vivo* be compensated for by other cytokines such as IL-4 or IL-15, which can support survival or proliferation of T cells *in vitro* [15].

The first hypothesis that was put forward to explain the phenotype of mice with disrupted IL-2 signaling postulated an essential role of IL-2 for activation induced cell death (AICD) [16]. AICD controls immune homeostasis through balancing the expansion phase of immune responses with a subsequent phase of contraction. Mechanistically, IL-2 was linked to AICD when it was shown that incubation of activated T cells with IL-2 *in vitro* can lead to up regulation of Fas ligand and tumor necrosis factor receptor (TNFR) as well as down regulation of c-FLIP, a caspase inhibitor [17, 18]. Until today, it has remained controversial in how far promotion of AICD indeed represents a non-redundant function of IL-2 signaling. For instance, contradictory results have been published as to whether clonal contraction and apoptosis of activated T cells *in vivo* is impaired in the absence of IL-2Rα or IL-2 [19–21].

In the mid 90's, several labs provided evidence arguing against a cell intrinsic function of IL-2 signaling for the maintenance of immune homeostasis through priming cells for AICD. Thus, mixed bone-marrow chimeras in which 30% of hematopoietic cells carried a functional IL-2 gene did not develop the lymphoproliferative disease observed in IL-2 deficient mice [13]. More importantly, similar experiments using IL-2Rβ deficient mice clearly showed that wild-type cells did not simply provide paracrine IL-2 in order to sensitize for AICD. Instead, these results implied that IL-2 signaling was required for the generation and/or maintenance of a regulatory T cell population.

Papiernik et al. were the first to suggest a role of IL-2 in the biology of $CD4^+CD25^+$ T_{reg} when they found that these cells were dramatically reduced in IL-2 deficient mice [22]. Constitutive expression of the high affinity IL-2R on the T_{reg} population strongly suggested a role for IL-2 signaling in the development and/or function of T_{reg} cells. Subsequently, it was also found that mice deficient in IL-2Rβ have reduced frequencies of $CD25^+$ T_{reg} in spleen and lymph nodes, and that adoptive transfer of $CD25^+$ T_{reg} rescued the autoimmunity of IL-2Rα and IL-2Rβ deficient mice [23, 24]. A large body of evidence now supports the prevailing view that the essential, non-redundant function of IL-2 is the maintenance of immune homeostasis through the action of IL-2 dependent T_{reg}. Deficiencies in components of the IL-2 signaling axis lead to a strong reduction in the number of these cells, and it is commonly accepted that IL-2 signaling is critical for the maintenance of these cells. While its role as a survival factor for T_{reg} is perhaps the most clearly documented function of IL-2, other, mutually not exclusive roles for IL-2 in T_{reg} development and function are controversial and less well explored. These aspects will be the topic of the present chapter.

IL-2 Signaling: An Overview

Despite the fact that the IL-2R lacks intrinsic kinase activity, its engagement leads to protein phosphorylation. This operates through protein kinases that can physically associate with the cytoplasmic tails of the IL-2Rβ and IL-2Rγ chain, such as the *src* family member *lck*, the Janus kinases (Jak) 1 and 3, and Syk [25]. Of note, Jak3, like the IL-2Rα chain, is not constitutively expressed in naïve T cells, but is only expressed upon activation. Moreover, Jak3 is downregulated in anergic T cells. Thus, Jak3 may be a critical gatekeeper of IL-2 signaling, and whether or not it contributes to IL-2R signaling could be important for the interpretation of signals leading to survival and/or proliferation.

In naïve T cells stimulated through the TCR in the presence of (autocrine) IL-2, Jak1 and Jak3, that are associated with the IL-2Rβ and IL-2Rγ [26, 27], respectively, activate *lck* and in concert phosphorylate tyrosine residues in the IL-2R subunits. It is believed that distinct tyrosine residues in the IL-2Rβ chain then determine the diversity, specificity and/or redundancy of downstream events [28, 29], as they represent docking sites for secondary signaling molecules via SH2 domains.

Perhaps the best known outcome of IL-2 signaling is the activation of signal transducers and activators of transcription (Stats) [30]. Stat5a and Stat5b, latent cytoplasmic transcription factors, associate with phosphotyrosine residues in the IL-2Rβ subunit. Within this signalosome, they become phosphorylated, leading to their dimerisation and subsequent translocation to the nucleus where they act as transcription factors for genes mostly associated with survival or proliferation, e.g. the G_1 cyclin D3 and c-*myc* [31]. Importantly, although Stat activity clearly contributes to cell cycle progression, it may not be absolutely essential for proliferation in response to IL-2.

Other transcription factors involved in IL-2R signaling are the AP-1 proteins *c-fos* and *c-jun*, acting as downstream effectors of the p38 mitogen activated protein kinase (MAPK)/JNK pathway [32]. Importantly,*c-fos* and *c-jun* are induced only when Jak3 is activated. This signaling cascade, most likely triggered through the non-receptor protein kinase Pyk2 in association with Jak3 [33], has mostly been implicated in driving proliferation of IL-2 stimulated cells.

Two IL-2 dependent signaling cascades emanate from the adaptor protein Shc that associates with the phosphorylated tyrosine 388 on the IL-2Rβ chain [34, 35]. First, phosphorylated Shc can recruit Grb2 and mSOS which will convert inactive Ras to active Ras, thus leading to activation of the MAP kinase pathway. Ultimately, this is translated into a proliferative response as a result of phosphorylation of Cdk2, which in complex with cyclin E translocates to the nucleus and drives cell cycle progression [36]. Second, IL-2 signaling through Shc can result in a survival/proliferation signal through the serine/threonine kinase AKT (protein kinase B). Activation of AKT occurs upon activation of the phosphatidylinositol 3-kinase (PI3K) pathway [37, 38]. The PI3K pathway may ultimately activate anti-apoptotic proteins such as Bcl-2, Bcl-x_L or the cellular Fas-associated death domain-like IL-1β converting enzyme inhibitory protein (c-FLIP) [39–41]. At the same time, it impinges on proliferation through activating the cell cycle regulators p70^{s6kinase} and E2F and down-regulating the cell cycle inhibitor p27^{kip} [42, 43].

Taken together, the signaling pathways activated upon binding of IL-2 to its high affinity receptor are extremely complex and not fully understood. It appears that efficient progression through the cell cycle as well as survival depend on the coordinated action of multiple signaling pathways. At present, we still lack a precise understanding of how signals emanating from the IL-2R mediate these cellular responses in a synergistic or eventually non-redundant manner.

Discrete IL-2 Signaling in CD4$^+$ CD25$^+$ Regulatory T Cells?

There can be little doubt that IL-2 signaling is essential for T_{reg} homeostasis *in vivo*, whereby IL-2 needs to be acquired in a paracrine fashion. This has been extensively documented in both IL-2- and IL-2R-deficient mice [22–24], where the frequency of T_{reg} is dramatically reduced, although their thymic generation appears "normal" [44, 45]. Similarly, adoptive transfer of peripheral T_{reg} into IL-2 deficient hosts [45] or IL-2 depletion experiments [46] clearly established this non-redundant function of IL-2 signaling. Consistent with this, in STAT5-deficient mice, the frequency of T_{reg} is dramatically reduced, while over-expression of active STAT5 in transgenic mice leads to an increase in T_{reg} frequency [47, 48].

What is less well defined is whether T_{reg} homeostasis requires "unique" IL-2 signaling that may differ from that in conventional T cells. Similarly, the above mentioned *in vivo* data may not allow distinguishing whether IL-2 provides *bona fide* survival signals or whether it may drive basal levels of proliferation that help to maintain the size of the T_{reg} pool *in vivo*. Bensinger et al. have recently addressed these questions in an *in vitro* system [49]. IL-2 was found to significantly increase the viability of *in vitro* cultured T_{reg} in the absence of any appreciable level of cycling of cells, consistent with the notion that IL-2 somehow maintains the metabolic fitness of T_{reg} [44]. The increased viability of T_{reg} in the presence of IL-2 involved signaling through the Jak/Stat pathway and upregulation of bcl-x_L, but not bcl-2. Strikingly, no detectable signaling through the PI3K/AKT pathway occured after IL-2 stimulation of T_{reg} *in vitro*, and inhibitors of PI3K did not revert the beneficial effect of IL-2 on the viability of T_{reg}. The absence of PI3K downstream signaling coincided with relatively high levels of the lipid phosphatase PTEN, a negative regulator of the PI3K pathway. Bensinger et al. interpreted their data to indicate that IL-2 signaling in T_{reg} would indeed be distinct from that in conventional T cells. Thus, PTEN would be critical to maintain the anergic state of T_{reg}, and it would do so by dissociating Jak/Stat promoted survival from PI3K/AKT mediated proliferation. Importantly, the same study showed that TCR ligation of T_{reg} down-modulated PTEN, thereby overcoming the proliferative block. Consistent with this, the *in vitro* anergy of T_{reg} can be broken by stimulation with antigen plus IL-2.

It appears highly unlikely that the *in vitro* anergy of T_{reg} reflects a physiological property of these cells in their *in vivo* environment. First, antigen specific T_{reg} readily proliferate upon antigen encounter *in vivo* [50, 51]. Second, the natural T_{reg} repertoire appears to be strongly biased in favor of autoreactive TCR-specificities [52, 53]. Thus, release of PTEN-mediated inhibition should be a common occurrence *in vivo*, so that it remains to be seen in how far the "distinct signaling model" is relevant for the *in vivo* behavior of T_{reg}.

A recent study elegantly demonstrated that T_{reg} may actually not have discrete IL-2 signaling requirements. Here, Yu and Malek tested whether transgenic expression of a chimeric receptor consisting of the extracellular domain of the IL-2Rβ chain and the cytoplasmic domain of the IL-7Rα could rescue the autoimmunity seen in IL-2Rβ-deficient mice [54]. Such a chimeric receptor is able to bind extracellular IL-2 like the IL-2R itself, while its cytoplasmic signaling modules, i.e. IL-7α and γc, are those of the IL-7 receptor. Importantly, signaling through the IL-7R overlaps with that of the IL-2R in that both activate the Jak/Stat5 and the PI3K/AKT pathways. It was found that "IL-2 dependent IL-7 signaling" indeed rescued the autoimmunity in IL-2Rβ-deficient mice and restored a nearly normal frequency of $CD25^+$ T_{reg}. This finding illustrates the striking redundancy between the signaling pathways of γc cytokines. Moreover, it shows that the non-redundant biological activities of cytokines may depend on their selective availability within certain "niches" *in vivo* or a narrow restriction of receptor expression rather than on discrete modes of intracellular signaling. Accordingly, Yu and Malek suggested that rather than unique IL-2 signaling *per se* being important for homeostasis of T_{reg}, it may be the selective compartmentalization of T_{reg} and IL-2 producing cells that renders IL-2 critical for T_{reg} survival.

What is the critical "niche" that provides IL-2 for Treg homeostasis? Data from the Sakaguchi lab showed that IL-2 mRNA was constitutively produced by $CD25^{intermediate}$ populations of non-regulatory T cells. These cells may represent one possible source or niche for T_{reg} to acquire the IL-2 required for their survival [46]. However, as previously mentioned, IL-2 has also been shown to be produced by various other cell types and it may be that *in vivo* these cells would provide IL-2 for T_{reg} homeostasis [2, 3, 55].

Finally, a discussion regarding IL-2 signaling in T_{reg} would not be complete without a note on the role of Foxp3. The transcription factor Foxp3 has been shown to be essential for T_{reg} development [56, 57]. Recent data indicate that Foxp3 may actually be responsible for repression of T_{reg} mediated production of IL-2 [58]. Here, the authors showed that Foxp3, in complex with NFAT, can repress expression of IL-2 in cells transduced with both of these transcription factors. Targeted mutagenesis of Foxp3 in the putative Foxp3/NFAT interface led to a disruption in the ability of these two factors to form a complex and to repress IL-2 production [58]. When further analyzed, the authors noticed that two sites in the IL-2 promoter could act as predicted NFAT/Foxp3 binding sites and using chromatin immunoprecipitation, they showed that NFAT and Foxp3 could both occupy the IL-2 promoter.

Taken together, the evidence for discrete signaling in T_{reg} upon IL-2 stimulation is rather sparce. It rather appears that it is the (at least in part Foxp3 mediated) inability of T_{reg} themselves to produce autocrine IL-2 in conjunction with their absolute dependence on paracrine IL-2 (perhaps not so co-incidentally produced by the cells that are controlled by T_{reg}) that renders IL-2 signaling unique for T_{reg} homeostasis.

IL-2 and *In Vitro* Suppression

The function of IL-2 signaling in $CD4^+CD25^+$ T_{reg} mediated suppression *in vitro* has been studied using the classic cell-culture suppression assay. Suppression by T_{reg} *in vitro* appears to be cell contact or close proximity dependent, as separation of the two cell types by a permeable membrane in transwell assays prevents suppression. This has led to the notion that *in vitro* suppression may be mediated by surface molecules on T_{reg}, and membrane bound TGF-β appeared to be a good candidate [59]. However, these results have been called into question by studies were cells with disrupted TGF-β signaling pathways remained responsive to *in vitro* suppression [60]. A commonly ignored alternative scenario that would explain absence of suppression when T_{reg} and naïve responders are separated in trans-well assays is competition for a factor derived from naïve responders. IL-2 certainly is a prime candidate as it is produced by naïve responders, but not by T_{reg}, and since T_{reg} possess an inherent competitive advantage for IL-2 due to their constitutive expression of the high affinity IL-2R.

What is the evidence for a competition mechanism of *in vitro* suppression? De la Rosa et al. investigated this possibility by first asking a very simple question. Can neutralization of IL-2 *in vitro* mimic the action of T_{reg}? They found that addition of anti-IL-2 antibody to an anti-CD3 stimulated monoculture of naïve responder cells indeed abrogated proliferation and up-regulation of CD25 [61]. Conversely, it was found that addition of exogenous IL-2 induced proliferation and further up-regulation of CD25 in a monoculture of anti-CD3 stimulated T_{reg}. Both these phenomena exactly mimicked the behavior of the two cell types when co-cultured, indicating that IL-2 is not completely suppressed in co-cultures, but is produced in low amounts by responders and taken up by T_{reg}. Using an elegant chimeric human/mouse suppression assay, they went on to show that blocking of the IL-2R specifically on T_{reg} during the assay abrogated suppression, which again would be consistent with competition for IL-2 being involved. Alternatively, T_{reg} may continuously require stimulation through IL-2R to exert their suppressive function through an unknown mechanism. The latter is of particular significance in face of the fact that exogenous IL-2 may simply serve to maintain the viability of T_{reg}.

It has been postulated that T_{reg} exert their suppressive function by inhibiting IL-2 mRNA expression of naïve T cells [62, 63]. Can this be explained by T_{reg} competing for responder-derived IL-2 and thus abrogating an autocrine enhancing feedback loop within responder cells? If so, addition of high doses of exogenous IL-2 should restore endogenous IL-2 production by responders in the presence of T_{reg}. De la Rosa et al. could show that this is indeed the case. By contrast, IL-4, IL-7 or IL-15 could not rescue IL-2 production by responder cells [61]. As a note of caution, one should mention that there is conflicting data with respect to the capacity of exogenous IL-2 to restore IL-2 production by responders in co-cultures. Thus, Thornton and Shevach observed by qPCR that the block in IL-2 transcription persists even in the presence of high amounts of exogenous IL-2 [63]. On the basis of this observation they argued that competitive absorption of IL-2 is unlikely to be a critical mechanism of *in vitro* suppression. It remains to be seen how these conflicting data can be reconciled. Of note, de la Rosa et al. used a surface capture assay to detect IL-2 production, while Thornton and Shevach used a PCR based approach.

Malek et al. showed that $CD25^+$ CD4 T cells lacking the IL-2Rβ chain were efficient suppressor T cells *in vitro* [23]. Similarly, we and others could show that $Foxp3^+$ T_{reg} genetically deficient for CD25 showed some, albeit in our hands reduced, suppressive capacity *in vitro* [44, 45]. This indicates that competition for IL-2 through the high affinity IL-2R cannot be the exclusive mechanism of suppression *in vitro*. Importantly, these data would also argue that T_{reg} do not need to be "activated" by IL-2 (at least not through the high affinity receptor) for efficient *in vitro* suppression.

IL-2 Signaling and *In Vivo* Suppression

It has been suggested that immune regulation may at least in part act through competition for growth factors and space [64]. $CD25^+$ regulatory T cells, that only poorly or not at all produce IL-2, may utilize IL-2 and other growth factors produced by neighboring cells for their own survival (and eventually) expansion and thus deplete these factors in the local microenvironment.

Due to the obvious interference of the role of IL-2 signaling for the survival of T_{reg} and any other potential function of IL-2, it is extremely difficult to test this model experimentally. For example, in a model of spontaneous experimental autoimmune encephalomyelitis (EAE), Furtado et al. could show that adoptive transfer of $CD4^+$ T cells from an IL-2-deficient mouse can protect against EAE development, while $CD4^+$ T cells transferred from a CD25-deficient animal were unable to do so [65]. At face value, these results would argue that T_{reg} can be generated in the absence of IL-2 signaling, a view supported by more recent experiments [44, 45], but rely on the expression of CD25 to exert their function *in vivo*, possibly through competition for IL-2. This interpretation, however, is complicated by the fact that IL-2 signaling may serve to maintain the viability of T_{reg} after transfer, as T_{reg} from an IL-2 deficient donor may readily survive and eventually expand in IL-2 expressing hosts, whereas CD25-deficient T_{reg} may quickly disappear.

We have used an adoptive transfer model to study the mutual interplay of T_{reg} and naïve CD4 T cells of identical antigen-specificity *in vivo* [51]. We found that in the presence of specific T_{reg}, the up-regulation of CD25 on responder cells following antigen-challenge was diminished, very similar to the *in vitro* observations by de la Rosa [61]. This may indicate IL-2 "starvation", as IL-2 regulates its high-affinity receptor via a feedback mechanism.

Taken together, it may be fair to say that conclusive evidence for or against a role of IL-2 competition in *in vivo* suppression is lacking, but this mechanism certainly cannot account for all aspects of regulation and does not appear to be essential. This is perhaps best illustrated by the fact that autoimmunity in IL2R deficient mice, where autoimmunity is obviously driven by cells that cannot use IL-2 in the first place, can be rescued through transfer of T_{reg} [23].

IL-2 and Induction of Peripheral Regulatory T Cells

A number of experiments suggested the possibility of peripheral conversion of naïve T cells into $Foxp3^+$ T_{reg} *in vivo*. Presently, it is mostly assumed that these cells indeed may be identical to thymus derived $Foxp3^+$ T_{reg}, however, this issue certainly deserves further investigation. Early reports on peripheral *de novo* induction of T_{reg} have met some criticism with regard to how conclusively expansion of a minute pre-existing $Foxp3^+$ T_{reg} cohort was excluded [66–69]. However, using cleaner "precursor" populations of naïve cells obtained from TCR transgenic models on a rag deficient background (where $Foxp3^+$ cells are absent) the von Boehmer lab has firmly established that this conversion can occur without the contribution of thymic selection events [70, 71]. The authors suggested a model whereby subimmunogenic encounter of antigen on not fully activated dendritic cells would favor conversion of naïve cells into T_{reg}. Based on the observation that the efficiency with which T_{reg} were generated inversely correlated with the propensity to proliferate, it was tested whether the absence of autocrine IL-2 would eventually favor the conversion into

T_{reg}. Indeed, when IL-2 deficient naïve cells were used, reduced proliferation and increased conversion into T_{reg} of the cells that had divided the least was observed. These results show that peripheral conversion of naïve cells into T_{reg} does not require autocrine IL-2, but that in fact deprivation of autocrine IL-2 may favor T_{reg} induction through limiting full activation. As paracrine IL-2 was available within the adoptive host, these experiments certainly do not exclude a function of IL-2 signaling other than that of a growth-factor. The cleanest experiment to test whether T_{reg} conversion can occur in an entirely IL-2 independent manner probably would be to use naïve CD4 T cells that are deficient in IL-2R components and assess their potential to convert into Foxp3$^+$ T_{reg} in the system depicted above.

Knoechel et al. used an adoptive transfer system to assess the function of paracrine IL-2 in the peripheral induction of Foxp3$^+$ T_{reg} [72]. TCR transgenic CD4 T cells specific for ovalbumin were transferred into mice that systemically expressed ovalbumin as a serum protein (sOva-transgenic) [72]. Transfer into rag$^{+/+}$ sOva recipients did not cause disease, but rather resulted in tolerance of the transferred T cells. Mechanistically, this tolerance appeared to be a mixture of anergy and deletion. By contrast, transfer of the same cells to sOVA recipients on a rag deficient background, i.e. lacking endogenous T and B cells, resulted in severe autoimmunity that resembled acute graft verus host disease (GvHD). Interestingly, a certain fraction of mice survived this acute autoimmunity, whereby the recovery correlated with the *de novo* emergence of a small population of Ova-specific T_{reg}. Obviously, a single auto-antigen could concomitantly induce either autoimmune effector cells or T_{reg} from a presumably homogenous precursor population. Knoechel et al. went ahead to address the potential role of auto- or paracrine IL-2 in this conundrum. When the experiments were repeated in the complete absence of IL-2 (i.e. transfer of IL-2 deficient cells into an IL-2 deficient host), a delayed onset of GvHD was observed. Importantly, however, all mice went on to develop a chronic and ultimately fatal progressive disease, and induction of T_{reg} was never observed. Strikingly, transfer of IL-2 deficient T cells into IL-2 sufficient hosts likewise did not result in T_{reg} induction, whereas transfer of IL-2 sufficient cells into IL-2 deficient hosts did so.

Are these observations at odds with the observation by Kretschmer et al. that ablation of endogenous IL-2 led to more efficient T_{reg} induction in the periphery? At face value, the data by Knoechel et al. may indicate that autocrine, but not paracrine IL-2 is necessary for the conversion of naïve T cells into T_{reg}. More precisely, however, these experiments do only show that autocrine IL-2 can be sufficient for the accumulation of T_{reg} at the end of the experiment, while in the complete absence of IL-2 this is not the case. Moreover, eventually available IL-2 from sources other than T cells (e.g. DC) in rag deficient hosts was not sufficient to support T_{reg} accumulation. In the setup used by Knoechel et al., it remains open whether IL-2 is indeed required for the peripheral conversion into T_{reg} as such, or whether it is necessary for the survival/expansion of few cells that may initially undergo this conversion in an IL-2 independent manner. Thus, the data by Kretschmer and Knoechel can be readily reconciled with each other when one assumes that peripheral conversion of naïve cells into T_{reg} by itself operates without IL-2, but that once generated, these cells are critically dependent on IL-2 (which is provided in a paracrine fashion in the system used by Kretschmer et al.). Notably, this discussion is very reminiscent of the controversy surrounding the requirements for IL-2 during generation of T_{reg} in the thymus.

Conclusions

The evidence that IL-2 signaling is important for T_{reg} survival and homeostasis is continually expanding. By contrast, it remains controversial in how far IL-2 signaling is essential for T_{reg} generation, irrespective of whether thymic or peripheral pathways are concerned. This is at least in part due to experimental limitations with regard to systems that would allow conclusively distinguishing between IL-2 being required for *de novo* generation as such or for the subsequent maintenance of cells that have differentiated in an IL-2 independent fashion. It may be fair to say that essentially all experimental observations that have been published to date could be explained by the latter scenario. Having said this, the case of IL-2 signaling in T_{reg} biology illustrates the complexities of one signaling pathway potentially being involved in multiple aspects of the function and homeostasis of a particular cell type. Straight knock-out models have provided some answers here, but certainly are of limited usefulness. In the future, more sophisticated technologies, such as gene replacement in conjunction with mutations that may affect one particular aspect of IL-2 signaling (i.e. survival, differentiation, proliferation), if feasible, may afford more conclusive answers.

References

1. Taniguchi, T., et al., Structure and expression of a cloned cDNA for human interleukin-2. Nature, 1983, 302(5906):305–10.
2. Yang-Snyder, J.A. and E.V. Rothenberg, Spontaneous expression of interleukin-2 in vivo in specific tissues of young mice. Dev Immunol, 1998, 5(4):223–45.
3. Granucci, F., et al., Inducible IL-2 production by dendritic cells revealed by global gene expression analysis. Nat Immunol, 2001, 2(9):882–8.
4. Sugamura, K., et al., The interleukin-2 receptor gamma chain: its role in the multiple cytokine receptor complexes and T cell development in XSCID. Annu Rev Immunol, 1996, 14: 179–205.
5. Nelson, B.H. and D.M. Willerford, Biology of the interleukin-2 receptor. Adv Immunol, 1998, 70:1–81.
6. Eklund, J.W. and T.M. Kuzel, A review of recent findings involving interleukin-2-based cancer therapy. Curr Opin Oncol, 2004, 16(6):542–6.
7. Pahwa, S. and M. Morales, Interleukin-2 therapy in HIV infection. AIDS Patient Care STDS, 1998, 12(3):187–97.
8. Waldmann, T.A., The biology of interleukin-2 and interleukin-15: implications for cancer therapy and vaccine design. Nat Rev Immunol, 2006, 6(8):595–601.
9. Schorle, H., et al., Development and function of T cells in mice rendered interleukin-2 deficient by gene targeting. Nature, 1991, 352(6336):621–4.

10. Willerford, D.M., et al., Interleukin-2 receptor alpha chain regulates the size and content of the peripheral lymphoid compartment. Immunity, 1995, 3(4):521–30.
11. Suzuki, H., et al., Abnormal development of intestinal intraepithelial lymphocytes and peripheral natural killer cells in mice lacking the IL-2 receptor beta chain. J Exp Med, 1997, 185(3):499–505.
12. Sadlack, B., et al., Ulcerative colitis-like disease in mice with a disrupted interleukin-2 gene. Cell, 1993, 75(2):253–61.
13. Kramer, S., A. Schimpl, and T. Hunig, Immunopathology of interleukin (IL) 2-deficient mice: thymus dependence and suppression by thymus-dependent cells with an intact IL-2 gene. J Exp Med, 1995, 182(6):1769–76.
14. Ma, A., et al., T cells, but not B cells, are required for bowel inflammation in interleukin 2-deficient mice. J Exp Med, 1995, 182(5):1567–72.
15. Van Parijs, L., et al., Functional responses and apoptosis of CD25 (IL-2R alpha)-deficient T cells expressing a transgenic antigen receptor. J Immunol, 1997, 158(8):3738–45.
16. Lenardo, M.J., Interleukin-2 programs mouse alpha beta T lymphocytes for apoptosis. Nature, 1991, 353(6347):858–61.
17. Zheng, L., et al., T cell growth cytokines cause the superinduction of molecules mediating antigen-induced T lymphocyte death. J Immunol, 1998, 160(2):763–9.
18. Refaeli, Y., et al., Biochemical mechanisms of IL-2-regulated Fas-mediated T cell apoptosis. Immunity, 1998, 8(5):615–23.
19. Suzuki, H., et al., Normal thymic selection, superantigen-induced deletion and Fas-mediated apoptosis of T cells in IL-2 receptor beta chain-deficient mice. Int Immunol, 1997, 9(9):1367–74.
20. Leung, D.T., S. Morefield, and D.M. Willerford, Regulation of lymphoid homeostasis by IL-2 receptor signals in vivo. J Immunol, 2000, 164(7):3527–34.
21. D'Souza, W.N., et al., Essential role for IL-2 in the regulation of antiviral extralymphoid CD8 T cell responses. J Immunol, 2002, 168(11):5566–72.
22. Papiernik, M., et al., Regulatory CD4 T cells: expression of IL-2R alpha chain, resistance to clonal deletion and IL-2 dependency. Int Immunol, 1998, 10(4):371–8.
23. Malek, T.R., et al., CD4 regulatory T cells prevent lethal autoimmunity in IL-2Rbeta-deficient mice. Implications for the nonredundant function of IL-2. Immunity, 2002. 17(2):167–78.
24. Almeida, A.R., et al., Homeostasis of peripheral CD4+ T cells: IL-2R alpha and IL-2 shape a population of regulatory cells that controls CD4+ T cell numbers. J Immunol, 2002, 169(9):4850–60.
25. Taniguchi, T., et al., IL-2 signaling involves recruitment and activation of multiple protein tyrosine kinases by the IL-2 receptor. Ann N Y Acad Sci, 1995, 766:235–44.
26. Witthuhn, B.A., et al., Involvement of the Jak-3 Janus kinase in signalling by interleukins 2 and 4 in lymphoid and myeloid cells. Nature, 1994, 370(6485):153–7.
27. Johnston, J.A., et al., Phosphorylation and activation of the Jak-3 Janus kinase in response to interleukin-2. Nature, 1994, 370(6485):151–3.
28. Van Parijs, L., et al., Uncoupling IL-2 signals that regulate T cell proliferation, survival, and Fas-mediated activation-induced cell death. Immunity, 1999, 11(3):281–8.
29. Gaffen, S.L., et al., Distinct tyrosine residues within the interleukin-2 receptor beta chain drive signal transduction specificity, redundancy, and diversity. J Biol Chem, 1996, 271(35):21381–90.
30. Moriggl, R., et al., Stat5 activation is uniquely associated with cytokine signaling in peripheral T cells. Immunity, 1999, 11(2):225–30.
31. Lin, J.X. and W.J. Leonard, The role of Stat5a and Stat5b in signaling by IL-2 family cytokines. Oncogene, 2000, 19(21):2566–76.
32. Graves, J.D., et al., The growth factor IL-2 activates p21ras proteins in normal human T lymphocytes. J Immunol, 1992, 148(8):2417–22.
33. Miyazaki, T., et al., Pyk2 is a downstream mediator of the IL-2 receptor-coupled Jak signaling pathway. Genes Dev, 1998, 12(6):770–5.

34. Friedmann, M.C., et al., Different interleukin 2 receptor beta-chain tyrosines couple to at least two signaling pathways and synergistically mediate interleukin 2-induced proliferation. Proc Natl Acad Sci USA, 1996, 93(5):2077–82.
35. Ravichandran, K.S., et al., Evidence for a role for the phosphotyrosine-binding domain of Shc in interleukin 2 signaling. Proc Natl Acad Sci USA, 1996, 93(11):5275–80.
36. Blanchard, D.A., et al., Cdk2 associates with MAP kinase in vivo and its nuclear translocation is dependent on MAP kinase activation in IL-2-dependent Kit 225 T lymphocytes. Oncogene, 2000, 19(36):4184–9.
37. Merida, I., E. Diez, and G.N. Gaulton, IL-2 binding activates a tyrosine-phosphorylated phosphatidylinositol-3-kinase. J Immunol, 1991, 147(7):2202–7.
38. Augustine, J.A., S.L. Sutor, and R.T. Abraham, Interleukin 2- and polyomavirus middle T antigen-induced modification of phosphatidylinositol 3-kinase activity in activated T lymphocytes. Mol Cell Biol, 1991, 11(9):4431–40.
39. Kelly, E., et al., IL-2 and related cytokines can promote T cell survival by activating AKT. J Immunol, 2002, 168(2):597–603.
40. Ahmed, N.N., et al., Transduction of interleukin-2 antiapoptotic and proliferative signals via Akt protein kinase. Proc Natl Acad Sci USA, 1997, 94(8):3627–32.
41. Jones, R.G., et al., Protein kinase B regulates T lymphocyte survival, nuclear factor kappaB activation, and Bcl-X(L) levels in vivo. J Exp Med, 2000, 191(10):1721–34.
42. Reif, K., B.M. Burgering, and D.A. Cantrell, Phosphatidylinositol 3-kinase links the interleukin-2 receptor to protein kinase B and p70 S6 kinase. J Biol Chem, 1997, 272(22):14426–33.
43. Brennan, P., et al., Phosphatidylinositol 3-kinase couples the interleukin-2 receptor to the cell cycle regulator E2F. Immunity, 1997, 7(5):679–89.
44. Fontenot, J.D., et al., A function for interleukin 2 in Foxp3-expressing regulatory T cells. Nat Immunol, 2005, 6(11):1142–51.
45. D'Cruz, L.M. and L. Klein, Development and function of agonist-induced CD25+Foxp3+ regulatory T cells in the absence of interleukin 2 signaling. Nat Immunol, 2005, 6(11):1152–9.
46. Setoguchi, R., et al., Homeostatic maintenance of natural Foxp3(+) CD25(+) CD4(+) regulatory T cells by interleukin (IL)-2 and induction of autoimmune disease by IL-2 neutralization. J Exp Med, 2005, 201(5):723–35.
47. Burchill, M.A., et al., Distinct effects of STAT5 activation on CD4+ and CD8+ T cell homeostasis: development of CD4+CD25+ regulatory T cells versus CD8+ memory T cells. J Immunol, 2003, 171(11):5853–64.
48. Antov, A., et al., Essential role for STAT5 signaling in CD25+CD4+ regulatory T cell homeostasis and the maintenance of self-tolerance. J Immunol, 2003, 171(7):3435–41.
49. Bensinger, S.J., et al., Distinct IL-2 receptor signaling pattern in CD4+CD25+ regulatory T cells. J Immunol, 2004, 172(9):5287–96.
50. Walker, L.S., et al., Antigen-dependent proliferation of CD4+ CD25+ regulatory T cells in vivo. J Exp Med, 2003, 198(2):249–58.
51. Klein, L., K. Khazaie, and H. von Boehmer, In vivo dynamics of antigen-specific regulatory T cells not predicted from behavior in vitro. Proc Natl Acad Sci USA, 2003, 100(15):8886–91.
52. Romagnoli, P., D. Hudrisier, and J.P. van Meerwijk, Preferential recognition of self antigens despite normal thymic deletion of CD4(+)CD25(+) regulatory T cells. J Immunol, 2002, 168(4):1644–8.
53. Hsieh, C.S., et al., Recognition of the peripheral self by naturally arising CD25+ CD4+ T cell receptors. Immunity, 2004, 21(2):267–77.
54. Yu, A. and T.R. Malek, Selective availability of IL-2 is a major determinant controlling the production of CD4+CD25+Foxp3+ T regulatory cells. J Immunol, 2006, 177(8):5115–21.
55. Cheng, L.E., et al., Enhanced signaling through the IL-2 receptor in CD8+ T cells regulated by antigen recognition results in preferential proliferation and expansion of responding CD8+ T cells rather than promotion of cell death. Proc Natl Acad Sci USA, 2002, 99(5):3001–6.

56. Fontenot, J.D., M.A. Gavin, and A.Y. Rudensky, Foxp3 programs the development and function of CD4+CD25+ regulatory T cells. Nat Immunol, 2003, 4(4):330–6.
57. Hori, S., T. Nomura, and S. Sakaguchi, Control of regulatory T cell development by the transcription factor Foxp3. Science, 2003, 299(5609):1057–61.
58. Wu, Y., et al., FOXP3 controls regulatory T cell function through cooperation with NFAT. Cell, 2006, 126(2):375–87.
59. Nakamura, K., A. Kitani, and W. Strober, Cell contact-dependent immunosuppression by CD4(+)CD25(+) regulatory T cells is mediated by cell surface-bound transforming growth factor beta. J Exp Med, 2001, 194(5):629–44.
60. Piccirillo, C.A., et al., CD4(+)CD25(+) regulatory T cells can mediate suppressor function in the absence of transforming growth factor beta1 production and responsiveness. J Exp Med, 2002, 196(2):237–46.
61. de la Rosa, M., et al., Interleukin-2 is essential for CD4+CD25+ regulatory T cell function. Eur J Immunol, 2004, 34(9):2480–8.
62. Thornton, A.M. and E.M. Shevach, CD4+CD25+ immunoregulatory T cells suppress polyclonal T cell activation in vitro by inhibiting interleukin 2 production. J Exp Med, 1998, 188(2):287–96.
63. Thornton, A.M., et al., Cutting edge: IL-2 is critically required for the in vitro activation of CD4+CD25+ T cell suppressor function. J Immunol, 2004, 172(11):6519–23.
64. Barthlott, T., G. Kassiotis, and B. Stockinger, T cell regulation as a side effect of homeostasis and competition. J Exp Med, 2003, 197(4):451–60.
65. Furtado, G.C., et al., Interleukin 2 signaling is required for CD4(+) regulatory T cell function. J Exp Med, 2002, 196(6):851–7.
66. Liang, S., et al., Conversion of CD4+ CD25- cells into CD4+ CD25+ regulatory T cells in vivo requires B7 costimulation, but not the thymus. J Exp Med, 2005, 201(1):127–37.
67. Thorstenson, K.M. and A. Khoruts, Generation of anergic and potentially immunoregulatory CD25+CD4 T cells in vivo after induction of peripheral tolerance with intravenous or oral antigen. J Immunol, 2001, 167(1):188–95.
68. Mahnke, K., et al., Induction of CD4+/CD25+ regulatory T cells by targting of antigens to immature dendritic cells. Blood, 2003, 101(12):4862–9.
69. Chen, W., et al., Conversion of peripheral CD4+CD25- naive T cells to CD4+CD25+ regulatory T cells by TGF-beta induction of transcription factor Foxp3. J Exp Med, 2003, 198(12):1875–86.
70. Kretschmer, K., et al., Making regulatory T cells with defined antigen specificity: role in autoimmunity and cancer. Immunol Rev, 2006, 212:163–9.
71. Kretschmer, K., et al., Inducing and expanding regulatory T cell populations by foreign antigen. Nat Immunol, 2005, 6(12):1219–27.
72. Knoechel, B., et al., Sequential development of interleukin 2-dependent effector and regulatory T cells in response to endogenous systemic antigen. J Exp Med, 2005, 202(10):1375–86.

Chapter 6
TGF-Beta and Regulatory T Cells

Yisong Y. Wan and Richard A. Flavell

Abstract Suppression of the immune system is critical in maintaining self-tolerance and immune homeostasis. Multiple types of cytokines and cell types actively suppress immune responses. Among them, the pleiotropic cytokine TGF-β, and naturally occurring regulatory T cells (Treg) are the best characterized. Dysregulation of either one leads to various immunopathologies under physiological conditions, demonstrating their essential roles in immune suppression. In addition, therapeutic strategies to treat aberrations of immune function by manipulating TGF-β and Treg functions have shown promising results. In this chapter, we will discuss the biologic functions of TGF-β and Treg, and the potential therapeutic effects on immune-related diseases through the manipulation of TGF-β and Treg function.

Introduction

Every day we are being exposed to foreign pathogens, including airborne and food borne viruses and bacteria. Complex immune strategies have evolved in mammals to defend against foreign pathogens to maintain healthy. Maintaining defense against invasion by ever changing pathogens is a daunting task for the immune system. However, by definition, it has evolved mechanisms to efficiently cope with these insults in the living species. The two fundamental arms of the immune response, namely adaptive and innate immunity, form together a defensive front against these pathogens. The adaptive immune responses mediated mostly by T and B cells that are highly antigen specific and which retain memory of exposure, while innate immunity, which is mediated by innate cells such as macrophages, is specific only to class of micro-organisms. T cells bearing T cell receptors (TCR) that recognize specific antigens are not only the executioners of the adaptive immunity, but also they modulate other components of the adaptive as well as innate immunity. Developing in the thymus, and by using quasi-random recombinational mechanisms, T cells can

---

R.A. Flavell
Section of Immunobiology, Yale University School of Medicine; Howard Hughes Medical Institute, 300 Cedar Street, New Haven, Connecticut 06520, USA
e-mail: richard.flavell@yale.edu

potentially generate infinite numbers of specificities against antigens from self or non-self. While broad specificity of the repertoire enables T cells to mount potent immune response against any possible foreign antigens, it can also be detrimental to the organism. Self-reactive T cells can cause autoimmune diseases in the host if they are not tightly controlled. Multiple processes therefore are deployed to suppress the generation or the function of self-reactive T cells. One mechanism includes selective processes in the thymus where self-reactive T cells fail to mature, or in the periphery where these cells are clonally deleted. Nonetheless, such elimination processes are incomplete resulting in small populations of mostly low-affinity self-reactive T cells in the periphery to potentially initiate an autoimmune response. Fortunately, active immune suppressive mechanisms exist to suppress the function of these autoreactive T cells. In addition to preventing auto-immunity, active immune suppression also plays important roles in regulating the contraction phase of normal immune responses to pathogens, thereby maintaining immune homeostasis. Aberrant down- or up-regulation of the active immune suppression can result in various disorders, such as autoimmunity, inflammatory diseases, as well as cancers. In recent years, great progress has been made in understanding the cellular and molecular components of immune suppression. Active immune suppression is mediated mostly through either cytokines or through specialized cells previously called suppressor cells [1], and now usually termed regulatory cells. Among the cytokines and regulatory cells suppressing the immune system, the pleiotropic cytokine, TGF-β, and the immunosuppressive cell, regulatory T cells (Treg), play critical roles in suppressing the immune response and will be discussed in this chapter.

TGF-β and Regulatory T Cell

TGF-β, A Ploietropic Cytokine, Mediates Immune Suppression

Transforming growth factor-β (TGF-β) consists of a family of pleiotropic cytokines regulating multi-faceted cellular functions, such as proliferation, differentiation, migration and survival, therein, effecting a broad spectrum of biological processes, such as development, carcinogenesis, fibrosis and wound healing [2]. TGF-β function in the immune system was first described in 1986 [3,4]. However, the critical roles of TGF-β in suppressing immune responses in vivo were not revealed until the generation of TGF-β deficient mice [5,6]. Later studies evaluating the components of TGF-β signaling networks, including their receptors and intracellular signaling molecule, confirmed the suppressive role for TGF-β in the immune system. Further studies exposed the role of TGF-β in regulating the adaptive immunity components, such as T cells, as well as the innate immunity components, such as natural killer (NK) cells [7–13].

TGF-β and its Signaling

In addition to the TGF-βs, the TGF-β superfamily comprises bone morphogenetic proteins (BMPs), activins and growth differentiation factors (GDFs) [14]. Three

TGF-β isoforms, TGF-β1, -β2 and -β3 have been identified in mammals with similar functions but disparate expression patterns [15]. TGF-β1 is the isoform predominantly expressed in the immune system. Unlike most cytokines, TGF-β is synthesized as an inactive form, pre-pro-TGF-β precursor. Through proteolytic processing, active TGF-β is produced in association with latency associated protein (LAP) or with latent-TGF-β-binding protein (LTBP), and this association keeps TGF-β inactive [16]. Additional stimuli are required to liberate active TGF-β enabling it to exert its function by binding to its receptor [17–21]. The active form of TGF-β exerts its biological functions in either a cell-surface bound form or a soluble form [22,23].

To initiate signal transduction, TGF-β binds to heterodimeric receptor complex consisting of a type I and II trans-membrane serine/threonine kinase subunits. Although more than 35 TGF family members have been identified, only 5 type I (activin like receptor kinase (ALK) family) and 7 type II receptors have been reported [14]. The active forms of TGF-β dimerize and bind to the tetrameric ALK5 and TGFβRII receptor complex [24]. TGF-β binding triggers the association of the type II receptor with the type I receptor, and subsequently the type II receptor phosphorylates and activates the type I receptor. Intracellular signal transduction of TGF-β is mediated to a great degree via Smad proteins [24,25]. The eight vertebrate Smads identified thus far are grouped into three categories: five receptor associated Smads (R-Smad1, 2, 3, 5, 8), one common Smad (Co-Smad4) and two inhibitory Smads (I-Smad6, 7). Upon TGF-β stimulation, activated ALK5 phosphorylates R-Smad-2 and -3. Phosphorylated R-Smads associate with Co-Smad4 and translocate into the nucleus to bind to DNA containing a Smad binding element (SBE) [26–29]. Unlike R-Smads, I-Smad7 is not phosphorylated upon TGF-β activation [30] and I-Smad7 suppresses TGF-β signaling. At least two mechanisms are used by I-Smad7 to inhibit TGF-signaling: the first is by competing with R-Smads for the binding to ALK5; and the second is by recruiting ubiquitin ligase complexes to degrade ALK5 via proteasome [31,32]. Smad-independent TGF-β signaling pathways have also been revealed [33,34]. Through mechanisms yet to be determined, rapid activation of Ras-Erk, TAK-MKK4-JNK, TAK-MKK3/6-p38, Rho-Rac-cdc42 MAPK and PI3 K-Akt pathways occurs when cells are treated with TGF-β [35]. MAPKs also found to crosstalk with Smads to modulate TGF-β responses [36–38]. Moreover, through direct protein binding, TGF-β receptors activate TRIP-1 and PP2A, and regulate translation initiation [39–41]. T cell specific target genes of TGF-β are largely unknown, however, the transcription of GATA3, T-bet, STAT4, IFN-γ and granzyme-B is suppressed by TGF-β [42–46].

Regulatory T cells and Immune Suppression

A seminal study by Sakaguchi et al. discovered a subset of T cells known as suppressor T cells or regulatory T cells (Treg) [47]. This finding rekindled interest in immune suppression, a concept proposed by Gershon et al. over 30 years ago [1,48,49]. Immune suppression has subsequently become one of the most actively researched areas in immunology. In recent years, substantial progress has been made in identifying different types of Treg cells, as well as in understanding how these cells are generated and function.

Multiple Types of Treg Cells Mediate Immune Suppression

Tregs consist of diverse populations sharing common features of being hyporesponsive to antigenic stimulation and possessing immunosuppressive activities. Based on cell surface markers or cytokine secretion profiles, Tregs can be grouped into naturally occurring Tregs (nTreg) and acquired Treg (aTreg).

Naturally Occurring Treg (nTreg)

nTreg include a subset of CD4 T cells that develops in the thymus and constitutively expresses cell surface IL-2 receptor α chain (CD25). $CD4^+CD25^+$ nTregs comprise approximately 10% of peripheral CD4 T cells in the mouse and humans. Surface molecules other than CD25, such as cytotoxic T lymphocyte antigen-4 (CTLA-4), glucocorticoid-induced tumor necrosis factor receptor family related gene (GITR) and lymphocyte activation antigen-3 (LAG-3) have also been used to differentiate nTreg from other T cells [50]. TGF-β is expressed at high levels as a cell-surface bound form [23,51]. These cells suppress immune responses in an antigen independent fashion in vitro and in vivo [50,52–54]. nTregs are critical for maintaining self-tolerance as disruption of thymic development or peripheral maintenance of these cells invariably results in the development of autoimmunity. Thymectomy of neonatal mice within 3 days after birth or anti-CD25 antibody mediated depletion of these cells has led to lethal autoimmunity due to the loss of peripheral nTreg in adult mice [50]. Foxp3, a Forkhead family transcription factor expressed specifically in nTreg among the lymphocyte populations, controls Treg thymic development [52,55,56]. Spontaneous mutation in *Foxp3* gene results in systemic autoimmunity in *Scurfy* mice and *IPEX* (X-linked neonatal diabetes mellitus, enteropathy and endocrinopathy syndrome) patients [57–61]. The disease manifested in *Scurfy* mice was attributed to Treg deficiency, although T cell extrinsic elements were also reported to contribute [62].

Aquired Treg (aTreg)

Non-regulatory T cells can acquire immune suppressive activities and become aTreg. Currently known aTreg includes Tr1 and TH3 cells. Tr1 is one of the aTreg subsets that are often found within the intestinal mucosa with a function to suppress immune reactions towards a variety of cognate antigens [63]. No particular surface markers have been associated with Tr1 cells. However, these cells produce increased levels of IL-10 and TGF-β [64]. Tr1 cells do not express Foxp3 [65], suggesting that it is a subset of Treg distinct from nTreg.

Th3 is another aTreg subset induced primarily from naïve CD4 T cells after ingestion of a foreign antigen via the oral route, thereby eliciting oral tolerance [66,67]. While no particular surface marker is associated with these cells, Foxp3 is also expressed in Th3 cells [68]. In addition, TGF-β is produced at elevated levels by Th3 cells [67]. Whether Th3 cells form a distinct aTreg subset or are activated nTreg remains unknown.

TGF-β and Regulatory T Cell in Action

TGF-β Regulates T Cell Proliferation and Effector Function

As a pleiotropic cytokine, TGF-β exerts its effects on multiple cell types. Its role in controlling T cell functions and immune responses has been studied extensively [69]. The anti-proliferation function of TGF-β on T cells was first documented by studies performed in vitro using activated human T cells [4]. One way TGF-β suppresses T cell proliferation is through the inhibition of the production of interleukin (IL)-2, a lymphokine known to potently activate T cells, NK cells and other types of cells of the immune system. Addition of exogenous IL-2 partially relieved TGF-β mediated suppression [4]. TGF-β suppresses IL-2 production in T cells potentially through direct inhibition of IL-2 promoter activity. A cis-acting enhancer DNA element was identified to be critical in suppressing IL-2 production by TGF-β [70]. In addition, R-Smad3 is critical in mediating TGF-β inhibited IL-2 production, as TGF-β failed to suppress IL-2 production in murine T cells that lack this gene [71]. Moreover, the Smad-binding element has been located upstream of the human IL-2 promoter, which is important for Smad mediated transcriptional-suppression of IL-2 [72]. TGF-β also inhibits T cell proliferation through means other than suppressing IL-2 production, as addition of exogenous IL-2 did not fully reconstitute T cell proliferation [4]. Further studies are warranted to investigate the detailed mechanisms of how TGF-β suppresses T cell proliferation.

The expression of cell cycle regulators is changed in response to TGF-β. Upon TGF-β treatment, cyclin-dependent kinase inhibitors (CKIs), such as p15, p21 and p27, are upregulated while cell cycle promoting factors, such as c-myc, cyclin D2, CDK2 and cyclin E, are decreased [73–78]. The functional significance of these expression changes in TGF-β induced inhibition and how TGF-β modulates these genes in T cells awaits further elucidation.

While TGF-β inhibits naïve T cell proliferation, it has minimal effect on activated T cells, which may be due to reduced TGF-β receptor II expression in these cells [79]. The inhibition of activated T cells by TGF-β was restored when TGFβ receptor II expression was upregulated by IL-10 stimulation [79], reinforcing that TGF-β functions in a manner dependent on both cell type and immunologic environment.

Following activation, naïve T cells differentiate into effector T cells to perform immune function [80]. Based on their cytokine production, CD4 effector T cells, also named T helper (Th), can be profiled as Th1, Th2 and Th17 cells [81,82]. Th1 cells produce interferon-γ (IFN-γ) and lymphotoxin (LT), Th2 cells secrete IL-4, IL-13 and IL-5, and Th17 cells express IL-17 and IL-22 [83]. TGF-β potently inhibits cytokine production by activated T cells and thus their differentiation into Th1 or Th2 effector cells [84]. Th1 polarizing condition promotes CD122 expression through T-bet [85], thereby enhancing the clonal expansion and the survival of Th1 cells [85]. Addition of TGF-β suppressed CD122 upregulation under Th1 skewing conditions. Therefore, TGF-β also limits Th1 effector cell numbers through inhibiting the upregulation of CD122. It was also noted that TGF-β inhibited T cell differentiation independent of T cell proliferation [86]. Thus, TGF-β potentially regulates T cell proliferation and effector functions through discrete mechanisms with the greatest effects on suppressing their differentiation. Interestingly, TGF-β was recently identified to be important for the induction of IL-17 producing cells under inflammatory conditions [87–90], further demonstrating the importance of environmental cues in influencing the function of TGF-β on T cells.

Further studies have revealed that TGF-β regulates effector T cell function through multiple mechanisms. Distinct sets of transcription factors are preferentially expressed in, and are important for Th cell differentiation. These include T-bet and Stat4 which specify Th1 cells, and Gata-3 and Stat6 which specify Th2 cells [80]. While the detailed mechanism remains unknown, T-bet and Gata-3 expression is inhibited by TGF-β [43,44,91,92] possibly, in the latter case, through a mechanism via blocking Itk kinase activity and calcium influx [42]. Interestingly, effector cytokine production by fully differentiated Th2 cells is unaffected by TGF-β, while Th1 cells remain susceptible to TGF-β suppression [93]. Therefore, TGF-β exerts most of its inhibitory effects on the establishment of effector cell functions. While TGF-β inhibits the production of pro-inflammatory cytokines, it promotes T cell production of IL-10, an anti-inflammatory cytokine, likely through direct activation of the IL-10 promoter via Co-Smad4 [94]. Besides regulating CD4 T cells, TGF-β controls CD8 T cell proliferation and effector functions. The expression of effector molecules by CD8 T cells, such as IFN-γ and perforin, is inhibited by TGF-β [95–98]. Recent studies showed that TGF-β is important for Treg induced inhibition of the exocytosis of granules and cytolytic function of CD8 T cells [99]. Under certain conditions, TGF-β could also enhance the proliferation of mouse $CD8^+$ cells [100] and to increase TNF-α production by both $CD4^+$ and $CD8^+$ cells [101]. TGF-β accelerates T cell death in some studies [102,103], while an anti-apoptotic role for TGF-β has also been documented [104,105].

The critical physiological role for TGF-β in regulating immune suppression has been clearly illustrated by studies of genetically modified mice. TGF-$β1^{-/-}$ mice develop a multifocal inflammatory disease associated with increased inflammatory cytokine production [5,6,106]. This phenotype is mediated through T cells, as depletion of $CD4^+$ T cells or crossing TGF-$β1^{-/-}$ mice onto an MHC class II null background prevented this inflammation [107]. However, it was not clear from these studies whether T cells are direct targets of TGF-β since TGF-β1 acts on multiple cell types. Therefore, mice expressing a dominant-negative form of TGFβRII from the CD4 promoter (CD4-dnTβRII) in T cells were generated. These mice developed an autoimmune inflammatory phenotype associated with uncontrolled $CD4^+$ T cell differentiation into effector cells [9]. In addition, deficiency of TGF-βRII caused mice to develop fatal autoimmune diseases similar to the TGF-$β1^{-/-}$ mice.

Further studies attributed this phenotype to hyper-activation and exaggerated effector functions of immune cells, especially T cells [85,108].

Treg Suppresses Immune Responses

The essential role of Treg in immune suppression is indisputable. However, the mechanisms by which Treg cells carry out their function remain ill-defined. Nevertheless, it is agreed that Treg suppress immune responses through multiple mechanisms. Several surface molecules preferentially expressed by Treg cells are proposed to be important for their function. For example, CD25, a high affinity IL-2 binding receptor, is highly expressed by the nTreg cells [109]. One possible mechanism of nTreg suppression of activation of conventional T cells is through competition for IL-2 binding [110,111]. CTLA-4, another surface molecule preferentially expressed by nTreg, is important for shutting down the immune system by competing for costimulatory ligands on T cells [112] as well as inhibiting the function of antigen presenting cells [113,114]. Thus, it is suggested that CTLA-4 is important for nTreg mediated immune suppression. However, genetic evidence has not supported the critical roles for CD25 and CTLA-4 in nTreg function as the function of nTreg deficient in CD25 or CTLA-4 appeared to be normal [115,116]. These experiments and others have suggested that multiple elements are involved in nTreg function. For example, recent studies have unraveled important roles of cytokines in nTreg function. TGF-β appears to be critical in mediating nTreg function as T cells from CD4-dnTβRII mice that are unresponsive to TGF-β are refractory to nTreg mediated suppression in vitro and in vivo [23,117,118]. Despite the fact that TGF-β mRNA is not elevated in nTreg cells, it is suggested that the membrane bound form of TGF-β is increased in nTreg cells and is important for their function [23,51]. IL-10 is another immune suppressive cytokine preferentially expressed in nTreg [119,120], and is important in mediating the functions of these cells [120,121].

TGF-β and Treg Connection

In contrast to its inhibitory effects on the proliferation and the function of conventional T cells, TGF-β promotes the generation of Treg cells and induces Foxp3 expression in these cells. Early studies demonstrated that TGF-β was necessary and sufficient to promote human CD8$^+$ T cells to acquire suppressive activities [122]. In addition, regulatory activity was induced in human naïve (CD45RA$^+$RO$^-$) CD4 T cells by TGF-β following stimulation. In this context, the initial induction of Treg but not their function requires the presence of TGF-β [123]. TGF-β was subsequently demonstrated to induce the expression of Foxp3 in CD4$^+$CD25$^-$ human T cells [124], and in activated murine CD4$^+$ and CD8$^+$ T cells as well [125]. In the presence of TGF-β1, Staphylococcus endotoxin-B (SEB) activated CD8 T cells inhibited the proliferation and effector functions of CD4$^+$ and CD8$^+$ T cells. This was accompanied by elevated levels of IL-10 and TGF-β1. However, there were undetectable levels of effector cytokines including IL-4 [126]. TGF-β was later demonstrated to convert mouse CD4$^+$CD25$^-$ into CD4$^+$CD25$^+$ T cells with elevated Foxp3 expression [124,127].

Additional cytokines are needed for TGF-β promoted Treg induction. The combination of IL-2 and TGF-β enhanced the suppressive effects of nTreg [128–130]. In addition, IL-2 has been shown to be essential for directing stimulated CD4$^+$CD25$^-$ cells into Treg, as neutralization of this cytokine abolishes TGF-β induced suppressive activity of these cells [128]. It is conceivable that IL-2 and TGF-β play different but complementary roles for Treg generation. IL-2 may be promoting the "metabolic fitness" of Treg cells and thus their survival and maintenance [116], while TGF-β maintains as well as promotes the "regulatory property" of these cells.

The previously discussed studies demonstrated that TGF-β is able to convert CD4$^+$CD25$^-$ non-Treg into CD4$^+$CD25$^+$ Treg cells, and this was accompanied with increased Foxp3 expression. However, a substantial portion of Foxp3$^+$ Treg cells are negative for CD25 [125,131]. Therefore, it is difficult to discern whether such conversion is due to preferential expansion/survival of the existing Foxp3$^+$CD25$^-$ population or due to de novo Foxp3 expression in the Foxp3$^-$CD25$^-$ population. Unequivocal evidence for TGF-β conversion of Foxp3$^-$ cells into Foxp3$^+$ cells came from one study using Foxp3-mRFP knockin mice, where Foxp3 expressing cells are marked by mRFP expression [125]. TGF-β induced de novo Foxp3 expression in Foxp3$^-$ CD4 T cells. Furthermore, only Foxp3$^+$CD4$^+$ cells but not Foxp3$^-$CD4$^+$ counterparts possessed regulatory activities [125].

Although TGF-β promotes Treg generation in vitro, it has been controversial whether TGF-β is involved in the generation or maintenance of Foxp3 expressing Treg under physiological conditions. Transient expression of TGF-β by a transgene specifically expressed in islets promotes the generation of CD4$^+$CD25$^+$ Treg in situ with high Foxp3 expression in diabetes-predisposed NOD mice [132]. This observation correlated with the suppression of diabetes. In addition, induced Treg cells suppressed the onset of diabetes following adoptive transfer of these cells in NOD mice [132]. These findings demonstrated that TGF-β is sufficient to promote the generation of Treg under physiological conditions. Conflicting results have been presented with regards to whether TGF-β is essential for the development and maintenance of nTreg. In one study, the CD4$^+$CD25$^+$ Treg population was shown to be decreased in adult mice transgenic for a dominant negative form of TGF-β receptor II [133] under the control of the CD2 promoter (hCD2-ΔkTβRII). Four days after being transferred into mice subjected to DSS induced colitis, hCD2-ΔkTβRII transgenic CD4$^+$CD25$^+$ cells proliferated poorly compared with wild-type CD4$^+$CD25$^+$ nTreg cells, thus suggesting that TGF-β signaling was required for the maintenance and expansion of CD4$^+$CD25$^+$ nTreg in vivo [134]. However, this result was challenged by studies using another transgenic model where a similar form of TGFDNR (dnTβRII) is expressed under the control of the CD4 promoter [9]. CD4$^+$CD25$^+$ nTreg cells in CD4-dnTβRII transgenic mice developed normally. A slightly increased number of nTreg were found in the periphery of these mice compared to their wild-type counterparts ([117] and our unpublished observation). Peripheral but not thymic nTreg were found reduced in 8-10-day old TGF-β1$^{-/-}$

mice [135], suggesting an essential function for endogenous TGF-β1 in the maintenance of the peripheral population of Treg. These results contrast that of an earlier study where no defect of Treg development or maintenance was observed in TGF-$\beta 1^{-/-}$ mice [136]. A more recent definitive study demonstrated that Foxp3 expressing nTreg cells that lack TGFβRII developed normally in the thymus, but were poorly maintained in the periphery [85]. Interestingly, in the same study, TGFβRII deficient Treg were found to proliferate faster than the WT counterparts in the periphery, suggesting that TGF-β signaling is required to promote the survival of peripheral Treg [85]. The reasons for these discrepancies are unknown but may be related to the different experimental systems and mouse genetic backgrounds used. In addition, because CD25 is also expressed by activated T cells, $CD4^+CD25^+$ Treg identified in these studies may have been contaminated by activated T cell to varying degrees in the earlier studies. With the development of EGFP-Foxp3 and Foxp3-mRFP knock-in mice, and Foxp3 intracellular staining, these potential complications can be circumvented by identifying Treg based on Foxp3 expression [85,125,131].

The downstream signaling through which TGF-β mediates the generation of Foxp3 expressing Treg remains unclear. R-$Smad3^{-/-}$ mice developed by two different groups showed distinct phenotypes in the immune system. While no apparent abnormality was observed when exon1 of R-$Smad3$ gene was knocked out in mice [137], impaired mucosal immunity associated with an activated phenotype in T cells was found when the exon8 of R-$Smad3$ gene was targeted [13]. However, there is no evidence to suggest that the phenotype found in the latter study is due to perturbed Treg compartments. A recent study proposed a cooperative role for TGF-β in upregulating CD25 expression following TCR stimulation, and this function of TGF-β appeared to be mediated through the R-Smad3/Co-Smad4 pathways [138]. Thus R-Smad3/Co-Smad4 regulation of the generation or maintenance of Treg may be through cooperation with IL-2 signaling. However, because the Treg population appeared to be normal when Co-Smad4 is deficient in T cells ([13,138] and our unpublished observation), it is questionable that Co-Smad4 and R-Smad dependent TGF-β signaling is critically involved in Treg generation under physiological conditions.

TGF-β, Regulatory T Cells and Diseases

Disruption of immune suppression contributes to various immune diseases, some of which are fatal. On the other hand, excessive immune-suppression is associated with tumor growth. Dysregulated function of TGF-β and Treg could account for many immune disorders found in murine models and patients with these diseases. Progress in the treatment of several clinical conditions, such as autoimmunity, infectious disease and cancer, might be greatly influenced by the development of agents which can down- or up-regulate the immune responses by manipulating the immune regulatory network through TGF-β and Treg.

Inflammatory bowel disease (IBD) is a category of auto-inflammatory diseases which is thought to result from a breach of tolerance to commensal bacteria in the gut. TGF-β and Treg are both crucial in maintaining this tolerance. Blockade of TGF-β signaling in T cells resulted in the spontaneous development of colitis in the mice [9]. Transferring $CD4^+CD45RB^{high}$ cells or $CD4^+Foxp3^-$ cells into $Rag^{-/-}$ mice induces colitis in the recipients [125,139]. Addition of $CD4^+CD25^+$ nTreg prevents as well as cures colitis in murine models [140]. The mechanism behind the curative effects of nTreg in colitis appear to be dependent on TGF-β, as experiments T cells refractory to TGF-β signaling fail to suppress this disease [117]. In addition, adoptive transfer of in vitro generated Treg cells also prevented/cured colitis [141].

Studies of Treg in diabetes models, such as the NOD mouse, demonstrate their important roles in preventing this autoimmune disease. Destruction of islet β cells by self-activated T cells leads to type I diabetes (T1D). Diabetes in the NOD mouse may have a similar etiology as in human T1D, and is therefore used as an animal model to study T1D. Transferring Treg prevents or delays the onset of diabetes in NOD mice [23,142], suggesting their role in preventing/delaying this disease. More importantly, TGF-β expression in the islets is able to promote the generation/proliferation of Treg in situ and prevent diabetes in the NOD mice [132]. Collectively, by promoting immune-suppression through manipulating TGF-β and Treg activities in vitro and in vivo, it may be possible to devise effective therapies to treat autoimmune/inflammatory diseases, such as IBD and T1D.

High levels of TGF-β are produced by many types of tumor cells. In addition, tumors can promote TGF-β production by the surrounding cells in the tumor microenvironment [143]. TGF-β not only fosters tumor growth and progression, but also allows them to evade immune surveillance [144–146]. The frequencies of Treg cells in the peripheral blood of tumor-bearing patients are increased [147] but decreased following cytotoxic chemotherapies [148]. One of the consequences of the accumulation of Tregs at the tumor site is to promote tumorigenesis by inhibiting anti-tumor immunity [149]. In that way, enhanced anti-tumor immunity may be achieved by abrogating Treg functions as demonstrated in a number of preclinical studies [150–152]. As TGF-β promotes Treg generation upon TCR stimulation, it is thus conceivable that tumors promote immune privilege by generating Treg through TGF-β production. By understanding the roles of TGF-β and Treg in promoting tumorigenesis, we may be able to design more effective immunotherapies augmenting the antitumor response in cancer patients.

Conclusion

Active immune suppression has been established as an essential component of the immune regulation network. Dysregulation of peripheral active immune suppression leads to various disease states. As a critical immune suppressive cytokine, TGF-β has pleiotropic functions in regulating T cell proliferation and effector functions in a context dependent manner. As a central cell type in mediating immune suppression, Treg controls the activities of self-reactive T cells in the periphery through multiple mechanisms. Further studies of the mechanisms and signaling pathways mediating TGF-β and Treg function will lead to a better understanding of how our immune system achieves self-tolerance and mounting effective immune responses against unsolicited foreign antigens. This knowledge may ultimately lead to the development of innovative therapeutic strategies to manipulate the immune system to treat both autoimmune diseases such as IBD or diabetes as well as diseases resulting from dampened immunity such as cancer.

Acknowledgments The research performed in our laboratory mentioned in this book chapter is supported by the NIH, American Diabetes Association (ADA) and Howard Hughes Medical Institute. R.A.F. is an investigator of the Howard Hughes Medical Institute. Y.Y.W. is supported by a postdoctoral fellowship from Cancer Research Institute (CRI). We are grateful to S. Wrzesinski for the critical reading of this manuscript and helpful comments. We thank F. Manzo for secretarial assistance.

References

1. R. K. Gershon. A disquisition on suppressor T cells. Transplant Rev, 1975 26, 170–85
2. G. C. Blobe; W. P. Schiemann; H. F. Lodish. Role of transforming growth factor beta in human disease. N Engl J Med, 2000 May 4, 342, 1350–8
3. J. H. Kehrl; A. B. Roberts; L. M. Wakefield; S. Jakowlew; M. B. Sporn; A. S. Fauci. Transforming growth factor beta is an important immunomodulatory protein for human B lymphocytes. J Immunol, 1986 Dec 15, 137, 3855–60
4. J. H. Kehrl; L. M. Wakefield; A. B. Roberts; S. Jakowlew; M. Alvarez-Mon; R. Derynck; M. B. Sporn; A. S. Fauci. Production of transforming growth factor beta by human T lymphocytes and its potential role in the regulation of T cell growth. J Exp Med, 1986 May 1, 163, 1037–50
5. A. B. Kulkarni; C. G. Huh; D. Becker; A. Geiser; M. Lyght; K. C. Flanders; A. B. Roberts; M. B. Sporn; J. M. Ward; S. Karlsson. Transforming growth factor beta 1 null mutation in mice causes excessive inflammatory response and early death. Proc Natl Acad Sci USA, 1993 Jan 15, 90, 770–4
6. M. M. Shull; I. Ormsby; A. B. Kier; S. Pawlowski; R. J. Diebold; M. Yin; R. Allen; C. Sidman; G. Proetzel; D. Calvin; et al. Targeted disruption of the mouse transforming growth factor-beta 1 gene results in multifocal inflammatory disease. Nature, 1992 Oct 22, 359, 693–9
7. B. B. Cazac; J. Roes. TGF-beta receptor controls B cell responsiveness and induction of IgA in vivo. Immunity, 2000 Oct, 13, 443–51
8. M. B. Datto; J. P. Frederick; L. Pan; A. J. Borton; Y. Zhuang; X. F. Wang. Targeted disruption of Smad3 reveals an essential role in transforming growth factor beta-mediated signal transduction. Mol Cell Biol, 1999 Apr, 19, 2495–504
9. L. Gorelik; R. A. Flavell. Abrogation of TGFbeta signaling in T cells leads to spontaneous T cell differentiation and autoimmune disease. Immunity, 2000 Feb, 12, 171–81
10. Y. Laouar; F. S. Sutterwala; L. Gorelik; R. A. Flavell. Transforming growth factor-beta controls T helper type 1 cell development through regulation of natural killer cell interferon-gamma. Nat Immunol, 2005 Jun, 6, 600–7
11. P. J. Lucas; S. J. Kim; S. J. Melby; R. E. Gress. Disruption of T cell homeostasis in mice expressing a T cell-specific dominant negative transforming growth factor beta II receptor. J Exp Med, 2000 Apr 3, 191, 1187–96

12. A. Nakao; S. Miike; M. Hatano; K. Okumura; T. Tokuhisa; C. Ra; I. Iwamoto. Blockade of transforming growth factor beta/Smad signaling in T cells by overexpression of Smad7 enhances antigen-induced airway inflammation and airway reactivity. J Exp Med, 2000 Jul 17, 192, 151–8
13. X. Yang; J. J. Letterio; R. J. Lechleider; L. Chen; R. Hayman; H. Gu; A. B. Roberts; C. Deng. Targeted disruption of SMAD3 results in impaired mucosal immunity and diminished T cell responsiveness to TGF-beta. Embo J, 1999 Mar 1, 18, 1280–91
14. H. Chang; C. W. Brown; M. M. Matzuk. Genetic analysis of the mammalian transforming growth factor-beta superfamily. Endocr Rev, 2002 Dec, 23, 787–823
15. R. Govinden; K. D. Bhoola. Genealogy, expression, and cellular function of transforming growth factor-beta. Pharmacol Ther, 2003 May, 98, 257–65
16. C. M. Dubois; M. H. Laprise; F. Blanchette; L. E. Gentry; R. Leduc. Processing of transforming growth factor beta 1 precursor by human furin convertase. J Biol Chem, 1995 May 5, 270, 10618–24
17. J. P. Annes; Y. Chen; J. S. Munger; D. B. Rifkin. Integrin alphaVbeta6-mediated activation of latent TGF-beta requires the latent TGF-beta binding protein-1. J Cell Biol, 2004 Jun 7, 165, 723–34
18. J. P. Annes; J. S. Munger; D. B. Rifkin. Making sense of latent TGFbeta activation. J Cell Sci, 2003 Jan 15, 116, 217–24
19. S. E. Crawford; V. Stellmach; J. E. Murphy-Ullrich; S. M. Ribeiro; J. Lawler; R. O. Hynes; G. P. Boivin; N. Bouck. Thrombospondin-1 is a major activator of TGF-beta1 in vivo. Cell, 1998 Jun 26, 93, 1159–70
20. J. S. Munger; X. Huang; H. Kawakatsu; M. J. Griffiths; S. L. Dalton; J. Wu; J. F. Pittet; N. Kaminski; C. Garat; M. A. Matthay; D. B. Rifkin; D. Sheppard. The integrin alpha v beta 6 binds and activates latent TGF beta 1: A mechanism for regulating pulmonary inflammation and fibrosis. Cell, 1999 Feb 5, 96, 319–28
21. T. Yehualaeshet; R. O'Connor; J. Green-Johnson; S. Mai; R. Silverstein; J. E. Murphy-Ullrich; N. Khalil. Activation of rat alveolar macrophage-derived latent transforming growth factor beta-1 by plasmin requires interaction with thrombospondin-1 and its cell surface receptor, CD36. Am J Pathol, 1999 Sep, 155, 841–51
22. J. J. Letterio; A. B. Roberts. Regulation of immune responses by TGF-beta. Annu Rev Immunol, 1998 16, 137–61
23. E. A. Green; L. Gorelik; C. M. McGregor; E. H. Tran; R. A. Flavell. CD4+CD25+ T regulatory cells control anti-islet CD8+ T cells through TGF-beta-TGF-beta receptor interactions in type 1 diabetes. Proc Natl Acad Sci USA, 2003 Sep 16, 100, 10878–83
24. J. Massague. TGF-beta signal transduction. Annu Rev Biochem, 1998, 67, 753–91
25. M. Huse; T. W. Muir; L. Xu; Y. G. Chen; J. Kuriyan; J. Massague. The TGF beta receptor activation process: An inhibitor- to substrate-binding switch. Mol Cell, 2001 Sep, 8, 671–82
26. G. J. Inman; F. J. Nicolas; C. S. Hill. Nucleocytoplasmic shuttling of Smads 2, 3, and 4 permits sensing of TGF-beta receptor activity. Mol Cell, 2002 Aug, 10, 283–94
27. K. Johnson; H. Kirkpatrick; A. Comer; F. M. Hoffmann; A. Laughon. Interaction of Smad complexes with tripartite DNA-binding sites. J Biol Chem, 1999 Jul 16, 274, 20709–16
28. Y. Shi; Y. F. Wang; L. Jayaraman; H. Yang; J. Massague; N. P. Pavletich. Crystal structure of a Smad MH1 domain bound to DNA: Insights on DNA binding in TGF-beta signaling. Cell, 1998 Sep 4, 94, 585–94
29. L. Zawel; J. L. Dai; P. Buckhaults; S. Zhou; K. W. Kinzler; B. Vogelstein; S. E. Kern. Human Smad3 and Smad4 are sequence-specific transcriptional activators. Mol Cell, 1998 Mar, 1, 611–7
30. A. Nakao; M. Afrakhte; A. Moren; T. Nakayama; J. L. Christian; R. Heuchel; S. Itoh; M. Kawabata; N. E. Heldin; C. H. Heldin; P. ten Dijke. Identification of Smad7, a TGFbeta-inducible antagonist of TGF-beta signalling. Nature, 1997 Oct 9, 389, 631–5
31. T. Ebisawa; M. Fukuchi; G. Murakami; T. Chiba; K. Tanaka; T. Imamura; K. Miyazono. Smurf1 interacts with transforming growth factor-beta type I receptor through Smad7 and induces receptor degradation. J Biol Chem, 2001 Apr 20, 276, 12477–80

32. P. Kavsak; R. K. Rasmussen; C. G. Causing; S. Bonni; H. Zhu; G. H. Thomsen; J. L. Wrana. Smad7 binds to Smurf2 to form an E3 ubiquitin ligase that targets the TGF beta receptor for degradation. Mol Cell, 2000 Dec, 6, 1365–75
33. M. E. Engel; M. A. McDonnell; B. K. Law; H. L. Moses. Interdependent SMAD and JNK signaling in transforming growth factor-beta-mediated transcription. J Biol Chem, 1999 Dec 24, 274, 37413–20
34. L. Yu; M. C. Hebert; Y. E. Zhang. TGF-beta receptor-activated p38 MAP kinase mediates Smad-independent TGF-beta responses. Embo J, 2002 Jul 15, 21, 3749–59
35. R. Derynck; Y. E. Zhang. Smad-dependent and Smad-independent pathways in TGF-beta family signalling. Nature, 2003 Oct 9, 425, 577–84
36. F. Blanchette; N. Rivard; P. Rudd; F. Grondin; L. Attisano; C. M. Dubois. Cross-talk between the p42/p44 MAP kinase and Smad pathways in transforming growth factor beta 1-induced furin gene transactivation. J Biol Chem, 2001 Sep 7, 276, 33986–94
37. M. Funaba; C. M. Zimmerman; L. S. Mathews. Modulation of Smad2-mediated signaling by extracellular signal-regulated kinase. J Biol Chem, 2002 Nov 1, 277, 41361–8
38. M. Kretzschmar; J. Doody; I. Timokhina; J. Massague. A mechanism of repression of TGF-beta/ Smad signaling by oncogenic Ras. Genes Dev, 1999 Apr 1, 13, 804–16
39. L. Choy; R. Derynck. The type II transforming growth factor (TGF)-beta receptor-interacting protein TRIP-1 acts as a modulator of the TGF-beta response. J Biol Chem, 1998 Nov 20, 273, 31455–62
40. I. Griswold-Prenner; C. Kamibayashi; E. M. Maruoka; M. C. Mumby; R. Derynck. Physical and functional interactions between type I transforming growth factor beta receptors and Balpha, a WD-40 repeat subunit of phosphatase 2A. Mol Cell Biol, 1998 Nov, 18, 6595–604
41. S. McGonigle; M. J. Beall; E. J. Pearce. Eukaryotic initiation factor 2 alpha subunit associates with TGF beta receptors and 14-3-3 epsilon and acts as a modulator of the TGF beta response. Biochemistry, 2002 Jan 15, 41, 579–87
42. C. H. Chen; C. Seguin-Devaux; N. A. Burke; T. B. Oriss; S. C. Watkins; N. Clipstone; A. Ray. Transforming growth factor beta blocks Tec kinase phosphorylation, Ca2+ influx, and NFATc translocation causing inhibition of T cell differentiation. J Exp Med, 2003 Jun 16, 197, 1689–99
43. L. Gorelik; S. Constant; R. A. Flavell. Mechanism of transforming growth factor beta-induced inhibition of T helper type 1 differentiation. J Exp Med, 2002 Jun 3, 195, 1499–505
44. L. Gorelik; P. E. Fields; R. A. Flavell. Cutting edge: TGF-beta inhibits Th type 2 development through inhibition of GATA-3 expression. J Immunol, 2000 Nov 1, 165, 4773–7
45. J. T. Lin; S. L. Martin; L. Xia; J. D. Gorham. TGF-beta1 uses distinct mechanisms to inhibit IFN-gamma expression in CD4+ T cells at priming and at recall: Differential involvement of Stat4 and T-bet. J Immunol, 2005 May 15, 174, 5950–8
46. D. A. Thomas; J. Massague. TGF-beta directly targets cytotoxic T cell functions during tumor evasion of immune surveillance. Cancer Cell, 2005 Nov, 8, 369–80
47. S. Sakaguchi; N. Sakaguchi; M. Asano; M. Itoh; M. Toda. Immunologic self-tolerance maintained by activated T cells expressing IL-2 receptor alpha-chains (CD25). Breakdown of a single mechanism of self-tolerance causes various autoimmune diseases. J Immunol, 1995 Aug 1, 155, 1151–64
48. R. K. Gershon; K. Kondo. Infectious immunological tolerance. Immunology, 1971 Dec, 21, 903–14
49. R. K. Gershon; K. Kondo. Cell interactions in the induction of tolerance: The role of thymic lymphocytes. Immunology, 1970 May, 18, 723–37
50. S. Sakaguchi. Naturally arising CD4+ regulatory t cells for immunologic self-tolerance and negative control of immune responses. Annu Rev Immunol, 2004, 22, 531-62
51. K. Nakamura; A. Kitani; W. Strober. Cell contact-dependent immunosuppression by CD4(+)CD25(+) regulatory T cells is mediated by cell surface-bound transforming growth factor beta. J Exp Med, 2001 Sep 3, 194, 629–44

52. S. Sakaguchi. Naturally arising Foxp3-expressing CD25+CD4+ regulatory T cells in immunological tolerance to self and non-self. Nat Immunol, 2005 Apr, 6, 345–52
53. E. M. Shevach. Regulatory T cells in autoimmmunity*. Annu Rev Immunol, 2000, 18, 423–49
54. E. M. Shevach. CD4+ CD25+ suppressor T cells: More questions than answers. Nat Rev Immunol, 2002 Jun, 2, 389–400
55. J. D. Fontenot; J. L. Dooley; A. G. Farr; A. Y. Rudensky. Developmental regulation of Foxp3 expression during ontogeny. J Exp Med, 2005 Oct 3, 202, 901–6
56. R. H. Schwartz. Natural regulatory T cells and self-tolerance. Nat Immunol, 2005 Apr, 6, 327–30
57. C. L. Bennett; J. Christie; F. Ramsdell; M. E. Brunkow; P. J. Ferguson; L. Whitesell; T. E. Kelly; F. T. Saulsbury; P. F. Chance; H. D. Ochs. The immune dysregulation, polyendocrinopathy, enteropathy, X-linked syndrome (IPEX) is caused by mutations of FOXP3. Nat Genet, 2001 Jan, 27, 20–1
58. M. E. Brunkow; E. W. Jeffery; K. A. Hjerrild; B. Paeper; L. B. Clark; S. A. Yasayko; J. E. Wilkinson; D. Galas; S. F. Ziegler; F. Ramsdell. Disruption of a new forkhead/winged-helix protein, scurfin, results in the fatal lymphoproliferative disorder of the scurfy mouse. Nat Genet, 2001 Jan, 27, 68–73
59. S. Hori; T. Nomura; S. Sakaguchi. Control of regulatory T cell development by the transcription factor Foxp3. Science, 2003 Feb 14, 299, 1057–61
60. R. Khattri; T. Cox; S. A. Yasayko; F. Ramsdell. An essential role for Scurfin in CD4+CD25+ T regulatory cells. Nat Immunol, 2003 Apr, 4, 337–42
61. R. S. Wildin; F. Ramsdell; J. Peake; F. Faravelli; J. L. Casanova; N. Buist; E. Levy-Lahad; M. Mazzella; O. Goulet; L. Perroni; F. D. Bricarelli; G. Byrne; M. McEuen; S. Proll; M. Appleby; M. E. Brunkow. X-linked neonatal diabetes mellitus, enteropathy and endocrinopathy syndrome is the human equivalent of mouse scurfy. Nat Genet, 2001 Jan, 27, 18–20
62. X. Chang; J. X. Gao; Q. Jiang; J. Wen; N. Seifers; L. Su; V. L. Godfrey; T. Zuo; P. Zheng; Y. Liu. The Scurfy mutation of FoxP3 in the thymus stroma leads to defective thymopoiesis. J Exp Med, 2005 Oct 17, 202, 1141–51
63. H. Groux; A. O'Garra; M. Bigler; M. Rouleau; S. Antonenko; J. E. de Vries; M. G. Roncarolo. A CD4+ T-cell subset inhibits antigen-specific T-cell responses and prevents colitis. Nature, 1997 Oct 16, 389, 737–42
64. M. G. Roncarolo; R. Bacchetta; C. Bordignon; S. Narula; M. K. Levings. Type 1 T regulatory cells. Immunol Rev, 2001 Aug, 182, 68–79
65. P. L. Vieira; J. R. Christensen; S. Minaee; E. J. O'Neill; F. J. Barrat; A. Boonstra; T. Barthlott; B. Stockinger; D. C. Wraith; A. O'Garra. IL-10-secreting regulatory T cells do not express Foxp3 but have comparable regulatory function to naturally occurring CD4+CD25+ regulatory T cells. J Immunol, 2004 May 15, 172, 5986–93
66. A. M. Faria; H. L. Weiner. Oral tolerance. Immunol Rev, 2005 Aug, 206, 232–59
67. H. L. Weiner. Induction and mechanism of action of transforming growth factor-beta-secreting Th3 regulatory cells. Immunol Rev, 2001 Aug, 182, 207–14
68. M. Stassen; S. Fondel; T. Bopp; C. Richter; C. Muller; J. Kubach; C. Becker; J. Knop; A. H. Enk; S. Schmitt; E. Schmitt; H. Jonuleit. Human CD25+ regulatory T cells: Two subsets defined by the integrins alpha 4 beta 7 or alpha 4 beta 1 confer distinct suppressive properties upon CD4+ T helper cells. Eur J Immunol, 2004 May, 34, 1303–11
69. M. O. Li; Y. Y. Wan; S. Sanjabi; A. K. Robertson; R. A. Flavell. Transforming growth factor-beta regulation of immune responses. Annu Rev Immunol, 2006, 24, 99–146
70. T. Brabletz; I. Pfeuffer; E. Schorr; F. Siebelt; T. Wirth; E. Serfling. Transforming growth factor beta and cyclosporin A inhibit the inducible activity of the interleukin-2 gene in T cells through a noncanonical octamer-binding site. Mol Cell Biol, 1993 Feb, 13, 1155–62
71. S. C. McKarns; R. H. Schwartz; N. E. Kaminski. Smad3 is essential for TGF-beta 1 to suppress IL-2 production and TCR-induced proliferation, but not IL-2-induced proliferation. J Immunol, 2004 Apr 1, 172, 4275–84

72. D. Tzachanis; G. J. Freeman; N. Hirano; A. A. van Puijenbroek; M. W. Delfs; A. Berezovskaya; L. M. Nadler; V. A. Boussiotis. Tob is a negative regulator of activation that is expressed in anergic and quiescent T cells. Nat Immunol, 2001 Dec, 2, 1174–82
73. R. J. Coffey, Jr.; C. C. Bascom; N. J. Sipes; R. Graves-Deal; B. E. Weissman; H. L. Moses. Selective inhibition of growth-related gene expression in murine keratinocytes by transforming growth factor beta. Mol Cell Biol, 1988 Aug, 8, 3088–93
74. M. B. Datto; Y. Li; J. F. Panus; D. J. Howe; Y. Xiong; X. F. Wang. Transforming growth factor beta induces the cyclin-dependent kinase inhibitor p21 through a p53-independent mechanism. Proc Natl Acad Sci USA, 1995 Jun 6, 92, 5545–9
75. G. J. Hannon; D. Beach. p15INK4B is a potential effector of TGF-beta-induced cell cycle arrest. Nature, 1994 Sep 15, 371, 257–61
76. K. Polyak; J. Y. Kato; M. J. Solomon; C. J. Sherr; J. Massague; J. M. Roberts; A. Koff. p27Kip1, a cyclin-Cdk inhibitor, links transforming growth factor-beta and contact inhibition to cell cycle arrest. Genes Dev, 1994 Jan, 8, 9–22
77. J. J. Ruegemer; S. N. Ho; J. A. Augustine; J. W. Schlager; M. P. Bell; D. J. McKean; R. T. Abraham. Regulatory effects of transforming growth factor-beta on IL-2- and IL-4-dependent T cell-cycle progression. J Immunol, 1990 Mar 1, 144, 1767–76
78. L. A. Wolfraim; T. M. Walz; Z. James; T. Fernandez; J. J. Letterio. p21Cip1 and p27Kip1 act in synergy to alter the sensitivity of naive T cells to TGF-beta-mediated G1 arrest through modulation of IL-2 responsiveness. J Immunol, 2004 Sep 1, 173, 3093–102
79. F. Cottrez; H. Groux. Regulation of TGF-beta response during T cell activation is modulated by IL-10. J Immunol, 2001 Jul 15, 167, 773–8
80. K. M. Murphy; S. L. Reiner. The lineage decisions of helper T cells. Nat Rev Immunol, 2002 Dec, 2, 933–44
81. T. R. Mosmann; R. L. Coffman. TH1 and TH2 cells: Different patterns of lymphokine secretion lead to different functional properties. Annu Rev Immunol, 1989, 7, 145–73
82. H. Park; Z. Li; X. O. Yang; S. H. Chang; R. Nurieva; Y. H. Wang; Y. Wang; L. Hood; Z. Zhu; Q. Tian; C. Dong. A distinct lineage of CD4 T cells regulates tissue inflammation by producing interleukin 17. Nat Immunol, 2005 Nov, 6, 1133–41
83. S. C. Liang; X. Y. Tan; D. P. Luxenberg; R. Karim; K. Dunussi-Joannopoulos; M. Collins; L. A. Fouser. Interleukin (IL)-22 and IL-17 are coexpressed by Th17 cells and cooperatively enhance expression of antimicrobial peptides. J Exp Med, 2006 Oct 2, 203, 2271–9
84. L. Gorelik; R. A. Flavell. Transforming growth factor-beta in T-cell biology. Nat Rev Immunol, 2002 Jan, 2, 46–53
85. M. O. Li; S. Sanjabi; R. A. Flavell. Transforming growth factor-beta controls development, homeostasis, and tolerance of T cells by regulatory T cell-dependent and -independent mechanisms. Immunity, 2006 Sep, 25, 455–71
86. S. Sad; T. R. Mosmann. Single IL-2-secreting precursor CD4 T cell can develop into either Th1 or Th2 cytokine secretion phenotype. J Immunol, 1994 Oct 15, 153, 3514–22
87. M. Veldhoen; R. J. Hocking; C. J. Atkins; R. M. Locksley; B. Stockinger. TGFbeta in the context of an inflammatory cytokine milieu supports de novo differentiation of IL-17-producing T cells. Immunity, 2006 Feb, 24, 179–89
88. P. R. Mangan; L. E. Harrington; D. B. O'Quinn; W. S. Helms; D. C. Bullard; C. O. Elson; R. D. Hatton; S. M. Wahl; T. R. Schoeb; C. T. Weaver. Transforming growth factor-beta induces development of the T(H)17 lineage. Nature, 2006 May 11, 441, 231–4
89. M. Veldhoen; R. J. Hocking; R. A. Flavell; B. Stockinger. Signals mediated by transforming growth factor-beta initiate autoimmune encephalomyelitis, but chronic inflammation is needed to sustain disease. Nat Immunol, 2006 Nov, 7, 1151–6
90. C. T. Weaver; L. E. Harrington; P. R. Mangan; M. Gavrieli; K. M. Murphy. Th17: An effector CD4 T cell lineage with regulatory T cell ties. Immunity, 2006 Jun, 24, 677–88
91. J. D. Gorham; M. L. Guler; D. Fenoglio; U. Gubler; K. M. Murphy. Low dose TGF-beta attenuates IL-12 responsiveness in murine Th cells. J Immunol, 1998 Aug 15, 161, 1664–70
92. V. L. Heath; E. E. Murphy; C. Crain; M. G. Tomlinson; A. O'Garra. TGF-beta1 down-regulates Th2 development and results in decreased IL-4-induced STAT6 activation and GATA-3 expression. Eur J Immunol, 2000 Sep, 30, 2639–49

93. B. R. Ludviksson; D. Seegers; A. S. Resnick; W. Strober. The effect of TGF-beta1 on immune responses of naive versus memory CD4+ Th1/Th2 T cells. Eur J Immunol, 2000 Jul, 30, 2101–11
94. A. Kitani; I. Fuss; K. Nakamura; F. Kumaki; T. Usui; W. Strober. Transforming growth factor (TGF)-beta1-producing regulatory T cells induce Smad-mediated interleukin 10 secretion that facilitates coordinated immunoregulatory activity and amelioration of TGF-beta1-mediated fibrosis. J Exp Med, 2003 Oct 20, 198, 1179–88
95. M. Ahmadzadeh; S. A. Rosenberg. TGF-beta1 attenuates the acquisition and expression of effector function by tumor antigen-specific human memory CD8 T cells. J Immunol, 2005 May 1, 174, 5215–23
96. H. Bonig; U. Banning; M. Hannen; Y. M. Kim; J. Verheyen; C. Mauz-Korholz; D. Korholz. Transforming growth factor-beta1 suppresses interleukin-15-mediated interferon-gamma production in human T lymphocytes. Scand J Immunol, 1999 Dec, 50, 612–8
97. G. E. Ranges; I. S. Figari; T. Espevik; M. A. Palladino, Jr. Inhibition of cytotoxic T cell development by transforming growth factor beta and reversal by recombinant tumor necrosis factor alpha. J Exp Med, 1987 Oct 1, 166, 991–8
98. M. J. Smyth; S. L. Strobl; H. A. Young; J. R. Ortaldo; A. C. Ochoa. Regulation of lymphokine-activated killer activity and pore-forming protein gene expression in human peripheral blood CD8+ T lymphocytes. Inhibition by transforming growth factor-beta. J Immunol, 1991 May 15, 146, 3289–97
99. T. R. Mempel; M. J. Pittet; K. Khazaie; W. Weninger; R. Weissleder; H. von Boehmer; U. H. von Andrian. Regulatory T cells reversibly suppress cytotoxic T cell function independent of effector differentiation. Immunity, 2006 Jul, 25, 129–41
100. H. M. Lee; S. Rich. Differential activation of CD8+ T cells by transforming growth factor-beta 1. J Immunol, 1993 Jul 15, 151, 668–77
101. J. D. Gray; T. Liu; N. Huynh; D. A. Horwitz. Transforming growth factor beta enhances the expression of CD154 (CD40L) and production of tumor necrosis factor alpha by human T lymphocytes. Immunol Lett, 2001 Sep 3, 78, 83–8
102. E. J. Chung; S. H. Choi; Y. H. Shim; Y. J. Bang; K. C. Hur; C. W. Kim. Transforming growth factor-beta induces apoptosis in activated murine T cells through the activation of caspase 1-like protease. Cell Immunol, 2000 Aug 25, 204, 46–54
103. H. K. Sillett; S. M. Cruickshank; J. Southgate; L. K. Trejdosiewicz. Transforming growth factor-beta promotes 'death by neglect' in post-activated human T cells. Immunology, 2001 Mar, 102, 310–6
104. W. Chen; W. Jin; H. Tian; P. Sicurello; M. Frank; J. M. Orenstein; S. M. Wahl. Requirement for transforming growth factor beta1 in controlling T cell apoptosis. J Exp Med, 2001 Aug 20, 194, 439–53
105. L. Genestier; S. Kasibhatla; T. Brunner; D. R. Green. Transforming growth factor beta1 inhibits Fas ligand expression and subsequent activation-induced cell death in T cells via downregulation of c-Myc. J Exp Med, 1999 Jan 18, 189, 231–9
106. L. A. Rudner; J. T. Lin; I. K. Park; J. M. Cates; D. A. Dyer; D. M. Franz; M. A. French; E. M. Duncan; H. D. White; J. D. Gorham. Necroinflammatory liver disease in BALB/c background, TGF-beta 1-deficient mice requires CD4+ T cells. J Immunol, 2003 May 1, 170, 4785–92
107. J. J. Letterio; A. G. Geiser; A. B. Kulkarni; H. Dang; L. Kong; T. Nakabayashi; C. L. Mackall; R. E. Gress; A. B. Roberts. Autoimmunity associated with TGF-beta1-deficiency in mice is dependent on MHC class II antigen expression. J Clin Invest, 1996 Nov 1, 98, 2109–19
108. P. Leveen; J. Larsson; M. Ehinger; C. M. Cilio; M. Sundler; L. J. Sjostrand; R. Holmdahl; S. Karlsson. Induced disruption of the transforming growth factor beta type II receptor gene in mice causes a lethal inflammatory disorder that is transplantable. Blood, 2002 Jul 15, 100, 560–8

109. A. M. Thornton; E. M. Shevach. CD4+CD25+ immunoregulatory T cells suppress polyclonal T cell activation in vitro by inhibiting interleukin 2 production. J Exp Med, 1998 Jul 20, 188, 287–96
110. T. Barthlott; H. Moncrieffe; M. Veldhoen; C. J. Atkins; J. Christensen; A. O'Garra; B. Stockinger. CD25+ CD4+ T cells compete with naive CD4+ T cells for IL-2 and exploit it for the induction of IL-10 production. Int Immunol, 2005 Mar, 17, 279–88
111. M. de la Rosa; S. Rutz; H. Dorninger; A. Scheffold. Interleukin-2 is essential for CD4+CD25+ regulatory T cell function. Eur J Immunol, 2004 Sep, 34, 2480–8
112. J. G. Egen; J. P. Allison. Cytotoxic T lymphocyte antigen-4 accumulation in the immunological synapse is regulated by TCR signal strength. Immunity, 2002 Jan, 16, 23–35
113. J. M. Slavik; J. E. Hutchcroft; B. E. Bierer. CD28/CTLA-4 and CD80/CD86 families: Signaling and function. Immunol Res, 1999 19, 1–24
114. R. J. Greenwald; G. J. Freeman; A. H. Sharpe. The B7 family revisited. Annu Rev Immunol, 2005 23, 515–48
115. E. Boden; Q. Tang; H. Bour-Jordan; J. A. Bluestone. The role of CD28 and CTLA4 in the function and homeostasis of CD4+CD25+ regulatory T cells. Novartis Found Symp, 2003 252, 55–63; Discussion 63–6, 106–14
116. J. D. Fontenot; J. P. Rasmussen; M. A. Gavin; A. Y. Rudensky. A function for interleukin 2 in Foxp3-expressing regulatory T cells. Nat Immunol, 2005 Nov, 6, 1142–51
117. L. Fahlen; S. Read; L. Gorelik; S. D. Hurst; R. L. Coffman; R. A. Flavell; F. Powrie. T cells that cannot respond to TGF-beta escape control by CD4(+)CD25(+) regulatory T cells. J Exp Med, 2005 Mar 7, 201, 737–46
118. M. L. Chen; M. J. Pittet; L. Gorelik; R. A. Flavell; R. Weissleder; H. von Boehmer; K. Khazaie. Regulatory T cells suppress tumor-specific CD8 T cell cytotoxicity through TGF-beta signals in vivo. Proc Natl Acad Sci USA, 2005 Jan 11, 102, 419–24
119. H. H. Uhlig; J. Coombes; C. Mottet; A. Izcue; C. Thompson; A. Fanger; A. Tannapfel; J. D. Fontenot; F. Ramsdell; F. Powrie. Characterization of Foxp3+CD4+CD25+ and IL-10-secreting CD4+CD25+ T cells during cure of colitis. J Immunol, 2006 Nov 1, 177, 5852–60
120. M. Kamanaka; S. T. Kim; Y. Y. Wan; F. S. Sutterwala; M. Lara-Tejero; J. E. Galan; E. Harhaj; R. A. Flavell. Expression of Interleukin-10 in Intestinal Lymphocytes Detected by an Interleukin-10 Reporter Knockin tiger Mouse. Immunity, 2006 Nov 27,
121. C. Asseman; S. Mauze; M. W. Leach; R. L. Coffman; F. Powrie. An essential role for interleukin 10 in the function of regulatory T cells that inhibit intestinal inflammation. J Exp Med, 1999 Oct 4, 190, 995–1004
122. J. D. Gray; M. Hirokawa; D. A. Horwitz. The role of transforming growth factor beta in the generation of suppression: An interaction between CD8+ T and NK cells. J Exp Med, 1994 Nov 1, 180, 1937–42
123. S. Yamagiwa; J. D. Gray; S. Hashimoto; D. A. Horwitz. A role for TGF-beta in the generation and expansion of CD4+CD25+ regulatory T cells from human peripheral blood. J Immunol, 2001 Jun 15, 166, 7282–9
124. M. C. Fantini; C. Becker; G. Monteleone; F. Pallone; P. R. Galle; M. F. Neurath. Cutting edge: TGF-beta induces a regulatory phenotype in CD4+CD25− T cells through Foxp3 induction and down-regulation of Smad7. J Immunol, 2004 May 1, 172, 5149–53
125. Y. Y. Wan; R. A. Flavell. Identifying Foxp3-expressing suppressor T cells with a bicistronic reporter. Proc Natl Acad Sci USA, 2005 Apr 5, 102, 5126–31
126. S. Rich; M. Seelig; H. M. Lee; J. Lin. Transforming growth factor beta 1 costimulated growth and regulatory function of staphylococcal enterotoxin B-responsive CD8+ T cells. J Immunol, 1995 Jul 15, 155, 609–18
127. W. Chen; W. Jin; N. Hardegen; K. J. Lei; L. Li; N. Marinos; G. McGrady; S. M. Wahl. Conversion of peripheral CD4+CD25− naive T cells to CD4+CD25+ regulatory T cells by TGF-beta induction of transcription factor Foxp3. J Exp Med, 2003 Dec 15, 198, 1875–86

128. D. A. Horwitz; S. G. Zheng; J. D. Gray. The role of the combination of IL-2 and TGF-beta or IL-10 in the generation and function of CD4+ CD25+ and CD8+ regulatory T cell subsets. J Leukoc Biol, 2003 Oct, 74, 471–8
129. M. K. Levings; R. Sangregorio; M. G. Roncarolo. Human cd25(+)cd4(+) t regulatory cells suppress naive and memory T cell proliferation and can be expanded in vitro without loss of function. J Exp Med, 2001 Jun 4, 193, 1295–302
130. P. A. Taylor; C. J. Lees; B. R. Blazar. The infusion of ex vivo activated and expanded CD4(+)CD25(+) immune regulatory cells inhibits graft-versus-host disease lethality. Blood, 2002 May 15, 99, 3493–9
131. J. D. Fontenot; J. P. Rasmussen; L. M. Williams; J. L. Dooley; A. G. Farr; A. Y. Rudensky. Regulatory T cell lineage specification by the forkhead transcription factor foxp3. Immunity, 2005 Mar, 22, 329–41
132. Y. Peng; Y. Laouar; M. O. Li; E. A. Green; R. A. Flavell. TGF-beta regulates in vivo expansion of Foxp3-expressing CD4+CD25+ regulatory T cells responsible for protection against diabetes. Proc Natl Acad Sci USA, 2004 Mar 30, 101, 4572–7
133. C. Schramm; M. Protschka; H. H. Kohler; J. Podlech; M. J. Reddehase; P. Schirmacher; P. R. Galle; A. W. Lohse; M. Blessing. Impairment of TGF-beta signaling in T cells increases susceptibility to experimental autoimmune hepatitis in mice. Am J Physiol Gastrointest Liver Physiol, 2003 Mar, 284, G525–35
134. S. Huber; C. Schramm; H. A. Lehr; A. Mann; S. Schmitt; C. Becker; M. Protschka; P. R. Galle; M. F. Neurath; M. Blessing. Cutting edge: TGF-beta signaling is required for the in vivo expansion and immunosuppressive capacity of regulatory CD4+CD25+ T cells. J Immunol, 2004 Dec 1, 173, 6526–31
135. J. C. Marie; J. J. Letterio; M. Gavin; A. Y. Rudensky. TGF-beta1 maintains suppressor function and Foxp3 expression in CD4+CD25+ regulatory T cells. J Exp Med, 2005 Apr 4, 201, 1061–7
136. M. Mamura; W. Lee; T. J. Sullivan; A. Felici; A. L. Sowers; J. P. Allison; J. J. Letterio. CD28 disruption exacerbates inflammation in Tgf-beta1−/− mice: In vivo suppression by CD4+CD25+ regulatory T cells independent of autocrine TGF-beta1. Blood, 2004 Jun 15, 103, 4594–601
137. Y. Zhu; J. A. Richardson; L. F. Parada; J. M. Graff. Smad3 mutant mice develop metastatic colorectal cancer. Cell, 1998 Sep 18, 94, 703–14
138. H. P. Kim; B. G. Kim; J. Letterio; W. J. Leonard. Smad-dependent cooperative regulation of interleukin 2 receptor alpha chain gene expression by T cell receptor and transforming growth factor-beta. J Biol Chem, 2005 Oct 7, 280, 34042–7
139. J. L. Coombes; N. J. Robinson; K. J. Maloy; H. H. Uhlig; F. Powrie. Regulatory T cells and intestinal homeostasis. Immunol Rev, 2005 Apr, 204, 184–94
140. K. J. Maloy; L. R. Antonelli; M. Lefevre; F. Powrie. Cure of innate intestinal immune pathology by CD4+CD25+ regulatory T cells. Immunol Lett, 2005 Mar 15, 97, 189–92
141. M. C. Fantini; C. Becker; I. Tubbe; A. Nikolaev; H. A. Lehr; P. Galle; M. F. Neurath. Transforming growth factor beta induced FoxP3+ regulatory T cells suppress Th1 mediated experimental colitis. Gut, 2006 May, 55, 671–80
142. E. A. Green; Y. Choi; R. A. Flavell. Pancreatic lymph node-derived CD4(+)CD25(+) Treg cells: Highly potent regulators of diabetes that require TRANCE-RANK signals. Immunity, 2002 Feb, 16, 183–91
143. H. L. Chang; N. Gillett; I. Figari; A. R. Lopez; M. A. Palladino; R. Derynck. Increased transforming growth factor beta expression inhibits cell proliferation in vitro, yet increases tumorigenicity and tumor growth of Meth A sarcoma cells. Cancer Res, 1993 Sep 15, 53, 4391–8
144. H. Fakhrai; O. Dorigo; D. L. Shawler; H. Lin; D. Mercola; K. L. Black; I. Royston; R. E. Sobol. Eradication of established intracranial rat gliomas by transforming growth factor beta antisense gene therapy. Proc Natl Acad Sci USA, 1996 Apr 2, 93, 2909–14

145. M. Stander; U. Naumann; L. Dumitrescu; M. Heneka; P. Loschmann; E. Gulbins; J. Dichgans; M. Weller. Decorin gene transfer-mediated suppression of TGF-beta synthesis abrogates experimental malignant glioma growth in vivo. Gene Ther, 1998 Sep, 5, 1187–94
146. G. Torre-Amione; R. D. Beauchamp; H. Koeppen; B. H. Park; H. Schreiber; H. L. Moses; D. A. Rowley. A highly immunogenic tumor transfected with a murine transforming growth factor type beta 1 cDNA escapes immune surveillance. Proc Natl Acad Sci USA, 1990 Feb, 87, 1486–90
147. L. A. Ormandy; T. Hillemann; H. Wedemeyer; M. P. Manns; T. F. Greten; F. Korangy. Increased populations of regulatory T cells in peripheral blood of patients with hepatocellular carcinoma. Cancer Res, 2005 Mar 15, 65, 2457–64
148. M. Beyer; M. Kochanek; K. Darabi; A. Popov; M. Jensen; E. Endl; P. A. Knolle; R. K. Thomas; M. von Bergwelt-Baildon; S. Debey; M. Hallek; J. L. Schultze. Reduced frequencies and suppressive function of CD4+ CD25high regulatory T cells in patients with chronic lymphocytic leukemia after therapy with fludarabine. Blood, 2005 May 24,
149. H. Nishikawa; T. Kato; I. Tawara; T. Takemitsu; K. Saito; L. Wang; Y. Ikarashi; H. Wakasugi; T. Nakayama; M. Taniguchi; K. Kuribayashi; L. J. Old; H. Shiku. Accelerated chemically induced tumor development mediated by CD4+CD25+ regulatory T cells in wild-type hosts. Proc Natl Acad Sci USA, 2005 Jun 16,
150. S. Onizuka; I. Tawara; J. Shimizu; S. Sakaguchi; T. Fujita; E. Nakayama. Tumor rejection by in vivo administration of anti-CD25 (interleukin-2 receptor alpha) monoclonal antibody. Cancer Res, 1999 Jul 1, 59, 3128–33
151. J. Shimizu; S. Yamazaki; S. Sakaguchi. Induction of tumor immunity by removing CD25+CD4+ T cells: A common basis between tumor immunity and autoimmunity. J Immunol, 1999 Nov 15, 163, 5211–8
152. R. P. Sutmuller; L. M. van Duivenvoorde; A. van Elsas; T. N. Schumacher; M. E. Wildenberg; J. P. Allison; R. E. Toes; R. Offringa; C. J. Melief. Synergism of cytotoxic T lymphocyte-associated antigen 4 blockade and depletion of CD25(+) regulatory T cells in antitumor therapy reveals alternative pathways for suppression of autoreactive cytotoxic T lymphocyte responses. J Exp Med, 2001 Sep 17, 194, 823–32

Chapter 7
TGF-β Regulates Reciprocal Differentiation of CD4$^+$CD25$^+$Foxp3$^+$ Regulatory T Cells and IL-17-Producing Th17 Cells from Naïve CD4$^+$CD25$^-$ T Cells

Wanjun Chen

Abstract CD4$^+$CD25$^+$ T cells expressing forkhead transcription factor Foxp3 are recognized as professional regulatory T cells (Tregs) and are instrumental in the induction and maintenance of immune tolerance. In addition to their development intrathymically, CD4$^+$ CD25$^+$Foxp3$^+$ Tregs may also be converted extrathymically from CD4$^+$CD25$^-$ naïve T cells through induction of Foxp3 by transforming growth factor–beta (TGF-β). Most recently, an interleukin-17 (IL-17)-producing pro-inflammatory T helper (Th) subset, designated Th17, has been identified. Unexpectedly, the differentiation of Th17 cells from CD4$^+$CD25$^-$ naïve T cells also requires TGF-β, but in the presence of the proinflammatory cytokine interleukin-6 (IL-6). The transcription factor that programs and directs Th17 cell differentiation includes retinoic-acid-related orphan receptor-γt (RORrt), which is induced by the combined signals from both IL-6 and TGF-β engagement. This chapter attempts to highlight the discovery and development of the TGF-β reciprocal regulation of Tregs and Th17 cells.

Introduction

A delicate balance between immune response (immunity) and immune tolerance is prerequisite to maintaining the well-being of the immune system, and consequently that of the human body. The search for the factors that maintain this balance has led to remarkable discoveries of phenotypically and especially functionally distinguishable subsets of CD4$^+$ T subsets of CD4$^+$ T helper cells. In chronological order, Th1 and Th2 effector T cell subsets were first discovered about 20 years ago [1,2], followed by a resurgence of thymic natural CD4$^+$CD25$^+$(IL-2 receptor α

W. Chen
Mucosal Immunology Unit, OIIB, National Institute of Dental and Craniofacial Research, National Institutes of Health, Bethesda, MD 20892, USA
e-mail: wchen@mail.nih.gov

chain) Tregs a decade ago [3,4]. Within the past three years, "induced" or adoptive CD4$^+$CD25$^+$ Foxp3$^+$Tregs [5–7] and the IL-17 producing Th17 subset were also identified [8–12].

The naïve CD4$^+$ T helper lymphocyte is traditionally thought to differentiate into Th1 and Th2 effector subsets [1,13,14]. In the context of TCR engagement, IL-12 and IFN-γ program and direct CD4$^+$ T cells to develop into cells of the Th1 effector subset through T-bet [15], whereas IL-4 is required for the differentiation of Th2 cells by Gata-3 [16,17]. Th1 and Th2 cells are involved in orchestrating the host responses of clearing pathogens and establishing long-term memory for recall responses. Persistent effector T responses, however, may trigger chronic inflammation such as autoimmune diseases and allergy [18].

The resurgence of interest in regulatory T cells (suppressor) came from a seminal observation of Sakaguchi and colleagues in 1995 [3]. In normal mice, there is small population of CD4$^+$ T cells constitutively expressing CD25 (IL-2 receptor alpha chain) and developed in the thymus as "natural" CD4$^+$CD25$^+$ Tregs mastered by transcription factor Foxp3 [19–21]. These CD4$^+$CD25$^+$ T cells are not only unresponsive (anergic) to TCR stimulation in vitro, but also more importantly suppress normal responder CD4$^+$ T cell proliferation in a co-culture assay [22], and thus were named Tregs. Subsequent studies have revealed that CD4$^+$CD25$^+$ Tregs also inhibit proliferation of CD8$^+$ and B lymphocytes as well as macrophages, dendritic cells (DCs), and NK cells [23–28]. Importantly, CD4$^+$CD25$^+$Foxp3$^+$ Tregs exist in humans [29,30], and they function and suppress antigen-driven T cell expansion in vivo [31]. It has been well accepted that CD4$^+$CD25$^+$ Tregs are instrumental in maintaining immune tolerance. Although CD4$^+$CD25$^+$ Tregs were originally thought to be generated and developed in the thymus as a defined lineage, recent evidence has demonstrated that CD4$^+$CD25$^+$ Tregs can also be converted from naïve CD4$^+$CD25$^-$ T cells in the periphery (Fig. 7.1). Importantly, these "induced" Tregs are phenotypically and functionally indistinguishable from the thymically derived "natural" Tregs, and their differentiation is independent of the presence of the thymus. Significantly, the conversion requires engagement of T cell receptor (TCR) and transforming growth factor (TGF)-β signaling through induction of Foxp3 in CD4$^+$CD25$^-$ Foxp3$^-$ naïve T cells [5]. In addition, there are also induced Th3 and Tr1 regulatory cells, which have been discussed in detail in several excellent reviews [32,33]

Another unexpected function of TGF-β signaling in Th cells is its indispensable role in the differentiation of Th17 cells [8,10,12]. Th17 cells are the most recently identified subset of CD4$^+$ Th cells and are characterized by production of a distinct profile of effector cytokines, including IL-17, IL-17A, and IL-6 [9,18,34]. Th17 cells have emerged as one of the cell populations that are primarily responsible for a variety of chronic inflammatory and autoimmune disorders, including experimental autoimmune encephalitis (EAE) (for human multiple sclerosis) and collagen-induced arthritis (CIA) (for human rheumatoid arthritis) [9,34], although their role in host defense against foreign pathogens still remains to be elucidated. Initially, IL-23 was thought to be the main driving force in the differentiation of IL-17 cells [35]. However, recent evidence has revealed that TGF-β, in the presence

7 Differentiation of CD4+CD25+Foxp3+ Tregs and IL-17-Producing Th17 Cells

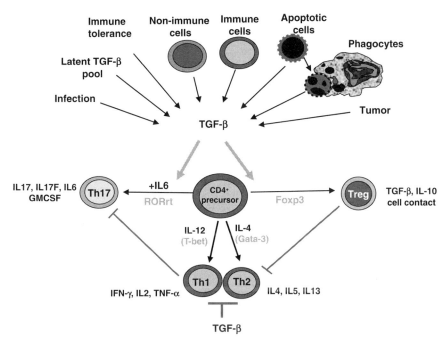

Fig. 7.1 TGF-β controls reciprocal differentiation of Tregs and Th17 cells. TGF-β is produced and secreted by almost all types of cells and from every tissues and organs. TGF-β is also activated and released from the pool of latent-TGF-β in the circulation in response to variety of physiological and pathological challenges. TGF-β plays critical roles in the regulation of immune responses. TGF-β inhibits T cell proliferation and Th1 and Th2 differentiation. TGF-β, in the context of TCR stimulation, may also induce Treg or Th17 cell differentiation from CD4+ T precursor depending on the absence or presence of IL-6, which are programmed and directed by Foxp3 and RORrt transcription factors respectively. Tregs suppress Th1 and Th2 responses, whereas Th1 and Th2 cells downregulate Th17 cell differentiation. *Green arrows* indicate positive regulation; *red lines* indicate negative regulation *(See also Color Insert)*

of proinflammatory cytokine IL-6, instigates the conversion of Th17 cells from the naïve CD4+ T precursors [18,34] (Fig. 7.1).

TGF-β belongs to a family of related molecules that affect almost all aspects of biological responses [36]. TGF-β1 is the prototype of the TGF-β family and mainly affects immune responses [37–41]. TGF-β is recognized primarily as an immunoregulatory cytokine, but its exact role in immune responses is heavily dependent on immune cell type, state, and microenvironment. The essential role of TGF-β in maintaining the balance of immunity and immune tolerance in vivo was established by TGF-β1 gene-deficient mice [42,43], because TGF-β1-null mice develop a rapid lethal inflammation with massive infiltrates in multiple tissues and organs. TGF-β initiates signaling by binding to TGF-β receptor II (TβRII) that phosphorylates the TGF-β receptor I (TβRI) kinase domain, which then propagates the signal through phosphorylation of the reactive Smads (Smad2/3) [44]. The classic paradigm emphasizes that phosphorylated Smad2/3(P-Smad2/3) then forms a complex with co-Smad (Smad4) to translocate into the nucleus to carry out TGF-β signaling. It has been known for a long time that TGF-β may down- or upregulate T cell responses and exerts both immunosuppressive and proinflammatory effects in several animal autoimmune disease models [37,39,45–47]. A clear picture of TGF-β as a " double agent" in immune response and especially in host defense is attributed to recent studies that TGF-β reciprocally regulates the differentiation of Tregs and Th17 cells from same naïve $CD4^+CD25^-$ precursors [5,7,8,10,12]. This review will focus on recent discoveries in the TGF-β-mediated differentiation of $Foxp3^+$ Tregs and Th17 cells from naïve $CD4^+CD25^-$ T cells, and discuss possible downstream molecular mechanisms and the implications of these findings for clinical settings.

Natural Versus Adoptive (Induced) $CD4^+CD25^+$ $Foxp3^+$ Regulatory T Cells

Current theory states that there are two populations of $CD4^+CD25^+Foxp3^+$ Tregs, natural vs. adoptive (induced). It is generally believed that "natural" Tregs are generated and developed in the thymus. Although the detailed molecular events and mechanisms remain elusive, it has been clearly demonstrated that Foxp3 is the master gene programming the generation and development of natural Tregs in the thymus [19–21,25]. The current theory suggests that $CD4^+CD25^+$ Tregs are derived from a defined lineage and somehow survived negative selection within a very small "window" with strong affinity to TCR engagement. The generation of $CD4^+CD25^+$ Tregs is believed to be dependent on the interaction and affinity between antigen and TCR on the thymocytes.

While the thymus is considered the key site or incubator for the generation and development of natural $CD4^+CD25^+$ Tregs, the detailed pathways through which the natural Tregs are developed remain largely unknown. The first significant question is at which stage of thymocyte development $CD4^+CD25^+Foxp3^+$ Tregs are generated. In a time-kinetic study, Rudensky and colleagues demonstrated that thymic $CD4^+CD25^+Foxp3^+$ Tregs appeared in $CD4^+CD8^-$ single positive thymocytes at about 3 days of age in mice, which is later than appearance of $CD4^+CD25^-$ or $CD4^+CD25^+Foxp3^-$ thymocytes [48]. However, once generated, the $CD4^+CD25^+Foxp3^+$ Tregs in the thymus will rapidly reach the level of an adult in a few days. This finding is important because it may offer an explanation for the deficiency of peripheral $CD4^+CD25^+$ Tregs in mice caused by thymectomy on day 3 of age. Regarding the cellular stage in the thymus, the same group of researchers has shown that the majority of $CD4^+CD25^+Foxp3^+$ Tregs are located in the medulla rather than the cortex of the thymus [25], where mature and single positive (SP) thymocytes reside. Significantly, the study also suggests that the thymic $CD4^+CD25^+Foxp3^+$ Tregs are unlikely to have differentiated from the $CD4^+CD8^+$ immature double positive (DP) thymocytes, although DP cells also contain a small number of cells expressing Foxp3 [25]. The data suggest that at a minimum a large number of natural $CD4^+CD25^+$ $Foxp3^+$ Tregs are dedifferentiated from mature or semimature $CD4^+$ thymocytes [25,49], although this hypothesis has not yet been proven experimentally. In support of this notion, however, CD4$^+$CD25$^-$Foxp3$^-$ single positive (SP) thymocytes can be converted into CD4$^+$CD25$^+$Foxp3$^+$ Tregs in the context of TCR stimulation and TGF-β treatment in culture [50] (discussed below). Similarly, human CD4$^+$CD25$^-$ Foxp3$^-$ thymocytes can also be converted into Foxp3$^+$ Tregs in vitro by thymic DCs pretreated with TLSP, a soluble factor secreted by Hassall's corpuscles in the thymus [51]. The biological significance and in vivo relevance of those in vitro findings, however, remains to be elucidated.

The second prominent issue is the molecular basis and pathways that program and direct the generation of thymic natural Tregs. It has been clearly established that Foxp3 is the master gene that governs the generation and development of natural Tregs, but the upstream molecules and pathways that switch on Foxp3 transcription in the thymic precursors are still far from clear. There is no doubt that TCR engagement is required in Treg generation and development. The questions of at which degree of engagement (affinity), with what kinds of co-stimulatory or accessory molecules or factors, and under which microenvironment TCR signaling leads to the generation of CD4$^+$CD25$^+$ Foxp3$^+$ Tregs rather than the development of CD4$^+$Foxp3$^-$ responder T cells remain unanswered. Several molecules have been implicated in the process, but none has yet been universally recognized. CD28 is the prototype and essential co-stimulatory molecule in CD4$^+$ T cell proliferation and differentiation [52]. Engagement of CD28 was originally believed to co-stimulate TCR signaling leading to T cell proliferation by producing IL-2 [53] and to protect from activation-induced T cell death by upregulating the antiapoptotic protein Bcl-XL [54]. Surprisingly, however, it has recently been found that CD28-null mice present a profound deficiency in CD4$^+$CD25$^+$ Tregs [55], linking the CD28/B7 costimulatory pathway with Treg homeostasis. Consequently, CD28-null NOD mice develop a more severe spontaneous diabetes [55]. In exploring the underlying molecular mechanisms of CD28 requirement and signaling, it has been demonstrated that generation of thymic natural CD4$^+$CD25$^+$ Tregs requires a motif that binds the tyrosine kinase lck, precisely the same motif that is required for CD28 costimulation of IL-2 production [49]. Nevertheless, CD28 engagement provides more than IL-2 to developing Tregs, as CD28 and TCR stimulation induces Foxp3 expression in CD4$^+$CD8$^+$DP thymocytes. This finding indicates that CD28 stimulation may directly program DP T cells toward Treg lineage. However, the unresolved questions include whether CD28 engagement presents a unique signal to DP to switch on Foxp3 transcription that directly develops CD25$^+$Foxp3$^+$ SP Tregs, or whether CD28 signaling helps mature DP into SP thymocytes that then differenciate toward Tregs. Furthermore, the biggest puzzle is under which condition or in which microenvironment CD28 engagement favors Treg development instead of promoting effector T cell proliferation, because CD28 is expressed in almost all murine T cells.

Another molecule implicated in the development and homeostasis of natural Tregs is IL-2. IL-2 was originally regarded as an essential cytokine in stimulating T cell growth and population expansion. Despite its critical role in Treg proliferation and function[56], IL-2 is dispensable for the induction of Foxp3 expression in thymocytes [57]. Using mice with the Foxp3^{gfp} knock-in allele that were genetically deficient in IL-2, Fontenot et al. clearly demonstrated that Foxp3^{gfp} IL-2–/– mice develop Foxp3$^+$ CD4$^+$ Tregs in the thymus, although the levels of Foxp3 expression are decreased. Using same techniques, they also showed that IL-2 receptors α (CD25) and β (CD122) are dispensable in Foxp3 induction in the thymus. However, deletion of IL-2Rγ almost abrogates Foxp3$^+$ Tregs in the thymus as well as in the periphery [57], indicating signaling through IL-2Rγ may be essential for the Foxp3 expression in natural Tregs. The upstream factors responsible for signaling IL-2Rγ to induce Foxp3 expression remain to be elucidated. IL-2 and IL-15 are unlikely candidates, because they bind both IL-2Rα and IL-2Rβ on T cells. In line with these observations, deletion of Stat5, a transcription factor involved in a variety of signaling transduction events also results in a deficiency of Foxp3$^+$ Tregs in the thymus [58]. Whether Stat5–/– would be sufficient to account for the absence of Foxp3 in IL-2Rγ–/– mice remains unknown [57]. Despite the positive correlation between IL-2Rγ or Stat5 deficiency and the absence of Foxp3$^+$ T cells in the thymus, it has been noticed that both types of null mice have a substantial deficiency and lower numbers of total CD4$^+$ T cells in the thymus, indicating that these molecules are required for CD4$^+$ T cell positive selection and development. Whether the Foxp3$^+$ Treg deficiency in those null mice is due to the direct role of those molecular pathways or is an indirect effect of the defect in CD4$^+$ T cell development needs to be determined.

TGF-β signaling in the generation and development of thymic CD4$^+$CD25$^+$Foxp3$^+$ Tregs is another area that is still being debated. The similar lethal inflammatory phenotype of TGF-β1-null mice and Foxp3 deficient mice has inevitably linked TGF-β1 signaling and Fxop3 expression. Indeed, in vitro and some in vivo models have clearly demonstrated that Foxp3 transcription in peripheral CD4$^+$CD25$^-$ T cells requires TGF-β1 signaling (discussed below). The relationship between TGF-β signaling and Foxp3 expression in thymic T cells in vivo, however, is obscure. Despite the massive inflammation, spontaneous T cell activation and proliferation, and decreased number of peripheral CD4$^+$CD25$^+$ Tregs in TGF-β1-null mice [59] (our unpublished data), CD4$^+$CD25$^+$ Tregs are generated and develop at a relatively normal level in the thymus. The interpretation of the results, though, is not simple; namely, one should not leap to the conclusion that TGF-β is completely uninvolved in thymic Treg development, for two reasons. First, in the TGF-β1-null mice, especially during their neonatal period before symptoms appear, maternal TGF-β1 may be passively carried over in their heterozygous mothers' milk [24]. Since Foxp3 expression and induction most likely begins at 3–4 days of age and the TGF-β receptors and subsequent mediators are intact in TGF-β1-null thymocytes, a role of TGF-β1 in Foxp3 expression in the thymus cannot be completely excluded. Second, the isoforms of the TGF-β family, TGF-β2 and TGF-β3, exist in the thymus in TGF-β1-null mice. TGF-β2 and TGF-β3 share the same receptors and mediators and deliver the same TGF-β signal as efficiently as TGF-β1 in the induction of Foxp3 in CD4$^+$CD25$^-$ T cells in vitro [60] (unpublished data). A compensatory effect of TGF-β2/3 in the thymus remains a significant possibility.

On the other hand, it has been reported that mice with T cell-specific negative dominant TGF-β receptor II transgene (dnTβRII) have the same levels of Foxp3$^+$ CD4$^+$ T cells as the control wild-type mice [61]. Since TβRII is the receptor to which TGF-β binds and initiates signaling, the data seem to argue against a further role of TGF-β signaling in Treg development. However, these transgenic mice develop rather a slow and milder progression of inflammation and diseases compared to TGF-β1-null mice. This may indicate a technical limitation, namely the insufficient and incomplete deletion of TβRII, resulting in only partial blockage of TGF-β signaling in the T cells in these mice. Supporting this argument is the most recent discovery that with T-cell-specific conditional knockout of TβRII using the Lck-cre or CD4-cre- lopflox system [41, 62], these conditional knockout mice, unlike dnTβRII transgenic mice, develop an aggressive lethal inflammation and 100% death as do TGF-β1-null mice. Interestingly, even in these T cell-specific TβRII knockout mice, the natural Tregs in the thymus also seem relatively normal [62] or even increased [41] in comparison with those of wild type mice. Significantly, however, these T cell-specific conditional knockout mice manifest a profound deficiency of CD4$^+$CD25$^+$Foxp3$^+$ T regs. In our own studies, we produced mice with T-cell-specific knockout of TGF-β receptor I (TβRI), an immediate downstream molecule of TβRII and an essential component of TGF-β signal transduction. The T-cell specific TβRI knockout mice also develop an aggressive lethal inflammation and die at 3-7 weeks of age (100%) (Liu et al., submitted). Strikingly, these TβRI conditional mice also exhibit dramatically decreased levels of CD4$^+$Foxp3$^+$ Tregs in the spleen and lymph nodes. Thus, a significant question is raised as to whether TGF-β signaling is indeed dispensable for the natural Tregs in the thymus, but required for the peripheral CD4$^+$Foxp3$^+$ Tregs. The data also raise the question of whether peripheral CD4$^+$CD25$^+$ Foxp3$^+$ Tregs are exclusively generated and migrate from the thymus or whether they can also be converted from peripheral CD4$^+$CD25$^-$ Foxp3$^-$ T cells. Recent evidence has started to favor both possibilites and reveal that TGF-β does indeed, in the context of TCR stimulation, convert CD4$^+$CD25$^-$ Foxp3$^-$ T cells to CD4$^+$CD25$^+$ Foxp3$^+$ Tregs through induction of Foxp3 [5,7]. The conversion also occurs in vivo in a TGF-dependent, but thymus-independent, manner [63]. It is now recognized that TGF-β converts CD4$^+$CD25$^+$ Foxp3$^+$ Tregs as a group of "induced" or "adoptive" Tregs.

Conversion of CD4$^+$CD25$^-$ Foxp3$^-$ Naïve T Cells into CD4$^+$CD25$^+$Foxp3$^+$ Tregs In Vitro

Stimulation of peripheral CD4$^+$CD25$^-$ T cells with TCR engagement in the presence of TGF-β converts them to a CD4$^+$CD25$^+$ regulatory T cell phenotype through induction of Foxp3 expression. This process requires collaboration between TCR and TGF-β signaling. Lack of either of these two signals causes Tregs not to be induced. In the absence of TGF-β, TCR stimulation with co-stimulatory molecules such as CD28 initiates a program to induce IL-2 gene transcription and protein production to drive T cell proliferation. The expanded and proliferated T cells express

CD25 and all other markers for effector T cells, but there is no Foxp3 gene transcription or protein expression [27,64,65]. Strikingly, addition of exogenous TGF-β into this culture induces Treg-specific gene Foxp3 expression [5–7]. Foxp3 can be induced to transcribe within 24 hours in the absence of obvious signs of T cell proliferation [50]. Then Foxp3 continues to increase until about 72–96 hours, when it plateaus. Using single-cell intracellular protein analysis with anti-Foxp3 antibody, it was revealed that by 72–96 hours, more than 50% of $CD4^+$ T cells were $CD25^+$ $Foxp3^+$ [8,50]. Without TCR stimulation, TGF-β fails to induce Foxp3 expression. The conversion is dependent on the dose of TGF-β. Although 2 ng/ml of TGF-β reaches the optimal expression of Foxp3 and higher levels of TGF-β fail to enhance more $Foxp3^+$ T cells, as little as 50 pg/ml of TGF-β induces about 10% of $Foxp3^+$ Tregs[50]. This is important, because it may have implications in physiological as well as pathological situations in vivo: 50 pg/ml of active TGF-β is not a difficult threshold to reach in tissues or in circulation. In a naïve mouse and a healthy human body, about 10–20 ng/ml of the latent form and a few hundred pg/ml of the active form of TGF-β is normally circulating in the blood. In contrast to TGF-β, the concentration of anti-TCR antibody seems less critical; 0.5 or 10 μg/ml anti-CD3 antibody elicited no significant difference in term of $Foxp3^+$ T cell conversion under the same optimal amount of TGF-β (2 ng/ml) (unpublished data). However, lower levels of anti-TCR antibodies have not been tested.

In addition to TCR, IL-2 and CD28 signalings are also involved in the TGF-β induction of Foxp3 and Tregs. Stimulation of $CD4^+CD25^-$ naïve T cells with anti-TCR and IL-2 is insufficient to induce Foxp3 expression; only in presence of TGF-β can Foxp3 be detected [5–7]. $CD4^+$ T cells derived from antigen-specific TCR transgenic mice in the background of Rag–/– or IL-2–/– lack $CD25^+Foxp3^+$ Tregs [27,48,66]. When $CD4^+$ $CD25^-$ T cells from 5C.C7 TCR transgenic mice that are Rag2–/– and IL-2–/– are stimulated with anti-CD3 and anti-CD28 (or with specific peptide and antigen-presenting cells [APCs]) in the presence of TGF-β, Foxp3 expression can be detected only in the cultures that include exogenous IL-2 (unpublished data), indicating a role of IL-2 in the process. Similar results have also been reported in recent studies [67]. Thus, IL-2 is important in the TGF-β induction of $Foxp3^+$ Tregs. However, it remains unclear whether IL-2 is directly involved in TGF-β induction of Foxp3 transcription or whether it indirectly maintains or promotes the proliferation and survival of TGF-β-converted Tregs.

CD28 signaling is also involved in the TGF-β induction of Tregs. First, CD28 co-stimulatory signaling may facilitate the Foxp3 expression induced by TCR and TGF-β stimulation in $CD4^+CD25^-$ T cells. In this regard, the presence of agonist anti-CD28 antibody in the culture of TCR stimulation with TGF-β enhances Foxp3 transcription [50]. Accordingly, $CD4^+CD25^-$ T cells in CD28-null mice show a transit defect, with lower levels of Foxp3 expression at the early phase of T cell culture compared to that in wild type T cells. This defect seems to be directly related to CD28 signaling rather than a deficiency in IL-2 production, because the addition of exogenous IL-2 into the cultures of CD28–/– T cells fails to correct the defect [50]. It has been observed, however, that the defect in Foxp3 expression in CD28–/–T cells is transit and limited, because expression is completely restored when the cultures are extended for more than 3 days. Consequently, TGF-β and TCR co-stimulation of CD28–/– CD4$^+$CD25$^-$ T cells can still induce CD25$^+$ Foxp3$^+$ Tregs that are functionally indistinguishable from those of wild-type mice. Strikingly, however, the total number of Tregs harvested from the CD28–/– cultures is dramatically lower than that of wild-type mice, suggesting that CD28 signaling is essential in safeguarding the expansion and/or survival of TGF-β-converted Tregs.

APCs appear dispensable in the TGF-β induction of Foxp3 in CD4$^+$CD25$^-$ T cells. Although TGF-β can induce Foxp3$^+$ Tregs in CD4$^+$CD25$^-$ T cell cultures in the presence of synegeic APCs, stimulation of the same CD25$^-$ T cells with anti-CD3 and anti-CD28 without APCs is equally efficient in inducing Foxp3 and differentiation of Tregs [5–8, 60]. This finding may have biological significance in the future induction of Tregs by TGF-β for therapeutic purposes where exclusion of APCs may facilitate clinical application.

In contrast to CD4$^+$CD25$^-$ T cells, TGF-β signaling in the context of TCR stimulation fails to further upregulate Foxp3 expression in natural CD4$^+$CD25$^+$Foxp3$^+$ Tregs [5]. Even in the presence of higher levels of exogenous IL-2, TGF-β is incapable of enhancing Foxp3 in natural Tregs. Rather, TGF-β signaling might suppress existing Foxp3 in natural Tregs. Our own preliminary data indicate addition of exogenous TGF-β to natural Treg cultures downregulates Foxp3 expression in vitro (unpublished data) and it may also occur in vivo [41]. The significance and molecular mechanisms, however, remain to be elucidated.

CD4$^+$CD25$^-$ T cells in mice consist of a minor population (2–3%) that express Foxp3$^+$ and exhibit immunosuppressive activity [25,66]. At one point, there was debate regarding whether the TGF-β upregulation of Foxp3 was attributable to TGF-β selective inhibition of Foxp3$^-$ cells within CD4$^+$CD25$^-$ T cells, which thus indirectly increased the relative frequency of the CD4$^+$CD25$^-$ Foxp3$^+$ subpopulation. Subsequently, however, several independent studies have demonstrated that TGF-β indeed induces Foxp3 transcription rather than selectively expanding the minor population of CD4$^+$CD25$^-$Foxp3$^+$ subsets [5,7,50]. First, TGF-β induces significant expression of Foxp3 in the wild type CD4$^+$CD25$^-$ T cell cultures within 24 hours in the absence of T cell proliferation [50]. Second, almost 50% of TGF-β treated cells express Foxp3$^+$CD25$^+$ by 72 hours and this increase represents about a 40–60-fold increase in total Foxp3$^+$ T cells compared to freshly isolated CD4$^+$CD25$^-$ T cells, yet it is impossible that TGF-β induces the original minor population of Foxp3$^+$CD25$^-$ to proliferate about 50-fold within three days [50]. Third, in TCR transgenic CD4$^+$ T cells known to lack Foxp3$^+$ Tregs against a background of Rag1–/– and IL-2–/–, TGF-β and TCR stimulation in the presence of IL-2 clearly induced significant Foxp3 expression (unpublished data). Finally, by using a mouse model in which a bicistronic reporter expressing a fluorescent protein has been knocked in to the endogenous Foxp3 locus, independent studies have clearly confirmed that TGF-β and TCR stimulation can convert GFP$^-$(RFP)(Foxp3$^-$) CD4$^+$ T cells into GFP$^+$(RFP$^+$)(Foxp3$^+$) Tregs [7,8]. Importantly, the TGF-β converted Foxp3$^+$ (GFP$^+$) T cells possess phenotype and functional activity similar to those of natural Tregs [7,8]. Thus, TGF-β, in the context of TCR stimulation, induces Foxp3 expression in CD4$^+$ Foxp3$^-$ naïve precursors and converts them into Tregs.

Phenotype and Function of TGF-β-Induced Tregs

CD4$^+$CD25$^+$Foxp3$^+$ Tregs induced by TGF-β are phenotypically and functionally indistinguishable from the thymically derived "natural" Tregs. As natural Tregs, TGF-β-induced Tregs express CD25 and Foxp3. Notably, the CD25 expression in TGF-β-converted Tregs is persistently high and distinctly different from the parallel T cells without TGF-β treatment, in which the expression of CD25 is at low levels [5,50]. Although how TGF-β upregulates CD25 remains unknown, it is possible TGF-β signaling may directly upregulate CD25 or prevent CD25 degradation and downregulation [50,68]. TGF-β-converted CD25$^+$Tregs express levels of Foxp3 similar to those of natural Tregs; one unresolved question, however, is whether Foxp3 in these induced Tregs is as stable and persistent as in natural Tregs. In addition to CD25 and Foxp3, TGF-β induced Tregs also express CD45RB$^{-/low}$, GITR$^+$, and intracellular CTLA-4$^+$ [5–8,60,69] similarly to natural Tregs. Although it is as yet unclear how TGF-β signaling affects expression of these molecules in T cells, previous studies have demonstrated that TGF-β is able to downregulate CD45RB [37,70,71]. TGF-β may have a role in the expression and/or stability of intracellular CTLA-4 in CD4$^+$ T cells [5,72]. Moreover, similar to natural Tregs [24,73–76], TGF-β-induced Foxp3$^+$ Tregs show cell membrane associated TGF-β [5,77], which may be associated with their regulatory activity. However, it is unclear how these TGF-β induced Tregs acquire cell membrane-bound TGF-β; it is conceivable that TGF-β enhances its own production and/or binds onto its cognate receptors on the Tregs, because TCR stimulation upregulates TGF-β receptor II (TβRII) in CD4$^+$ T cells [24,72–75,78,79].

Functionally, TGF-β-induced Tregs exhibit suppressive activity equal in potency to that of natural Tregs in vitro, and inhibit antigen-specific CD4$^+$ T cell expansion and prevent and suppress pathogenic immune responses in vivo [5–8,60,80,81]. TGF-β induced Tregs exhibit anergic features and produce no Th1 or Th2 cytokines upon TCR restimulation [5]. Exogenous IL-2 can reverse their anergy and restore their cytokine production. When added into a co-culture assay, TGF-β-converted Foxp3$^+$ Tregs exhibit potency equal to that of natural CD4$^+$CD25$^+$ Tregs in suppressing TCR-drive T cell proliferation in normal CD4$^+$ T cells. The immunosuppression mediated by TGF-β-induced Tregs also requires cell-cell contact in vitro, and exogenous IL-2 is capable of reversing their suppressive activity. Importantly, the TGF-β-induced Foxp3 Tregs also suppress T cell expansion in vivo. In a classic adoptive transfer model of chicken oval egg albumin (OVA) peptide transgenic CD4$^+$ T cells into normal synegeneic BALB/c mice, TGF-β-converted OVA peptide-specific Tregs effectively suppressed OVA antigen-driven CD4$^+$ T cell proliferation in vivo [5]. As natural Tregs [31,82], OVA-specific TGF-β-induced CD4$^+$CD25$^+$Foxp3$^+$ Tregs proliferate to their specific antigen in vivo and express high levels of CD25. The immunoregulatory effect was further demonstrated in a host dust mite (HDM)-induced asthmatic model. Co-administration of TGF-β-induced Tregs not only greatly reduced/prevented HDM-induced inflammation in the lungs, but also almost completely abrogated mucin production in the airways induced by HDM [5]. In another study, CD4$^+$CD25$^+$ Foxp3$^+$ Tregs induced by TGF-β2, an isoform of the TGF-β family, inhibited innate inflammatory responses to syngeneic transplanted pancreatic islets and enhanced islet transplant survival, suggesting a role in transplantation tolerance [60]. Moreover, TGF-β-induced Tregs also significantly suppressed Th1-mediated colitis on CD4$^+$CD62L$^+$ T cell transfer in vivo [83]. Thus, TGF-β-induced CD4$^+$CD25$^+$Foxp3$^+$ Tregs are phenotypically and functionally indistinguishable from the natural T regs.

Evidence of TGF-β Requirement for Converting CD4$^+$CD25$^-$ T Cells to CD4$^+$CD25$^+$ Tregs In Vivo

The successful conversion of CD4$^+$CD25$^+$ Foxp3$^+$ Tregs by TGF-β in vitro has raised the question of whether it also occurs in vivo. Despite early skepticism about peripheral generation of CD4$^+$CD25$^+$Foxp3$^+$ Tregs in vivo, evidence has gradually accumulated in support of the notion that peripheral CD4$^+$CD25$^-$ naïve T cells can be converted into Foxp3$^+$ Tregs. Significantly, this conversion may require TGF-β signaling. Using several experimental systems, independent studies have shown that peripheral CD4$^+$CD25$^-$ T cells can be converted de novo into CD25$^+$ Tregs. By transferring splenic CD4$^+$ HA specific (6.5$^+$) naïve T cells isolated from HA-TCR transgenic mice with Rag1–/– into sublethally irradiated syngeneic mice expressing agonist antigen HA (Ig-HA), it has been found that CD4$^+$6.5$^+$CD25$^-$ T cells could be converted into CD25$^+$ regulatory T cells. This de novo conversion is independent of other tutor natural Tregs or the presence of thymus [28,31,84,85]. In similar studies, adoptive transfer of CD45.1$^+$CD4$^+$CD25$^-$ naïve T cells into congeneic CD45.2$^+$ mice led to conversion of CD45.1$^+$CD25$^+$CD4$^+$ Tregs [86]. Again, this conversion in vivo was independent of thymus, but required co-stimulatory B7 molecules. The in vivo instruction of suppressor commitment in naïve T cells was further tested in a system in which a prolonged subcutaneous infusion of low doses of peptide by means of osmotic pumps transformed mature T cells into CD4$^+$CD25$^+$ regulatory T cells [87]. The induced CD4$^+$CD25$^+$Tregs can persist for long periods of time in the absence of antigen and confer specific immunologic tolerance upon challenge with antigen. Although the aforementioned studies clearly demonstrated the peripheral conversion of CD4$^+$CD25$^+$ Tregs from naïve CD4$^+$CD25$^-$ T cells by variety of means, the role of TGF-β in the process was not addressed.

To investigate this question, several independent studies have gradually demonstrated that TGF-β is indeed involved in the conversion of Tregs from naïve peripheral CD4$^+$CD25$^-$ T cells in vivo. In an animal model for transplantation tolerance induced by injection of non-depleting CD4 and CD8 or CD154 monoclonal antibodies, it has been shown that the tolerance is dependent on the generation of CD4$^+$Foxp3$^+$ regulatory T cells [88]. Importantly, the induction of Foxp3$^+$ Tregs requires the presence of TGF-β in vivo; elimination of TGF-β with anti-TGF-β neutralization antibody prevents induction of Foxp3$^+$ Tregs and abolishes the transplant tolerance. In a type 1 diabetic mouse model [89], systemic delivery of TGF-β gene therapy blocks islet destructive autoimmunity with an increase in CD4$^+$CD25$^+$foxp3$^+$ Tregs in the islet. Transient doxycycline-controlled expression of TGF-β in pancreatic β-cell islets inhibits the onset of diabetes by increasing the "intra-islet" Foxp3$^+$CD4$^+$CD25$^+$ regulatory T cells through proliferation of already established Tregs or de novo conversion of such cells [90]. More over, in an EAE model, neurons possess B7 costimulatory molecules and produce TGF-β. The interaction between neurons and T cells leads to conversion of encephalitogenic T cells to CD25$^+$TGF-β1$^+$Foxp3$^+$ Tregs, and this process is dependent on TGF-β in vivo [77]. More direct evidence that TGF-β plays a critical role in peripheral CD4$^+$CD25$^+$ Foxp3$^+$ Treg conversion in vivo came from a study by Kretschmer et al [63]. In this study, Thy1.1$^+$ BALB/c mice were injected with CFSE-labeled Thy1.2$^+$CD4$^+$CD25$^-$6.5$^+$(HA TCR transgenic) T cells from dominant negative TGF-β receptor II (dnTβRII) or wild-type mice, and injected mice were immunized with specific antigen (anti-DEC-HA) the next day. After two weeks, it was found that the percentage of dnTβRII$^+$CD4$^+$CD25$^+$Foxp3$^+$Thy1.2$^+$ Tregs was diminished considerably and accompanied by much more extensive proliferation than that of wild-type CD4$^+$6.5$^+$ T cells. Despite of vigorous T cell proliferation, the total number of dnTβRII CD4$^+$CD25$^+$Foxp3$^+$ Tregs was significantly lower than that in TβRII wild-type mice. In contrast, CD4$^+$CD25$^-$6.5$^+$IL-2-/- T cells gives rise to significantly more CD4$^+$CD25$^+$Foxp3$^+$ regulatory cells in the same experimental system. The study provides strong evidence to support a critical role of TGF-β in the de novo conversion of Treg in the periphery. It was noted, however, that the low but detectable levels of dnTβRII T cells could be still be converted into CD4$^+$CD25$^+$Foxp3$^+$ T cells, which might be attributed to incomplete deletion of TGF-β signaling in dnTβRII mice [91]. To circumvent the problem of incomplete deletion in dnTβRII transgenic mice, we injected Rag1-/- mice with CD4$^+$CD25$^-$Foxp3$^-$ T cells from our T cell-specific TβRI knockout mice in which TGF-β signaling is completely abrogated (manuscript in preparation). After 6–8 weeks, we found that the CD4$^+$CD25$^-$ T cells derived from the wild-type mice give rise to about 3% CD4$^+$Foxp3$^+$ T regs, but CD4$^+$CD25$^-$ T cells from TβRI knockout mice failed to result in any detectable Foxp3$^+$CD4$^+$ T cells in the spleen and lymph nodes of Rag1-/- mice. Taken together, the evidence has established the essential role of TGF-β signaling in converting CD4$^+$CD25$^-$ naïve T cells into Tregs in vitro and in vivo.

While the TGF-β-induced conversion of Tregs in vivo has been established, the biological significance of this conversion in physiological and pathological conditions needs to be discussed. Although it is generally believed that in the steady state, the natural Tregs play a dominant role in maintaining immune tolerance, a role for the induced Tregs cannot be completely excluded. In this regard, the mucosal immune system may serve as a typical example. Both respiratory and gastric mucosal tissues produce a large quantity of TGF-β; daily inhalation of a variety of foreign antigens, digestion of food proteins, and residual commensal bacteria provide the antigens for TCR engagement. In a microenvironment where the two basic prerequisites of conversion exist, it is inconceivable that the delicate balance of immunity and tolerance can be impeccably maintained without the instant induction of Tregs in the mucosal systems. The TGF-β induction of Foxp3+ Tregs may also occur under pathological conditions. It has been gradually recognized that tumor tissues harbor large numbers of CD4+CD25+Foxp3+ Tregs that exhibit potent suppressive activity toward anti-tumor immunity [92]. The etiology and source of these intratumor Tregs, however, are still being debated. The currently dominant view is that natural CD4+CD25+Foxp3+ Tregs are attracted by tumor cells through chemokine-associated molecules into the tumor tissues [92]. Recent evidence, however, has indicated that the intratumor Foxp3+ Tregs can also be de novo generated locally from CD4+Foxp3− T cells, although the role of TGF-β in the process has not been thoroughly investigated [93]. Our preliminary data indicate that TGF-β produced and secreted from human oral carcinoma is responsible for the local generation and conversion of intratumor Foxp3+ Tregs (unpublished data). In HIV infection, the sudden and rapid apoptosis of CD4+ T cells killed by HIV virus not only exhausts T cell immune responses, but also greatly influences the subsequent generation and differentiation of CD4+ T effector cells. TGF-β secreted from the phagocytes after uptake of apoptotic CD4+T cells and/or released from the nonphagocytosed apoptotic T cells [94] may switch the newly generated CD4+ T cells into Foxp3+ Tregs, which are not necessarily HIV-infected CD4+ T cells. The large accumulation of Foxp3+ Tregs may then be a driving force decreasing immune responses and further worsening the immune deficiency. It should be noted, however, that the exact role for Tregs in HIV infection and disease development still remains to be elucidated.

Molecular Mechanisms and Pathways in TGF-β Induction of Foxp3

Despite recognition of the TGF-β conversion of Tregs, the underlying molecular mechanism by which TGF-β induces Foxp3 expression is largely unknown. Considering that both TCR and TGF-β signals are required for Foxp3 transcription, the interaction and cross-talk of downstream mediators between the TCR and TGF-β pathways might be the key to deciphering the mystery. It remains unclear whether the transcription of Foxp3 by TGF-β treatment is attributable to the direct effects of the TGF-β-Smad pathway, to an indirect influence of TCR signal transduction by TGF-β, or to both. Our own preliminary study has indicated that several Smad bind sequences (SBS) exist in the promoters of both mouse and human Foxp3 genes (unpublished data), but the evidence of a direct role of Smad is still lacking and under intense investigation. Nevertheless, TGF-β signaling is essential, as deletion of TGF-β receptor I, which abolishes TGF-β signal transduction, completely abrogates conversion of CD4+CD25− T cells into CD4+CD25+ Tregs (Liu and Chen, manuscript in preparation). The roles of Smad2/3, however, remain to be elucidated. Deletion of Smad3 in T cells seems to be ineffective in blocking TGF-β-induced Foxp3 expression (unpublished data), suggesting a redundancy of Smad2 and/or that other Smad-independent pathways may exist. It has been reported that the inhibitory Smad7 protein that is normally induced by TGF-β and limits TGF-β signaling is strongly downregulated by Foxp3 at the transcriptional level [6]. Foxp3-mediated downregulation of Smad7 subsequently renders CD4+CD25− T cells highly susceptible to the morphogenic and regulatory effects of TGF-β signaling via Smad3/4. However, the inhibition of Smad 7 is a subsequent effect of Foxp3 expression; thus, the role of Smad7 in the TGF-β induction of Foxp3 is still unclear. Given the fact that T cells deficient in LAT of TCR signaling [95], in Stat5 [58], or in IL-2Rγ [57] also lack Foxp3 transcription, the possibility that TGF-β signaling interacts with those pathways is still open to investigation.

TGF-β Induction of Tregs in Human T Cells

As in mouse T cells, TGF-β also converts human CD4+CD25− Foxp3− T cells into CD4+Foxp3+ regulatory T cells. Stimulation of CD4+CD25− Foxp3− T cells with anti-CD3 and CD28 antibodies in the presence of TGF-β induces expression of Foxp3 mRNA and also protein [6] (our unpublished data). The TGF-β converted Tregs are anergic to TCR stimulation and potently suppress T cell proliferation and cytokine production driven by TCR. The TGF-β-induced Tregs are also phenotypically and functionally indistinguishable from human natural Tregs. In contrast to mouse T cells, however, it has been reported that stimulation of human CD4+CD25− Foxp3− T cells with TCR and anti-CD28 in the absence of exogenous TGF-β also induces Foxp3+ Tregs [96], although whether the induced Foxp3+ T cells are immunosuppressive is still in debate [97]. These findings have raised a critical issue, namely how to reconcile this observation (induction of Foxp3 and Tregs) with the established paradigm that the primary goal of T cell activation by TCR and CD28 is to induce T cell proliferation and differentiation to mount specific T cell immunity [98]. Since the underlying events responsible for Foxp3 induction by TCR and CD28 stimulation had not previously been addressed, we investigated this question and demonstrated that Foxp3 induction by TCR and CD28 stimulation in human CD4+CD25− T cells is in great part attributable to the endogenous secretion and activation of latent TGF-β by activated T cells (unpublished data). Nevertheless, the finding that TGF-β converts Tregs opens a new avenue for inducing antigen-specific regulatory T cells in vitro as needed, which may help re-build or balance immune responses in vivo in clinical settings.

TGF-β Induces Differentiation of Th17 Cells from CD4+CD25− Naïve T Cells

The recent finding that TGF-β, in the presence of IL-6, is essential for the differentiation of Th17 cells from naïve CD4+ T precursors has further highlighted a non-redundant role for TGF-β in the regulation of immune responses. Th17 cells are a recently identified new lineage of Th cells that are distinguishable from Th1 and Th2 cells [9,18,34]. In contrast to Th1 and Th2 cells, Th17 cells produce a distinct profile of effector cytokines, including 17(IL-17A), IL-6, IL17F, GM-CSF, and chemokine CXCL1. The initial indication that Th17 cells might represent a new lineage of Th cells distinct from Th1 and Th2 came from studies of murine models of autoimmunity, including EAE and CIA. In contrast to previous suspicions that IL-12-mediated Th1 cells play a dominant role in the development of these autoimmune diseases, a number of independent studies have revealed that it is IL-23, which shares the P40-subunit of IL-12 [99], rather than IL-12 that plays a critical role in EAE and CIA. Indeed, data from mice deficient in IFN-γ are even more susceptible to EAE and CIA, further weakening the Th1 or IL-12-IFN-γ cytokine axis link [18]. Subsequent studies from Chua and coworkers have resolved this paradox and established the vital role of IL-23, not IL-12, in EAE and CIA using mice deficient in IL-12, IL-23, or both [11,100]. The disease development was abolished in IL-23-, but not IL-12-null mice.

Indications that IL-23 is involved in IL-17-producing T cells came from studies which found that production of IL-17, rather than IFN-γ, has been consistently observed in T cells in EAE mice and the disease development was ablated in with IL-23- or IL-23 subunit-deficient mice. Deletion of IL-17 by neutralization antibodies or IL-17−/− mice improves the diseases of EAE and CIA [101–103]. Importantly, IL-17+ and IFN-γ+ T cells show a reciprocal pattern. A positive correlation was established between the availability of IL-23 and IL-17 and a negative correlation was established between IL-12- and IFN-γ-producing Th1 cells and disease development [18]. It was also reported that IL-23 promoted IL-17 production from CD4+ effector and memory cells [35]. Although Th-17 cells express the receptor for IL-23 and require that key cytokine for their effector function, survival, and proliferation, subsequent studies have discovered that IL-23 is not necessary for the differentiation of Th17 cells from naïve CD4+ T cells.

Instead, three independent studies have demonstrated the surprising finding that TGF-β is the essential factor in the differentiation of Th17 cells from naïve CD4+ precursors [8,10,12]. Unexpectedly, the proinflammatory cytokine IL-6 is a key co-stimulatory factor in TGF-β-induced Th17 commitment. Neither TGF-β nor IL-6 alone is sufficient to direct CD4+ T cell precursors to differentiate into Th17 cells; only the combination of the two accomplishes the task.

Stockinger and colleagues have revealed that Tregs, through TGF-β secretion, induce Th17 differentiation in CD4+CD25− precursors stimulated with TCR engagement in the presence of IL-6 produced from LPS-treated DCs [12,104]. When naïve CD4+ T cells were co-cultured with Tregs in the presence of DCs stimulated with LPS, it was found that IL-2, IL-4, and IFN-γ were still suppressed despite the fact that T cell proliferation was restored to normal levels. Surprisingly, a substantial portion of the CD4+ T cells in the co-cultures produced and secreted IL-17 [12]. They further identified that TGF-β, derived from Tregs, is the key in differentiation of Th-17 cells in the co-cultures, since neutralization of TGF-β with anti-TGF-β antibody abrogated IL-17 production. To positively correlate TGF-β as a key component in Th17 differentiation, the authors replaced Tregs with exogenous TGF-β1 in LPS-DC-stimulated CD4+ naïve T cells and Th17 cells were induced. IL-6 produced from DC treated with LPS is the key co-stimulatory cytokine with

TGF-β in Th17 differentiation, because only anti-IL-6, but not anti-IL23 or anti-IL12, ablated the IL-17-producing cells. The requirement for TGF-β and IL-6 in the differentiation of Th17 cells was further verified by use of pure CD4$^+$ T cell cultures stimulated with TCR antibody in the presence of exogenous IL-6 and TGF-β1. Furthermore, the study demonstrated that Th17 cells failed to express either Th1 or Th2 transcriptions, including T-bet, HIX, and Gata-3, and also did not increase inhibitory Smad7 of the TGF-β signaling pathway. Another interesting observation was that Th17 cells induced by TGF-β and IL-6 might require IL-23 for survival or maintenance, since the presence of IL-23 in the secondary culture further increased the number of Th17 cells, whereas IL-2 abolished IL-17 and switched the cells into IFN-γ-producing Th1 cells. Thus, TGF-β derived from Tregs or other sources and IL-6 produced by LPS-treated DCs induced differentiation of Th17 cells from naïve CD4$^+$ precursors.

Using different approaches with a Foxp3-EGFP knock in a reporter mouse model, Kuchroo and co-workers have independently drawn the same conclusion that TGF-β and IL-6 co-stimulation induces differentiation of Th17 cells [8]. Consistent with the original finding in wild-type mice[5], stimulation of CD4$^+$GFP$^-$ (Foxp3$^-$) T cells with immobilized anti-CD3 and CD28 antibodies in the presence of TGF-β induces Foxp3$^+$(GFP$^+$) expression in about 50% of T cells [8]. A panel of pro- or anti-inflammatory cytokines was then added into the cultures to assess which one might modulate TGF-β-induced Foxp3 expression. Surprisingly, only IL-6, but not IL-1, IL-2, IL-7, IL-10, IL-11, IL-12, IL-13, IL-15, IL-18 or TNF-α, strongly inhibited Foxp3$^+$ Treg conversion [8]. Importantly, this IL-6 treatment in TGF-β-stimulated CD4$^+$ T cells results in large amounts of IL-17 production. IL-23 is excluded as a critical factor in the differentiation of Th17 cells by the fact that stimulation of CD4$^+$ T cells in the presence of either IL-23 alone, with neutralization antibodies to IL-4 and IFN-γ, or together with IL-6 or TGF-β failed to induce Th17 cells. Significantly, the study unequivocally demonstrated the reciprocal differentiation of Foxp3$^+$ Tregs and Th17 cells from same CD4$^+$ precursors. When CD4$^+$CD62L$^+$GFP$^-$ naïve T cells were cultured with TCR stimulation alone, only IFN-γ-producing T cells were induced; addition of TGF-β to the culture resulted in Foxp3$^+$ Tregs, but not IFN-γ or IL-17 production. However, addition of TGF-β plus IL-6 to T cells during differentiation abrogated Foxp3 and IFN-γ, but led to concomitant expression of IL-17 from these T cells. Finally, in an EAE model, transgenic mice with T cell-specific expression of TGF-β that produce higher levels of TGF-β developed much more severe clinical disease scores. T cells from the spinal cord of TGF-β transgenic mice produced higher amounts of IL-17, further strengthening the evidence for TGF-β affecting IL-17-producing T cells in vivo. The critical role of TGF-β signaling in Th17 cell differentiation in the initiation of EAE was further verified in another study using dnTβRII transgenic mice, in which TGF-β signaling was severely blocked and the mice exhibited much less severe EAE disease [105].

In another independent study, Weaver and co-workers also identified TGF-β as an essential cytokine in the induction of Th17 cells [10]. Addition of TGF-β in TCR-stimulated CD4$^+$ primary T cells in the presence of anti-IFN-γ and anti-IL-4 maximally induced IL-17-producing T cell commitment. Accordingly, in IFN-γ- or IFN-γ receptor-deficient CD4$^+$ T cells, exogenous TGF-β induced even greater Th17 cell development, independent of IL-23. Importantly, Weaver et al. have shown that TGF-β1-null mice exhibit much lower levels of IL-17 in the circulation and fewer IL-17$^+$CD4$^+$ T cells in the mesenteric lymph nodes and in the lamina propria, where a higher frequency of Th17 T cells is found in normal mice[10]. They went one step further and showed that TGF-β-treated CD4$^+$ T cells had increased IL-23 receptors on their surface, providing a molecular basis for the role of IL-23 in Th17 cell growth and survival. Using a mouse model infected with orally administered *Citrobacter Rodentium*, a natural rodent pathogen, against which the IL-1-IL-23 axis seems to be essential for host protection, the critical role of TGF-β in Th17 cells in mucosal immunity in vivo was further confirmed. IL-23-null mice treated with anti-TGF-β antibody developed severe ulcerative and hemorrhagic intestinal lesions with gross bacterial invasion, neither of which was found in IL-23-null mice treated with isotype control antibody in vivo. Taken together, these independent studies have clearly established that TGF-β is the key in the reciprocal differentiation of Foxp3$^+$ Tregs and Th17 cells, which is dependent on the absence or presence of IL-6.

TGF-β and IL-6 Induce Orphan Nuclear Receptor RORgammat to Direct Th17 Differentiation

A few months after the identification of TGF-β as an essential factor in the induction of Th17 cells, Littman and his co-workers discovered that RORgammat is a critical transcription factor in programming Th17 cell differentiation from naïve CD4$^+$ T cells [106], RORgammat induces transcription of the genes encoding IL-17 and the related cytokine IL-17F in naïve CD4$^+$ T helper cells. Importantly, RORgammat is required for IL-17 expression in response to TGF-β and IL-6. In vitro treatment of naïve CD4$^+$ T cells with TCR stimulation in the presence of IL-6 and TGF-β maximally induced RORgammat expression, for which mRNA peaked at 16 hours, ahead of peak IL-17 production at 48 hours [106]. TGF-β and IL-6 costimulation failed to induce Th17 cells in RORgammat–/– CD4$^+$ T cells. Further proof of RORgammat's essential role was obtained by enforced expression of RORgammat in the absence of other stimulation; this induced IL-17 in naïve CD4$^+$ T cells. Significantly, RORgammat–/– mice lack Th17 cells in mucosal lymphoid tissues and have reduced severity of EAE when immunized with MOG antigen. Paradoxically, TGF-β and IL-6 can individually induce expression of RORgammat, but neither alone can induce IL-17. Accordingly, CD4$^+$ T cells in the lamina propria of IL-6–/– mice did not express RORgammat and significantly reduced IL-17$^+$ T cells. Nonetheless, a critical role of RORgammat transcription factor has been revealed in the programming of Th17 cells, which requires the participation of TGF-β and IL-6.

Conclusion

CD4$^+$CD25$^+$Foxp3$^+$ Tregs and Th17 cells are recently identified new members of the CD4$^+$ Th cell family. The unexpected finding that TGF-β plays key roles in the differentiation of these phenotypically and functionally distinguished T cell subsets further strengthens the long-standing argument that TGF-β is essential in the regulation of immune responses (Fig. 7.1). In a steady state, when naïve CD4$^+$ T precursors are presented with their specific antigen in the presence of dominant TGF-β without pathological invaders, T cells have the tendency to differentiate into Foxp3$^+$ Tregs, which in turn safeguard and maintain peripheral immune tolerance. The benefit of TGF-β induction of Tregs may also be demonstrated in tolerance induction in the mucosal system, in the recovery mechanisms of some acute inflammations, and in the remission phase of certain chronic inflammatory and autoimmune diseases. This same process of TGF-β conversion of Tregs can also be detrimental in the circumstances of anti-tumor immunity against cancers. Tumor cells might escape immune surveillance or antagonize anti-tumor immunity by converting CD4$^+$(CD8$^+$) T cells toward a Treg phenotype by producing and secreting large amounts of TGF-β. The same detrimental effects of TGF-β induction of Tregs may also occur in the late stage of HIV infection, in which TGF-β produced by phagocytes digesting apoptotic T cells that were released by massive CD4$^+$ T cell apoptosis leads CD4$^+$ or even CD8$^+$ T cells towards an anergic/suppressor phenotype, which may subsequently inhibit immune responses against HIV or other pathogenic or oncologic factors, leading to an unavoidable collapse of the immune system.

On the other hand, when CD4$^+$ helper T cells are confronted with their specific antigen in the presence of IL-6 and TGF-β, such as pathogen infection and chronic inflammation, CD4$^+$ T cells are directed to differentiate into Th17 cells (Fig. 7.1). If the regulatory mechanisms are defective, or the Th17 response is overwhelming, the regulatory mechanism can no longer manage the situation. A flow or even storm of accumulating neutrophils and other inflammatory cells could persist, and consequent damage of the tissue could occur. Chronic inflammation and autoimmune disease may ensue. Thus, emphasis in the field should be on the critical factors controlling the balance between Treg and Th17 cell differentiation, in order to identify possible targets for clinical manipulation.

Another significant issue is the molecular pathways and points of interaction between TGF-β and IL-6 signalings during the process of Treg versus Th17 cell differentiation. For instance, where is the turning point, and what are the molecule(s) or step(s) that determine the direction of naïve CD4$^+$ precursors toward Tregs or Th17 cells in response to TGF-β or/and IL-6? By refocusing our investigations on these underlying molecular events and interactions, it may become possible to design strategies to increase the limited and/or inadequate numbers of antigen-specific CD4$^+$CD25$^+$Foxp3$^+$ regulatory cells as needed, while eliminating the production and differentiation of Th17 cells in autoimmune disease and chronic inflammation, by manipulating TGF-β and its influencing factors—in other words, keep the good but control the bad and eliminate the ugly of TGF-β [39,78].

Acknowledgments This research was supported by the Intramural Research Program of the NIH, National Institute for Dental and Craniofacial Research.

References

1. Mosmann, T. R., H. Cherwinski, M. W. Bond, M. A. Giedin, and R. L. Coffman. 1986. Two types of murine helper T cell clone. I. Definition according to profiles of lymphokine activities and secreted proteins. *J Immunol 136:2348.*
2. Seder, R. A., and W. E. Paul. 1994. Acuisition of lymphokine-producing phenotype by CD4+ T cells. *Annu Rev Immunol 12:635.*
3. Sakaguchi, S., N. Sakaguchi, M. Asano, M. Itoh, and M. Toda. 1995. Immunologic self-tolerance maintained by activated T cells expressing IL-2 receptor alpha-chains (CD25). Breakdown of a single mechanism of self-tolerance causes various autoimmune diseases. *J Immunol 155:1151.*
4. Shevach, E. M. 2001. Certified professionals: CD4(+)CD25(+) suppressor T cells. *J Exp Med 193:F41.*
5. Chen, W., W. Jin, N. Hardegen, K. J. Lei, L. Li, N. Marinos, G. McGrady, and S. M. Wahl. 2003. Conversion of peripheral CD4+CD25– naive T cells to CD4+CD25+ regulatory T cells by TGF-beta induction of transcription factor Foxp3. *J Exp Med 198:1875.*
6. Fantini, M. C., C. Becker, G. Monteleone, F. Pallone, P. R. Galle, and M. F. Neurath. 2004. Cutting edge: TGF-beta induces a regulatory phenotype in CD4+CD25– T cells through Foxp3 induction and down-regulation of Smad7. *J Immunol 172:5149.*
7. Wan, Y. Y., and R. A. Flavell. 2005. Identifying Foxp3-expressing suppressor T cells with a bicistronic reporter. *Proc Natl Acad Sci USA 102:5126.*
8. Bettelli, E., Y. Carrier, W. Gao, T. Korn, T. B. Strom, M. Oukka, H. L. Weiner, and V. K. Kuchroo. 2006. Reciprocal developmental pathways for the generation of pathogenic effector TH17 and regulatory T cells. *Nature 441:235.*
9. Dong, C. 2006. Diversification of T-helper-cell lineages: Finding the family root of IL-17-producing cells. *Nat Rev Immunol 6:329.*
10. Mangan, P. R., L. E. Harrington, D. B. O'Quinn, W. S. Helms, D. C. Bullard, C. O. Elson, R. D. Hatton, S. M. Wahl, T. R. Schoeb, and C. T. Weaver. 2006. Transforming growth factor-beta induces development of the T(H)17 lineage. *Nature 441:231.*
11. Murphy, C. A., C. L. Langrish, Y. Chen, W. Blumenschein, T. McClanahan, R. A. Kastelein, J. D. Sedgwick, and D. J. Cua. 2003. Divergent pro- and antiinflammatory roles for IL-23 and IL-12 in joint autoimmune inflammation. *J Exp Med 198:1951.*
12. Veldhoen, M., R. J. Hocking, C. J. Atkins, R. M. Locksley, and B. Stockinger. 2006. TGFbeta in the context of an inflammatory cytokine milieu supports de novo differentiation of IL-17-producing T cells. *Immunity 24:179.*
13. Murphy, K., A. Heimberger, and D. Loh. 1990. Induction by antigen of intrathymic apoptosis of CD4+CD8+TCRlo thymocytes in vivo. *Science 250:1720.*
14. Murphy, K. M., and S. L. Reiner. 2002. The lineage decisions of helper T cells. *Nat Rev Immunol 2:933.*
15. Szabo, S. J., S. T. Kim, G. L. Costa, X. Zhang, C. G. Fathman, and L. H. Glimcher. 2000. A novel transcription factor, T-bet, directs Th1 lineage commitment. *Cell 100:655.*
16. Ouyang, W., S. H. Ranganath, K. Weindel, D. Bhattacharya, T. L. Murphy, W. C. Sha, and K. M. Murphy. 1998. Inhibition of Th1 development mediated by GATA-3 through an IL-4-independent mechanism. *Immunity 9:745.*
17. Zheng, W., and R. A. Flavell. 1997. The transcription factor GATA-3 is necessary and sufficient for Th2 cytokine gene expression in CD4 T cells. *Cell 89:587.*
18. Weaver, C. T., L. E. Harrington, P. R. Mangan, M. Gavrieli, and K. M. Murphy. 2006. Th17: An effector CD4 T cell lineage with regulatory T cell ties. *Immunity 24:677.*

19. Fontenot, J. D., M. A. Gavin, and A. Y. Rudensky. 2003. Foxp3 programs the development and function of CD4(+)CD25(+) regulatory T cells. *Nat Immunol 4:330*.
20. Hori, S., T. Takahashi, and S. Sakaguchi. 2003. Control of autoimmunity by naturally arising regulatory CD4+ T cells. *Adv Immunol 81:331*.
21. Khattri, R., T. Cox, S. A. Yasayko, and F. Ramsdell. 2003. An essential role for Scurfin in CD4(+)CD25(+) T regulatory cells. *Nat Immunol 4:337*.
22. Thornton, A. M., and E. M. Shevach. 1998. CD4+CD25+ immunoregulatory T cells suppress polyclonal T cell activation in vitro by inhibiting interleukin 2 production. *J Exp Med 188:287*.
23. Bluestone, J. A., and A. K. Abbas. 2003. Natural versus adaptive regulatory T cells. *Nat Rev Immunol 3:253*.
24. Chen, W., and S. M. Wahl. 2003. TGF-beta: The missing link in CD4(+)CD25(+) regulatory T cell-mediated immunosuppression. *Cytokine Growth Factor Rev 14:85*.
25. Fontenot, J. D., J. P. Rasmussen, L. M. Williams, J. L. Dooley, A. G. Farr, and A. Y. Rudensky. 2005. Regulatory T cell lineage specification by the forkhead transcription factor foxp3. *Immunity 22:329*.
26. Sakaguchi, S. 2000. Regulatory T cells: Key controllers of immunologic self-tolerance. *Cell 101:455*.
27. Shevach, E. M. 2002. CD4+ CD25+ suppressor T cells: More questions than answers. *Nat Rev Immunol 2:389*.
28. von Boehmer, H. 2005. Mechanisms of suppression by suppressor T cells. *Nat Immunol 6:338*.
29. Baecher-Allan, C., J. A. Brown, G. J. Freeman, and D. A. Hafler. 2001. CD4+CD25 high regulatory cells in human peripheral blood. *J Immunol 167:1245*.
30. Sakaguchi, S. 2002. Immunologic tolerance maintained by regulatory T cells: Implications for autoimmunity, tumor immunity and transplantation tolerance. *Vox Sang 1(Suppl 83):151*.
31. Klein, L., K. Khazaie, and H. Von Boehmer. 2003. In vivo dynamics of antigen-specific regulatory T cells not predicted from behavior in vitro. *Proc Natl Acad Sci USA 100:8886*.
32. Roncarolo, M. G., R. Bacchetta, C. Bordignon, S. Narula, and M. K. Levings. 2001. Type 1 T regulatory cells. *Immunol Rev 182:68*.
33. Weiner, H. L. 2001. Induction and mechanism of action of transforming growth factor-beta-secreting Th3 regulatory cells. *Immunol Rev 182:207*.
34. Cua, D. J., and R. A. Kastelein. 2006. TGF-beta, a 'double agent' in the immune pathology war. *Nat Immunol 7:557*.
35. Aggarwal, S., N. Ghilardi, M. H. Xie, F. J. de Sauvage, and A. L. Gurney. 2003. Interleukin-23 promotes a distinct CD4 T cell activation state characterized by the production of interleukin-17. *J Biol Chem 278:1910*.
36. Shi, Y., and J. Massague. 2003. Mechanisms of TGF-beta signaling from cell membrane to the nucleus. *Cell 113:685*.
37. Chen, W., and S. M. Wahl. 2002. TGF-beta: Receptors, signaling pathways and autoimmunity. *Curr Dir Autoimmun 5:62*.
38. Strober, W., B. Kesall, I. Fuss, T. Marth, B. Ludviksson, R. Ehrhardt, and M. Neurath. 1997. Reciprocal IFN-γ and TGF-β response regulate the occurrence of mucosal inflammation. *Immunol Today 18:61*.
39. Wahl, S. M. 1994. Transforming growth factor: The good, the bad, and the ugly. *J Exp Med 180:1587*.
40. Weiner, H. L. 1997. Oral tolerance: Immune mechanisms and treatment of autoimmune diseases. *Immunol Today 18:335*.
41. Li, M. O., S. Sanjabi, and R. A. Flavell. 2006. Transforming growth factor-beta controls development, homeostasis, and tolerance of T cells by regulatory T cell-dependent and -independent mechanisms. *Immunity 25:455*.
42. Kulkarni, A. B., C.-H. Huh, D. Becker, A. Gerser, M. Lyght, K. C. Flanders, A. B. Roberts, M. B. Sporn, J. M. Ward, and S. Karlsson. 1993. Transforming growth factor-β null mutation

in mice causes excessive inflammatory response and early death. *Proc Natl Acad Sci USA* 90:770.
43. Shull, M. M., I. Ormsby, A. B. Kier, S. Pawlowski, R. J. Diebold, M.Yin, R.Allen, C. Sidman, B. Proetzel, D. calvin, N. Annuniziata, and T. Doeschman. 1992. Targeted disruption of the mouse transforming growth factor-β1 gene results in multifocal inflammatory disease. *Nature* 359:693.
44. Massague, J. 1998. TGF-β signal transduction. *Annu Rev Biochem* 67:753.
45. Cerwenka, A., D. Bevec, O. Majdic, W. Knapp, and W. Holter. 1994. TGF-β1 is a potent inducer of human effector T cells. *J Immunol* 153:4367.
46. Chen, W., W. Jin, H. Tian, P. Sicurello, M. Frank, J. M. Orenstein, and S. M. Wahl. 2001. Requirement for Transforming Growth Factor beta1 in Controlling T Cell Apoptosis. *J Exp Med* 194:439.
47. Kehrl, J. H., L. M. Wakefield, A. B. Roberts, S. Jakowelw, M. Alvarez-Mon, R. Derynck, M. B. Sporn, and A. S. Fauci. 1986. Production of transforming growth factor β by human T lymphocytes and its potential role in the regulation of T cell growth. *J Exp Med* 163:1037.
48. Fontenot, J. D., J. L. Dooley, A. G. Farr, and A. Y. Rudensky. 2005. Developmental regulation of Foxp3 expression during ontogeny. *J Exp Med* 202:901.
49. Tai, X., M. Cowan, L. Feigenbaum, and A. Singer. 2005. CD28 costimulation of developing thymocytes induces Foxp3 expression and regulatory T cell differentiation independently of interleukin 2. *Nat Immunol* 6:152.
50. Liu, Y., S. Amarnath, and W. Chen. 2006. Requirement of CD28 signaling in homeostasis/survival of TGF-beta converted CD4+CD25+ Tregs from thymic CD4+CD25– single positive t cells. *Transplantation* 82:953.
51. Watanabe, N., Y. H. Wang, H. K. Lee, T. Ito, Y. H. Wang, W. Cao, and Y. J. Liu. 2005. Hassall's corpuscles instruct dendritic cells to induce CD4+CD25+ regulatory T cells in human thymus. *Nature* 436:1181.
52. Schwartz, R. H. 1996. Models of T cell anergy is there a common molecular mechanism. *J Exp Med* 184:1.
53. Schwartz, R. H. 2002. T Cell Anergy. *Annu Rev Immunol* 21:305.
54. Thompson, C. B., and J. P. Allison. 1997. The emerging role of CTLA-4 as an immune attenuator. *Immunity* 7:445.
55. Salomon, B., D. J. Lenschow, L. Rhee, N. Ashourian, B. Singh, A. Sharpe, and J. A. Bluestone. 2000. B7/CD28 costimulation is essential for the homeostasis of the CD4+CD25+ immunoregulatory T cells that control autoimmune diabetes. *Immunity* 12:431.
56. Malek, T. R., and A. L. Bayer. 2004. Tolerance, not immunity, crucially depends on IL-2. *Nat Rev Immunol* 4:665.
57. Fontenot, J. D., J. P. Rasmussen, M. A. Gavin, and A. Y. Rudensky. 2005. A function for interleukin 2 in Foxp3-expressing regulatory T cells. *Nat Immunol* 6:1142.
58. Antov, A., L. Yang, M. Vig, D. Baltimore, and L. Van Parijs. 2003. Essential role for STAT5 signaling in CD25+CD4+ regulatory T cell homeostasis and the maintenance of self-tolerance. *J Immunol* 171:3435.
59. Marie, J. C., J. J. Letterio, M. Gavin, and A. Y. Rudensky. 2005. TGF-beta1 maintains suppressor function and Foxp3 expression in CD4+CD25+ regulatory T cells. *J Exp Med* 201:1061.
60. Fu, S., N. Zhang, A. C. Yopp, D. Chen, M. Mao, D. Chen, H. Zhang, Y. Ding, and J. S. Bromberg. 2004. TGF-beta induces Foxp3 + T-regulatory cells from CD4 + CD25 – precursors. *Am J Transplant* 4:1614.
61. Fahlen, L., S. Read, L. Gorelik, S. D. Hurst, R. L. Coffman, R. A. Flavell, and F. Powrie. 2005. T cells that cannot respond to TGF-beta escape control by CD4(+)CD25(+) regulatory T cells. *J Exp Med* 201:737.
62. Marie, J. C., D. Liggitt, and A. Y. Rudensky. 2006. Cellular mechanisms of fatal early-onset autoimmunity in mice with the T cell-specific targeting of transforming growth factor-beta receptor. *Immunity* 25:441.

63. Kretschmer, K., I. Apostolou, D. Hawiger, K. Khazaie, M. C. Nussenzweig, and H. von Boehmer. 2005. Inducing and expanding regulatory T cell populations by foreign antigen. *Nat Immunol 6:1219*.
64. Sakaguchi, S. 2003. Control of immune responses by naturally arising CD4+ regulatory T cells that express toll-like receptors. *J Exp Med 197:397*.
65. Schwartz, R. H. 2005. Natural regulatory T cells and self-tolerance. *Nat Immunol 6:327*.
66. Sakaguchi, S. 2003. The origin of FOXP3-expressing CD4+ regulatory T cells: Thymus or periphery. *J Clin Invest 112:1310*.
67. Horwitz, D. A., S. G. Zheng, J. D. Gray, J. H. Wang, K. Ohtsuka, and S. Yamagiwa. 2004. Regulatory T cells generated ex vivo as an approach for the therapy of autoimmune disease. *Semin Immunol 16:135*.
68. Kim, H. P., B. G. Kim, J. Letterio, and W. J. Leonard. 2005. Smad-dependent cooperative regulation of interleukin 2 receptor alpha chain gene expression by T cell receptor and transforming growth factor-beta. *J Biol Chem 280:34042*.
69. Zheng, S. G., J. H. Wang, J. D. Gray, H. Soucier, and D. A. Horwitz. 2004. Natural and induced CD4+CD25+ cells educate CD4+CD25– cells to develop suppressive activity: The role of IL-2, TGF-beta, and IL-10. *J Immunol 172:5213*.
70. Chen, W., and S. M. Wahl. 1999. Manipulation of TGF-beta to control autoimmune and chronic inflammatory diseases. *Microbes Infect 1:1367*.
71. Swain, S. L. 1994. Generation and in vivo persistence of polarized Th1 and Th2 memory cells. *Immunity 1:543*.
72. Oida, T., L. Xu, H. L. Weiner, A. Kitani, and W. Strober. 2006. TGF-beta-mediated suppression by CD4+CD25+ T cells is facilitated by CTLA-4 signaling. *J Immunol 177:2331*.
73. Nakamura, K., A. Kitani, I. Fuss, A. Pedersen, N. Harada, H. Nawata, and W. Strober. 2004. TGF-beta 1 plays an important role in the mechanism of CD4+CD25+ regulatory T cell activity in both humans and mice. *J Immunol 172:834*.
74. Nakamura, K., A. Kitani, and W. Strober. 2001. Cell contact-dependent immunosuppression by CD4(+)CD25(+) regulatory T cells is mediated by cell surface-bound transforming growth factor beta. *J Exp Med 194:629*.
75. Oida, T., X. Zhang, M. Goto, S. Hachimura, M. Totsuka, S. Kaminogawa, and H. L. Weiner. 2003. CD4+CD25– T cells that express latency-associated peptide on the surface suppress CD4+CD45RBhigh-induced colitis by a TGF-beta-dependent mechanism. *J Immunol 170:2516*.
76. Oldenhove, G., M. de Heusch, G. Urbain-Vansanten, J. Urbain, C. Maliszewski, O. Leo, and M. Moser. 2003. CD4+ CD25+ regulatory T cells control T helper cell type 1 responses to foreign antigens induced by mature dendritic cells in vivo. *J Exp Med 198:259*.
77. Liu, Y., I. Teige, B. Birnir, and S. Issazadeh-Navikas. 2006. Neuron-mediated generation of regulatory T cells from encephalitogenic T cells suppresses EAE. *Nat Med 12:518*.
78. Chen, W. 2006. Dendritic cells and (CD4++)CD25+ T regulatory cells: Crosstalk between two professionals in immunity versus tolerance. *Front Biosci 11:1360*.
79. Zhang, X., L. Izikson, L. Liu, and H. L. Weiner. 2001. Activation of CD25(+)CD4(+) regulatory T cells by oral antigen administration. *J Immunol 167:4245*.
80. Cacoub, P., L. Musset, P. Hausfater, P. Ghillani, F. L. Fabiani, F. Charlotte, E. Angevin, P. Opolon, T. Poynard, J. C. Piette, and B. Autran. 1998. No evidence for abnormal immune activation in peripheral blood T cells in patients with hepatitis C virus (HCV) infection with or without cryoglobulinaemia. Multivirc Group. *Clin Exp Immunol 113:48*.
81. Gui, S. Y., W. Wei, H. Wang, L. Wu, W. Y. Sun, W. B. Chen, and C. Y. Wu. 2006. Effects and mechanisms of crude astragalosides fraction on liver fibrosis in rats. *J Ethnopharmacol 103:154*.
82. Walker, L. S., A. Chodos, M. Eggena, H. Dooms, and A. K. Abbas. 2003. Antigen-dependent Proliferation of CD4+ CD25+ Regulatory T Cells In Vivo. *J Exp Med 198:249*.

83. Fantini, M. C., C. Becker, I. Tubbe, A. Nikolaev, H. A. Lehr, P. Galle, and M. F. Neurath. 2006. Transforming growth factor beta induced FoxP3+ regulatory T cells suppress Th1 mediated experimental colitis. *Gut 55:671.*
84. von Boehmer, H., I. Aifantis, F. Gounari, O. Azogui, L. Haughn, I. Apostolou, E. Jaeckel, F. Grassi, and L. Klein. 2003. Thymic selection revisited: How essential is it? *Immunol Rev 191:62.*
85. Apostolou, I., A. Sarukhan, L. Klein, and H. von Boehmer. 2002. Origin of regulatory T cells with known specificity for antigen. *Nat Immunol 3:756.*
86. Liang, S., P. Alard, Y. Zhao, S. Parnell, S. L. Clark, and M. M. Kosiewicz. 2005. Conversion of CD4+ CD25– cells into CD4+ CD25+ regulatory T cells in vivo requires B7 costimulation, but not the thymus. *J Exp Med 201:127.*
87. Apostolou, I., and H. von Boehmer. 2004. In vivo instruction of suppressor commitment in naive T cells. *J Exp Med 199:1401.*
88. Cobbold, S. P., R. Castejon, E. Adams, D. Zelenika, L. Graca, S. Humm, and H. Waldmann. 2004. Induction of foxP3+ regulatory T cells in the periphery of T cell receptor transgenic mice tolerized to transplants. *J Immunol 172:6003.*
89. Luo, X., H. Yang, I. S. Kim, F. Saint-Hilaire, D. A. Thomas, B. P. De, E. Ozkaynak, T. Muthukumar, W. W. Hancock, R. G. Crystal, and M. Suthanthiran. 2005. Systemic transforming growth factor-beta1 gene therapy induces Foxp3+ regulatory cells, restores self-tolerance, and facilitates regeneration of beta cell function in overtly diabetic nonobese diabetic mice. *Transplantation 79:1091.*
90. Peng, Y., Y. Laouar, M. O. Li, E. A. Green, and R. A. Flavell. 2004. TGF-beta regulates in vivo expansion of Foxp3-expressing CD4+CD25+ regulatory T cells responsible for protection against diabetes. *Proc Natl Acad Sci USA 101:4572.*
91. Li, M. O., Y. Y. Wan, S. Sanjabi, A. K. Robertson, and R. A. Flavell. 2006. Transforming growth factor-beta regulation of immune responses. *Annu Rev Immunol 24:99.*
92. Zou, W. 2006. Regulatory T cells, tumour immunity and immunotherapy. *Nat Rev Immunol 6:295.*
93. Valzasina, B., S. Piconese, C. Guiducci, and M. P. Colombo. 2006. Tumor-induced expansion of regulatory T cells by conversion of CD4+CD25– lymphocytes is thymus and proliferation independent. *Cancer Res 66:4488.*
94. Fadok, V. A., D. L. Bratton, A. Konowal, P. W. Freed, J. Y. Westcott, and P. M. Henson. 1998. Macrophages that have ingested apoptotic cells in vitro inhibit proinflammatory cytokine production through autocrine/paracrine mechanisms involving TGF-beta, PGE2, and PAF. *J Clin Invest 101:890.*
95. Koonpaew, S., S. Shen, L. Flowers, and W. Zhang. 2006. LAT-mediated signaling in CD4+CD25+ regulatory T cell development. *J Exp Med 203:119.*
96. Walker, M. R., D. J. Kasprowicz, V. H. Gersuk, A. Benard, M. Van Landeghen, J. H. Buckner, and S. F. Ziegler. 2003. Induction of FoxP3 and acquisition of T regulatory activity by stimulated human CD4+CD25– T cells. *J Clin Invest 112:1437.*
97. Gavin, M. A., T. R. Torgerson, E. Houston, P. DeRoos, W. Y. Ho, A. Stray-Pedersen, E. L. Ocheltree, P. D. Greenberg, H. D. Ochs, and A. Y. Rudensky. 2006. Single-cell analysis of normal and FOXP3-mutant human T cells: FOXP3 expression without regulatory T cell development. *Proc Natl Acad Sci USA 103:6659.*
98. Schwartz, R. H. 1989. Acquisition of immunologic self-tolerance. *Cell 57:1073.*
99. Oppmann, B., R. Lesley, B. Blom, J. C. Timans, Y. Xu, B. Hunte, F. Vega, N. Yu, J. Wang, K. Singh, F. Zonin, E. Vaisberg, T. Churakova, M. Liu, D. Gorman, J. Wagner, S. Zurawski, Y. Liu, J. S. Abrams, K. W. Moore, D. Rennick, R. de Waal-Malefyt, C. Hannum, J. F. Bazan, and R. A. Kastelein. 2000. Novel p19 protein engages IL-12p40 to form a cytokine, IL-23, with biological activities similar as well as distinct from IL-12. *Immunity 13:715.*
100. Cua, D. J., J. Sherlock, Y. Chen, C. A. Murphy, B. Joyce, B. Seymour, L. Lucian, W. To, S. Kwan, T. Churakova, S. Zurawski, M. Wiekowski, S. A. Lira, D. Gorman, R. A. Kastelein,

and J. D. Sedgwick. 2003. Interleukin-23 rather than interleukin-12 is the critical cytokine for autoimmune inflammation of the brain. *Nature 421:744*.
101. Nakae, S., A. Nambu, K. Sudo, and Y. Iwakura. 2003. Suppression of immune induction of collagen-induced arthritis in IL-17-deficient mice. *J Immunol 171:6173*.
102. Nakae, S., S. Saijo, R. Horai, K. Sudo, S. Mori, and Y. Iwakura. 2003. IL-17 production from activated T cells is required for the spontaneous development of destructive arthritis in mice deficient in IL-1 receptor antagonist. *Proc Natl Acad Sci USA 100:5986*.
103. Koenders, M. I., E. Lubberts, B. Oppers-Walgreen, L. van den Bersselaar, M. M. Helsen, F. E. Di Padova, A. M. Boots, H. Gram, L. A. Joosten, and W. B. van den Berg. 2005. Blocking of interleukin-17 during reactivation of experimental arthritis prevents joint inflammation and bone erosion by decreasing RANKL and interleukin-1. *Am J Pathol 167:141*.
104. Pasare, C., and R. Medzhitov. 2003. Toll Pathway-Dependent Blockade of CD4+CD25+ T Cell-Mediated Suppression by Dendritic Cells. *Science 299:1033*.
105. Veldhoen, M., R. J. Hocking, R. A. Flavell, and B. Stockinger. 2006. Signals mediated by transforming growth factor-beta initiate autoimmune encephalomyelitis, but chronic inflammation is needed to sustain disease. *Nat Immunol 7:1151*.
106. Ivanov, II, B. S. McKenzie, L. Zhou, C. E. Tadokoro, A. Lepelley, J. J. Lafaille, D. J. Cua, and D. R. Littman. 2006. The orphan nuclear receptor RORgammat directs the differentiation program of proinflammatory IL-17+ T helper cells. *Cell 126:1121*.

Chapter 8
Molecular Signalling in T Regulatory Cells

Natasha R. Locke, Natasha K. Crellin and Megan K. Levings

Abstract T regulatory (Treg) cells contribute to immune homeostasis and maintain peripheral tolerance by actively inhibiting the expansion and function of conventional T cells. Their importance in controlling a wide range of immune responses, including autoimmune disease, cancer and transplantation tolerance is well established. Treg cells differ from T effector cells in many aspects, including their capacity to proliferate, produce cytokines, and express a unique complement of proteins. This unique phenotype suggests that Treg cells may have alterations in molecular signalling pathways. This review summarizes recent progress in defining the distinct intracellular signalling events in Treg cells that may underlie their phenotype and suppressive function. We focus on changes in events downstream of the T cell receptor and the IL-2 receptor, and discuss how costimulation events may contribute to altered signalling events. Much more research is required to better define the mechanistic basis for theses changes in Treg cells and to define their functional significance.

Introduction

It is well-recognised that active suppression of immune responses by T regulatory (Treg) cells is a key mechanism for the induction and maintenance of peripheral tolerance. Evidence from mouse models and human diseases has shown that Treg cells are essential for the maintenance of peripheral tolerance to self-antigens as well as commensal intestinal flora, and that they can modulate immune responses to pathogens, tumor antigens and alloantigens. Although T cells with a regulatory/suppressor function exist within all major subsets, in both humans and rodents

M.K. Levings
Department of Surgery, University of British Columbia and Immunity and Infection Research Centre, Vancouver Coastal Health Research Institute, 2660 Oak St. Vancouver, B.C., V6H 3Z6, Canada
e-mail: mlevings@interchange.ubc.ca

most attention has been focused on CD4$^+$ Treg cell subsets, including CD4$^+$CD25$^+$ Treg, type 1 Treg (Tr1) and T helper type 3 (Th3) cells [1–3].

CD4$^+$CD25$^+$ Treg cells were originally characterised by their high constitutive expression of the IL-2 receptor α chain (CD25), a marker traditionally used to identify activated T cells. More recently, a more definitive marker for this population was found to be the FoxP3 transcription factor [4,5]. In mice, FOXP3 appears to be not only critical for the development of CD4$^+$CD25$^+$ Treg cells, but also sufficient to confer a Treg phenotype [5]. In contrast, in humans, FOXP3 does not appear to be an exclusive lineage factor for Treg cells as it can also be expressed in activated T effector cells [6–8]. Many in vitro studies have demonstrated that CD4$^+$CD25$^+$ Treg cells strongly suppress the function of both naive and memory CD4$^+$ T effector cells, as well as CD8$^+$ T cells, antigen presenting cells and B cells [9,10]. Their suppressive activity directed towards T cells is related to their ability to inhibit IL-2 production and promote cell-cycle arrest. Although their mechanism of action remains to be fully defined, it may involve membrane-bound TGF-β [11] and/or granzyme A-mediated perforin-independent cytotoxicity [12]. The in vivo relevance of these cells has been demonstrated in several mouse models of human disease. In the absence of functional CD4$^+$CD25$^+$ Treg cells mice develop lethal autoimmunity, whereas adoptive transfer can prevent autoimmunity, allergies, graft rejection, anti-tumor, anti-viral and anti-parasite immunity [13–16].

In contrast to CD4$^+$CD25$^+$ Treg cells, which can either arise in the thymus or differentiate in the periphery, Th3 and Tr1 subsets are exclusively generated in the periphery in the presence of TGF-β and IL-10, respectively [1,17]. Currently no specific cell surface markers for Tr1 or Th3 cells have been defined, and they can only be defined based on their unique cytokine production profiles. Tr1 cells are IL-10$^+$, IL-4$^-$, IL-2^{low}, IFN-γ$^+$ and suppress both naive and memory Th1 or Th2 cells via an IL-10-dependent mechanism [1]. Th3 cells are defined by their production of high levels of TGF-β1, which is crucial for their generation and suppressive function [18].

There is a great deal of hope that therapeutic approaches based on modulating the numbers and/or function of Treg cells will expand the treatment options for many immune-mediated diseases. For example, trials are already underway to test whether enhancing the numbers of polyclonal or antigen-specific populations of Treg cells can ameliorate graft versus host disease and type 1 diabetes [19,20]. A more detailed knowledge, however, of how Treg cells differ from effector T cells at the molecular level is essential to develop better ways to monitor their function, elucidate their mechanism of action, and design strategies to specifically target different Treg cell subsets in vivo. Recently there have been several advances in defining how the intracellular signalling events in Treg cells differ from those in Teffector cells. This review will focus on evidence that Treg cells have differences in several of the major signalling pathways activated by the T cell receptor (TCR) and that they may also respond differentially to co-stimulatory signals and cytokines. Since the majority of studies have focused on CD4$^+$CD25$^+$ Treg cells, we will primarily focus on this subset, and hereafter refer to them as Treg cells.

TCR-Mediated Signalling in T Effector Cells

Signalling cascades activated by the TCR have been extensively studied in T cell lines and primary cells. This topic has been reviewed extensively [21–26], and here we will simply highlight some of the major molecular changes that coincide with T cell activation. It is clear that in order for T cells to be fully activated, proliferate and acquire effector functions, they must receive signal 1 via the T cell receptor (TCR) and signal 2 by one of several possible co-stimulatory molecules expressed by antigen presenting cells (APCs) [27]. Provision of both signals in conventional T cells results in IL-2 production and proliferation, while provision of signal 1 alone results in T-cell anergy [28]. In conventional T cells, coligation of the TCR and CD4 by peptide-MHC complexes results in phosphorylation of tyrosine residues in immunoreceptor tyrosine-based activation motifs (ITAMs) in the TCR/CD3

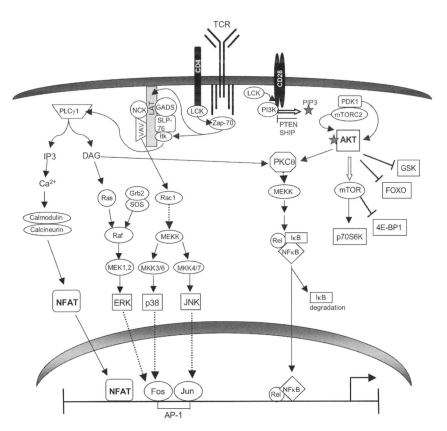

Fig. 8.1 *Overview of TCR signalling.* Coligation of the TCR and CD4 in the presence of CD28 costimulation results in the recruitment of kinases and adaptor molecules to the plasma membrane, allowing complex formation and clustering to occur. The major downstream consequences are the nuclear translocation of transcription factors and the modulation of proteins controlling cell growth and survival

complex by the src family kinases $p56^{lck}$ or $p59^{fyn}$. This leads to recruitment and phosphorylation of ZAP-70 [29]. Activated ZAP-70 then phosphorylates the adaptor protein, linker for activation of T cells (LAT), which forms a platform that connects the major intracellular signalling pathways. The signalling components juxtaposed on this platform include Gads/SLP-76, Grb-2/SOS; the Tec family kinase Itk; phospholipase Cγ1 (PLCγ1), Rac1, Rac2, the adaptor Nck and RhoG guanine nucleotide exchanger Vav-1 (reviewed in [30]). These initial events initiate multiple signalling cascades, resulting in the activation of the phosphatidylinositol 3-kinase (PI3'K) pathway, the mitogen-activated protein kinases (MAPKs), calcium mobilization and release of NFAT to the nucleus (summarized in Fig. 8.1). CD28 ligation by B7 further enhances phosphorylation and activation of Lck, Vav-1, and PI3 K. Disruption in any of these major pathways often leads to changes in thymic development and in the function of peripheral T cells, highlighting that each of these pathways has a unique and essential role in T effector cells.

Signalling Downstream of the TCR in Treg Cells

Knowledge of how Treg cells differ from conventional T cells in terms of intracellular signalling cascades largely remains to be determined. These studies have been hampered by limiting cell numbers, difficulties in isolating pure populations due to the lack of reliable cell surface markers, and their relatively poor proliferation in vitro. Despite these difficulties, there have been several reports suggesting that intracellular signalling events in Treg cells differ from those in conventional T cells.

Changes in the Ras/MAPK Pathway

Based on the fact that Treg cells are hyporesponsive to TCR-mediated stimulation, many groups hypothesized that like classical anergic T cells, they may have defects in the Ras/MAPK signalling pathways and/or calcium mobilization [31]. One report found that human Treg cells have a defect in phosphorylation of CD3ζ ITAM residues [32], resulting in a lack of ZAP-70 recruitment and activation, and phosphorylation of SLP-76; key events necessary for activation of MAPKs. In agreement with this finding, mouse Treg cells were reported to have reduced PLC-γ1 phosphorylation and activity, and consequently reduced intracellular diacylglycerol (DAG) levels and activation of PKC and Ras compared with T effector cells [33]. The ability of LAT to recruit PLC-γ1 is also necessary for the development of Treg cells [34].

Our own experiments, with ex vivo human Treg cells found that they do not have a defect in activation of p38 MAPK or ERK1/2 compared to effector T cells [35]. In contrast, expanded Treg cell lines from human cord blood showed reduced activation of Ras, MAP kinase kinase (MEK1/2) and ERK1/2 [36]. These discrepancies may be explained by the different methods employed. In order to analyze events in ex vivo Tregs, and to overcome the limitations of using CD25 as a Treg-specific marker, we used flow cytometric analysis to analyze phosphorylation of ERK specifically in CD4+CD25+FOXP3+ Treg cells [35]. Although our findings in pure populations of FOXP3+ cells are therefore more likely to reflect the true capacity of Treg cells to signal via this pathway, further research will be required to reconcile these differences.

Stimulation with PMA and the calcium ionophore, ionomycin, activates signalling pathways downstream of Ras and PKC, bypassing the requirement for TCR engagement. Notably, upon stimulation with PMA/ionomycin, the proliferative capacity of human and mouse CD4+CD25+ Treg cells is restored [32,37], and under these stimulatory conditions, Treg cells display normal levels of p38 MAPK and ERK1/2 phosphorylation and AP-1 complex activity [37]. In contrast, murine Tregs have been reported to display reduced JNK activation following stimulation with PMA/ionomycin, although this change did not appear to be related to their function [37]. There is evidence that calcium mobilization upon TCR engagement is impaired in murine and human Treg cells [33,38]. The function, however, of the calcium channels which are directly responsible for calcium influx, appears to be normal in Treg cells, indicating that impaired calcium mobilization is caused by a more proximal signalling defect [33].

The Role of NFAT and Interactions with FOXP3

Anergic T cell clones generated by stimulation with ionomycin primarily display defects in the Ras/MAPK signalling cascade, resulting in a deficiency of AP-1 formation [39]. T cells lacking NFAT are resistant to anergy induction, whereas infection of T cells with a retrovirus expressing a constitutively active form of NFAT that is unable to interact with AP-1 results in anergy induction [39]. These data indicate that NFAT induces T cell anergy if prevented from interacting with AP-1. One candidate for an NFAT binding partner that leads to the induction of anergy is the transcription factor Egr-2 [40]. Interestingly, expression of Egr-2 is high in Treg cells, and its presence is correlated with upregulation of the cell cycle inhibitors $p21^{cip1}$ and $p27^{kip}$, as well as suppressive capacity [41]. Thus, altered levels of AP-1 in Treg cells may alter the function of NFAT to act as a negative rather than positive regulator of gene transcription.

Interestingly, FoxP3 has also been shown to form heterodimeric complexes with NFAT, and this complex can bind NFAT-AP-1 consensus sequences and suppress transcription [42,43]. The introduction of mutations into FoxP3 that disrupt its interaction with NFAT, impair the ability of FOXP3 to repress expression of IL-2 and to upregulate expression of the Treg markers CTLA-4 and CD25 [42], indicating the biological importance of this interaction. Therefore FoxP3/NFAT complexes act to suppress transcription of cytokine promoters and promote the transcription of Treg-associated molecules. Furthermore, this FoxP3/NFAT association is required for the suppressive function of Treg cells in a murine model of autoimmune diabetes [42]. Together these data suggest that one of the major roles of FOXP3 is to alter the downstream consequences of NFAT activation.

Influence of the PI3′K Pathway on Treg Cell Differentiation and Function

There is substantial evidence that PI3′K is a critical regulator of tolerance. Expression of a constitutively active mutant of either PI3′K or AKT under the control of a T-cell specific promoter results in autoimmunity [44,45]. Conversely, treatment of human T cells with wortmannin, a natural inhibitor of PI3′K, induces alloantigen-specific tolerance [46]. Finally, rapamycin, an inhibitor of PI3′K pathway component mTOR, can block cell-cycle progression from the G1 to S phase in activated T cells, resulting in the induction of tolerance [47]. Rapamycin forms a complex with FK506-binding protein-12 (FKBP12), and then binds to and specifically inhibits the function of the mammalian target of Rapamycin complex 1 (mTORC1), the activation of which is essential for protein synthesis and cell cycle progression. Repeated antigen-specific stimulation of murine T cells in the presence of rapamycin gives rise to a cell-contact-dependent suppressive population enriched for $CD4^+CD25^+$ cells with enhanced FoxP3 expression in vitro [48]. Experiments with human T cells argue that these effects of rapamycin are not due to the selective expansion of Treg cells, but rather to the capacity of rapamycin to induce suppressive function in conventional $CD4^+$ T cells [49]. In contrast, in vivo treatment of mice with rapamycin in combination with IL-10 expands functional Tr1 cells, but not Treg cells [50]. Taken together, these data indicate that inhibition of the PI3′K pathway by rapamycin may act by inhibiting effector T cell proliferation while allowing the expansion of multiple subsets of Treg cells.

In line with the hypothesis that altered activity of the PI3′K pathway may be important for Treg development and/or function, we found that human Treg cells have a defect in activation of AKT downstream of the TCR [35]. We have recently confirmed this finding in murine cells (Fig. 8.2). Compared to effector CD4+ T cells, Treg cells displayed reduced phosphorylation of AKT on Ser473 after stimulation via the TCR and CD28. This observation is consistent with the finding that murine $CD4^+CD25^+$ Treg cells have reduced AKT activation upon IL-2R stimulation [51]. In addition, a similar defect in AKT^{Ser473} phosphorylation has been described in CD46-induced Tr1-like cells [52]. These cells were shown to be more susceptible to activation-induced cell death due to defective expression of the anti-apoptotic protein Survivin.

AKT is phosphorylated within the activation loop at threonine 308 by PDK-1 and the C-terminus at Serine 473 [53]. There has been long-standing disagreement on whether phosphorylation of Ser473 is a prerequisite for Thr308 phosphorylation or whether these events are independent. Two recent studies in which components of mTOR complex 2 (mTORC2), SIN1 and rictor, were genetically deleted have identified mTORC2 as the previously unknown Ser473 kinase and shown that Thr308 phosphorylation is independent of Ser473 phosphorylation [54,55]. Importantly, the kinase activity of AKT in SIN and rictor deficient cells was largely intact, and the only AKT-dependent phosphorylation targets affected were the Forkhead transcription factors FoxO1 and FoxO3 [55]. These data suggest that Ser473 phosphorylation affects AKT substrate specificity rather than absolute activity and have important

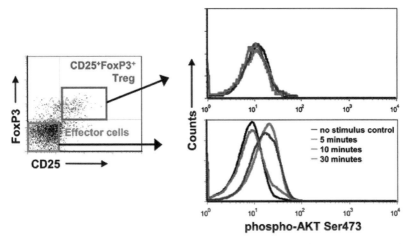

Fig. 8.2 *Mouse Treg cells fail to phosphorylate AKT at Ser473.* Splenic and lymph node CD4$^+$ T cells were purified from 8-week-old C57Bl/6 mice by negative selection and rested for 2 hours and either left unstimulated or stimulation with anti-CD3 (5 μg/ml) and anti-CD28 (1 μg/ml) mAbs and Fc crosslinker antibody (20 μg/ml). Cells were fixed in FOXP3 Fix/Perm buffer after 5, 10 and 30 min of stimulation, stained for CD4, CD25, FoxP3 and phospho-AKT and examined by flow cytometry. The mean flourescence intensity of AKT phosphorylation at Ser473 is compared in CD4$^+$CD25^{hi}FoxP3+ Treg (*top*) and CD4$^+$CD25$^-$ effector (*bottom*) cells *(See also Color Insert)*

implications for our understanding of the effects of rapamycin, which inhibits the kinase activity of mTORC1 but not mTORC2. Differential AKT substrate specificity determined by the balance or activation state of mTORC1/mTORC2 may provide an explanation for the known anti-apoptotic effects of AKT and yet lack of activation-induced cell death induced by blockade of AKT activation by rapamycin [48] (which blocks only mTORC1).

The biological significance of altered activation of the PI3'K pathway in Treg cells was directly demonstrated by restoration of AKT activity via lenti-viral mediated expression of an inducibly-active form of the kinase [35]. Surprisingly, these experiments revealed that reduced activity of the PI3'K pathway was necessary for the in vitro suppressive function of CD4$^+$CD25$^+$ Treg cells but was not required for the maintenance of their hypoproliferative state. It is important to note that the inducibly-active AKT used in these experiments is likely phosphorylated at both Ser473 and Thr308 [56]. Additional experiments will be required to further dissect the molecular control of Treg suppressive capacity. In particular, it remains to be determined whether AKT^{Ser473}-dependent (FoxO) or AKT^{Thr308}-dependent (S6 kinase, GSK-3, TSC2 and mTOR) phosphorylation events are required for inhibition of Treg suppressive capacity.

Confirmation that PI3 K is critical for Treg suppressive capacity comes from the demonstration that Treg cells lacking the main PI3 K isoform activated downstream of TCR stimulation, p110δ, lack suppressive function [57]. This is consistent with the known role of CD28 costimulation in Treg cell development and minimal enhancement of the PI3 K pathway downstream of CD28 signalling. Intriguingly, in vitro assays demonstrated that p110δ–deficient Treg cells stimulated with anti-CD3 and anti-CD28 coated beads lacked suppressive capacity while those stimulated with anti-CD3 coated APC exhibited approximately 50% reduced suppressive capacity. It is tempting to speculate that this difference reflects the contribution of PI3 K pathway-independent signalling through inhibitory ligands expressed by Treg cells, such as CTLA-4 or GITR. Nevertheless, although p110δ–deficient Treg cells may retain some suppressive function, they are unable to prevent the spontaneous development of colitis [57].

Negative Regulators of the PI3'K Pathway

PI3'K activity is regulated primarily by two phosphastases, the phosphatase and tensin homologue deleted on chromosome 10 (PTEN) and SH2 containing inositol 5-phosphatase (SHIP). These phosphatases hydrolyse the key second messenger in the PI3 K pathway, PIP3, and thus negatively regulate the phosphorylation and activity of AKT. Therefore it was hypothesized that increased activity or concentrations of PTEN or SHIP may be required to maintain reduced PI3 K pathway activation in Treg cells. Despite indications that PTEN levels remain elevated after TCR signalling in Treg cells compared to effector T cells [51], analyses of mice with PTEN-deficient T cells revealed that PTEN is not vital for the development or in vivo function of $CD4^+CD25^+$ Treg cells [58]. However, T cells deficient in PTEN are rendered resistant to suppression by Treg cells due to enhanced AKT activity, resulting in lymphoproliferative disorders and autoimmunity [59]. These data highlight the critical balance between PI3'K and PTEN in regulating appropriate T cell activation.

A role for SHIP in promoting Treg development has also recently been documented [60]. SHIP acts on the PI3 K product, PIP_3, to dephosphorylate the 5' position of the inositol ring. Deletion of SHIP results in a dramatic increase in the percentage of $CD4^+CD25^+FoxP3^+$ Treg cells and double deletion of SHIP and the adaptor protein downstream of kinase (Dok-1) further increases Treg numbers [60]. However, the role of grossly expanded $Mac1^+Gr1^+$ cells in these knockout mice in Treg lineage promotion has not been addressed. Importantly, SHIP and SHIP/Dok-1-deficient Treg cells are functional [60], indicating that these pathways are not directly required for their in vitro suppressive capacity.

Dok-1 and Dok-2 have also been shown to negatively regulate AKT activation [61,62]. However, in addition to the role of SHIP in counteracting the PI3 K pathway, new evidence suggests that SHIP may also act as a chaperone/adaptor molecule, bringing Dok-1 or Dok-2 to the TCR complex upon TCR stimulation. Dok-1 and Dok-2 double deficient mice exhibit enhanced lymphocyte proliferation and spontaneously develop chronic myelogenous leukaemia and lymphomas [62,63]. Interestingly, combined inhibition of Dok-1 and Dok-2 by siRNA knockdown results in enhanced TCR-induced IL-2 secretion and ZAP-70 and AKT^{Ser473} phosphorylation, indicating that Dok-1 and Dok-2 negatively control TCR signal initiation [64]. The mechanism of Dok signal inhibition has not yet been elucidated but it is possible that competitive binding to Grb-2 SH2-domains by SHIP/Dok-2 prevents the association of other positive signalling molecules. Alternatively, Dok-1 has been demonstrated to attenuate signalling in mouse embryonic fibroblasts upon epidermal growth factor stimulation by recruiting C-terminal Src kinase (Csk) [65], a potent negative regulator of Src family kinase members [66]. Thus, recruitment of Csk to the immunological synapse by Dok-1 or Dok-2 may attenuate TCR signalling by inhibiting p56^{lck} and p59^{fyn} phosphorylation. Given the evidence for altered regulation of ZAP-70 and PI3 K pathways in Treg cells, investigation of whether this negative feedback loop may be enhanced in Treg cells is warranted.

NF-κB

An important consequence of AKT activation is the induction of NF-κB nuclear translocation following TCR and CD28 signal transduction [67]. Inducible degradation of the inhibitory protein IκB, which normally sequesters NF-κB in the cytoplasm, results in NFκB nuclear translocation. PKC (activated by the TCR) and AKT (activated downstream of CD28 costimulation) appear to act in concert to phosphorylate CARMA-1, leading to downstream formation of the IKK complex and IκB degradation [68]. It is currently not known, however, whether phosphorylation at Ser473 and/or Thr308 is required for this function of AKT. Blockade of NF-κB nuclear translocation has been demonstrated to result in the production of antigen-specific CD4$^+$ Treg cells [69]. It is interesting to note that FoxP3 has been shown to associate with and negatively regulate the transcriptional activity of NFAT and NF-κB [43]. Thus it appears that the inhibition of NF-κB nuclear translocation and transcriptional activity is an important regulatory mechanism in Treg cells. It would be of interest to investigate whether the nuclear translocation of NF-κB in CD4$^+$CD25$^+$ Treg cells is altered in light of their demonstrated failure to phosphorylate AKT at Ser473.

Mechanistic Basis for Changes in TCR-Mediated Signalling

A number of mechanisms, including reduced duration/strength of TCR engagement, poor immunological synapse formation and degradation of key second messenger molecules, may result in these changes in TCR-mediated signalling in Treg cells. With regard to the latter, a report investigating the signalling defects in ionomycin-induced anergic T cells found that PKC and PLC-γ1 are specifically targeted for degradation upon TCR engagement [70]. This does not, however, appear to be the case in Treg cells as the levels of PLCγ1, Ras and ERK proteins appear equivalent in unstimulated and CD3-stimulated Treg cells [33]. Upon TCR stimulation, receptors involved in TCR signalling such as CD4, TCR and CD3 complexes are recruited to the actin cytoskeleton via a process, termed actin cap formation, which increases the effective concentration of signalling molecules and guides the formation of the immunological synapse. One of the key proteins involved in this process is VAV, a guanine nuclear exchange factor that regulates the activity of Rho and Rac1, proteins that regulate actin remodeling [71]. Consistent with a lack of VAV recruitment, $CD4^+CD25^+$ Treg cells were shown to exhibit defective actin cap formation [32]. A failure of actin cap formation and subsequent lack of sustained $p56^{lck}$ activation would be predicted to further downmodulate proximal TCR signalling. Further research into this topic is required to elucidate the precise mechanisms that result in the altered signalling observed in Treg cells.

Costimulatory Molecules Which May Contribute to Unique Signalling Cascades in Treg Cells

On naïve T cells, the CD28 receptor provides the primary costimulatory signal following interaction with B7-1(CD80) or B7-2 (CD86) on APCs. Since the cytoplasmic tail of CD28 lacks intrinsic catalytic activity, the intracellular signals are mediated by associating enzymes. SRC family kinases, including Lck and Fyn, and the TEC family kinase ITK (IL-2- inducible T-cell kinase), have been shown to phosphorylate CD28 molecules at four tyrosine residues within the intracellular portion of the receptor [72,73]. CD28 signalling through PI3 K results in the recruitment of PKC to the TCR complex, subsequent activation of NF-kB and induction of IL-2 transcription. CD28 costimulation also enhances and stabilizes TCR-induced phosphorylation by directing the clustering of lipid raft-associated intracellular kinases at the site of TCR engagement [74,75] and thus is thought to reduce the interaction time and number of TCRs that need to be engaged. In addition to augmenting TCR-mediated signals, CD28 also activates independent signalling pathways, such as upregulation of the anti-apoptotic protein $BCL-X_L$ [76].

Treg cells do not require CD28 ligation for their suppressive function, as evidenced by the presence of functional Treg cells in CD28-deficient mice [77]. However, mice deficient in either CD28 or B7 have notably reduced numbers of Treg cells in both the thymus and periphery [78,79]. Elegant experiments involving mutation of the CD28 cytoplasmic tail at kinase binding motifs have demonstrated that Treg cell development from thymocyte precursors requires CD28 costimulation, and is specifically dependent on an Lck-binding motif in the CD28 cytosolic tail [80]. CD28 costimulation was also shown to induce Foxp3 expression and to upregulate GITR and CTLA-4 expression, indicating that this molecule is required at an early stage of Treg cell differentiation, before the expression of CD25 or acquisition of regulatory function. Importantly, this requirement for CD28 is independent of CD28-induced IL-2 production by effector T cells as provision of IL-2-competent cells does not rescue Treg development in these mice.

In addition to positive costimulatory molecules such as CD28, inhibitory receptors such as Glucocortocoid-induced tumor necrosis factor receptor (GITR) and CTLA-4 are highly expressed on Treg cells [77,81,82]. CD28 and CTLA-4 are closely related members of the immunoglobulin supergene family and bind the same ligands, CD80 and CD86, although CTLA-4 has a higher binding avidity. CTLA-4 ligation raises the threshold for T cell activation, arrests cell cycle progression, and therefore acts as a potent negative regulator of T-cell responses and in the induction of self-tolerance [76]. Many groups have attempted to define the role of CTLA-4 in Treg cells using antibody-mediated blockade studies. It should be noted, however, that this protein is predominantly intracellular and antibody blockade is therefore relatively ineffective [83,84]. Analysis of Treg cells in CTLA-4-deficient mice has also been challenging as they develop a severe lymphoproliferative disorder and die from an autoimmune-like disease within 3–5 weeks of birth [85].

Despite the critical role of CTLA-4 in maintaining T cell homeostasis, the molecular mechanisms involved remain largely unknown. CTLA-4 may mediate negative regulation in various ways; by competing for CD80/CD86 ligand binding and thereby attenuating CD28-mediated costimulatory signals, by delivering a negative signal and/or by enhancing suppression by Treg cells [86]. The demonstration that Treg cells from CTLA-4-deficient mice have normal suppressive activity in vitro [87,88] argues against the latter possibility. Interestingly, CTLA-4-deficient Treg cells were shown to develop a compensatory TGF-β-dependent suppressive function [88]. Studies have shown that CTLA-4 can remove the TCRζ chain from lipid rafts [89] and inhibit signalling via a mechanism involving PP2A. In contrast, ligation of PD-1 can prevent CD28-mediated PI3 K signalling through AKT [90]. Although further research is needed, it is tempting to speculate that one or both of these proteins may have a key role in mediating the unique intracellular signalling events in Treg cells.

It is well know that at least in vitro, Treg cells mediate regulatory function directly via cell-cell interactions. There is growing evidence that proteins of the Notch family receptors and their ligands, such as Jagged in humans and Delta and Serrate in mice, may have a role in this process [91]. The binding of Notch ligands induces 2 cleavage events in the intracellular portion of the Notch receptor, releasing the intracellular domain of Notch, which moves into the nucleus and binds to the transcription factor CBF1. Treg cells have been shown to modify APC such that they can induce tolerance in subsequently encountered naïve cells [92]. Notch plays a central role in this phenomenon as murine APC overexpressing the notch ligand Serrate induced naïve T cells to differentiate into Treg cells capable of transferring antigen-specific tolerance and preventing streptozotocin-induced autoimmune diabetes [93]. These findings suggest that activation of the Notch pathway regulates the de novo induction of Treg cells from naïve precursors.

IL-2R Signalling in Treg Cells

Although stimulation of Treg cells via the TCR is required to induce their suppressive function, they are hyporesponsive, as measured by the lack of DNA replication and IL-2 secretion in vitro. This inability of Treg cells to produce IL-2 following activation appears to result from a lack of chromatin remodeling of the IL-2 promoter region [37]. Despite their inability to produce IL-2, Treg cells are highly dependent on production from exogenous sources [94]. Several studies have shown that signalling via the IL-2R to STAT5 appears to be normal in Treg cells [51], and essential for their development and survival in vivo [95,96]. This requirement for IL-2 can be replaced by IL-7 if it is provided to Treg cells in culture or if a chimeric receptor consisting of the extracellular portion of IL-2Rα and the intracellular portion if IL-7Rα is expressed on Treg cells [97]. These data indicate that it is cytokine availability rather than distinct receptor signalling that determines the absolute requirement of Treg cells for IL-2. Interestingly, in line with our finding that signalling via the PI3'K pathway is deficient downstream of the TCR, this also appears to be the case downstream of the IL-2R [51]. Notably, the anti-apoptotic effect of IL-2 to upregulate BCL-X_L is maintained despite the lack of AKT activation. These studies also revealed that the inability of Treg cells to proliferate in response to IL-2 is related to their inability to downregulate PTEN expression since deletion of PTEN restored their capacity to expand in response to IL-2 [58].

Conclusion

There is now widespread consensus that Treg cells have a major role in establishing and maintaining tolerance. Currently, a major focus is to better define the molecular mechanisms that control their development and function. It is clear that the intracellular signalling mechanisms operational in Tregs are distinct from those in Teffector cells and anergic T cells (summarized in Fig. 8.3), but much more research is required to fully understand the mechanistic basis for these changes and their biological relevance. For example, we have shown that altered activity of the

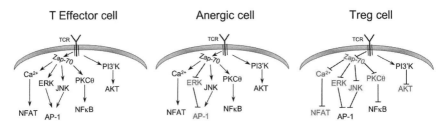

Fig. 8.3 *Differences between T effector, anergic and Treg cells in the major TCR-mediated signalling pathways.* Upon stimulation of a T effector cell through its TCR and co-receptors, the calcium/NFAT, Ras/MAPK, NFκB and PI3'K signalling pathways are activated (*left*). In contrast, cells rendered anergic by incomplete stimulation are characterised by defects in the Ras/MAPK signalling cascade resulting in a lack of AP-1 nuclear formation (*middle*). Changes documented in TCR-mediated signalling pathways in Treg cells (*right*) include defective phosphorylation of ITAMs within the CD3ζ chain, resulting in a lack of Zap-70 recruitment and reduced phosphorylation of downstream substrates LAT and PLCγ1 [32], reduced calcium mobilisation and subsequent reductions in NFAT nuclear translocation and in PKC phosphorylation [33,38], a failure to phosphorylate AKT at Ser^{473} downstream of PI3'K signalling [35], modest reductions in ERK phosphorylation [36] and reduced JNK activation following PMA/ionomycin stimulation [37]. Signalling components whose activation/phosphorylation has been demonstrated to be defective in Treg cells are depicted in *red (See also Color Insert)*

PI3'K pathway is necessary for the suppressive function of Tregs, but currently do not know what upstream changes lead to a deficit in AKT phosphorylation. The functional consequences of changes in other signalling pathways remain unclear. A major challenge has been to perform traditional biochemical studies in pure populations of Treg cells. Many past studies were based on analyses of signalling events in populations of cells defined on CD25 expression, which, particularly in humans, is not a reliable marker [6,7]. Recent advances in intracellular staining methodology for detecting phospho-epitopes by flow cytometry permits the detection of many phosphorylation events in individual cells. Importantly, this technique also allows for FoxP3 co-staining so that signalling events can be followed in more accurately defined Treg cell populations. Ultimately, we can foresee that manipulation of Treg-specific signalling events may offer the potential to manage Treg cell numbers and/or function. The discovery of unique molecular pathways in Treg cells will reveal natural mechanisms that limit T-cell growth and highlight potential targets for the development of drugs which could be used to manipulate a diverse array of immune responses by Treg cells.

Acknowledgments The authors' own work is supported by grants from the Canadian Institutes for Health Research (CIHR) (MOP57834 and MOP127506) and BC Transplant Society. MKL holds a Canada Research Chair in Transplantation and is a Michael Smith Foundation for Health Research (MSFHR) Scholar. NRL holds a CIHR/MSFHR Training program postdoctoral fellowship. NKC holds a MSFHR Senior Graduate Studentship award.

References

1. Roncarolo, M G; Gregori, S; Battaglia, M; Bacchetta, R; Fleischhauer, K; Levings, M K. Interleukin-10-secreting type 1 regulatory T cells in rodents and humans. Immunol Rev, 2006 Aug, 212, 28–50.
2. Sakaguchi, S; Ono, M; Setoguchi, R; Yagi, H; Hori, S; Fehervari, Z; Shimizu, J; Takahashi, T; Nomura, T. Foxp3+ CD25+ CD4+ natural regulatory T cells in dominant self-tolerance and autoimmune disease. Immunol Rev, 2006 Aug, 212, 8–27.
3. Shevach, E M; DiPaolo, R A; Andersson, J; Zhao, D M; Stephens, G L; Thornton, A M. The lifestyle of naturally occurring CD4+ CD25+ Foxp3+ regulatory T cells. Immunol Rev, 2006 Aug, 212, 60–73.
4. Ziegler, S F. FOXP3: of mice and men. Annu Rev Immunol, 2006, 24, 209–26.
5. Fontenot, J D; Rudensky, A Y. A well adapted regulatory contrivance: regulatory T cell development and the forkhead family transcription factor Foxp3. Nat Immunol, 2005 Apr, 6(4), 331–7.
6. Allan, S E; Crome, S Q; Crellin, N K; Passerini, L; Steiner, T S; Bacchetta, R; Roncarolo, M G; Levings, M K. Activation-induced FOXP3 in human T effector cells does not suppress proliferation or cytokine production. Int Immunol, 2007 Apr, 19(4), 345–54.
7. Ziegler, S F. FOXP3: Not just for regulatory T cells anymore. Eur J Immunol, 2007 Jan, 37(1), 21–3.
8. Wang, J; Ioan-Facsinay, A; van der Voort, E I; Huizinga, T W; Toes, R E. Transient expression of FOXP3 in human activated nonregulatory CD4(+) T cells. Eur J Immunol, 2007 Jan, 37(1), 129–38.
9. Levings, M K; Sangregorio, R; Sartirana, C; Moschin, A L; Battaglia, M; Orban, P C; Roncarolo, M G. Human CD25+CD4+ T suppressor cell clones produce TGF-b, but not IL-10 and are distinct from type 1 T regulatory cells. J Exp Med, 2002, 196, 1335–46.

10. Levings, M K; Sangregorio, R; Roncarolo, M G. Human CD25+CD4+ T regulatory cells suppress naive and memory T-cell proliferation and can be expanded in vitro without loss of function. J Exp Med, 2001, 193, 1295–302.
11. Oida, T; Zhang, X; Goto, M; Hachimura, S; Totsuka, M; Kaminogawa, S; Weiner, H L. CD4+CD25- T cells that express latency-associated peptide on the surface suppress CD4+CD45RBhigh-induced colitis by a TGF-beta-dependent mechanism. J Immunol, 2003 Mar 1, 170(5), 2516–22.
12. Gondek, D C; Lu, L F; Quezada, S A; Sakaguchi, S; Noelle, R J. Cutting edge: contact-mediated suppression by CD4+CD25+ regulatory cells involves a granzyme B-dependent, perforin-independent mechanism. J Immunol, 2005 Feb 15, 174(4), 1783–6.
13. Waldmann, H; Cobbold, S. Regulating the immune response to transplants: a role for CD4+ regulatory cells? Immunity, 2001, 14, 399–406.
14. Sakaguchi, S. Naturally arising CD4+ regulatory T cells for immunologic self-tolerance and negative control of immune responses. Annu Rev Immunol, 2004, 22, 17–32.
15. Shevach, E M. Regulatory T cells in autoimmmunity. Annu Rev Immunol, 2000, 18, 423–49.
16. Liu, H; Leung, B P. CD4+CD25+ regulatory T cells in health and disease. Clin Exp Pharmacol Physiol, 2006 May–Jun, 33(5–6), 519–24.
17. Fu, S; Zhang, N; Yopp, A C; Chen, D; Mao, M; Zhang, H; Ding, Y; Bromberg, J S. TGF-beta induces Foxp3 + T-regulatory cells from CD4 + CD25 – precursors. Am J Transplant, 2004 Oct, 4(10), 1614–27.
18. Faria, A M; Weiner, H L. Oral tolerance and TGF-beta-producing cells. Inflamm Allergy Drug Targets, 2006 Sep, 5(3), 179–90.
19. Rezvani, K; Mielke, S; Ahmadzadeh, M; Kilical, Y; Savani, B N; Zeilah, J; Keyvanfar, K; Montero, A; Hensel, N; Kurlander, R; Barrett, A J. High donor FOXP3-positive regulatory T-cell (Treg) content is associated with a low risk of GVHD following HLA-matched allogeneic SCT. Blood, 2006 Aug 15, 108(4), 1291–7.
20. Filippi, C; Bresson, D; von Herrath, M. Antigen-specific induction of regulatory T cells for type 1 diabetes therapy. Int Rev Immunol, 2005 Sep–Dec, 24(5–6), 341–60.
21. Altman, A; Villalba, M. Protein kinase C-theta (PKCtheta): it's all about location, location, location. Immunol Rev, 2003 Apr, 192, 53–63.
22. Palacios, E H; Weiss, A. Function of the Src-family kinases, Lck and Fyn, in T-cell development and activation. Oncogene, 2004 Oct 18, 23(48), 7990–8000.
23. Dustin, M L. A dynamic view of the immunological synapse. Semin Immunol, 2005 Dec, 17(6), 400–10.
24. Mittelstadt, P R; Salvador, J M; Fornace, A J, Jr.; Ashwell, J D. Activating p38 MAPK: new tricks for an old kinase. Cell Cycle, 2005 Sep, 4(9), 1189–92.
25. Kabouridis, P S. Lipid rafts in T cell receptor signalling. Mol Membr Biol, 2006 Jan–Feb, 23(1), 49–57.
26. Weil, R Israel, A. Deciphering the pathway from the TCR to NF-kappaB. Cell Death Differ, 2006 May, 13(5), 826–33.
27. Lafferty, K J. Cunningham, A J. A new analysis of allogeneic interactions. Aust J Exp Biol Med Sci, 1975 Feb, 53(1), 27–42.
28. Schwartz, R H. T cell anergy. Annu Rev Immunol, 2003, 21, 305–34.
29. Chan, A C; Dalton, M; Johnson, R; Kong, G H; Wang, T; Thoma, R; Kurosaki, T. Activation of ZAP-70 kinase activity by phosphorylation of tyrosine 493 is required for lymphocyte antigen receptor function. Embo J, 1995 Jun 1, 14(11), 2499–508.
30. Aguado, E; Martinez-Florensa, M; Aparicio, P. Activation of T lymphocytes and the role of the adapter LAT. Transpl Immunol, 2006 Dec, 17(1), 23–6.
31. Appleman, L J; Tzachanis, D; Grader-Beck, T; van Puijenbroek, A A; Boussiotis, V A. Helper T cell anergy: from biochemistry to cancer pathophysiology and therapeutics. J Mol Med, 2001, 78(12), 673–83.
32. Tsang, J Y; Camara, N O; Eren, E; Schneider, H; Rudd, C; Lombardi, G; Lechler, R. Altered proximal T cell receptor (TCR) signalling in human CD4+CD25+ regulatory T cells. J Leukoc Biol, 2006 Jul, 80(1), 145–51.

33. Hickman, S P; Yang, J; Thomas, R M; Wells, A D; Turka, L A. Defective activation of protein kinase C and Ras-ERK pathways limits IL-2 production and proliferation by CD4+CD25+ regulatory T cells. J Immunol, 2006 Aug 15, 177(4), 2186–94.
34. Koonpaew, S; Shen, S; Flowers, L; Zhang, W. LAT-mediated signalling in CD4+CD25+ regulatory T cell development. J Exp Med, 2006 Jan 23, 203(1), 119–29.
35. Crellin, N K; Garcia, R V; Levings, M K. Altered activation of AKT is required for the suppressive function of human CD4+CD25+ T regulatory cells. Blood, 2007 Mar 1, 109(5), 2014–22.
36. Li, L; Godfrey, W R; Porter, S B; Ge, Y; June, C H; Blazar, B R; Boussiotis, V A. CD4+CD25+ regulatory T-cell lines from human cord blood have functional and molecular properties of T-cell anergy. Blood, 2005 Nov 1, 106(9), 3068–73.
37. Su, L; Creusot, R J; Gallo, E M; Chan, S M; Utz, P J; Fathman, C G; Ermann, J. Murine CD4+CD25+ regulatory T cells fail to undergo chromatin remodeling across the proximal promoter region of the IL-2 gene. J Immunol, 2004 Oct 15, 173(8), 4994–5001.
38. Gavin, M A; Clarke, S R; Negrou, E; Gallegos, A; Rudensky, A. Homeostasis and anergy of CD4(+)CD25(+) suppressor T cells in vivo. Nat Immunol, 2002 Jan, 3(1), 33–41.
39. Macian, F; Garcia-Cozar, F; Im, S H; Horton, H F; Byrne, M C; Rao, A. Transcriptional mechanisms underlying lymphocyte tolerance. Cell, 2002 Jun 14, 109(6), 719–31.
40. Harris, J E; Bishop, K D; Phillips, N E; Mordes, J P; Greiner, D L; Rossini, A A; Czech, M P. Early growth response gene-2, a zinc-finger transcription factor, is required for full induction of clonal anergy in CD4+ T cells. J Immunol, 2004 Dec 15, 173(12), 7331–8.
41. Anderson, P O; Manzo, B A; Sundstedt, A; Minaee, S; Symonds, A; Khalid, S; Rodriguez-Cabezas, M E; Nicolson, K; Li, S; Wraith, D C; Wang, P. Persistent antigenic stimulation alters the transcription program in T cells, resulting in antigen-specific tolerance. Eur J Immunol, 2006 Jun, 36(6), 1374–85.
42. Wu, Y; Borde, M; Heissmeyer, V; Feuerer, M; Lapan, A D; Stroud, J C; Bates, D L; Guo, L; Han, A; Ziegler, S F; Mathis, D; Benoist, C; Chen, L; Rao, A. FOXP3 controls regulatory T cell function through cooperation with NFAT. Cell, 2006 Jul 28, 126(2), 375–87.
43. Bettelli, E; Dastrange, M; Oukka, M. Foxp3 interacts with nuclear factor of activated T cells and NF-kappa B to repress cytokine gene expression and effector functions of T helper cells. Proc Natl Acad Sci USA, 2005 Apr 5, 102(14), 5138–43.
44. Borlado, L R; Redondo, C; Alvarez, B; Jimenez, C; Criado, L M; Flores, J; Marcos, M A; Martinez, A C; Balomenos, D; Carrera, A C. Increased phosphoinositide 3-kinase activity induces a lymphoproliferative disorder and contributes to tumor generation in vivo. Faseb J, 2000 May, 14(7), 895–903.
45. Parsons, M J; Jones, R G; Tsao, M S; Odermatt, B; Ohashi, P S; Woodgett, J R. Expression of active protein kinase B in T cells perturbs both T and B cell homeostasis and promotes inflammation. J Immunol, 2001 Jul 1, 167(1), 42–8.
46. Taub, D D; Murphy, W J; Asai, O; Fenton, R G; Peltz, G; Key, M L; Turcovski-Corrales, S; Longo, D L. Induction of alloantigen-specific T cell tolerance through the treatment of human T lymphocytes with wortmannin. J Immunol, 1997 Mar 15, 158(6), 2745–55.
47. Powell, J D; Lerner, C G; Schwartz, R H. Inhibition of cell cycle progression by rapamycin induces T cell clonal anergy even in the presence of costimulation. J Immunol, 1999 Mar 1, 162(5), 2775–84.
48. Battaglia, M; Stabilni, A; Roncarolo, M G. Rapamycin selectively expands CD4+CD25+FOXP3+ regulatory T cells. Blood, 2005 Jun 15, 105(12), 4743–8.
49. Valmori, D; Tosello, V; Souleimanian, N E; Godefroy, E; Scotto, L; Wang, Y; Ayyoub, M. Rapamycin-mediated enrichment of T cells with regulatory activity in stimulated CD4+ T cell cultures is not due to the selective expansion of naturally occurring regulatory T cells but to the induction of regulatory functions in conventional CD4+ T cells. J Immunol, 2006 Jul 15, 177(2), 944–9.
50. Battaglia, M; Stabilini, A; Draghici, E; Gregori, S; Mocchetti, C; Bonifacio, E; Roncarolo, M G. Rapamycin and interleukin-10 treatment induces T regulatory type 1 cells that mediate antigen-specific transplantation tolerance. Diabetes, 2006 Jan, 55(1), 40–9.

51. Bensinger, S J; Walsh, P T; Zhang, J; Carroll, M; Parsons, R; Rathmell, J C; Thompson, C B; Burchill, M A; Farrar, M A; Turka, L A. Distinct IL-2 receptor signalling pattern in CD4+CD25+ regulatory T cells. J Immunol, 2004 May 1, 172(9), 5287–96.
52. Meiffren, G; Flacher, M; Azocar, O; Rabourdin-Combe, C; Faure, M. Cutting edge: abortive proliferation of CD46-induced Tr1-like cells due to a defective Akt/Survivin signalling pathway. J Immunol, 2006 Oct 15, 177(8), 4957–61.
53. Cantley, L C; Neel, B G. New insights into tumor suppression: PTEN suppresses tumor formation by restraining the phosphoinositide 3-kinase/AKT pathway. Proc Natl Acad Sci USA, 1999 Apr 13, 96(8), 4240–5.
54. Shiota, C; Woo, J T; Lindner, J; Shelton, K D; Magnuson, M A. Multiallelic disruption of the rictor gene in mice reveals that mTOR complex 2 is essential for fetal growth and viability. Dev Cell, 2006 Oct, 11(4), 583–9.
55. Jacinto, E; Facchinetti, V; Liu, D; Soto, N; Wei, S; Jung, S Y; Huang, Q; Qin, J; Su, B. SIN1/MIP1 maintains rictor-mTOR complex integrity and regulates Akt phosphorylation and substrate specificity. Cell, 2006 Oct 6, 127(1), 125–37.
56. Kohn, A D; Barthel, A; Kovacina, K S; Boge, A; Wallach, B; Summers, S A; Birnbaum, M J; Scott, P H; Lawrence, J C, Jr.; Roth, R A. Construction and characterization of a conditionally active version of the serine/threonine kinase Akt. J Biol Chem, 1998 May 8, 273(19), 11937–43.
57. Patton, D T; Garden, O A; Pearce, W P; Clough, L E; Monk, C R; Leung, E; Rowan, W C; Sancho, S; Walker, L S; Vanhaesebroeck, B; Okkenhaug, K. Cutting edge: the phosphoinositide 3-Kinase p110{delta} is critical for the function of CD4+CD25+Foxp3+ regulatory T cells. J Immunol, 2006 Nov 15, 177(10), 6598–602.
58. Walsh, P T; Buckler, J L; Zhang, J; Gelman, A E; Dalton, N M; Taylor, D K; Bensinger, S J; Hancock, W W; Turka, L A. PTEN inhibits IL-2 receptor-mediated expansion of CD4+ CD25+ Tregs. J Clin Invest, 2006 Sep, 116(9), 2521–31.
59. Suzuki, A; Yamaguchi, M T; Ohteki, T; Sasaki, T; Kaisho, T; Kimura, Y; Yoshida, R; Wakeham, A; Higuchi, T; Fukumoto, M; Tsubata, T; Ohashi, P S; Koyasu, S; Penninger, J M; Nakano, T; Mak, T W. T cell-specific loss of Pten leads to defects in central and peripheral tolerance. Immunity, 2001 May, 14(5), 523–34.
60. Kashiwada, M; Cattoretti, G; McKeag, L; Rouse, T; Showalter, B M; Al-Alem, U; Niki, M; Pandolfi, P P; Field, E H; Rothman, P B. Downstream of tyrosine kinases-1 and Src homology 2-containing inositol 5'-phosphatase are required for regulation of CD4+CD25+ T cell development. J Immunol, 2006 Apr 1, 176(7), 3958–65.
61. Van Slyke, P; Coll, M L; Master, Z; Kim, H; Filmus, J; Dumont, D J. Dok-R mediates attenuation of epidermal growth factor-dependent mitogen-activated protein kinase and Akt activation through processive recruitment of c-Src and Csk. Mol Cell Biol, 2005 May, 25(9), 3831–41.
62. Yasuda, T; Shirakata, M; Iwama, A; Ishii, A; Ebihara, Y; Osawa, M; Honda, K; Shinohara, H; Sudo, K; Tsuji, K; Nakauchi, H; Iwakura, Y; Hirai, H; Oda, H; Yamamoto, T; Yamanashi, Y. Role of Dok-1 and Dok-2 in myeloid homeostasis and suppression of leukemia. J Exp Med, 2004 Dec 20, 200(12), 1681–7.
63. Niki, M; Di Cristofano, A; Zhao, M; Honda, H; Hirai, H; Van Aelst, L; Cordon-Cardo, C; Pandolfi, P P. Role of Dok-1 and Dok-2 in leukemia suppression. J Exp Med, 2004 Dec 20, 200(12), 1689–95.
64. Dong, S; Corre, B; Foulon, E; Dufour, E; Veillette, A; Acuto, O; Michel, F. T cell receptor for antigen induces linker for activation of T cell-dependent activation of a negative signalling complex involving Dok-2, SHIP-1, and Grb-2. J Exp Med, 2006 Oct 30, 203(11), 2509–18.
65. Zhao, M; Janas, J A; Niki, M; Pandolfi, P P; Van Aelst, L. Dok-1 independently attenuates Ras/mitogen-activated protein kinase and Src/c-myc pathways to inhibit platelet-derived growth factor-induced mitogenesis. Mol Cell Biol, 2006 Apr, 26(7), 2479–89.

66. Okada, M; Nada, S; Yamanashi, Y; Yamamoto, T; Nakagawa, H. CSK: a protein-tyrosine kinase involved in regulation of src family kinases. J Biol Chem, 1991 Dec 25, 266(36), 24249–52.
67. Jones, R G; Parsons, M; Bonnard, M; Chan, V S; Yeh, W C; Woodgett, J R; Ohashi, P S. Protein kinase B regulates T lymphocyte survival, nuclear factor kappaB activation, and Bcl-X(L) levels in vivo. J Exp Med, 2000 May 15, 191(10), 1721–34.
68. Narayan, P; Holt, B; Tosti, R; Kane, L P. CARMA1 is required for Akt-mediated NF-kappaB activation in T cells. Mol Cell Biol, 2006 Mar, 26(6), 2327–36.
69. Martin, E; O'Sullivan, B; Low, P; Thomas, R. Antigen-specific suppression of a primed immune response by dendritic cells mediated by regulatory T cells secreting interleukin-10. Immunity, 2003 Jan, 18(1), 155–67.
70. Heissmeyer, V; Macian, F; Im, S H; Varma, R; Feske, S; Venuprasad, K; Gu, H; Liu, Y C; Dustin, M L; Rao, A. Calcineurin imposes T cell unresponsiveness through targeted proteolysis of signalling proteins. Nat Immunol, 2004 Mar, 5(3), 255–65.
71. Michel, F; Mangino, G; Attal-Bonnefoy, G; Tuosto, L; Alcover, A; Roumier, A; Olive, D; Acuto, O. CD28 utilizes Vav-1 to enhance TCR-proximal signalling and NF-AT activation. J Immunol, 2000 Oct 1, 165(7), 3820–9.
72. Raab, M; Cai, Y C; Bunnell, S C; Heyeck, S D; Berg, L J; Rudd, C E. p56Lck and p59Fyn regulate CD28 binding to phosphatidylinositol 3-kinase, growth factor receptor-bound protein GRB-2, and T cell-specific protein-tyrosine kinase ITK: implications for T-cell costimulation. Proc Natl Acad Sci USA, 1995 Sep 12, 92(19), 8891–5.
73. King, P D; Sadra, A; Teng, J M; Xiao-Rong, L; Han, A; Selvakumar, A; August, A; Dupont, B. Analysis of CD28 cytoplasmic tail tyrosine residues as regulators and substrates for the protein tyrosine kinases, EMT and LCK. J Immunol, 1997 Jan 15, 158(2), 580–90.
74. Viola, A; Schroeder, S; Sakakibara, Y; Lanzavecchia, A. T lymphocyte costimulation mediated by reorganization of membrane microdomains. Science, 1999 Jan 29, 283(5402), 680–2.
75. Tavano, R; Gri, G; Molon, B; Marinari, B; Rudd, C E; Tuosto, L; Viola, A. CD28 and lipid rafts coordinate recruitment of Lck to the immunological synapse of human T lymphocytes. J Immunol, 2004 Nov 1, 173(9), 5392–7.
76. Alegre, M L; Frauwirth, K A; Thompson, C B. T-cell regulation by CD28 and CTLA-4. Nat Rev Immunol, 2001 Dec, 1(3), 220–8.
77. Takahashi, T; Tagami, T; Yamazaki, S; Uede, T; Shimizu, J; Sakaguchi, N; Mak, T W; Sakaguchi, S. Immunologic self-tolerance maintained by CD25(+)CD4(+) regulatory T cells constitutively expressing cytotoxic T lymphocyte-associated antigen 4. J Exp Med, 2000 Jul 17, 192(2), 303–10.
78. Salomon, B; Lenschow, D J; Rhee, L; Ashourian, N; Singh, B; Sharpe, A; Bluestone, J A. B7/CD28 costimulation is essential for the homeostasis of the CD4+CD25+ immunoregulatory T cells that control autoimmune diabetes. Immunity, 2000 Apr, 12(4), 431–40.
79. Tang, Q; Henriksen, K J; Boden, E K; Tooley, A J; Ye, J; Subudhi, S K; Zheng, X X; Strom, T B; Bluestone, J A. Cutting edge: CD28 controls peripheral homeostasis of CD4+CD25+ regulatory T cells. J Immunol, 2003 Oct 1, 171(7), 3348–52.
80. Tai, X; Cowan, M; Feigenbaum, L; Singer, A. CD28 costimulation of developing thymocytes induces Foxp3 expression and regulatory T cell differentiation independently of interleukin 2. Nat Immunol, 2005 Feb, 6(2), 152–62.
81. Shimizu, J; Yamazaki, S; Takahashi, T; Ishida, Y; Sakaguchi, S. Stimulation of CD25(+)CD4(+) regulatory T cells through GITR breaks immunological self-tolerance. Nat Immunol, 2002 Feb, 3(2), 135–42.
82. Annunziato, F; Cosmi, L; Liotta, F; Lazzeri, E; Manetti, R; Vanini, V; Romagnani, P; Maggi, E; Romagnani, S. Phenotype, localization, and mechanism of suppression of CD4(+)CD25(+) human thymocytes. J Exp Med, 2002 Aug 5, 196(3), 379–87.
83. Linsley, P S; Bradshaw, J; Greene, J; Peach, R; Bennett, K L; Mittler, R S. Intracellular trafficking of CTLA-4 and focal localization towards sites of TCR engagement. Immunity, 1996 Jun, 4(6), 535–43.

84. Pentcheva-Hoang, T; Egen, J G; Wojnoonski, K; Allison, J P. B7-1 and B7-2 selectively recruit CTLA-4 and CD28 to the immunological synapse. Immunity, 2004 Sep, 21(3), 401–13.
85. Tivol, E A; Borriello, F; Schweitzer, A N; Lynch, W P; Bluestone, J A; Sharpe, A H. Loss of CTLA-4 leads to massive lymphoproliferation and fatal multiorgan tissue destruction, revealing a critical negative regulatory role of CTLA-4. Immunity, 1995 Nov, 3(5), 541–7.
86. Sansom, D M; Walker, L S. The role of CD28 and cytotoxic T-lymphocyte antigen-4 (CTLA-4) in regulatory T-cell biology. Immunol Rev, 2006 Aug, 212, 131–48.
87. Kataoka, H; Takahashi, S; Takase, K; Yamasaki, S; Yokosuka, T; Koike, T; Saito, T. CD25(+)CD4(+) regulatory T cells exert in vitro suppressive activity independent of CTLA-4. Int Immunol, 2005 Apr, 17(4), 421–7.
88. Tang, Q; Boden, E K; Henriksen, K J; Bour-Jordan, H; Bi, M; Bluestone, J A. Distinct roles of CTLA-4 and TGF-beta in CD4+CD25+ regulatory T cell function. Eur J Immunol, 2004 Nov, 34(11), 2996–3005.
89. Chikuma, S; Imboden, J B; Bluestone, J A. Negative regulation of T cell receptor-lipid raft interaction by cytotoxic T lymphocyte-associated antigen 4. J Exp Med, 2003 Jan 6, 197(1), 129–35.
90. Parry, R V; Chemnitz, J M; Frauwirth, K A; Lanfranco, A R; Braunstein, I; Kobayashi, S V; Linsley, P S; Thompson, C B; Riley, J L. CTLA-4 and PD-1 receptors inhibit T cell activation by distinct mechanisms. Mol Cell Biol, 2005 Nov, 25(21), 9543–53.
91. Hoyne, G F; Dallman, M J; Lamb, J R. Linked suppression in peripheral T cell tolerance to the house dust mite derived allergen Der p 1. Int Arch Allergy Immunol, 1999, 118(2–4), 122–4.
92. Taams, L S; Wauben, M H. Anergic T cells as active regulators of the immune response. Hum Immunol, 2000, 61(7), 633–9.
93. Hoyne, G F; Le Roux, I; Corsin-Jimenez, M; Tan, K; Dunne, J; Forsyth, L M; Dallman, M J; Owen, M J; Ish-Horowicz, D; Lamb, J R. Serrate1-induced notch signalling regulates the decision between immunity and tolerance made by peripheral CD4(+) T cells. Int Immunol, 2000, 12(2), 177–85.
94. Thornton, A M; Donovan, E E; Piccirillo, C A; Shevach, E M. Cutting edge: IL-2 is critically required for the in vitro activation of CD4+CD25+ T cell suppressor function. J Immunol, 2004 Jun 1, 172(11), 6519–23.
95. Snow, J W; Abraham, N; Ma, M C; Herndier, B G; Pastuszak, A W; Goldsmith, M A. Loss of tolerance and autoimmunity affecting multiple organs in STAT5A/5B-deficient mice. J Immunol, 2003 Nov 15, 171(10), 5042–50.
96. Burchill, M A; Goetz, C A; Prlic, M; O'Neil, J J; Harmon, I R; Bensinger, S J; Turka, L A; Brennan, P; Jameson, S C; Farrar, M A. Distinct effects of STAT5 activation on CD4+ and CD8+ T cell homeostasis: development of CD4+CD25+ regulatory T cells versus CD8+ memory T cells. J Immunol, 2003 Dec 1, 171(11), 5853–64.
97. Yu, A; Malek, T R. Selective availability of IL-2 is a major determinant controlling the production of CD4+CD25+Foxp3+ T regulatory cells. J Immunol, 2006 Oct 15, 177(8), 5115–21.

Part II
Regulatory T Cells in Disease and Clinical Application

Chapter 9
$CD4^+Foxp3^+$ Regulatory T Cells in Immune Tolerance

Ciriaco A. Piccirillo

Abstract Peripheral regulatory T cell networks suppress immune responses in various inflammatory contexts and ultimately assure peripheral tolerance. Naturally-occurring $CD4^+$ Treg cells (nTreg) represent a major lymphocyte population maintaining dominant self-tolerance and controlling a variety of pathological immune responses. These nTreg cells specifically express Foxp3, a transcription factor that plays a critical role in their development and function. Functional abrogation of Foxp3-expressing nTreg cells *in vivo,* or genetic defects that affect their development or function, unequivocally predisposes animals and humans to the onset of multi-organ autoimmune and inflammatory diseases. $CD4^+$ nTreg cells differentiate in the thymus as a functionally distinct subset of T cells, although their post-thymic differentiation can occur from conventional T cells. These cells bear a broad T cell receptor repertoire endowing these cells with the capacity to recognize a wide spectrum of self and non-self Ag specificities. These nTreg cells are dependent on IL-2 for their fitness in the periphery. In addition, deficiency or functional alteration of other surface molecules expressed on immune cells, may affect the development/function of nTreg cells, and consequently favor the onset of autoimmunity. Recent studies have shed light in our understanding of the cellular and molecular basis of $CD4^+$ nTreg cell–mediated active maintenance of self-tolerance immune regulation, and will facilitate both our understanding of the pathogenetic mechanism of autoimmune disease and the development of novel methods of autoimmune disease prevention and treatment via enhancing and re-establishing Treg-mediated dominant control over autoreactive T cells. In this chapter, we discuss the contribution of $CD4^+$ nTreg cells in the induction of immunologic self-tolerance in animal models and humans.

C.A. Piccirillo
Departments of Microbiology and Immunology, McGill Centre for the Study of Host Resistance, McGill University Health Center, Montreal, QC, H3A 2B4, Canada
e-mail: ciro.piccirillo@mcgill.ca

Introduction

Dominant Regulation of Peripheral T Cell Tolerance

Immune tolerance relies on a homeostatic and regulated balance between maintaining peripheral tolerance to self antigens while generating protective immunity against invading pathogens [1]. In an attempt to reach a balance between these two different immunological outcomes, a peripheral network of induced (i) and naturally occurring (n) $CD4^+$ regulatory T cells (Treg) cells exists to maintain this balance [2]. Thus, Treg cells can simultaneously suppress autoreactive T cells that escape thymic negative selection, maintain normal intestinal immunity towards commensal bacteria, and dampen the anti-pathogen effector mechanisms from inducing immune pathology [3].

Initial studies of the 1970's indicated that suppressor T cells with regulatory function could be induced following antigen stimulation and could subsequently downregulate antigen-specific, T cell responses [4,5]. However, these seminal studies led to inconsistent results, hindering the attempts to characterize the lineage, phenotype and molecular pathway needed for the suppressor mechanism. Today, the existence of T cell-mediated suppression in the regulation of immune responses to self and non-self antigens is undisputed. $CD4^+$ T cells endowed with the capacity to control the activation and function of other T cells exist in two general categories: induced and naturally-occurring (Fig. 9.1). Induced $CD4^+CD25^+$ Treg (iTreg) cells whose suppressive activity and CD25 expression develops as a consequence of *in vivo* or *ex vivo* activation of nTreg-depleted $CD4^+$ T cells, without intrinsic regulatory potential, under unique, differentiation signals in the periphery [6–8]. In the last 2 decades, an array of iTreg cells capable of assuring an efficient control of peripheral T cell responses at multiple levels has been described, and their definition has largely been based on their potential to produce certain signature cytokines after antigen challenge. Some of these iTreg cells include counter-regulatory IFN-γ producing-Th1 cells and IL-4 producing Th2 cells, IL-10-producing Tr1 cells, and TGF-β-secreting Th3 cells induced in the gut following oral ingestion of antigen [6–8]. In contrast, nTreg cells develop in the normal and naïve T cell repertoire, are intrinsically regulatory in nature, and may cooperate with iTreg cells to modulate immune responses [9–15]. In contrast to iTreg cells, they are therefore already specialized for suppressive function before antigen encounter [16]. This makes them distinct from iTreg cells that differentiate from naïve T cells following antigen exposure in the periphery, although some Treg cells that are phenotypically and functionally similar to thymus-produced $CD4^+CD25^+$ nTreg cells may be able to differentiate in the periphery from naïve T cells under certain conditions [17]. A great deal of confusion currently exists in the literature regarding the relative roles of iTreg vesus nTreg cells since in many cases it is not possible to trace the origins of regulatory activity due to the lack of unique lineage markers. In this chapter, we discuss recent findings on the role of $CD4^+$ nTreg cells in mediating dominant self-tolerance, and their contribution to the control of immunopathology and immune responses to non-self antigens. We will focus our discussion on Foxp3 and IL-2

9 CD4+Foxp3+ Regulatory T Cells in Immune Tolerance

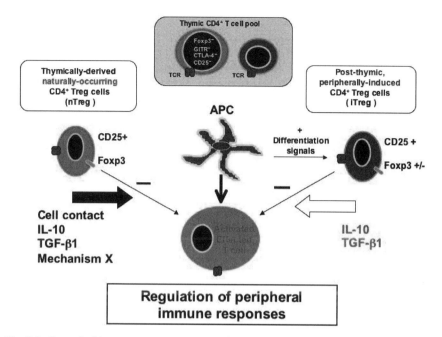

Fig. 9.1 Control of immune responses by CD4+ regulatory T cells Naturally-occurring (*red*) and induced (*blue*) CD4+ regulatory T cell subsets downregulate the function of activated effector T cells (*green*) in several types of peripheral immune responses. While CD4+CD25+Foxp3+ nTreg cells differentiate in the thymus and are found in the normal, naïve CD4+ T cell repertoire, multiple iTreg cell subsets, possibly expressing CD25 and Foxp3, originate from the activation and differentiation of conventional CD4+ T cells in the periphery under unique stimulatory conditions. The relative contribution of each subset in the overall regulation of immune responses is unclear but both conceivably can synergize to achieve this outcome. (*See also Color Insert*)

since both are essential for the development, function, and/or fitness of nTreg cells, and because mutations or genetic polymorphisms in the coding sequences of these molecules are indeed causative of and predisposing to autoimmune disease in both rodents and humans. In addition, use of Foxp3 expression enables to discriminate nTreg cells from other T cells and allows for the determination of the extent and spectrum of autoimmunity that could potentially evolve abrogating of nTreg ell function. We will also discuss potential genetic and immunological factors, which may influence their development, function and/or peripheral homeostasis, create an imbalance between nTreg cells and autoreactive T cells, and ultimately provoke the spontaneous onset of autoimmune disease.

Maintenance of Immune Self-Tolerance by CD4+ nTreg cells

Naturally-Occurring CD4+ Regulatory T Cells

The mechanisms underlying immune non-responsiveness to self antigens, and how a breakdown in these mechanisms can provoke autoimmune disease is unclear. In addition to the mechanisms of clonal deletion and anergy that physically eliminate or functionally inactivate potentially hazardous autoreactive lymphocytes, there is accumulating evidence that Treg actively suppress the activation and expansion of autoreactive T cells, and onset of autoimmune disease [1,18,19]. In 1995, the laboratory of Shimon Sakaguchi observed that a unique subset of $CD4^+$ T cells expressing the IL-2 receptor (R) alpha (α) chain, termed CD25, in normal animals display potent immunoregulatory functions *in vitro* and *in vivo* [10]. nTreg cells develop as a functionally mature and distinct T cell subpopulation during the normal thymopoesis, survive in the periphery poised for normal surveillance of self-Ags, and prevent potential autoimmune responses by an as of yet undefined mechanism [9,20,21]. nTreg cells represent 1–10% of total $CD4^+$ T cells in thymus, peripheral blood and lymphoid tissues, and, at least *in vitro*, are a hypoproliferative (anergic) and produce little no T cell derived cytokines. Currently, CD25 continues to be the most useful cell surface marker for nTreg cells in the normal T cell repertoire, although several other markers may allow more accurate identification and/or sorting of specific subsets of nTreg. In particular, the Foxp3 transcription factor is likely a more specific marker than CD25, and it is generally viewed that both should be used in conjunction [22].

Experiments involving depletion or functional abrogation of nTreg cells from the periphery of normal rodents leads to the spontaneous development of various organ-specific autoimmune diseases, such as gastritis, thyroiditis, orchitis, oophoritis, type 1 diabetes (T1D), and inflammatory bowel disease (IBD) [23,24]. Prophylactic infusion of normal $CD4^+CD25^+$ T cells prevents these disorders [23,24]. Furthermore, depletion of $CD4^+CD25^+$ T cells provokes effective tumor immunity to tumor antigens in otherwise non-responsive hosts, enhances immune responses to invading or commensal microbes, provokes allergic responses to innocuous environmental antigens, and breaks feto-maternal tolerance during pregnancy (reviewed in [25]). Thus, these studies conclusively show that nTreg cells are essential for the regulation of organ-specific autoimmunity, suppress a variety of inflammatory responses to a wide spectrum of non-self antigens, as well as dampen any possible immunopathological consequences of over-zealous immune responses. It can be easily envisaged nTreg cells can be manipulated to establish immunologic tolerance to non-self antigens, as in organ transplantation, and to treat various autoimmune and chronic inflammatory conditions.

The antigens driving suppression and immunity in many of the above-described models are ill-defined. The TCR repertoire of nTreg cells is as polyclonal and diverse as naïve T cells, although the former bears TCRs of higher affinity for the MHC class II/self-peptide ligands positively selecting them in the thymus thus endowing them to be preferentially autoreactive in the periphery [26,27]. These nTreg cells are anergic upon TCR triggering, do not produce pro-inflammatory cytokines upon antigenic stimulation and thus are seemingly not pathogenic despite the higher reactivity for self antigens. Once activated, they potently suppress the activation, proliferation, and effector functions of $CD4^+$ and $CD8^+$ T cells, NK cells, NKT cells, B cells, and dendritic cells, although by an as of yet undefined mechanism [11,28–33]. These unique immunological characteristics of $CD4^+$

nTreg cells endow them with ability to control pathogenic autoreactive T cells efficiently if self-antigens should be aberrantly or excessively presented in the immune system.

Foxp3: Biological Marker and Master Switch of nTreg Cells Development and Function

Several studies clearly show that Foxp3, a forkhead winged helix family transcriptional factor, represents an essential molecular switch for the genetic programming of nTreg cell development and function [34–39]. In mice, the majority of Foxp3$^+$ cells are CD4$^+$CD25$^+$, and a genetic deficiency in Foxp3 results in an autoimmune pathology secondary to a loss of CD4$^+$CD25$^+$ nTreg cells [34,40,41]. Interestingly, Foxp3-overexpressing mice have an increased development of nT$_{reg}$ cells, and these cells can potently suppress the development of autoimmune disease [36]. *In vitro* activation of naïve CD4$^+$CD25$^-$ cells, even under Th1 or Th2 differentiation conditions fails to induce Foxp3 expression in mice [42]. Interestingly, overexpression of Foxp3 in non-regulatory murine CD4$^+$CD25$^-$ or CD8$^+$ T cells induces potent suppressive functions *in vitro*, and suppresses the development of the lymphoproliferative syndrome in CTLA-4 deficient mice [36]. Similarly, Foxp3-transduced T cells can inhibit T1D in animal models [43]. Peripheral and thymic CD4$^+$CD25$^+$ nTreg cells account for the vast majority of Foxp3 expression, and natural mutations of the *foxp3* gene result in a loss in Foxp3 protein and nTreg cells, hyperactivation of T cells reactive with a variety of self and non-self antigens, and the development of various spontaneous and early onset organ-specific autoimmune pathologies due to hyper-activation of CD4$^+$ T cells including T1D in *Scurfy* mice, a natural, X-linked recessive mutant with lethality in hemizygous males within a month after birth, exhibiting hyperactivation of CD4$^+$ T cells, and overproduction of proinflammatory cytokines [34,36]. A similar phenotype is seen in mice depleted of nT$_{reg}$ cells or with *foxp3* gene deficiency, and mutations of the human Foxp3 gene were subsequently found to be the cause of immune dysregulation, polyendocrinopathy, enteropathy, X-linked syndrome (IPEX), which is a rare X-linked immunodeficiency syndrome that develops organ-specific autoimmune disease (including T1D), IBD, allergic dermatitis, food allergy, hematological disorders, hyperimmunoglobulinemia E, and serious infections [34,36,40,44–49]. Transfer of CD4$^+$CD25$^+$ nTreg cells, but not CD4$^+$CD25$^-$ T cells, into Foxp3$^{-/-}$ or *Scurfy* mice prevents the fatal *Scurfy* syndrome suggesting that Foxp3$^+$ nTreg cells are critical for preventing autoimmunity in mice [34,36,42]. Bone marrow chimera experiments have shown that Foxp3 functions in CD4$^+$ nTreg cells in a cell-autonomous fashion, although Foxp3 is also reportedly expressed in some thymic stromal cells in mice in turn affecting thymic selection events [50,51]. Recent analysis of Foxp3-reporter mice and intracellular staining of the Foxp3 protein revealed a correlation between the ontogeny of CD25-expressing nTreg cells and Foxp3-expressing T cells ([40], unpublished data). This is consistent with the previous finding that CD4$^+$CD25$^+$ nTreg cells ontogenically become detectable in the periphery of normal mice a few days after birth and that day 3 thymectomy after birth in select strains of mice abrogates the thymic production of nTreg cells thus permitting potentially autoreactive T cells that have seeded the periphery prior to thymectomy to become aberrantly primed, and cause autoimmune disease [23].

Although Foxp3 is highly expressed in both mouse and human nTreg cells, and is likely key for their development, its molecular mechanism of action and role in their suppressive function remain unknown. A number of over-expression, knock-out and molecular studies show that Foxp3 can directly influence several T cell functional parameters, including their capacity to express cell-surface markers, proliferate, and secrete produce cytokines [22,34,36,52,53]. Foxp3 binds DNA, at least in part, through its conserved forkhead domain (FHD) [54]. The presence of classical protein-protein interaction domains, such as leucine zipper and zinc-finger regions, suggest that Foxp3 may act as homo- or hetero-dimer to repress transcription of genes including IL-2 and other NFAT-regulated cytokines [55,56]. Recent data suggest there may also be a direct interation between Foxp3, NFAT and/or NF-KB, strongly suggesting that Foxp3 negatively regulate the effects of these critical transcription factors [57,58].

Like mice, human nTreg cells also express high levels of FOXP3, and the devastating autoimmunity that results in patients with FOXP3 mutations demonstrates its fundamental importance in immune homeostasis [34,36,40,44–49]. In contrast to mice, however, humans express two different splice variants of FOXP3, with the smaller form lacking exon 2 [52]. Allen et al. has shown that both variants independently possess the abililty to render $CD4^+$ T cells anergic. It is unknown whether the two variants are similtaneously co-expressed at the single cell level, and/or if they may cooperate at the molecular level. There is currently conflicting views as to whether retroviral-mediated over-expression of Foxp3 is sufficient to induce potent suppressive activity [52]. Importantly, in contrast to murine cells, human $CD4^+CD25^-$ T cells upregulate expression of FOXP3 upon activation, and non-suppressive, $FOXP3^+$ T cell clones have been described. Thus, in humans, expression of FOXP3 may not be exclusively linked to suppression, and this molecule may have an important role outside nTreg cells. Foxp3 is currently the most specific, reliable molecular marker, irrespective of CD25 expression, for thymic or peripheral nTreg cells in rodents and humans [21]. Overall, these findings indicate that Foxp3 is a key control switch for the development or function of $CD4^+CD25^+$ nTreg cells.

Lessons from Scurfy Mice and IPEX Patients

Foxp3 deficiency in humans and rodents leads to severe multi-organ autoimmunity [21,54]. The incidence of autoimmune disease in IPEX patients is almost 100%, with the majority of subjects suffering from T1D within the first 6–12 month of life [44]. It is known that besides $CD4^+CD25^+$ T cells, the $CD4^+CD25^-$ T cells contains a low frequency of Foxp3-expressing cells presumably with regulatory activity [40]. A recent study by the Sakaguchi group addressed whether complete deficiency of $CD4^+Foxp3^+$ T cells, including $CD25^+$ and $CD25^-$ Treg cells, would merely increase the severity of the autoimmune diseases that can be induced by $CD25^+$ nTreg cell depletion or would evoke a wider spectrum of autoimmune diseases, including those usually unobserved in $CD25^+$ cell-depleted animals. Multiparametric flow cytometric studies indicate that about 10–15% of $CD4^+$ T cells are $Foxp3^+$, and 70–80% of $Foxp3^+$ $CD4^+$ T cells are $CD25^+$, in the periphery of normal naïve mice [59]. In the thymus, approximately 3–5% of $CD4^+CD8^-$ mature thymocytes are $Foxp3^+$, and approximately 70% of $Foxp3^+$ cells are $CD25^+$ [59]. Notably, Foxp3-expressing $CD4^+$ T cells in both the thymus and the periphery are all $GITR^{high}$. When $GITR^{high}$ cells were depleted from splenocytes of normal BALB/c mice and the remaining cells transferred to BALB/c nude mice, the recipients developed a wider spectrum and more severe organ specific autoimmune diseases than the transfer of $CD4^+CD25^-$ T cells [59]. Most mice receiving $GITR^{low}$ cells die within 8 weeks post transfer, while recipients of $CD4^+CD25^-$ T cells survive and develop nonlethal autoimmune diseases such as gastritis, thyroiditis, and oophoritis [59]. The immediate cause of death in the recipients of $GITR^{low}$ cells was heart failure due to the development of autoantibodies specific for cardiac myosin and ensuing severe autoimmune myocarditis [59].

These experimental observations collectively indicate several important conclusions about nTreg cell-mediated control of self-tolerance. First, $CD4^+Foxp3^+$ T cells, irrespective of CD25 surface expression, engage in the maintenance of immunological self-tolerance, and the suppressive activity of $CD4^+CD25^+$ T cells shown in various rodent models of autoimmunity is likely attributed to $Foxp3^+CD4^+CD25^+$ nTreg cells (20–30% of total Foxp3-expressing T cells) present within this population. Second, the degree of $Foxp3^+$ nTreg cell functional deficiency has a direct effect on the extent of autoimmunity. In addition to provoking autoimmunity that can be elicited by CD25 depletion, a more profound deficiency of $Foxp3^+$ nTreg cells can activate even weak/rare autoreactive T cells thus inducing a wider spectrum autoimmune/inflammatory diseases that are not induced by a less severe deficiency of nTreg cells. Third, the susceptibility to particular autoimmune diseases is seemingly genetically determined in each host strain of mice [24,25,60]. For instance, BALB/c and NOD mice develop overlapping but also different autoimmune diseases possibly due to an apparent similarity in their degree of nTreg deficiency. The longer or more severe the reduction of $CD4^+$ nTreg cells, the higher the incidence of a particular autoimmune disease, and the wider the spectrum of clinical autoimmune diseases, which manifest themselves in a genetically determined hierarchical pattern in different strains of mice. Furthermore, many autoimmune or chronic inflammatory diseases in mice that are induced by severe nTreg depletion resemble their human clinically and immunologically. For example, autoimmune myocarditis in mice has similar immunopathological features to human giant cell myocarditis, and dilated cardiomyopathy, both of which are frequently associated with autoimmune thyroiditis and T1D [25,59]. Therefore, a combination of a profound nTreg deficiency and a particular host genetic background is able to produce an autoimmune/inflammatory disease in mice that corresponds to its human equivalent.

Homeostatic Functions of IL-2 for nTreg Cells

The cytokine signals required for nTreg cell development and function are not completely defined. In the resting, naïve immune system, CD25 is a useful surface marker for discriminating effector and nTreg cells, although every T cell expresses CD25 upon TCR engagement. There is an increasing body of evidence showing that CD25 (high-affinity IL-2 receptor) is also an essential component of the IL-2R pathway. Based on this, many have reasoned that IL-2 is an important molecular switch for the development, peripheral survival, suppression function and fitness of nTreg cells [61]. Mice deficient for IL-2, IL-2Rα, IL-2Rβ or STAT5A/5B have a drastically reduced pool of nTreg cells and die prematurely from a severe lymphoproliferative and autoimmune syndrome (inflammatory bowel disease (IBD) and lymphocytic infiltration of multiple organs) [61,62]. Genetic deficiency of CD25 in humans also displays similar clinical and pathological manifestation [63]. In IL-2Rβ$^{-/-}$ mice with a IL-2Rβ transgene that is expressed predominantly in the thymus, CD4$^+$CD25$^+$ T cell development is restored and lymphoadenopathy and autoimmunity is prevented, indicating that an intact IL-2/IL-2R pathway is also required for thymic generation of nTreg [64]. The adoptive transfer of WT nTreg cells can only prevent autoimmunity in mice lacking a functional IL-2R, in contrast to IL-2$^{-/-}$ mice, and WT Treg cells transferred into neonatal IL-2Rβ$^{-/-}$ mice, undergo rapid and extensive IL-2 dependent proliferation in lymph nodes and spleen [65,66]. These studies imply that the lack of nTreg cells contributes to the autoimmune phenotype of IL-2$^{-/-}$ and IL-2R$^{-/-}$ mice, and that IL-2 is a critical growth and differentiation factor for Treg cells. Similarly, CD4$^+$ T cells from IL-2$^{-/-}$ mice can protect mice from spontaneous experimental autoimmune encephalomyelitis (EAE), however CD25$^{-/-}$ CD4$^+$ T cells cannot protect from EAE [67]. Consistently, systemic IL-2 neutralization induces autoimmune gastritis in BALB/c mice, provokes spontaneous autoimmune neuropathy and exacerbates diabetes in NOD mice [68]. In humans with cancer after chemotherapy-induced lymphopenia, peripheral expansion of Treg cells is augmented by recombinant human IL-2 therapy. Collectively, these studies show that IL-2 may be required in the production of nTreg, and that early IL-2 production in sites of inflammation may drive CD4$^+$CD25$^+$ Treg cell-mediated suppression of T cell responses, and possible alterations in this pathway may block Treg cell development and provoke autoimmunity.

A few lines of evidence, however, do not entirely support a mandatory role for IL-2 in the genesis of Foxp3$^+$ Treg cell function. Recently, Fontenot et al. examined whether IL-2 was required for the development of Foxp3$^+$ nTreg cells [69]. In this study, mice containing a Foxp3-GFP knock-in allele encoding a GFP-Foxp3 fusion protein were used, thus enabling the identification of Foxp3$^+$ Treg cells by GFP fluorescence [40]. From the analysis of Foxp3^{gfp} mice genetically deficient in IL-2 or CD25, IL-2 signaling does not appear to be required for the induction of Foxp3 expression in developing thymocyte, and both IL-2$^{-/-}$ and IL-2Rα$^{-/-}$ Treg cells are unexpectedly able to suppress T cell proliferation *in vitro*, despite a substantially reduced frequency of these cells *in vivo* [69]. Interestingly, this study showed that IL-2 signaling was nonetheless necessary for maintaining the expression of genes involved in controlling cell growth and metabolism, possibly indicating that IL-2 signaling may be required for sustaining the homeostasis and competitive fitness of nTreg cells *in vivo* [69]. Consistently, Bcl-2 deficiency does not affect CD4$^+$ CD25$^+$ Treg homeostasis and transgenic expression of Bcl-2 does not rescue CD4$^+$CD25$^+$ Treg number in IL-2$^{-/-}$ or STAT5$^{-/-}$ mice, suggesting that abrogation of IL-2 signaling in Treg cells does not merely disrupt an essential survival pathway [70]. The reasons for the detection of Treg activity in IL-2- and IL-2Rα-deficient mice are not thoroughly understood but could be explained by the inability to enrich for Treg cells in the midst of the large peripheral pool of pre-activated CD4$^+$ T cell pool. The few CD25$^+$ T cells that could be purified from these mice were likely effector T cells, and thus non-suppressive *in vitro*. In Fontenot et al., the apparent *in vitro* regulatory function in Foxp3$^+$ Treg cells from IL-2- and IL-2Rα-deficient mice could be explained by a preferential enrichment of Foxp3$^+$ CD4$^+$ T cells (based on GFP expression, and not CD25) from the total CD4$^+$ T cell population, and thus an exclusion of potentially pre-activated effector T cells from the final preparation [69]. Analysis of an agonist-induced population of TCR-transgenic Foxp3$^+$Treg cells in IL-2$^{-/-}$ or CD25$^{-/-}$ mice has shown that intrathymic nTreg development is not IL-2 dependent, and peripheral survival is IL-2 dependent, although in stark opposition to the findings of Malek et al. [64,71]. These observations are also consistent with the regulatory function seen in CD25$^+$ Treg cells transferred into lymphopenic hosts, and whose CD25 is down-regulated post-transfer, although sufficient IL-2 signaling may have been given prior to CD25 down-regulation in the lymphopenic environment [72]. Alternatively, other lymphopenia-associated signaling may have compensated or overcome the need for IL-2 in this system. Lastly, the peripheral detection of nTreg function in IL-2Rβ deficient mice after thymic reconstitution of IL-2Rβ, despite the lack of IL-2Rβ in peripheral Treg cells is difficult to interpret, but could be explained by increased IL-2 responsiveness in developing nTreg cells in the thymus as a result of forced expression of the IL-2Rβ chain, in turn possibly circumventing the requirement for this critical signal in the periphery [73]. This apparent discrepancy in the functional requirement of IL-2 in thymic development, Foxp3 expression, function acquisition, and peripheral fitness of nTreg cells may be explained by a number of reasons, including the MHC and non-MHC genetic background, co-stimulatory burden, and degree of lymphopenia in each of the mouse models used [73]. Variation in these factors may ultimately influence IL-2 availability *in vivo* and the IL-2 dependency of Treg cells possibly by influencing the antigenic repertoire, TCR avidity, and activation thresholds in developing T cells [73]. It also remains unknown whether compensatory signals, including cytokines, may actually drive nTreg development or function in the functional absence of IL-2, as shown with other γ_c signaling cytokines such as IL-4 which can compensate for IL-2 *in vitro* and *in vivo*. Interestingly, the finding that complete deficiency of Foxp3$^+$ nTreg cells in mice deficient of the common γ-chain, in contrast with IL-2R α-chain- or IL-2-deficiency, indicates that the binding of other cytokines like IL-4, IL-7, IL-9, and IL-15 to the receptors sharing the common γ-chain is also involved in the maintenance of nTreg cells [69]. Overall, these findings show that IL-2 is indispensable for the peripheral survival of nTreg cells and that IL-2 deficiency, for a limited duration in time, is sufficient to elicit T-cell-mediated autoimmune disease in otherwise normal animals.

Co-Stimulatory Signals Favoring nTreg Cell Development and Homeostasis

The role of B7/CD28 interactions in thymic development and peripheral homeostasis of the nTreg cells has recently been highlighted. In addition to proper co-stimulation to developing immature thymic precursors, CD28 engagement might be required to sustain a stable peripheral pool of nTreg cells by promoting their survival and self-renewing potential possibly through the expression anti-apoptotic molecules, or through cytokines that function as growth, survival or suppressor activity maintenance factors [74]. CD28 seems to play a key role in the generation of $CD4^+CD25^+$ nTreg cells in the thymus and presumably in their survival in the periphery because $CD28^{-/-}$ mice develop a substantially reduced number of $CD4^+CD25^+$ nTreg cells in the thymus and periphery [75,76]. Interestingly, the abrogation of the B7/CD28 co-stimulatory pathway results in quantitative and qualitative defects in $CD4^+CD25^+$ nTreg cells in lymphoid tissues, and a consequential induction of autoimmunity, as illustrated by $CD28^{-/-}$ and $B7.1$-$B7.2^{-/-}$ NOD mice, which develop an exacerbated form of diabetes compared to their NOD control littermates [75]. Similarly, $CD4^+CD25^+CD45RB^{low}$ T cells are reduced in $CD40^{-/-}$ mice, and transfer of $CD40^{-/-}$ T cells into nude mice induces autoimmune disease. However, blocking of CD40/CD40L interaction does not affect nTreg suppressive activity *in vivo* [77]. IL-2 administration to $CD40^{-/-}$ mice normalizes nTreg number by promoting both their survival and homeostatic proliferation suggesting that the CD40/CD40L pathway is also needed for the genesis of nTreg cells [78]. Efficient generation of nTreg cells in the thymus requires CD28 costimulation since expression of Foxp3 and the induction of the nTreg cell differentiation program are specifically induced in DP thymocytes signaled *in vitro* by simultaneous co-engagement of TCR and CD28 surface molecules [79]. However CD28 signaling is essential for IL-2 production, so it has not been cleared whether reduction of nTreg in CD28-deficient may be IL-2 independent. Generation of nTreg cells and production of IL-2 require an identical Lck-binding motif in the CD28 cytosolic domain. Tai et al. shows that the function of CD28 in nTreg cell generation and in IL-2 production were independent of each other [79]. While critical for nTreg cell development. B7-CD28 interactions do not seem to be required for the *in vitro* activation of suppressor function in peripheral Treg cells. Thus, as B7/CD28 engagement appears to be required for thymic development and peripheral survival, the activation of nTreg cell function is CD28-independent.

Expression and Cellular Sources of IL-2 In Vivo

As $CD4^+CD25^+$ nTreg cells suppress T cell responses by down-regulating IL-2 synthesis, and paradoxically require IL-2 for their function and fitness, the dynamics of IL-2 production and suppression remains enigmatic *in vivo*. What cells produce IL-2 to stimulate nTreg cells is unclear. However, IL-2 is likely to originate from activated T cells. Setoguchi et al. showed that the active expression of IL-2 mRNA and protein was a predominant feature of CD4$^+$CD25^{low} T cells isolated from normal naïve mice [68]. These seemingly activated CD4$^+$CD25^{low} T cells were also found in DO11.10 expressing a transgenic TCR specific for an ovalbumin peptide but not in those DO11.10/RAG$^{-/-}$ mice suggesting that these cells may be kept in a physiological active state by self-antigens or commensal bacteria [68]. In addition to CD4$^+$CD25^{low} T cells, other lymphocytes, including CD8$^+$, NK, and NKT cells, showed low-level transcription of the IL-2 gene in normal naïve mice, possibly contributing to the maintenance and activation of nTreg cells [68]. DC are also thought to be possible candidates since matured DCs promote nTreg cell proliferation, although whether this effect is dependent on IL-2 is controversial [80,81]. No differences are shown in nTreg proliferation activity between WT and IL-2$^{-/-}$ matured DCs [80,81]. In contrast, CD40$^{-/-}$ DCs are impaired in nTreg proliferation, but additional of exogenous IL-2 is able to restore the normal number of nTreg cells in CD40$^{-/-}$ mice suggesting that DC-derived IL-2 plays a critical role in nTreg cell activation [77]. Watanabe et al. has also shown that thymic stromal lymphopoietin (TSLP) expressed by human Hassall's corpuscles activates DCs maturation, and these TSLP-conditioned DCs induces the proliferation and differentiation of CD4$^+$CD8$^-$CD25$^-$ thymic T cells into CD4$^+$CD25$^+$Foxp3$^+$ nTreg cells [82].

The Role of IL-2 in Self-Tolerance and Autoimmunity

The above findings have the following implications for the roles of the IL-2/IL-2R system in nTreg-mediated self-tolerance and its abnormality in autoimmune disease. First, IL-2 is not only a cytokine promoting adaptive immunity but also a mediator of negative feedback control by which activated nonregulatory T cells contribute to the maintenance and activation of CD4$^+$CD25$^+$ nTreg cells, which in turn limit the expansion of the former. This may occur locally where both nTreg cells and effector T cells are recruited to an antigen-presenting cell (APC) presenting a particular self or non-selfantigen, thus contributing to the accumulation and expansion of nTreg cells at the site of inflammation, as observed in the organ/tissue affected in organ-specific autoimmune disease. Second, these results suggest that genetic mutations or variations that modify the half-life, synthetic rate or function of IL-2 can be a predisposing factor to organ-specific autoimmunity by potentially abrogating the cellular frequency, function or peripheral homeostasis of nTreg cells. This provides a possible biological explanation for the IL-2 gene polymorphism at the *Idd3* locus that is linked with the susceptibility to T1D and EAE in NOD mice (see below) [83–85]. This polymorphism might impact nTreg cell function by modulating IL-2 expression or via altered glycosylation of the IL-2 molecule in NOD mice [85]. Similar to the variation at the IL-2 gene locus in NOD mice, there is significant linkage of an allelic variation at the CD25 gene locus with human T1D [63]. This could be attributed to the altered function of the CD25 molecule in the maintenance of nTreg cells, as revealed by the severe autoimmunity in CD25-deficient individuals. Genetic polymorphisms of the genes encoding other IL-2R chains, or signal transduction molecules like STAT5, might also contribute to determining the susceptibility to T1D and other autoimmune diseases [86]. Third, reduction of IL-2 for therapeutic purposes may elicit autoimmunity under certain circumstances. For example, cyclosporine A, an immunosuppressant that blocks IL-2 formation in T cells, causes organ-specific autoimmune diseases similar to those produced by neonatal IL-2 neutralization when the drug is administered to normal mice during the first week after birth, but not later [87]. Finally, IL-2 can be used therapeutically for expanding and activating nTreg cells such that *ex vivo* stimulation with specific antigen in the presence of high dose of IL-2 can expand antigen-specific Treg cells for the eventual cellular therapy of autoimmune disease [88,89]. Thus, IL-2 can be good targets for controlling *in vivo* immune responses via the modulation of the number and function of nTreg cells.

A Model for IL-2-mediated T Cell Regulation: Inflammation Drives Regulation

The attribution of IL-2 as an essential molecular determinant for nTreg suppressive activity has 2 important implications [73]. It is known that TCR triggering alone does not induce full suppressive activity in nTreg cells. This likely represents a fail-safe system since most nTreg probably recognize autoantigens with high affinity, would constantly be activated by antigen *in vivo*, and would likely result in permanent systemic suppression. Thus, productive nTreg activation of function may depend on another co-activating signal, such as IL-2, provided by another cell type since nTreg cells do not produce it themselves. Under physiological conditions, IL-2 is largely produced by activated $CD4^+$ T cells and to a smaller extent $CD8^+$ T cells [62]. Some studies have also reported DC as being a possible IL-2 source [77]. Thus, the activation status of effector T cells may modulate the induction of T cell suppression, such that the level of secreted IL-2 in inflammatory microenvironments may have a functional impact on nTreg cells. In normal non-inflamed mice, low-level IL-2 production resulting from low-affinity TCR stimulation within the heterogeneous T cell repertoire, may determine nTreg cell peripheral homeostasis without necessarily inducing overt nTreg cell effector functions. Under such conditions, nTreg suppressive activity may not be required since the lack of co-stimulation prevents efficient effector T cell activation. On the other hand, inflammation may result in higher IL-2 levels within inflammatory settings, subsequently permitting efficient activation of responder T cells, and simultaneously allowing for the induction of nTreg suppressor activity. This model would suggest that the temporal lag between effector T cell activation and induction of $Foxp3^+$ nTreg cell suppressive activity avoids hindering T cell priming but would permit control of activated effector T cells within inflamed tissues. It remains to be seen whether selective tissues alter the magnitude or kinetic of IL-2 production. However, the activated nTreg state is restricted only to the short time-window of active immunity such that nTreg cells are induced to suppress by IL-2 only transiently and when the immune response is terminated, and external IL-2 levels declines, the activity of nTreg cells wanes thus preventing systemic immunosuppression. It remains to be determined whether IL-2 drives various nTreg effector functions, including secretion of suppressive cytokines like IL-10.

Contribution of CTLA-4 and TGF-β1 Downregulatory Signals in the Development and Function of nTreg Cell

CTLA-4 and TGF-β1 have often been ascribed to nTreg cell function, however, their respective contribution in the development and/or function of these cells is unclear.

CTLA-4

CTLA-4 is expressed upon T cell activation, and transduces a negative signal to activated T cells, thereby attenuating T cell responses [90]. The functional role of CTLA-4 in $CD4^+CD25^+$ T cell-mediated suppression remains ill-defined, and several lines of evidence indicate that CTLA-4 could potentially play a critical role in nTreg cell–mediated suppression [91]. A lethal lymphoproliferative autoimmune syndrome that spontaneously develops in $CTLA-4^{-/-}$ mice is not T cell autonomous but can be inhibited by WT T cells, indicating that CTLA-4-deficiency may lead to impaired regulation [92]. Takehashi et al. reported that, in contrast to naïve T cells, intracellular CTLA-4 was present in resting $CD4^+CD25^+$ cells and antibodies to CTLA-4 could reverse $CD4^+CD25^+$ mediated suppression, concluding that CTLA-4 signaling was required for the activation of $CD4^+CD25^+$ cells [93]. Anti-CTLA-4 mAb treatment in normal mice over a short period induced autoimmune diseases similar to those produced by depletion of $CD4^+CD25^+$ nTreg cells, without reducing the absolute number of $CD4^+CD25^+$ T cells [93]. Similarly, Read et al. showed that administration of anti-CTLA-4 mAb abolished the protective activity of $CD4^+CD25^+$ nTreg in the murine IBD model [91,94]. *In vitro* blockade of CTLA-4 by Fab fragments of anti-CTLA-4 mAb neutralized the $CD4^+CD25^+$ nTreg cell-mediated suppression of $CTLA-4^{-/-}$ $CD4^+CD25$-T cells upon anti-CD3 stimulation [93]. Although CTLA-4 engagement may deliver an activating signal in this system, it is more likely that CTLA-4 engagement on $CD4^+CD25^+$ cells leads to an inhibition of signals required for induction of suppressor function, a feature consistent with the widely accepted down-regulatory role of CTLA-4. Recently, another study showed that when $CD4^+CD25^+$ cells were pre-activated with either WT or $B7.1^{-/-}/B7.2^{-/-}$ APC or in the presence of CTLA-4-Ig, $CD4^+CD25^+$ T cells retained potent suppressor activity under each stimulatory condition [95]. $CD4^+CD25^+$ cells were fully capable of inhibiting the remaining response despite the lack of CD28 or CTLA-4 co-stimulation, although co-stimulation through CD28-like molecules is conceivable [95]. Interestingly, one study has shown that nTreg cells initiated tryptophan catabolism in DC cells in a CTLA-4 dependent fashion [96]. Furthermore, CD4$^+$CD25$^+$ cells from CTLA-4$^{-/-}$ mice also exhibit a significant suppressive activity *in vitro* [97]. The effects of *in vivo* CTLA-4 blockade to exacerbate autoimmunity, enhance rejection of transplanted organs, or provoke tumor immunity can be attributed to the downregulatory role of CTLA-4. The effects could also be attributed to the blockade of CTLA-4 on CD4$^+$CD25$^+$ nTreg cells and consequent abrogation of nTreg cell–mediated suppression. It remains to be examined at molecular levels how the balance between signals through CTLA-4 and CD28, both of which interact with B7.1 and B7.2 APC, contribute to the tuning of the regulatory activity of nTreg cells. It was also suggested that CTLA-4 on nTreg cells may ligate CD80, and to a lesser extent CD86, on activated T cells and directly transduce a negative signal to responder T cells [98].

There are several lines of evidence indicating that CTLA-4 may not be needed for the suppressive function of nTreg cells. CD4$^+$CD25$^+$Foxp3$^+$ T cells are found in CTLA-4$^{-/-}$ mice but these mice still develop fatal autoimmunity [97]. Transgenic expression of Foxp3 in CTLA-4$^{-/-}$ mice ameliorates their debilitating systemic inflammation and prevents early death [36]. This difference between the *in vivo* and *in vitro* activity of CTLA-4$^{-/-}$ nTreg cells could be attributed to a different mechanism of suppression. For example, CTLA-4$^{-/-}$ nTreg cells reportedly secrete a large amount of TGF-β when stimulated *in vitro* [97].

In humans, it known that a particular polymorphic variant of the CTLA-4 gene is linked with a variety of autoimmune diseases, including T1D, Graves' disease, Hashimoto's thyroiditis, and Addison's disease (reviewed in [99]). A recent genetic study showed a significant correlation between the presence of this variant and mRNA levels of soluble CTLA-4 and also the number of CD4$^+$CD25$^+$ nTreg cells in healthy control subjects. Consistently, compared with T1D-susceptible NOD mice, nTreg cells of T1D-resistant NOD mice congenic at the *Idd5.1* locus (from C57BL/6 mice), which contains the CTLA-4 gene, were reported to express a higher level of ligand-independent CTLA-4. The reasons for this are unclear but this CTLA-4 variant, which lacks the CD80/CD86-binding domain, can result in attenuated TCR signals in the thymus or periphery and favor nTreg function (reviewed in [84,100,101]).

The Pleiotropic Functions of TGF-β1

The potential effector role of TGF-β1 in CD4$^+$CD25$^+$ nTreg cell-mediated suppression *in vitro* is controversial [73]. TGF-β1 is a logical candidate because it plays a critical role in the downregulation of immune responses, as illustrated by the development of a severe autoimmune-like syndrome in TGF-β1$^{-/-}$ mice characterized by the spontaneous and progressive multi-organ infiltration of mononuclear cells and pathogenic autoantibodies [102]. This is further corroborated by the observation that genetic disruption of TGF-β1 signaling in T cells by overexpression of a dominant-negative TGF-β type II receptor [DNRIITg], conditional deletion of the TGF-β type II receptor in hematopoetic progenitors, or inactivation of the gene encoding the receptor-activated Smad3, alters the sensitivity of T cells to the inhibitory effects of TGF-β, and lead to aberrant T cell responses [21,103–107]. One study reported that activated CD4$^+$CD25$^+$ T cells express an inactive form of TGF-β1 complexed to its latency associated peptide [108], that is retained as a membrane-bound complex by an undefined surface receptor, a complex hypothesized to account for the enhanced suppressive capacity of activated CD4$^+$CD25$^+$ T cells *in vitro* [109]. Nonetheless, most murine and human *in vitro* studies conclude that neither secreted nor membrane-bound forms of active or latent TGF-β1 are responsible for contact-dependent suppression mediated by resting or activated CD4$^+$CD25$^+$ T cells (reviewed in [20,73,110]). DNRIITg and Smad3$^{-/-}$ T cells, which are resistant to the growth inhibitory effects of exogenous TGF-β1, remain susceptible to suppression by CD4$^+$CD25$^+$ T cells [111]. More importantly, CD4$^+$CD25$^+$ T cells isolated from neonatal TGF-β1$^{-/-}$ mice are anergic to TCR signals and display comparable suppressive activity to WT nTreg cells *in vitro*, and in contrast to their CD4$^+$CD25$^-$ counterparts, are positive for CTLA-4, GITR, CD103 and Foxp3, a phenotype consistent with WT CD4$^+$CD25$^+$ nTreg cells [111–113].

A consensus view on the role of TGF-β1 in CD4$^+$CD25$^+$ nTreg cell function *in vivo* is currently lacking. Some reports have suggested that secretion of TGF-β1 by CD4$^+$CD25$^+$ nTreg cells is required to protect SCID mice from IBD induced by CD4$^+$CD45RB^{high} effector T cells as treatment of recipients of CD4$^+$CD45RB^{high} and CD4$^+$CD25$^+$ nTreg cells with neutralizing anti-TGF-β antibody reversed suppression [114]. Similarly, one recent study reported a requirement for TGF-β1 in CD4$^+$CD25$^+$ nTreg cell-mediated control of CD8$^+$ T cell anti-tumor activity [115]. Administration of anti-TGF-β antibody neutralized suppressive activity of CD4$^+$CD45RC^{low} T cells in rat T1D and thyroiditis produced by adult thymectomy and irradiations. However, the cellular source of the bioactive TGF-β1 was not determined in these experiments and remains largely unknown. The regulatory role for TGF-β1 in CD4$^+$CD25$^+$ T cell-mediated suppression is complicated by the fact that a variety of cell types can produce TGF-β1 and might include nTreg cells themselves, activated effector T cells or, possibly induced by nTreg cells in non-lymphoid target tissues that are inflamed or in the process of healing.

One study has concluded that CD4$^+$CD25$^+$LAP$^+$ T cells expressing a membrane-bound form of TGF-β1 was responsible for the control of IBD induced by CD4$^+$CD45RB^{high} effector T cells [116]. In contrast, Oida et al. reported that TGF-β1-dependent suppression was observed in the LAP$^+$CD4$^+$CD25$^-$ T cell subset [117]. In addition, some studies have suggested a correlation between nTreg effector function and the apparent preferential expression of a latent, membrane-bound form of TGF-β1 on CD4$^+$CD25$^+$ nTreg cells isolated from inflamed pancreatic lymphoid tissues [118]. However, these studies did not conclusively show functional evidence for a direct effect of nTreg cell-derived TGF-β1 on responder T cells. Furthermore, additional studies have suggested that cell surface TGF-β1, from autocrine or paracrine sources, may mediate its effects by actively signaling in CD4$^+$CD25$^+$ nTreg cells themselves, possibly by maintaining their survival, differentiation, expansion, or suppressive effector mechanism [119–121]. Furthermore, a recent study has shown that TGF-β1 signaling in nTreg cells may promote Foxp3 expression and subsequent nTreg function *in vitro* and *in vivo* [122]. We have recently demonstrated that CD4$^+$CD25$^+$ nTreg cells from either TGF-β1$^{+/+}$ or neonatal TGF-β1$^{-/-}$ mice can suppress the incidence and severity of IBD as well as colonic IFN-γ mRNA expression induced by WT CD4$^+$CD25$^-$ effector T cells [111–113]. Furthermore, TGF-β-resistant Smad3$^{-/-}$ CD4$^+$CD25$^+$ nTreg cells are equivalent to WT nTreg cells in their capacity to suppress disease induced by either WT and Smad3$^{-/-}$ CD4$^+$CD25$^-$ effector T cells [111–113]. Although CD4$^+$CD25$^+$ T cells from Smad3$^{-/-}$ mice appeared to function as efficiently as WT CD4$^+$CD25$^+$ T cells in our colitis model, it remains possible that TGF-β could play a stimulatory role in the induction of CD4$^+$CD25$^+$ suppressor activity, perhaps in a Smad3-independent fashion [111–113]. Thus, CD4$^+$CD25$^+$ nTreg cells are able to suppress intestinal inflammation by a mechanism not requiring nTreg cell-derived TGF-β1 or effector T cell/Treg cell Smad3-dependent responsiveness to TGF-β.

These findings strongly suggest that TGF-β plays a critical role in the suppression of disease, but that the TGF-β is derived from host non-T cells or non-lymphoid cells or even from the effector T cells themselves. Furthermore, these results support the view that tissue/context dependent factors may influence the mechanism of immune suppression by CD4$^+$CD25$^+$ nTreg cells *in vivo*. In autoimmune gastritis, CD4$^+$CD25$^+$ nTreg cells do not need suppressor cytokines, and possibly suppress disease by resorting to contact-dependent mechanism, a finding similar to what is observed *in vitro* [11,123]. However, in the milieu of a bacteria-driven inflammation in the intestine, the contact-dependent pathway may not be sufficient to mediate disease protection and must be supplemented by suppressor cytokines, like IL-10 by nTreg cells and by production of TGF-β by non-nTreg cells. It is also possible that the CD4$^+$CD25$^+$ nTreg cells may facilitate the induction of TGF-β production by host cells that subsequently promotes the induction and differentiation of Foxp3$^+$ nTreg cells from CD4$^+$CD25$^-$ T cell precursors. We have recently demonstrated that TGF-β1 can selectively promote the differentiation of IL-10 secreting, Foxp3$^+$CD4$^+$ Treg cells from CD4$^+$CD45RB^{low}CD25$^-$ T cell precursors (M.Pyzik and C.A. Piccirillo, unpublished). Thus, inflammatory cytokines, like TGF-β1, may enhance immunosuppression by sustaining regulatory networks via the post-thymic development of cytokine-secreting Foxp3$^+$iT$_{reg}$ cells, and maintaining nT$_{reg}$ cell peripheral homeostasis by bolstering Foxp3 expression.

Determinants Affecting the Balance Between Regulatory and Autoreactive T Cells and the Onset of Autoimmune Disease

The onset of autoimmune disease in a given host is determined by a complex array of environmental, genetic, and immune factors, which can synergize to variable degrees to impact the incidence and severity of disease. As CD4$^+$ nTreg cells play a central role in the induction of self-tolerance, an unresolved issue pertains to the robustness and persistence of nTreg cell activity during the onset of clinical autoimmunity. The development of spontaneous autoimmunity may be the resultant of 2 non-mutually exclusive immunological outcomes: (1) The onset of disease in mice and humans may result from the overriding of normal CD4$^+$ nTreg cell-mediated regulation by an uncontrollable activation of autoreactive T cells; (2) or the developmental/functional deficiencies in CD4$^+$ nTreg cells may promote the occurrence of a dysregulated immune system and consequently tip the balance towards the priming of autoreactive T cells, and disease onset [22].

Impaired Development or Function Within the nTreg Cell Compartment

The etiological basis of any given autoimmune disease is multi-factorial, and the nature and dose of the etiologic agent may vary between afflicted individuals and diseases. Genetic or environmental insults can conceivably provoke a developmental or functional deficiency in nTreg cells, alter the delicate balance between nTreg/autoreactive T cells, and ultimately provoke autoimmunity. In various rodent systems, it has been shown that physical, chemical, and biological agents or genetic abrogation can indeed cause autoimmune disease by either affecting thymic production or the size of the peripheral pool of nTreg cells [9,22,25]. Indeed, administration of cyclosporine A, T-cell-tropic viral infection, low-dose irradiation, or changes in TCR gene expression caused autoimmunity (e.g.; gastritis, thyroiditis or IBD) in mice, a clinical outcome also seen in similar to those induced by depletion of CD4$^+$CD25$^+$ nTreg or day 3 thymectomy in similar hosts [9,10,23,25,124–126]. These treatments are not nTreg-specific, but nTreg cells are likely more sensitive to these treatments because of their unique cellular properties: they possess a high radiosensitivity because of their active proliferation in the physiological state, are highly dependent on exogenous IL-2 for their survival, and are relatively scarce early in life, hence easier depletion of them in young mice. As CD4$^+$CD25$^+$ nTreg continuously cycle in response to self-antigen recognition in the periphery, it remains possible that they are inherently more sensitive than conventional T cells to certain environmental/genetic triggers throughout life, particularly during gestational life, which in turn may affect developing nTreg cells and thereby trigger autoimmunity.

An array of MHC and non-MHC genes in the host determines the onset, specificity, severity, and duration of autoimmune responses. Polymorphisms in the MHC II locus can obviously influence the repertoire of nTreg cells. Genetic alterations of other critical non-MHC immune genes such as Foxp3, Aire, CTLA-4, TGF-β1, CD40, IL-2, CD25 (IL-2Rα), and CD122 (IL-2Rβ) elicit fulminant autoimmune diseases [90,102,127–130]. Thus, it is reasonable that evolutionary conservation of various defects or polymorphisms of certain genes critical for nTreg cell development may also contribute to determining the overall genetic susceptibility to autoimmune disease. In addition, unless an abnormality in nTreg cells is present in a given host, predisposing genes, on their own, may be unable to induce autoimmune disease. However, the mere deficiency of nTreg cells on its own cannot determine which tissues will be targeted by autoreactive T cells, as illustrated by the day 3 thymectomy and CD25-depletion studies where the incidence, onset, and severity of many autoimmune diseases varies according to the age, sex and genetic background of the host [10,23,131–133]. As discussed in section Lessons from Scurfy Mice and IPEX Patients, the degree and duration of Foxp3$^+$ nTreg cell functional deficiency represents a critical factor determining the manifestation and extent of a given autoimmune disease. In addition to provoking autoimmunity that can be elicited by CD25 depletion, a more profound deficiency of Foxp3$^+$ nTreg cells can reduce the activation thresholds of seemingly weaker/rarer autoreactive T cells thus inducing a wider spectrum of autoimmune/inflammatory diseases, which would otherwise not be induced if the deficiency of nTreg cells was less drastic and severe. The susceptibility to particular autoimmune diseases is also genetically determined in each host as shown in BALB/c, C57BL/6 and NOD mice which develop overlapping and distinct autoimmune diseases with a similar degree of nTreg deficiency induced by CD25 depletion or day 3 thymectomy [10,23,131–133]. It is thought that the longer or more severe the reduction of CD4$^+$ nTreg cells, the higher the incidence and severity of a particular autoimmune disease, and the spectrum of autoimmune diseases generally occurs following a genetically-determined hierarchy. Unlike C57BL/6 or NOD mice, the incidence of disease in BALB/c.*scid* mice recipients of splenocytes depleted of CD4$^+$CD25$^+$ nTreg cells is AIG, oophoritis, thyroiditis, adrenalitis, and T1D [10,23,131–133]. Furthermore, many autoimmune diseases in mice that are induced by severe nTreg depletion resemble their human equivalents clinically and immunologically suggesting that a combination of a profound Treg deficiency and a particular host genetic background is able to produce an autoimmune/inflammatory disease in mice that corresponds to its human equivalent. It must be noted that a functional deficiency in nTreg cells might not be evident as a reduction in the cellular frequency in peripheral tissues, and might be resultant to selective gaps in antigen-specific TCR specificities or various genetic polymorphisms potentially modulating various aspects of nTreg cell function: activation, effector function, survival, or trafficking. Although a reduction in nTreg cellular frequency or function is a critical predisposing factor to autoimmunity, it may be insufficient at times, and the nature and magnitude of the inflammatory context that could accompany a nTreg deficiency may also be a critical parameter. This is corroborated by the fact that a mere deficiency of nTreg cells was insufficient to cause AIG in Balb/c mice unless this deficiency is accompanied by inflammatory signals triggered by immunization with autoantigen and adjuvant [133].

How do genetic and environmental factors synergize to reveal the pathogenesis of any given autoimmune disease? In contrast to Scurfy mice or IPEX patients where the degree of nTreg cell depletion is severe and complete, most genetically determined Treg anomalies, if at all, in common autoimmune diseases, like T1D in humans may be more subtle. If an environmental trigger affects nTreg cells in any way, it would likely be for a brief time period as in viral infections. However, this mild/transient imbalance between nTreg and autoreactive effector T cells may be sufficient to trigger the activity of the latter and thus cause autoimmune disease if the affected individuals bear other susceptible genes, including MHC and non-MHC genes, as shown in animal models. One etiologic agent affecting nTreg cells may lead to the occurrence of different autoimmune diseases in one host. Alternatively, different causal agents may produce the same autoimmune disease in genetically susceptible individuals through a common mechanism. It is possible that most autoimmune diseases have a common mechanism, and not necessarily a specific etiology for each autoimmune disease. One can speculate that a genetically induced imbalance between nTreg and autoreactive T cells could be a key causative event in these spontaneous models of autoimmunity. Thus, there appears to be several common elements among various animal models in which the primary cause of autoimmune disease is either genetic or environmental. One such element can be an imbalance between nTreg cells and autoreactive T cells, especially due to a deficiency or dysfunction of nTreg cells even for a limited period of time, produced by genetic anomalies or environmental insults. In addition, although nTreg cells may be quantitatively or functionally normal at disease onset, it does not exclude the possibility that a given nTreg anomaly may still have impacted disease pathogenesis in more downstream phases of the evolution of disease.

Can Abnormal Central Tolerance Increase the Peripheral Pool of Autoreactive T Cells, Over-Ride nTreg Cell Function, and Provoke Autoimmunity?

CD4$^+$ nTreg cells assure a means of assuring dominant suppression of potentially pathogenic, autoreactive T cells that escaped thymic negative selection. It is currently unknown if/how these central and peripheral mechanisms synchronize their efforts to maintain self-tolerance. Can abnormal production of autoreactive T cells with moderate to high affinity TCRs for self during thymopoesis be sufficient to provoke autoimmunity, despite a normal functional pool of circulating nTreg cells in the periphery? Recent studies in mice harboring point mutations in the LAT or ZAP-70 genes suggest that intrinsic defects of TCR signaling may impact negative selection of autoreactive T cells and production of nTreg cells in the thymus, and the ensuing autoimmune disease in these mice can be prevented by the reconstitution of normal nTreg cells [25,134]. It is likely that mutations or genetic variations of critical genes involved in T cell signaling molecules, in particular those operating in TCR proximal signals, may also alter the peripheral balance between nTreg and autoreactive T cells toward autoimmunity. In contrast, studies in *Aire*-deficient mice suggest that thymic *Aire*-deficient stromal cells allow for a greater thymic output of potentially pathogenic, autoreactive T cells may overcome normal Treg-mediated peripheral suppression, and consequently trigger organ-specific autoimmunity [127,135,136]. It was noted that other genes may preferentially impact nTreg cells as shown by studies with TRAF6, CD40, or NF-κB-inducing kinase deficient thymic stromal cells, which predominantly block nTreg cell development [78,137,138]. It also remains to be determined whether these defects are also operative in the periphery of these mice.

Can Autoreactive T Cells be Resistant to nTreg Cell Mediated Suppression?

In addition to molecular abnormalities affecting nTreg cells, deficiency of certain molecules may predominantly affect responder T cells and render the former resistant to suppression by nTreg cells. For example, mice deficient in Cbl-b, an E3 ubiquitin ligase and a key adapter protein in CD28 signalling in T cells, show CD28-independent T cell activation, hyperreactivity, and spontaneous autoimmunity [139]. Cbl-b-deficient $CD4^+CD25^-$ T cells proliferate, express increased amounts of IL-2 upon TCR stimulation, and are resistant to suppression by either Cbl-b-deficient or wildtype $CD4^+CD25^+$ T cells [139]. Mice deficient for the NFATc2 and NFATc3 transcription factors also exhibit massive lymphadenopathy, splenomegaly, and autoimmune-like responses [140]. Although they develop $CD4^+CD25^+$ nTreg cells their $CD4^+CD25^-$ T cells are unresponsive to Treg suppression in *in vitro* coculture with double-deficient or wildtype $CD4^+CD25^+$ T cells [140]. It is likely that spontaneously activated double-deficient $CD4^+CD25^+$ T cells secrete a large amount of IL-2, which might abrogate the suppression. Why this increased IL-2 production does not drive regulation in these autoimmune-prone mice is unclear. It was shown that IL-6 secreted by DCs, together with an uncharacterized mediator, is able to render effector T cells resistant to nTreg-mediated suppression [141]. Thus, in addition to Treg abnormalities, resistance of autoreactive T cells to Treg-mediated suppression may trigger the development of autoimmune disease through affecting the balance between the two cellular populations.

Role of nTreg Cells in Animal Models of Autoimmunity

The NOD Mouse Model of Spontaneous Diabetes

There are several animal models that spontaneously develop T-cell-mediated organ-specific autoimmune diseases immunopathologically similar to those produced by Treg depletion or dysfunction. Recent genetic studies provide some insights into how pathogenic autoreactive T cells are generated and activated to cause autoimmune disease in these models. In addition, there are animal models in which environmental insults are responsible for causing similar autoimmune diseases in otherwise normal animals by affecting nTreg cells.

The non-obese diabetic mouse represents a prototypic model of immune dysregulation since it spontaneously develops several autoimmune diseases including type 1 diabetes (T1D) [100,142]. The extended time lag between the initial immune cell infiltration of β-islets, termed insulitis (checkpoint 1), and the onset of overt T1D (checkpoint 2), suggests that regulatory mechanisms in the periphery control self-reactivity and disease progression in prediabetic NOD mice [18]. An array of different types of Treg cells, but particularly $Foxp3^+$ nTreg cells, have been found to represent a central control point in tolerance induction in NOD mice [18].

An unresolved question is whether the onset of spontaneous T1D in NOD mice results from a simple decline in nTreg cell function over time, consequently tipping the balance towards the activation, expansion and recruitment of diabetogenic T cells, and clinical T1D. If this is indeed the case, therapeutic potentiation of their numbers and/or function could shift disease progression towards tolerance and protection from T1D.

Is nTreg cell function normal in NOD mice? As NOD mice spontaneously develop autoimmunity, it begs this question. Studies have implicated nTreg cells in the control of T1D, and that reduced CD4$^+$ nTreg cell frequencies or function in NOD mice, represent a primary predisposing factor to spontaneous autoimmunity [143–147]. Transfer of CD25-depleted NOD splenocytes into NOD.*scid* hosts leads to a quicker onset of T1D than total splenocytes [75]. A disruption of B7/CD28 or CD40/CD40L pathways in NOD mice alters the thymic development and peripheral homeostasis of nTreg cells, and leads to an accelerated T1D onset, which can be prevented by infusion of wild-type nTreg cells [75,76]. Other studies have also reported a significant expansion of CD25-expressing CD4$^+$ T cells with apparent regulatory activity, sometimes correlating with the expression of a membrane-bound form of TGF-β1, in the inflamed pancreatic lymph nodes of adult, pre-diabetic, and insulitic NOD mice [118]. Transfer of such cells after long-term *in vitro* expansion has been shown to be protective, although it is unknown whether *in vitro* conditioning of T$_{reg}$ cells may have adversely affected their physiological role [89,148,149]. It is unclear from these studies whether these CD4$^+$CD25$^+$ T cells are iTreg cells from CD4$^+$CD25$^-$ progenitors during pancreatic inflammation, or whether they emerge from the thymus-derived, Foxp3$^+$ nTreg cell pool. A recent study by Chen et al. used NOD mice harbouring the scurfy mutation (FoxP3^{sf}) to examine the functional contribution of this defined subset of nTreg cells to tolerance induction [150]. NOD.FoxP3^{sf} displayed significantly increased incidence and earlier onset of T1D compared to normal NOD mice, implying that Foxp3$^+$ nTreg cells control T1D pathogenesis [150]. This study did not address whether the injection of Treg cells, which rescues from T1D, compensates for the primary deficit in nTreg cells believed to underlie T1D pathogenesis in these mice, or whether such injection was actually suppressing the global inflammation/pathology that likely arose as a secondary consequence of Foxp3 deficiency. Furthermore, this study did not exclude the possibility that NOD.FoxP3^{sf} mice abnormally present autoAgs, and in conjunction with increased co-stimulation in these mice, may reduce activation thresholds for effector T cells. Indeed, Chang et al. recently demonstrated that a T cell extrinsic defect may contribute to the development of the *Scurfy* disease and IPEX, since the Foxp3^{sf} mutation in thymic stromal cells leads to defective thymopoiesis [51]. Collectively, these studies imply a critical role for Foxp3$^+$ nTreg cells in the control of T1D.

It is currently unknown whether a quantitative or qualitative deficiency in CD4$^+$ nTreg cells leads to a failure to control the onset of T1D. Thymic and peripheral CD4$^+$CD25$^+$ nTreg cells from 10d old neonatal NOD mice are fully functional *invitro*, as they potently suppress the *in vitro* proliferation of activated T cells in a contact-dependent and cytokine independent fashion (M. Tritt and C.A. Piccirillo, unpublished observation). In most studies of T1D models, the CD25 surface marker is used to monitor nTreg cell frequencies in pre-diabetic, and adult NOD mice. CD25 is only a specific marker for resting $CD4^+CD25^+$ nTreg cells in neonatal lymphoid environments, and becomes unreliable during immune responses as conventional $CD4^+$ T cells upregulate CD25 upon activation [2,142]. The frequency of $Foxp3^+CD4^+$ nTreg cells, irrespective of CD25 expression, represent a stable pool within total $CD4^+$ T cells from thymus, LN or spleen in neonatal and adult NOD mice, and is comparable to T1D-resistant B6 mice, thus refuting the view that a quantitative nTreg deficiency underlies T1D onset in NOD mice (M. Tritt and C.A. Piccirillo, unpublished observation). $CD4^+CD25^-$ T cells from prediabetic NOD mice are diabetogenic when transferred alone in NOD.*scid* mice [75,132] (M. Tritt and C.A. Piccirillo, unpublished observation). However, when $CD4^+CD25^+$ T cells from neonatal NOD mice are co-transferred, T1D development is dramatically halted, even when T1D is mediated by $CD4^+$ T cells from diabetic mice [75,132] (M. Tritt and C.A. Piccirillo, unpublished observation). The cellular potency of nTreg cells is fully operative in neonatal mice, but declines with age, despite a stable cellular frequency of $Foxp3^+$ nTreg cells in primary and secondary lymphoid tissues (M. Tritt and C.A. Piccirillo, unpublished observation). In a BDC2.5 TCR Tg $CD4^+$ T cell transfer model of T1D in $NOD.TCR\alpha^{-/-}$ mice, $CD4^+CD25^+Foxp3^+$ nTreg cells from 2–4 week old mice were fully functional *in vivo*. Interestingly, $CD4^+CD25^+$ nTreg cells isolated from adult 6–8 week old BDC2.5 mice seemingly fail to control T1D, unless their frequency is increased in the periphery suggesting that their functional potency may decline with age (M. Tritt, and C.A. Piccirillo, unpublished data). These results do not exclude the possibility that a functional deficiency in nTreg cells in NOD mice may not be manifested as a sudden decline in the cellular frequency of these cells in peripheral tissues, but rather by subtle variations in effector and regulatory functions [151]. Thus, qualitative, and not quantitative, changes in $CD4^+Foxp3^+$ nTreg cells over time may be a critical determinant in T1D onset.

The protective role of nTreg cells *in vivo* has been well established in a variety of autoimmune diseases, and may involve induction of alternate migration, activation, differentiation, and/or clonal expansion of effector T cells in lymphoid and non-lymphoid tissues [22]. Whether nTreg cells prevent priming of effector T cells is unclear. While Tang et al. has shown that priming is suppressed in pancLN, other studies in BDC2.5 mice have shown that the initial activation of effector T cells in draining lymph nodes is unaffected in the presence of nTreg cells suggesting that Ag presentation, and proximal TCR signals are not inhibited by nTreg cells [148,150]. This is also in agreement with the observation that nTreg cells suppress disease transfer mediated by primed T cells from diabetic mice, which likely traffic directly to islets, and circumventing priming in the pancreatic lymph node (pancLN) [149] (M. Tritt, and C.A. Piccirillo, unpublished data). In addition, the frequency of proliferating diabetogenic $CD4^+$ T cells in the pLN, either in the presence of absence of nTreg cells remains unchanged, suggesting that Ag-induced cell division of autoreactive T cells is not directly affected by nTreg cells [150]. In contrast, the frequency of effector T cells is dramatically increased in the absence of nTreg cell function, suggesting that regulation may effect later stages of T cell clonal expansion, survival and/or homing (M. Tritt, and C.A. Piccirillo, unpublished data). These results do not exclude the possibility that nTreg cells alter effector T cells to become less pathogenic only when localized in islets. A similar observation was also made in studies involving the transfer of BDC2.5 T cells into thymectomized NOD.B7-2$^{-/-}$ recipients, in conjunction with *in vivo* depletion of CD25$^+$ T cells, which resulted in an increased accumulation of BDC2.5 T cells in the pLN compared to control mice [152]. Furthermore, there is also evidence suggesting that suppression of APC functions may also be, in part, responsible for nTreg cell-mediated protection. Serra et al. found that nTreg cells control the pathogenicity of islet-specific, CD8$^+$ effector T cells by inhibiting DC maturation in the pancLN [153]. Similarly, *in vitro* expanded CD25$^+$ Treg cells seemingly control diabetogenic responses by disrupting the interaction between BDC2.5 CD4$^+$ T effector cells and DCs in the pLN.

The location where nTreg cells mediate tolerance induction occurs *in vivo* is currently unknown. Studies show the majority of nTreg cells are actively suppressing in the pancreas, rather than the pLN where the initial priming of the autoreactive T cells response is presumably occurring. In T1D models, nTreg cells preferentially home to and expand within inflamed pLN and islets of T1D-protected mice, suggesting that these local tissues are the sites of control [148,150,153,154]. The preferential accumulation of nTreg cells in the inflamed pancreas also suggests that nTreg cells only become activated or mediate their suppressor function once inflammation has been initiated. A recent study from Chen et al. found that the gene expression profile of nTreg cells within the insulitic infiltrate differs from nTreg cells residing in the pLN, suggesting that stimuli in the target tissue initiate a unique transcriptional program, possibly thereby increasing their ability to regulate immune responses in these sites [148,150,153,154]. It remains to be determined, however, whether these distinct nTreg cell gene signatures occur as a result of their tissue localization, or are merely the consequence of their own suppression. It is also remains to be seen whether nTreg cells are functionally and phenotypically heterogeneous in nature, and whether each nTreg cell subset can potentially control autoreactive T cells via various mechanisms, and in tissue-specific fashions [132]. The reasons behind this context-dependent mode of T cell regulation are unknown, but target organs may confer unique regulatory pressures on infiltrating effector T cells, and may shape the type of regulation needed for disease resolution.

Immunogenetic Determinants of Autoimmunity

The etiology of T1D is complex and mediated by multiple genetic and environmental factors [84]. T1D susceptibility is inherited through multiple genes, with a strong predisposition from genes affecting immune responses to β islet cell antigens [84,100]. Family studies in humans and genomic mapping of congenic NOD strains, which harbor defined genetic intervals contributed from T1D-resistant mouse strains, have shown that at least 20 regions of the NOD genome collectively contribute to disease susceptibility [84]. These T1D-linked regions termed insulin-dependent diabetes (Idd) are essential for susceptibility although no single gene is both necessary and sufficient [84,100]. Although the presence of resistance alleles at a single *Idd* locus can decrease T1D incidence, particular combinations of *Idd* loci can provide nearly complete protection from disease [84,100]. The *Idd1* locus encodes the disease-prone class II MHC I-A^{g7} molecule, which remains the strongest genetic contributor to the disease. While expression of the MHC I-A^{g7} molecule makes NOD mice susceptible to autoimmunity, introduction of a different class II MHC haplotype onto the NOD background renders the strain resistant to T1D, For example, MHC-congenic NOD.2 H4 mice, which express I-A^k, spontaneously develop autoimmune thyroiditis [155]. I-A^{g7} thus appears to promote the thymic selection and/or peripheral activation of diabetogenic T cell clones in NOD mice [156]. Further study is required to understand how MHC skewed production/priming of diabetogenic T cells, in addition to possible genetic aberrations of nTreg cells, affects the peripheral balance between immunity and tolerance in NOD mice.

As discussed above, NOD mice show an age-dependent decline in the function of nTreg cells to suppress T1D although adult NOD mice harbor a normal frequency of CD4$^+$Foxp3$^+$ T cells with normal *in vitro* suppressive activity (unpublished observations). A number of immune molecules encoded by the prime candidate genes include CTLA-4 (*Idd5.1*), NRAMP1 (*Idd5.2*), and 4-1BB (*Idd9.3*), and each can affect nTreg cell function directly or indirectly [84,157–160]. Another major *Idd* loci for T1D susceptibility includes *Idd3* [161]. The *Idd3* locus, which maps to a 0.15 CM interval in the proximal region of chr.3, imparts potent resistance to T1D as NOD mice congenic for the B6 *Idd3* locus (NOD.*Idd3*^{B6}) and show a markedly reduced incidence and delayed T1D onset compared to control NOD.*Idd3*^{WT} mice [161]. Fine mapping studies show that the *Idd3* locus contains at least 3 genes: *Il21*, *Fgf-2*, and *Il2*, which represents the strongest candidate for protection in NOD.*Idd3*^{B6} mice [83,162,163]. The "susceptible" and "resistant" IL-2 variants possess no polymorphisms within regulatory elements of the IL-2 promoter, but do differ in their N-terminal sequence, correlating with their differential glycosylation states [85]. This suggests that IL-2 variants may be functionally distinct between these strains. Although it is unknown how these genetic polymorphisms modify the pathophysiology of T1D in NOD mice, expression or functional variations of IL-2 may, at least in part, may affect the development, function or survival of nTreg cells. The highly protective NOD susceptibility region *Idd3*, harboring the strong *Il2* candidate gene, confers T1D protection by supporting nTreg cell homeostasis and regulation of self-specific CD4$^+$ T cells *in vivo* (A. Albanese, E. Sgouroudis and C. A. Piccirillo, unpublished observations). *Idd3*^{B6} controls the amount of IL-2 mRNA and protein produced by diabetogenic CD4$^+$ T cells, and favors the suppressive function of CD4$^+$Foxp3$^+$ nTreg cells *in vitro*. In contrast to controls, CD4$^+$ T cells from BDC2.5.*Idd3*^{B6} mice expand less efficiently in pancreatic LN, are less pathogenic, and abrogation of nT$_{reg}$ cells unleashes BDC2.5.*Idd3*^{B6} T cell proliferation, IL-2/IFN-γ secretion, and disease potential *in vivo*. Although the systemic frequency of Foxp3$^+$ nTreg cells is not augmented, the protective *Idd3*^{B6} allele affords resistance to T1D by, in part, by heightening the function/homeostasis of nTreg cells locally in the pancreas. Interestingly, the protective Idd3 allele may also act outside of the T cell compartment since mature $Idd3^{B6}$ DC, in contrast to controls, can potentiate nT_{reg} cell function *in vitro* and *in vivo*. Thus, the protective *Idd3* allele helps create an local environment, which tips the balance in favor of nTreg cell function and consequential tolerance induction. Collectively, the activation status of pathogenic, autoreactive T cells may modulate the induction of T cell suppression via its production of IL-2, such that the level of secreted IL-2 in inflammatory microenvironments may have a functional impact on nTreg cells.

Many genetic studies of human T1D have revealed that several candidate IDDM susceptibility genes, such as those encoding class II MHC, insulin, CTLA-4, the protein tyrosine phosphatase PTPN22, and CD25 (reviewed in [84,164]). Some genes like CTLA-4 and PTPN22 in humans are also common susceptibility genes to other autoimmune diseases, including Hashimoto's thyroiditis, Graves' disease, Addison's disease, and rheumatoid arthritis [101,157,164]. It is possible that the PTPN22 mutation (variant R620 W) might affect the thymic generation and peripheral activation of autoreactive T cells as well as nTreg cells as observed with the ZAP-70 or LAT mutation in mice [134,165–168]. Disease-susceptible variants of class II MHC, CTLA-4, and CD25 might well affect the balance between nTreg cells and autoreactive T cells toward autoimmunity.

Dysfunctional $CD4^+$ nTreg Cells in Human Autoimmunity

Scurfy and $Foxp3^{-/-}$ mice fail to develop fully functional nTreg cells and suffer from a severe autoimmune syndrome, and as such provide direct evidence for the capacity of nTreg cells to prevent autoimmunity [34,36]. A similar syndrome, known as IPEX develops in humans with mutations in FOXP3, and represents the clearest example in humans that an abnormality in nTreg cell development or function can be a primary cause in the induction of autoimmune disease, IBD, and allergy [44,48]. Although IPEX is a rare disease, its clinical picture and underlying causal mechanism for nTreg cell developmental and/or functional deficiencies has several important implications for the role of nTreg cells in immune self-tolerance in humans and also for the pathogenic mechanism of human autoimmunity. Clinical symptoms of the disease vary widely, but IPEX patients frequently develop not only T1D and thyroiditis but also various other autoimmune diseases, including gastritis, hemolytic anemia, thrombocytopenia, and hepatitis, within a few years after birth. Unlike Scurfy mice, IPEX patients have a wide spectrum of mutations (>20 have been documented), and variable clinical outcomes [44,54]. This variability is due to the fact that whereas Scurfy mice fail to translate any protein, IPEX patients may either have a null or a point mutation that results in a fully translated, but functionally impaired, FOXP3 protein. The most severe disease typically arises in IPEX patients who have a null mutation or a point mutation in the DNA-binding forkhead domain. In contrast, children with mutations outside of the forkhead domain often present with mild or late-onset disease [44]. Remarkably, although patients with IPEX have autoimmunity that resembles Scurfy mice, $CD4^+CD25^+$ T cells with suppressive activity are detectable in the circulation of children with point mutations in FOXP3, albeit with less potent suppressor activity in comparison to cells from age-matched controls [169]. Interestingly, a patient with a mutation in the initiating codon of FOXP3 and consequential loss in FOXP3 expression, had severely reduced numbers of $CD4^+CD25^+$ cells that were non-functional *in vitro* [169]. These data indicate that the type of mutation, and its location in the FOXP3 gene, may dictate the function of nTreg cells and therefore be predictive of the clinical outcomes of these patients. In addition, evidence that siblings with the same FOXP3 mutation can have drastically different phenotypes, suggests that other genetic and environmental factors ultimately influence the manifestation of autoimmune disease. Interestingly, children with IPEX also have a major defect in effector T cell functions [169]. Thus, unlike the hyperactive phenotype of T cells in Scurfy mice, mutations in FOXP3 lead to a significant defect in the capacity of non-Treg cells to proliferate and produce cytokines in humans suggesting that FOXP3 may also operate also outside of the nTreg compartment. Overall, although analysis of the Scurfy mouse has provided invaluable insight into the link between nTreg cells and autoimmunity, further studies are required to unravel the exact mechanistic basis of autoimmune pathology believed to arise due to aberrant nTreg development and/or function in humans.

IPEX patients have provided us with a salient example that nTreg cells indeed represent a dominant mechanism of self-tolerance in humans. Furthermore, due to the random inactivation of the X-chromosome in individual nTreg cells, it is conceivable that some hemizygous females may have functional and dysfunctional $CD4^+$ nTreg cells, and remain disease-free, similar to Scurfy mice reconstituted with normal $CD4^+CD25^+$ T cells [25]. This observation further reinforces the notion that normal nTreg cells actively control autoreactive T cells in hemizygous females and in most normal individuals harboring pathogenic autoreactive T cells. A recent study has shown that T1D can rapidly develop even in IPEX patients bearing the T1D-protective HLA-DR2 haplotype, suggesting that an overwhelming nTreg cell deficiency may potentially unleash even rare diabetogenic T cells in subjects with a protective HLA haplotype [49,53]. These findings confirm that autoreactive T cells potentially pathogenic for various autoimmune diseases circulate in most individuals, and that nTreg cell-mediated suppression is a critical mechanism of peripheral self-tolerance. These findings also suggest that a monogenic mutation affecting nTreg cells can evoke several autoimmune diseases in rodents and humans, thus suggesting that any genetic defect or environmental agent could be a predisposing factor to a variety of autoimmune diseases, granted it affects the development, function or fitness of nTreg cells. Evidently, it is very likely that a developmental or functional deficiency in nTreg cells cannot, on its own, represent the sole cause of many autoimmune diseases, and other mechanisms probably contribute to the manifestation of clinical autoimmunity, such as aberrant thymic production of autoreactive cells, or excessive priming of such cells in the periphery. It remains to be seen to what extent a defective nTreg cell compartment contributes to the various human autoimmune disorders, and to what extent nTreg cells can represent a common therapeutic target for all of these diseases.

The triggers that initiate and exacerbate autoimmune disease have remained largely elusive. While it is difficult to establish a direct causative link between human diseases and nTreg cells, recent studies suggest that a decrease in number and/or function of these cells is associated with autoimmunity, and likely a critical determining factor in disease development. For example, a recent study of systemic lupus erythrematosus (SLE) patients demonstrated that patients have a reduced frequency of peripheral nTreg cells, although the functional activity of these cells was normal *in vitro* [146]. Lee et al. also found a correlation between a decrease in nTreg cell numbers and active SLE and increased autoantibody levels [170]. In contrast, Liu et al. failed to find such a connection [171]. Studies of patients with multiple sclerosis have revealed that patients have normal numbers of circulating nTreg cells, but a functional defect in their suppressive capacity was detected *in vitro* [147,172] A similar trend was also observed in a study of patients with myasthenia gravis, in which thymic nTreg cells showed a striking defect in suppressive capacity in comparison to healthy controls [173]. Defects in nTreg cell function were also detected in patients with psoriasis [174], diabetes [175], and autoimmune polyglandular syndromes [176]. In the case of rhuematoid arthritis, evidence suggests that the immunological defect involves changes to both T effector cells and nTreg cells within affected joints, and that inflammation and pathology results from an imbalance of the two populations [177–179].

While the evidence for decreased nTreg number or function as a factor in autoimmunity is increasing, the reasons for why and how these abnormalities arise remain elusive. It is also puzzling that the reported global defects in peripheral nTreg cell function do not lead to widespread autoimmunity, suggesting that a tissue-specific imbalance may also predispose individuals to a given disease. There is also the possibility that nTreg cell defects are indicative, but not causative of autoimmunity. However, some studies have shown that some clinically effective treatments actually reverse nTreg cell functional defects [25,179,180], supporting the concept that nTreg cell dysfunction is invovled in disease pathology, and that therapeutic manipulation of these cells will be an effective treatment for a variety of autoimmune diseases.

Regulation of Immune Responses to Microbes

An emerging notion in the immunology of infectious diseases is that persistent, chronic infections may indicate a compromise between the pathogen and host [181,182]. While the host attempts to maintain strong effector immunity against re-infection and limit pathology due to an uncontrolled inflammatory response, pathogens strive to subvert host immunity to actively promote immunosuppression, and pathogen persistence. In addition to regulating immune responses to "self", $CD4^+$ nTreg cells are also engaged in suppressing excessive immune responses to various microbial pathogens. There is increasing evidence suggesting that nTreg also control host immune responses to a variety of pathogenic microorganisms. Many studies show that persistent pathogens establish chronic infections in immunocompetent hosts by engaging nTreg cells which function to suppress host immunity, and control excessive effector immune responses [183,184]. Depletion of CD4$^+$CD25$^+$ nTreg cells can enhance protective immune responses against invading microbes including intracellular and extracellular bacteria, viruses, fungi, and intracellular parasites, leading to their eradication from hosts [183,185–187].

One the most characterized models of nTreg cell-mediated control of immune responses towards microbes is the IBD mouse model of bacterial-driven, T cell-mediated intestinal inflammation [188,189]. In this model, the transfer of normal CD4$^+$CD45Rb^{high} T cells into immunodeficient BALB/c.*scid* mice induces a form of IBD, which is characterized by the activation of bacterial-specific Th1 cells and subsequent histopathology in the gut. Interestingly, co-transfer of CD4$^+$CD45Rb^{high} T cells with CD4$^+$CD25$^+$ nTreg cells prevents primary or established disease. Interestingly, similar experiments performed in germ-free SCID mice fail to develop the disease suggesting that induction and suppression of intestinal inflammation is bacterial-driven [188,189]. Alterations in the nTreg cell balance, in favor of effector T cells, can provoke a loss of T cell tolerance to bacterial antigens and consequential induction of dysregulated, bacterial-driven, Th1 colitogenic responses in the gut. Thus, nTreg cells play a central role in this immunological balancing act as they establish and maintain tolerance to enteric bacteria and preserve normal intestinal homeostasis [3].

The establishment of chronicity and pathogen persistence at sites of infection is dependent on a tight equilibrium between effector and nTreg cells within these sites. For example, nTreg cells are essential for the development and maintenance of chronic cutaneous infection in *L. major* resistant C57BL/6 mice [183]. In this model, disease is initiated by an acute phase hallmarked by pathogen replication, and then followed by the establishment of a chronic phase, characterized by the stable maintenance of a low number of parasites at the site of primary challenge, the absence of overt pathology and resistance to re-infection [182,190]. A function for IL-10 in promoting parasite persistence and susceptibility to *L. major* infection in both susceptible and resistant strains is well documented. The absence or inhibition of nTreg function or IL-10 production will promote complete clearance of the parasite, while depletion of effector cells or pro-inflammatory cytokines like IFN-γ will promote disease reactivation and parasite elimination [183,191–193]. IL-10, which is in part produced by nTreg cells contributes directly to parasite persistence and concomitant immunity. The mechanism by which IL-10 allows parasite growth and survival are not fully defined, but it is likely due to its potent deactivating role of infected APCs that would become unresponsive to activation by IFN-γ. During the chronic phase of disease, IL-10-producing nTreg dominate over effector mechanisms resulting in IFN-γ suppression [183]. This immunosuppressive role for IL-10 is also applicable to other animal models of infection, in which the levels of a deactivating cytokine such as IL-10 are highly predictive of the outcome of the clinical course infection. Interestingly, the control of pathogen persistence in sites of infection occurred independently of nTreg-derived IL-10 in the initial phase of disease establishment and appeared to be IL-10 dependent in chronic phases of disease, confirming the notion that nTreg can adapt to their inflammatory microenvironments for efficient and timely disease control [183]. This apparent requirement for IL-10 in chronic stages of infection might reflect possible roles of IL-10 in nTreg activity, or in the induction of IL-10 from other sources. The depletion of nTreg may thus lead to complete pathogen clearance but may advertently prevent induction of long-term immunity because of insufficient maintenance of memory T cells owing to lack of microbe persistence. $CD4^+$ nTreg cells may be required for the maintenance of balanced symbiosis between the host and microbes.

Chemokine Directed Homing of nTreg Cells to Sites of Inflammation

A topic that has recently been explored is the relative capacity of nTreg versus effector cells to home to lymphoid organs and sites of inflammation. Lymphocyte homing and trafficking in inflamed lymphoid and non-lymphoid tissues is controlled by the expression of distinct sets of chemokine receptors, which provide directional cues for the migration and recruitment of T cells into sites of inflammation [194]. Currently, the chemokines involved in directing nTreg cells to traffic to sites of inflammation are not clearly understood. *In vitro* studies have shown that a large fraction of nTreg cells from human peripheral blood selectively express CCR4 and CCR8 and show a strong chemotactic response to CCR4 ligands [195,196]. Another study showed that nTreg cells in prediabetic NOD mice express high levels of CCR7, giving them the capacity to migrating towards the lymphoid-derived chemokines CCL19 and CCL21 [197]. Recently, Kleinewietfeld et al. found that CCR6 is expressed on a distinct subset of mouse and human nTreg cells and that these $CCR6^+$ nTreg cells are enriched in the peripheral blood, have a high turnover rate, rapidly produce IL-10 following Ag stimulation, and accumulate in the central nervous system after induction of EAE [198]. Bystry et al. demonstrated that murine nTreg cells express CCR5 and respond *in vitro* to CCR5-activating chemokines including MIP-1α, MIP-1β, or RANTES [199]. In addition, other studies have indicated that CCR5 may modulate nTreg cell activity in graft versus host disease and human tumors [200,201]. Overall, there is substantial evidence indicating that expression of a variety of chemokine receptors on nTreg cells may endow them with a competitive advantage over effector T cells, and allow them migrate more efficiently to inflammatory sites and prevent immune responses.

$CD4^+$ nTreg cells rapidly accumulate at sites of infection, where they suppress anti-pathogen $CD4^+$ T cell responses, favor the persistence of a small number of parasites within cutaneous lesions, and consequently control concomitant immunity as mice lacking nTreg cells achieve sterile cure and lose immunity to a secondary infection [183]. Therefore, a plausible explanation would be that both T cell populations expand, but that their pattern of trafficking dictated by distinct chemokine receptor expression profiles may differ, as previously proposed in various models. Recently, the CCR5 chemokine receptor regulates critical aspects of immunity in *L. major* infection. $CD4^+CD25^+$ nTreg cells preferentially express CCR5 and respond very efficiently to the CCR5 ligands MIP-1α, MIP-1β, or RANTES *in vitro*. [202] In a mouse model of *L. major* infection, CCR5 was shown to direct the homing of

IL-10 producing nTreg cells into sites of infection where they suppress effector T cell expansion and IFN-γ production, thus promoting the establishment of infection and ensuring the long-term survival of the parasite in the immune host [202]. $CCR5^{-/-}$ $CD4^+CD25^+$ nTreg cells, in contrast to WT nTreg cells, migrate less efficiently to infected dermal sites, and fail to suppress the magnitude of parasite-specific, IFN-γ-producing $CD4^+$ T cells, thus resulting in a dramatic reduction in parasite numbers and potent resistance to infection [202]. Consistently, mice deficient for CCR5 are resistant to *L. major* infection in contrast to their WT counterparts, and resistance is associated with a significant increased infiltration and sustained accumulation of IFN-γ producing $CD4^+$ effector T cells compared to the infected sites of control mice [202]. IFN-γ could contribute indirectly to the resolution of *L. major* infection in $CCR5^{-/-}$ mice by prolonging/sustaining the inflammatory response in these mice. Intrestingly, CCR5 deficiency actually augmented IFN-γ production by $CD4^+$ T cells further supporting the notion that disease resistance is likely caused by a lack of regulation mediated by nTreg cells in infected sites [202]. CCR5 deficiency results in a drastic reduction in the production of IL-10 by dermal $CD4^+$ T cells, an observation consistent with a reduction of $CD4^+CD25^+$ nTreg function in infectious sites [202]. Moreover, nTreg cell development in $CCR5^{-/-}$ mice is similar to that observed in WT mice, thus excluding the possibility that inherent differences in nTreg development or maturation in WT and $CCR5^{-/-}$ mice used in this study underlies our findings. However, these results do not exclude the possibility that CCR5 may in fact have a direct functional impact on the expansion, survival and function of nTreg cells in inflammatory settings, nor does it argue against a possible role of other $CCR5^+$ immune cells within the infected host capable of affecting nTreg or effector cell activity in these sites. Although CCR5 deficiency decreases the cellular frequency of nTreg cells and consequential IL-10 production in dermal sites, these results do not exclude the possibility that CCR5 signals may actually be required for the terminal differentiation of IL-10 producing $CD4^+CD25^+$ nTreg cells within sites of inflammation. The selective recruitment and migration of effector or nTreg cells to sites of infection would result in distinct immunological outcomes: in the presence of nTreg cells, effector T cells, and infected APCs would be deactivated in their functions thus favoring parasite replication; in the absence of nTreg cells, the activation and expansion of effector T cells could more efficiently activate local APCs leading to an amplification of the effector immune response, such as IFN-γ production, and parasite elimination. The inflammatory signals impacting nTreg cell homing, and influencing the functional dynamics of nTreg cells within infected tissue remain elusive. CCR5 chemokines are actively induced following *L. major* infection, and are temporally correlated with the recruitment of nTreg cells in infected dermal sites. The cellular sources of the CCR5 chemokines MIP-1α, MIP-1β, or RANTES these are unknown, and may be produced by a variety of cell types throughout the course of an immune response, including activated effector T cells, infected macrophages and dendritic cells, fibroblasts and endothelial cells. It is plausible that the initial inflammatory events following pathogen infection results in chemokine production by these cell types, which subsequently attracts nTreg cells to dampen local immune responses.

Thus, the expression of CCR5 chemokine receptor is a critical checkpoint influencing the balance between regulatory and effector T cells in infectious sites. CCR5 directs the homing of CD4$^+$ nTreg cells to *L. major* infected dermal sites where they promote the establishment of chronic infection and parasite persistence in the immune host by dampening anti-pathogen effector mechanisms, and controlling the intensity of the inflammatory response within these sites [202].

Several infectious diseases such as AIDS, tuberculosis, and leishmaniasis are caused by pathogens that also result in chronic infections with severe consequences on human health. Persistent infections may represent a compromise between pathogen and host, allowing the host to maintain strong effector immunity against re-infection and limit the potential of immunopathology. Experimental evidence that pathogens subvert the mammalian immune system to actively recruit nTreg cells in order to promote immunosuppression and pathogen persistence, would cause a paradigm shift in our understanding of the host-pathogen network. The cellular and molecular mechanisms underlying the establishment, maintenance and disruption of the homeostatic balance between host and pathogen remain poorly understood. CD4$^+$ nTreg can be exploited to tune the intensity of antimicrobial immune responses in acute and chronic infections, and to develop novel immunotherapeutic strategies for various infectious diseases including transient CCR5 blockade to abrogate nTreg function and elicit potent anti-pathogen immunity.

Conclusion

It is likely that most normal individuals harbor potentially pathogenic autoreactive T cells capable of causing a particular autoimmune disease. There is now compelling evidence in rodents, non-human primates and humans that abrogation of nTreg cell development and/or peripheral function may permit the preferential activation, expansion and differentiation of such autoreactive T cell in peripheral blood of normal individuals and ultimately cause autoimmune disease *in vivo*. These alterations in nTreg cell activity may also simultaneously unleash immunity to pathogens and tumors in the same host. This indicates that CD4$^+$Foxp3$^+$ nTreg cells actively engage in the maintenance of dominant self-tolerance in humans and functional aberrations in the nTreg compartment may induce autoimmune disease. Alterations in thymic T cell selection processes may cause an imbalance in the effector or nTreg cells produced and populating the periphery. Several autoimmune susceptibility genes and various environmental triggers may also exert their effects by affecting this balance (Fig. 9.2). Genetic analysis, detection and functional testing of antigen-specific nTreg cells with specific and stable markers, such as Foxp3, will facilitate our understanding at the cellular and molecular levels of how nTreg cells control autoreactive T cells in health, how such mechanism fails in instances of autoimmune disease. Experimentation along these lines will undeniably shed light into the development of novel therapeutic strategies, which are destined to either potentiate or abrogate nTreg functions in cases of autoimmunity and tumor/pathogen immunity, respectively. Alternatively, such studies may also allow us to determine if the

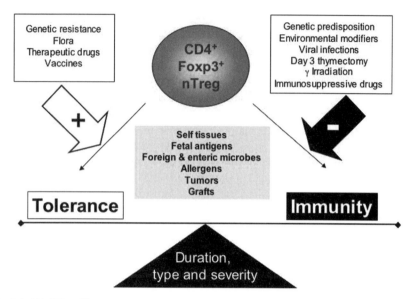

Fig. 9.2 CD4+Foxp3+ regulatory T cells: master-switch of peripheral tolerance CD4+Foxp3+ nTreg cells represent a central master-switch of peripheral T cell tolerance as abrogation of nTreg development or function can provoke autoimmunity, and also increase immunity to tumor, allergens, grafts and various pathogens. Thus, nTreg cells play a determining role in the balance between tolerance and immunity, and alterations in their development or function provoked by physical, chemical, environmental or genetic triggers, may represent a determining variable in disease resistance or susceptible. The duration, type and severity of nTreg cell dysfunction may also affect this balance, and ultimately determine degree of tolerance or immunity to self and non-self antigens. *(See also Color Insert)*

resistance or susceptibility of various effector T cells to nTreg cell function may be modified in order to direct the immune response desired. The use of antigen-specific nTreg cells may be a future strategy for the treatment and prevention of autoimmune disease through the reestablishment of dominant self-tolerance.

Acknowledgments I would like to thank the core members of my team: Alexander Albanese, Evridiki Sgouroudis, Michal Pyzik, Michael Tritt, Ekaterina Yurchenko, Eva d'Hennezel, Jessica St-Pierre and Valerie Hay for endless discussions and hard work. I acknowledge the financial support of the Canadian Institutes for Health Research, Canadian Diabetes Association, and Canada Research Chair program.

References

1. Abbas, A.K., J. Lohr, B. Knoechel, and V. Nagabhushanam. 2004. T cell tolerance and autoimmunity. *Autoimmun Rev* 3:471–475.
2. Piccirillo, C.A., and A.M. Thornton. 2004. Cornerstone of peripheral tolerance: naturally occurring CD4+CD25+ regulatory T cells. *Trends Immunol* 25:374–380.
3. Powrie, F. 2004. Immune regulation in the intestine: A Balancing act between effector and regulatory T cell responses. *Ann NY Acad Sci* 1029:132–141.

4. Gershon, R.K., and K. Kondo. 1970. Cell interactions in the induction of tolerance: the role of thymic lymphocytes. *Immunology* 18:723–737.
5. Gershon, R.K., and K. Kondo. 1971. Infectious immunological tolerance. *Immunology* 21:903–914.
6. Levings, M.K., and M.G. Roncarolo. 2005. Phenotypic and functional differences between human CD4+CD25+ and type 1 regulatory T cells. *Curr Top Microbiol Immunol* 293:303–326.
7. O'Garra, A., and F.J. Barrat. 2003. In vitro generation of IL-10-producing regulatory CD4+ T cells is induced by immunosuppressive drugs and inhibited by Th1- and Th2-inducing cytokines. *Immunol Lett* 85:135–139.
8. Roncarolo, M.G., S. Gregori, M. Battaglia, R. Bacchetta, K. Fleischhauer, and M.K. Levings. 2006. IL-10 secreting type 1 regulatory T cells in rodents and humans. *Immunol Rev* 212:28–50.
9. Sakaguchi, S. 2004. Naturally arising CD4+ regulatory t cells for immunologic self-tolerance and negative control of immune responses. *Annu Rev Immunol* 22:531–562.
10. Sakaguchi, S., N. Sakaguchi, M. Asano, M. Itoh, and M. Toda. 1995. Immunologic self-tolerance maintained by activated T cells expressing IL-2 receptor alpha-chains (CD25). Breakdown of a single mechanism of self-tolerance causes various autoimmune diseases. *J Immunol* 155:1151–1164.
11. Thornton, A.M., and E.M. Shevach. 1998. CD4+CD25+ immunoregulatory T cells suppress polyclonal T cell activation in vitro by inhibiting interleukin 2 production. *J Exp Med* 188:287–296.
12. Baecher-Allan, C., J.A. Brown, G.J. Freeman, and D.A. Hafler. 2001. CD4+CD25high regulatory cells in human peripheral blood. *J Immunol* 167:1245–1253.
13. Baecher-Allan, C., J.A. Brown, G.J. Freeman, and D.A. Hafler. 2003. CD4+CD25+ regulatory cells from human peripheral blood express very high levels of CD25 ex vivo. *Novartis Found Symp* 252:67–88; discussion 88–91, 106–114.
14. Levings, M.K., R. Sangregorio, and M.G. Roncarolo. 2001. Human CD25+CD4+ T regulatory cells suppress naive and memory T-cell proliferation and can be expanded *in vitro* without loss of function. *J Exp Med* 193:1295–1302.
15. Levings, M.K., R. Sangregorio, C. Sartirana, A.L. Moschin, M. Battaglia, P.C. Orban, and M.G. Roncarolo. 2002. Human CD25+CD4+ T suppressor cell clones produce transforming growth factor beta, but not interleukin 10, and are distinct from type 1 T regulatory cells. *J Exp Med* 196:1335–1346.
16. Sakaguchi, S. 2005. Naturally arising Foxp3-expressing CD25+CD4+ regulatory T cells in immunological tolerance to self and non-self. *Nat Immunol* 6:345–352.
17. Chen, W., W. Jin, N. Hardegen, K.J. Lei, L. Li, N. Marinos, G. McGrady, and S.M. Wahl. 2003. Conversion of peripheral CD4+CD25- naive T cells to CD4+CD25+ regulatory T cells by TGF-beta induction of transcription factor Foxp3. *J Exp Med* 198:1875–1886.
18. Bach, J.F. 2003. Regulatory T cells under scrutiny. *Nat Rev Immunol* 3:189–198.
19. Shevach, E.M. 2002. CD4+ CD25+ suppressor T cells: more questions than answers. *Nat Rev Immunol* 2:389–400.
20. Piccirillo, C.A., and E.M. Shevach. 2004. Naturally-occurring CD4+CD25+ immunoregulatory T cells: central players in the arena of peripheral tolerance. *Semin Immunol* 16:81–88.
21. Rudensky, A. 2005. Foxp3 and dominant tolerance. *Philos Trans R Soc Lond B Biol Sci* 360:1645–1646.
22. Levings, M.K., S. Allan, E. d'Hennezel, and C.A. Piccirillo. 2006. Functional dynamics of naturally occurring regulatory T cells in health and autoimmunity. *Adv Immunol* 92:119–155.
23. Asano, M., M. Toda, N. Sakaguchi, and S. Sakaguchi. 1996. Autoimmune disease as a consequence of developmental abnormality of a T cell subpopulation. *J Exp Med* 184:387–396.
24. Sakaguchi, S. 2002. Immunologic tolerance maintained by regulatory T cells: implications for autoimmunity, tumor immunity and transplantation tolerance. *Vox Sang* 83 Suppl 1:151–153.

25. Sakaguchi, S., M. Ono, R. Setoguchi, H. Yagi, S. Hori, Z. Fehervari, J. Shimizu, T. Takahashi, and T. Nomura. 2006. Foxp3+ CD25+ CD4+ natural regulatory T cells in dominant self-tolerance and autoimmune disease. *Immunol Rev* 212:8–27.
26. Hsieh, C.S., and A.Y. Rudensky. 2005. The role of TCR specificity in naturally arising CD25+ CD4+ regulatory T cell biology. *Curr Top Microbiol Immunol* 293:25–42.
27. Hsieh, C.S., Y. Zheng, Y. Liang, J.D. Fontenot, and A.Y. Rudensky. 2006. An intersection between the autoreactive regulatory and nonregulatory T cell receptor repertoires. *Nat Immunol* 7:401–410.
28. Cederbom, L., H. Hall, and F. Ivars. 2000. CD4+CD25+ regulatory T cells down-regulate co-stimulatory molecules on antigen-presenting cells. *Eur J Immunol* 30:1538–1543.
29. Hunig, T., and A. Schimpl. 1997. Systemic autoimmune disease as a consequence of defective lymphocyte death. *Curr Opin Immunol* 9:826–830.
30. Piccirillo, C.A., and E.M. Shevach. 2001. Cutting edge: control of CD8+ T cell activation by CD4+CD25+ immunoregulatory cells. *J Immunol* 167:1137–1140.
31. Seo, S.J., M.L. Fields, J.L. Buckler, A.J. Reed, L. Mandik-Nayak, S.A. Nish, R.J. Noelle, L.A. Turka, F.D. Finkelman, A.J. Caton, and J. Erikson. 2002. The impact of T helper and T regulatory cells on the regulation of anti-double-stranded DNA B cells. *Immunity* 16:535–546.
32. Trzonkowski, P., E. Szmit, J. Mysliwska, A. Dobyszuk, and A. Mysliwski. 2004. CD4+CD25+ T regulatory cells inhibit cytotoxic activity of T CD8+ and NK lymphocytes in the direct cell-to-cell interaction. *Clin Immunol* 112:258–267.
33. Azuma, T., T. Takahashi, A. Kunisato, T. Kitamura, and H. Hirai. 2003. Human CD4+ CD25+ regulatory T cells suppress NKT cell functions. *Cancer Res* 63:4516–4520.
34. Fontenot, J.D., M.A. Gavin, and A.Y. Rudensky. 2003. Foxp3 programs the development and function of CD4+CD25+ regulatory T cells. *Nat Immunol* 4:330–336.
35. Hori, S., and S. Sakaguchi. 2004. Foxp3: a critical regulator of the development and function of regulatory T cells. *Microbes Infect* 6:745–751.
36. Khattri, R., T. Cox, S.A. Yasayko, and F. Ramsdell. 2003. An essential role for Scurfin in CD4+CD25+ T regulatory cells. *Nat Immunol* 4:337–342.
37. Gavin, M.A., J.P. Rasmussen, J.D. Fontenot, V. Vasta, V.C. Manganiello, J.A. Beavo, and A.Y. Rudensky. 2007. Foxp3-dependent programme of regulatory T-cell differentiation. *Nature* 445:771–775.
38. Williams, L.M., and A.Y. Rudensky. 2007. Maintenance of the Foxp3-dependent developmental program in mature regulatory T cells requires continued expression of Foxp3. *Nat Immunol* 8:277–284.
39. Kim, J.M., and A. Rudensky. 2006. The role of the transcription factor Foxp3 in the development of regulatory T cells. *Immunol Rev* 212:86–98.
40. Fontenot, J.D., J.P. Rasmussen, L.M. Williams, J.L. Dooley, A.G. Farr, and A.Y. Rudensky. 2005. Regulatory T cell lineage specification by the forkhead transcription factor foxp3. *Immunity* 22:329–341.
41. Kim, J.M., J.P. Rasmussen, and A.Y. Rudensky. 2007. Regulatory T cells prevent catastrophic autoimmunity throughout the lifespan of mice. *Nat Immunol* 8:191–197.
42. Hori, S., T. Nomura, and S. Sakaguchi. 2003. Control of regulatory T cell development by the transcription factor Foxp3. *Science* 299:1057–1061.
43. Jaeckel, E., H. von Boehmer, and M.P. Manns. 2005. Antigen-specific FoxP3-transduced T-cells can control established type 1 diabetes. *Diabetes* 54:306–310.
44. Ochs, H.D., S.F. Ziegler, and T.R. Torgerson. 2005. FOXP3 acts as a rheostat of the immune response. *Immunol Rev* 203:156–164.
45. Bennett, C.L., J. Christie, F. Ramsdell, M.E. Brunkow, P.J. Ferguson, L. Whitesell, T.E. Kelly, F.T. Saulsbury, P.F. Chance, and H.D. Ochs. 2001. The immune dysregulation, polyendocrinopathy, enteropathy, X-linked syndrome (IPEX) is caused by mutations of FOXP3. *Nat Genet* 27:20–21.

46. Gambineri, E., T.R. Torgerson, and H.D. Ochs. 2003. Immune dysregulation, polyendocrinopathy, enteropathy, and X-linked inheritance (IPEX), a syndrome of systemic autoimmunity caused by mutations of FOXP3, a critical regulator of T-cell homeostasis. *Curr Opin Rheumatol* 15:430–435.
47. Owen, C.J., C.E. Jennings, H. Imrie, A. Lachaux, N.A. Bridges, T.D. Cheetham, and S.H. Pearce. 2003. Mutational analysis of the FOXP3 gene and evidence for genetic heterogeneity in the immunodysregulation, polyendocrinopathy, enteropathy syndrome. *J Clin Endocrinol Metab* 88:6034–6039.
48. Tommasini, A., S. Ferrari, D. Moratto, R. Badolato, M. Boniotto, D. Pirulli, L.D. Notarangelo, and M. Andolina. 2002. X-chromosome inactivation analysis in a female carrier of FOXP3 mutation. *Clin Exp Immunol* 130:127–130.
49. Wildin, R.S., S. Smyk-Pearson, and A.H. Filipovich. 2002. Clinical and molecular features of the immunodysregulation, polyendocrinopathy, enteropathy, X linked (IPEX) syndrome. *J Med Genet* 39:537–545.
50. Smyk-Pearson, S.K., A.C. Bakke, P.K. Held, and R.S. Wildin. 2003. Rescue of the autoimmune scurfy mouse by partial bone marrow transplantation or by injection with T-enriched splenocytes. *Clin Exp Immunol* 133:193–199.
51. Chang, X., J.X. Gao, Q. Jiang, J. Wen, N. Seifers, L. Su, V.L. Godfrey, T. Zuo, P. Zheng, and Y. Liu. 2005. The Scurfy mutation of FoxP3 in the thymus stroma leads to defective thymopoiesis. *J Exp Med* 202:1141–1151.
52. Allan, S.E., L. Passerini, R. Bacchetta, N. Crellin, M. Dai, P.C. Orban, S.F. Ziegler, M.G. Roncarolo, and M.K. Levings. 2005. The role of 2 FOXP3 isoforms in the generation of human CD4+ Tregs. *J Clin Invest* 115:3276–3284.
53. Yagi, H., T. Nomura, K. Nakamura, S. Yamazaki, T. Kitawaki, S. Hori, M. Maeda, M. Onodera, T. Uchiyama, S. Fujii, and S. Sakaguchi. 2004. Crucial role of FOXP3 in the development and function of human CD25+CD4+ regulatory T cells. *Int Immunol* 16: 1643–1656.
54. Ziegler, S.F. 2006. FOXP3: Of Mice and Men. *Annu Rev Immunol* 24:209–226.
55. Schubert, L.A., E. Jeffery, Y. Zhang, F. Ramsdell, and S.F. Ziegler. 2001. Scurfin (FOXP3) acts as a repressor of transcription and regulates T cell activation. *J Biol Chem* 276: 37672–37679.
56. Zheng, Y., S.Z. Josefowicz, A. Kas, T.T. Chu, M.A. Gavin, and A.Y. Rudensky. 2007. Genome-wide analysis of Foxp3 target genes in developing and mature regulatory T cells. *Nature* 445:936–940.
57. Rudensky, A.Y., M. Gavin, and Y. Zheng. 2006. FOXP3 and NFAT: partners in tolerance. *Cell* 126:253–256.
58. Bettelli, E., M. Dastrange, and M. Oukka. 2005. Foxp3 interacts with nuclear factor of activated T cells and NF-kappa B to repress cytokine gene expression and effector functions of T helper cells. *Proc Natl Acad Sci USA* 102:5138–5143.
59. Ono, M., J. Shimizu, Y. Miyachi, and S. Sakaguchi. 2006. Control of autoimmune myocarditis and multiorgan inflammation by glucocorticoid-induced TNF receptor family-related protein(high), Foxp3-expressing CD25+ and CD25- regulatory T cells. *J Immunol* 176: 4748–4756.
60. Sakaguchi, S. 2003. The origin of FOXP3-expressing CD4+ regulatory T cells: thymus or periphery. *J Clin Invest* 112:1310–1312.
61. Malek, T.R., and A.L. Bayer. 2004. Tolerance, not immunity, crucially depends on IL-2. *Nat Rev Immunol* 4:665–674.
62. Nelson, B.H. 2004. IL-2, regulatory T cells, and tolerance. *J Immunol* 172:3983–3988.
63. Vella, A., J.D. Cooper, C.E. Lowe, N. Walker, S. Nutland, B. Widmer, R. Jones, S.M. Ring, W. McArdle, M.E. Pembrey, D.P. Strachan, D.B. Dunger, R.C. Twells, D.G. Clayton, and J.A. Todd. 2005. Localization of a type 1 diabetes locus in the IL2RA/CD25 region by use of tag single-nucleotide polymorphisms. *Am J Hum Genet* 76:773–779.

64. Malek, T.R., A. Yu, V. Vincek, P. Scibelli, and L. Kong. 2002. CD4 regulatory T cells prevent lethal autoimmunity in IL-2Rbeta-deficient mice. Implications for the nonredundant function of IL-2. *Immunity* 17:167–178.
65. Bayer, A.L., A. Yu, D. Adeegbe, and T.R. Malek. 2005. Essential role for interleukin-2 for CD4(+)CD25(+) T regulatory cell development during the neonatal period. *J Exp Med* 201:769–777.
66. Wolf, M., A. Schimpl, and T. Hunig. 2001. Control of T cell hyperactivation in IL-2-deficient mice by CD4(+)CD25(–) and CD4(+)CD25(+) T cells: evidence for two distinct regulatory mechanisms. *Eur J Immunol* 31:1637–1645.
67. Furtado, G.C., D. Olivares-Villagomez, M.A. Curotto de Lafaille, A.K. Wensky, J.A. Latkowski, and J.J. Lafaille. 2001. Regulatory T cells in spontaneous autoimmune encephalomyelitis. *Immunol Rev* 182:122–134.
68. Setoguchi, R., S. Hori, T. Takahashi, and S. Sakaguchi. 2005. Homeostatic maintenance of natural Foxp3(+) CD25(+) CD4(+) regulatory T cells by interleukin (IL)-2 and induction of autoimmune disease by IL-2 neutralization. *J Exp Med* 201:723–735.
69. Fontenot, J.D., J.P. Rasmussen, M.A. Gavin, and A.Y. Rudensky. 2005. A function for interleukin 2 in Foxp3-expressing regulatory T cells. *Nat Immunol* 6:1142–1151.
70. Antov, A., L. Yang, M. Vig, D. Baltimore, and L. Van Parijs. 2003. Essential role for STAT5 signaling in CD25+CD4+ regulatory T cell homeostasis and the maintenance of self-tolerance. *J Immunol* 171:3435–3441.
71. D'Cruz, L.M., and L. Klein. 2005. Development and function of agonist-induced CD25+Foxp3+ regulatory T cells in the absence of interleukin 2 signaling. *Nat Immunol* 6:1152–1159.
72. Gavin, M., and A. Rudensky. 2003. Control of immune homeostasis by naturally arising regulatory CD4+ T cells. *Curr Opin Immunol* 15:690–696.
73. Toda, A., and C.A. Piccirillo. 2006. Development and function of naturally occurring CD4+CD25+ regulatory T cells. *J Leukoc Biol* 80:458–470.
74. Salomon, B., and J.A. Bluestone. 2001. Complexities of CD28/B7: CTLA-4 costimulatory pathways in autoimmunity and transplantation. *Annu Rev Immunol* 19:225–252.
75. Salomon, B., D.J. Lenschow, L. Rhee, N. Ashourian, B. Singh, A. Sharpe, and J.A. Bluestone. 2000. B7/CD28 costimulation is essential for the homeostasis of the CD4+ CD25+ immunoregulatory T cells that control autoimmune diabetes. *Immunity* 12:431–440.
76. Tang, Q., K.J. Henriksen, E.K. Boden, A.J. Tooley, J. Ye, S.K. Subudhi, X.X. Zheng, T.B. Strom, and J.A. Bluestone. 2003. Cutting edge: CD28 controls peripheral homeostasis of CD4+CD25+ regulatory T cells. *J Immunol* 171:3348–3352.
77. Guiducci, C., B. Valzasina, H. Dislich, and M.P. Colombo. 2005. CD40/CD40L interaction regulates CD4(+)CD25(+) T reg homeostasis through dendritic cell-produced IL-2. *Eur J Immunol* 35:557–567.
78. Kumanogoh, A., X. Wang, I. Lee, C. Watanabe, M. Kamanaka, W. Shi, K. Yoshida, T. Sato, S. Habu, M. Itoh, N. Sakaguchi, S. Sakaguchi, and H. Kikutani. 2001. Increased T cell autoreactivity in the absence of CD40-CD40 ligand interactions: a role of CD40 in regulatory T cell development. *J Immunol* 166:353–360.
79. Tai, X., M. Cowan, L. Feigenbaum, and A. Singer. 2005. CD28 costimulation of developing thymocytes induces Foxp3 expression and regulatory T cell differentiation independently of interleukin 2. *Nat Immunol* 6:152–162.
80. Fehervari, Z., and S. Sakaguchi. 2004. Control of Foxp3+ CD25+CD4+ regulatory cell activation and function by dendritic cells. *Int Immunol* 16:1769–1780.
81. Yamazaki, S., T. Iyoda, K. Tarbell, K. Olson, K. Velinzon, K. Inaba, and R.M. Steinman. 2003. Direct expansion of functional CD25+ CD4+ regulatory T cells by antigen-processing dendritic cells. *J Exp Med* 198:235–247.
82. Tarbell, K.V., S. Yamazaki, and R.M. Steinman. 2006. The interactions of dendritic cells with antigen-specific, regulatory T cells that suppress autoimmunity. *Semin Immunol* 18:93–102.

83. Encinas, J.A., L.S. Wicker, L.B. Peterson, A. Mukasa, C. Teuscher, R. Sobel, H.L. Weiner, C.E. Seidman, J.G. Seidman, and V.K. Kuchroo. 1999. QTL influencing autoimmune diabetes and encephalomyelitis map to a 0.15-cM region containing Il2. *Nat Genet* 21:158–160.
84. Maier, L.M., and L.S. Wicker. 2005. Genetic susceptibility to type 1 diabetes. *Curr Opin Immunol* 17:601–608.
85. Podolin, P.L., M.B. Wilusz, R.M. Cubbon, U. Pajvani, C.J. Lord, J.A. Todd, L.B. Peterson, L.S. Wicker, and P.A. Lyons. 2000. Differential glycosylation of interleukin 2, the molecular basis for the NOD Idd3 type 1 diabetes gene? *Cytokine* 12:477–482.
86. Snow, J.W., N. Abraham, M.C. Ma, B.G. Herndier, A.W. Pastuszak, and M.A. Goldsmith. 2003. Loss of tolerance and autoimmunity affecting multiple organs in STAT5A/5B-deficient mice. *J Immunol* 171:5042–5050.
87. Nacsa, J., Y. Edghill-Smith, W.P. Tsai, D. Venzon, E. Tryniszewska, A. Hryniewicz, M. Moniuszko, A. Kinter, K.A. Smith, and G. Franchini. 2005. Contrasting Effects of Low-Dose IL-2 on Vaccine-Boosted Simian Immunodeficiency Virus (SIV)-Specific CD4+ and CD8+ T Cells in Macaques Chronically Infected with SIVmac251. *J Immunol* 174: 1913–1921.
88. Bluestone, J.A., and Q. Tang. 2004. Therapeutic vaccination using CD4+CD25+ antigen-specific regulatory T cells. *Proc Natl Acad Sci USA* 101 Suppl 2:14622–14626.
89. Tang, Q., K.J. Henriksen, M. Bi, E.B. Finger, G. Szot, J. Ye, E.L. Masteller, H. McDevitt, M. Bonyhadi, and J.A. Bluestone. 2004. In vitro-expanded antigen-specific regulatory T cells suppress autoimmune diabetes. *J Exp Med* 199:1455–1465.
90. Greenwald, R.J., G.J. Freeman, and A.H. Sharpe. 2004. The B7 Family Revisited. *Annu Rev Immunol* 23:515–548.
91. Read, S., R. Greenwald, A. Izcue, N. Robinson, D. Mandelbrot, L. Francisco, A.H. Sharpe, and F. Powrie. 2006. Blockade of CTLA-4 on CD4+CD25+ regulatory T cells abrogates their function in vivo. *J Immunol* 177:4376–4383.
92. Bachmann, M.F., G. Kohler, B. Ecabert, T.W. Mak, and M. Kopf. 1999. Cutting edge: lymphoproliferative disease in the absence of CTLA-4 is not T cell autonomous. *J Immunol* 163:1128–1131.
93. Takahashi, T., T. Tagami, S. Yamazaki, T. Uede, J. Shimizu, N. Sakaguchi, T.W. Mak, and S. Sakaguchi. 2000. Immunologic self-tolerance maintained by CD25(+)CD4(+) regulatory T cells constitutively expressing cytotoxic T lymphocyte-associated antigen 4. *J Exp Med* 192:303–310.
94. Rao, P.E., A.L. Petrone, and P.D. Ponath. 2005. Differentiation and expansion of T cells with regulatory function from human peripheral lymphocytes by stimulation in the presence of TGF-{beta}. *J Immunol* 174:1446–1455.
95. Thornton, A.M., C.A. Piccirillo, and E.M. Shevach. 2004. Activation requirements for the induction of CD4+CD25+ T cell suppressor function. *Eur J Immunol* 34:366–376.
96. Fallarino, F., U. Grohmann, K.W. Hwang, C. Orabona, C. Vacca, R. Bianchi, M.L. Belladonna, M.C. Fioretti, M.L. Alegre, and P. Puccetti. 2003. Modulation of tryptophan catabolism by regulatory T cells. *Nat Immunol* 4:1206–1212.
97. Tang, Q., E.K. Boden, K.J. Henriksen, H. Bour-Jordan, M. Bi, and J.A. Bluestone. 2004. Distinct roles of CTLA-4 and TGF-beta in CD4+CD25+ regulatory T cell function. *Eur J Immunol* 34:2996–3005.
98. Paust, S., L. Lu, N. McCarty, and H. Cantor. 2004. Engagement of B7 on effector T cells by regulatory T cells prevents autoimmune disease. *Proc Natl Acad Sci USA* 101:10398–10403.
99. Boden, E., Q. Tang, H. Bour-Jordan, and J.A. Bluestone. 2003. The role of CD28 and CTLA4 in the function and homeostasis of CD4+CD25+ regulatory T cells. *Novartis Found Symp* 252:55–63; discussion 63–56, 106–114.
100. Solomon, M., and N. Sarvetnick. 2004. The pathogenesis of diabetes in the NOD mouse. *Adv Immunol* 84:239–264.
101. Ueda, H., J.M. Howson, L. Esposito, J. Heward, H. Snook, G. Chamberlain, D.B. Rainbow, K.M. Hunter, A.N. Smith, G. Di Genova, M.H. Herr, I. Dahlman, F. Payne, D. Smyth, C. Lowe, R.C. Twells, S. Howlett, B. Healy, S. Nutland, H.E. Rance, V. Everett, L.J. Smink,

A.C. Lam, H.J. Cordell, N.M. Walker, C. Bordin, J. Hulme, C. Motzo, F. Cucca, J.F. Hess, M.L. Metzker, J. Rogers, S. Gregory, A. Allahabadia, R. Nithiyananthan, E. Tuomilehto-Wolf, J. Tuomilehto, P. Bingley, K.M. Gillespie, D.E. Undlien, K.S. Ronningen, C. Guja, C. Ionescu-Tirgoviste, D.A. Savage, A.P. Maxwell, D.J. Carson, C.C. Patterson, J.A. Franklyn, D.G. Clayton, L.B. Peterson, L.S. Wicker, J.A. Todd, and S.C. Gough. 2003. Association of the T-cell regulatory gene CTLA4 with susceptibility to autoimmune disease. *Nature* 423:506–511.

102. Kulkarni, A.B., C.G. Huh, D. Becker, A. Geiser, M. Lyght, K.C. Flanders, A.B. Roberts, M.B. Sporn, J.M. Ward, and S. Karlsson. 1993. Transforming growth factor beta 1 null mutation in mice causes excessive inflammatory response and early death. *Proc Natl Acad Sci USA* 90:770–774.

103. Lucas, P.J., S.J. Kim, S.J. Melby, and R.E. Gress. 2000. Disruption of T cell homeostasis in mice expressing a T cell-specific dominant negative transforming growth factor beta II receptor. *J Exp Med* 191:1187–1196.

104. Yang, X., J.J. Letterio, R.J. Lechleider, L. Chen, R. Hayman, H. Gu, A.B. Roberts, and C. Deng. 1999. Targeted disruption of SMAD3 results in impaired mucosal immunity and diminished T cell responsiveness to TGF-beta. *Embo J* 18:1280–1291.

105. Ashcroft, G.S., X. Yang, A.B. Glick, M. Weinstein, J.L. Letterio, D.E. Mizel, M. Anzano, T. Greenwell-Wild, S.M. Wahl, C. Deng, and A.B. Roberts. 1999. Mice lacking Smad3 show accelerated wound healing and an impaired local inflammatory response. *Nat Cell Biol* 1:260–266.

106. Datto, M.B., J.P. Frederick, L. Pan, A.J. Borton, Y. Zhuang, and X.F. Wang. 1999. Targeted disruption of Smad3 reveals an essential role in transforming growth factor beta-mediated signal transduction. *Mol Cell Biol* 19:2495–2504.

107. Marie, J.C., D. Liggitt, and A.Y. Rudensky. 2006. Cellular mechanisms of fatal early-onset autoimmunity in mice with the T cell-specific targeting of transforming growth factor-beta receptor. *Immunity* 25:441–454.

108. Blois, S., M. Tometten, J. Kandil, E. Hagen, B.F. Klapp, R.A. Margni, and P.C. Arck. 2005. Intercellular adhesion molecule-1/LFA-1 cross talk is a proximate mediator capable of disrupting immune integration and tolerance mechanism at the feto-maternal interface in murine pregnancies. *J Immunol* 174:1820–1829.

109. Nakamura, K., A. Kitani, and W. Strober. 2001. Cell contact-dependent immunosuppression by CD4(+)CD25(+) regulatory T cells is mediated by cell surface-bound transforming growth factor beta. *J Exp Med* 194:629–644.

110. Shevach, E.M., C.A. Piccirillo, A.M. Thornton, and R.S. McHugh. 2003. Control of T cell activation by CD4+CD25+ suppressor T cells. *Novartis Found Symp* 252:24–36; discussion 36–44, 106–114.

111. Piccirillo, C.A., J.J. Letterio, A.M. Thornton, R.S. McHugh, M. Mamura, H. Mizuhara, and E.M. Shevach. 2002. CD4(+)CD25(+) regulatory T cells can mediate suppressor function in the absence of transforming growth factor beta1 production and responsiveness. *J Exp Med* 196:237–246.

112. Kullberg, M.C., V. Hay, A.W. Cheever, M. Mamura, A. Sher, J.J. Letterio, E.M. Shevach, and C.A. Piccirillo. 2005. TGF-beta1 production by CD4+ CD25+ regulatory T cells is not essential for suppression of intestinal inflammation. *Eur J Immunol* 35:2886–2895.

113. Mamura, M., W. Lee, T.J. Sullivan, A. Felici, A.L. Sowers, J.P. Allison, and J.J. Letterio. 2004. CD28 disruption exacerbates inflammation in Tgf-beta1-/- mice: in vivo suppression by CD4+CD25+ regulatory T cells independent of autocrine TGF-beta1. *Blood* 103: 4594–4601.

114. Read, S., V. Malmstrom, and F. Powrie. 2000. Cytotoxic T lymphocyte-associated antigen 4 plays an essential role in the function of CD25(+)CD4(+) regulatory cells that control intestinal inflammation. *J Exp Med* 192:295–302.

115. Chen, M.L., M.J. Pittet, L. Gorelik, R.A. Flavell, R. Weissleder, H. von Boehmer, and K. Khazaie. 2005. Regulatory T cells suppress tumor-specific CD8 T cell cytotoxicity through TGF-beta signals in vivo. *Proc Natl Acad Sci USA* 102: 419–424.

116. Nakamura, K., A. Kitani, I. Fuss, A. Pedersen, N. Harada, H. Nawata, and W. Strober. 2004. TGF-beta 1 plays an important role in the mechanism of CD4+CD25+ regulatory T cell activity in both humans and mice. *J Immunol* 172:834–842.
117. Oida, T., X. Zhang, M. Goto, S. Hachimura, M. Totsuka, S. Kaminogawa, and H.L. Weiner. 2003. CD4+CD25- T cells that express latency-associated peptide on the surface suppress CD4+CD45RBhigh-induced colitis by a TGF-beta-dependent mechanism. *J Immunol* 170:2516–2522.
118. Green, E.A., L. Gorelik, C.M. McGregor, E.H. Tran, and R.A. Flavell. 2003. CD4+CD25+ T regulatory cells control anti-islet CD8+ T cells through TGF-beta-TGF-beta receptor interactions in type 1 diabetes. *Proc Natl Acad Sci USA* 100:10878–10883.
119. Huber, S., C. Schramm, H.A. Lehr, A. Mann, S. Schmitt, C. Becker, M. Protschka, P.R. Galle, M.F. Neurath, and M. Blessing. 2004. Cutting edge: TGF-beta signaling is required for the in vivo expansion and immunosuppressive capacity of regulatory CD4+CD25+ T cells. *J Immunol* 173:6526–6531.
120. Peng, Y., Y. Laouar, M.O. Li, E.A. Green, and R.A. Flavell. 2004. TGF-beta regulates in vivo expansion of Foxp3-expressing CD4+CD25+ regulatory T cells responsible for protection against diabetes. *Proc Natl Acad Sci USA* 101:4572–4577.
121. Yamagiwa, S., J.D. Gray, S. Hashimoto, and D.A. Horwitz. 2001. A role for TGF-beta in the generation and expansion of CD4+CD25+ regulatory T cells from human peripheral blood. *J Immunol* 166:7282–7289.
122. Marie, J.C., J.J. Letterio, M. Gavin, and A.Y. Rudensky. 2005. TGF-beta1 maintains suppressor function and Foxp3 expression in CD4+CD25+ regulatory T cells. *J Exp Med* 201:1061–1067.
123. Suri-Payer, E., and H. Cantor. 2001. Differential cytokine requirements for regulation of autoimmune gastritis and colitis by CD4(+)CD25(+) T cells. *J Autoimmun* 16:115–123.
124. Morse, S.S., N. Sakaguchi, and S. Sakaguchi. 1999. Virus and autoimmunity: induction of autoimmune disease in mice by mouse T lymphotropic virus (MTLV) destroying CD4+ T cells. *J Immunol* 162:5309–5316.
125. Barrett, S.P., B.H. Toh, F. Alderuccio, I.R. van Driel, and P.A. Gleeson. 1995. Organ-specific autoimmunity induced by adult thymectomy and cyclophosphamide-induced lymphopenia. *Eur J Immunol* 25:238–244.
126. Sakaguchi, S., and N. Sakaguchi. 1989. Organ-specific autoimmune disease induced in mice by elimination of T cell subsets. V. Neonatal administration of cyclosporin A causes autoimmune disease. *J Immunol* 142:471–480.
127. Anderson, M.S., E.S. Venanzi, Z. Chen, S.P. Berzins, C. Benoist, and D. Mathis. 2005. The cellular mechanism of Aire control of T cell tolerance. *Immunity* 23:227–239.
128. Gorelik, L., and R.A. Flavell. 2000. Abrogation of TGFbeta signaling in T cells leads to spontaneous T cell differentiation and autoimmune disease. *Immunity* 12:171–181.
129. Schimpl, A., I. Berberich, B. Kneitz, S. Kramer, B. Santner-Nanan, B. Wagner, M. Wolf, and T. Hunig. 2002. IL-2 and autoimmune disease. *Cytokine Growth Factor Rev* 13:369–378.
130. Girvin, A.M., M.C. Dal Canto, L. Rhee, B. Salomon, A. Sharpe, J.A. Bluestone, and S.D. Miller. 2000. A critical role for B7/CD28 costimulation in experimental autoimmune encephalomyelitis: a comparative study using costimulatory molecule-deficient mice and monoclonal antibody blockade. *J Immunol* 164:136–143.
131. Sakaguchi, S., M. Toda, M. Asano, M. Itoh, S.S. Morse, and N. Sakaguchi. 1996. T cell-mediated maintenance of natural self-tolerance: its breakdown as a possible cause of various autoimmune diseases. *J Autoimmun* 9:211–220.
132. Alyanakian, M.A., S. You, D. Damotte, C. Gouarin, A. Esling, C. Garcia, S. Havouis, L. Chatenoud, and J.F. Bach. 2003. Diversity of regulatory CD4+T cells controlling distinct organ-specific autoimmune diseases. *Proc Natl Acad Sci USA* 100:15806–15811.
133. McHugh, R.S., and E.M. Shevach. 2002. Cutting edge: depletion of CD4+CD25+ regulatory T cells is necessary, but not sufficient, for induction of organ-specific autoimmune disease. *J Immunol* 168:5979–5983.

134. Sakaguchi, N., T. Takahashi, H. Hata, T. Nomura, T. Tagami, S. Yamazaki, T. Sakihama, T. Matsutani, I. Negishi, S. Nakatsuru, and S. Sakaguchi. 2003. Altered thymic T-cell selection due to a mutation of the ZAP-70 gene causes autoimmune arthritis in mice. *Nature* 426: 454–460.
135. Anderson, M.S., E.S. Venanzi, L. Klein, Z. Chen, S.P. Berzins, S.J. Turley, H. von Boehmer, R. Bronson, A. Dierich, C. Benoist, and D. Mathis. 2002. Projection of an immunological self shadow within the thymus by the aire protein. *Science* 298:1395–1401.
136. Ramsey, C., O. Winqvist, L. Puhakka, M. Halonen, A. Moro, O. Kampe, P. Eskelin, M. Pelto-Huikko, and L. Peltonen. 2002. Aire deficient mice develop multiple features of APECED phenotype and show altered immune response. *Hum Mol Genet* 11:397–409.
137. Akiyama, T., S. Maeda, S. Yamane, K. Ogino, M. Kasai, F. Kajiura, M. Matsumoto, and J. Inoue. 2005. Dependence of self-tolerance on TRAF6-directed development of thymic stroma. *Science* 308:248–251.
138. Kajiura, F., S. Sun, T. Nomura, K. Izumi, T. Ueno, Y. Bando, N. Kuroda, H. Han, Y. Li, A. Matsushima, Y. Takahama, S. Sakaguchi, T. Mitani, and M. Matsumoto. 2004. NF-kappa B-inducing kinase establishes self-tolerance in a thymic stroma-dependent manner. *J Immunol* 172:2067–2075.
139. Wohlfert, E.A., M.K. Callahan, and R.B. Clark. 2004. Resistance to CD4+CD25+ regulatory T cells and TGF-beta in Cbl-b-/- mice. *J Immunol* 173:1059–1065.
140. Bopp, T., A. Palmetshofer, E. Serfling, V. Heib, S. Schmitt, C. Richter, M. Klein, H. Schild, E. Schmitt, and M. Stassen. 2005. NFATc2 and NFATc3 transcription factors play a crucial role in suppression of CD4+ T lymphocytes by CD4+ CD25+ regulatory T cells. *J Exp Med* 201:181–187.
141. Pasare, C., and R. Medzhitov. 2003. Toll pathway-dependent blockade of CD4+CD25+ T cell-mediated suppression by dendritic cells. *Science* 299:1033–1036.
142. Piccirillo, C.A., M. Tritt, E. Sgouroudis, A. Albanese, M. Pyzik, and V. Hay. 2005. Control of Type 1 Autoimmune Diabetes by Naturally Occurring CD4+CD25+ Regulatory T Lymphocytes in Neonatal NOD Mice. *Ann NY Acad Sci* 1051:72–87.
143. Pop, S.M., C.P. Wong, D.A. Culton, S.H. Clarke, and R. Tisch. 2005. Single cell analysis shows decreasing FoxP3 and TGFbeta1 coexpressing CD4+CD25+ regulatory T cells during autoimmune diabetes. *J Exp Med* 201:1333–1346.
144. Gregg, R.K., R. Jain, S.J. Schoenleber, R. Divekar, J.J. Bell, H.H. Lee, P. Yu, and H. Zaghouani. 2004. A sudden decline in active membrane-bound TGF-beta impairs both T regulatory cell function and protection against autoimmune diabetes. *J Immunol* 173: 7308–7316.
145. Gregori, S., N. Giarratana, S. Smiroldo, and L. Adorini. 2003. Dynamics of pathogenic and suppressor T cells in autoimmune diabetes development. *J Immunol* 171:4040–4047.
146. Miyara, M., Z. Amoura, C. Parizot, C. Badoual, K. Dorgham, S. Trad, D. Nochy, P. Debre, J.C. Piette, and G. Gorochov. 2005. Global natural regulatory T cell depletion in active systemic lupus erythematosus. *J Immunol* 175:8392–8400.
147. Viglietta, V., C. Baecher-Allan, H.L. Weiner, and D.A. Hafler. 2004. Loss of functional suppression by CD4+CD25+ regulatory T cells in patients with multiple sclerosis. *J Exp Med* 199:971–979.
148. Tang, Q., J.Y. Adams, A.J. Tooley, M. Bi, B.T. Fife, P. Serra, P. Santamaria, R.M. Locksley, M.F. Krummel, and J.A. Bluestone. 2006. Visualizing regulatory T cell control of autoimmune responses in nonobese diabetic mice. *Nat Immunol* 7:83–92.
149. Tarbell, K.V., S. Yamazaki, K. Olson, P. Toy, and R.M. Steinman. 2004. CD25+ CD4+ T cells, expanded with dendritic cells presenting a single autoantigenic peptide, suppress autoimmune diabetes. *J Exp Med* 199:1467–1477.
150. Chen, Z., A.E. Herman, M. Matos, D. Mathis, and C. Benoist. 2005. Where CD4+CD25+ T reg cells impinge on autoimmune diabetes. *J Exp Med* 202:1387–1397.
151. You, S., M. Belghith, S. Cobbold, M.A. Alyanakian, C. Gouarin, S. Barriot, C. Garcia, H. Waldmann, J.F. Bach, and L. Chatenoud. 2005. Autoimmune diabetes onset results from

qualitative rather than quantitative age-dependent changes in pathogenic T-cells. *Diabetes* 54:1415–1422.
152. Bour-Jordan, H., B.L. Salomon, H.L. Thompson, G.L. Szot, M.R. Bernhard, and J.A. Bluestone. 2004. Costimulation controls diabetes by altering the balance of pathogenic and regulatory T cells. *J Clin Invest* 114:979–987.
153. Serra, P., A. Amrani, J. Yamanouchi, B. Han, S. Thiessen, T. Utsugi, J. Verdaguer, and P. Santamaria. 2003. CD40 ligation releases immature dendritic cells from the control of regulatory CD4+CD25+ T cells. *Immunity* 19:877–889.
154. Herman, A.E., G.J. Freeman, D. Mathis, and C. Benoist. 2004. CD4+CD25+ T regulatory cells dependent on ICOS promote regulation of effector cells in the prediabetic lesion. *J Exp Med* 199:1479–1489.
155. Balasa, B., and N. Sarvetnick. 1998. Cytokines and IDDM: Implications for etiology and therapy. *Drug News Perspect* 11:356–360.
156. Chen, Z., C. Benoist, and D. Mathis. 2005. How defects in central tolerance impinge on a deficiency in regulatory T cells. *Proc Natl Acad Sci USA* 102:14735–14740.
157. Wicker, L.S., J. Clark, H.I. Fraser, V.E. Garner, A. Gonzalez-Munoz, B. Healy, S. Howlett, K. Hunter, D. Rainbow, R.L. Rosa, L.J. Smink, J.A. Todd, and L.B. Peterson. 2005. Type 1 diabetes genes and pathways shared by humans and NOD mice. *J Autoimmun* 25 Suppl: 29–33.
158. Greve, B., L. Vijayakrishnan, A. Kubal, R.A. Sobel, L.B. Peterson, L.S. Wicker, and V.K. Kuchroo. 2004. The diabetes susceptibility locus Idd5.1 on mouse chromosome 1 regulates ICOS expression and modulates murine experimental autoimmune encephalomyelitis. *J Immunol* 173:157–163.
159. Hill, N.J., P.A. Lyons, N. Armitage, J.A. Todd, L.S. Wicker, and L.B. Peterson. 2000. NOD Idd5 locus controls insulitis and diabetes and overlaps the orthologous CTLA4/IDDM12 and NRAMP1 loci in humans. *Diabetes* 49:1744–1747.
160. Lyons, P.A., W.W. Hancock, P. Denny, C.J. Lord, N.J. Hill, N. Armitage, T. Siegmund, J.A. Todd, M.S. Phillips, J.F. Hess, S.L. Chen, P.A. Fischer, L.B. Peterson, and L.S. Wicker. 2000. The NOD Idd9 genetic interval influences the pathogenicity of insulitis and contains molecular variants of Cd30, Tnfr2, and Cd137. *Immunity* 13:107–115.
161. Wicker, L.S., J.A. Todd, J.B. Prins, P.L. Podolin, R.J. Renjilian, and L.B. Peterson. 1994. Resistance alleles at two non-major histocompatibility complex-linked insulin-dependent diabetes loci on chromosome 3, Idd3 and Idd10, protect nonobese diabetic mice from diabetes. *J Exp Med* 180:1705–1713.
162. Denny, P., C.J. Lord, N.J. Hill, J.V. Goy, E.R. Levy, P.L. Podolin, L.B. Peterson, L.S. Wicker, J.A. Todd, and P.A. Lyons. 1997. Mapping of the IDDM locus Idd3 to a 0.35-cM interval containing the interleukin-2 gene. *Diabetes* 46:695–700.
163. Lyons, P.A., N. Armitage, F. Argentina, P. Denny, N.J. Hill, C.J. Lord, M.B. Wilusz, L.B. Peterson, L.S. Wicker, and J.A. Todd. 2000. Congenic mapping of the type 1 diabetes locus, Idd3, to a 780-kb region of mouse chromosome 3: identification of a candidate segment of ancestral DNA by haplotype mapping. *Genome Res* 10:446–453.
164. Wicker, L.S., C.L. Moule, H. Fraser, C. Penha-Goncalves, D. Rainbow, V.E. Garner, G. Chamberlain, K. Hunter, S. Howlett, J. Clark, A. Gonzalez-Munoz, A.M. Cumiskey, P. Tiffen, J. Howson, B. Healy, L.J. Smink, A. Kingsnorth, P.A. Lyons, S. Gregory, J. Rogers, J.A. Todd, and L.B. Peterson. 2005. Natural genetic variants influencing type 1 diabetes in humans and in the NOD mouse. *Novartis Found Symp* 267:57–65; discussion 65–75.
165. Bottini, N., L. Musumeci, A. Alonso, S. Rahmouni, K. Nika, M. Rostamkhani, J. MacMurray, G.F. Meloni, P. Lucarelli, M. Pellecchia, G.S. Eisenbarth, D. Comings, and T. Mustelin. 2004. A functional variant of lymphoid tyrosine phosphatase is associated with type I diabetes. *Nat Genet* 36:337–338.
166. Bottini, N., T. Vang, F. Cucca, and T. Mustelin. 2006. Role of PTPN22 in type 1 diabetes and other autoimmune diseases. *Semin Immunol* 18:207–213.

167. Steck, A.K., S.Y. Liu, K. McFann, K.J. Barriga, S.R. Babu, G.S. Eisenbarth, M.J. Rewers, and J.X. She. 2006. Association of the PTPN22/LYP gene with type 1 diabetes. *Pediatr Diabetes* 7:274–278.
168. Vang, T., M. Congia, M.D. Macis, L. Musumeci, V. Orru, P. Zavattari, K. Nika, L. Tautz, K. Tasken, F. Cucca, T. Mustelin, and N. Bottini. 2005. Autoimmune-associated lymphoid tyrosine phosphatase is a gain-of-function variant. *Nat Genet* 37:1317–1319.
169. Bacchetta, R., L. Passerini, E. Gambineri, M. Dai, S.E. Allan, L. Perroni, F. Dagna-Bricarelli, C. Sartirana, S. Matthew, A. Lawitschka, C. Azzari, S.F. Ziegler, M.K. Levings, and M.G. Roncarolo. 2006. Defective regulatory and effector T cell functions in patients with FOXP3 mutations. *J Clin Invest* 116:1713–1722.
170. Lee, J.H., L.C. Wang, Y.T. Lin, Y.H. Yang, D.T. Lin, and B.L. Chiang. 2006. Inverse correlation between CD4+ regulatory T-cell population and autoantibody levels in paediatric patients with systemic lupus erythematosus. *Immunology* 117:280–286.
171. Liu, M.F., C.R. Wang, L.L. Fung, and C.R. Wu. 2004. Decreased CD4+CD25+ T cells in peripheral blood of patients with systemic lupus erythematosus. *Scand J Immunol* 59: 198–202.
172. Haas, J., A. Hug, A. Viehover, B. Fritzsching, C.S. Falk, A. Filser, T. Vetter, L. Milkova, M. Korporal, B. Fritz, B. Storch-Hagenlocher, P.H. Krammer, E. Suri-Payer, and B. Wildemann. 2005. Reduced suppressive effect of CD4+CD25high regulatory T cells on the T cell immune response against myelin oligodendrocyte glycoprotein in patients with multiple sclerosis. *Eur J Immunol* 35:3343–3352.
173. Balandina, A., S. Lecart, P. Dartevelle, A. Saoudi, and S. Berrih-Aknin. 2005. Functional defect of regulatory CD4(+)CD25+ T cells in the thymus of patients with autoimmune myasthenia gravis. *Blood* 105:735–741.
174. Sugiyama, H., R. Gyulai, E. Toichi, E. Garaczi, S. Shimada, S.R. Stevens, T.S. McCormick, and K.D. Cooper. 2005. Dysfunctional blood and target tissue CD4+CD25high regulatory T cells in psoriasis: mechanism underlying unrestrained pathogenic effector T cell proliferation. *J Immunol* 174:164–173.
175. Lindley, S., C.M. Dayan, A. Bishop, B.O. Roep, M. Peakman, and T.I. Tree. 2005. Defective suppressor function in CD4(+)CD25(+) T-cells from patients with type 1 diabetes. *Diabetes* 54:92–99.
176. Kriegel, M.A., T. Lohmann, C. Gabler, N. Blank, J.R. Kalden, and H.M. Lorenz. 2004. Defective suppressor function of human CD4+ CD25+ regulatory T cells in autoimmune polyglandular syndrome type II. *J Exp Med* 199:1285–1291.
177. Cao, D., V. Malmstrom, C. Baecher-Allan, D. Hafler, L. Klareskog, and C. Trollmo. 2003. Isolation and functional characterization of regulatory CD25brightCD4+ T cells from the target organ of patients with rheumatoid arthritis. *Eur J Immunol* 33:215–223.
178. de Kleer, I.M., L.R. Wedderburn, L.S. Taams, A. Patel, H. Varsani, M. Klein, W. de Jager, G. Pugayung, F. Giannoni, G. Rijkers, S. Albani, W. Kuis, and B. Prakken. 2004. CD4+CD25bright regulatory T cells actively regulate inflammation in the joints of patients with the remitting form of juvenile idiopathic arthritis. *J Immunol* 172:6435–6443.
179. Valencia, X., G. Stephens, R. Goldbach-Mansky, M. Wilson, E.M. Shevach, and P.E. Lipsky. 2006. TNF down-modulates the function of human CD4+CD25hi T regulatory cells. *Blood* 108:253–261.
180. Ehrenstein, M.R., J.G. Evans, A. Singh, S. Moore, G. Warnes, D.A. Isenberg, and C. Mauri. 2004. Compromised function of regulatory T cells in rheumatoid arthritis and reversal by anti-TNFalpha therapy. *J Exp Med* 200:277–285.
181. Rouse, B.T., and S. Suvas. 2004. Regulatory cells and infectious agents: detentes cordiale and contraire. *J Immunol* 173:2211–2215.
182. Belkaid, Y. 2003. The role of CD4(+)CD25(+) regulatory T cells in Leishmania infection. *Expert Opin Biol Ther* 3:875–885.
183. Belkaid, Y., C.A. Piccirillo, S. Mendez, E.M. Shevach, and D.L. Sacks. 2002. CD4+CD25+ regulatory T cells control Leishmania major persistence and immunity. *Nature* 420:502–507.

184. Suvas, S., A.K. Azkur, B.S. Kim, U. Kumaraguru, and B.T. Rouse. 2004. CD4+CD25+ regulatory T cells control the severity of viral immunoinflammatory lesions. *J Immunol* 172:4123–4132.
185. Manigold, T., E.C. Shin, E. Mizukoshi, K. Mihalik, K.K. Murthy, C.M. Rice, C.A. Piccirillo, and B. Rehermann. 2006. Foxp3+CD4+CD25+ T cells control virus-specific memory T cells in chimpanzees that recovered from hepatitis C. *Blood* 107:4424–4432.
186. Hesse, M., C.A. Piccirillo, Y. Belkaid, J. Prufer, M. Mentink-Kane, M. Leusink, A.W. Cheever, E.M. Shevach, and T.A. Wynn. 2004. The pathogenesis of schistosomiasis is controlled by cooperating IL-10-producing innate effector and regulatory T cells. *J Immunol* 172:3157–3166.
187. Suvas, S., U. Kumaraguru, C.D. Pack, S. Lee, and B.T. Rouse. 2003. CD4+CD25+ T cells regulate virus-specific primary and memory CD8+ T cell responses. *J Exp Med* 198:889–901.
188. Powrie, F., S. Read, C. Mottet, H. Uhlig, and K. Maloy. 2003. Control of immune pathology by regulatory T cells. *Novartis Found Symp* 252:92–98; discussion 98–105, 106–114.
189. Singh, B., S. Read, C. Asseman, V. Malmstrom, C. Mottet, L.A. Stephens, R. Stepankova, H. Tlaskalova, and F. Powrie. 2001. Control of intestinal inflammation by regulatory T cells. *Immunol Rev* 182:190–200.
190. Belkaid, Y., S. Mendez, R. Lira, N. Kadambi, G. Milon, and D. Sacks. 2000. A natural model of Leishmania major infection reveals a prolonged "silent" phase of parasite amplification in the skin before the onset of lesion formation and immunity. *J Immunol* 165:969–977.
191. Mendez, S., S.K. Reckling, C.A. Piccirillo, D. Sacks, and Y. Belkaid. 2004. Role for CD4(+)CD25(+) regulatory T cells in reactivation of persistent leishmaniasis and control of concomitant immunity. *J Exp Med* 200:201–210.
192. Suffia, I.J., S.K. Reckling, C.A. Piccirillo, R.S. Goldszmid, and Y. Belkaid. 2006. Infected site-restricted Foxp3+ natural regulatory T cells are specific for microbial antigens. *J Exp Med* 203:777–788.
193. Belkaid, Y., K.F. Hoffmann, S. Mendez, S. Kamhawi, M.C. Udey, T.A. Wynn, and D.L. Sacks. 2001. The role of interleukin (IL)-10 in the persistence of Leishmania major in the skin after healing and the therapeutic potential of anti-IL-10 receptor antibody for sterile cure. *J Exp Med* 194:1497–1506.
194. Zlotnik, A., O. Yoshie, and H. Nomiyama. 2006. The chemokine and chemokine receptor superfamilies and their molecular evolution. *Genome Biol* 7:243.
195. D'Ambrosio, D., F. Sinigaglia, and L. Adorini. 2003. Special attractions for suppressor T cells. *Trends Immunol* 24:122–126.
196. Iellem, A., M. Mariani, R. Lang, H. Recalde, P. Panina-Bordignon, F. Sinigaglia, and D. D'Ambrosio. 2001. Unique chemotactic response profile and specific expression of chemokine receptors CCR4 and CCR8 by CD4(+)CD25(+) regulatory T cells. *J Exp Med* 194:847–853.
197. Szanya, V., J. Ermann, C. Taylor, C. Holness, and C.G. Fathman. 2002. The subpopulation of CD4+CD25+ splenocytes that delays adoptive transfer of diabetes expresses L-selectin and high levels of CCR7. *J Immunol* 169:2461–2465.
198. Kleinewietfeld, M., F. Puentes, G. Borsellino, L. Battistini, O. Rotzschke, and K. Falk. 2005. CCR6 expression defines regulatory effector/memory-like cells within the CD25(+)CD4+ T-cell subset. *Blood* 105:2877–2886.
199. Bystry, R.S., V. Aluvihare, K.A. Welch, M. Kallikourdis, and A.G. Betz. 2001. B cells and professional APCs recruit regulatory T cells via CCL4. *Nat Immunol* 2:1126–1132.
200. Curiel, T.J., G. Coukos, L. Zou, X. Alvarez, P. Cheng, P. Mottram, M. Evdemon-Hogan, J.R. Conejo-Garcia, L. Zhang, M. Burow, Y. Zhu, S. Wei, I. Kryczek, B. Daniel, A. Gordon, L. Myers, A. Lackner, M.L. Disis, K.L. Knutson, L. Chen, and W. Zou. 2004. Specific recruitment of regulatory T cells in ovarian carcinoma fosters immune privilege and predicts reduced survival. *Nat Med* 10:942–949.

201. Wysocki, C.A., Q. Jiang, A. Panoskaltsis-Mortari, P.A. Taylor, K.P. McKinnon, L. Su, B.R. Blazar, and J.S. Serody. 2005. Critical role for CCR5 in the function of donor CD4+ CD25+ regulatory T cells during acute graft-versus-host disease. *Blood* 106:3300–3307.
202. Yurchenko, E., M. Tritt, V. Hay, E.M. Shevach, Y. Belkaid, and C.A. Piccirillo. 2006. CCR5-dependent homing of naturally occurring CD4+ regulatory T cells to sites of Leishmania major infection favors pathogen persistence. *J Exp Med* 203:2451–2460.

Chapter 10
Regulatory T Cell Control of Autoimmune Diabetes and Their Potential Therapeutic Application

Qizhi Tang and Jeffrey A. Bluestone

Abstract Extensive research in animal models and human patients in the past ten years since the mid-1990s has established that regulatory T cells (Tregs) are essential in maintaining normal immune homeostasis. Apart from this fundamental role, both natural and adaptive Tregs function to control various autoimmune disorders including Type 1 diabetes (T1D). These autoantigen-specific Tregs suppress the activation of a wide variety of cells including pathogenic T cells, B cells, and cells of the innate immune system such as dendritic cells, macrophages and granulocytes. The mechanisms of Treg control of autoreactive T cells in T1D are likely two-fold. First, in steady state, Tregs prevent activation of autoreactive T cells in the lymph nodes by limiting their access to dendritic cells and thus their expansion and acquisition of effector functions. These activities are largely mediated by thymus-derived natural Tregs. Second, when immune homeostasis is perturbed and inflammation erupts in the tissues, both natural Tregs and cytokine-induced adaptive Tregs traffic to the site of inflammation and curtail the functions of fully differentiated pathogenic effector T cells in the target tissue. Thus, the ability of Tregs to actively control unwanted immunity even after the onset of pathological manifestation makes them an attractive candidate of immunotherapy for autoimmune diseases such as T1D.

Introduction

Since their discovery as a key component of peripheral tolerance, regulatory T cells (Tregs) have been known for their function in controlling autoimmunity. Absence of Tregs due to mutations in the forkhead transcription factor, Foxp3, which is essential for Treg development, leads to multi-organ autoimmunity in mice and humans [1–5]. The vital function of Foxp3$^+$ T cells in maintaining normal immune homeostasis is further demonstrated by the fatal consequence of depletion of these cells in

---

J.A. Bluestone
UCSF Diabetes Center, Department of Medicine and Department of Pathology, University of California, San Francisco, HSW 1118, 513 Parnassus Avenue, Box 0540, San Francisco, CA 94143-0540, USA
e-mail: jbluest@diabetes.ucsf.edu

normal adult mice [6]. Similarly, defects in IL-2 production or IL-2 receptor expression interfere with the peripheral survival of Tregs and lead to loss of self tolerance [7–9]. In addition, CD28 is essential to the thymic development and peripheral homeostasis of Tregs. In the absence of CD28, the autoimmune-prone non-obese diabetic (NOD) mice develop exacerbated autoimmune diseases that affect multiple organs including pancreatic islets, exocrine pancreas, salivary glands, thyroid and peripheral nervous tissues (Tang and Bluestone, unpublished observation). Thus, Tregs are essential to the maintenance of self tolerance and prevention of autoimmune diseases. Importantly, the Treg subsets include both the thymically-derived Tregs as well as those induced during the course of immune responses. For example, therapies, such as anti-CD3, can induce Foxp3$^+$ Tregs even in natural Treg-deficient CD28-deficient mice [10].

Many challenges and questions remain in clinical application of Tregs. First of all, the mechanism of Treg function remains to be elucidated. Ample experimental evidence demonstrate that Tregs can suppress expansion and differentiation of pathogenic T cells during lymph node (LN) priming, and they can also halt the progressive tissue destruction at the site of inflammation. It is not clear how Tregs exert such broad effects in steady state and on an ongoing immune response. Second, the dynamics between Tregs and pathogenic T cells during the chronic course of an autoimmune response such as Type 1 diabetes (T1D) has not been delineated. It is not clear whether autoimmune diseases develop as a consequence of reduced Treg number and function or heightened resistance of pathogenic T cells to regulation. Third, to-date there have been no unique cell surface markers that distinguish Tregs from other T cell subsets, however, Tregs express a unique constellation of surface molecules that make their isolation and expansion possible. These markers include CD25, CTLA-4, HLA-DR, CD58, folate receptor-4, neuropilin 1 and reduced expression of IL-7Rα, CD127 [11,12]. Yes, it remains unclear whether these markers can be adapted to clinical efforts to select cells for immunotherapy. Fourth, Tregs represent less than 2% of the T cells in the peripheral blood; and obtaining enough Tregs to achieve a therapeutic effect in vivo remains a daunting task. This is especially challenging given the wealth of experimental data suggesting that the in vivo function of the Tregs depend on its antigen specificity. Finally, it remains unclear how successful isolation of antigen-specific Tregs will be in the clinical setting or whether this will be essential to avoid non-specific suppressive activity of adoptively-transferred cells. In this chapter, we will summarize the efforts made in the past several years in understanding of Treg function in vivo and in developing protocols to expand and enrich antigen-specific Tregs ex vivo for therapeutic use. These efforts have made the goal of defining Treg phenotype and treatment regimen to specifically control and reverse the course of autoimmune diabetes more attainable.

Dynamics of Tregs and Pathogenic T Cells During the Progression of Autoimmune Diabetes

Autoimmune diseases develop as a consequence of defects in central and/or peripheral tolerance. It has been known for quite some time that thymic deletion does not completely eliminate all autoreactive T cells and healthy individuals harbor an autoimmune repertoire [13]; however, most people and animals do not develop autoimmune diseases because tolerogenic mechanisms in the periphery such as deletion, anergy and Tregs keep autoimmunity in check. The study of animal models of autoimmune diabetes suggests that failure of Tregs to maintain peripheral tolerance is one of the underlying causes of the disease. For example, in the NOD mouse, leukocytic infiltrates in islets are observed as early as 2–3 weeks of age in all male and female mice. However, β cell destruction and diabetes develop in 70–90% of the females and 20–50% of males within 7–8 months of age. Reasons for the incomplete penetrance and protracted disease course are not clear. One likely explanation, among several others, is that β cell destruction is held in check by Tregs. This notion is supported by results from several independent lines of investigation. First, disruption of the CD28/B7 pathway impairs Treg development and homeostasis, and in NOD mice, exacerbates diabetes so that 100% of the female and male mice develop diabetes between 8–12 weeks of age [14]. Similarly in NOD mice, Treg depletion early in life by IL-2 neutralization results in accelerated diabetes onset [15]. Lastly, $TCR\alpha^{-/-}$, $RAG^{-/-}$ or scurfy (Foxp3-deficient) NOD.BDC2.5 T cell receptor (TCR) transgenic (BDC2.5) mice, which T cells are specific for an islet antigen and are completely devoid of Tregs, displayed no delay between onset of insulitis and overt diabetes [16–18]. Furthermore, three independent investigations suggested that progression to diabetes is due to an age-dependent decline in Treg numbers and their suppressive activities in the NOD mice accompanied by an associated emergence of Treg-resistant pathogenic T cells [19–21].

To further investigate the cause of Treg and effector T cells (Teff) imbalance in diabetic NOD mice, we compared the numbers of Foxp3-expressing cells in the pancreatic LNs and pancreatic islets of young prediabetic mice and mice within one week of disease onset. Surprisingly, diabetes onset was accompanied by an increase of Tregs in the pancreatic LN in both the absolute numbers and in proportion to the $CD4^+$ T cells. The increase in $Foxp3^+$ cells correlated with reduced proliferation of $Foxp3^-$ cells, suggesting elevated Treg function in the pancreatic LN at the time of diabetes onset (QT, Jason Adams, Cristina Penaranda, and JAB unpublished results). In contrast, in the islets, a precipitous loss of Tregs was observed as Teff population expands. The loss of Tregs was not a consequence of differential proliferation of Teffs and Tregs in vivo but likely a result of a defect in intra-islet Treg survival due to reduced expression of CD25. Thus, diabetes onset is associated with heightened regulatory function in the pancreatic LN and progressive loss of Tregs in inflamed islets suggesting critical role of intra-islet Treg function in regulating diabetes pathogenesis during disease progression (Fig. 10.1).

Several reports have been published analyzing peripheral blood Treg frequency and function in patients with T1D (Table 10.1). While some found reduced Treg frequency [22] or function [23,24], and others did not find any changes when comparing diabetics to healthy controls [25]. Several factors may have contributed to the variability in these investigations. First, as suggested above, the absence of unique Treg markers, especially in humans, makes elucidation of Treg numbers and function problematic. Although CD25 and Foxp3 are excellent markers for Tregs in mice [26], in human peripheral blood mononuclear cells, there is no clear

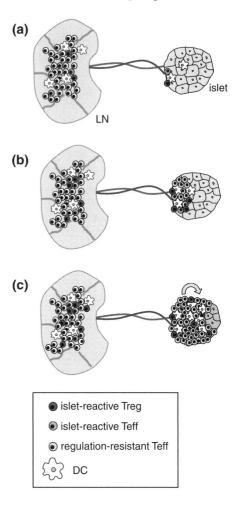

Fig. 10.1 Dynamics of Tregs and pathogenic T cells during the progression of autoimmune diabetes. (**a**) At three weeks of age, β cell antigen shedding initiated by program cell death during developmental islet tissue remodeling leads to priming of islet-reactive T cells in the pancreatic LN. The primed Teff cells and Tregs infiltrate islets. (**b**) T cells mediated β cell death leads to more antigen shedding and activation and expansion of Tregs in pancreatic LN. Tregs limit the priming of Teff cells. Over the long disease course, Treg-resistant Teff cells emerge and become activated and infiltrate and accumulate in islet tissue despite the presence of functional Tregs in the LN. (**c**) More islet damage and antigen shedding leads to more Treg activation and expansion in pancreatic LN. Treg-resistant Teff cells expand and differentiate in islets and eventually destroy β cells and leads to diabetes *(See also Color Insert)*

demarcation of $CD25^+$ and $CD25^-$ $CD4^+$ T cells [27], thus making the quantification of Treg levels among the peripheral blood mononuclear cells somewhat a guess work. The availability of anti-Foxp3 antibody has taken some of the uncertainty out of the guess work. However, Foxp3 expression can be induced in human T cells upon T cell receptor activation [28], and mere expression of Foxp3 does not ensure a regulatory phenotype [29]. Moreover, since Foxp3 is a transcription factor localized in the nucleus, T cells have to be fixed and permeabilized to permit Foxp3 staining therefore prohibiting the isolation of live $Foxp3^+$ cells for functional studies.

In addition to the uncertainty in the markers used to identify Tregs, the method by which the Tregs are isolated may also contribute to the variation in the findings. Magnetic column-isolated cells are more likely to be contaminated with more cells that express low to intermediate level of CD25 that have no regulatory activity than cells purified by flow cytometry. Moreover, it is not clear whether the levels of Treg in peripheral blood reflect the number of Treg in regional LN and sites of

Table 10.1 Publications on Tregs and human T1D

Publication	Treg number	Treg Function	Treg identification	Treg isolation	Patient population
Ref 22	Reduced in T1D	Not tested	CD4$^+$CD25$^+$	Not done	Mixed new-onset and chronic T1D
Ref 23	No difference	decreased in T1D	CD4$^+$CD25$^+$	MACS	New onset T1D
Ref 24	older patients had more CD4$^+$CD25$^+$ cells	decreased in T1D	CD4$^+$CD25$^+$	MACS	Mixed new-onset and chronic T1D
Ref 25	No difference	No difference	CD4$^+$CD25^{high}	FACS	Chronic T1D

inflammation, where their action matters most. Thus, future experiments using Treg-specific markers to examine the most relevant tissues will help to clarify the field. In this regard, we recently identified CD127 as a useful marker to isolate and study human Tregs. Utilizing CD127 as a marker, we observed that virtually all CD25$^+$Foxp3$^+$ T cells expressed low levels of CD127. Moreover, selection of human cells based solely on the CD4$^+$ and CD127^{low} cell surface phenotype resulted in a highly suppressive population of T cells in vitro in spite of the fact that less than 50% of the resulting population was Foxp3$^+$. These results suggested that this marker may be necessary and sufficient to purify large numbers of human Tregs for analysis and therapeutic expansion [12].

Mechanisms of Regulatory T Cell Function *In Vivo*

The mechanism of in vivo Treg function is an area of extensive ongoing investigation. In vitro suppression experiments demonstrated that Tregs suppress proliferation of other T cells by shutting down IL-2 transcription. In vivo experimental data thus far from many laboratories suggest that Tregs can control various aspects of an immune response including proliferation, effector cytokine production, tissue trafficking, and effector functions within the target tissue. It is not clear how Tregs mediate such diverse effects in vivo and if all of these controls are operational in the autoimmune settings. In the following sections, we will summarize findings from our research efforts in the NOD model of T1D. Our approach in analyzing Treg function in vivo has been to compare pathogenic autoreactive T cell response in mice with varying level of Tregs, namely NOD, Treg-deficient NOD.CD28$^{-/-}$, NOD and CD28$^{-/-}$ mice reconstituted with Tregs isolated from BDC2.5 mice. The NOD.CD28$^{-/-}$ mice have the lowest number of Tregs, which correlates with increased severity and incidence (100%). In contrast, NOD.CD28$^{-/-}$ mice reconstituted with ex vivo expanded BDC2.5 Tregs show complete protection against diabetes. Tracking the activities of adoptively transferred or endogenous diabetogenic T cells in these mice allows a determination of where and how Tregs exert their effect on the autoimmune process.

This model provided an opportunity to study the in vivo function of Tregs both in the lymphoid compartments and the pancreatic tissue.

Effect of Tregs on T Cell Priming in the LN

Our studies and others suggest that Tregs function at various stages of autoreactive T cells activation, differentiation and egress-specific events in the tissue antigen-draining LNs. Naive autoreactive T cells circulate through the body, enter LN through high endothelial venules, and home to T cell zones in response to the chemokines, CCL19 and CCL21 [30]. In the presence of cognate antigen, T cells accumulate in an area bordering the paracortical T cell zone and B cell follicles where they become activated by tissue emigrant dendritic cell (DCs) [31,32], which lead to the expansion of the autoreactive population. In the NOD mice, pancreatic LNs are essential during the early phase of disease course when initial priming of the diabetogenic T cells is first detected [33]. Splenic T cells from mice without pancreatic LN from birth do not possess sufficient numbers of diabetogenic T cells to transfer disease [34]. Moreover, deficiency in co-stimulatory molecule B7-2 results in marked reduction in LN priming of islet reactive T cells and protection against diabetes in NOD mice [35,36]. Together these findings suggest that priming in the pancreatic LN is essential for diabetes pathogenesis by expanding the population of low frequency autoreactive T cells and facilitating autoreactive T cell differentiation including the emergence of IFNγ,TNFα and/or IL-17 producing pathogenic cells as well as cytotoxic T cells [37–41]. Another important function of LN priming is the acquisition of adhesion molecule and chemokine receptor expression that facilitates LNs egress and target tissue invasion [42]. This last step in LN priming is an elaborate process that is dictated by the nature of T cell and DC interactions. For example, DCs in the subcutaneous LN induce T cells to express P-selectin and E-selectin ligands and chemokine receptor CCR4 and CCR10 for skin trafficking after leaving LNs, whereas DCs in gut-associated lymphoid organs direct T cells to infiltrate the intestine through induction of integrin $\alpha 4\beta 7$ and chemokine receptor CCR9 [43–46]. In the NOD mice, $\alpha 4$ integrin interactions with mucosal addressin cell adhesion molecule 1 ligands are critical for T cell infiltration into islets [47] while L-selectin is dispensable for diabetes induction [48,49]. In addition, chemokine CCL22 and CXCL10 are induced in inflamed islets, and the expression of their receptor, CCR4 and CXCR3, on autoreactive T cells has been implicated in diabetes pathogenesis [50,51].

Direct examination of function of Tregs in controlling T cell priming have shown varying effect on proliferation in the LN [52,53], however, the studies have uniformly observed inhibition on T cell differentiation [54–57]. These disparities are likely due to kinetic and specificity differences and the fact that large numbers of antigen-specific Tregs are introduced by adoptive transfer in most system. In the NOD model, we found that the proliferation of Teff cells inversely correlated with the level of antigen-specific Tregs in the LN [58]. The most robust proliferation of BDC2.5 Teff cells was observed in Treg-deficient $CD28^{-/-}$ or $B7^{-/-}$ mice. In contrast, significantly fewer Teff cells entered cell cycle in NOD mice that harbor intermediate level of Tregs. Teff proliferation was completely abrogated in NOD or $CD28^{-/-}$ mice that had been reconstituted with antigen-specific Tregs prior to the Teff cell transfer. In contrast, no inhibition of Teff proliferation was observed when Treg reconstitution of $NOD.CD28^{-/-}$ was performed concomitantly with Teff cell transfer. However, the in vivo IFNγ expression was markedly suppressed. Thus, inhibition of proliferation requires the presence of antigen-specific Tregs prior to the arrival of Teff cells, whereas delayed arrival of Tregs can still control T cell differentiation without affecting proliferation. Similar observations have been made in other models. For instance, although antigen-specific Tregs abrogated anti-tumor immunity in an ectopic hemagglutinin (HA)-expressing tumor model, the proliferation of HA-specific $CD8^+$ cells in the draining LN of a HA-expressing tumor-bearing mouse was not affected. In fact, in this model, IFNγ expression by the $CD8^+$ cells was not suppressed; however, cytotoxicity as measured by in vivo killing of HA bearing cells was inhibited by the Tregs [55]. Therefore, Tregs can control T cell proliferation, differentiation, and effector development during LN priming in vivo, yet, which process is regulated varies depending on the timing of Treg versus Teff cell encounter with antigen, and the relative activity of the Tregs and effector T cells. In this regard, in vitro analysis demonstrated that both TCR engagement and IL-2 signaling are required to promote Treg suppression, and it takes 1–3 days to fully activate Treg function [59,60]. Thus, when Tregs are transferred at the same time as the Teff cells, they may not be able to exert their maximal suppressive effect for several days *after* Teff cells have received sufficient stimulation to enter cell cycle. In contrast, differentiation of Teff cells requires repeated and prolonged stimulation and can be inhibited at later time points after antigen encounter. Similarly, Tregs with lower affinity TCR may not be sufficiently activated to suppress the proliferation of high affinity Teff cells. However, even suboptimal suppression may be able to inhibit differentiation and effector function development due to higher and more protracted activation requirements.

Effect of Tregs in the Peripheral Tissues

The presence of Tregs in inflamed tissues such as pancreatic islets, synovia of arthritic joints, colitis lesions, asthmatic lung, skin, intestinal tissue of graft versus host disease (GvHD) patients, cardiac and renal allografts, various tumors, and virus and parasite infected tissues suggests the potential for Tregs to act at the site of inflammation (reviewed in [61]). Some studies have demonstrated that these tissue-dwelling Tregs have suppressive activity. For example, in the skin allograft setting, the Treg-infiltrated allograft was capable of transferring antigen-specific tolerance when re-transplanted to a new host suggesting that the tolerant skin graft harbored functional Tregs [62]. Two separate studies also demonstrated that Tregs within tumor and allograft tissues were necessary for immunosuppression. In one study, depletion of Tregs within the tumor was sufficient to promote productive anti-tumor responses and induce tumor regression [63]. By comparison, in an allogeneic cardiac transplant model, CCR4-deficiency was associated with a lack of intra-graft Foxp3 expression and failure in maintaining tolerance even though Tregs were present at normal levels in the LN [64]. More recently, it has been shown that Foxp3 mRNA level in urine samples inversely correlate with serum creatinine levels in patients with acute renal graft rejection. In fact, higher levels of Foxp3 mRNA in urine predicted reversal of acute rejection while lower levels of Foxp3 mRNA level identified patients at risk of graft failure [65]. Together these results suggest that Tregs can function within the peripheral tissue to suppress inflammation and tissue damage.

Using the BDC2.5 mouse model system, Benoist and colleagues observed that Treg-deficient scurfy BDC2.5 or BDC.RAG$^{-/-}$ mice succumb to destructive insulitis and diabetes at four weeks of age, whereas almost all Foxp3-sufficient BDC2.5 mice never develop diabetes even though extensive insulitis was evident early in life [16–18]. No difference in LN priming was detected in the Treg-deficient and Treg-sufficient BDC2.5 mice and the only discernable change was the aggressiveness of the insulitis. $CD4^+CD25^+CD69^-$ are detected in the islets of Treg-sufficient mice that appear to function via an ICOS-dependent mechanism [16]. Moreover, adoptive transfer of splenic $CD4^+$ cells to the Treg-deficient BDC2.5 mice protected them from diabetes by suppressing inflammation in the islets [17]. These results suggest that Tregs function within the insulitic lesions to prevent β cell destruction. In the conventional NOD mice exhibiting a polyclonal T cell repertoire, we found that diabetes progression was associated with enhanced regulation in the pancreatic LN and reduced survival of Tregs in the islets (QT, Cristina Penaranda, and JAB unpublished observation). These findings emphasize the central importance of both pancreatic LN and intra-islet Treg function in regulating disease pathogenesis.

Mechanism of Treg Suppression of LN Priming

T cell priming in vivo is initiated by their cognate interaction with antigen-bearing DCs in the LN. In vitro studies demonstrated that productive T cell activation requires a persistent engagement with antigen presenting cells (APC) [66–68]. Live tissue imaging of in vivo T cell priming in intact LN revealed a sequential multistage process during cognate interaction between T cells and DCs. The two cell types initially probe each other for cognate ligands by repeated transient engagements, during which T cells appear to be swarming around DCs. Such dynamic interactions are followed by arrest of T cells and long-lasting stable conjugations between DCs and T cells, which lead to onset of T cell division and resumption of rapid motility [69–73]. Islet-antigen-specific BDC2.5 Teff cells responding to an autoantigen in Treg-deficient NOD.CD28$^{-/-}$ mice also swarm and arrest in a manner similar to that described for T cells responding to foreign antigens. However, limited swarming and no cellular arrest was observed in the NOD recipients suggesting that endogenous NOD Tregs prevent cellular arrest and stable conjugation between DCs and autoreactive T cells. The swarming-only phenotype was recapitulated by reconstitution of CD28$^{-/-}$ mice with polyclonal NOD Tregs. Furthermore, reconstitution of NOD or NOD.CD28$^{-/-}$ mice with antigen-specific BDC2.5 Treg cells completely

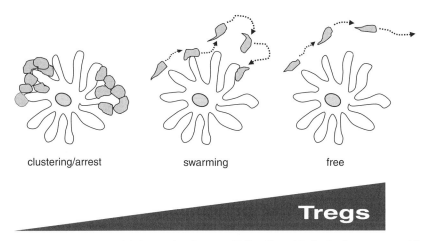

Fig. 10.2 Tregs inhibit stable interaction between Teff cells and DCs. In the presence of low level of Tregs, Teff cells form long stable conjugates with antigen-bearing DCs as indicated by their clustering and arrest on the DCs. With increase Tregs, the interaction between Teff and DCs becomes transient evident from the swarming behavior of the Teff cells. Presence of high level of Tregs completely abolish Teff and DC interaction and Teff cells moves freely in the LN as if no cognate antigen is present *(See also Color Insert)*

abolished both cell arrest and the swarming of autoreactive Teff cells [58]. Similarly, by monitoring myelin basic protein (MBP)-specific Teff cells in LN of Treg sufficient and deficient recipients, Dustin and Lafaille's group found that Teff cells failed to arrest in response to MBP peptide in the presence of host-derived Tregs [74]. Thus, Tregs can control the very first step of T cell priming and subsequent T cell/DCs interactions during differentiation by preventing persistent T cell conjugation with the DCs (Fig. 10.2).

A number of in vitro studies have suggested that Treg cells function in a cytokine independent, cell-cell contact dependent manner. In fact, it has been demonstrated that Treg cells suppress via a direct T-T interaction. The mechanisms of suppression include Treg surface-bound transforming growth factor (TGFβ) [75], direct engagement of B7-1 expressed on responder T cells by CTLA-4 expressed on Tregs[76], and direct killing of responder T cells by Tregs [77]. However, the majority of in vivo models of suppression provide strong evidence for dependence on interleukin 10 or TGFβ leading to bystander suppression [10,78–81]. These results did not exclude a role for direct T cell-T cell interactions in vivo. Thus, we examined this question using 2-photon microscopy. We observed that both Tregs and Teff cells homed to the T cell zone of the LN and are enriched at the T-B boundary in the presence of antigen in the draining LN. However, live dynamic imaging revealed no stable contacts between the two T cell types. Instead, Tregs were found to form stable conjugations with DCs antigen-bearing bearing prior to the establishment of suppression [58] suggesting that DCs are essential to Treg function in vivo either as direct Treg targets and/or by activating Tregs to express suppressive functions (Fig. 10.3).

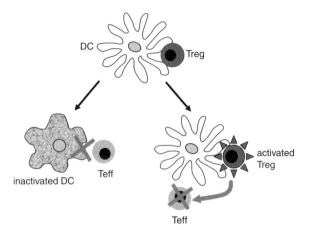

Fig. 10.3 Role of DCs in Treg function in vivo. Treg engagement of DCs may lead to direct inactivation of DCs so that they can not activate Teff cells. Alternatively, DC may activate Tregs to produce soluble inhibitory factors, which in turn prevent activation of Teff cells *(See also Color Insert)*

The functional importance of the complexity of Treg/DC interplay in an immune response has just begun to be appreciated. Among all APC, mature DCs have the unique ability to promote natural Treg expansion [82,83], whereas immature DCs are able to induce the generation of adaptive Tregs [84–87]. In fact, recent work by Travis et al. (Mark A. Travis, Boris Reizis, Emma Masteller, Qizhi Tang, John Proctor, Yanli Wang, Xin Bernstein, Xiazhou Huang, Lou Reichardt, Jeffrey Bluestone, Dean Sheppard, submitted) suggest that $CD11c^+$ DCs express an integrin, $\alpha v \beta 8$, on their cell surface responsible for cleavage of TGFβ and the induction of $Foxp3^+$ adaptive Tregs in vivo. Intriguingly, in vitro T cell responses stimulated by DCs matured with toll-like receptor stimulation are resistant to Treg suppression [83,88], however, inflammatory-cytokine matured DCs can still be suppressed by Tregs [89]. Thus, DCs in the draining LN or inflamed tissue in an autoimmune setting promote Treg expansion without overriding their suppressive functions.

Some in vitro and in vivo evidence suggests that Tregs can modify DCs to down-regulate the expression of MHC class II, costimulatory molecules, and IL-12 [90–94]. However, it remains to be seen whether these affects are primarily responsible for functional Treg activity in vivo or whether DCs are often merely activators of Tregs which leads to bystander suppressive activity.

A Unifying Model

We propose that Tregs have two modes of actions in vivo, homeostatic control in the steady state and damage control at the site of inflammation. In the steady state, Tregs efficiently home to LN, expand, and accumulate where cognitive antigens are present. With the TCR repertoire skewed toward self recognition [95] and more TCR capable of sensing low level of ligands [96], Treg have a higher probability to engage activated self antigen-bearing DCs than pathogenic cells and are, thus, well-positioned to prevent priming of autoreactive T cells and maintain immune homeostasis. Thymically-derived natural Tregs are largely responsible for this homeostatic activity. These cells are likely to function in a cell-cell contact dependent manner most likely interacting with autoantigen-bearing DCs providing ongoing tonic signals that stimulate the Tregs leading to direct regulation of DC expansion and immunogenic APC function. We hypothesize that natural Tregs directly interfere with cellular functions by controlling a key step in this cascade of events – *the antigen-presenting capability of DC*. In vitro and in vivo studies have established that long-lasting stable engagement with APC during the first antigenic exposure is essential to the "fitness" of the T cells when proliferation, cytokine production, and survival are measured. Thus, T cell activation is not a binary switch, but instead a programming process, in which the duration of the T-APC engagement dictates the outcome of the stimulation. Long stable engagement allows the T cells to progress through the complete program including proliferation, differentiation, and expression of survival molecules to develop into full-fledged effector T cells capable of invading tissue and cause tissue destruction. However, more transient engagement with APC permits only partial progression through this developmental program and only cell functions that require less antigenic stimulation such as proliferation are manifested. Recent imaging experiments showed that stable T-DC conjugations, as indicted by T cell arrest and clustering, were only observed in mice with low numbers of endogenous Tregs, which can be progressively reduced to transient engagements (swarming) to apparently no interactions with increasing numbers of antigen-specific Tregs. Thus, in the normal homeostatic state, natural Tregs function in vivo to interrupt the earliest stages of the effector development through limiting the autoreactive T cell access to DCs. By controlling this vital step in LN priming, Tregs limit the chance for a T cell to initiate their differentiation entire program and become pathogenic effector cells. The balance between the strength of the immunogenic stimulation and potency of the regulation force determines how far a T cell progresses along this developmental axis. Consistent with this model, islet infiltration of transferred T cells, indicative of completion of the effector T cell programming in the LN, was only observed in the low Treg setting (QT and JAB unpublished observation) whereas in the presence of increasing numbers of islet antigen-specific Tregs, IFNγ expression and proliferation was progressively inhibited. Since the program requires continuous contact between T cells and DCs over a 10-hour period, there is a wide window of opportunity for therapeutic interventions using Tregs.

Unfortunately, ongoing homeostasis can be disrupted by a number of genetic and environmental challenges. In autoimmune-prone individuals, the genetic make-up leads to dysregulation at various levels including apoptosis, central deletion and DC function. Environmental exposure to viral or bacterial challenges or other foreign proteins can lead to DC maturation and inadvertent activation of the dormant autoreactive T cells. These effects can "overwhelm" the innate control processes through induction of pro-inflammatory cytokines, TLR ligation on DCs or Tregs and potentially cross-presentation of shared antigenic epitopes. The result is that autoreactivity moves from quiescent to pathogenic with increased inflammation, T effector differentiation, tissue invasion into tissue sites and appearance of self-generating lymphoid-like structures within the target tissues (e.g. pancreas). In an

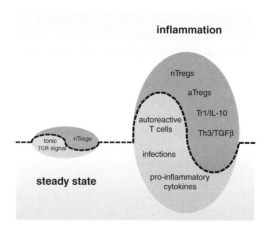

Fig. 10.4 Function of Treg in steady state and during inflammation. In steady state, thymus-derived natural Tregs (nTreg) with a repertoire skewed toward recognition of self antigens maintains normal immune homeostasis by interacting with DCs and preventing their activation of Teff cells. When normal immune homeostasis is perturbed by the presence of highly autoreactive T cells or infections, inflammation erupts. In response, nTregs expand and adaptive Tregs (aTregs) including IL-10 producing Tr1 cells and TGFβ expressing Th3 cells develop to control the immune response *(See also Color Insert)*

attempt to control autoimmune pathogenesis, natural Tregs expand and a second subset of IL-10- or TGFβ-producing adaptive Tregs develop in response to inflammation well as the recruitment of specialized DCs that accumulate in the lesions (Fig. 10.4).

The emergence of regulation resistant clones or toll-like receptor engagement during an infection allows Teff cells to escape regulation and tissue destruction occurs. Local inflammatory signals and increasing antigen availability promotes further expansion and activation of Tregs in the draining LN enabling the Tregs to acquire chemokine receptors and homing molecules to traffic to the site of inflammation, where they would exert damage control to limit further tissue destruction. Pathogenic cells require reactivation in the tissue to exert their effector functions and tissue DCs have been shown to be critical in sustaining local inflammation [97,98]. Thus in this scenario, Tregs function in tissues in much the same way as they do in the LN where they control the ability of DCs to stimulate effector responses. Although these counter-regulatory processes can results in disease remission short term (in diseases such as SLE and MS and maybe even in T1D), usually, the effector response resurfaces, the Tregs are overwhelmed and ultimately complete tissue destruction is achieved.

Therapeutic Use of Regulatory T Cells

The notion of using $CD4^+$ Tregs as therapeutics has been around for over two decades. Sakaguchi and colleagues observed that the adoptive transfer of T cells depleted of $CD4^+CD25^+$ cells induced multi-organ autoimmunity in the recipient animals [99,100]. Adoptive transfer of Tregs into autoimmune prone animals has profound effect on disease incidence and progression in animal models of T1D, experimental autoimmune encephalitis, gastritis, thyroiditis [101], oophoritis, inflammatory bowel disease, GvHD, arthritis, and systemic lupus erythematosus (SLE) [14,78,99,102–115]. In organ transplant setting, cells from tolerant animals can be transferred to naïve recipients, suppressing not only the original antigen specificities but other antigens linked through the same APC [116]. This concept of "infectious tolerance" is instrumental in providing a context for understanding Tregs as a potential therapy [116,117]. However, only recently has the field progressed to the point that these cells may likely become therapeutics for the treatment of the increasing number of human diseases. This section of the chapter will focus on efforts to develop Treg-based therapies for treating autoimmune diseases. Many of the current therapies are non-specific and as yet untested in humans but provide a strong rationale for moving into the clinic.

In Vivo *Efficacy of Adoptive Treg Therapy: The Issue of Antigen-Specificity*

Initial preclinical and clinical studies will likely be focused on the use of polyclonal Tregs as therapeutics since there is a body of literature that show, in certain disease settings, Tregs can function in vivo without enriching for antigen-specific subpopulations. For example, in a mouse model of SLE, adoptive transfer of polyclonal Tregs reduced the rate of renal disease development similar to that observed in T1D using high numbers ($>20 \times 10^6$) of Tregs. A second transfer of Tregs after the treated animals had developed proteinuria further slowed the progression of renal disease and significantly improved survival [115]. The protection required the transfer of 5×10^6 Tregs. It is not completely clear why polyclonal Tregs are more efficacious in SLE than in the autoimmune diabetes model. It is possible that antigens for SLE, such as double-stranded DNA and cell nuclear antigens, are ubiquitously present through out the body; thus Tregs specific for these antigens are maintained in all peripheral LNs and in the spleen, whereas islet antigen-specific Tregs are only maintained in the pancreatic LN. It is conceivable that Tregs for SLE antigens are at much higher frequency than islet-reactive Tregs in population of pooled peripheral LN and spleens, which are used to isolate Tregs for all these studies. Thus, for systemic autoimmune disease such as SLE, when antigen-specific Tregs are likely present at sufficiently high level in the circulation, polyclonal Tregs can be a viable therapeutic when sufficient numbers can be applied.

In mouse GvHD models, protection is clearly evident with the use of polyclonal Tregs [107], whereas Tregs enriched for allospecificity showed only moderately improved efficacy [118]. In these models, a variety of reasons may contribute to the lack of complete dependency on antigen-specific Tregs. First of all, these experiments relied on the use of lymphopenic hosts. Tregs has been shown to regulate lymphopenia driven homeostatic proliferation in an antigen nonspecific manner, which may contribute to their inhibitory effect by limiting the repopulation of the co-transferred Teff cells [119,120]. Additionally, a highly skewed ratio of as many as three Tregs to one Teff cell was used that may exaggerate the non-antigen-specific effect of homeostatic control by Tregs. Moreover, frequency of alloantigen-specific T cells is estimated to be as high as 1–10% [121] as opposed to 0.05–0.0005% for autoreactive T cells [122]. If alloreactive Tregs are also present at such high frequency, polyclonal pool may afford sufficient protection without enrichment for antigen-specific cells, especially in a lymphopenic host. It should be pointed out, however, that for solid organ allogeneic transplantation, polyclonal Tregs do not delay graft rejection in lymphopenic hosts even when very high Treg to Teff ratios are used unless the Tregs come from mice that are conditioned to be tolerant to the graft. This suggests that other factors such as trafficking and site of action are likely to also affect the efficacy of polyclonal Tregs.

In the NOD model of autoimmune diabetes, islet-antigen-specific Tregs were more effective at preventing diabetes than polyclonal Tregs as islet antigen-specific Tregs were at least 50-fold more potent than polyclonal Tregs [103]. Similarly, Tarbell et al observed that BDC2.5 Tregs were at least 20 fold more potent than polyclonal NOD Tregs in preventing diabetes transfer to NOD.SCID hosts [104]. It is noteworthy to point out that the BDC2.5 Tregs used in these studies are specific for a single antigen but was able to suppress complex T cell responses to an entire organ, suggesting that these Tregs are capable of mediating bystander suppression and controlling the activation of T cells with distinct antigen-specificity in vivo. These results are encouraging for potential therapeutic application as it is feasible to control multifaceted autoimmune response using Tregs specific for a limited subset of antigens.

The reduced efficacy of polyclonal Tregs is not due to their intrinsic deficit in suppressive activity. When stimulated with anti-CD3, polyclonal Tregs are equally suppressive as BDC2.5 Tregs in vitro. However, in vivo suppression of diabetes requires Treg activation by endogenous islet antigens, and polyclonal Tregs may not contain sufficient numbers of islet-reactive Tregs to control disease. In vitro experiments demonstrated that the specificity of Treg function is dictated by the specificity of the TCR they express. Once activated by its cognate ligand, the activated Treg suppress in an antigen non-specific manner [123]. In vivo, the efficacy of Tregs in preventing diabetes correlates with their ability to become activated in the draining LN [103]. In this regard, it has been shown that enriching for insulin-specific Treg by in vivo immunization and in vitro restimulation with insulin B9-23 peptide enhances the in vivo efficacy of the Tregs in preventing diabetes transfer to lymphodeficient hosts [124]. Studies using mice in which autoimmune diabetes is induced by transgenic expression of TNFα and CD80 in the islets demonstrate that Treg that are extremely potent in preventing diabetes expand and accumulate preferentially in the pancreatic LNs and not the spleen or other LNs [125]. In a rat model of autoimmune thyroiditis, the expansion and/or survival of thyroid-specific Treg depends on the presence of thyroid antigens since Tregs from mice whose thyroids are ablated in utero were unable to prevent disease upon adoptive transfer [126]. Recently, Tung and colleagues have elegantly demonstrated, in a mouse oophoritis model, that ovarian antigen-specific Tregs are enriched in the ovary draining LN.

Ovarian disease is completely inhibited by the transfer of large numbers of polyclonal Tregs of normal adult mice into 5-day-old day 3 thymectomized neonates. Transferred Treg are observed to distribute and proliferate equally in ovarian LNs and non-draining LNs. However, only the input Treg re-isolated from the ovarian LNs are capable of suppressing autoimmune ovarian disease when transferred in small numbers to second group of thymectomized mice; whereas Tregs re-isolated from the ovarian LNs are not able to suppress autoimmunity of the lachrymal gland while Tregs re-isolated from combined LNs that including those draining the lachrymal glands prevent autoimmunity of the lachrymal gland. These results suggest that protection depends on organ-specific Tregs present at higher frequency in the draining LN. Thus, in organ specific autoimmune diseases such as T1D, autoimmune thyroiditis, and autoimmune ovarian disease, tissue antigen specific Tregs in the regional LN of the target organs are superior therapeutic options when compared to polyclonal Tregs. However, it should be pointed out that although the use of antigen-specific Tregs is a laudable goal, current protocols for the expansion of human antigen-specific Tregs are problematic. Thus, it is likely that polyclonal Tregs, especially those with an adaptive phenotype will be tested in organ-specific autoimmune settings such as T1D due to source constraints and the possibility that in combination with other Teff "de-bulking" therapies will have a positive outcome. More importantly, recent data suggesting that polyclonal Tregs are not likely to have a pan immunosuppressive effect due to the ability of viruses and bacteria to circumvent Treg function and thus likely will be safe adjunct therapies.

Phenotype of Tregs and Mode of Their Actions **In Vivo**

Characterizing Treg phenotypes associated with better suppression in vivo will help to improve efficacy and lessen the requirements for high cell numbers. Investigations in this regard have identified numerous cell surface markers, in addition to CD4 and CD25, to enrich for more potent Tregs. These markers include CD62L, CD45Rb^{lo}, CD38, CD103, Dx5, active form of CD44, CD27, and chemokine receptors CCR4, CCR6, CCR8, and reduced CD127 expression [12,17,78,127–137]. Interestingly, the majority of these markers are associated with T cell activation status and tissue tropism, suggesting that Treg efficacy in vivo is linked to their ability to traffic to relevant tissue and become activated. However, expression of some of these markers, such as CD103 and CD62L, are mutually exclusive, making one wonder how both can be associated with enhanced Treg activity. One way to unify these findings is to consider that Treg efficacy is dictated by the tissue types and aggressiveness of the autoimmune immune response [132]. For example, studies from several laboratories demonstrate that CD62L$^+$ Tregs exhibit higher suppressive activity in vivo in preventing diabetes and GvHD [128,129,138–140]. One report further demonstrate that CD62L$^+$ Tregs express higher level of CCR7 [138], which is critical for the cells response to lymphoid homing chemokine CCL19. In models of chronic autoimmune diseases such as T1D and GvHD, effector T cells need to expand and differentiate first before infiltrating tissue and causing tissue damage.

Thus, Tregs that can migrate to LN and control the expansion of pathogenic T cells would achieve better therapeutic outcome. In contrast, in acute arthritis and delayed type hypersensitivity models, αE integrin $CD103^+$ $CD62L^{low}$ Tregs are found more efficacious. These Tregs migrated toward proinflammatory chemokines CCL20 and CXCL9 rather than LN homing chemokine, CCL19 [53,130]. CD103 expression also promotes the infiltration and retention of Tregs in peripheral tissue [141,142]. In the acute inflammation models used in these studies, the enhanced efficacy of $CD103^+$ Tregs may be due to their ability to efficiently and quickly migrate to the site of inflammation. Together, these results are consistent with the notion that Treg can control immune response both in the lymphoid organs (homeostatic control) and during an inflammatory response in the peripheral tissue (inflammation control). In this scenario, the LN-homing $CD62L^+$ Tregs are more efficacious at suppressing LN priming and clonal expansion thus exhibited superior protection before onset of tissue injury while tissue-seeking $CD103^+$ Tregs might offer immediate alleviation of tissue inflammation by directly trafficking to the inflamed tissue.

In Vivo *Efficacy of Adoptive Treg Therapy: The Issue of Numbers*

Multiple parallel efforts have been undertaken to use polyclonal Tregs therapeutically in mice. Several groups have shown that co-transplant of purified $CD4^+CD25^+$ Tregs in a murine model of bone marrow transplantation results in a significant reduction of GvHD while retaining potent graft versus leukemia responses [107–109]. Similarly, there has been multiple mouse studies reported in which Tregs have been shown to be efficacious in vivo following adoptive transfer into mice or rats either before or during clinical manifestations of the disease [103,105,106,115]. Although the efficacy of the studies is without dispute, the ability to adopt this therapy to the human clinical setting will ultimately depend on techniques to isolate and transfer adequate numbers.

The number of Tregs needed to achieve therapeutic effect in vivo is likely to vary depending on the disease in question. Many dose titration experiments described in the literature rely on the use of adoptive transfer systems in which the number of pathogenic T cell is known. In most of these experiments, a large number of Tregs, up to 3:1 Tregs to T effector cell ratio, are needed unless antigen-specific Tregs are used. In intact animals and patients, it is nearly impossible to estimate the total numbers of pathogenic T cells in LN and tissues, thus the therapeutic dosage needs to be determined empirically. In our experience, a large number of polyclonal Tregs ($>20 \times 10^6$/mouse) afforded partial protection against diabetes in the non-obese diabetic (NOD) mice [14]. The numbers are easier to estimate in the GvHD setting since the number of donor T cells is known. In mouse models of MHC haplotype mismatched bone marrow transplantation, ratios of 3:1–1:2 Treg to naïve T cells are required to protect against GvHD despite the likely presence of high frequency of antigen-specific (alloreactive) Tregs in the inoculums. In human bone marrow transplantation, approximately 5×10^8 T cells are transfused along with the bone marrow graft, thus, in principle, 2.5–15.0×10^8 Tregs would be needed to control GvHD. It is possible that less Tregs will be efficacious since most of the bone marrow transplants in clinical setting have at least partial MHC matching. However, the degree of homeostatic proliferation of the potential T effectors, or Tregs for that matter is not clear, complicating estimates of therapeutic dosing. Thus, even under the most generous estimates, it would seem unlikely that one could purify sufficient Tregs from an individual for transfer. For instance, assuming 1:1 Treg to effector T cell ratio is used, to isolate 5×10^8 Tregs from donor peripheral blood, which contain less than 2% of $CD4^+CD25^+Foxp3^+$ Tregs, one would need to sort from $2.5-5.0 \times 10^{10}$ peripheral blood mononuclear cells, which would amount to 25–50 liters of blood! However, if Tregs can be expanded 100-fold, one would need to sort cells from 250 to 500 ml of blood, the task becomes much more manageable.

Tregs were described to be anergic initially due to their inability to proliferate in vitro to antigens or anti-CD3 stimulation [96]. However, it is now known that Tregs replicate quite effectively in vivo [143–145]. Moreover, given the critical role for TCR occupancy, CD28 co-stimulation, and IL-2 in Treg activation and survival, several groups have used these agents to push the cells into cycle in vitro. Taking advantage of these observations, many laboratories have been able to expand mouse and human Tregs in vitro [27,82,103,104,108,146,147]. For instance, we have been able to use anti-CD3- and anti-CD28-coated paramagnetic beads and IL-2 to expand mouse Tregs more than 300-fold in less than two weeks. Such expansion is critically dependent on a large dose, e.g. 1000–2000 IU/ml, of recombinant IL-2 in addition to signaling through TCR and CD28. Polyclonal Tregs from BALB/c, C57BL/6, and NOD mice have been successfully expanded using this protocol. In addition, by isolating Tregs from TCR transgenic mice such as NOD.BDC2.5 and BALB/c DO11.10, antigen-specific Tregs can also be expanded for experimental use. Using a similar approach, human Tregs isolated from peripheral blood or cord blood can be repeatedly stimulated and expanded to more than 100-fold in three to four weeks while maintaining their suppressive functions [27,146].

Mature DCs (as well as artificial APCs) are also capable of overcoming the anergic phenotype of Tregs and drive them into proliferation without addition of anti-CD28. In fact, it has been postulated that DCs contribute to self-tolerance by maintaining Treg homeostasis in steady state [82]. Taking advantage of this unique property of the DCs, investigators were able to expand Tregs in vitro. Tarbell et al reported that using bone marrow-derived LPS-matured DCs, Tregs isolated from BDC2.5 mice were expanded 10-fold in one week using either anti-CD3 or an agonist peptide [104]. The ability to process and present antigens gives the DC-based protocols an added advantage in expanding antigen-specific Tregs (discussed in more details below). Instead of DCs or antibody-coated beads, Edinger and colleagues used modified L cells as artificial APC to stimulate human Tregs. They show that human Fcγ receptor II transfected murine L cells together with anti-CD3, anti-CD28, and IL-2 can be used to expand human Tregs up to 40,000 fold [147]. In the same study, these investigators reported that extensive expansion can also be achieved in a helper cell-free culture by using anti-CD3 and anti-CD28-coated beads. Similar results have been reported by June and colleagues in humans using artificial APC ([148]; unpublished observations). June and collaborators are adapting a T cell expansion protocol used successfully to grow bulk CD4+ T cells in a recently concluded clinical trial to expand Tregs [146,149,150]. The expanded Tregs will be used in an adoptive Treg therapy trial (Bruce Blazar, principle investigator, University of Minnesota) scheduled to start in 2007 for preemptive therapy of GvHD in adults at high risk for GvHD following high dose chemotherapy and stem cell transplantation.

In vitro expanded Tregs maintain their quintessential lineage markers and phenotype and are in general more suppressive than the freshly isolated Tregs [103,147]. Expanded Tregs remain to be $CD25^{hi}$ when compared to similarly expanded Teff cells, and typically greater than 90% of the cells express high level of Foxp3 after expansion. Other molecules implicated in Treg function such as CTLA-4 and PD-1 remains highly expressed on the cell surface of expanded Tregs. In some settings, but not all, enhanced production of TGFβ and IL-10 was observed after expansion [103,149]. Importantly, secretion of IL-2 and IFNγ by the cells remains low, suggesting minimal outgrowth and contamination of potentially pathogenic Teff cells. However, it should be noted that in the human setting, the inability to effectively isolate a "pure" Treg population can lead to overgrowth of non-Tregs. This problem has been approached in two ways. First, we have observed that CD127 is an excellent marker to distinguish Treg and memory cells in humans. Thus, using this additional marker assures a purer Treg isolation [12] In addition, recent studies have shown that the addition of rapamycin to the in vitro culture selectively inhibits Teff expansion and assures outgrowth of the Tregs [151,152]. Thus, a combination of cell surface markers, culture conditions and regulatory agents will be need in development of expanded Tregs for clinical use.

Importantly, even after extensive expansion, Tregs continue to express CD62L and CCR7, and maintain their ability to home to peripheral LNs and further expand in vivo [103,146,147]. Despite prolonged in vitro stimulation during expansion, in vitro suppression by the expanded Tregs require reactivation through their TCR demonstrating that the expanded Tregs are not constitutively suppressive. Thus, when applied in vivo, functions of expanded Tregs will depend on their encounter with cognate antigens making them unlikely to be pan-immunosuppressive. Although, there is some evidence that Tregs can function in an antigen non-specific manner by competing for space [119], we believe that the concerns related to non-specific immunosuppressive effects of Tregs on infectious disease are limited as investigators have demonstrated that effective pathogen-specific immune responses are Treg-resistant [88]. In summary, it has been clearly established that the anergic phenotype of the Tregs can be overcome in vitro through a varieties of means. In fact, Tregs show enormous proliferative potential when properly stimulated demonstrating that they are not terminally differentiated end-stage cells. Most importantly, expanded Tregs are almost indistinguishable from freshly isolated Tregs in terms of their phenotypic markers and suppressive activities. These findings affirm the technical feasibility of expanding these cells for therapeutic use in vivo and suggest that the polyclonal approach to autoimmune therapy is a good starting point to test this cell therapy clinically.

In Vitro *Expansion of Antigen-Specific Tregs*

The use of Tregs from BDC2.5 mice provided the proof-of-principal experimental data that in vivo efficacy of Treg-based therapy for type I diabetes relies critically on the specificity of the Tregs. In order to apply the therapy to patients, a protocol must be developed to selectively expand antigen-specific Tregs from a polyclonal source. Expansion of antigen-specific Treg from a pool of polyclonal Treg is technically challenging. The frequency of antigen-specific Treg is currently unknown but is likely similar to that for antigen-specific T cells in the $CD4^+$ effector pool, which are estimated to be present in the range of 1 in 2,000–1 in 200,000 or less [122]. Despite the expected low frequency, expansion of antigen-specific Tregs from polyclonal populations has been achieved. We generated recombinant soluble NOD MHC class II, IA^{g7}, linked to a peptide mimetope for the BDC2.5 TCR, and immobilized the MHC-peptide complex together with anti-CD28 mAb onto paramagnetic beads. These beads were used to stimulate polyclonal NOD Tregs. By seeding a tracer population of BDC2.5 Tregs in the polyclonal pool, we demonstrated that such an approach would selectively expand antigen-specific Tregs when the precursor frequency was as low as 1 in 100,000. Indeed, using this method we were able to expand antigen-specific Tregs from the polyclonal population. The resulting expanded Tregs showed enhanced in vivo activity in suppressing diabetes when compared to anti-CD3-expanded Tregs [153]. However, the islet peptide-mimic-reactive Tregs expanded from polyclonal source were not as efficient as the BDC2.5 Tregs in controlling disease. The reason for this is currently unknown. One possibility is that the Tregs expanded from polyclonal source may not recognize cognate antigen with the same high avidity as the BDC2.5 Treg. Tregs expanded with BDC2.5 mimetope were found to be oligoclonal as determined by TCR β usage and likely to bind to peptide-MHC class II multimers with varying avidity. In addition, it is likely that only a subset of islet peptide-mimic expanded Treg responded to naturally occurring islet antigen in vivo. It is also possible that the different cytokine milieu generated under the two conditions account for the difference in their suppressive functions in vivo. In conditions where essentially the entire population of BDC2.5 Tregs was activated, high concentrations of IL-10 and potentially TGFβ were produced, both of which have been reported to promote Treg production and function [103]. This is in contrast to the cytokine milieu expected for cultures of polyclonal Tregs stimulated with a single specificity MHC-peptide complex where only a few antigen-specific cells are activated and capable of producing IL-10 and TGFβ. Thus, careful screening of antigenic peptides and supplementing cultures with factors that promote Treg development may help to enhance the in vivo suppressive function of the expanded Tregs.

DCs have also been used to expand antigen-specific Tregs. The ability of DCs to process and present multiple epitopes from whole protein endows these cells with the possibility of generating oligoclonal tissue-specific Treg. In particular, Tregs with direct allogeneic antigen-specificities are readily expanded with allogeneic DCs [154]. Tregs enriched for alloantigen specificity have been shown to be more efficacious in preventing GvHD in bone marrow transplant recipients. The relative ease in expanding alloreactive cells is likely due to their extraordinarily high frequency comparing to autoantigen-specific Tregs. The use of DCs for expanding rare antigen-specific Treg from polyclonal pool has been demonstrated using transgenic mice expressing HA in pancreatic islets (ins-HA). Starting with Treg from ins-HA mice where HA is expressed as a neo-self pancreatic antigen, DCs presenting HA peptide expanded Treg with HA-dependent suppressive function [155]. Furthermore, although bulk DCs processed and presented specific HA epitopes from lysates of HA-producing islet cells, $CD8^+$ lymphoid DCs were more efficient in inducing the expansion of antigen-specific Treg due to the decreased non-specific background proliferation of polyclonal Treg as compared to $CD8^-$ myeloid DCs.

Complementary Approaches to Treg-Based Therapies

Given the challenges in isolating Tregs for cell-based therapeutics, a variety of protocols have been developed to augment Treg activities in vivo using antigen [156,157], cytokines [158], monoclonal antibodies [10,159], DCs [160], and immunosuppressive drugs such as rapamycin [161]. It is conceivable that these therapies can be used to condition Treg donors to provide an enriched source of Tregs. Alternatively, they can be combined with Treg-based therapies to enhance treatment efficacy especially when Treg numbers are limiting. Since these areas are subjects of other reviews in this issue, we will limit our discussion to anti-CD3 therapy.

One of the most promising interventions for T1D is the application of Fc receptor (FCR) non-binding anti-CD3 mAb [159,162–168]. Preclinical studies in mice with autoimmune diabetes induced with multiple low doses of streptozocin and in the NOD mouse showed that FCR non-binding anti-CD3 mAb prevented and even reversed autoimmune diabetes [167,169,170]. Chatenoud et al found that treatment of NOD mice at the time of diabetes onset with anti-CD3 mAb 145-2C11 restored normal glucose homeostasis in 80% of the mice. More importantly, the mice remain normoglycemic for the duration of the study in the absence of any ongoing immunotherapy. Based on these animal studies, a humanized mouse anti-human CD3 (OKT3 human IgG1 with mutations at positions 234 and 235 to eliminate FCR binding) was tested in patients with recent diabetes onset. Patients were treated within 6 weeks of diagnosis for two weeks with the FCR non-binding anti-CD3 mAb. At one year post therapy, the majority of patients showed sustained insulin secretion, as indicated by serum C-peptide levels, indicating increased preservation of remaining β-cell mass as a consequence of the tolerogenic therapy. Recent two-year data suggests that the affects of the FCR non-binding anti-CD3 therapy is long-term [171]. Similar results were noted in a subsequent blinded phase II trial with the additional observations that a significant percentage of patients were taking a homeopathic dose of insulin consistent with the therapy resulting in insulin independence [172]. Mechanistic studies of the treated patients and mice suggested that the effect of the antibody was multifold. On the one hand, short term, in vivo treatment with FCR non-binding anti-CD3 preferentially deleted activated

CD4$^+$ effector T cells (QT and JAB, unpublished observation, [173]), induced unresponsiveness of pathogenic cells [174,175], and eliminated the T cell infiltrate in the pancreas. In addition, there are long-term consequences of the immunotherapy. Analysis of mice in diabetes remission showed that the FCR non-binding anti-CD3 treatment augmented tolerogenic mechanisms including the induction of Tregs [10,176].

The effects of FCR non-binding anti-CD3 therapy on Tregs are complex. First of all, thymic-derived CD4$^+$CD25$^+$ "natural Tregs" are more resistant to anti-CD3 induced cell death than their CD25$^-$ counterpart. Thus, while the absolute number of the Tregs is not significantly affected by FCR non-binding anti-CD3 treatment; the relative proportion of Treg to auto-aggressive T cells is greatly enhanced due to selective depletion of CD25$^-$ cells (QT and JAB, unpublished observations). In addition, FCR non-binding anti-CD3 treatment also induces adaptive Tregs that function in a cytokine-dependent manner. Belgith et al. demonstrated that the FCR non-binding anti-CD3 therapy induced prolonged elevation of TGFβ expressing CD4 T cells in the draining LN and long-term diabetes remission was TGFβ dependent [10]. In humans, production of IL-5 and IL-10 was significantly increased in the patients' sera treated with FCR non-binding anti-CD3, while IL-2 and IFN-γ were infrequently detected [168]. The production of these suppressive cytokines most likely explains the long-lasting effects of the monoclonal antibodies in murine models and in patients. Additional support for the induction of adaptive Tregs comes from experiments in the natural Treg-deficient NOD.CD28$^{-/-}$ mice. Diabetes in these mice can also be reversed with FCR non-binding anti-CD3 treatment given at the time of disease onset. The protection was associated with the appearance of CD4$^+$CD25^{low} T cells that exhibit TGFβ dependent suppressive activity [177]. Lastly, in vivo and in vitro FCR non-binding anti-CD3 treatment also induces CD8$^+$CD25$^+$ that express CTLA-4 and Foxp3. These CD8$^+$ Tregs are anergic and suppress CD4$^+$ T cell proliferation in a contact dependent manner [168]. Taken together, these data favor the model that FCR non-binding anti-CD3 function by purging autoreactive cells and inducing active regulation in both autoimmune prone mice and humans.

Engineering Tregs with Desired Specificity

One alternative to expanding antigen-specific Tregs is to redirect polyclonal Tregs to recognize particular autoantigen of choice through a gene therapy approach [178,179]. In these studies, the Tregs were modified to express a chimeric receptor that links the myelin basic protein peptide 89-101 to its restricting MHC I-A^s with the cytoplasmic domain of TCR ζ chain. This chimeric receptor serves as a surrogate TCR and transduces an activation signal upon engaging their cognate TCR expressed on the surface of I-A^s MBP 89-101-specific autoreactive T cells. In vivo infusion of 1×10^6 of the receptor-modified Tregs prevented and reversed experimental autoimmune encephalitis. The receptor-modified Tregs was efficacious even after the antigen epitopes have spread to proteolipid protein, distinct from the

MBP 89-101 used for the immunization, suggesting that they mediated bystander suppression [179]. The receptor-modified Tregs appeared to suppress disease by converting endogenous MBP-reactive cells into IL-10-producing adaptive Tregs [178]. It is important to point out that unlike the conventional Tregs that are activated by APC; the receptor-modified Tregs directly interact with self-reactive T cells. Thus, the function of these receptor-modified Tregs may not resemble the mode of action of conventional Tregs. Redirecting T cell by receptor modification has been explored as a therapeutic avenue for enhancing anti-tumor immunity [180]. In majority of these studies, one of two means was used to redirect the T cell specificity. First, forced expression of a TCR specific for a tumor peptide and its presenting MHC has been shown to confer the desired specificity. Second, chimeric receptor with an extracellular domain of an immunoglobulin specific for an antigen of interest linked to the intracellular domain of TCR ζ chain has been shown to behave like a surrogate TCR upon ligation with the specified antigen in its unprocessed native form. It would be interesting to determine if these approaches can be applied to redirect Tregs for therapeutic use in autoimmune settings. Lastly, one novel protocol employed by Wood and colleagues to generate Tregs of desired in vivo efficacy was to acutely activate Tregs with antigen in vitro before adoptive transfer. These activated "effector" Tregs were able to exert their suppressive effect without further in vivo activation, thus obviate the need for antigen-specific Tregs. In a skin transplant model, the acutely activated antigen-irrelevant Tregs were able to prevent graft rejection through the induction of dominant infectious tolerance [181].

Conclusion

Ample experimental evidence demonstrates that Tregs can be generated in sufficient quantity in vitro and can be an effective treatment option for autoimmune diseases. Over the past several years, various protocols have been developed to overcome many technical challenges in generating Tregs in sufficient number with desired properties for therapeutic applications. Better understanding of Treg functions in vivo and dynamics between Tregs and pathogenic T cells in autoimmune diseases will help us to improve the design of Treg-based therapeutics. Investigations in this regard in the past decade since the rebirth of Tregs/suppressor cells have provided valuable information that allows us to sketch out a design for Treg-based therapies. Such therapy represents a novel form of individualized medicine that allows physicians to tailor the treatment for the specific condition of individual patient to achieve optimal therapeutic outcome. Furthermore, data are emerging to show that conventional therapies already in clinical use, such as FCR non-binding anti-CD3 and rapamycin, may synergize with Tregs in controlling autoimmune responses. Thus, our hopes are high that Tregs holds the promise of halting autoimmune attacks and restoring long-term self-tolerance.

References

1. Khattri R, Cox T, Yasayko SA, Ramsdell F: An essential role for Scurfin in CD4+CD25+ T regulatory cells. Nat Immunol 2003 Apr, 4:337–42.
2. Fontenot JD, Gavin MA, Rudensky AY: Foxp3 programs the development and function of CD4+CD25+ regulatory T cells. Nat Immunol 2003 Apr, 4:330–6.
3. Hori S, Nomura T, Sakaguchi S: Control of regulatory T cell development by the transcription factor Foxp3. Science 2003 Feb 14, 299:1057–61.
4. Wildin RS, Ramsdell F, Peake J, Faravelli F, Casanova JL, Buist N, Levy-Lahad E, Mazzella M, Goulet O, Perroni L, et al.: X-linked neonatal diabetes mellitus, enteropathy and endocrinopathy syndrome is the human equivalent of mouse scurfy. Nat Genet 2001 Jan, 27:18–20.
5. Bennett CL, Christie J, Ramsdell F, Brunkow ME, Ferguson PJ, Whitesell L, Kelly TE, Saulsbury FT, Chance PF, Ochs HD: The immune dysregulation, polyendocrinopathy, enteropathy, X-linked syndrome (IPEX) is caused by mutations of FOXP3. Nat Genet 2001 Jan, 27:20–1.
6. Kim J, Rasmussen J, Rudensky A: Regulatory T cells prevent catastrophic autoimmunity throughout the lifespan of mice. Nat Immunol 2006 Nov 30.
7. Malek TR, Yu A, Vincek V, Scibelli P, Kong L: CD4 regulatory T cells prevent lethal autoimmunity in IL-2Rbeta- deficient mice. Implications for the nonredundant function of IL-2. Immunity 2002, 17:167–78.
8. Fontenot JD, Rasmussen JP, Gavin MA, Rudensky AY: A function for interleukin 2 in Foxp3-expressing regulatory T cells. Nat Immunol 2005 Nov, 6:1142–51.
9. D'Cruz LM, Klein L: Development and function of agonist-induced CD25+Foxp3+ regulatory T cells in the absence of interleukin 2 signaling. Nat Immunol 2005 Nov, 6:1152–9.
10. Belghith M, Bluestone JA, Barriot S, Megret J, Bach JF, Chatenoud L: TGF-beta-dependent mechanisms mediate restoration of self-tolerance induced by antibodies to CD3 in overt autoimmune diabetes. Nat Med 2003 Sep, 9: –8.
11. Sakaguchi S: Naturally arising CD4+ regulatory t cells for immunologic self-tolerance and negative control of immune responses. Annu Rev Immunol 2004, 22:531–62.
12. Liu W, Putnam AL, Xu-Yu Z, Szot GL, Lee MR, Zhu S, Gottlieb PA, Kapranov P, Gingeras TR, Fazekas de St Groth B, et al.: CD127 expression inversely correlates with FoxP3 and suppressive function of human CD4+ T reg cells. J Exp Med 2006 Jul 10, 203:1701–11.
13. Moudgil KD, Sercarz EE: The self-directed T cell repertoire: its creation and activation. Rev Immunogenet 2000, 2:26–37.
14. Salomon B, Lenschow DJ, Rhee L, Ashourian N, Singh B, Sharpe A, Bluestone JA: B7/CD28 costimulation is essential for the homeostasis of the CD4+CD25+ immunoregulatory T cells that control autoimmune diabetes. Immunity 2000, 12:431–40.
15. Setoguchi R, Hori S, Takahashi T, Sakaguchi S: Homeostatic maintenance of natural Foxp3(+) CD25(+) CD4(+) regulatory T cells by interleukin (IL)-2 and induction of autoimmune disease by IL-2 neutralization. J Exp Med 2005 Mar 7, 201:723–35.
16. Herman AE, Freeman GJ, Mathis D, Benoist C: CD4+CD25+ T regulatory cells dependent on ICOS promote regulation of effector cells in the prediabetic lesion. J Exp Med 2004 Jun 7, 199:1479–89.
17. Gonzalez A, Andre-Schmutz I, Carnaud C, Mathis D, Benoist C: Damage control, rather than unresponsiveness, effected by protective DX5+ T cells in autoimmune diabetes. Nat Immunol 2001 Dec, 2:1117–25.
18. Chen Z, Herman AE, Matos M, Mathis D, Benoist C: Where CD4+CD25+ T reg cells impinge on autoimmune diabetes. J Exp Med 2005 Nov 21, 202:1387–97.
19. You S, Belghith M, Cobbold S, Alyanakian MA, Gouarin C, Barriot S, Garcia C, Waldmann H, Bach JF, Chatenoud L: Autoimmune diabetes onset results from qualitative rather than quantitative age-dependent changes in pathogenic T-cells. Diabetes 2005 May, 54:1415–22.

20. Pop SM, Wong CP, Culton DA, Clarke SH, Tisch R: Single cell analysis shows decreasing FoxP3 and TGFbeta1 coexpressing CD4+CD25+ regulatory T cells during autoimmune diabetes. J Exp Med 2005 Apr 18, 201:1333–46.
21. Gregori S, Giarratana N, Smiroldo S, Adorini L: Dynamics of pathogenic and suppressor T cells in autoimmune diabetes development. J Immunol 2003 Oct 15, 171:4040–7.
22. Kukreja A, Cost G, Marker J, Zhang C, Sun Z, Lin-Su K, Ten S, Sanz M, Exley M, Wilson B, et al.: Multiple immuno-regulatory defects in type-1 diabetes. J Clin Invest 2002 Jan, 109:131–40.
23. Lindley S, Dayan CM, Bishop A, Roep BO, Peakman M, Tree TI: Defective suppressor function in CD4(+)CD25(+) T-cells from patients with type 1 diabetes. Diabetes 2005 Jan, 54:92–9.
24. Brusko TM, Wasserfall CH, Clare-Salzler MJ, Schatz DA, Atkinson MA: Functional defects and the influence of age on the frequency of CD4+ CD25+ T-cells in type 1 diabetes. Diabetes 2005 May, 54:1407–14.
25. Putnam AL, Vendrame F, Dotta F, Gottlieb PA: CD4+CD25high regulatory T cells in human autoimmune diabetes. J Autoimmun 2005 Feb, 24:55–62.
26. Fontenot JD, Dooley JL, Farr AG, Rudensky AY: Developmental regulation of Foxp3 expression during ontogeny. J Exp Med 2005 Oct 3, 202:901–6.
27. Earle KE, Tang Q, Zhou X, Liu W, Zhu S, Bonyhadi ML, Bluestone JA: In vitro expanded human CD4+CD25+ regulatory T cells suppress effector T cell proliferation. Clin Immunol 2005 Apr, 115:3–9.
28. Walker MR, Kasprowicz DJ, Gersuk VH, Benard A, Van Landeghen M, Buckner JH, Ziegler SF: Induction of FoxP3 and acquisition of T regulatory activity by stimulated human CD4+CD25- T cells. J Clin Invest 2003 Nov, 112:1437–43.
29. Allan SE, Passerini L, Bacchetta R, Crellin N, Dai M, Orban PC, Ziegler SF, Roncarolo MG, Levings MK: The role of 2 FOXP3 isoforms in the generation of human CD4+ Tregs. J Clin Invest 2005 Nov, 115:3276–84.
30. Cyster JG: Lymphoid organ development and cell migration. Immunol Rev 2003 Oct, 195:5–14.
31. Itano AA, Jenkins MK: Antigen presentation to naive CD4 T cells in the lymph node. Nat Immunol 2003 Aug, 4:733–9.
32. Itano AA, McSorley SJ, Reinhardt RL, Ehst BD, Ingulli E, Rudensky AY, Jenkins MK: Distinct dendritic cell populations sequentially present antigen to CD4 T cells and stimulate different aspects of cell-mediated immunity. Immunity 2003 Jul, 19:47–57.
33. Gagnerault MC, Luan JJ, Lotton C, Lepault F: Pancreatic lymph nodes are required for priming of beta cell reactive T cells in NOD mice. J Exp Med 2002 Aug 5, 196:369–77.
34. Levisetti MG, Suri A, Frederick K, Unanue ER: Absence of lymph nodes in NOD mice treated with lymphotoxin-beta receptor immunoglobulin protects from diabetes. Diabetes 2004 Dec, 53:3115–9.
35. Bour-Jordan H, Salomon BL, Thompson HL, Szot GL, Bernhard MR, Bluestone JA: Costimulation controls diabetes by altering the balance of pathogenic and regulatory T cells. J Clin Invest 2004 Oct, 114:979–87.
36. Yadav D, Judkowski V, Flodstrom-Tullberg M, Sterling L, Redmond WL, Sherman L, Sarvetnick N: B7-2 (CD86) controls the priming of autoreactive CD4 T cell response against pancreatic islets. J Immunol 2004 Sep 15, 173:3631–9.
37. Rabinovitch A, Suarez-Pinzon WL: Role of cytokines in the pathogenesis of autoimmune diabetes mellitus. Rev Endocr Metab Disord 2003 Sep, 4:291–9.
38. Segal BM: CNS chemokines, cytokines, and dendritic cells in autoimmune demyelination. J Neurol Sci 2005 Feb 15, 228:210–4.
39. Bour-Jordan H, Thompson HL, Bluestone JA: Distinct effector mechanisms in the development of autoimmune neuropathy versus diabetes in nonobese diabetic mice. J Immunol 2005 Nov 1, 175:5649–55.

40. Langrish CL, Chen Y, Blumenschein WM, Mattson J, Basham B, Sedgwick JD, McClanahan T, Kastelein RA, Cua DJ: IL-23 drives a pathogenic T cell population that induces autoimmune inflammation. J Exp Med 2005 Jan 17, 201:233–40.
41. Nakae S, Saijo S, Horai R, Sudo K, Mori S, Iwakura Y: IL-17 production from activated T cells is required for the spontaneous development of destructive arthritis in mice deficient in IL-1 receptor antagonist. Proc Natl Acad Sci USA 2003 May 13, 100:5986–90.
42. Luster AD, Alon R, von Andrian UH: Immune cell migration in inflammation: present and future therapeutic targets. Nat Immunol 2005 Dec, 6:1182–90.
43. Svensson M, Johansson-Lindbom B, Wurbel MA, Malissen B, Marquez G, Agace W: Selective generation of gut-tropic T cells in gut-associated lymphoid tissues: requirement for GALT dendritic cells and adjuvant. Ann NY Acad Sci 2004 Dec, 1029:405–7.
44. Johansson-Lindbom B, Svensson M, Wurbel MA, Malissen B, Marquez G, Agace W: Selective generation of gut tropic T cells in gut-associated lymphoid tissue (GALT): requirement for GALT dendritic cells and adjuvant. J Exp Med 2003 Sep 15, 198:963–9.
45. Mora JR, Bono MR, Manjunath N, Weninger W, Cavanagh LL, Rosemblatt M, Von Andrian UH: Selective imprinting of gut-homing T cells by Peyer's patch dendritic cells. Nature 2003 Jul 3, 424:88–93.
46. Dudda JC, Lembo A, Bachtanian E, Huehn J, Siewert C, Hamann A, Kremmer E, Forster R, Martin SF: Dendritic cells govern induction and reprogramming of polarized tissue-selective homing receptor patterns of T cells: important roles for soluble factors and tissue microenvironments. Eur J Immunol 2005 Apr, 35:1056–65.
47. Michie SA, Sytwu HK, McDevitt JO, Yang XD: The roles of alpha 4-integrins in the development of insulin-dependent diabetes mellitus. Curr Top Microbiol Immunol 1998 231: 65–83.
48. Friedline RH, Wong CP, Steeber DA, Tedder TF, Tisch R: L-selectin is not required for T cell-mediated autoimmune diabetes. J Immunol 2002 Mar 15, 168:2659–66.
49. Mora C, Grewal IS, Wong FS, Flavell RA: Role of L-selectin in the development of autoimmune diabetes in non-obese diabetic mice. Int Immunol 2004 Feb, 16:257–64.
50. Kim SH, Cleary MM, Fox HS, Chantry D, Sarvetnick N: CCR4-bearing T cells participate in autoimmune diabetes. J Clin Invest 2002 Dec, 110:1675–86.
51. Frigerio S, Junt T, Lu B, Gerard C, Zumsteg U, Hollander GA, Piali L: Beta cells are responsible for CXCR3-mediated T-cell infiltration in insulitis. Nat Med 2002 Dec, 8: 1414–20.
52. Apostolou I, Sarukhan A, Klein L, von Boehmer H: Origin of regulatory T cells with known specificity for antigen. Nat Immunol 2002, 3:756–63.
53. Siegmund K, Feuerer M, Siewert C, Ghani S, Haubold U, Dankof A, Krenn V, Schon MP, Scheffold A, Lowe JB, et al.: Migration matters: regulatory T cell compartmentalization determines suppressive activity in vivo. Blood 2005 Jul 12.
54. McHugh RS, Shevach EM: Cutting edge: depletion of CD4+CD25+ regulatory T cells is necessary, but not sufficient, for induction of organ-specific autoimmune disease. J Immunol 2002, 168:5979–83.
55. Chen ML, Pittet MJ, Gorelik L, Flavell RA, Weissleder R, von Boehmer H, Khazaie K: Regulatory T cells suppress tumor-specific CD8 T cell cytotoxicity through TGF-beta signals in vivo. Proc Natl Acad Sci USA 2005 Jan 11, 102:419–24.
56. Lohr J, Knoechel B, Jiang S, Sharpe AH, Abbas AK: The inhibitory function of B7 costimulators in T cell responses to foreign and self-antigens. Nat Immunol 2003 Jul, 4: 664–9.
57. Sarween N, Chodos A, Raykundalia C, Khan M, Abbas AK, Walker LS: CD4+CD25+ cells controlling a pathogenic CD4 response inhibit cytokine differentiation, CXCR-3 expression, and tissue invasion. J Immunol 2004 Sep 1, 173:2942–51.
58. Tang Q, Adams JY, Tooley AJ, Bi M, Fife BT, Serra P, Santamaria P, Locksley RM, Krummel MF, Bluestone JA: Visualizing regulatory T cell control of autoimmune responses in nonobese diabetic mice. Nat Immunol 2006 Jan, 7:83–92.

59. Thornton AM, Donovan EE, Piccirillo CA, Shevach EM: Cutting edge: IL-2 is critically required for the in vitro activation of CD4+CD25+ T cell suppressor function. J Immunol 2004 Jun 1, 172:6519–23.
60. Thornton AM, Piccirillo CA, Shevach EM: Activation requirements for the induction of CD4+CD25+ T cell suppressor function. Eur J Immunol 2004 Feb, 34:366–76.
61. Huehn J, Siegmund K, Hamann A: Migration rules: functional properties of naive and effector/memory-like regulatory T cell subsets. Curr Top Microbiol Immunol 2005 293: 89–114.
62. Graca L, Cobbold SP, Waldmann H: Identification of regulatory T cells in tolerated allografts. J Exp Med 2002 Jun 17, 195:1641–6.
63. Yu P, Lee Y, Liu W, Krausz T, Chong A, Schreiber H, Fu YX: Intratumor depletion of CD4+ cells unmasks tumor immunogenicity leading to the rejection of late-stage tumors. J Exp Med 2005 Mar 7, 201:779–91.
64. Lee I, Wang L, Wells AD, Dorf ME, Ozkaynak E, Hancock WW: Recruitment of Foxp3+ T regulatory cells mediating allograft tolerance depends on the CCR4 chemokine receptor. J Exp Med 2005 Apr 4, 201:1037–44.
65. Muthukumar T, Dadhania D, Ding R, Snopkowski C, Naqvi R, Lee JB, Hartono C, Li B, Sharma VK, Seshan SV, et al.: Messenger RNA for FOXP3 in the urine of renal-allograft recipients. N Engl J Med 2005 Dec 1, 353:2342–51.
66. Iezzi G, Scotet E, Scheidegger D, Lanzavecchia A: The interplay between the duration of TCR and cytokine signaling determines T cell polarization. Eur J Immunol 1999 Dec, 29:4092–101.
67. Gett AV, Sallusto F, Lanzavecchia A, Geginat J: T cell fitness determined by signal strength. Nat Immunol 2003 Apr, 4:355–60.
68. Huppa JB, Gleimer M, Sumen C, Davis MM: Continuous T cell receptor signaling required for synapse maintenance and full effector potential. Nat Immunol 2003 Aug, 4:749–55.
69. Stoll S, Delon J, Brotz TM, Germain RN: Dynamic imaging of T cell-dendritic cell interactions in lymph nodes. Science 2002 Jun 7, 296:1873–6.
70. Mempel TR, Henrickson SE, Von Andrian UH: T-cell priming by dendritic cells in lymph nodes occurs in three distinct phases. Nature 2004 Jan 8, 427:154–9.
71. Miller MJ, Wei SH, Parker I, Cahalan MD: Two-photon imaging of lymphocyte motility and antigen response in intact lymph node. Science 2002 Jun 7, 296:1869–73.
72. Miller MJ, Safrina O, Parker I, Cahalan MD: Imaging the single cell dynamics of CD4+ T cell activation by dendritic cells in lymph nodes. J Exp Med 2004 Oct 4, 200:847–56.
73. Bousso P, Robey E: Dynamics of CD8+ T cell priming by dendritic cells in intact lymph nodes. Nat Immunol 2003 Jun, 4:579–85.
74. Tadokoro CE, Shakhar G, Shen S, Ding Y, Lino AC, Maraver A, Lafaille JJ, Dustin ML: Regulatory T cells inhibit stable contacts between CD4+ T cells and dendritic cells in vivo. J Exp Med 2006 Mar 13.
75. Nakamura K, Kitani A, Strober W: Cell contact-dependent immunosuppression by CD4(+)CD25(+) regulatory T cells is mediated by cell surface-bound transforming growth factor beta. J Exp Med 2001, 194:629–44.
76. Paust S, Lu L, McCarty N, Cantor H: Engagement of B7 on effector T cells by regulatory T cells prevents autoimmune disease. Proc Natl Acad Sci USA 2004 Jul 13, 101:10398–403.
77. Grossman WJ, Verbsky JW, Barchet W, Colonna M, Atkinson JP, Ley TJ: Human T regulatory cells can use the perforin pathway to cause autologous target cell death. Immunity 2004 Oct, 21:589–601.
78. Asseman C, Mauze S, Leach MW, Coffman RL, Powrie F: An essential role for interleukin 10 in the function of regulatory T cells that inhibit intestinal inflammation. J Exp Med 1999 Oct 4, 190:995–1004.
79. Read S, Malmstrom V, Powrie F: Cytotoxic T lymphocyte-associated antigen 4 plays an essential role in the function of CD25(+)CD4(+) regulatory cells that control intestinal inflammation. J Exp Med 2000, 192:295–302.

80. Kingsley CI, Karim M, Bushell AR, Wood KJ: CD25+CD4+ regulatory T cells prevent graft rejection: CTLA-4- and IL-10-dependent immunoregulation of alloresponses. J Immunol 2002 Feb 1, 168:1080–6.
81. Belkaid Y, Piccirillo CA, Mendez S, Shevach EM, Sacks DL: CD4+CD25+ regulatory T cells control Leishmania major persistence and immunity. Nature 2002 Dec 5, 420:502–7.
82. Yamazaki S, Iyoda T, Tarbell K, Olson K, Velinzon K, Inaba K, Steinman RM: Direct expansion of functional CD25+ CD4+ regulatory T cells by antigen-processing dendritic cells. J Exp Med 2003 Jul 21, 198:235–47.
83. Kubo T, Hatton RD, Oliver J, Liu X, Elson CO, Weaver CT: Regulatory T cell suppression and anergy are differentially regulated by proinflammatory cytokines produced by TLR-activated dendritic cells. J Immunol 2004 Dec 15, 173:7249–58.
84. Kared H, Masson A, Adle-Biassette H, Bach JF, Chatenoud L, Zavala F: Treatment with granulocyte colony-stimulating factor prevents diabetes in NOD mice by recruiting plasmacytoid dendritic cells and functional CD4(+)CD25(+) regulatory T-cells. Diabetes 2005 Jan, 54:78–84.
85. Bilsborough J, George TC, Norment A, Viney JL: Mucosal CD8alpha+ DC, with a plasmacytoid phenotype, induce differentiation and support function of T cells with regulatory properties. Immunology 2003 Apr, 108:481–92.
86. Cong Y, Konrad A, Iqbal N, Hatton RD, Weaver CT, Elson CO: Generation of antigen-specific, Foxp3-expressing CD4+ regulatory T cells by inhibition of APC proteosome function. J Immunol 2005 Mar 1, 174:2787–95.
87. Bruder D, Westendorf AM, Hansen W, Prettin S, Gruber AD, Qian Y, von Boehmer H, Mahnke K, Buer J: On the edge of autoimmunity: T-cell stimulation by steady-state dendritic cells prevents autoimmune diabetes. Diabetes 2005 Dec, 54:3395–401.
88. Pasare C, Medzhitov R: Toll pathway-dependent blockade of CD4+CD25+ T cell-mediated suppression by dendritic cells. Science 2003 Feb 14, 299:1033–6.
89. Oldenhove G, de Heusch M, Urbain-Vansanten G, Urbain J, Maliszewski C, Leo O, Moser M: CD4+ CD25+ regulatory T cells control T helper cell type 1 responses to foreign antigens induced by mature dendritic cells in vivo. J Exp Med 2003 Jul 21, 198:259–66.
90. Cederbom L, Hall H, Ivars F: CD4+CD25+ regulatory T cells down-regulate co-stimulatory molecules on antigen-presenting cells. Eur J Immunol 2000 Jun, 30:1538–43.
91. Serra P, Amrani A, Yamanouchi J, Han B, Thiessen S, Utsugi T, Verdaguer J, Santamaria P: CD40 ligation releases immature dendritic cells from the control of regulatory CD4+CD25+ T cells. Immunity 2003 Dec, 19:877–89.
92. Misra N, Bayry J, Lacroix-Desmazes S, Kazatchkine MD, Kaveri SV: Cutting edge: human CD4+CD25+ T cells restrain the maturation and antigen-presenting function of dendritic cells. J Immunol 2004 Apr 15, 172:4676–80.
93. Sato K, Tateishi S, Kubo K, Mimura T, Yamamoto K, Kanda H: Downregulation of IL-12 and a novel negative feedback system mediated by CD25+CD4+ T cells. Biochem Biophys Res Commun 2005 Apr 29, 330:226–32.
94. Lewkowich IP, Herman NS, Schleifer KW, Dance MP, Chen BL, Dienger KM, Sproles AA, Shah JS, Kohl J, Belkaid Y, et al.: CD4+CD25+ T cells protect against experimentally induced asthma and alter pulmonary dendritic cell phenotype and function. J Exp Med 2005 Dec 5, 202:1549–61.
95. Hsieh CS, Liang Y, Tyznik AJ, Self SG, Liggitt D, Rudensky AY: Recognition of the peripheral self by naturally arising CD25+ CD4+ T cell receptors. Immunity 2004 Aug, 21:267–77.
96. Takahashi T, Kuniyasu Y, Toda M, Sakaguchi N, Itoh M, Iwata M, Shimizu J, Sakaguchi S: Immunologic self-tolerance maintained by CD25+CD4+ naturally anergic and suppressive T cells: induction of autoimmune disease by breaking their anergic/suppressive state. Int Immunol 1998, 10:1969–80.
97. Nikolic T, Geutskens SB, van Rooijen N, Drexhage HA, Leenen PJ: Dendritic cells and macrophages are essential for the retention of lymphocytes in (peri)-insulitis of the nonobese diabetic mouse: a phagocyte depletion study. Lab Invest 2005 Apr, 85:487–501.

98. van Rijt LS, Jung S, Kleinjan A, Vos N, Willart M, Duez C, Hoogsteden HC, Lambrecht BN: In vivo depletion of lung CD11c+ dendritic cells during allergen challenge abrogates the characteristic features of asthma. J Exp Med 2005 Mar 21, 201:981–91.
99. Sakaguchi S, Takahashi T, Nishizuka Y: Study on cellular events in post-thymectomy autoimmune oophoritis in mice. II. Requirement of Lyt-1 cells in normal female mice for the prevention of oophoritis. J Exp Med 1982 Dec 1, 156:1577–86.
100. Sakaguchi S: Regulatory T cells: key controllers of immunologic self-tolerance. Cell 2000 May 26, 101:455–8.
101. Mason D, Powrie F: Control of immune pathology by regulatory T cells. Curr Opin Immunol 1998 Dec, 10:649–55.
102. Bach JF, Boitard C, Yasunami R, Dardenne M: Control of diabetes in NOD mice by suppressor cells. J Autoimmun 1990 Apr, 3(Suppl 1):97–100.
103. Tang Q, Henriksen KJ, Bi M, Finger EB, Szot G, Ye J, Masteller EL, McDevitt H, Bonyhadi M, Bluestone JA: In vitro-expanded antigen-specific regulatory T cells suppress autoimmune diabetes. J Exp Med 2004 Jun 7, 199:1455–65.
104. Tarbell KV, Yamazaki S, Olson K, Toy P, Steinman RM: CD25+ CD4+ T cells, expanded with dendritic cells presenting a single autoantigenic peptide, suppress autoimmune diabetes. J Exp Med 2004 Jun 7, 199:1467–77.
105. Zhang X, Koldzic DN, Izikson L, Reddy J, Nazareno RF, Sakaguchi S, Kuchroo VK, Weiner HL: IL-10 is involved in the suppression of experimental autoimmune encephalomyelitis by CD25+CD4+ regulatory T cells. Int Immunol 2004 Feb, 16:249–56.
106. Kohm AP, Carpentier PA, Anger HA, Miller SD: Cutting edge: CD4+CD25+ regulatory T cells suppress antigen-specific autoreactive immune responses and central nervous system inflammation during active experimental autoimmune encephalomyelitis. J Immunol 2002 Nov 1, 169:4712–6.
107. Edinger M, Hoffmann P, Ermann J, Drago K, Fathman CG, Strober S, Negrin RS: CD4+CD25+ regulatory T cells preserve graft-versus-tumor activity while inhibiting graft-versus-host disease after bone marrow transplantation. Nat Med 2003 Sep, 9:1144–50.
108. Trenado A, Charlotte F, Fisson S, Yagello M, Klatzmann D, Salomon BL, Cohen JL: Recipient-type specific CD4+CD25+ regulatory T cells favor immune reconstitution and control graft-versus-host disease while maintaining graft-versus-leukemia. J Clin Invest 2003 Dec, 112:1688–96.
109. Taylor PA, Lees CJ, Blazar BR: The infusion of ex vivo activated and expanded CD4(+)CD25(+) immune regulatory cells inhibits graft-versus-host disease lethality. Blood 2002 May 15, 99:3493–9.
110. Frey O, Petrow PK, Gajda M, Siegmund K, Huehn J, Scheffold A, Hamann A, Radbruch A, Brauer R: The role of regulatory T cells in antigen-induced arthritis: aggravation of arthritis after depletion and amelioration after transfer of CD4+CD25+ T cells. Arthritis Res Ther 2005, 7:R291–301.
111. Suri-Payer E, Cantor H: Differential cytokine requirements for regulation of autoimmune gastritis and colitis by CD4(+)CD25(+) T cells. J Autoimmun 2001 Mar, 16:115–23.
112. DiPaolo RJ, Glass DD, Bijwaard KE, Shevach EM: CD4+CD25+ T cells prevent the development of organ-specific autoimmune disease by inhibiting the differentiation of autoreactive effector T cells. J Immunol 2005 Dec 1, 175:7135–42.
113. Sakaguchi S, Sakaguchi N, Asano M, Itoh M, Toda M: Immunologic self-tolerance maintained by activated T cells expressing IL-2 receptor alpha-chains (CD25). Breakdown of a single mechanism of self-tolerance causes various autoimmune diseases. J Immunol 1995 Aug 1, 155:1151–64.
114. Tung KS, Setiady YY, Samy ET, Lewis J, Teuscher C: Autoimmune ovarian disease in day 3-thymectomized mice: the neonatal time window, antigen specificity of disease suppression, and genetic control. Curr Top Microbiol Immunol 2005 293:209–47.

115. Scalapino KJ, Tang Q, Bluestone JA, Bonyhadi ML, Daikh DI: Suppression of disease in New Zealand Black/New Zealand White lupus-prone mice by adoptive transfer of ex vivo expanded regulatory T cells. J Immunol 2006 Aug 1, 177:1451–9.
116. Cobbold S, Waldmann H: Infectious tolerance. Curr Opin Immunol 1998 Oct, 10:518–24.
117. Gershon RK, Kondo K: Infectious immunological tolerance. Immunology 1971 Dec, 21:903–14.
118. Trenado A, Sudres M, Tang Q, Maury S, Charlotte F, Gregoire S, Bonyhadi M, Klatzmann D, Salomon BL, Cohen JL: Ex Vivo-Expanded CD4+CD25+ Immunoregulatory T Cells Prevent Graft-versus-Host-Disease by Inhibiting Activation/Differentiation of Pathogenic T Cells. J Immunol 2006 Jan 15, 176:1266–73.
119. Stockinger B, Barthlott T, Kassiotis G: T cell regulation: a special job or everyone's responsibility? Nat Immunol 2001 Sep, 2:757–8.
120. Barthlott T, Kassiotis G, Stockinger B: T cell regulation as a side effect of homeostasis and competition. J Exp Med 2003 Feb 17, 197:451–60.
121. Suchin EJ, Langmuir PB, Palmer E, Sayegh MH, Wells AD, Turka LA: Quantifying the frequency of alloreactive T cells in vivo: new answers to an old question. J Immunol 2001 Jan 15, 166:973–81.
122. Novak EJ, Masewicz SA, Liu AW, Lernmark A, Kwok WW, Nepom GT: Activated human epitope-specific T cells identified by class II tetramers reside within a CD4high, proliferating subset. Int Immunol 2001 Jun, 13:799–806.
123. Thornton AM, Shevach EM: Suppressor effector function of CD4+CD25+ immunoregulatory T cells is antigen nonspecific. J Immunol 2000 Jan 1, 164:183–90.
124. Mukherjee R, Chaturvedi P, Qin HY, Singh B: CD4+CD25+ regulatory T cells generated in response to insulin B:9-23 peptide prevent adoptive transfer of diabetes by diabetogenic T cells. J Autoimmun 2003 Nov, 21:221–37.
125. Green EA, Choi Y, Flavell RA: Pancreatic lymph node-derived CD4(+)CD25(+) Treg cells: highly potent regulators of diabetes that require TRANCE-RANK signals. Immunity 2002 Feb, 16:183–91.
126. Seddon B, Mason D: Peripheral autoantigen induces regulatory T cells that prevent autoimmunity. J Exp Med 1999, 189:877–82.
127. Kuniyasu Y, Takahashi T, Itoh M, Shimizu J, Toda G, Sakaguchi S: Naturally anergic and suppressive CD25(+)CD4(+) T cells as a functionally and phenotypically distinct immunoregulatory T cell subpopulation. Int Immunol 2000, 12:1145–55.
128. Herbelin A, Gombert JM, Lepault F, Bach JF, Chatenoud L: Mature mainstream TCR alpha beta+CD4+ thymocytes expressing L-selectin mediate "active tolerance" in the nonobese diabetic mouse. J Immunol 1998, 161:2620–8.
129. Lepault F, Gagnerault MC: Characterization of peripheral regulatory CD4+ T cells that prevent diabetes onset in nonobese diabetic mice. J Immunol 2000 Jan 1, 164:240–7.
130. Huehn J, Siegmund K, Lehmann JC, Siewert C, Haubold U, Feuerer M, Debes GF, Lauber J, Frey O, Przybylski GK, et al.: Developmental stage, phenotype, and migration distinguish naive- and effector/memory-like CD4+ regulatory T cells. J Exp Med 2004 Feb 2, 199: 303–13.
131. Firan M, Dhillon S, Estess P, Siegelman MH: Suppressor activity and potency among regulatory T cells is discriminated by functionally active CD44. Blood 2006 Jan 15, 107: 619–27.
132. Alyanakian MA, You S, Damotte D, Gouarin C, Esling A, Garcia C, Havouis S, Chatenoud L, Bach JF: Diversity of regulatory CD4+T cells controlling distinct organ-specific autoimmune diseases. Proc Natl Acad Sci USA 2003 Dec 23, 100:15806–11.
133. Freeman CM, Chiu BC, Stolberg VR, Hu J, Zeibecoglou K, Lukacs NW, Lira SA, Kunkel SL, Chensue SW: CCR8 is expressed by antigen-elicited, IL-10-producing CD4+CD25+ T cells, which regulate Th2-mediated granuloma formation in mice. J Immunol 2005 Feb 15, 174:1962–70.

134. Iellem A, Mariani M, Lang R, Recalde H, Panina-Bordignon P, Sinigaglia F, D'Ambrosio D: Unique chemotactic response profile and specific expression of chemokine receptors CCR4 and CCR8 by CD4(+)CD25(+) regulatory T cells. J Exp Med 2001 Sep 17, 194:847–53.
135. Kleinewietfeld M, Puentes F, Borsellino G, Battistini L, Rotzschke O, Falk K: CCR6 expression defines regulatory effector/memory-like cells within the CD25(+)CD4+ T-cell subset. Blood 2005 Apr 1, 105:2877–86.
136. Sanchez-Ramon S, Navarro AJ, Aristimuno C, Rodriguez-Mahou M, Bellon JM, Fernandez-Cruz E, de Andres C: Pregnancy-induced expansion of regulatory T-lymphocytes may mediate protection to multiple sclerosis activity. Immunol Lett 2005 Jan 31, 96:195–201.
137. Koenen HJ, Fasse E, Joosten I: CD27/CFSE-based ex vivo selection of highly suppressive alloantigen-specific human regulatory T cells. J Immunol 2005 Jun 15, 174:7573–83.
138. Szanya V, Ermann J, Taylor C, Holness C, Fathman CG: The subpopulation of CD4+CD25+ splenocytes that delays adoptive transfer of diabetes expresses L-selectin and high levels of CCR7. J Immunol 2002 Sep 1, 169:2461–5.
139. Ermann J, Hoffmann P, Edinger M, Dutt S, Blankenberg FG, Higgins JP, Negrin RS, Fathman CG, Strober S: Only the CD62L+ subpopulation of CD4+CD25+ regulatory T cells protects from lethal acute GVHD. Blood 2005 Mar 1, 105:2220–6.
140. Taylor PA, Panoskaltsis-Mortari A, Swedin JM, Lucas PJ, Gress RE, Levine BL, June CH, Serody JS, Blazar BR: L-Selectin(hi) but not the L-selectin(lo) CD4+25+ T-regulatory cells are potent inhibitors of GVHD and BM graft rejection. Blood 2004 Dec 1, 104:3804–12.
141. Suffia I, Reckling SK, Salay G, Belkaid Y: A role for CD103 in the retention of CD4+CD25+ Treg and control of Leishmania major infection. J Immunol 2005 May 1, 174:5444–55.
142. Huehn J, Hamann A: Homing to suppress: address codes for Treg migration. Trends Immunol 2005 Dec, 26:632–6.
143. Tang Q, Henriksen KJ, Boden EK, Tooley AJ, Ye J, Subudhi SK, Zheng XX, Strom TB, Bluestone JA: Cutting edge: CD28 controls peripheral homeostasis of CD4+CD25+ regulatory T cells. J Immunol 2003 Oct 1, 171:3348–52.
144. Walker LS, Chodos A, Eggena M, Dooms H, Abbas AK: Antigen-dependent proliferation of CD4+ CD25+ regulatory T cells in vivo. J Exp Med 2003 Jul 21, 198:249–58.
145. Fisson S, Darrasse-Jeze G, Litvinova E, Septier F, Klatzmann D, Liblau R, Salomon BL: Continuous activation of autoreactive CD4+ CD25+ regulatory T cells in the steady state. J Exp Med 2003 Sep 1, 198:737–46.
146. Godfrey WR, Ge YG, Spoden DJ, Levine BL, June CH, Blazar BR, Porter SB: In vitro-expanded human CD4(+)CD25(+) T-regulatory cells can markedly inhibit allogeneic dendritic cell-stimulated MLR cultures. Blood 2004 Jul 15, 104:453–61.
147. Hoffmann P, Eder R, Kunz-Schughart LA, Andreesen R, Edinger M: Large-scale in vitro expansion of polyclonal human CD4(+)CD25high regulatory T cells. Blood 2004 Aug 1, 104:895–903.
148. Thomas AK, Maus MV, Shalaby WS, June CH, Riley JL: A cell-based artificial antigen-presenting cell coated with anti-CD3 and CD28 antibodies enables rapid expansion and long-term growth of CD4 T lymphocytes. Clin Immunol 2002 Dec, 105:259–72.
149. Godfrey WR, Spoden DJ, Ge YG, Baker SR, Liu B, Levine BL, June CH, Blazar BR, Porter SB: Cord blood CD4(+)CD25(+)-derived T regulatory cell lines express FoxP3 protein and manifest potent suppressor function. Blood 2005 Jan 15, 105:750–8.
150. Rapoport AP, Stadtmauer EA, Aqui N, Badros A, Cotte J, Chrisley L, Veloso E, Zheng Z, Westphal S, Mair R, et al.: Restoration of immunity in lymphopenic individuals with cancer by vaccination and adoptive T-cell transfer. Nat Med 2005 Nov, 11:1230–7.
151. Strauss L, Whiteside TL, Knights A, Bergmann C, Knuth A, Zippelius A: Selective survival of naturally occurring human CD4+CD25+Foxp3+ regulatory T cells cultured with rapamycin. J Immunol 2007 Jan 1, 178:320–9.
152. Battaglia M, Stabilini A, Migliavacca B, Horejs-Hoeck J, Kaupper T, Roncarolo MG: Rapamycin promotes expansion of functional CD4+CD25+FOXP3+ regulatory T cells of both healthy subjects and type 1 diabetic patients. J Immunol 2006 Dec 15, 177:8338–47.

153. Masteller EL, Warner MR, Tang Q, Tarbell KV, McDevitt H, Bluestone JA: Expansion of functional endogenous antigen-specific CD4+CD25+ regulatory T cells from nonobese diabetic mice. J Immunol 2005 Sep 1, 175:3053–9.
154. Trenado A, Fisson S, Braunberger E, Klatzmann D, Salomon BL, Cohen JL: Ex vivo selection of recipient-type alloantigen-specific CD4(+)CD25(+) immunoregulatory T cells for the control of graft-versus-host disease after allogeneic hematopoietic stem-cell transplantation. Transplantation 2004 Jan 15, 77:S32–4.
155. Fisson S, Djelti F, Trenado A, Billiard F, Liblau R, Klatzmann D, Cohen JL, Salomon BL: Therapeutic potential of self-antigen-specific CD4+ CD25+ regulatory T cells selected in vitro from a polyclonal repertoire. Eur J Immunol 2006 Apr, 36:817–27.
156. Jordan MS, Boesteanu A, Reed AJ, Petrone AL, Holenbeck AE, Lerman MA, Naji A, Caton AJ: Thymic selection of CD4+CD25+ regulatory T cells induced by an agonist self-peptide. Nat Immunol 2001, 2:301–6.
157. Apostolou I, von Boehmer H: In vivo instruction of suppressor commitment in naive T cells. J Exp Med 2004 May 17, 199:1401–8.
158. Roncarolo MG, Bacchetta R, Bordignon C, Narula S, Levings MK: Type 1 T regulatory cells. Immunol Rev 2001 Aug, 182:68–79.
159. Herold KC, Hagopian W, Auger JA, Poumian-Ruiz E, Taylor L, Donaldson D, Gitelman SE, Harlan DM, Xu D, Zivin RA, et al.: Anti-CD3 monoclonal antibody in new-onset type 1 diabetes mellitus. N Engl J Med 2002, 346:1692–8.
160. Mahnke K, Qian Y, Knop J, Enk AH: Induction of CD4+/CD25+ regulatory T cells by targeting of antigens to immature dendritic cells. Blood 2003 Jun 15, 101:4862–9.
161. Battaglia M, Stabilini A, Roncarolo MG: Rapamycin selectively expands CD4+CD25+FoxP3+ regulatory T cells. Blood 2005 Jun 15, 105:4743–8.
162. Chatenoud L: Immunotherapy of type 1 diabetes mellitus. Curr Dir Autoimmun 2001 4: 333–50.
163. Hirsch R, Gress RE, Pluznik DH, Eckhaus M, Bluestone JA: Effects of in vivo administration of anti-CD3 monoclonal antibody on T cell function in mice. II. In vivo activation of T cells. J Immunol 1989, 142:737–43.
164. Campos HH, Bach JF, Chatenoud L: Devising murine models to better adapt clinical protocols: sequential low-dose treatment with anti-CD3 and anti-CD4 monoclonal antibodies to prevent fully mismatched allograft rejection. Transplant Proc 1993 Feb, 25:798–9.
165. Chatenoud L, Ferran C, Bach JF: In-vivo anti-CD3 treatment of autoimmune patients. Lancet 1989 Jul 15, 2:164.
166. Chatenoud L, Thervet E, Primo J, Bach JF: Remission of established disease in diabetic NOD mice induced by anti-CD3 monoclonal antibody. C R Acad Sci III 1992, 315:225–8.
167. Chatenoud L, Thervet E, Primo J, Bach JF: Anti-CD3 antibody induces long-term remission of overt autoimmunity in nonobese diabetic mice. Proc Natl Acad Sci USA 1994, 91:123–7.
168. Herold KC, Burton JB, Francois F, Poumian-Ruiz E, Glandt M, Bluestone JA: Activation of human T cells by FcR nonbinding anti-CD3 mAb, hOKT3gamma1(Ala-Ala). J Clin Invest 2003 Feb, 111:409–18.
169. Herold KC, Bluestone JA, Montag AG, Parihar A, Wiegner A, Gress RE, Hirsch R: Prevention of autoimmune diabetes with nonactivating anti-CD3 monoclonal antibody. Diabetes 1992, 41:385–91.
170. Chatenoud L, Primo J, Bach JF: CD3 antibody-induced dominant self tolerance in overtly diabetic NOD mice. J Immunol 1997, 158:2947–54.
171. Herold KC, Gitelman SE, Masharani U, Hagopian W, Bisikirska B, Donaldson D, Rother K, Diamond B, Harlan DM, Bluestone JA: A single course of anti-CD3 monoclonal antibody hOKT3gamma1(Ala-Ala) results in improvement in C-peptide responses and clinical parameters for at least 2 years after onset of type 1 diabetes. Diabetes 2005 Jun, 54:1763–9.
172. Keymeulen B, Vandemeulebroucke E, Ziegler AG, Mathieu C, Kaufman L, Hale G, Gorus F, Goldman M, Walter M, Candon S, et al.: Insulin needs after CD3-antibody therapy in new-onset type 1 diabetes. N Engl J Med 2005 Jun 23, 352:2598–608.

173. Yu XZ, Anasetti C: Enhancement of susceptibility to Fas-mediated apoptosis of TH1 cells by nonmitogenic anti-CD3epsilon F(ab')2. Transplantation 2000 Jan 15, 69:104–12.
174. Woodle ES, Xu D, Zivin RA, Auger J, Charette J, O'Laughlin R, Peace D, Jollife LK, Haverty T, Bluestone JA, et al.: Phase I trial of a humanized, Fc receptor nonbinding OKT3 antibody, huOKT3gamma1(Ala-Ala) in the treatment of acute renal allograft rejection. Transplantation 1999, 68:608–16.
175. Kohm AP, Williams JS, Bickford AL, McMahon JS, Chatenoud L, Bach JF, Bluestone JA, Miller SD: Treatment with nonmitogenic anti-CD3 monoclonal antibody induces CD4+ T cell unresponsiveness and functional reversal of established experimental autoimmune encephalomyelitis. J Immunol 2005 Apr 15, 174:4525–34.
176. Chatenoud L: CD3 antibody treatment stimulates the functional capability of regulatory Tcells. Novartis Found Symp 2003, 252:279–86; discussion 86–90.
177. You S, Leforban B, Garcia C, Bach JF, Bluestone JA, Chatenoud L: Adaptive TGF-beta-dependent regulatory T cells control autoimmune diabetes and are a privileged target of anti-CD3 antibody treatment. Proc Natl Acad Sci USA. 2007 Apr 10, 104:6335–40.
178. Mekala DJ, Alli RS, Geiger TL: IL-10-dependent infectious tolerance after the treatment of experimental allergic encephalomyelitis with redirected CD4+CD25+ T lymphocytes. Proc Natl Acad Sci USA 2005 Aug 16, 102:11817–22.
179. Mekala DJ, Geiger TL: Immunotherapy of autoimmune encephalomyelitis with redirected CD4+CD25+ T lymphocytes. Blood 2005 Mar 1, 105:2090–2.
180. Kershaw MH, Teng MW, Smyth MJ, Darcy PK: Supernatural T cells: genetic modification of T cells for cancer therapy. Nat Rev Immunol 2005 Dec, 5:928–40.
181. Karim M, Feng G, Wood KJ, Bushell AR: CD25+CD4+ regulatory T cells generated by exposure to a model protein antigen prevent allograft rejection: antigen-specific reactivation in vivo is critical for bystander regulation. Blood 2005 Jun 15, 105:4871–7.

Chapter 11
CD4+CD25+ Regulatory T Cells as Adoptive Cell Therapy for Autoimmune Disease and for the Treatment of Graft-Versus-Host Disease

Swati Acharya and C. Garrison Fathman

Abstract Researchers in the field of T cell biology, pharmacology, molecular biology and genetic strategies have, over the past several years, systematically developed therapeutic approaches to tackle systemic and organ-specific autoimmune diseases, and treatment with general immunosupressants may soon be replaced by novel therapies based upon this research. Recently, interest in the field of regulatory T cells along with accumulated information that these cells play significant roles in preventing autoimmunity, may revolutionize the field of autoimmune disease therapy. Numerous research groups worldwide have combined forces to develop successful adoptive transfer methods so that regulatory T cells expanded *in vivo* or *ex vivo*, as antigen specific or polyclonal, can be used to treat autoimmune diseases. This chapter discusses the evolution of therapeutics for the treatment of autoimmune diseases with particular focus on regulatory T cell therapy using adoptive cell transfer and its treatment for graft-versus-host disease.

Introduction

The primary function of the body's immune system is to eliminate non-self pathogens without attacking the body's own tissues. However, autoimmune diseases result from tissue and organ damage that arise as a consequence of a pathologic immune attack against self. Since the immune system has developed efficient mechanisms to ensure self-tolerance, autoimmune diseases are of relatively infrequent occurrence. Strongly self reactive B and T cells are eliminated during development in the bone marrow or thymus by the process of negative selection (central tolerance), however some immune cells that have the potential to behave as autoreactive cells, occasionally escape into the systemic circulation. These potentially

---

C.G. Fathman
School of Medicine, Stanford University, Palo Alto, CA 94305, USA
e-mail: cfathman@stanford.edu

autoreactive cells are normally controlled by processes of peripheral tolerance that are brought about by three well-understood mechanism; anergy, apoptosis (activated cell death), and immune regulation mediated by subsets of antigen presenting cells (APCs) as well as a specialized class of T cells called regulatory T cells.

Although autoimmune diseases share common underlying pathophysiological principles, these diseases vary in terms of their demographic profile and primary clinical manifestations. For clinicians, autoimmune diseases are either systemic or organ–specific. The most common autoimmune diseases are Rheumatoid Arthritis (RA), Insulin Dependant or Type One Diabetes Mellitus (T1D), Multiple Sclerosis (MS) and Systemic Lupus Erythematosus (SLE). According to the latest NIH Autoimmune Disease report, based on the August 2004 Census Bureau figures, about 5–8% of the U.S. population, corresponding to between 14.7 and 23.5 million people are afflicted by at least one of the autoimmune diseases. Jacobsen et al. reviewed more that 130 published studies in order to estimate prevalence of autoimmune disease. More recently, Cooper and Stroela [1] extended the Jacobsen analysis and observed an increased incidence of autoimmune disease prevalence amongst women as well as an increased incidence of T1D worldwide over the last 40 years. For some autoimmune diseases such as thyroiditis, scleroderma, SLE and Sjogren's syndrome, more than 80% of the patients are female. This disparity is less significant in MS, myasthenia gravis and inflammatory bowel disease. T1D doesn't display gender disparity, while ankylosing spondylitis occurs with higher frequency in men, demonstrating an involvement of gender in the pathophysiology of autoimmune diseases.

In recent years, substantial progress has been made in better understanding the factors that lead to the development of autoimmune diseases and to their improved diagnosis, treatment and prevention. Many common human autoimmune diseases have well characterized experimental counterparts in animal models that provide biologic and genetic models for the study of autoimmune diseases including MS, RA, T1D, SLE, Orchitis, Uveitis, Thyroiditis, Colitis and many others. These animal models are used to study human disease pathophysiology and potential therapeutic interventions.

Although a large amount of research on novel therapeutics has been generated in such animal models of autoimmune disease over the past few years, suppression of the immune system, which is often associated with severe adverse side affects, is still the most common current clinical approach to therapy. However, novel approaches to adoptive cellular gene therapy of autoimmune diseases have evolved from understanding the pathophysiological mechanisms of autoimmune diseases and seem to hold promise for specific targeted treatment with minimal off target side effects. Additionally, novel proteomic technologies have been used recently in a search for new biomarkers of disease etiology, or surrogate markers of therapeutic intervention.

This chapter focuses on (1) Generic and selective therapies for autoimmune diseases. (2) Development of gene transfer and adoptive transfer strategies. (3) Regulatory T cells in Autoimmune Disease. (4) Adoptive transfer of regulatory T cells for therapy.

Generic and Selective Therapies for Autoimmune Diseases

Although there is variability in autoimmune disease manifestations, there exist shared pathophysiological immune mechanisms that respond to similar immunosuppressive treatment strategies, but the use of immunosuppressive drugs to ameliorate the autoimmune disease can frequently give rise to adverse side effects. One successful therapeutic strategy involves repairing or replacing the damaged tissue or products of the damaged tissue. Notable examples of this strategy are insulin replacement in patients with T1D (since insulin is not produced following autoimmune destruction of the insulin producing beta cells) and thyroid hormone replacement in patients with autoimmune thyroiditis. A damaged organ may also be replaced; kidney transplantation in patients who lose kidney function following autoimmune-glomerulonephritis. Advances in research are underway for islet transplants as a replacement for beta cell loss in T1D and stem cell therapies for future treatment strategies of other autoimmune diseases.

Generic Strategies

Most current treatment of human autoimmune diseases uses non-selective immunosuppressants such as corticosteroids, cyclophosphamide, methotrexate or a slightly more selective approach using monoclonal antibodies to lymphocyte cell surface markers such as CD3, CD4, and CD20. Side effects of this approach arise from the fact that these therapeutic measures do not discriminate between pathogenic and beneficial immune responsive lymphocytes. Additional somewhat selective therapeutics including glatiramer acetate [2], natalizumab [3] and mitoxantrone [4] in MS and the use of inhibition of tumor necrosis factor (TNF) in RA [5] were developed in experimental animal models. However, many therapies that are successful in animal models [6] are not effective in treating the human disease, as seen in studies on MS [7], adverse outcome of treatment with Interferon-γ [8], development of progressive multifocal leukoencepahlopathy in recipients of an adhesion molecule blocker in MS [9,10], and massive T cell activation in volunteers treated with anti-CD28 antibodies [11]. Due to the heterogeneity of human autoimmune disease manifestations and our incomplete understanding of tolerizing therapies (oral tolerance, [12]), T-cell receptor peptide immunization [13], co-stimulation blockade [14], DNA vaccination ([15,16] and administration of altered peptide ligands, APL [7], recipients of these therapies may actually have an immune induction effect resulting in disease relapse or progression.

Selective Strategies

In the last decades, due to advances in microarray technology [17], *in vivo* imaging [18] and proteomics [19], more selective therapies have been developed to replace global immunosuppression to treat autoimmune diseases. Potentially autoantigenic CD4+ T cells that escape thymic selection and circulate in the immune system may occasionally encounter peptide presented by a professional APC in an activating mode, which leads to full activation and differentiation of this effector T cell and enhances its migratory potential to the non-lymphoid target tissue. Activation of this cell leads to immune–mediated tissue destruction and loss of function that eventually manifests as autoimmune disease symptoms. The ideal approach to the development of selective therapy would be to identify a pathogenic autoantigen and block the response of the CD4+ T cell at the point of antigen recognition. Blockade of pathogenic clones of lymphocytes instead of blocking the entire immune system, would lack toxicity, an important features of selective antigen specific therapies.

There is no difference between an immune response to a foreign antigen and an immune response to a self antigen, thus many molecules participating in the inflammatory processes such as cytokines, chemokines and their receptors, the signaling pathways in APCs, T cells and B cells, and the molecules orchestrating the regulatory phase of the immune response, are similar in an appropriate immune response to infection and an autoimmune response. RA, ankylosing spondylitis and Crohn's disease have been treated successfully with TNF inhibition, even though stopping treatment is associated with the relapse of the disease [5] thus demonstrating TNF [20] plays a major role in these diseases. However the most selective treatments will be those that can block autoantigen response or induce regulatory T cells that recognize the autoantigenic T cells. Selectivity in immunotherapy could be achieved by using the autoantigen to manipulate the immune system to dampen its reactivity or induce tolerance. Myelin basic protein [21], collagen type II [22] and insulin [23] have been administered orally or nasally as effective treatment in experimental animal models including autoimmune encepahlomeylitis (EAE), collagen induced arthritis, and T1D, respectively. Based on positive results in animals, human oral tolerance trials were initiated but with much less impressive results [24]. In some cases tolerance induction protocols using antigen via the oral route may have resulted in immune activation instead of immunesuppression, leading to exacerbation of the autoimmune disease [25]. Additional ways to induce antigen specific tolerance in experimental animals include injection of autoantigens into the thymus [26], co-stimulation blockade [27], and DNA vaccination, all of which remain to be tested in humans. An alternative approach to autoantigen tolerance induction strategies was to use the reactive T cell [28,29] or its receptor [30] to immunize, in the hope that anti-idiotypic responses would arise following this T cell vaccination (TCV). The TCV model was successfully used in several animal models of autoimmune disease [31]. TCV has been used in human trials mostly in MS and RA by several groups in Europe [32], US [33] and Israel [34] although the overall performance of this approach in treating human disease has not demonstrated effectiveness.

Development of Gene Transfer and Adoptive Transfer Strategies

Gene Therapy

Gene therapy has been developed over the last decade as a particularly interesting and promising approach to the treatment of autoimmune diseases. It offers the unique chance of providing long-term expression of immune-modulating molecules *in vivo* that can antagonize the chronic inflammatory processes in autoimmune diseases, long-term efficacy and reduced adverse side effects, giving it an edge over conventional generic immunosuppressants.

Gene therapy involves introduction of DNA into a host cell for therapeutic purposes. DNA insertion into a host cell followed by the expression of the gene products of interest with the intention of achieving one of three effects, a) targeting a known gene defect, b) delivery of immune modulating molecules, such as cytokines (IL4, IL-10, TGF-β), cytokine antagonists (IL-1R antagonist and IL-12p40), soluble cytokine receptors or blocking antibodies and c) interference with signaling processes involved in autoimmune reactions, for example, TCR signaling and co-stimulation or apoptosis pathways. Naked DNA, DNA complexed with liposomes and viral vectors are amongst the common approaches used to introduce DNA into host cells. Viral vectors bearing the gene of interest can be either injected systemically or locally into the host tissues or can be used to transduce host cells *in vitro* which can then be used as adoptively transferred transgenic T cells [35–37], fibroblasts [38–40] or dendritic cells [41–43] in an attempt to deliver regulatory proteins to the site of autoimmune inflammation. Even though autoimmune diseases are polygenic disorders, studies have revealed that gene therapy can replace a single gene product with therapeutic success. Fibroblasts transduced with a retroviral construct encoding a genetically modified human proinsulin, cleavable into insulin in non-beta cells followed by transplantation of these fibroblasts into diabetic mice reversed hyperglycemia [44]. Also, a single chain insulin analog, SIA, (encoded by a recombinant adeno-associated viral vector with a rat L type pyruvate kinase (LPK)) instead of proinsulin was able to treat streptozotocin induced inflammatory diabetes in non–obese diabetic (NOD) mice resulting in long term remission [45]. The delivery of immune modulating molecules by various means represents another well-investigated strategy for gene therapy in autoimmune disease. The anti-inflammatory cytokines IL-4 [43,46–49], IL4-IgG1 chimeric protein 1 L-1, IL 10 [50,51], Epstein-Barr virus encoded vIL-10 [52–54] and IL-13 [55] have been confirmed to be beneficial in disease prevention in MS, T1D and Autoimmune Thyroiditis in animal models [43,46,48,51,55–57]. Soluble TNFR molecules [40,54,58], IFNγR-IgG fusion molecules [59,60], and IL-12R antagonist IL-12p40 [37,61] have been used successfully in various studies in order to antagonize the pro-inflammatory cytokines such as TNF, IFNγ, and IL-12.

The onset of autoimmune disease in humans is difficult to establish since all biomarkers still need to be validated, thus experimental therapeutics are easier to evaluate in established disease. Successful treatment of established disease using single cytokines has yielded variable results as reported in studies with autoimmune encephalomyelitis (EAE) using intra-thecal injection of HSV-1 derived vectors encoding IL-4 [47] whereas IL-4 treatment of rat adjuvant–induced arthritis (AIA) and murine collagen induced arthritis (CIA) has been reported to be therapeutic [46,49]. Adenovirus encoding rat IL-4 injected into inflamed AIA joints reduced the severity of clinical arthritis. Also almost complete disease suppression of CIA was seen using intravenous adoptive transfer of DCs expressing IL-4 after adenoviral gene transfer. However, no improvement of clinical disease was noted when IL-4 was administered by intra-articular injection of recombinant human type 5 adenovirus encoding IL-4 [62]. IL-10 has been used successfully in treatment of murine autoimmune thryroiditis [50] resulting in reduced lymphocytic thyroid infiltration, IFN-γ levels and anti-thyroglobulin antibody response, by surgical instillation of liposome and poly-L-Lysine complexed DNA encoding IL-10 into the inflamed murine thyroid glands. In the case of CIA, adenoviral gene transfer of viral IL-10 in an attempt to treat established disease had only minimal effects [54] despite effective disease prevention when treated at the disease onset. In yet another study [63] adenovirus encoded bivalent human p55 soluble TNFR-Ig murine IgG1 fusion protein had no effect on treating murine CIA. However, the dimeric p75TNFR was able to ameliorate both acute and relapsing disease when immortalized fibroblasts expressing this product were adoptively transferred.

Adoptive Transfer

Adoptive cell transfer is a form of passive immunotherapy whereby previously sensitized immune reactive cells are transferred to recipients as a form of therapy. Since gene therapy offers the unique chance of providing long-term expression of immune-modulating molecules *in vivo* that can antagonize the chronic inflammatory processes in autoimmune diseases with low side effects while specifically targeting known pathogenic mechanisms, our laboratory has developed the approach of adoptive cellular gene therapy. This *ex vivo* approach utilizes antigen-specific T cells or dendritic cells (DCs) for the targeted and safe delivery of immune-regulatory molecules to sites of autoimmune inflammation after retro or lentiviral transduction of the vehicle cells *in vitro* and adoptive transfer by intravenous or intraperoneal injection.

Viral Vectors: Amongst various approaches used to introduce DNA into host cells, adenoviruses have commonly been used for gene delivery. However, they afford only transient expression in infected cells and they are immunogenic and have been associated with fatality in a recent clinical trial. Retro or lentiviral-mediated gene transfer into T cells and DCs allows stable integration of the transgene with resultant long-term expression. Furthermore, retroviral gene delivery has been used safely to treat human disease and has avoided the viral protein immunogenicity that occurs with the use of adenoviral gene delivery. Gene transfer into target cells via retroviruses utilizes the ability of murine retroviruses to bind cell surface molecules in the initial stages of the infection process. After binding and fusion, the retroviral core and genome enter the cell being infected (packaging cell lines). Cell lines used for packaging the retrovirus are capable of producing all of the necessary proteins required for packaging, processing, reverse transcription, and integration of recombinant genomes. Here, retroviral reverse transcription occurs with subsequent integration into the genome after cell division. The producer cells will encapsulate or "package" only the transcripts containing the psi packaging signal and release the viral particles into the culture supernatant, which is then harvested and used for transduction of target cells.

Even though, retroviral vectors have been extensively characterized and used in gene therapy, the inability of these vectors to infect non-dividing cells limits their usefulness in clinical gene therapy applications. Lentiviral vectors are able to infect non-dividing and terminally differentiated cells. Additional accessory elements such as the Tat and Rev proteins in lentiviruses allow efficient viral gene expression. Even though retroviruses can enter non-dividing cells, the viral genome is excluded from the cell nucleus by the nuclear envelope. The pre-integration complex of lentiviruses contains nuclear localization signals that permit its active transport through nuclear pores into the nucleus during interphase of the non-dividing cell. Development of lentiviral vectors for such experimental protocols have enhanced prospects of adoptive cell transfer in clinical therapy by conserving time and reagents spent in intermediate protocols associated with adenoviral and retroviral vector mediated approaches. Efforts in our lab have established protocols for delivery of genes via lentiviral vectors aimed towards gene therapy coupled with *in vivo* imaging studies.

Vehicle cells: Antigen-specific CD4+ T cells and antigen-presenting dendritic cells (DCs) are important mediators in the pathogenesis of autoimmune disease and thus are ideal candidates for adoptive cellular gene therapy. Using retrovirally transduced cells and luciferase bioluminescence, it has been demonstrated that primary T cells, T cell hybridomas, and DCs rapidly and preferentially home to the sites of inflammation in animal models of multiple sclerosis, arthritis, and diabetes. T cells and their cytokines play a central role in the initiation and perpetuation of organ-specific autoimmune disease. Numerous studies have demonstrated successful prevention or amelioration of autoimmune diseases by blocking Th1 cytokines with specific antagonists or by counteracting the inflammatory response with immune-regulatory cytokines such as TGF-β, (IL-1Ra), IL-4, IL-10, IL-12p40, or anti-TNF which serve as effective therapies in models of autoimmune disease such as CIA, EAE, and NOD. These protocols appear to work by polarizing T cells away from Th1 differentiation pathways. Several studies have used antigen-specific CD4+ T cells and T cell hybridomas to deliver regulatory cytokines and cytokine inhibitors to autoimmune lesions, showing that T cells are suitable vehicles for targeted immunotherapy. Our group originally described amelioration of EAE with IL-4 produced by retrovirally transduced myelin reactive T cell hybridomas. Effective prevention of EAE, and of CIA by adoptive cellular gene therapy using autoantigen-specific T cells and T cell hybridomas retrovirally transduced to express the IL-12 receptor blocker IL-12p40, the regulatory cytokine IL-4 [64] and TNF-antagonizing anti-TNF scFv [65]. Besides, therapeutic potential, these studies demonstrated good safety profile of our adoptive cellular gene therapy approach in terms of low systemic side effects.

DCs not only provide a common set of signals to initiate clonal expansion of T cells, but also provide T cells with selective signals driving differentiation that lead to either Th1 or Th2 immunity. Transfer of TNF-treated, incompletely matured DCs (semi-mature DCs) induces peptide-specific IL-10-producing T cells *in vivo* and prevents EAE. Similarly, injecting immature DCs, infected with IL-4-encoding adenovirus into mice with established CIA resulted in almost complete suppression of disease with no disease recurrence for up to 4 weeks post-treatment. We have found that DCs transduced to express either IL-12p40 or IL-10 are effective in suppressing CIA. These experiments raised the exciting possibility of using DCs for adoptive cellular gene therapy of autoimmune disease. The adoptive transfer of IL-4-expressing DCs led to suppression of Th1-type immune responses in the lymph nodes and spleen and diminished the associated humoral immune responses and the therapeutic DCs migrated to the lymphoid tissues and modulated T cell immune responses by expression of the regulatory cytokine IL-4 through specific DC-T cell interactions. Studies in our lab with bone marrow-derived DC migration in CIA using bioluminescence imaging suggested that injected DCs not only home to lymphoid organs, but also accumulate in inflamed joints and therefore could be used to deliver anti-inflammatory molecules directly to the site of inflammation. Taken together, these results indicate that the use of genetically engineered DCs is a very promising approach for adoptive cellular gene therapy of autoimmune disease.

In vivo imaging: Specific homing to sites of autoimmune inflammation is the central principle of adoptive cellular gene therapy. In order to ascertain that the vehicle cells would exhibit the assumed homing behavior *in vivo*, whole-body bioluminescence imaging was used. This novel and powerful technique allows tracking of cells *in vivo* in real time. For our purposes, T cell hybridomas and DCs were transduced with a retrovirus encoding luciferase and transferred into the recipients. The mice were anesthetized and received the substrate luciferin by intraperitoneal injection. The enzymatic reaction between luciferase and luciferin causes emission of photons within the luciferase positive cells detected by a cooled charge-coupled device camera. A pseudo-color image representing light intensity of the emission is superimposed on a gray-scale body-surface reference image collected under weak illumination, and the data are acquired and analyzed using appropriate software to allow trafficking of the luciferase positive cells. Transfer of luciferase expressing T cell hybridomas and DCs into arthritic CIA mice, EAE-affected mice, and NOD mice confirmed the assumed homing behavior. Three days after the cell transfer, photons emitted from the cells were detected in arthritic joints from all mice tested in the CIA model, while five of six mice with EAE demonstrated luciferase-positive cells in the brain within three days of adoptive transfer. Similarly, cellular homing to the area of the pancreas was found in the NOD mouse [66]. No luciferase-positive cells were detected in the paws, CNS, or pancreas, respectively, in naïve control mice receiving luciferase transduced vehicle cells at any time.

Currently efforts to establish protocols for Lentiviral mediated gene therapy in animal models are being modified for use in adoptive cell transfer for the treatment of autoimmune diseases in man.

Role of T Regulatory Cells in Autoimmune Disease

As defined by CD4 and CD25 expression, CD25+CD4+ regulatory T cells comprise about 5–10% of the mature CD4+ helper T cell subpopulation in mice and about 1–2% CD4+ helper T cells in humans. The forkhead transcription factor,

Foxp3, has been shown to be an important molecular marker of T regulatory cells, expression of which is required for the developmental fate of regulatory T cells (Tregs). To designate Tregs, the establishment of two cell surface molecules, CD4 and CD25, occurred prior to the identification of Foxp3, hence the designation CD4+CD25+ Tregs. The large majority of Foxp3+ Tregs are found within the MHC class II restricted CD4 expressing cell populations. In addition to the Foxp3-expressing CD4+CD25+, there exists a minor population of MHC class I restricted $CD8^+$ Foxp3-expressing Tregs. High levels of CTLA-4 (cytotoxic T-lymphocyte associated molecule-4) and GITR (glucocorticoid-induced TNF receptor) are also expressed on regulatory T cells and their function significance is currently being analyzed. Interest in identifying cell surface markers that are uniquely and specifically expressed on all Foxp3-expressing Tregs by several research groups including our own is underway [67].

The term "naturally-occurring" CD4+CD25+ regulatory T cells (nTregs) is used to distinguish Tregs from *in vitro* generated "suppressor" T cells and is a subset of the total Foxp3-expressing Treg population. The regulatory T cell field has also been reported to include additional populations such as Tr1, CD8+CD28-, and Qa-1 restricted T cells. Genetic mutations in the gene encoding Foxp3 have been identified in both humans and mice based on the heritable disease caused by these mutations. Humans with mutations in Foxp3 suffer from a severe and rapidly fatal autoimmune disorder known as **I**mmune dysregulation, **P**olyendocrinopathy, **E**nteropathy **X**-linked (**IPEX**) syndrome, characterized by the development of overwhelming systemic autoimmunity in the first year of life resulting in the commonly observed triad of watery diarrhea, eczematous dermatitis, and endocrinopathy seen most commonly as insulin-dependent. The majority of affected males die within the first year of life of either metabolic derangements or sepsis. The analogous disease observed in a spontaneous Foxp3 mutant mouse is known as scurfy.

The molecular mechanism by which Tregs exert their suppressor and immune regulatory function is an interesting subject of research today. Over the past few years, many investigators have shown the enormous potential of regulatory T cells in murine models to suppress pathological immune responses in autoimmune diseases, transplantation, and graft-versus-host disease. The role of Tregs is important in infectious diseases and cancer as well. Cell surface markers CD4, CD25, CTLA-4, and GITR, have enabled investigators to selectively purify the Foxp3+ T cells in order to study the function of T regulatory cells *in vivo* and *in vitro*. The $CD4^+CD25^+$ Tregs arise within the thymus early in development. Adaptive Tregs develop throughout the course of the immune response *in vivo*. TCR engagement of conventional $CD25^-$ $CD4^+$ helper T (Th) cells by small amounts of antigen can induce Foxp3 and differentiate these cells into a regulatory population. Adaptive Tregs (also termed Th3 or Tr1) express a conventional cell surface phenotype, produce cytokines such as IL-10 and TGF, and express a repertoire broader than the repertoire of the thymus-derived $CD25^+$ Tregs, in some cases, inducing TGF-β related Foxp3 expression, hence, making the various subsets indistinguishable. *In vitro* assays of Treg activity have been widely used to characterize Treg-mediated suppression that is cell–cell contact dependent. It has been shown by some that anti-TGF-β antibodies reverse Treg function *in vitro*. Some reports suggest that contact of nTreg with other T cells (antigen presenting cells (APCs) or Th cells) induces a second population of Tregs, which produce suppressive cytokines, implying a role for both contact-dependent and contact-independent effects.

It has been established that thymectomy before day three of life, or transfer of mature thymocytes depleted of CD4+CD25+ cells to mice lacking T cells, leads to development of severe autoimmune disease. These observations reaffirm both the presence of peripheral "non-negative selected pathogenic autoreactive cells" as well as regulatory function of CD4+CD25+T cells to control the exported pool of autoreactive cells. If antigen selection in the thymus follows the same rules as a recall response to an immunizing foreign antigen, then the tolerance inducing or CD25+ Treg pool should be positively selected on dominant epitopes and contain a transcriptome (set of expressed genes) that prevents apoptosis of Tregs during what would be negative selection processes for conventional T cells. These possibilities have resulted in our hypothesis on the manner in which Tregs control autoimmunity: tolerance-inducing Tregs see dominant self-epitopes on the surface of antigen presenting cells (APC or specialized epithelial cells) and regulate peripheral autoreactive T cells by one of two mechanisms. Either a dominant regulatory response results from the display of major antigenic determinants by the APC or specialized epithelial cell that recruits high affinity CD25+ cells (Tregs) that have not been negatively selected by high affinity self antigen recognition in the thymus. These Tregs could simply physically out compete cells that see minor determinants with low avidity, those CD4+CD25- T cells that have survived negative selection (potentially pathogenic autoreactive T cells). Alternatively, it is possible that a few CD25+ cells (considering that they are rare cells in the periphery) engage APC or specialized epithelial cells displaying major antigenic determinants, prolonging the Treg interaction time with the APC and thereby inhibiting target autoreactive cells by direct contact as has been demonstrated in models of suppression by CD25+ cells *in vitro*. Experiments tracking CD25+ Tregs *in vivo* in models of autoimmunity or allo-transplantation suggest that the Tregs accumulate and perform their immunosuppressive function in the draining lymph nodes [L-selectin high (CD62L) Tregs function *in vivo* whereas both L-selectin high and low work equally effectively *in vitro*] rather than in the target tissue itself [68,69] reinforcing the hypothesis put forward above. Despite studies that have robust end points as readouts, the molecular mechanisms underlying the functions of Tregs remain largely unknown.

Numerous recent reports have convincingly proven that Tregs are able to reduce the incidence of autoimmunity. With recent advances in protocols of adoptive cell transfer, a modern day therapy using regulatory T cells is logically possible using adoptive cell transfer. Sakaguchi and colleagues observed that the adoptive transfer of T cells depleted of CD4+CD25+ cells induces multiorgan autoimmunity in recipient animals. Antigen–specificity of regulatory T cells is an important factor in harnessing these cells for treatment of specific tissues. TCR expression on Tregs has been implicated in the development of antigen-specific Tregs, while some studies find no evidence of the requirement for antigen-specific Tregs. Conventional CD4+CD25⁻ TCR transgenic T cells can effectively protect recipient mice against inflammatory bowel disease. T cells from tolerant animals can be transferred to naïve recipients resulting in the suppression of T cell responses not only to the original antigen specificities but to other antigens presented by the same APC. Treg function *in vivo* is IL-10 and/or TGF-β?dependent in several systems, and this suppression can be abrogated by antibodies against IL-10 and/or TGF-β? Systemic autoimmune response characterized by inflammatory infiltration of auto-reactive T cells followed by tissues destruction results from a loss or elimination of Tregs within the first few days of birth. In non-obese diabetic (NOD) models of T1D, islet antigen-specific BDC2.5 Tregs have been demonstrated to be more effective in treating diabetes than polyclonal Tregs from NOD mice in both lymphopenic and lymphoreplete systems.

Convincing studies by several groups have shown that Tregs function *in vivo* through direct and indirect effects on dendritic cells (DCs), and become activated following encounter with DCs or specialized epithelial cells. Islet antigen-specific nTregs cluster around antigen-bearing DCs in the draining lymph nodes (LNs) of prediabetic NOD mice. In addition, the suppression of T cell responses to antigen-pulsed mature DCs *in vivo* is regulated by nTregs. Selective DC subsets have the unique ability to induce aTregs, mature DCs can stimulate nTregs to expand and suppress. *In vitro*, robust expansion of Tregs is seen with DCs either matured with GM-CSF stimulation or isolated from draining LNs after challenge with complete Freund's adjuvant. DCs may also be the target of Treg activity. Multiple studies *in vitro* have shown that IL-10 and TGF-β can downregulate DC function by altering DC maturation or modulating cell surface expression of molecules that are critical for T cell activation, such as B7, CD40 and IL-12 through cell contact or cytokines. Thus, during the course of an immune response, the development of a Treg is directly controlled by the interaction of conventional Th cells with 'tolerogenic' DCs or mature DCs in a tolerogenic setting whereas the activation and suppressive properties of Tregs is promoted by mature DCs.

Emerging experimental evidence suggest that Tregs control autoimmunity at multiple levels, In the steady state, Tregs efficiently home to LNs, expand and accumulate where cognitive antigens are present. This basal level of activity prevents priming of autoreactive T cells, and facilitates the maintenance of immune homeostasis. Failure in this first line of defense results in tissue destruction and inflammation, Tregs further differentiate into effector Tregs, traffic to affected tissue, and control the inflammatory response at the local site. Evidence suggests that an *in vivo* immune response is dictated by the interplay between Tregs and DCs. Increased DC activation during ongoing autoimmunity leads to heightened Treg function, which feeds back to limit the immune response. Breakdown of the balance of these immune cells leads to insufficiency of functional Tregs, leading to autoimmunity. Therapeutic intervention using Tregs of relevant specificity is thus likely to restore normal immune homeostasis and prevent further tissue destruction.

Adoptive Transfer of T Regulatory Cells

Recent research and increasing evidence of Treg involvement in autoimmune diseases has brought to focus the potential of using these cells for therapy. Several noteworthy concerns serve as limiting factors for the use of these cells by adoptive transfer for the treatment of autoimmune diseases. The small number of cells available and the potential requirement for antigen specific Tregs for therapy are two primary concerns. To address the first concern, several protocols have been developed and are in the process of development to allow expansion of Tregs in such a way that they maintain their regulatory properties. It is known that Tregs can proliferate well *in vivo* but become anergic during *in vitro* proliferation. Even though antigen-specific Tregs would be the magic bullet of choice in treatment of autoimmune disease, polyclonal Tregs may be a practical option. In support of treatment with polyclonal Tregs, it has been shown that they can be used to block unwanted immunity in mouse models of graft-versus-host disease (GVHD) and colitis when co-transferred with effector T cells into lymphopenic hosts.

Antigen-Non-Specific Clonal Expansion

One disease where Tregs are probably the most useful therapeutically is Graft-versus-Host – Disease (GVHD). GVHD, a frequent complication of allogeneic Bone Marrow Transplant (BMTs), is an inflammatory disease characterized by inflammatory cell recruitment to the skin, gastrointestinal tract and other organs and leads to a systemic chronic debilitating condition. Efforts to reduce GVHD by removing T cells from the donated bone marrow have resulted in poor engraftment and recurrence of the primary disease. Identification, characterization and potential uses of CD4+CD25+ regulatory T cells (Tregs) in the prevention of autoimmune/inflammatory diseases such as GVHD [70], is an important therapeutic approach. Polyclonal Tregs can support allogeneic haematopoietic stem cell (progenitor cell) engraftment without inducing GVHD in a mouse model of fully MHC-mismatched BMT [71]. Tregs expanded *ex vivo* [74] or co-transplantation of purified Tregs in a mouse model of BMT, show significant reduction in GVHD while retaining graft-versus-leukaemia responses [72,73]. The basis for this potent and selective effect might be the extensive homeostatic-proliferation properties of Tregs compared with effector T cells or a relatively high precursor frequency of alloreactive cells in the polyclonal Treg population. Tregs expanded *in vitro* using CD3 and CD28 specific antibodies are relatively efficient at suppressing human allogeneic mixed lymphocyte reactions *in vitro* [75]. The clinical application of antigen non-specific expanded Tregs that retain alloreactivity, homing and expansion properties is an ideal approach to treat GVHD. Several groups are developing approaches to increase the suppressive activity of the polyclonal regulatory T cells that have marked allogeneic crossreactivity. Human CD4+ T cells can be activated by autologous APCs in the presence of IL-10 and type I interferons to expand IL-10 producing CD4+ Treg clones [76]. These cells have been shown to suppress the *in vitro* alloantigen-specific proliferation of CD4+ T cells. T cells primed in the presence of TGF-β developed into contact-dependent and, in some cases, cytokine-producing Tregs. These studies have led to novel clinical trials. Although the relative frequency of allospecific Tregs that are present in cultures expanded with IL-2 plus beads coated with CD3- and CD28-specific antibodies remains unknown, in one mouse model, alloantigen or polyclonally expanded Tregs suppressed GVHD in a manner comparable with a similar number of freshly isolated Tregs. However, the cells that were expanded with alloantigen were more effective in preventing some of the graft-versus-host pathology. So, although antigen non-specific expansion of Tregs might work in the case of BMT, owing to the high frequency of alloreactive Tregs, antigen-specific expansion of the relevant cells might be advantageous while perhaps compromising engrafment. Polyclonal Tregs could be used to treat systemic autoimmunity such as SLE where Tregs have been reported to be reduced in numbers. Treatment of such patients with *ex vivo* polyclonally expanded, autologous Tregs at the time of relapse might halt the relapse. Similar treatment could be used for IBD, another autoimmune disease that is characterized by multiple relapses.

Antigen-Specific Clonal Expansion

In vitro experiments indicate that suppressive functions require the activation of Tregs through their TCR suggesting that the *in vivo* activation and function of Tregs is controlled by the specificity of the TCR. Autoantigen and alloantigen-specific Tregs have been from the natural Treg repertoire. In the non-obese diabetic (NOD) mouse model of T1D, islet-specific Tregs from TCR-transgenic mice expanded using beads coated with CD3 and CD28-specific antibodies were markedly more effective at suppressing disease than polyclonal Tregs. Peptide or antigen-pulsed DCs (or other APCs) and the use of peptide–MHC tetramers have been described to expand Tregs in an antigen-specific manner. Antigen-specific Tregs from the islet-antigen-specific TCR-transgenic mice have also been expanded using DCs from NOD mice pulsed with specific autoantigen *in vitro* [77]. IL-2 and polystyrene beads conjugated to both CD28-specific monoclonal antibodies and MHC dimers presenting an islet-antigen can selectively expand islet-antigen-specific Tregs from the polyclonal Treg population. An 'artificial' APC, using a human APC-like tumor line, K562, transfected with genes encoding several co-stimulatory molecules that induce the rapid and effective expansion of peripheral T cells, has been described [78].

As the problems of limited numbers and the use of antigen specificity are being tackled by ongoing research, the development of protocols for the therapeutic use of Tregs is now a major focus of research in this field. In an attempt to study the potential therapeutic uses of Tregs, our lab settled on a murine model of *Graft Versus Host Disease* (GVHD). Tregs are potent modulators of alloimmune responses. Mature donor T cells cause GVHD, however, they are also the main mediators of the beneficial graft-versus-tumor (GVT) activity of allogeneic bone marrow transplantation.

Suppression of GVHD with maintenance of GVT activity is a desirable outcome for clinical transplantation.

We initially demonstrated that donor-derived Tregs inhibited lethal GVHD after allogeneic bone marrow transplantation across MHC class I, and MHC class II barriers in mice [70]. In collaboration with Dr. Negrin, we demonstrated in host mice with leukemia and lymphoma, that Tregs suppressed the early expansion of alloreactive donor T cells, their IL-2-receptor α chain expression (CD25), and their capacity to induce GVHD without abrogating their GVT effector function that was mediated primarily by the perforin lysis pathway [72]. Thus, CD4+CD25+ T cells are potent regulatory cells that can separate GVHD from GVT activity mediated by conventional donor T cells. We also examined the differential effect of CD62L+ and CD62L- subsets of Tregs on T1D transfer and on aGVHD-related mortality. Tregs had been previously implicated in the control of diabetes suggesting that the inflamed islets of Langerhans in pre-diabetic NOD mice are under peripheral immune surveillance. We also demonstrated that CD4+CD25+ splenocytes inhibited diabetes in co-transfer with islet-infiltrating cells [68]. Furthermore, CD62L expression was necessary for this disease delaying effect of Tregs *in vivo*, but not for their suppressor function *in vitro*. We demonstrated that the CD62L+ Tregs expressed CCR7 at high levels, and migrated towards SLC and ELC, "lymphoid chemokines," whereas CD62L- Tregs preferentially expressed CCR2, CCR4 and CXCR3 and migrated towards the corresponding "inflammatory chemokines." These data demonstrated that CD62L+Tregs but not CD62L-Tregs delayed diabetes transfer, and that Tregs are comprised of at least two sub-populations that behave differently in co-transfer *in vivo* and express distinct chemokine receptor and chemotactic response profiles, despite demonstrating equivalent suppressor functions *in vitro*. Similarly, in the GVHD mouse model, although both L selectin positive and L selectin negative subpopulations showed the characteristic features of Tregs *in vitro*, in co-transfer with donor CD4 + CD25- T cells, only the CD62L+ subset of Tregs prevented severe tissue damage to the colon, and protected recipients from lethal aGVHD [70]. Early after transplantation, a higher number of donor-type Tregs could be recovered from host mesenteric lymph nodes (LNs) and spleen when CD62L+ Tregs were transferred as compared to the CD62L- subset. Importantly, CD62L + Tregs showed a significantly higher capacity than their CD62L- counterpart to inhibit the expansion of donor CD4+ CD25- T cells. These findings, obtained in collaboration with Dr. Negrin, suggest that the ability of Tregs to efficiently enter secondary lymphoid organs is a prerequisite for their protective function in aGVHD [69]. A follow up on this observation indicates that by studying the homing receptors L-Selectin and $\alpha_4\beta_7$ integrin facilitate entry of T cells into the gut associated organized lymphoid tissues such as the mesenteric lymph nodes and Peyer's patches [79]. We studied the impact of inactivation of genes encoding these receptors on the ability of purified donor CD4$^+$ T cells to induce acute lethal graft versus host disease (GVHD) associated with severe colitis in irradiated MHC-mismatched mice. Whereas lack of expression of a single receptor had no significant impact on the severity of colitis and GVHD, the lack of expression of both receptors markedly ameliorated both colitis and the early deaths observed with wild-type (WT) T cells.

The changes in colitis and GVHD were reflected in a marked reduction in the early accumulation of donor T cells in the mesenteric lymph nodes, and subsequently in the colon. The purified WT donor CD4+ T cells did not accumulate early in the Peyer's patches and failed to induce acute injury to the small intestine. These data suggest that the combination of CD62L and β_7 integrin are required to induce acute colitis and facilitate entry of CD4$^+$ donor T cells in the mesenteric nodes associated with lethal GVHD in allogeneic hosts.

In a mechanistic approach to understanding Tregs, we asked why Tregs maintained an anergy phenotype when activated *in vitro* under conditions that would induce IL-2 production and proliferation in conventional T cells. We compared signaling events in freshly isolated murine Tregs compared to conventional CD4+CD25- T cells, following stimulation with PMA and Ionomycin. Both p38 MAPK and ERK1/2 protein kinases were phosphorylated with similar kinetics in both populations. Although we observed diminished activation of JNK in the Tregs, retrovirally mediated reconstitution of the JNK pathway in Tregs did not abrogate the inhibition of IL-2 synthesis by these cells under fully activating conditions. Furthermore, we did not observe any perturbation in the formation of AP-1, NF-AT, and NF-kB transcriptional complexes in Tregs compared to conventional CD4+ CD25- T cells. Using a PCR based chromatin accessibility assay (CHART-PCR) we found that the minimal IL-2 promoter region of Tregs, unlike conventional T cells,

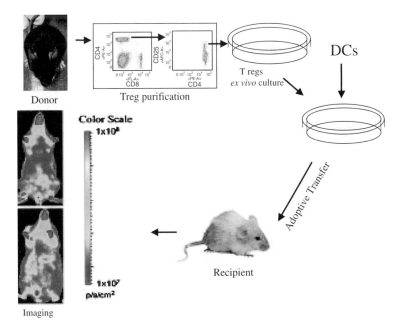

Fig. 11.1 Schematic representation of basic Treg Adoptive transfer Protocol. Treg cells from donor mice are used either for expansion *ex vivo* or expanded *in vivo* (not represented here) together with DCs and adoptively transferred to another experimental recipient mouse. *In vivo* luciferase base imaging is used to track cells homing to sites of inflammation *(See also Color Insert)*

remained in a closed chromatin configuration after stimulation suggesting that the anergy phenotype of Tregs is controlled by epigenetic mechanisms [80]. (Figs. 11.1 and 11.2)

In collaboration with Dr. Chris Contag, we have utilized bioluminescent-based imaging (BLI) to gain unique insights into the spatial and temporal events in the induction of GVHD. BLI is based upon the concept that light penetrates tissues and by introducing an internal light source into cellular populations of interest non-invasive imaging can be performed which guides subsequent analysis. We have utilized the luciferase (*luc*) gene from the North American Firefly which emits light upon introduction of the substrate luciferin in an ATP, O_2 dependent reaction. In collaboration with Dr. Contag, novel fusion constructs with *gfp* and *yfp* have been developed with *luc* and introduced into either lenti or retroviral vectors that can then be used to transduce cells of interest. Initial studies were focused on tumor models to explore residual malignancy after either chemotherapy or HCT. Remarkably BLI is extremely sensitive and as few as 100–1,000 luc^+ cells can be non-invasively and quantitatively measured in living animals [72,81]. Since the animals

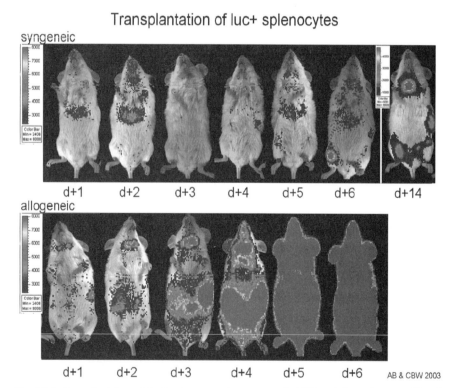

Fig. 11.2 Transplantation of *luc*$^+$ splenocytes into syngeneic FVB recipients (*top panels*) with imaging at indicated time points vs allogeneic BALB/C recipients (*bottom panels*). The dramatic proliferation of alloreactive T cells can be readily visualized. These images are representative of large number of animals studied. The allogeneic animals die from GVHD at approximately day 14 (*See also Color Insert*)

do not need to be sacrificed re-imaging is possible to gain insight into the temporal sequence of complex *in vivo* biological processes. This strategy was effective, however, the limitations of viral transduction of primary T cells was highly variable when imaging of GVHD induction was attempted. To overcome this limitation Dr Contag's group has recently developed a novel transgenic strain of animals in the FVB ($H-2^q$) background, termed L2G85 that constitutively express both *gfp* and *luc* in all hematopoietic cells. These animals have been utilized to study hematopoietic stem cell engraftment [82] and GVHD pathology [83].

Using L2G85 animals we have been able to serially follow alloreactive T cells *in vivo* that allowed a more detailed analysis. As shown by data presented in Fig. 11.2, following ablative radiation (800 rads), animals were transplanted with 5×10^6 T cell depleted (TCD) bone marrow (BM) from FVB wt animals and 10^6 luc^+ splenocytes from L2G85 animals. Recipient animals included allogeneic BALB/c ($H-2^d$) and syngeneic FVB control animals. Initially BLI demonstrated primarily a signal in the lung followed by infiltration of cervical lymph nodes and sites within the gut within 1–3 days. In the allogeneic but not syngeneic animals, massive T cell proliferation occurred by day 3–6, with infiltration of the gut and skin (note ear signal). By day 10–14 massive T cell proliferation was observed throughout the animal along with typical signs of GVHD including hair loss, diarrhea, weight loss and eventually animal death [83]. Detailed analysis revealed that specific structures within the GI tract infiltrated by 2–3 days included individual Peyer's patches and mesenteric lymph nodes as well as the spleen and other nodal sights prior to infiltration of GVHD target organs such as the gut, skin and liver. As discussed above the addition of Tregs markedly attenuated the induction of GVHD. The advantages of this model are the rapid quantitative readout after approximately 7 days of Treg function, the requirement for smaller numbers of animals per group to reach meaningful endpoints with statistical power and the reproducibility of the assay.

Conclusion

Modern day research using novel tools including microarrays, proteomics platforms, *in vivo* imaging and using mouse models, has paved the way for the emergence of novel therapies for the treatment of autoimmune diseases. Recent studies have shown that it may be possible to harness antigen-specific T cells and DCs for adoptive cell therapy. Clinical use of adoptive transfer as a cell based therapy can be predicted to arise from further corroboration of homing capabilities and novel engineered chemokine receptor-bearing cells that can more specifically follow chemokine gradients to inflammatory lesions. Furthermore, knowing the genetic signature of regulatory T cells as well as their ontogeny will provide us with information to develop better methods of Treg expansion for therapy. Future research should also allow combination therapy in order to achieve therapeutic effectiveness and reduced side effects. Improved adoptive cellular gene therapy delivering optimized combinations of immune-regulatory molecules will undoubtedly be informative in characterizing the underlying immune mechanisms in organ-specific autoimmune diseases and will potentially lead to new therapeutic options for treating human autoimmune diseases. A major challenge in the field of adoptive cell therapy will be to control the number of Tregs for therapy as well as post therapeutic proliferation *in vivo* that might lead to oncogenesis. Regardless of the success of relevant adoptive transfer based therapy in mouse models, enormous compliance issues will have to be dealt with before consideration for human clinical trials. Development of current protocols for human use would be expensive as this therapy is in general individual patient based. A major need in this field would be to establish a genetic signature of human regulatory T cells to identify them as well as detect them pre- and post adoptive transfer based therapy for autoimmune disease. Additionally, surrogate markers of effect will need to be identified to follow the effectiveness of therapy.

References

1. Cooper GS, Stroehla BC. The epidemiology of autoimmune diseases. Autoimmunity Reviews 2003; 2:119–125
2. Sela M, Teitelbaum D. Glatiramer acetate in the treatment of multiple sclerosis. Expert Opinion on Pharmacotherapy 2001; 2:1149–1165
3. Steinman L. Blocking adhesion molecules as therapy for multiple sclerosis: natalizumab. Nature Reviews 2005; 4:510–518
4. Gonsette RE. Mitoxantrone immunotherapy in multiple sclerosis. Multiple sclerosis (Houndmills, Basingstoke, England) 1996; 1:329–332
5. Maini RN, Brennan FM, Williams R, et al. TNF-alpha in rheumatoid arthritis and prospects of anti-TNF therapy. Clinical and Experimental Rheumatology 1993; 11 Suppl 8:S173-S175
6. Shoda LK, Young DL, Ramanujan S, et al. A comprehensive review of interventions in the NOD mouse and implications for translation. Immunity 2005; 23:115–126
7. Bielekova B, Goodwin B, Richert N, et al. Encephalitogenic potential of the myelin basic protein peptide (amino acids 83–99) in multiple sclerosis: results of a phase II clinical trial with an altered peptide ligand. Nature Medicine 2000; 6:1167–1175
8. Panitch HS, Hirsch RL, Haley AS, Johnson KP. Exacerbations of multiple sclerosis in patients treated with gamma interferon. Lancet 1987; 1:893–895
9. Bartt RE. Multiple sclerosis, natalizumab therapy, and progressive multifocal leukoencephalopathy. Current Opinion in Neurology 2006; 19:341–349
10. Ransohoff RM. Natalizumab and PML. Nature Neuroscience 2005; 8:1275
11. Suntharalingam G, Perry MR, Ward S, et al. Cytokine storm in a phase 1 trial of the anti-CD28 monoclonal antibody TGN1412. The New England Journal of Medicine 2006; 355:1018–1028
12. Blanas E, Carbone FR, Allison J, Miller JF, Heath WR. Induction of autoimmune diabetes by oral administration of autoantigen. Science 1996; 274:1707–1709
13. Desquenne-Clark L, Esch TR, Otvos L, Jr., Heber-Katz E. T-cell receptor peptide immunization leads to enhanced and chronic experimental allergic encephalomyelitis. Proceedings of the National Academy of Sciences of the United States of America 1991; 88:7219–7223
14. Perrin PJ, Scott D, June CH, Racke MK. B7-mediated costimulation can either provoke or prevent clinical manifestations of experimental allergic encephalomyelitis. Immunologic Research 1995; 14:189–199
15. Robinson WH, Fontoura P, Lee BJ, et al. Protein microarrays guide tolerizing DNA vaccine treatment of autoimmune encephalomyelitis. Nature Biotechnology 2003; 21:1033–1039

16. Roffe E, Souza AL, Caetano BC, et al. A DNA vaccine encoding CCL4/MIP-1beta enhances myocarditis in experimental Trypanosoma cruzi infection in rats. Microbes and infection/Institut Pasteur 2006; 8:2745–2755
17. Miura K, Bowman ED, Simon R, et al. Laser capture microdissection and microarray expression analysis of lung adenocarcinoma reveals tobacco smoking- and prognosis-related molecular profiles. Cancer Research 2002; 62:3244–3250
18. Stosiek C, Garaschuk O, Holthoff K, Konnerth A. In vivo two-photon calcium imaging of neuronal networks. Proceedings of the National Academy of Sciences of the United States of America 2003; 100:7319–7324
19. Medh RD. Microarray-based expression profiling of normal and malignant immune cells. Endocrine Reviews 2002; 23:393–400
20. Rajan AJ, Gao YL, Raine CS, Brosnan CF. A pathogenic role for gamma delta T cells in relapsing-remitting experimental allergic encephalomyelitis in the SJL mouse. Journal of Immunology 1996; 157:941–949
21. El Behi M, Dubucquoi S, Lefranc D, et al. New insights into cell responses involved in experimental autoimmune encephalomyelitis and multiple sclerosis. Immunology Letters 2005; 96:11–26
22. Thorbecke GJ, Schwarcz R, Leu J, Huang C, Simmons WJ. Modulation by cytokines of induction of oral tolerance to type II collagen. Arthritis and Rheumatism 1999; 42:110–118
23. Maron R, Blogg NS, Polanski M, Hancock W, Weiner HL. Oral tolerance to insulin and the insulin B-chain: cell lines and cytokine patterns. Annals of the New York Academy of Sciences 1996; 778:346–357
24. Weiner HL, Mackin GA, Matsui M, et al. Double-blind pilot trial of oral tolerization with myelin antigens in multiple sclerosis. Science 1993; 259:1321–1324
25. Blaha P, Bigenzahn S, Koporc Z, et al. The influence of immunosuppressive drugs on tolerance induction through bone marrow transplantation with costimulation blockade. Blood 2003; 101:2886–2893
26. Khoury SJ, Hancock WW, Weiner HL. Oral tolerance to myelin basic protein and natural recovery from experimental autoimmune encephalomyelitis are associated with downregulation of inflammatory cytokines and differential upregulation of transforming growth factor beta, interleukin 4, and prostaglandin E expression in the brain. The Journal of Experimental Medicine 1992; 176:1355–1364
27. Wekerle T, Kurtz J, Bigenzahn S, Takeuchi Y, Sykes M. Mechanisms of transplant tolerance induction using costimulatory blockade. Current Opinion in Immunology 2002; 14:592–600
28. Mor F. Preparation of lymphocytes for autolymphocyte therapy in metastatic renal carcinoma. Lancet 1990; 336:62
29. Ben-Nun A, Cohen IR. Vaccination against autoimmune encephalomyelitis (EAE): attenuated autoimmune T lymphocytes confer resistance to induction of active EAE but not to EAE mediated by the intact T lymphocyte line. European Journal of Immunology 1981; 11: 949–952
30. Vandenbark AA, Hashim G, Offner H. Immunization with a synthetic T-cell receptor V-region peptide protects against experimental autoimmune encephalomyelitis. Nature 1989; 341: 541–544
31. Cohen IR, Quintana FJ, Mimran A. Tregs in T cell vaccination: exploring the regulation of regulation. The Journal of Clinical Investigation 2004; 114:1227–1232
32. Hellings N, Raus J, Stinissen P. T-cell vaccination in multiple sclerosis: update on clinical application and mode of action. Autoimmunity Reviews 2004; 3:267–275
33. Hafler DA, Cohen I, Benjamin DS, Weiner HL. T cell vaccination in multiple sclerosis: a preliminary report. Clinical Immunology and Immunopathology 1992; 62:307–313
34. Achiron A, Kishner I, Sarova-Pinhas I, et al. Intravenous immunoglobulin treatment following the first demyelinating event suggestive of multiple sclerosis: a randomized, double-blind, placebo-controlled trial. Archives of Neurology 2004; 61:1515–1520
35. Tuohy VK, Mathisen PM. T cell design for therapy in autoimmune demyelinating disease. Journal of Neuroimmunology 2000; 107:226–232

36. Setoguchi K, Misaki Y, Araki Y, et al. Antigen-specific T cells transduced with IL-10 ameliorate experimentally induced arthritis without impairing the systemic immune response to the antigen. Journal of Immunology 2000; 165:5980–5986
37. Nakajima A, Seroogy CM, Sandora MR, et al. Antigen-specific T cell-mediated gene therapy in collagen-induced arthritis. The Journal of Clinical Investigation 2001; 107: 1293–1301
38. Rabinovich GA, Daly G, Dreja H, et al. Recombinant galectin-1 and its genetic delivery suppress collagen-induced arthritis via T cell apoptosis. The Journal of Experimental Medicine 1999; 190:385–398
39. Dreja H, Annenkov A, Chernajovsky Y. Soluble complement receptor 1 (CD35) delivered by retrovirally infected syngeneic cells or by naked DNA injection prevents the progression of collagen-induced arthritis. Arthritis and Rheumatism 2000; 43:1698–1709
40. Croxford B, Tham KW, Young A, Oreszczyn T, Wyon D. A study of local electrostatic filtration and main pre-filtration on airborne and surface dust levels in air-conditioned office premises. Indoor Air 2000; 10:170–177
41. Fiehn C, Wettschureck N, Krauthoff A, Haas R, Ho AD. Bone marrow-derived cells as carriers of recombinant immunomodulatory cytokine genes to lymphoid organs. Cancer Gene Therapy 2000; 7:1105–1112
42. Kim SH, Kim S, Evans CH, et al. Effective treatment of established murine collagen-induced arthritis by systemic administration of dendritic cells genetically modified to express IL-4. Journal of Immunology 2001; 166:3499–3505
43. Morita Y, Yang J, Gupta R, et al. Dendritic cells genetically engineered to express IL-4 inhibit murine collagen-induced arthritis. The Journal of Clinical Investigation 2001; 107: 1275–1284
44. Falqui L, Martinenghi S, Severini GM, et al. Reversal of diabetes in mice by implantation of human fibroblasts genetically engineered to release mature human insulin. Human Gene Therapy 1999; 10:1753–1762
45. Lee HC, Kim SJ, Kim KS, Shin HC, Yoon JW. Remission in models of type 1 diabetes by gene therapy using a single-chain insulin analogue. Nature 2000; 408:483–488
46. Woods JM, Katschke KJ, Volin MV, et al. IL-4 adenoviral gene therapy reduces inflammation, proinflammatory cytokines, vascularization, and bony destruction in rat adjuvant-induced arthritis. Journal of Immunology 2001; 166:1214–1222
47. Martino G, Furlan R, Brambilla E, et al. Cytokines and immunity in multiple sclerosis: the dual signal hypothesis. Journal of Neuroimmunology 2000; 109:3–9
48. Cottard V, Mulleman D, Bouille P, et al. Adeno-associated virus-mediated delivery of IL-4 prevents collagen-induced arthritis. Gene Therapy 2000; 7:1930–1939
49. Kim SH, Evans CH, Kim S, et al. Gene therapy for established murine collagen-induced arthritis by local and systemic adenovirus-mediated delivery of interleukin-4. Arthritis Research 2000; 2:293–302
50. Batteux F, Trebeden H, Charreire J, Chiocchia G. Curative treatment of experimental autoimmune thyroiditis by in vivo administration of plasmid DNA coding for interleukin-10. European Journal of Immunology 1999; 29:958–963
51. Koh JJ, Ko KS, Lee M, et al. Degradable polymeric carrier for the delivery of IL-10 plasmid DNA to prevent autoimmune insulitis of NOD mice. Gene Therapy 2000; 7:2099–2104
52. Lechman ER, Jaffurs D, Ghivizzani SC, et al. Direct adenoviral gene transfer of viral IL-10 to rabbit knees with experimental arthritis ameliorates disease in both injected and contralateral control knees. Journal of Immunology 1999; 163:2202–2208.
53. Whalen JD, Lechman EL, Carlos CA, et al. Adenoviral transfer of the viral IL-10 gene periarticularly to mouse paws suppresses development of collagen-induced arthritis in both injected and uninjected paws. Journal of Immunology 1999; 162:3625–3632
54. Kim KN, Watanabe S, Ma Y, et al. Viral IL-10 and soluble TNF receptor act synergistically to inhibit collagen-induced arthritis following adenovirus-mediated gene transfer. Journal of Immunology 2000; 164:1576–1581

55. Bessis N, Honiger J, Damotte D, et al. Encapsulation in hollow fibres of xenogeneic cells engineered to secrete IL-4 or IL-13 ameliorates murine collagen-induced arthritis (CIA). Clinical and Experimental Immunology 1999; 117:376–382
56. Chang Y, Prud'homme GJ. Intramuscular administration of expression plasmids encoding interferon-gamma receptor/IgG1 or IL-4/IgG1 chimeric proteins protects from autoimmunity. The Journal of Gene Medicine 1999; 1:415–423
57. Piccirillo CA, Prud'homme GJ. Prevention of experimental allergic encephalomyelitis by intramuscular gene transfer with cytokine-encoding plasmid vectors. Human Gene Therapy 1999; 10:1915–1922
58. Quattrocchi E, Walmsley M, Browne K, et al. Paradoxical effects of adenovirus-mediated blockade of TNF activity in murine collagen-induced arthritis. Journal of Immunology 1999; 163:1000–1009
59. Prud'homme GJ, Chang Y. Prevention of autoimmune diabetes by intramuscular gene therapy with a nonviral vector encoding an interferon-gamma receptor/IgG1 fusion protein. Gene Therapy 1999; 6:771–777
60. Lawson BR, Prud'homme GJ, Chang Y, et al. Treatment of murine lupus with cDNA encoding IFN-gammaR/Fc. The Journal of Clinical Investigation 2000; 106:207–215
61. Costa GL, Sandora MR, Nakajima A, et al. Adoptive immunotherapy of experimental autoimmune encephalomyelitis via T cell delivery of the IL-12 p40 subunit. Journal of Immunology 2001; 167:2379–2387
62. Lubberts E, Joosten LA, Chabaud M, et al. IL-4 gene therapy for collagen arthritis suppresses synovial IL-17 and osteoprotegerin ligand and prevents bone erosion. The Journal of Clinical Investigation 2000; 105:1697–1710
63. Zhang HG, Fleck M, Kern ER, et al. Antigen presenting cells expressing Fas ligand downmodulate chronic inflammatory disease in Fas ligand-deficient mice. The Journal of Clinical Investigation 2000; 105:813–821
64. Tarner IH, Nakajima A, Seroogy CM, et al. Retroviral gene therapy of collagen-induced arthritis by local delivery of IL-4. Clinical Immunology (Orlando, Fla) 2002; 105:304–314
65. Smith R, Tarner IH, Hollenhorst M, et al. Localized expression of an anti-TNF single-chain antibody prevents development of collagen-induced arthritis. Gene Therapy 2003; 10:1248–1257
66. Urbanek-Ruiz I, Ruiz PJ, Paragas V, et al. Immunization with DNA encoding an immunodominant peptide of insulin prevents diabetes in NOD mice. Clinical Immunology (Orlando, Fla) 2001; 100:164–171
67. Marson A, Kretschmer K, Frampton GM, et al. Foxp3 occupancy and regulation of key target genes during T-cell stimulation. Nature 2007; 445:931–935.
68. Szanya V, Ermann J, Taylor C, Holness C, Fathman CG. The subpopulation of CD4+CD25+ splenocytes that delays adoptive transfer of diabetes expresses L-selectin and high levels of CCR7. Journal of Immunology 2002; 169:2461–2465
69. Ermann J, Hoffmann P, Edinger M, et al. Only the CD62L+ subpopulation of CD4+CD25+ regulatory T cells protects from lethal acute GVHD. Blood 2005; 105:2220–2226
70. Hoffmann P, Ermann J, Edinger M, Fathman CG, Strober S. Donor-type CD4(+)CD25(+) regulatory T cells suppress lethal acute graft-versus-host disease after allogeneic bone marrow transplantation. The Journal of Experimental Medicine 2002; 196:389–399
71. Hanash AM, Levy RB. Donor CD4+CD25+ T cells promote engraftment and tolerance following MHC-mismatched hematopoietic cell transplantation. Blood 2005; 105:1828–1836
72. Edinger M, Hoffmann P, Ermann J, et al. CD4+CD25+ regulatory T cells preserve graft-versus-tumor activity while inhibiting graft-versus-host disease after bone marrow transplantation. Nature Medicine 2003; 9:1144–1150
73. Trenado A, Charlotte F, Fisson S, et al. Recipient-type specific CD4+CD25+ regulatory T cells favor immune reconstitution and control graft-versus-host disease while maintaining graft-versus-leukemia. The Journal of Clinical Investigation 2003; 112:1688–1696

74. Taylor PA, Lees CJ, Blazar BR. The infusion of ex vivo activated and expanded CD4(+)CD25(+) immune regulatory cells inhibits graft-versus-host disease lethality. Blood 2002; 99:3493–3499
75. Godfrey WR, Ge YG, Spoden DJ, et al. In vitro-expanded human CD4(+)CD25(+) T-regulatory cells can markedly inhibit allogeneic dendritic cell-stimulated MLR cultures. Blood 2004; 104:453–461
76. Roncarolo MG, Bacchetta R, Bordignon C, Narula S, Levings MK. Type 1 T regulatory cells. Immunological Reviews 2001; 182:68–79
77. Tarbell KV, Yamazaki S, Olson K, Toy P, Steinman RM. CD25+ CD4+ T cells, expanded with dendritic cells presenting a single autoantigenic peptide, suppress autoimmune diabetes. The Journal of Experimental Medicine 2004; 199:1467–1477
78. Maus MV, Thomas AK, Leonard DG, et al. Ex vivo expansion of polyclonal and antigen-specific cytotoxic T lymphocytes by artificial APCs expressing ligands for the T-cell receptor, CD28 and 4-1BB. Nature Biotechnology 2002; 20:143–148
79. Dutt S, Ermann J, Tseng D, et al. L-selectin and beta7 integrin on donor CD4 T cells are required for the early migration to host mesenteric lymph nodes and acute colitis of graft-versus-host disease. Blood 2005; 106:4009–4015
80. Su L, Creusot RJ, Gallo EM, et al. Murine CD4+CD25+ regulatory T cells fail to undergo chromatin remodeling across the proximal promoter region of the IL-2 gene. Journal of Immunology 2004; 173:4994–5001
81. Sweeney TJ, Mailander V, Tucker AA, et al. Visualizing the kinetics of tumor-cell clearance in living animals. Proceedings of the National Academy of Sciences of the United States of America 1999; 96:12044–12049
82. Cao D, van Vollenhoven R, Klareskog L, Trollmo C, Malmstrom V. CD25brightCD4+ regulatory T cells are enriched in inflamed joints of patients with chronic rheumatic disease. Arthritis Research and Therapy 2004; 6:R335–R346
83. Beilhack A, Schulz S, Baker J, et al. In vivo analyses of early events in acute graft-versus-host disease reveal sequential infiltration of T-cell subsets. Blood 2005; 106:1113–1122

Chapter 12
Natural CD4$^+$CD25$^+$ Regulatory T Cells in Regulation of Autoimmune Disease

Adam P. Kohm and Stephen D. Miller

Abstract An essential characteristic of thymic T cell development is the generation of TCR diversity enabling T cells to respond to an unlimited number of foreign antigens. However, one inevitable consequence of TCR diversity is the generation of self-reactive TCRs creating the potential for autoimmune disease. To balance this, the immune system has developed the processes of central and peripheral tolerance to deter the generation of self-reactive T cells as well as to regulate the function of autoreactive T cells that persist in the mature T cell repertoire, respectively. We'll discuss here one critical mediator of peripheral tolerance, CD4$^+$CD25$^+$ regulatory or suppressor T cells, and their role in regulating the various processes of autoimmune disease.

Introduction

As critical members of the adaptive immune system, CD4$^+$ T cells are key mediators in multiple phases of the protective immune response by recognizing foreign antigens via their antigen-specific T cell receptor (TCR) complex during cognate interactions with antigen-presenting cells (APCs) displaying peptide/MHC class II complexes. Thus, an essential characteristic of thymic T cell development is the generation of TCR diversity enabling T cells to respond to an unlimited number of foreign antigens. However, one inevitable consequence of TCR diversity is the generation of self-reactive TCRs creating the potential for development of autoimmune disease.

To balance this, the immune system has developed regulatory checkpoints that govern lymphocyte development which includes the biphasic processes of central tolerance which only permits the generation of T cells with a functional TCR while deleting populations of T cells which express TCRs specific for self-peptides

---

S.D. Miller
Department of Microbiology-Immunology and Interdepartmental Immunobiology Center, Feinberg School of Medicine, Northwestern University, 303 E. Chicago Avenue, Chicago, IL 60611 USA
e-mail: s-d-miller@northwestern.edu

expressed in the thymus. When properly functioning, the process of central tolerance ensures the selective generation of functional, non-self-reactive T cells. However, many tissue antigens are not expressed at sufficient levels in the thymus to ensure culling the repertoire of all self-reactive T cells, thus central tolerance is not absolute. Therefore, autoreactive T cells persist in the mature T cell repertoire with the potential of being activated by various means to mediate autoimmune disease creating a requirement for additional endogenous peripheral regulatory mechanisms.

A critical peripheral immune regulatory mechanism consists of the activity of $CD4^+CD25^+Foxp3^+$ regulatory T cells (T-regs). One fundamental question is whether $CD4^+CD25^+$ T-reg cells influence the onset of autoimmune disease, and more importantly, does T-reg cell dysfunction/deficiency lead to increased susceptibility to autoimmune disease? These are important and hotly contested questions. There is a growing body of evidence indicating that $CD4^+CD25^+$ T-regs play a key role in the maintenance of self-tolerance. Examples of this critical responsibility can be gleaned from experimental models in which T-regs are either naturally inactive or artificially depleted [29, 49]. In both cases, T-reg dysfunction directly correlates with fatal lymphoproliferative disease and eventual death. Such regulatory responsibilities suggest a significant role for T-regs in modulating the development of autoimmunity [48, 55]. This review will discuss the role of T-regs in regulating the induction and expression of a number of autoimmune diseases.

T-Regs in EAE/Multiple Sclerosis

Multiple sclerosis (MS) is an immune-mediated disease of the central nervous system (CNS) characterized by perivascular $CD4^+$ T cell and mononuclear cell infiltration with subsequent primary demyelination of axonal tracks leading to progressive paralysis [57]. MS is generally considered to be an autoimmune disease characterized by T cell responses to MBP, PLP, and/or myelin-oligodendrocyte glycoprotein (MOG) [5, 9, 45], however a clear-cut cause-effect relationship between myelin reactivity and disease pathology has yet to be demonstrated. In addition to a significant genetic component [12], epidemiological studies provide strong circumstantial evidence for an environmental trigger, most likely viral, in the induction of MS [28, 44, 56]. CNS pathology may therefore result from bystander myelin damage mediated via T cell targeting of virus persisting in the CNS; *and/or* from direct activation of autoreactive T cells secondary to an encounter with a pathogen via *molecular mimicry* [18, 43, 59], or indirectly by *epitope spreading* resulting from the release of sequestered antigens secondary to virus-specific T cell-initiated myelin damage [37, 53]. Despite years of intensive research, the inducing antigen(s) and precise immunologic mechanisms involved in the induction and chronic course of MS are still poorly understood and there are limited therapeutic options available for managing this disease.

There is a substantial body of evidence supporting a role for $CD4^+CD25^+$ T-reg cells in regulating autoimmune T cell function during EAE in mice and potentially MS in humans. Early observations noted that the occurrence of spontaneous EAE in myelin basic protein (MBP) TCR transgenic mice can be efficiently prevented via the transfer of purified, syngeneic CD4$^+$ T cells [reviewed in [19]]. However, the sub-population of CD4$^+$ T cells responsible for protection in these studies remains unclear as protection was obtained by the transfer of either CD4$^+$CD25$^+$ and CD4$^+$CD25$^-$ T cells.

More recently, studies using the actively induced models of both chronic and relapsing-remitting EAE (R-EAE) provide strong support for a potential contribution of CD4$^+$CD25$^+$ T-reg cells in regulating the onset, progression, and spontaneous recovery from EAE [23, 24]. At a basic level, supplementation of the normal population of CD4$^+$CD25$^+$ T-reg cells via adoptive transfer attenuates the clinical progression of EAE. One of the most common approaches to determine if T-reg cells influence disease progression is to "deplete" this population of cells at various times either prior to or following disease induction [36]. To achieve this, animals are injected with one of the two most common clones of anti-CD25 mAb (PC61 – rIgG1 or 7D4 – rIgM). Injection of mice with either clone leads to what appears to be a rapid depletion of T-reg cells as determined by flow cytometry using the other clone as a detecting antibody. It is important to note that while anti-CD25 mAb injection leads to the rapid depletion of T-reg cells, the T-reg cell population rapidly recovers and returns to normal levels within 7–14 days following depletion. Thus, interpretations from previous data resulting from anti-CD25 mAb injection and presumed CD4$^+$CD25$^+$ T-reg cell depletion may need to be re-evaluated. Accordingly, T-reg cell inactivation by the injection of anti-CD25 mAb exacerbates clinical EAE induced by immunization with a sub-optimal amount of myelin peptide [25]. These findings are in agreement with others that observe that the clinical disease course of EAE is exacerbated in mice previously depleted of CD4$^+$CD25$^+$ T-reg cells and that passive transfer of CD4$^+$CD25$^+$ T-reg cells prevents EAE [35]. In addition, injection of anti-CD25 mAb to deplete CD4$^+$CD25$^+$ T-reg cells converts mice that are normally resistant to EAE (B10.S) to susceptible to disease induction [47]. Thus, these basic findings demonstrate the powerful regulatory role of CD4$^+$CD25$^+$ T-reg cells in controlling the clinical disease course of EAE.

One important consideration is the possibility that the regulatory deficiency in EAE/MS may not lie directly in the T-reg cell population, but instead, in a supporting population. Recent findings suggest a role for CD11c$^+$ dendritic cells in supporting the functional capacity of T-reg cells [10], but in a dynamic manner such that alterations in the levels of costimulatory molecule expression may permit CD11c$^+$ dendritic cells to both promote autoimmune-linked inflammation as well as T-reg cell function and remission from active disease. Thus, when examining the potential role that T-regs may play in mediating both the onset and remission from autoimmune disease, it is important to consider potential contributions of supporting populations.

Another question that arises with any model of autoimmune disease regulated by CD4$^+$CD25$^+$ T-reg cells is the site of suppressive action. The two obvious possibilities for the site of regulations are the secondary lymphoid organs (wherein activation of autoimmune effector cells would be regulated) and/or the target organ (wherein the effector functions of autoimmune effector cells could be suppressed).

Interestingly, the passive transfer of T-reg cells does not appear to result in significant effects on the peripheral immune response during EAE. However, the possibility exists that CD4$^+$CD25$^+$ T-reg cells mediate their effects in the target organ. As discussed above, supplementation of CD4$^+$CD25$^+$ T-reg cells attenuates clinical progression of clinical symptoms of EAE. Of interest, CD4$^+$CD25$^+$ T-reg cells can be found within the CNS [25, 35] and appear to be localized directly within CNS lesions. The presence of CD4$^+$CD25$^+$ T-reg cells within the lesions suggests that regulatory T cells may decrease the level of CNS inflammation by suppressing the continued activation of autoreactive CD4$^+$ responder T cells within the target organ. This may explain the observations that CD4$^+$CD25$^+$ T-reg cells appear to prevent clinical disease progression in the absence of observable effects on the peripheral autoreactive immune response [23].

Findings from clinical studies provide additional support for the potential contribution of CD4$^+$CD25$^+$ T-reg cells in preventing autoimmune disease in humans. Most notably, it appears that CD4$^+$CD25$^+$ T reg cells isolated from the peripheral blood of MS patients possess a reduced capacity to suppress responder CD4$^+$ T cell effector function [4, 54]. There also appears to be a lower frequency of CD4$^+$CD25$^+$ T-reg cell in sick versus healthy individuals, arguing against the potential alternative explanation that the CD4$^+$CD25$^+$ T cells isolated from sick individuals were in fact activated responder CD4$^+$ T cells as opposed to T-reg cells. However, this is a legitimate concern as a high percentage of circulating CD4$^+$ T cells in autoimmune patients are previously activated and therefore may express elevated levels of CD25. Thus, similar to animals studies, research concerning the role of human CD4$^+$CD25$^+$ T-reg cells in regulating autoimmune disease would also significantly benefit from the discovery of a definitive phenotypic marker specifying the regulatory cell population.

Of great interest is the potential role of CD4$^+$CD25$^+$ T-reg cells in mediating spontaneous remission in EAE (or other autoimmune diseases). Due the increased frequency of T-reg cells that express self-reactive TCR, it is conceivable that CD4$^+$CD25$^+$ T-reg cells represent an endogenous mechanism by which the immune system discourages progression self-directed immune responses. Our recent preliminary studies strongly indicate a critical role for T-regs in mediating remission in the SJL/J mouse R-EAE model. Current studies in our laboratory are actively investigating how T-regs are activated in response to ongoing self tissue destruction and their role in regulating spontaneous EAE in a TCR transgenic model of disease. These studies should provide additional information concerning the role of CD4$^+$CD25$^+$ T-reg cells in affecting various components of the pathophysiology of MS including susceptibility, progression, and spontaneous remission.

T-Regs in Type 1 Diabetes

Insulin-dependent diabetes mellitus or type 1 diabetes (T1D) is an autoimmune disease resulting in the progressive loss of insulin-producing β cells in the pancreatic islets of Langerhans [58]. Tissue destruction appears to be the direct result of a breakdown in central tolerance and the subsequent activation of autoreactive CD4$^+$ and CD8$^+$ T cells. In light of the increasing incidence of disease, as well as the contribution of antigen-specific CD4$^+$ and CD8$^+$ T cells to disease pathogenesis, the NOD model of diabetes is widely employed for the general study of autoimmunity and treatment strategy design. Since, IDDM is believed to be the direct result of breakdown of self tolerance, tolerance restoration either via the depletion, inactivation, and/or alteration of the effector function of β-cell antigen-specific autoreactive T cells are all efficient techniques to relieve clinical symptoms. In light of this, a significant effort has been made to understand the endogenous mechanisms that regulate CD4$^+$ T cell effector function and dysfunction.

Since it is believed that the autoreactive immune responses during T1D are directed against islet antigens [22, 42], one natural mechanism for preventing diabetes would the deletion of islet-reactive T cells during negative selection. However, we now know that negative selection is not either dependable or that deficiencies in the protective process are ultimate indicators of who will develop diabetes since islet-reactive T cells are found in the circulation of healthy individuals [33]. Thus, there appears to be a role for active regulatory mechanisms in preventing the activation of these autoreactive T cells and that it is a loss of central tolerance that leads to IDDM onset and progression. At the center of these investigations is the role of T-reg cells in regulating IDDM onset and progression.

Numerous studies have investigated the role of CD4$^+$CD25$^+$ T-reg cells in regulating the onset and progression of IDDM [reviewed in [30]]. Not surprisingly, these studies have reached the general consensus that supplementation of the endogenous CD4$^+$CD25$^+$ T-reg cell population confers resistance to disease, whereas injection of anti-CD25 mAb exacerbates disease progression [2, 3, 51].

Interestingly, unlike in EAE [26], TGF-β production by CD4$^+$CD25$^+$ T-reg cells appears to play a major role in the mechanisms by which this regulatory T cell population influences the progression of IDDM. For example, TGF-β induces the expansion of T-reg cells that prevent diabetes [46] and TGF-β can increase the CD4$^+$CD25$^+$ T-reg cell population to approximately 50% of all CD4$^+$ T cells in the pancreatic islets. This accumulation of T-reg cells is believed to be a direct result of TGF-β induced proliferation of CD4$^+$CD25$^+$FoxP3$^+$ T-reg cells. However, another possibility is the TGF-β induced conversion of CD4$^+$CD25$^-$ T cells to CD4$^+$FoxP3$^+$CD25$^+$ T cells [16, 34]. Regardless of whether TGF-β mediates its effects through T-reg cell expansion and/or conversion, it appears to be a critical component to CD4$^+$CD25$^+$ T-reg cell-mediated regulation of IDDM and this regulation may include both CD4$^+$ and CD8$^+$ autoreactive T cells [20].

It is not surprising that supplementation of the CD4$^+$CD25$^+$Foxp3$^+$ T-reg cell population confers protection against IDDM onset and progression in the animal model systems, as this has been demonstrated in virtually every model of autoimmune disease. However, the more important question is what role do endogenous CD4$^+$CD25$^+$FoxP3$^+$ T-reg cells play in the natural onset and progression of diabetes in IDDM patients? While the number of reports concerning T-reg cell manipulation and its effects on IDDM appears never-ending, there is also a scarcity of data reporting deficits in T-reg cells in these same models of autoimmune disease. It is well-accepted that enhancing the number of T-reg cells confers protection against islet infiltration and damage, but it is also important to note that the number and function of the endogenous $CD4^+CD25^+FoxP3^+$ T-reg cells appears normal in these animals even in the face of impending autoimmune disease [38]. This raises that possibility that the regulatory dysfunction in IDDM is not in the $CD4^+CD25^+FoxP3^+$ T-reg cell compartment, but that supplementing this population compensates for other defunct regulatory processes. One potential explanation may indeed involve $CD4^+CD25^+FoxP3^+$ T-reg cells, but the deficit may originate on the other side of the immunological synapse in the effector cell population. More recent studies have now begun a much need analysis of changes in effector cell susceptibility to regulation and future studies may reveal that the true regulatory deficits lie within the target, not regulatory, cell population.

T-Regs in Rheumatoid Arthritis

Rheumatoid arthritis (RA) is debilitating autoimmune disease of unknown etiology characterized by systemic inflammation of the joints and the ensuing destruction of cartilage and bone. While the exact destructive mechanism(s) of RA is unknown, clinical symptoms are paralleled by the production of pro-inflammatory cytokines and immune cell activation within the joints [40]. It is believed that a fundamental breakdown in the processes of self-tolerance is responsible for the initiation of RA and subsequent destruction of joint tissue.

At first, it seems reasonable to assume that if a deficiency in T-regs is, in part, involved in the initiation of RA, then one might expect lower numbers of T-regs to be present in RA patients. However, the number of T-regs as defined by the number or $CD25^{high}$ CD4+ T cells in the peripheral blood does appear to be decreased in RA patients [31]. Certainly more interesting is the observed increased in T-reg cell numbers in the synovial fluid of RA patients [6], a finding that obviously is contradictory to initial assumptions. This finding leads to the next obvious possibility which is that T-reg cells in RA patients may be non-functional. This latter hypothesis currently appears correct, as T-reg cells isolated from the synovial fluid of RA patients appear to be functional inactive in regards to their ability to suppress pro-inflammatory cytokine production [13].

While the exact mechanism responsible for this deficit in T-reg cell function is unknown, a number of possible explanations exist. First and foremost, it is possible that the level of the local immune response may simply exceed that which may be inhibited by T-reg cells. It is generally accepted that the level of $CD4^+$ T cell activation inversely correlates with the ability of T-reg cells to block their effector function. Thus, the inflammatory processes may outpace the regulatory capacity of the local T-reg cell population regardless of the enhanced numbers of these cells in the autoimmune inflammatory environment. Alternately, the costimulatory environment also plays a significant role in the determination of a successful T-reg cell response. High expression of costimulatory molecules by the supporting antigen-presenting cell populations may also block T-reg cell inhibitory function. Lastly, active mechanisms may directly inhibit T-reg cell function. For example, treatment with anti-TNF-α resulted in increased numbers of T-reg cells in RA patients [41] and clinical remission suggesting an active inhibitor role for TNF-α on T-reg cell function in vivo. Anti-TNF-α treatment, in the form of infliximab, is now an accepted treatment for RA [32].

Findings from both animal models and the clinic clearly demonstrate a role for T-reg cells in regulating the progression of RA and current treatments exhibit known effects on the T-reg cell population to augment both their numbers and effector function. Regardless, the mechanisms responsible for the initial deficits in the capacity for T-reg cells to maintain immune homeostasis within the joints of RA patients is still unclear.

T-Regs in Lupus Erythematosus

Lupus erythematosus (LE) clinically manifests in a variety of forms ranging from localized cutaneous lesions (CLE) to a broad systemic phenotype (SLE). LE is characterized by the presence of autoantibodies directed against a variety of antigens including DNA, nuclear proteins and cell surface antigens [7, 39] that are believed to be the result of hyperactive B cell function and/or dysfunctional self-tolerance. While the Ag depot appears to be linked to the inefficient or defective clearance of apoptotic cells and their contents, these deficiencies are believed to be exaggerated by dysfunctional B and/or T cell regulatory mechanisms creating a potential for abnormal T-reg function to contribute to disease progression.

Supporting this hypothesis is the observation of increased titers of anti-dsDNA antibodies in animals lacking functional T-regs [27]. These findings from animal models have been translated into the clinic. Studies investigating the role of T-regs in LE patients reveal decreased numbers of circulating $CD4^+CD25^+$ T-regs in both SLE [8] and CLE [17] patients suggesting that developmental abnormalities in the T-reg linage, in combination with inefficient clearance of apoptotic cells, may contribute to the onset of LE.

T-Regs in Wiskott-Aldrich Syndrome

Wiskott-Aldrich syndrome (WAS) is an X-linked primary immunodeficiency characterized by opportunistic infections, thrombocytopenia, and eczema, as well as autoimmune disease. The primary deficiency in WAS is linked to the WAS protein, which is a critical regulator of actin polymerization. It is believed that this deficit is responsible for the ensuing abnormal lymphocyte function, mostly localized to the T cell compartment, that characterizes this disease. Of importance to the discussion at hand is that observation that as many as 70% of WAS patients display symptoms of various autoimmune disorders [11]. The combination of an affected T cell compartment with the prevalence of autoimmune symptoms in WAS patients lends significant support to the hypothesis that WAS protein deficiency negatively affects $CD4^+CD25^+$ T-reg cell function

A number of studies have investigated the effect of WAS protein deficiency on $CD4^+CD25^+$ T-reg cell development and function. Interestingly, WAS protein deficiency does not appear to block thymic development of T-reg cells [21], yet the numbers of T-regs in the periphery are decreased in WAS subjects. One explanation of this may be effects of WAS protein deficiency on the "competitive fitness" of the $CD4^+CD25^+$ T-reg cell population. WAS protein is known to play a critical role in T cell activation, thus providing a mechanism whereby T-reg cells may not receive the required level of stimulation for prolonged survival in the periphery. However, T-reg cell survival may not be the sole explanation for the aberrant immunoregulation in WAS patients, as WAS protein deficiency also appears to negatively influence $CD4^+CD25^+$ T-reg cell function. T-regs isolated from WASP patients displayed impaired suppressive function in vitro, however, this impairment could partially be rescued by exogenous IL-2 treatment [1]. Taken together, these findings suggest that $CD4^+CD25^+$ T-reg cell dysfunction may contribute to the clinical symptoms of WAS, however, it is likely that WAS protein deficiency globally affects all T cells and that it is this effect on multiple T cell populations that is responsible for the wide variety of symptoms, including autoimmunity, in these patients.

T-Reg Cells as Treatments for Autoimmune Disease

Due to the massive recent interest in T-regs and their contribution to the regulatory mechanisms that govern autoimmune disease onset and progression, these cells have been proposed for use in cell-based therapy for both autoimmune disease and organ/tissue transplantation [50]. As discussed above, many animal models have demonstrated that the transfer of an antigen-specific population of T-regs appears to be an effective deterrent to autoimmune disease onset and/or progression. Thus, the autologous transfer of enriched T-regs would appear to be a viable treatment option for clinical use. However perhaps the greatest challenge to this type of therapy is not the technical processes of T-reg cell isolation, activation and transfer, but instead the correct identification of the appropriate antigens to target during autoimmune disease and concern about the antigen non-specific effector arm of T-reg-medicated suppression.

In many of the autoimmune disease models discussed in this chapter, significant progress has been made towards the identification of the immunodominant epitopes that are responsible for both the initiation and progression of disease. Unfortunately, most of these studies were performed in transgenic mice in which only a single MHC haplotype is present thus significantly simplifying epitope determination. This is not the case in the clinic where patients express many more possible haplotype combinations which significantly complicates the identification and selection of the appropriate Ag targets. However, in the future it is likely that a successful combination of haplotype screening with the referencing of an extensive peptide library of immunodominant epitopes may correctly predict which T-reg cell population should be expanded for treatment in specific diseases.

Conclusions

In summary, it appears that $CD4^+CD25^+$ T-reg cells play a significant role in the body's defense against and control of autoimmune disease. As discussed above, $CD4^+CD25^+$ T-reg cells appear to play major roles in regulating both EAE/MS and IDDM, and importantly, T-reg cells have also been reported to influence a number of other (auto)immune diseases such as colitis [15] and graft versus host disease [14, 52]. By preferentially expressing TCRs that recognize self-peptide/MHC complexes and by efficiently quenching both $CD4^+$ and $CD8^+$ T cell effector function upon activation, $CD4^+CD25^+$ T-reg cells appear to be a well-suited defense against the rogue activation of self-reactive T cells. However, at this point, there appears to be more questions than answers in regards to the origin, phenotype, and the mechanism of action of this regulatory cell population. To date, despite an intensive effort, identification of a definitive surface phenotypic surface marker of $CD4^+CD25^+$ T-reg cells remains elusive. However, the realization that multiple sub-populations of regulatory T cells appear to cooperate in regulating the immune response may contribute to future success in defining both the phenotype and mechanism of action of each of these regulatory cell populations. In addition, the future ability to isolate and propagate self antigen-specific T-regs may provide a therapeutic option for the therapy of a variety of autoimmune diseases.

References

1. Adriani, M., Aoki, J., Horai, R., Thornton, A. M., Konno, A., Kirby, M., Anderson, S. M., Siegel, R. M., Candotti, F., and Schwartzberg, P. L. (2007). Impaired in vitro regulatory T cell function associated with Wiskott-Aldrich syndrome. *Clin. Immunol.* 124:41–48.
2. Alyanakian, M. A., You, S., Damotte, D., Gouarin, C., Esling, A., Garcia, C., Havouis, S., Chatenoud, L., and Bach, J. F. (2003). Diversity of regulatory CD4+T cells controlling distinct organ-specific autoimmune diseases. *Proc. Natl. Acad. Sci. U.S.A.* 100:15806–15811.
3. Anderson, M. S., and Bluestone, J. A. (2005). The NOD mouse: A model of immune dysregulation. *Ann. Rev. Immunol.* 23:447–485.
4. Baecher-Allan, C., Brown, J. A., Freeman, G. J., and Hafler, D. A. (2001). CD4+CD25high regulatory cells in human peripheral blood. *J. Immunol.* 167:1245–1253.
5. Bernard, C. C., and de Rosbo, N. K. (1991). Immunopathological recognition of autoantigens in multiple sclerosis. *Acta Neurol.* 13:171–178.
6. Cao, D., Malmstrom, V., Baecher-Allan, C., Hafler, D., Klareskog, L., and Trollmo, C. (2003). Isolation and functional characterization of regulatory CD25brightCD4+ T cells from the target organ of patients with rheumatoid arthritis. *Eur. J. Immunol.* 33:215–223.
7. Cohen, P. L. (1993). T- and B-cell abnormalities in systemic lupus. *J. Invest. Dermatol.* 100:69S–72S.
8. Crispin, J. C., Martinez, A., and Alcocer-Varela, J. (2003). Quantification of regulatory T cells in patients with systemic lupus erythematosus. *J. Autoimmun.* 21:273–276.

9. de Rosbo, N. K., Hoffman, M., Mendel, I., Yust, I., Kaye, J., Bakimer, R., Flechter, S., Abramsky, O., Milo, R., Karni, A., and Ben-Nun, A. (1997). Predominance of the autoimmune response to myelin oligodendrocyte glycoprotein (MOG) in multiple sclerosis: Reactivity to the extracellular domain of MOG is directed against three main regions. *Eur. J. Immunol.* 27:3059–3069.
10. Deshpande, P., King, I. L., and Segal, B. M. (2007). Cutting edge: CNS CD11c+ cells from mice with encephalomyelitis polarize Th17 cells and support CD25+CD4+ T cell-mediated immunosuppression, suggesting dual roles in the disease process. *J. Immunol.* 178: 6695–6699.
11. Dupuis-Girod, S., Medioni, J., Haddad, E., Quartier, P., Cavazzana-Calvo, M., Le Deist, F., de Saint Basile, G., Delaunay, J., Schwarz, K., Casanova, J. L., et al. (2003). Autoimmunity in Wiskott-Aldrich syndrome: Risk factors, clinical features, and outcome in a single-center cohort of 55 patients. *Pediatrics* 111:e622–627.
12. Ebers, G. C., Sadovnick, A. D., and Risch, N. J. (1995). A genetic basis for familial aggregation in multiple sclerosis. *Nature* 377:150–151.
13. Ehrenstein, M. R., Evans, J. G., Singh, A., Moore, S., Warnes, G., Isenberg, D. A., and Mauri, C. (2004). Compromised function of regulatory T cells in rheumatoid arthritis and reversal by anti-TNFalpha therapy. *J. Exp. Med.* 200:277–285.
14. Ermann, J., Hoffmann, P., Edinger, M., Dutt, S., Blankenberg, F. G., Higgins, J. P., Negrin, R. S., Fathman, C. G., and Strober, S. (2005). Only the CD62L+ subpopulation of CD4+CD25+ regulatory T cells protects from lethal acute GVHD. *Blood* 105:2220–2226.
15. Fahlen, L., Read, S., Gorelik, L., Hurst, S. D., Coffman, R. L., Flavell, R. A., and Powrie, F. (2005). T cells that cannot respond to TGF-beta escape control by CD4(+)CD25(+) regulatory T cells. *J. Exp. Med.* 201:737–746.
16. Fantini, M. C., Becker, C., Monteleone, G., Pallone, F., Galle, P. R., and Neurath, M. F. (2004). Cutting edge: TGF-beta induces a regulatory phenotype in CD4+CD25- T cells through Foxp3 induction and down-regulation of Smad7. *J. Immunol.* 172:5149–5153.
17. Franz, B., Fritzsching, B., Riehl, A., Oberle, N., Klemke, C. D., Sykora, J., Quick, S., Stumpf, C., Hartmann, M., Enk, A., et al. (2007). Low number of regulatory T cells in skin lesions of patients with cutaneous lupus erythematosus. *Arthritis Rheum.* 56:1910–1920.
18. Fujinami, R. S., and Oldstone, M. B. (1985). Amino acid homology between the encephalitogenic site of myelin basic protein and virus: Mechanism for autoimmunity. *Science* 230: 1043–1045.
19. Furtado, G. C., Olivares-Villagomez, D., Curotto de Lafaille, M. A., Wensky, A. K., Latkowski, J. A., and Lafaille, J. J. (2001). Regulatory T cells in spontaneous autoimmune encephalomyelitis. *Immunol. Rev.* 182:122–134.
20. Green, E. A., Gorelik, L., McGregor, C. M., Tran, E. H., and Flavell, R. A. (2003). CD4+CD25+ T regulatory cells control anti-islet CD8+ T cells through TGF-beta-TGF-beta receptor interactions in type 1 diabetes. *Proc. Natl. Acad. Sci. U.S.A.* 100:10878–10883.
21. Humblet-Baron, S., Sather, B., Anover, S., Becker-Herman, S., Kasprowicz, D. J., Khim, S., Nguyen, T., Hudkins-Loya, K., Alpers, C. E., Ziegler, S. F., et al. (2007). Wiskott-Aldrich syndrome protein is required for regulatory T cell homeostasis. *J. Clin. Invest.* 117: 407–418.
22. Kent, S. C., Chen, Y., Bregoli, L., Clemmings, S. M., Kenyon, N. S., Ricordi, C., Hering, B. J., and Hafler, D. A. (2005). Expanded T cells from pancreatic lymph nodes of type 1 diabetic subjects recognize an insulin epitope. *Nature* 435:224–228.
23. Kohm, A. P., Carpentier, P. A., Anger, H. A., and Miller, S. D. (2002). Cutting Edge: CD4(+)CD25(+) regulatory T cells suppress antigen-specific autoreactive immune responses and central nervous system inflammation during active experimental autoimmune encephalomyelitis. *J. Immunol.* 169:4712–4716.
24. Kohm, A. P., Carpentier, P. A., and Miller, S. D. (2003). Regulation of experimental autoimmune encephalomyelitis (EAE) by CD4$^+$CD25$^+$ regulatory T cells. *Novartis Found. Symp.* 252:45–52.

25. Kohm, A. P., McMahon, J. S., Podojil, J. R., Smith Begolka, W., DeGutes, M., Kasprowicz, D. J., Ziegler, S. F., and Miller, S. D. (2006). Cutting Edge: Anti-CD25 mAb injection results in the functional inactivation, not depletion of CD4+CD25+ Treg cells. *J. Immunol.* 176:3301–3305.
26. Kohm, A. P., Williams, J. S., Bickford, A. L., McMahon, J. S., Chatenoud, L., Bach, J. F., Bluestone, J. A., and Miller, S. D. (2005). Treatment with nonmitogenic anti-CD3 monoclonal antibody induces CD4+ T cell unresponsiveness and functional reversal of established experimental autoimmune encephalomyelitis. *J. Immunol.* 174:4525–4534.
27. Koonpaew, S., Shen, S., Flowers, L., and Zhang, W. (2006). LAT-mediated signaling in CD4+CD25+ regulatory T cell development. *J. Exp. Med.* 203:119–129.
28. Kurtzke, J. F. (1993). Epidemiologic evidence for multiple sclerosis as an infection. *Clin. Microbiol. Rev.* 6:382–427.
29. Lahl, K., Loddenkemper, C., Drouin, C., Freyer, J., Arnason, J., Eberl, G., Hamann, A., Wagner, H., Huehn, J., and Sparwasser, T. (2007). Selective depletion of Foxp3+ regulatory T cells induces a scurfy-like disease. *J. Exp. Med.* 204:57–63.
30. Lan, R. Y., Ansari, A. A., Lian, Z. X., and Gershwin, M. E. (2005). Regulatory T cells: Development, function and role in autoimmunity. *Autoimmun. Rev.* 4:351–363.
31. Leipe, J., Skapenko, A., Lipsky, P. E., and Schulze-Koops, H. (2005). Regulatory T cells in rheumatoid arthritis. *Arthritis Res. Ther.* 7:93.
32. Lipsky, P. E., van der Heijde, D. M., St Clair, E. W., Furst, D. E., Breedveld, F. C., Kalden, J. R., Smolen, J. S., Weisman, M., Emery, P., Feldmann, M., et al. (2000). Infliximab and methotrexate in the treatment of rheumatoid arthritis. Anti-tumor necrosis factor trial in rheumatoid arthritis with concomitant therapy study group. *N. Engl. J. Med.* 343:1594–1602.
33. Lohmann, T., Leslie, R. D., and Londei, M. (1996). T cell clones to epitopes of glutamic acid decarboxylase 65 raised from normal subjects and patients with insulin-dependent diabetes. *J. Autoimmun.* 9:385–389.
34. Luo, X., Tarbell, K. V., Yang, H., Pothoven, K., Bailey, S. L., Ding, R., Steinman, R. M., and Suthanthiran, M. (2007). Dendritic cells with TGF-{beta}1 differentiate naive CD4+CD25– T cells into islet-protective Foxp3+ regulatory T cells. *Proc. Natl. Acad. Sci. U.S.A.* 104: 2821–2826.
35. McGeachy, M. J., Stephens, L. A., and Anderton, S. M. (2005). Natural recovery and protection from autoimmune encephalomyelitis: Contribution of CD4+CD25+ regulatory cells within the central nervous system. *J. Immunol.* 175:3025–3032.
36. McHugh, R. S., and Shevach, E. M. (2002). Cutting edge: Depletion of CD4+CD25+ regulatory T cells is necessary, but not sufficient, for induction of organ-specific autoimmune disease. *J. Immunol.* 168:5979–5983.
37. McRae, B. L., Vanderlugt, C. L., Dal Canto, M. C., and Miller, S. D. (1995). Functional evidence for epitope spreading in the relapsing pathology of experimental autoimmune encephalomyelitis. *J. Exp. Med.* 182:75–85.
38. Mellanby, R. J., Thomas, D., Phillips, J. M., and Cooke, A. (2007). Diabetes in non-obese diabetic mice is not associated with quantitative changes in CD4+ CD25+ Foxp3+ regulatory T cells. *Immunology* 121:15–28.
39. Mills, J. A. (1994). Systemic lupus erythematosus. *N. Engl. J. Med.* 330:1871–1879.
40. Muller-Ladner, U., Pap, T., Gay, R. E., Neidhart, M., and Gay, S. (2005). Mechanisms of disease: The molecular and cellular basis of joint destruction in rheumatoid arthritis. *Nature Clin. Pract. Rheumatol.* 1:102–110.
41. Nadkarni, S., Mauri, C., and Ehrenstein, M. R. (2007). Anti-TNF-alpha therapy induces a distinct regulatory T cell population in patients with rheumatoid arthritis via TGF-beta. *J. Exp. Med.* 204:33–39.
42. Nakayama, M., Abiru, N., Moriyama, H., Babaya, N., Liu, E., Miao, D. M., Yu, L. P., Wegmann, D. R., Hutton, J. C., Elliott, J. F., and Eisenbarth, G. S. (2005). Prime role for an insulin epitope in the development of type 1 diabetes in NOD mice. *Nature* 435: 220–223.

43. Olson, J. K., Croxford, J. L., Calenoff, M., Dal Canto, M. C., and Miller, S. D. (2001a). A virus-induced molecular mimicry model of multiple sclerosis. *J. Clin. Invest.* 108:311–318.
44. Olson, J. K., Croxford, J. L., and Miller, S. D. (2001b). Virus-induced autoimmunity: Potential role of viruses in initiation, perpetuation, and progression of T cell-mediated autoimmune diseases *Viral Immunol.* 14 227–250.
45. Ota, K., Matsui, M., Milford, E. L., Mackin, G. A., Weiner, H. L., and Hafler, D. A. (1990). T-cell recognition of an immunodominant myelin basic protein epitope in multiple sclerosis. *Nature* 346:183–187.
46. Peng, Y., Laouar, Y., Li, M. O., Green, E. A., and Flavell, R. A. (2004). TGF-beta regulates in vivo expansion of Foxp3-expressing CD4+CD25+ regulatory T cells responsible for protection against diabetes. *Proc. Natl. Acad. Sci. U.S.A.* 101:4572–4577.
47. Reddy, J., Illes, Z., Zhang, X., Encinas, J., Pyrdol, J., Nicholson, L., Sobel, R. A., Wucherpfennig, K. W., and Kuchroo, V. K. (2004). Myelin proteolipid protein-specific CD4+CD25+ regulatory cells mediate genetic resistance to experimental autoimmune encephalomyelitis. *Proc. Natl. Acad. Sci. U.S.A.* 101:15434–15439.
48. Sakaguchi, S. (2005). Naturally arising Foxp3-expressing CD25+CD4+ regulatory T cells in immunological tolerance to self and non-self. *Nat. Immunol.* 6:345–352.
49. Sakaguchi, S., Ono, M., Setoguchi, R., Yagi, H., Hori, S., Fehervari, Z., Shimizu, J., Takahashi, T., and Nomura, T. (2006). Foxp3+ CD25+ CD4+ natural regulatory T cells in dominant self-tolerance and autoimmune disease. *Immunol. Rev.* 212:8–27.
50. Tang, Q., and Bluestone, J. A. (2006). Regulatory T-cell physiology and application to treat autoimmunity. *Immunol. Rev.* 212:217–237.
51. Tarbell, K. V., Petit, L., Zuo, X., Toy, P., Luo, X., Mqadmi, A., Yang, H., Suthanthiran, M., Mojsov, S., and Steinman, R. M. (2007). Dendritic cell-expanded, islet-specific CD4+ CD25+ CD62L+ regulatory T cells restore normoglycemia in diabetic NOD mice. *J. Exp. Med.* 204:191–201.
52. Taylor, P. A., Panoskaltsis-Mortari, A., Swedin, J. M., Lucas, P. J., Gress, R. E., Levine, B. L., June, C. H., Serody, J. S., and Blazar, B. R. (2004). L-Selectin(hi) but not the L-selectin(lo) CD4+25+ T-regulatory cells are potent inhibitors of GVHD and BM graft rejection. *Blood* 104:3804–3812.
53. Vanderlugt, C. L., and Miller, S. D. (2002). Epitope spreading in immune-mediated diseases: Implications for immunotherapy. *Nat. Rev. Immunol.* 2:85–95.
54. Viglietta, V., Baecher-Allan, C., Weiner, H. L., and Hafler, D. A. (2004). Loss of functional suppression by CD4+CD25+ regulatory T cells in patients with multiple sclerosis. *J. Exp. Med.* 199:971–979.
55. von Boehmer, H. (2005). Mechanisms of suppression by suppressor T cells. *Nat. Immunol.* 6:338–344.
56. Waksman, B. H. (1995). Multiple sclerosis: More genes versus environment. *Nature* 377:105–106.
57. Wekerle, H. (1991). Immunopathogenesis of multiple sclerosis. *Acta Neurol.* 13:197–204.
58. Wong, F. S., and Janeway, C. A. (1999). Insulin-dependent diabetes mellitus and its animal models. *Curr. Opin. Immunol.* 11:643–647.
59. Wucherpfennig, K. W., and Strominger, J. L. (1995). Molecular mimicry in T cell-mediated autoimmunity: Viral peptides activate human T cell clones specific for myelin basic protein. *Cell* 80:695–705.

Chapter 13
Multiple Sclerosis and Regulatory T Cells

Jonathon Hutton, Clare Baecher-Allan, and David A. Hafler

Abstract Multiple sclerosis (MS) is a complex genetic disease associated with inflammation in the central nervous system (CNS) white matter. This disease thought to be mediated by an autoimmune processes, involves autoreactive T cells and the clonal expansion of B cells and their antibody products. Consistent with this hypothesis, MS is associated with major histocompatibility complex genes, the occurrence of inflammatory white matter infiltrates, and can be treated with immunomodulatory and immunosuppressive therapies. The underlying disease pathology is thought to be caused by autoreactive myelin-specific effector T cells that enter into the CNS. Whilst autoreactive T cells are present in the periphery of healthy individuals, other regulatory mechanisms exist to prevent autoreactive T cells from causing immune disorders. Active suppression by regulatory T (Treg) cells plays a key role in the control of self-antigen-reactive T cells and the induction of peripheral tolerance in vivo. In particular, the importance of antigen-specific Treg cells in conferring genetic resistance to organ specific autoimmunity and in limiting autoimmune tissue damage has been documented in many disease models including MS. The current consensus suggests that the frequency of Tregs in MS patients is unchanged from controls, but their function measured in vitro may be diminished, correlating with impaired inhibitory activity in vivo. This chapter discusses the immunopathology of MS with particular focus given to regulatory T cells and their potential for the development of new therapies to treat this disease.

Work was supported by the NIH grants: UO1DK6192601, RO1NS2424710, PO1AI39671, and PO1NS38037; and grants from the National Multiple Sclerosis Society: RG2172C9 and RG3308A10, and from the 2004 FOCIS Centers of Excellence Amgen Award.

D.A. Hafler
Division of Molecular Immunology, Center for Neurologic Diseases, Brigham and Women's Hospital, Harvard Medical School, 77 Ave Louis Pasteur, Boston, MA. 02115, USA
e-mail: hafler@broad.mit.edu

Introduction

Multiple sclerosis (MS) was first described in 1868 noting the accumulation of inflammatory cells in a perivascular distribution within the brain and spinal cord white matter of patients with intermittent episodes of neurologic dysfunction. This led to the term *sclerose en plaques disseineeś* or MS. The more recent observation in 1948 by Elvin Kabat [1] of increases in oligoclonal immunoglobulin in the cerebro spinal (CSF) of patients with MS provided further evidence of an inflammatory nature to the disease [1]. In the past half-century, several large population-based MS twin studies demonstrated a strong genetic basis to this clinical-pathologic entity [2]. Lastly, the demonstration of an autoimmune, at times demyelinating, disease in mammals with immunization of central nervous system (CNS) myelin [experimental autoimmune encephalomyelitis (EAE)], first made by Thomas Rivers [3] at the Rockefeller Institute in 1933 with the repeated injection of rabbit brain and spinal cord into primates, has led to the generally accepted hypothesis that MS is secondary to an autoimmune response to self-antigens in a genetically susceptible host. It should be pointed out that although the inflammation found in the CNS of patients with MS is thought to represent an autoimmune response, this belief is based on negative results with the inability to consistently isolate a microbial agent from the tissue of diseased patients. Nevertheless, primary viral infections in the CNS may induce an autoimmune response [4], and the recurring lesson from the EAE model is that the minimal requirement for inducing inflammatory, autoimmune CNS-demyelinating disease is the activation of myelinreactive T cells in the peripheral immune system [5].

Immunopathophysiology of MS

A critical lesson from the EAE model is that of epitope spreading, first observed by Eli Sercarz [6]. With the injection of a single myelin protein epitope into mice with subsequent development of EAE, it was observed that T cells became activated against other epitopes of the same protein; this was followed by T-cell activation to other myelin proteins that become capable of adoptively transferring the disease to naive mice. The epitope spreading requires costimulation with B7/CD28, suggesting that with tissue damage in the CNS, an adjuvant is created in the CNS with the expression of high amounts of B7.1 costimulatory molecules associated with antigen release [7]. Moreover, we have recently observed that a transgenic mouse expressing DR2 (DRB1*1501) and a T-cell receptor (TCR) (Ob1A12) cloned from the blood of a patient with MS-recognizing myelin basic protein (MBP) 85–99 spontaneously developed EAE with epitope spreading to a number of antigens implicated in MS, including a-crystalline and proteolipid protein (PLP). As we have observed high expression of B7.1 costimulatory molecules in the CNS white matter of patients with MS [7] and most patients exhibit T-cell reactivity to a number of myelin antigens [8], it is likely that by the time a patient develops clinical MS there has been epitope spreading, with reactivity to multiple myelin epitopes.

However, the presence of clonally-expanded T cells in the CSF and brain tissue of patients with the disease raises the issue that there may be clonal reactivity to just a few myelin antigens. Single cell cloning of T cells from the inflamed CNS tissue screened against combinatorial and protein libraries may allow new insights into pathophysiology of MS.

Data from a number of laboratories combined with experimental data in the EAE model where myelin antigen is injected with adjuvant into mammals indicate that there are autoreactive T cells recognizing myelin antigens in the circulation of mammals. It appears that the activation of these T cells is the critical event in inducing autoimmune disease. We and others first demonstrated over a decade ago that T-cell clones isolated from the blood of patients with MS frequently exhibit exquisite specificity for the immunodominant p85–99 epitope of MBP [8–10]. However, while the TCR appears to be highly specific in recognizing this peptide, altering the peptide ligand can change the TCR conformation to yield a higher degree of T-cell cross-reactivity [11]. Using combinatorial chemistry, even a greater degree of cross-reactivity could be demonstrated, and a number of viral epitopes were identified that could trigger autoreactive T-cell clones in a manner that would not be predicted by simple algorithms [12]. Indeed, one MBP-reactive T-cell clone recognized an epitope of an entirely different self-protein, the myelin oligodendrocyte glycoprotein. Hence, a significant degree of functional degeneracy exists in the recognition of self-antigens by T cells. This finding is consistent with the hypothesis that MS is triggered by autoreactive T cells activated by microbial antigens cross-reactive with myelin [13]. The high frequency of activated, myelin-reactive T cells in the circulation and CSF of patients with MS is further consistent with this hypothesis.

CNS-Specific Regulation of Inflammation

A variety of peripheral mechanisms regulate CNS autoimmunity, whether it be editing of the autoreactive T-cell repertoire, limiting activation of autoreactive T cells, or suppressing autoreactive T-cell activity with a variety of regulatory T-cell populations. Nevertheless, the unique anatomy and cell types of the CNS together provide additional mechanisms that limit CNS inflammation and autoimmunity. Recent advances in neurobiology and immunology, aided by technical advances in gene expression analysis and in vivo imaging, have begun to provide greater insight into the molecular events occurring within and around MS plaques, and this knowledge should soon help elucidate the relative roles resident cells in the CNS play in regulating inflammation.

Cellular Events Occurring Within and Around MS Plaques

Gross examination of MS brain tissue reveals multiple sharply demarcated plaques in the CNS white matter with a predilection to the optic nerves and white matter tracts of the periventricular regions, brain stem, and spinal cord. As was recognized early on and so elegantly investigated in more recent studies, substantial axonal injuries with axonal transactions are abundant throughout active MS lesions [14]. The inflammatory cell profile of active lesions is characterized by perivascular infiltration of oligoclonal T cells [15] consisting of CD4$^+$/CD8 α/β [16, 17] and $\gamma\delta$ [18] T cells and monocytes with occasional B cells and infrequent plasma cells [19]. Lymphocytes may be found in normal appearing white matter beyond the margin of active demyelination [20]. Macrophages are most prominent in the center of the plaques and are seen to contain myelin debris, while oligodendrocyte counts are reduced. In chronic active lesions, the inflammatory cell infiltrate is less prominent and may be largely restricted to the rim of the plaque, suggesting the presence of ongoing inflammatory activity along the lesion edge. Recently, four pathologic categories of the disease were described related to potentially different pathophysiologic disease mechanisms, though this has yet to be demonstrated at a molecular level [21]. It was of interest that the pattern of pathology tended to be the same in multiple lesions from any single individual with MS.

Regulation of CNS Inflammation by Resident Cells of the CNS

For well over a decade, there has been considerable debate about the relative influences microglia and astrocytes have on CNS inflammation associated with MS. Many studies have characterized the relative abilities of astrocytes and microglia to process and present antigen or to influence Th-cell activation, differentiation, or apoptosis [22, 23]. Increasingly, focus is shifting to the roles of glial cells in regulation of CNS inflammation. Glial cells may respond in predetermined manners to inflammatory cues, with the glial responses having indirect but profound consequences on neurons and infiltrating lymphocytes. For example, lipopolysaccharide (LPS), a molecular component of Gram-negative bacteria, has been shown to induce neuronal and axonal loss both in vitro and in vivo due to activation of microglia through Toll-like receptor 4 [24]. In another study, peripheral administration of LPS to female rats resulted in a 240% increase in the density of activated microglia in the dentate gyrus of the hippocampus, which correlated with a 35% decrease in hippocampal neurogenesis [25]. Co-culture of neural progenitor cells with LPS-stimulated but not resting microglia inhibited neurogenesis in vitro by approximately 50%, an effect that was mediated by microglial secretion of IL-6. The effect of peripheral LPS administration on neurogenesis was completely blocked by systemic administration of the non-steroidal anti-inflammatory drug indomethacin. Similar results have been obtained in another study, which documented an inverse correlation between the number of activated microglia and new neurons in the hippocampus after CNS delivery of LPS [26]. Again, suppression of microglia activation, this time with systemic administration of minocycline, reduced the number of activated microglia and increased the number of new neurons in the hippocampus.

With respect to MS lesion formation, IL-6 is upregulated in active lesions relative to inactive lesions [27, 28] and could contribute to axonal destruction. Indeed, IL-6-deficient mice are completely resistant to EAE [29]. While both microglia and astrocytes become activated in response to CNS inflammation, they may respond in different ways to inflammatory stimuli (45). For example, microglia but not astrocytes are responsible for production of the Th1-promoting cytokine IL-12 [22]. Aloisi et al. [30] have demonstrated that astrocytes inhibit IL-12 secretion by in vitro activated microglia, while others have demonstrated using primary astrocyte/microglia co-cultures that astrocyte-derived IL-10 inhibits LPS-induced secretion of the proinflammatory molecules nitric oxide (NO), IL-6, tumor necrosis factor-a (TNF-α), and IL-1b [31]. Orian and colleagues [32] have recently demonstrated that depending on the mouse strain used (NOD/Lt or C57Bl/6), injection with encephalitogenic myelin oligodendrocyte glycoprotein (MOG) peptide 35–55 induces disease that resembles relapsing/remitting or primary progressive forms of MS. Subsequently, they have found that in the relapsing/remitting form of disease induced in NOD/Lt mice, there is evidence of both astrocyte and microglial activation prior to the first clinical sign of disease [33]. In contrast, microglia but not astrocytes appear activated in the primary progressive form of disease induced in C57Bl/6 mice. While preliminary, the data suggest that differences in microglial and astrocytic responses to inflammation could influence the pathogenesis of MS. Astrocytes may also prove to indirectly suppress CNS inflammation by induction of regulatory populations of T cells. A recent report demonstrated that primary rat astrocytes (or astrocytoma cell lines) cultured for 24 h with primary T cells inhibited T-cell secretion of interferon (IFN)-γ in a cell contact-dependent manner [34]. T cells cultured with astrocytes in turn acquired a regulatory phenotype in that they could inhibit Con A-induced T-cell proliferation in vitro and ameliorate the severity of EAE induced with rat spinal cord homogenate in vivo. Using ex vivo human malignant glioma tumor specimens, we have recently demonstrated that astrocyte-derived tumor cells are notable for their secretion of the immunosuppressive cytokines IL-10 and TGF-β and that $CD4^+CD25^+$ effector cells that infiltrate the tumors are notable for their secretion of IL-10 in the near absence of IFN-γ. Due to their propensity to secrete IL-10, it is tempting to speculate that astrocytes may naturally promote generation of IL-10-secreting type 1 T-regulatory cells that could help suppress CNS inflammation [35]. It remains to be seen whether nontransformed reactive astrocytes can similarly induce regulatory populations of lymphocytes and in which CNS inflammatory diseases this may arise.

Regulatory T Cells in MS

Clonal deletion of self-reactive T cells in the thymus and induction of T-cell anergy alone do not explain the maintenance of immunologic self-tolerance, as potentially pathogenic autoreactive T cells are present in the periphery of healthy individuals [8, 36]. Thus, other regulatory mechanisms exist to prevent autoreactive T cells from causing immune disorders. Active suppression by regulatory T cells plays a key role in the control of self-antigen-reactive T cells and the induction of peripheral tolerance in vivo [37, 38]. Seminal experiments performed by Sakaguchi et al. [39] have shown that depletion of $CD4^+CD25^+$ suppressor cells results in the onset of systemic autoimmune diseases in mice. Furthermore, co-transfer of these cells with $CD4^+CD25^-$ cells prevents the development of experimentally induced autoimmune diseases such as colitis, gastritis, insulin-dependent autoimmune diabetes, and thyroiditis [40].

We and others described a population of $CD4^+CD25^{hi}$ regulatory T cells in human peripheral blood and thymus [41, 42]. Human $CD4^+CD25^{hi}FoxP3^+$ natural regulatory T cells (nTreg), similar to the mouse $CD4^+CD25^+$ suppressor cells, are anergic to in vitro antigenic stimulation and strongly suppress the proliferation of responder T cells upon coculture. $CD4^+CD25^+FoxP3^+$ T cells are among the best-characterized immune regulatory subsets shown to prevent activation and effector function of activated responder T cells [43]. While autoreactive T cells are present in healthy individuals and patients with autoimmune disorders, autoreactive T cells found in patients with autoimmune disease are more easily activated as compared to those from normal subjects [43]. This finding led us to hypothesize that either deficient generation or reduced effector function of $CD4^+CD25^{hi}$ regulatory T cells plays a role in the autoimmune state of patients with MS. Thus, we recently compared the frequency and function of $CD4^+CD25^{hi}$ T-regulatory cells derived from a group of untreated patients who have relapsing/remitting MS with those from age-matched healthy control subjects [44].

$CD4^+CD25^{hi}$ T Cells are Present with the Same Frequency in Healthy Donors and Patients with MS

We stained whole mononuclear cells from freshly drawn human blood with different combination of anti-CD4-CyChrome, anti-CD25-phosphatidylethanolamine, and a cocktail of fluorescein isothiocyanate (FITC) labeled anti-CD14, anti-CD32, and anti-CD116. The cells were gated on lymphocytes via their forward- and side-scatter features, and all FITC-labeled cells were negatively selected during sorting. Human peripheral blood contains a heterogeneous population of $CD4^+CD25^+$ T cells that express either moderate levels of CD25 consisting of non-regulatory T cells or high levels of CD25 that exhibit regulatory function [41]. As there are no other known cell-surface makers able to identify regulatory T cells ex vivo, we used CD25 expression to discriminate regulatory T cells in humans. We analyzed the mean fluorescent intensity of the $CD25^+$ population in both patients with MS and control subjects and found no differences between the two groups. Approximately 10% of $CD4^+$ T cells express the a-chain of IL-2 receptor CD25, but only 1–2% are $CD25^{hi}$. No differences in the frequency of $CD4^+$ $CD25^{hi}$ T cells were found between patients and healthy controls. Vandenbark and coworkers [45] showed that there were decreases in FoxP3 levels in the Tregs in patients with MS, and this decrease correlated with their loss of function. While Venken et al., has reported that relapsing-remitting, in contrast to secondary progressive MS patients show lower levels of FoxP3 expression [46]. However, it is not yet clear from these studies whether lower FoxP3 expression in MS patients is due to a reduced frequency of FoxP3 expressing cells in the $CD4^+CD25^{hi}$ T-cell population of these

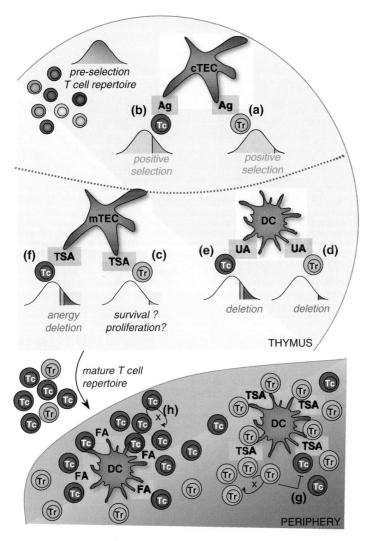

Fig. 3.1 Thymic Selection Positive selection of Treg-precursors (Tr) (**a**) requires higher avidity interactions with cortical thymic epithelial cells (cTEC) than positive selection of conventional T-cell precursors (Tc) (**b**). Medullary thymic epithelial cells (mTEC) somehow secondarily positively select Tr specific for autoantigens (**c**). Thus, a two-step positive selection process appears to substantially contribute to the generation of a Tr repertoire specific for autoantigens. Negative selection prunes the developing Tr and Tc repertoires of cells specific for ubiquitous antigens (UA) expressed by dendritic cells (DC) (**d, e**), neutralizes Tc specific for tissue-specific antigen (TSA) expressed by mTEC (**f**), but spares TSA-specific Tr (**c**). The Tr repertoire is thus exquisitely shaped to prevent autoimmune attacks on tissues (**g**) and to allow development of useful immune responses against foreign antigen (FA) (**h**). The bell-shaped curves indicate the assumed normal distribution of the avidity of pre-selection thymocytes' TCR for self MHC/peptide ligands

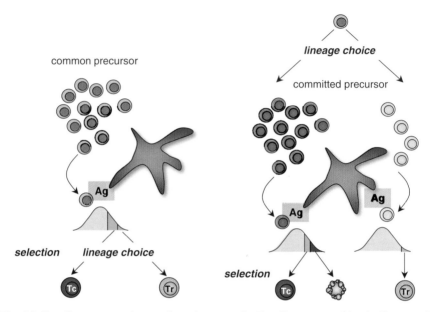

Fig. 3.2 Treg lineage commitment Commitment to the Treg lineage may either be "instructed" by high affinity interaction of common precursors with thymic stroma (**a**) or be caused by any other factor (**b**). In the instructive model, positive selection would therefore take place concomitant with lineage commitment. In the stochastic model for lineage commitment, positive selection (rather than lineage commitment) would depend on high avidity interaction of the thymocyte's TCR with self MHC class II/peptide ligand. It appears at present more likely that precursors choose the Treg lineage independently of the TCR's specificity

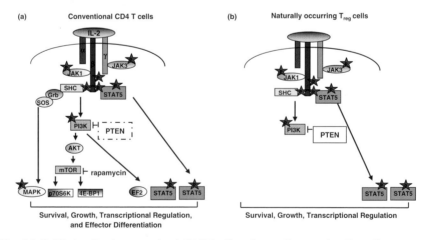

Fig. 4.1 IL-2R signaling in conventional CD4 T cells and naturally occurring T_{reg} cells

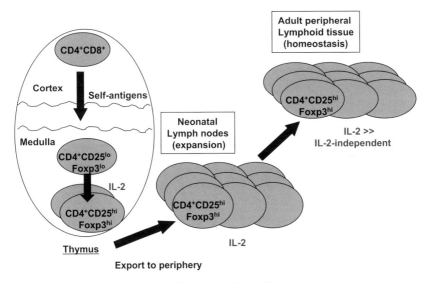

Fig. 4.4 A model for the contribution of IL-2 during T_{reg} cell development and maintenance

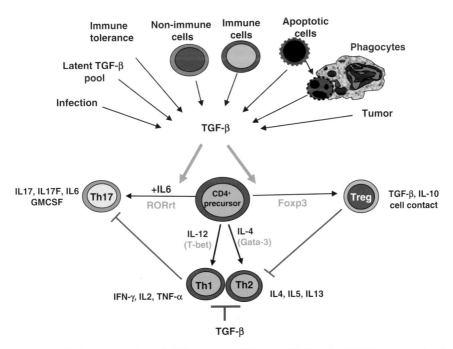

Fig. 7.1 TGF-β controls reciprocal differentiation of Tregs and Th17 cells. TGF-β is produced and secreted by almost all types of cells and from every tissues and organs. TGF-β is also activated and released from the pool of latent-TGF-β in the circulation in response to variety of physiological and pathological challenges. TGF-β plays critical roles in the regulation of immune responses. TGF-β inhibits T cell proliferation and Th1 and Th2 differentiation. TGF-β, in the context of TCR stimulation, may also induce Treg or Th17 cell differentiation from CD4+ T precursor depending on the absence or presence of IL-6, which are programmed and directed by Foxp3 and RORrt transcription factors respectively. Tregs suppress Th1 and Th2 responses, whereas Th1 and Th2 cells downregulate Th17 cell differentiation. *Green arrows* indicate positive regulation; *red lines* indicate negative regulation

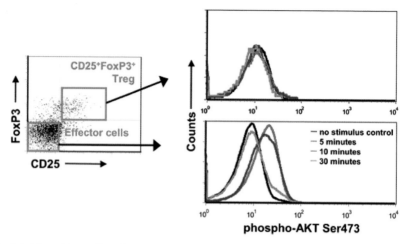

Fig. 8.2 *Mouse Treg cells fail to phosphorylate AKT at Ser473.* Splenic and lymph node CD4+ T cells were purified from 8-week-old C57Bl/6 mice by negative selection and rested for 2 hours and either left unstimulated or stimulation with anti-CD3 (5 μg/ml) and anti-CD28 (1 μg/ml) mAbs and Fc crosslinker antibody (20 μg/ml). Cells were fixed in FOXP3 Fix/Perm buffer after 5, 10 and 30 min of stimulation, stained for CD4, CD25, FoxP3 and phospho-AKT and examined by flow cytometry. The mean flourescence intensity of AKT phosphorylation at Ser473 is compared in CD4+CD25^{hi}FoxP3+ Treg (*top*) and CD4+CD25− effector (*bottom*) cells

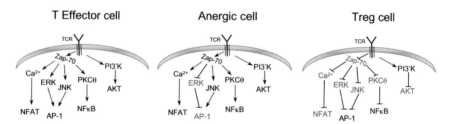

Fig. 8.3 *Differences between T effector, anergic and Treg cells in the major TCR-mediated signalling pathways.* Upon stimulation of a T effector cell through its TCR and co-receptors, the calcium/NFAT, Ras/MAPK, NFκB and PI3'K signalling pathways are activated (*left*). In contrast, cells rendered anergic by incomplete stimulation are characterised by defects in the Ras/MAPK signalling cascade resulting in a lack of AP-1 nuclear formation (*middle*). Changes documented in TCR-mediated signalling pathways in Treg cells (*right*) include defective phosphorylation of ITAMs within the CD3ζ chain, resulting in a lack of Zap-70 recruitment and reduced phosphorylation of downstream substrates LAT and PLCγ1 [32], reduced calcium mobilisation and subsequent reductions in NFAT nuclear translocation and in PKC phosphorylation [33,38], a failure to phosphorylate AKT at Ser^{473} downstream of PI3'K signalling [35], modest reductions in ERK phosphorylation [36] and reduced JNK activation following PMA/ionomycin stimulation [37]. Signalling components whose activation/phosphorylation has been demonstrated to be defective in Treg cells are depicted in *red*

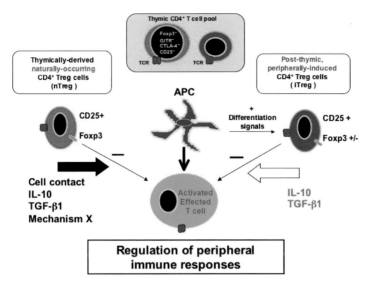

Fig. 9.1 Control of immune responses by CD4⁺ regulatory T cells Naturally-occurring (*red*) and induced (*blue*) CD4⁺ regulatory T cell subsets downregulate the function of activated effector T cells (*green*) in several types of peripheral immune responses. While CD4+CD25+Foxp3+ nTreg cells differentiate in the thymus and are found in the normal, naïve CD4+ T cell repertoire, multiple iTreg cell subsets, possibly expressing CD25 and Foxp3, originate from the activation and differentiation of conventional CD4⁺ T cells in the periphery under unique stimulatory conditions. The relative contribution of each subset in the overall regulation of immune responses is unclear but both conceivably can synergize to achieve this outcome.

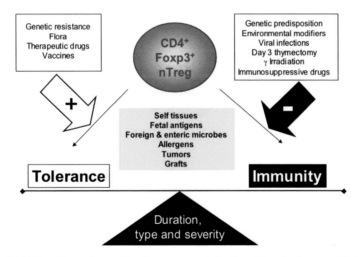

Fig. 9.2 CD4⁺Foxp3⁺ regulatory T cells: master-switch of peripheral tolerance CD4⁺Foxp3⁺ nTreg cells represent a central master-switch of peripheral T cell tolerance as abrogation of nTreg development or function can provoke autoimmunity, and also increase immunity to tumor, allergens, grafts and various pathogens. Thus, nTreg cells play a determining role in the balance between tolerance and immunity, and alterations in their development or function provoked by physical, chemical, environmental or genetic triggers, may represent a determining variable in disease resistance or susceptible. The duration, type and severity of nTreg cell dysfunction may also affect this balance, and ultimately determine degree of tolerance or immunity to self and non-self antigens.

Fig. 10.1 Dynamics of Tregs and pathogenic T cells during the progression of autoimmune diabetes. (**a**) At three weeks of age, β cell antigen shedding initiated by program cell death during developmental islet tissue remodeling leads to priming of islet-reactive T cells in the pancreatic LN. The primed Teff cells and Tregs infiltrate islets. (**b**) T cells mediated β cell death leads to more antigen shedding and activation and expansion of Tregs in pancreatic LN. Tregs limit the priming of Teff cells. Over the long disease course, Treg-resistant Teff cells emerge and become activated and infiltrate and accumulate in islet tissue despite the presence of functional Tregs in the LN. (**c**) More islet damage and antigen shedding leads to more Treg activation and expansion in pancreatic LN. Treg-resistant Teff cells expand and differentiate in islets and eventually destroy β cells and leads to diabetes

- islet-reactive Treg
- islet-reactive Teff
- regulation-resistant Teff
- DC

clustering/arrest   swarming   free

Tregs

Fig. 10.2 Tregs inhibit stable interaction between Teff cells and DCs. In the presence of low level of Tregs, Teff cells form long stable conjugates with antigen-bearing DCs as indicated by their clustering and arrest on the DCs. With increase Tregs, the interaction between Teff and DCs becomes transient evident from the swarming behavior of the Teff cells. Presence of high level of Tregs completely abolish Teff and DC interaction and Teff cells moves freely in the LN as if no cognate antigen is present

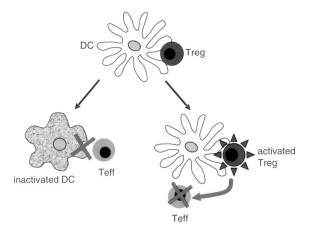

Fig. 10.3 Role of DCs in Treg function in vivo. Treg engagement of DCs may lead to direct inactivation of DCs so that they can not activate Teff cells. Alternatively, DC may activate Tregs to produce soluble inhibitory factors, which in turn prevent activation of Teff cells

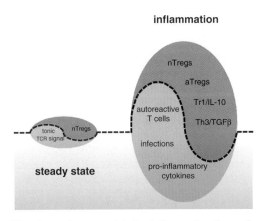

Fig. 10.4 Function of Treg in steady state and during inflammation. In steady state, thymus-derived natural Tregs (nTreg) with a repertoire skewed toward recognition of self antigens maintains normal immune homeostasis by interacting with DCs and preventing their activation of Teff cells. When normal immune homeostasis is perturbed by the presence of highly autoreactive T cells or infections, inflammation erupts. In response, nTregs expand and adaptive Tregs (aTregs) including IL-10 producing Tr1 cells and TGFβ expressing Th3 cells develop to control the immune response

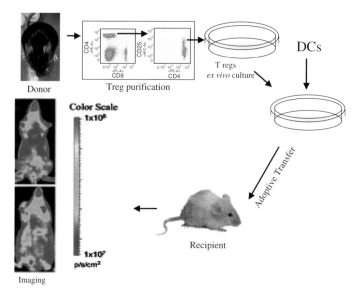

Fig. 11.1 Schematic representation of basic Treg Adoptive transfer Protocol. Treg cells from donor mice are used either for expansion *ex vivo* or expanded *in vivo* (not represented here) together with DCs and adoptively transferred to another experimental recipient mouse. *In vivo* luciferase base imaging is used to track cells homing to sites of inflammation

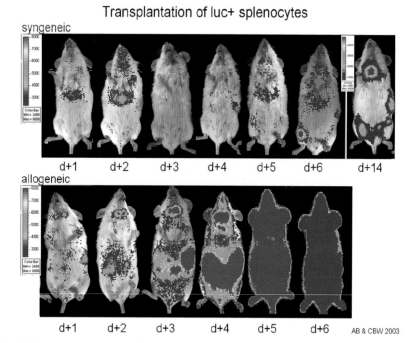

Fig. 11.2 Transplantation of luc^+ splenocytes into syngeneic FVB recipients (*top panels*) with imaging at indicated time points vs allogeneic BALB/C recipients (*bottom panels*). The dramatic proliferation of alloreactive T cells can be readily visualized. These images are representative of large number of animals studied. The allogeneic animals die from GVHD at approximately day 14

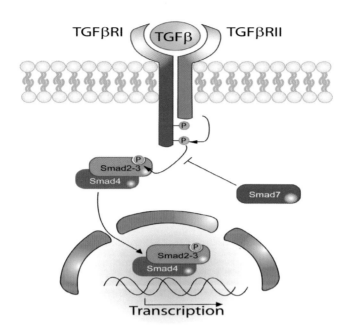

Fig. 14.1 TGF-beta pathway. TGF-beta interaction with the TGF-beta receptor II (TGFβRII) causes morpho-structural changes leading the TGF-beta/TGF-betaRII complex to interact with the TGF-beta receptor I. This interaction in turn activates the kinase activity of the TGF-betaR I intracellular domain which phosphorylates Smad2/3. Phospho-Smad2/4 couple with the CoSmad Smad4 and together translocate into the nucleus inducing the expression of TGF-beta-dependent genes. The inhibitory molecule Smad7 prevents the activation of the Smad-dependent TGF-beta signal transduction thus dampening most of the TGF-beta-mediated effects

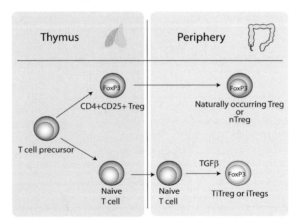

Fig. 14.2 Peripheral induction of Treg (i.e. iTreg, TiTregs). nTregs originate from the thymus after the first days of post-natal life and migrate in the periphery where they are thought to play a pivotal role in the maintenance of tolerance towards self-antigens. In the periphery, CD4+ naïve T cells may differentiate in suppressive- FoxP3 expressing-cells in the presence of TGF-beta. TiTregs (or nTregs) might be important for tolerance towards antigens contained in the gut lumen either produced by the intestinal flora or introduced with the diet

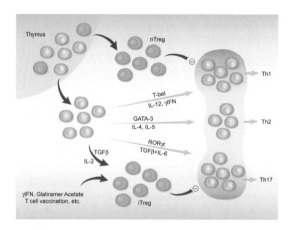

Fig. 15.1 Unlike naturally occurring Tregs (nTreg) that are differentiated in and migrate out of thymus, inducible Tregs are induced or converted from peripheral CD4+CD25− T cell pool by TGF-beta, gamma-interferon and other conditions that are reviewed in this chapter. Tregs of both types work in concert to keep various effectors (Th1, Th2 and Th17) in check

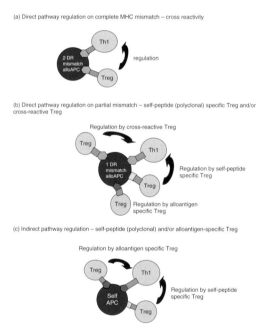

Fig. 18.1 Pathways of allorecognition and regulation. (**a**) In the case of complete mismatched direct pathway alloresponse, only those Tregs with cross-reactivity for intact alloantigen ● can regulate. There are likely to be few of these because Tregs represent less than 10% of circulating $CD4^+$ cells and Tregs are selected for high avidity to self-peptides. (**b**) With partial matching of the direct pathway, Tregs with any combination of specificity can regulate: those specific to self peptides will be able to be ligated by the matched MHC:peptide complex ◇; indirect pathway, alloantigen-specific Tregs can be ligated by the matched MHC in combination with processed allopeptide ●; cross-reactive direct pathway Tregs can suppress as before. (**c**) With self-APC, only those Tregs with self-peptide or allopeptide specificity can suppress. As discussed, most Tregs will fall into these two categories

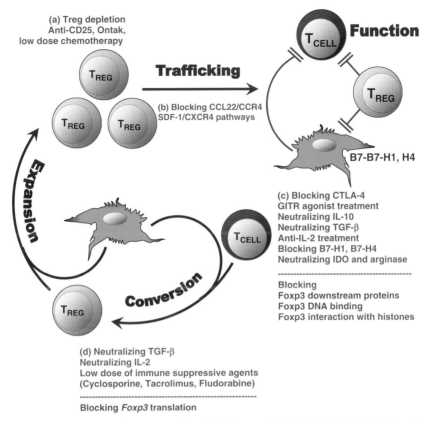

Fig. 20.2 Targeting Tregs in cancer (a) Treg depletion: anti-CD25, ONTAK. (b) Blocking Treg tumour trafficking. (c) Reducing Treg function. (d) Blocking Treg expansion and conversion

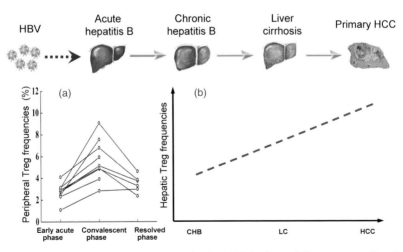

Fig. 21.1 Upper panel: Treg in the pathogenesis of HBV infection and disease progression. Down panel: (a) Longitudinal Treg alteration throughout the resolution of acute HBV infection; (b) Schematic alteration of the increased trend of hepatic Treg infiltration from CHB, to LC and finally to primary HCC

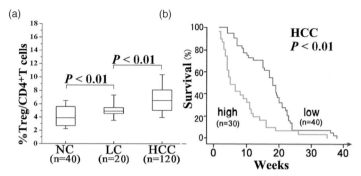

Fig. 21.2 There is an increased Treg frequency in peripheral blood, which is associated with the disease progression of primary HCC. (**a**) Treg frequency is significantly increased compared with NC and LC patients. (**b**) HCC patients with higher Treg frequency showed shorter survival time. Actual overall survival rates were analyzed by the Kaplan Meier method and survival was measured in weeks from diagnosis to death. The log-rank test was applied to compare between the groups. Multivariate analysis of prognostic factors for overall survival was performed using the Cox proportional hazards model (Modified from Fu et al. Gastroenterology, 2007, 132: 2328–2339)

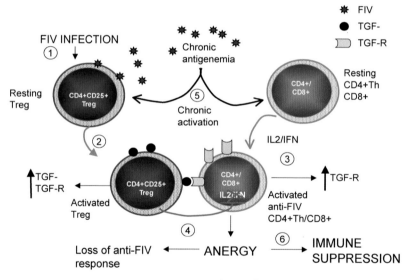

Fig. 22.1 Potential role of immunosuppressive $CD4^+CD25^+$ T reg cells in the immunopathogenesis of FIV: A model for persistent viral replication and immunosuppression FIV productively infects Treg cells (1), resulting in Treg cell activation characterized by expression of membrane TGF-β (mTGF-β) (2). Infection activates virus specific $CD4^+$ Th and $CD8^+$ T cells characterized by expression of TGF-βRII and secretion of IL2 and IFN-γ (3). TGF-β on the activated Treg cell is able to bind TGF-β RII on the virus activated $CD4^+$ and $CD8^+$ cells, transducing a signal for down-regulation of IL2 and IFN-γ gene transcription, resulting in anergy and loss of anti-FIV immune responses (4). This loss of the antiviral immune response contributes to continued virus replication, chronic antigenemia, and chronic activation of Treg cells (5). These activated Treg cells are then capable of suppressing not only anti-viral immune responses, but immune responses to other antigens, thus contributing to the global immune suppression associated with FIV infection (6)

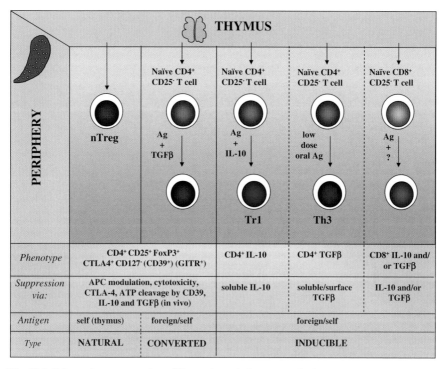

Fig. 23.1 Schematic representation of Treg subsets in humans and mice

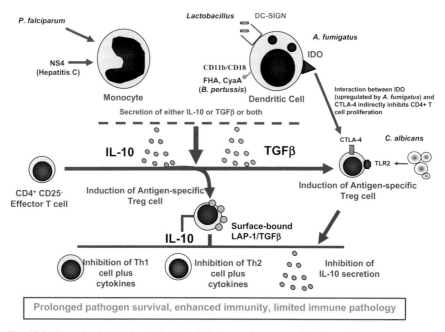

Fig. 23.2 The role of TGF-β and IL-10 in inducing pathogen-specific Treg cells

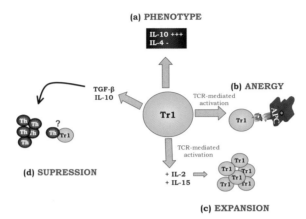

Fig. 24.1 Main features of human CD4$^+$ Tr1 cells CD4$^+$ Tr1 cells are distinguished by their ability to produce IL-10 in the absence of IL-4 (**a**). Autocrine production of IL-10 render these cells highly anergic upon TCR-mediated activation (**b**), but addition of exogenous cytokines, such as IL-2 and IL-15, can revert this anergic phenotype allowing their in vitro expansion (**c**). Upon TCR engagement, CD4$^+$ Tr1 cells exert their suppressive function in a non antigen-specific manner by secreting TGF-β and IL-10, although some cell-cell contact mechanisms cannot be excluded (**d**)

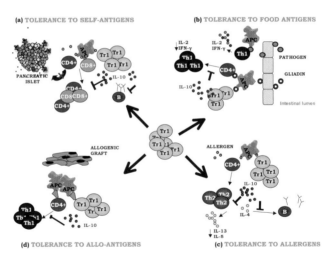

Fig. 24.2 Tolerance and Tr1 cells (**a**) Auto-agressive T and B cells escaping central deletion may get activated by self antigens (such as pancreas-related antigens) and develop an auto-aggressive immune response that leads to self-tissues destruction (such as destruction of insulin-producing cells, in type 1 diabetes) and autoimmunity. IL-10 produced by Tr1 cells can efficiently keep under control the auto-aggressive reaction and prevent disease development. (**b**) Tr1 cells present in the intestinal lumen can control undesired immune responses to non-self non harmful antigens such as gliadin, through IL-10 production. Despite this activity, Tr1 cells permit active immune responses versus pathogens, which are very abundant in the mucosal system. (**c**) The Th2-mediated immune response to allergens and the consequent IgE production can be actively controlled by Tr1 cells. (**d**) Tr1 cells avoid the immunological reactions activated by allo-antigens introduced by transplantation. Importantly, in all conditions depicted in the figure, Tr1 cells need to be first activated by their specific antigens in order to exert their suppressive function

Fig. 25.1 Anterior chamber inoculation After 2 µl of aqueous humor is removed, the antigen is delivered into the chamber through a custom-made glass needle inserted in the hole made by the extraction needle. Injected eyes are followed for three days for signs of inflammation and used only if the cornea and eye are clear

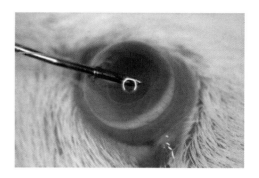

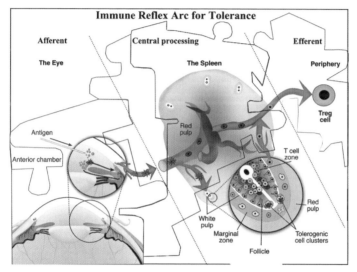

Fig. 25.2 Immune reflex arc for tolerance Illustration of the traffic of eye-derived F4/80+ cells (*red cells*) showing it path through the blood to the marginal zone of the T cell areas in the spleen. Five to seven days post a.c. inoculation, antigen specific Treg cells are in the peripheral tissue and can be assessed from by testing the cells in the spleen for its ability to suppress Th1 and Th2 responses

Fig. 25.3 Illustration of a Mouse Eye The diagram shows the anatomy of the eye with symbols representing each of the various immunosuppressive factors that collaborate to establish an immunosuppressive environment

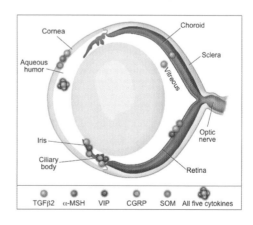

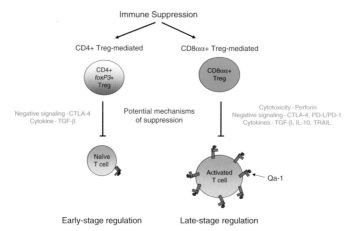

Fig. 26.1 Targets and mechanisms of CD4+CD25+FOXP3+ and CD8αα+TCRαβ+ Treg cell-mediated suppression. See text for details

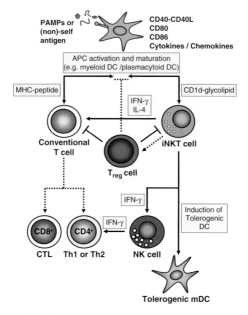

Fig. 27.2 Schematic model of iNKT cell activation and regulation iNKT cells upon activation with endogenous or exogenous glycolipid can potentiate immune responses. Upon interaction with glycolipid-CD1d complexes expressed on dendritic cells (DC), iNKT cells can mature DC to activate conventional T cells, which can be further activated with soluble signals derived from both DC and iNKT cells. During cell mediated immune responses, iNKT cells can activate NK cells and direct the activation of conventional T cells towards Th1 or Th2 immunity to antigen-specific targets. During prevention from autoimmune disease, activated iNKT cells can induce the activation and recruitment of tolerogenic DC or otherwise tolerize autoreactive pathogenic T cells through cell-cell contact dependent mechanisms. CD4+CD25+Foxp3+ T_{reg} cells at the centre of this activation schema can regulate iNKT cell or conventional T cell responses via iNKT-DC and/or T-cell-DC interactions, or during downstream iNKT mediated responses

patients, or is due to decreased expression at a cellular level. Haas and colleagues recently reported on a Treg subset of CD4$^+$CD25$^+$CD45A$^+$CD45RO$^-$FOXP3$^+$ cells co-expressing CD31 which has a significantly reduced frequency in MS patients [47]. They demonstrated that reduced de novo generation of naive Treg cells was compensated for by higher proportions of memory Treg cells, resulting in an apparent overall stable population.

CD4$^+$CD25^{hi} Regulatory T Cells Display Impaired Function in Patients with MS

As it was critical to examine the regulatory T-cell function, we isolated highly pure CD4$^+$CD25^{hi} regulatory and CD4$^+$CD25$^-$ responder cell populations by high speed flow cytometric sorting. CD4$^+$CD25$^-$ responder cells from both patients with MS and healthy individuals responded similarly in a dose-dependent fashion to varying concentration of plate-bound anti-CD3 monoclonal antibody (mAb). CD4$^+$CD25^{hi} T cells isolated from both groups were anergic to stimulation at all doses of platebound anti-CD3, indicating that CD4$^+$CD25^{hi} T cells isolated from patients with MS do exhibit this regulatory property. To quantitate their regulatory function, CD4$^+$CD25^{hi} T cells were co-cultured with autologous responder cells (2.5×10^3/well) at different ratios (responder/suppressor ratio 1 : 1, 1 : 1/2, 1 : 1/4, and 1 : 1/8). As previously reported in healthy individuals, CD4$^+$CD25^{hi} T cells consistently suppressed proliferation at a 1 : 1 ratio. Increasing the ratio of responder/suppressor T cells resulted in less suppression. In striking contrast, the regulatory CD4$^+$CD25^{hi} T cells isolated from the circulation of patients with MS, while normal in frequency as compared to healthy controls, poorly inhibited responder CD4$^+$CD25$^-$ T-cell proliferation. As we have previously shown that increased strength of signal inhibits regulation [48], we stimulated co-cultures of regulatory and responder T cells from patients with MS and healthy controls with a maximal concentration (2.5 mg/ml) of platebound anti-CD3mAb. As predicted, the CD4$^+$CD25^{hi} regulatory T cells no longer suppressed the proliferation of responder T cells. We also examined the production of cytokines in all the cultures and the ability to inhibit their secretion by CD4$^+$CD25^{hi} T cells co-cultured with CD4$^+$CD25$^-$ responder cells. The secretion of the Th1 cytokine IFN-γ, resulting from the activation of destructive autoreactive T cells, was suppressed by CD4$^+$CD25^{hi} T-cell co-cultures from healthy controls but not in co-cultures derived from patients with MS. IL-10 was variably secreted, predominantly by the CD4$^+$CD25$^-$ T cells. The secretion of IL-10 was often reduced upon co-culture with CD4$^+$CD25^{hi}T cells, excluding a potential role of this cytokine in mediating this regulatory suppressor function. In this ex vivo model of suppression, blocking IL-10 or TGF-β does not result in loss of suppressor function by CD4$^+$CD25$^+$ regulatory T cells.

It was important to examine whether the loss of regulatory function was due to a decrease in CD4$^+$CD25^{hi} T-cell function or an increase in the resistance of activated CD4$^+$CD25$^-$ responder T cells to inhibition. Thus, we performed mixing experiments in which patient and control regulatory CD4$^+$CD25^{hi} T cells were co-cultured with the autologous and the converse target cells isolated from either healthy subjects or patients with MS. Regulatory T cells from patients with MS could not suppress the proliferative response of target responder T cells from either patients or healthy controls (suppression 23%). In the reciprocal experiments, regulatory $CD4^+CD25^{hi}$ T cells from healthy controls suppressed the proliferative response of target $CD4^+CD25^-$ T cells derived from both controls and patients with MS (suppression 78%). These data indicate that the primary regulatory defect is in the function of $CD4^+CD25^{hi}$ T cells isolated from the circulation of patients with MS. Consistent with this data, we have also reported functional abnormalities in CD46-mediated Tr1 Treg cells in patients with MS [49]. CD46 is a newly defined costimulatory molecule that can induce a Tr1 Treg cell phenotype, capable of secreting the anti-inflammatory cytokine IL-10. We observed striking defects in the induction of Tr1 cells with CD46 co-stimulation, in their ability to secrete IL-10 (but not INF-γ) in patients with MS compared with healthy subjects [49].

CD62L Expression on $CD4^+CD25^{hi}$ Regulatory T Cells

Although there were no differences in the frequency of $CD4^+CD25^{hi}$ T cells or in their proliferative or cytokine secretion in response to different stimuli between healthy subjects and patients with MS, it was important to determine whether an increase in the frequency of activated $CD4^+$ T cells in the circulation was diluting the regulatory $CD4^+CD25^{hi}$ T cells. Therefore, we used CD62L expression to further purify regulatory from the activated T cells, because l-selectin expression is down-regulated upon activation. We isolated $CD4^+CD25^{hi}$ $CD62L^+$ and total $CD4^+CD25^{hi}$ regulatory T cells from healthy subjects and patients with MS. Whereas in the healthy controls both populations were able to suppress the proliferative response to anti-CD3 stimulation, the $CD4^+CD25^{hi}$ regulatory cells isolated from patients with MS, although further depleted of the potentially activated $CD62L^-$ T cells, were still unable to inhibit the proliferation of the $CD4^+CD25^-$ responder population. These data strongly confirm a defect in the highly purified regulatory subset in patients with MS.

Discussion of $CD4^+CD25^+$ T Cells in Patients with MS

An important aspect of these investigations was the measurement of regulatory T-cell function as opposed to simple phenotypic measurement of $CD4^+CD25^+$ T-cell frequency. Unlike 6-week-old mice raised in clean facilities, whose total $CD4^+CD25^+$ T-cell population manifests regulatory properties, humans are exposed to a myriad of infections and show a significant population of $CD4^+CD25$medium/low T cells that do not exhibit regulatory function [41]. Thus, it was critical to examine the functional state of the regulatory T cells expressing high levels of CD25. We have previously demonstrated that the strength of signal delivered through the TCR of target T cells is one factor determining whether regulatory

$CD4^+CD25^{hi}$ T cells can suppress the responder T-cell proliferation [41]. Thus, to properly examine the function of regulatory $CD4^+CD25^{hi}$ T cells, we used a number of different strengths of stimulatory signals in these experiments. We observed that the strong signal provided by maximal concentration of plate-bound anti-CD3 mAb similarly abrogated suppression in both patient and control co-cultures. In contrast, lower concentrations of plate-bound anti-CD3 delivered a signal that resulted in the appearance of a significant defect in the suppressive function of this subset of regulatory cells derived from patients with MS. The use of different stimulatory conditions allowed us to reveal alterations in the regulatory function of $CD4^+CD25^{hi}$ T cells while still demonstrating that they are $CD25^+$ regulatory T cells as opposed to activated responder cells expressing CD25. Stimulation of cultures with soluble anti-CD3 and anti- CD28, which has previously shown to be the most permissive for enabling co-culture suppression, gave equivalent levels of suppression in patients and control subjects when co-cultured at 1 : 1 ratios. In contrast, the stimulation provided by platebound anti-CD3 at 0.1 and 0.5 mg/ml resulted in a threefold decrease of suppression by $CD4^+CD25^{hi}$ cells derived from patients with MS as compared to normal controls. Previous experiments showed that delivering a low strength TCR signal to responder T cells, such as that provided by self-antigens as compared to microbial antigens, resulted in a greater sensitivity to suppression. Thus, the present findings may help explain defects in suppression of autoreactive T cells in autoimmune patients as compared to T cells stimulated by microbial antigens during infections. An important control to note in all these experiments is the anergy or lack of thymidine incorporation resulting from stimulation of $CD4^+CD25^{hi}$ T cells cultured alone. This anergy indicates that $CD4^+CD25^{hi}$ T cells isolated from patients with MS are not $CD25^+$ activated T cells; such cells would not exhibit regulatory activity but rather enhance proliferation. It was critical to determine whether the decrease in T-cell regulatory function observed in patients with MS was due to a defect in the $CD4^+CD25^{hi}$ T-cell subset or whether the responder $CD4^+CD25^-$ T cells were refractory to suppression. By performing co-mixing experiments, we could clearly demonstrate that the defect lies in the $CD4^+CD25^{hi}$ T-cell function, as opposed to enhanced responder T-cell resistance in patients with MS. Although in vitro measurement of biologic function in patients with autoimmune diseases will always be correlative, these in vitro experiments, based on in vitro and in vivo experimentation in mouse models of autoimmune disease, provided the first definitive evidence for a defect in regulatory T-cell function in a human autoimmune disease. Ultimately, monitoring the effects of immunomodulatory drugs on this regulatory T-cell subset will help define their pathogenic role in MS and other human autoimmune diseases.

Regulatory T Cells and MS Therapy

The majority of patients present with relapsing/remitting MS, often with a single clinical event. Unlike in other auto- immune diseases, such as diabetes where the majority of the pancreatic islet cells are destroyed with clinical presentation, patients with MS are now identified relatively early in the disease, allowing manipulation of the immune system in disease prevention. Earlier in this chapter, we discussed in detail how the cells of the immune system are involved in damaging the CNS, resulting in the varied and debilitating symptoms of MS. An ideal therapy seeks to prevent the initiation of such attacks, to halt such attacks if in progress, and optimally, to reverse the damage done. Because the disease is the result of a dysregulated immune system, rational therapy would be aimed at correcting the underlying dysfunction. Immunotherapy refers to therapy seeking to alter the immune response to prevent or treat the disease in question. Thus, unlike treatment of other diseases, such as some types of cancer, in the case of MS almost any promising therapy will be termed an immunotherapy, due to the autoimmune nature of the disease. The potential for Treg cells to control CNS autoimmunity has been well documented in experimental models. Treg cells administered to mice can significantly reduce EAE severity [50], and have been shown to accumulate within the CNS during the recovery [51]. In relapsing-remitting EAE models, not only does depletion of Treg cells increase acute-phase severity, but it also prevents remission [52, 53]. The key role that Treg cells play in the EAE model identifies this cell type as a major potential target for human immunotherapy.

The origin of the observed defect of suppressive function of Tregs in relapsing-remitting MS patients has not yet been elucidated. We have recently reported using large-scale genomic screens, that certain alleles of the genes encoding for IL2α and IL7Rα chains are associated with increased risk for MS [54]. This supports the idea that polymorphisms within genes related to the regulation of the immune response, and to Treg cells in particular (which express high levels of CD25) are important factors in MS. While this data suggests that the functional alterations of Treg cells in these patient groups could be linked to genetic factors, as with any in vitro observation, clinical trials targeting the function or frequency of Tregs and any association with clinical responses, be it autoimmune disease, HIV, or with tissue transplantation must be performed. Moreover, the lack of a role for cytokine in these in vitro studies of these $CD4^+CD25^{hi}$ Treg cells should not exclude a critical role for IL-10, TGF-β, or other cytokines in the combined in vivo mechanisms of these regulatory cells. The limitations of in vitro analyses enhance the difficulty of understanding human disease, as the assays are likely to reflect only single in vivo events rather than the culmination of all biologic mechanisms that would operate in vivo; that is, these models only observe what is asked of them. Nevertheless, the apparently parallel observations of $CD4^+CD25^+$ Tregs in animal models and humans indicate that suppressor cells act as central regulators of immune responses in human diseases.

We have shown that the regulatory function of the $CD4^+CD25^{hi}$ T-cell subpopulation is defective in patients with MS [44]. These cells are thought to have a role in immune homeostasis, that is, in tempering the immune response and preventing postimmune response inflammation, etc. Therefore, a functional lack of $CD4^+CD25^{hi}$ T cells in MS patients is likely to contribute to the pathophysiology of the disease. Thus, it follows from this, and from experimental models of MS, that reversing this lack of functional regulatory cells would be a viable and productive treatment avenue. However, effective long-term treatment is likely to be more complex. It has been recently shown, at least in mouse models, that myelin-specific functional Treg cells accumulate in the CNS but fail to control autoimmune inflammation [55]. The authors speculate that in order for Tregs to effectively control autoimmune reactions in the CNS, it may be necessary to control tissue inflammation. Unfortunately, while a great deal is known about the requirements and characteristics of these cells in mice, much less is known about them in humans. Once we have a more complete understanding of the regulatory T-cell subsets in humans, this area will be an attractive one for intervention or immunotherapy with a goal of increasing the number and/or the functionality of these T cells in MS patients.

References

1. Kabat, E.A., M. Glusman, and V. Knaub, Quantitative Estimation of the Albumin and Gamma-Globulin in Normal and Pathologic Cerebrospinal Fluid by Immunochemical Methods. *American Journal of Medicine*, 1948, **4**(5):653–662.
2. Mackay, R.P. and Myrianthopoulos, N.C, Multiple Sclerosis in Twins and Their Relatives – Final Report. *Archives of Neurology*, 1966, **15**(5):449–462.
3. Rivers, T.M., D.H. Sprunt, and G.P. Berry, Observations on attempts to produce acute disseminated encephalomyelitis in monkeys. *Journal of Experimental Medicine*, 1933, **58**(1):39–U58.
4. Hafler, D.A., The distinction blurs between an autoimmune versus microbial hypothesis in multiple sclerosis. *Journal of Clinical Investigation*, 1999, **104**(5):527–529.
5. Goverman, J., et al., Transgenic Mice That Express a Myelin Basic Protein-Specific T-Cell Receptor Develop Spontaneous Autoimmunity. *Cell*, 1993, **72**(4):551–560.
6. Lehmann, P.V., et al., Spreading of T-Cell Autoimmunity to Cryptic Determinants of an Autoantigen. *Nature*, 1992, **358**(6382):155–157.
7. Windhagen, A., et al., Expression of Costimulatory Molecules B7-1 (Cd80), B7-2 (Cd86), and Interleukin-12 Cytokine in Multiple-Sclerosis Lesions. *Journal of Experimental Medicine*, 1995, **182**(6):1985–1996.
8. Ota, K., et al., T-Cell Recognition of an Immunodominant Myelin Basic-Protein Epitope in Multiple-Sclerosis. *Nature*, 1990, **346**(6280):183–187.
9. Pette, M., et al., Myelin Basic Protein-Specific Lymphocyte-T Lines from Ms Patients and Healthy-Individuals. *Neurology*, 1990, **40**(11):1770–1776.
10. Martin, R., et al., Fine Specificity and Hla Restriction of Myelin Basic Protein-Specific Cytotoxic T-Cell Lines from Multiple-Sclerosis Patients and Healthy-Individuals. *Journal of Immunology*, 1990, **145**(2):540–548.
11. Ausubel, L.J., et al., Complementary mutations in an antigenic peptide allow for crossreactivity of autoreactive t-cell clones. *Proceedings of the National Academy of Sciences of the United States of America*, 1996, **93**(26):15317–15322.
12. Hemmer, B., et al., Identification of high potency microbial and self ligands for a human autoreactive class II-restricted T cell clone. *Journal of Experimental Medicine*, 1997, **185**(9):1651–1659.
13. Wucherpfennig, K.W. and J.L. Strominger, Molecular Mimicry in T-Cell-Mediated Autoimmunity – Viral Peptides Activate Human T-Cell Clones Specific for Myelin Basic-Protein. *Cell*, 1995, **80**(5):695–705.
14. Trapp, B.D., et al., Axonal transection in the lesions of multiple sclerosis. *New England Journal of Medicine*, 1998, **338**(5):278–285.
15. Wucherpfennig, K.W., et al., T-Cell Receptor V-Alpha-V-Beta Repertoire and Cytokine Gene-Expression in Active Multiple-Sclerosis Lesions. *Journal of Experimental Medicine*, 1992, **175**(4):993–1002.

16. Traugott, U., E.L. Reinherz, and C.S. Raine, Multiple-Sclerosis – Distribution of T-Cell Subsets within Active Chronic Lesions. *Science*, 1983, **219**(4582):308–310.
17. Hauser, S.L., et al., Immunohistochemical Analysis of the Cellular Infiltrate in Multiple-Sclerosis Lesions. *Annals of Neurology*, 1986, **19**(6):578–587.
18. Wucherpfennig, K.W., et al., Gamma-Delta T-Cell Receptor Repertoire in Acute Multiple-Sclerosis Lesions. *Proceedings of the National Academy of Sciences of the United States of America*, 1992, **89**(10):4588–4592.
19. Prineas, J.W. and R.G. Wright, Macrophages, Lymphocytes, and Plasma-Cells in Perivascular Compartment in Chronic Multiple-Sclerosis. *Laboratory Investigation*, 1978, **38**(4):409–421.
20. Prineas, J., Pathology of Early Lesion in Multiple-Sclerosis. *Human Pathology*, 1975, **6**(5):531–554.
21. Lucchinetti, C.F., et al., Distinct patterns of multiple sclerosis pathology indicates heterogeneity in pathogenesis. *Brain Pathology*, 1996, **6**(3):259–274.
22. Becher, B., et al., Soluble tumor necrosis factor receptor inhibits interleukin 12 production by stimulated human adult microglial cells in vitro. *Journal of Clinical Investigation*, 1996, **98**(7):1539–1543.
23. Dong, Y.S. and E.N. Benveniste, Immune function of astrocytes. *Glia*, 2001, **36**(2):180–190.
24. Lehnardt, S., et al., Activation of innate immunity in the CNS triggers neurodegeneration through a Toll-like receptor 4-dependent pathway. *Proceedings of the National Academy of Sciences of the United States of America*, 2003, **100**(14):8514–8519.
25. Monje, M.L., H. Toda, and T.D. Palmer, Inflammatory blockade restores adult hippocampal neurogenesis. *Science*, 2003, **302**(5651):1760–1765.
26. Ekdahl, C.T., et al., Inflammation is detrimental for neurogenesis in adult brain. *Proceedings of the National Academy of Sciences of the United States of America*, 2003, **100**(23): 13632–13637.
27. Baranzini, S.E., et al., Transcriptional analysis of multiple sclerosis brain lesions reveals a complex pattern of cytokine expression. *Journal of Immunology*, 2000, **165**(11): 6576–6582.
28. Mycko, M.P., et al., Microarray gene expression profiling of chronic active and inactive lesions in multiple sclerosis. *Clinical Neurology and Neurosurgery*, 2004, **106**(3):223–229.
29. Eugster, H.P., et al., IL-6-deficient mice resist myelin oligodendrocyte glycoprotein-induced autoimmune encephalomyelitis. *European Journal of Immunology*, 1998, **28**(7):2178–2187.
30. Aloisi, F., et al., IL-12 production by central nervous system microglia is inhibited by astrocytes. *Journal of Immunology*, 1997, **159**(4):1604–1612.
31. Ledeboer, A., et al., Interleukin-10, interleukin-4, and transforming growth factor-beta differentially regulate lipopolysaccharide-induced production of pro-inflammatory cytokines and nitric oxide in co-cultures of rat astroglial and microglial cells. *Glia*, 2000, **30**(2):134–142.
32. Onuki, I., et al., Axonal degeneration is an early pathological feature in autoimmune-mediated demyelination in mice. *Microscopy Research and Technique*, 2001, **52**(6):731–739.
33. Ayers, M.M., et al., Early glial responses in murine models of multiple sclerosis. *Neurochemistry International*, 2004, **45**(2–3):409–419.
34. Trajkovic, V., et al., Astrocyte-induced regulatory T cells mitigate CNS autoimmunity. *Glia*, 2004, **47**(2):168–179.
35. Roncarolo, M.G., et al., Type 1 T regulatory cells. *Immunological Reviews*, 2001, **182**:68–79.
36. Pullen, A.M., P. Marrack, and J.W. Kappler, Evidence That Mls-2 Antigens Which Delete V-Beta-3+ T-Cells Are Controlled by Multiple Genes. *Journal of Immunology*, 1989, **142**(9):3033–3037.
37. Sakaguchi, S., Regulatory T cells: Key controllers of immunologic self-tolerance. *Cell*, 2000, **101**(5):455–458.
38. Shevach, E.M., et al., Control of T-cell activation by CD4(+) CD25(+) suppressor T cells. *Immunological Reviews*, 2001, **182**:58–67.
39. Sakaguchi, S., et al., Organ-Specific Autoimmune-Diseases Induced in Mice by Elimination of T-Cell Subset .1. Evidence for the Active Participation of T-Cells in Natural Self-Tolerance –

Deficit of a T-Cell Subset as a Possible Cause of Autoimmune-Disease. *Journal of Experimental Medicine*, 1985, **161**(1):72–87.
40. Read, S., V. Malmstrom, and F. Powrie, Cytotoxic T lymphocyte-associated antigen 4 plays an essential role in the function of CD25(+)CD4(+) regulatory cells that control intestinal inflammation. *Journal of Experimental Medicine*, 2000, **192**(2):295–302.
41. Baecher-Allan, C., et al., CD4+CD25(high) regulatory cells in human peripheral blood. *Journal of Immunology*, 2001, **167**(3):1245–1253.
42. Dieckmann, D., et al., Ex vivo isolation and characterization of CD4(+)CD25(+) T cells with regulatory properties from human blood. *Journal of Experimental Medicine*, 2001, **193**(11):1303–1310.
43. Reijonen, H., et al., Detection of GAD65-specific T-cells by major histocompatibility complex class II tetramers in type 1 diabetic patients and at-risk subjects. *Diabetes*, 2002, **51**(5): 1375–1382.
44. Viglietta, V., et al., Loss of functional suppression by CD4+CD25+ regulatory T cells in patients with multiple sclerosis. *The Journal of Experiment Medicine*, 2004, **199**(7):971–9.
45. Huan, J., et al., Decreased FOXP3 levels in multiple sclerosis patients. *Journal of Neuroscience Research*, 2005, **81**(1):45–52.
46. Venken, K., et al., Secondary progressive in contrast to relapsing-remitting multiple sclerosis patients show a normal CD4+CD25+ regulatory T-cell function and FOXP3 expression. *Journal of Neuroscience Research*, 2006, **83**(8):1432–46.
47. Haas, J., et al., Prevalence of newly generated naive regulatory T cells (Treg) is critical for Treg suppressive function and determines Treg dysfunction in multiple sclerosis. *Journal of Immunology*, 2007, **179**(2):1322–30.
48. Baecher-Allan, C., V. Viglietta, and D.A. Hafler, Inhibition of human CD4(+)CD25(+high) regulatory T cell function. *Journal of Immunology*, 2002, **169**(11):6210–6217.
49. Astier, A.L., et al., Alterations in CD46-mediated Tr1 regulatory T cells in patients with multiple sclerosis. *The Journal of Clinical Investigation*, 2006, **116**(12):3252–7.
50. Kohm, A.P., et al., Cutting edge: CD4+CD25+ regulatory T cells suppress antigen-specific autoreactive immune responses and central nervous system inflammation during active experimental autoimmune encephalomyelitis. *Journal of Immunology*, 2002, **169**(9):4712–6.
51. McGeachy, M.J., L.A. Stephens, and S.M. Anderton, Natural recovery and protection from autoimmune encephalomyelitis: contribution of CD4+CD25+ regulatory cells within the central nervous system. *Journal of Immunology*, 2005, **175**(5):3025–32.
52. Gartner, D., et al., CD25 regulatory T cells determine secondary but not primary remission in EAE: impact on long-term disease progression. *Journal of Neuroimmunology*, 2006, **172** (1–2):73–84.
53. Zhang, X., et al., Recovery from experimental allergic encephalomyelitis is TGF-beta dependent and associated with increases in CD4+LAP+ and CD4+CD25+ T cells. *International Immunology*, 2006, **18**(4):495–503.
54. Hafler, D.A., et al., Risk alleles for multiple sclerosis identified by a genomewide study. *The New England Journal of Medicine*, 2007, **357**(9):851–62.
55. Korn, T., et al., Myelin-specific regulatory T cells accumulate in the CNS but fail to control autoimmune inflammation. *Nature Medicine*, 2007, **13**(4):423–31.

Chapter 14
CD4+CD25+ Regulatory T Cells and TGF-Beta in Mucosal Inflammation

M. Fantini and Markus F. Neurath

Abstract Transforming growth factor-beta (TGF-beta) is an anti-inflammatory cytokine which plays a key role in the maintenance of the immune system homeostasis. Indeed the abrogation of the TGF-beta signaling in immune cells leads to autoimmunity and inflammation in several organs including the gut. TGF-beta acts at multiple levels to maintain the immune system in check. However, TGF-beta has been recently shown to play a key role in the peripheral generation and function of CD4+CD25+ regulatory T cells, a subset of suppressive lymphocytes involved in the control of effector T cell activation and proliferation. Consistently abrogation of Tregs maturation as observed in different systems leads to a phenotype resembling that caused by the impairment of the TGF-beta signaling thus linking Tregs and TGF-beta activities. Here we review data generated in the last years at support of this link.

Introduction

Mucosae are complex structures covering the largest surface of the human body. The largest mucosal system is located in the gut, although mucosae are also present in the airways and urinary tract. In the gut, the main function of the mucosa is to absorb nutrients and to maintain the hydroelectolitic balance. To accomplish to these tasks, the intestinal epithelium extends on a surface of around 400 m^2 representing the widest area of contact with the external environment. Moreover, a tight interaction between intestinal epithelial cells and bacteria, which form the physiologic intestinal flora, is required to efficiently absorb nutrients introduced with the diet. Therefore, the intestinal mucosa is characterized in normal conditions by a homeostatic balance between the antigens contained in the intestinal lumen and the mucosal immune system. Indeed the gut associated-lymphoid tissue which is formed by lymphocytes scattered all along the intestinal tract and specialized lymphoid structures

---

M.F. Neurath
Laboratoty of Mucosal Immunology, Johannes Gutenberg University of Mainz, Mainz, Germany
e-mail: neurath@1-med.klinik.uni-mainz.de

localized in the small intestine and the colon (i.e. Payers' patches and lymphoid follicles respectively) are able to recognize harmless antigens (i.e. dietary antigens and antigens produced by the normal intestinal bacterial flora) from harmful antigens deriving from pathogens that may occasionally infect the intestine. From the capacity to tolerate the presence of harmless antigens while specifically reacting against the pathogen-associated ones, derives the intestinal immune homeostasis which results impaired, for example, in the inflammatory bowel disease.

Crohn's disease (CD) and ulcerative colitis (UC) are the two major forms of inflammatory bowel diseases (IBDs), characterized by the chronic inflammation of the gut. IBDs are thought to be the consequence of a loss of tolerance towards antigens normally contained in the intestinal lumen. Indeed an abnormal accumulation of activated lymphocytes with the consequent tissue damage characterizes these diseases. Despite no specific defects of the mucosal immune system have been thus far identified to be responsible for the development of these diseases, many lines of evidence suggest that defects of the counter regulatory systems, which normally operate to prevent an abnormal activation of the mucosal immune system might contribute to the development of IBDs [1].

Alteration of the transforming growth factor-beta (TGF-beta) activity, an immuomodulatory cytokine produced by lymphoid and non lymphoid cells, as well as defects in the generation and/or activity of regulatory T cells, a subset of T cells endowed with immunosuppressive capacity have been claimed to contribute the abnormal activation of the mucosal immune system observed in the IBDs.

Transforming Growth Factor-Beta (TGF-Beta) and Gut Inflammation

In mammals, TGF-beta family of cytokines plays a pivotal role in the immune system and in the morphogenesis of several organs. TGF-beta exist includes three isoforms: TGF-beta1, 2 and 3. TGF-beta1 is the isoform preferentially expressed by hematopoietic cells, while the TGF-beta2 and 3 are produced in negligible amount by these cells. In the immune system, TGF-beta1 has been shown to play a pivotal role in the maintenance of the immune system homeostasis and tolerance towards self antigens. Indeed, early studies showed that the specific deletion of the gene encoding for the TGF-beta1 on a mixed genetic background, induces in mice an autoimmune phenotype characterized by the development of a multiorgan autoimmune disease including myocarditis, hepatitis, vasculitis and lmphoadenopathy leading mice to death few weeks after birth [2,3]. These data indicate that TGF-beta1 plays a major immunosuppressive effect on the immune system. Similar data were obtained in mice overexpressing a dominant negative form of the TGF-beta receptor II under control of the CD4 (CD4-DNRII) [4,5]. These mice overexpress a non functional form of the TGF-beta receptor II exclusively on the CD4+ cell surface thus rendering these cells hyporesponsive to the TGF-beta stimulation. Similarly to the data obtained in TGF-beta1 knockout mice, the CD4-DNRII mice developed spontaneous autoimmune disease. However n contrast to TGF-beta1 knockout mice, in CD4-DNRII mice, activated T cells accumulated especially in the mucosal system of lungs and gut thus causing cytokine release and tissue damage, indicating that the CD4-specifc TGF-beta pathway is required for the maintenance of the mucosal immune system homeostasis. Since TGF-beta is ubiquitously expressed, in order to identify the cellular source of TGF-beta important for the control the immune system at the level of mucosae, a CD4 specific TGF-beta1 conditional knockout was generated [6]. These mice are characterized by the specific inactivation of the TGF-beta1 gene in CD4+ T cells. Results from this mouse model have shown that the loss of CD4-specific TGF-beta1 expression is sufficient to generate an autoimmune phenotype characterized by the development of airways inflammation and spontaneous colitis. Therefore the autocrine expression of TGF-beta1 by T cells is crucial for the maintenance of the mucosal immune system homeostasis.

Functionally TGF-beta1 prevents T cell activation [7] and proliferation by inhibiting IL-2 expression [8,9]. Moreover in vitro data have shown that TGF-beta1 signaling can directly interfere with the Th1 and Th2 differentiation process. Indeed TGF-beta1 has been shown to directly affect the expression of T-bet and GATA-3, the transcription factors which play pivotal role in the differentiation of Th1 and Th2 cells respectively [10,11].

Summing up these data, despite the wide expression of TGF-beta1 in different tissues by numerous cell types, the autocrine expression of TGF-beta1 by T cells seems to play a key role in the TGF-beta1-mediated control of the immune system activation. Therefore alteration of the TGF-beta signaling may contribute to the development and maintenance of the chronic inflammation observed in the IBDs. Accordingly, a block of the TGF-beta signaling has been identified in inflammatory conditions such as IBDs and *H.Pylori*-related gastritis [12,13].

The TGF-beta signaling starts with the specific binding of homodimeric TGF-beta with the TGF-betaR II. The interaction between TGF-beta and the TGF-beta RII, determines, morpho-structural changes which promote the interaction between the TGF-beta/TGF-betaRII complex and the TGF-betaR I (Fig. 14.1) [14]. The specific interaction of the TGF-beta with its receptor complex activates the tyrosine kinase activity of the TGF-beta receptor II-intracellular domain leading to the phosphorylation and activation of the TGF-beta receptor I which in turn phosphorylates and activates members of the SMAD family of intracellular factors. Eight members of the SMAD proteins have been identified so far each of them belonging to distinct subfamilies based on their function. Smad 1, 2, 3, 5 and 8, the receptor-SMADs (rSMAD), directly bind the intracellular domain of the activated TGF-beta receptor complex and translocate into the nucleus upon phosphorylation. Smad4, belonging to the Co-SMAD subfamily, forms a heterodimer with the activated R-SMAD and allows the translocation of the SMADs complex into the nucleus where it binds the DNA on specific consensous sequences thus determining gene expression. Smad7 and Smad9 are inhibitory molecules which downregulate the TGF-beta signal transduction by preventing the activation of the R-SMADs. Smad7 expression, which is specifically involved in the negative modulation of the TGF-beta1 signaling, has been shown to be induced by proinflamatory cytokines (i.e. TNF-alpha, IFN-gamma and IL-6) thus suggesting that during inflammation the massive release of these

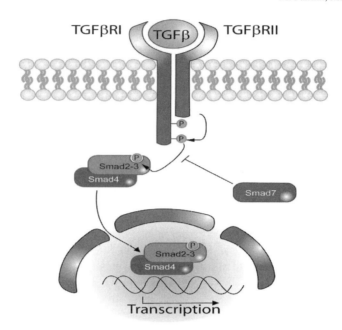

Fig. 14.1 TGF-beta pathway. TGF-beta interaction with the TGF-beta receptor II (TGFβRII) causes morpho-structural changes leading the TGF-beta/TGF-betaRII complex to interact with the TGF-beta receptor I. This interaction in turn activates the kinase activity of the TGF-betaR I intracellular domain which phosphorylates Smad2/3. Phospho-Smad2/4 couple with the CoSmad Smad4 and together translocate into the nucleus inducing the expression of TGF-beta-dependent genes. The inhibitory molecule Smad7 prevents the activation of the Smad-dependent TGF-beta signal transduction thus dampening most of the TGF-beta-mediated effects *(See also Color Insert)*

cytokines contribute to the inhibition of the immunosuppressive signal generated by TGF-beta1 [15–17]. Indeed in both CD and UC but also in H.Pylori-related gastritis, Smad7 upregulation has been described, suggesting that the Smad7-mediated inhibition of the TGF-beta signaling may contribute to sustain mucosal inflammation [12,13]. In these studies the downregulation of Smad7 expression obtained ex vivo by RNA silencing in human organ cultures, showed a dramatic reduction of the T cell infiltration, indicating that the reactivation of the TGF-beta pathway, impaired by the expression of the natural inhibitor Smad7 is sufficient to control chronic inflammation.

Naturally Occurring CD4+CD25+ Reguatory T Cells (Tregs) and Mucosal Inflammation

CD4+CD25+ Regulatory T cells (Tregs) is a class of lymphocytes characterized by the capacity to suppress activation and proliferation of other T cells as shown both in vitro and in vivo. Tregs were initially identified as a subset of CD4 memory

T cells characterized by the constitutive high expression of the activation marker CD25, the alpha subunit of the IL-2 receptor. However, CD25 is readily upregulated also on effector cells upon activation, rendering difficult the use of this marker to discriminate Tregs from recently activated effector T cells. Therefore CD25 should not be considered a specific marker of Tregs. The use of CD25 as marker of Tregs resulted particularly misleading in studies performed in humans where the fraction of non Tregs/memory cells or recently activated T cells is bigger in compare to mice. Indeed while murine CD4+CD25+ contain almost 90% of Tregs while in humans this faction is more variable. In contrast to CD25, FoxP3, an intracellular transcription factor belonging to the family of the winged helix of transcription factor, is specifically expressed by Tregs cells. Moreover FoxP3 is responsible for the lineage commitment of Tregs during their development. Consistent with this role, the forced expression of FoxP3 in naïve CD4+ T cells was sufficient to induce a suppressive phenotype in naïve CD4+ T cells, thus suggesting that FoxP3 might be sufficient to direct the specific genetic program leading to the development of Tregs in mice [18].

Deletion or loss of function mutations of the FoxP3 gene leads in both humans and mice to the development of autoimmunity. Indeed, loss of function mutations of FoxP3 in humans give rise to the immunodisregulation, polyendocrinopathy, enteropathy, X-linked syndrome (IPEX) a severe autoimmune disease developing early after birth. Similarly, an autoimmune disease involving different organs characterizes the phenotype of the *Scurfy* mice, in which FoxP3 gene results deleted [19,20]. The autoimmunity developing both in mice and humans carrying mutation of FoxP3 has been linked to the absence of circulating Tregs [21]. Therefore, the loss of FoxP3 expression or the expression of defective forms of FoxP3 leads to the absence of Tregs and to an uncontrolled activation of self-reactive T cells which, at the level of the gut, causes a massive infiltration of the mucosa and tissue damage.

Numerous lines of evidence sustain that FoxP3-expressing regulatory T cells may originate either from the thymus during the T cells positive and negative selection process or in the periphery resulting from the differentiation of naïve T cells. Thymus-derived Tregs, also indicated as naturally occurring Tregs (nTregs), originate from the thymus during the first days of postnatal life. Indeed while thymectomy performed on mice at birth (0Tx) resulted in the complete loss of peripheral T cells, thymectomy performed at day three (3Tx) after birth resulted in a specific loss of peripheral Tregs, indicating that nTregs depend on the thymus for their development and that they are released in periphery late during the immune system ontogenesis [22]. Interestingly, 3Tx mice developed an autoimmune phenotype characterized by autoimmune gastritis and high titers of anti parietal cells antibodies indicating that the loss of nTregs is sufficient to alter the mucosal immune system homeostasis. Moreover the same autoimmune phenotype was induced in immunodeficient BALBc$^{nu/nu}$ mice receiving CD4+CD25+-depleted T cells, thus indicating that in the absence of Tregs, autoimmunity can develop during the homeostatic expansion of naïve T cells in a lymphopenic host. Indeed the homeostatic expansion of T cells in the absence of Tregs leads to an uncontrolled expansion and activation of self-reactive T cell clones which are responsible for inflammation and inflammation-related tissue damage. The role of CD4+CD25+ nTregs depletion in the development of autoimmunity was further demonstrated by the observation that the cotransfer of CD4+CD25+ Tregs isolated from a syngenic mouse together with cells isolated from a 3Tx mice in a lymphopenic recipients or the early administration of CD4+CD25+ cells to 3Tx mice was sufficient to prevent the disease, confirming the pivotal role of nTregs in the control of self-reactive T cells [23].

The role of nTregs in the control of the immune system homeostasis in the intestinal mucosa has been also investigated in murine models of colitis based on the adoptive transfer of naïve T cells in immunodeficient mice. In these models either RAG1 knockout or SCID mice, characterized by the absence of T cells, develop colitis upon adoptive transfer of naïve T cells isolated from a syngenic wild type mouse [24]. Similarly to the gastritis model, colitis developing in these mice is thought to be dependent on an uncontrolled expansion, in an immunodeficient host, of autoreactive clones deprived of the Tregs control. Accordingly with the role played by Tregs, the cotransfer of naïve cells with Tregs, was sufficient to prevent colitis development without affecting the homeostatic expansion of naïve T cells. Interestingly, the same experiments performed in animals grown in germ free conditions failed to develop colitis indicating that T cells clones specific for antigens contained in the normal bacterial flora expand during the homeostatic expansion occurring in the immunodeficient host and that in the absence of Tregs these cells react against intestinal bacteria thus leading to colitis [25].

Although the depletion of nTregs is required to generate autoimmunity, in the animal models described, the homeostatic expansion of naïve T cells was required to generate autoimmune gastritis and colitis. Indeed gastritis developed in 3Tx mice which are characterized by a state of immunodepletion secondary to the early thymectomy. Analogously, colitis developed in lymphopenic hosts upon adoptive transfer of Treg–depleted T cells. Consistent with the role of homeostatic expansion in the generation of autoimmunity, mice depleted of Tregs by using an anti-CD25 antibody do not develop colitis while lymphopenic mice receiving cells isolated from depleted mice do [26]. This observation would lead to the notion that Tregs are able to control reactive T cells exclusively in the context of a homeostatic expansion. However, the adoptive transfer of sorted CD4+CD25+ nTregs in wild type mice was able to prevent colitis induced by dextrane sulfate sodium, a chemical which induces colitis in a homeostatic expansion-independent system [27]. In another animal model, colitis is induced in imunodefecent RAG1 knockout mice upon infection with *H. Hepaticus*. Colitis developing in these mice is secondary to the activation of the innate immunity with no role played by T cells. In this context, the adoptive transfer of Tregs was able to suppress colitis [28]. These data indicate the immunsuppressive effect of Tregs in vivo is more profound then the simple block of T cell activation and proliferation observed in vitro, negatively regulating the activity of non-T cells. Therefore Tregs prevent mucosal inflammation dampening both the acquired and innate immunity.

In humans, the role of CD4+CD25+ Tregs in the pathogenesis of mucosal inflammation has been investigated in IBDs. In one report the suppressive capacity of CD4+CD25+ Treg cells isolated from the gut of IBD and non-IBD patients were compared. Results from these experiments showed no difference in terms of suppressive capacity of Tregs, indicating that gut inflammation occurring in UC and CD is not secondary to a functional defect of Tregs [29].

The absence of specific defects in the suppressive capacity of Tregs isolated from IBD patients, however, did not rule out the possibility that a defective accumulation of Tregs in the lamina propria of these patients could contribute to the uncontrolled activation of the intestinal mucosal immune system. Indeed, in another study, the number of intestinal FoxP3 expressing cells was lower in IBD patients in compare to those affected by diverticulitis, an acute inflammatory condition of the colon not related to IBD, suggesting that an insufficient accumulation of Tregs in the gut mucosa might contribute to the maintenance of a chronic inflammation observed in IBDs [30].

TGF-Beta and the Peripheral Induction of Tregs

The observation that FoxP3-expressing cells may accumulate in the mucosa and that these cells may contribute to the maintenance of the mucosal immune system homeostasis, questioned whether Tregs expand from a restricted pool of cells generated in the thymus or if, alternatively, they can originate as result of a differentiation process from the pool of circulating naïve T cells. With this regard, data indicate that cells characterized by the expression of FoxP3 and endowed with suppressive capacity can be generated from a pool of naïve cells.

In addition to the aforementioned role of TGF-beta in suppressing naïve T cell activation and Th1/Th2 differentiation, TGF-beta has been shown to drive the differentiation of regulatory T cells from a population of naïve cells [31,32]. Indeed naïve cells activated in the presence of TGF-beta differentiate towards a regulatory phenotype. The differentiation of naïve cells into Tregs has been demonstrated in different models. In initial studies, CD4+CD25+-depleted CD4+ T cells, activated in the presence of TGF-beta acquired suppressive capacities and this phenotype was also characterized by the expression of FoxP3. Similar results were obtained using cells isolated from TCR-transgenic mice on a Rag1 knockout background which have only naïve cells as result of an abnormal positive and negative T cell selection occurring in the thymus [33]. Experiments performed using T cells isolated from these mice confirmed that TGF-beta does induce FoxP3 in activated naïve T cells. Finally, cells isolated from a FoxP3-GFP knockin mouse in which the FoxP3 expression in T cells is associated with green fluorescence, demonstrated that TGF-beta is able to induce FoxP3 in FoxP3 negative T cells (Fig. 14.2) [34].

The TGF-beta-dependent induction of peripheral Tregs is tightly controlled by proinflammatory stimuli. IL-6 [35–37] and IL-21 [38,39], two cytokines expressed at high levels during inflammation, have been shown to suppress iTregs generation thus promoting the induction of Th17 cells, a novel class of effector T cells involved in the inflammatory process affecting several organs. These data suggest that in physiologic conditions pathogen-elicited proinflammatory stimuli downregulate the

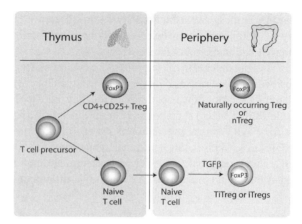

Fig. 14.2 Peripheral induction of Treg (i.e. iTreg, TiTregs). nTregs originate from the thymus after the first days of post-natal life and migrate in the periphery where they are thought to play a pivotal role in the maintenance of tolerance towards self-antigens. In the periphery, CD4+ naïve T cells may differentiate in suppressive- FoxP3 expressing-cells in the presence of TGF-beta. TiTregs (or nTregs) might be important for tolerance towards antigens contained in the gut lumen either produced by the intestinal flora or introduced with the diet *(See also Color Insert)*

anti-inflammatory signal provided by Tregs in order to mount an efficient immune response.

Functionally, TGF-beta induced Tregs (TiTregs or iTregs) appear undistinguishable from the nTregs in that they suppress the activation and cytokine expression of naïve T cells in vitro. In vivo iTregs suppress inflammation in an adoptive transfer model of colitis and in allergen-induced asthma model thus indicating that the peripheral induction of FoxP3 in naïve T cells generate a class of differentiated regulatory T cells which are able to suppress mucosal inflammatory diseases [31,40].

In humans the TGF-beta-mediated induction of Tregs is controversial. Initial studies showed that FoxP3 is readily expressed by activated CD4+ T cells, suggesting that FoxP3 induction in humans is independent from the TGF-beta signaling [41]. Nevertheless the same Authors showed that FoxP3 expression is associated with Treg differentiation of human CD4+CD25- cells. In contrast to these data, studies have demonstrated that the induction of FoxP3 in CD4+CD25- T cells is indeed TGF-beta dependent but that the induction of this transcription factor is not sufficient to confer suppressive capacity to human naïve T cells [42]. Therefore, in contrast to mice, supplementary signals, other than TGF-beta, are required for the peripheral generation of Tregs in humans.

TGF-Beta as Tregs Effector Molecule

Although with discordant results, TGF-beta has been associated with the Treg-mediated T cell suppression. Indeed, studies showed that neutralization of membrane-bound TGF-beta by means of either an anti-TGF-beta or a recombinant latency associated peptide (rLAP) abrogated the Tregs-mediated suppressive capacity indicating that TGF-beta expressed by Tregs might be responsible at least in part for their suppressive capacity [43,44]. In contrast with this concept are evidences produced in TGF-beta knockout mice. Although these animals develop autoimmune disease early in life, the development of thymus derived nTregs in not compromised making possible the isolation of CD4+CD25+ regulatory T cells. CD4+CD25+ Tregs isolated from TGF-beta deficient mouse retained the suppressive capacity in vitro thus bringing into question the role of TGF-beta as Treg effector molecule [45]. One possible explanation for the discordant results obtained in different systems might rely on the adoptive process undergoing in the TGF-beta knockout Tregs in which other suppressive mechanisms become upregulated thus making up for the TGF-beta deficiency.

In vivo, the role of TGF-beta as mediator of Tregs suppression of mucosal inflammation seems to be more defined. Indeed Tregs isolated from wild type mice failed to suppress colitis induced upon the adoptive transfer in SCID mice of naïve CD4+ T cells overexpressing the dominant negative form of the TGF-beta receptor II indicating that TGF-beta signaling on colitogenic cells is required to suppress mucosal inflammation by Tregs [46]. Although these data clearly implicate the T cell-specific TGF-beta pathway is required for Tregs-mediated colitis suppression they do not identify the source of TGF-beta. To address this issue, similar studies have been performed using CD4-TGF-beta conditional knockout mice, in which TGF-beta gene is selectively inactivated in CD4+ cells. By using the same model of experimental colitis, the authors showed that Tregs require TGF-beta expression to suppress colitis induced by the adoptive transfer of naïve CD4+ T cells [6]. However is worth noting that in this study SCID mice reconstituted with colitogenic CD4+ naïve T cells and Tregs both isolated from CD4+TGF-beta knockout mice showed the worst colitis, indicating that sources of TGF-beta different from Tregs may contribute to the mucosal immune homeostasis.

Manipulation of TGF-Beta Pathway and Tregs as Therapy of Mucosal Inflammation

Accumulating evidences suggest that administration of Tregs and/or manipulation of the TGF-beta pathway could be of potential interest for the therapy of mucosal inflammatory disorders. Concerning the use of Tregs for therapy, data produced in mice indicate that CD4+CD25+ Tregs can be potentially useful for therapy. Indeed, Tregs were able to be curative in mice affected by established colitis obtained by the adoptive transfer of naïve T cells into lymphopenic hosts. Moreover CD4+CD25+ nTegs suppressed colitis induced by *H.Hepaticus*. However, some considerations on the use of these cells for therapy need to be made. First of all the magnitude of the immunosuppression induced by the in vivo administration of Tregs is currently unknown. In fact, different levels of immunosuppression are known to be linked to the development of opportunistic infections and cancer development. Moreover the suppressive activity of Tregs is not antigen specific. Indeed, once activated in an antigen specific manner, Tregs may suppress the activation of T cells specific for different antigens (bystander suppression) thus potentially blocking the immune response directed against pathogens or cancer cells. Since the use of Tregs as therapy is likely to follow a pharmacokinetic which is deeply different from common drugs, the possible long standing immunosuppressive effect of Tregs on the mucosal immune system should be carefully evaluated. Nevertheless the powerful immunosuppressive effect of Tregs observed in different animal models makes mandatory to explore the possibility to use these cells for therapy.

Beyond the problem of safety, also the exiguous number of circulating Tregs renders the use of naturally occurring Tregs for therapy a challenging task. Treg cells represent around 10% of circulating CD4+ cells. Many protocols, including IL-2, a Treg growth factor, have been developed in order to expand Tregs in vitro. This approach would require a conspicuous effort to generate a sufficient number of Tregs to be inoculated back into the donor. The evidence that Treg can be induced in vitro from naïve T cells by TGF-beta seems to offer a solution to this problem, however, while in mice iTregs are clearly induced y TGF-beta, data obtained in humans are less clear.

A different approach to modulate mucosal inflammation is represented by the manipulation of the TGF-beta pathway. In mucosal inflammation, like that observed in inflammatory bowel disease, despite the large amount of TGF-beta present in the gut mucosa, the TGF-beta pathway results impaired by the Smad7 expression. Therefore, to reestablish the TGF-beta signaling by down-regulating Smad7 seems to be a feasible approach. This feasibility of this approach is supported by data showing that intrarectal administration of Smad7 oligoantisense, which silence the Smad7 expression at the RNA level, thus reducing the level of Smad7 protein, is able to prevent TNBS and oxazolone-mediated experimental colitis. Studies on humans will show the efficacy of such a treatment [47].

Conclusions

The bulk of knowledge produced in the last years has shown that regulatory T cells and TGF-beta are tightly connected and both are required for the maintenance of the mucosal immune system homeostasis. Alteration of one or both these counter regulatory systems leads to an uncontrolled activation of T cells which produce tissue damage like that observed in the gut of patients affected by IBDs. Therefore the understanding of the mechanisms governing the TGF-beta/Tregs immunosuppressive system is crucial to design new therapeutic approaches aimed at modulate the immune system thus avoiding the damage produced by a long lasting mucosal inflammation.

References

1. Strober, W., I. Fuss, and P. Mannon. 2007. The fundamental basis of inflammatory bowel disease. *J Clin Invest 117:514.*

2. Shull, M. M., I. Ormsby, A. B. Kier, S. Pawlowski, R. J. Diebold, M. Yin, R. Allen, C. Sidman, G. Proetzel, D. Calvin, et al. 1992. Targeted disruption of the mouse transforming growth factor-beta 1 gene results in multifocal inflammatory disease. *Nature 359:693.*
3. Kulkarni, A. B., C. G. Huh, D. Becker, A. Geiser, M. Lyght, K. C. Flanders, A. B. Roberts, M. B. Sporn, J. M. Ward, and S. Karlsson. 1993. Transforming growth factor beta 1 null mutation in mice causes excessive inflammatory response and early death. *Proc Natl Acad Sci USA 90:770.*
4. Lucas, P. J., S. J. Kim, S. J. Melby, and R. E. Gress. 2000. Disruption of T cell homeostasis in mice expressing a T cell-specific dominant negative transforming growth factor beta II receptor. *J Exp Med 191:1187.*
5. Gorelik, L., and R. A. Flavell. 2000. Abrogation of TGFbeta signaling in T cells leads to spontaneous T cell differentiation and autoimmune disease. *Immunity 12:171.*
6. Li, M. O., Y. Y. Wan, and R. A. Flavell. 2007. T cell-produced transforming growth factor-beta1 controls T cell tolerance and regulates Th1- and Th17-cell differentiation. *Immunity 26:579.*
7. Kehrl, J. H., L. M. Wakefield, A. B. Roberts, S. Jakowlew, M. Alvarez-Mon, R. Derynck, M. B. Sporn, and A. S. Fauci. 1986. Production of transforming growth factor beta by human T lymphocytes and its potential role in the regulation of T cell growth. *J Exp Med 163:1037.*
8. Morris, D. R., C. A. Kuepfer, L. R. Ellingsworth, Y. Ogawa, and P. S. Rabinovitch. 1989. Transforming growth factor-beta blocks proliferation but not early mitogenic signaling events in T-lymphocytes. *Exp Cell Res 185:529.*
9. Stoeck, M., S. Miescher, H. R. MacDonald, and V. Von Fliedner. 1989. Transforming growth factors beta slow down cell-cycle progression in a murine interleukin-2 dependent T-cell line. *J Cell Physiol 141:65.*
10. Gorelik, L., S. Constant, and R. A. Flavell. 2002. Mechanism of transforming growth factor beta-induced inhibition of T helper type 1 differentiation. *J Exp Med 195:1499.*
11. Gorelik, L., P. E. Fields, and R. A. Flavell. 2000. Cutting edge: TGF-beta inhibits Th type 2 development through inhibition of GATA-3 expression. *J Immunol 165:4773.*
12. Monteleone, G., G. Del Vecchio Blanco, G. Palmieri, P. Vavassori, I. Monteleone, A. Colantoni, S. Battista, L. G. Spagnoli, M. Romano, M. Borrelli, T. T. MacDonald, and F. Pallone. 2004. Induction and regulation of Smad7 in the gastric mucosa of patients with Helicobacter pylori infection. *Gastroenterology 126:674.*
13. Monteleone, G., A. Kumberova, N. M. Croft, C. McKenzie, H. W. Steer, and T. T. MacDonald. 2001. Blocking Smad7 restores TGF-beta1 signaling in chronic inflammatory bowel disease. *J Clin Invest 108:601.*
14. Shi, Y., and J. Massague. 2003. Mechanisms of TGF-beta signaling from cell membrane to the nucleus. *Cell 113:685.*
15. Jenkins, B. J., D. Grail, T. Nheu, M. Najdovska, B. Wang, P. Waring, M. Inglese, R. M. McLoughlin, S. A. Jones, N. Topley, H. Baumann, L. M. Judd, A. S. Giraud, A. Boussioutas, H. J. Zhu, and M. Ernst. 2005. Hyperactivation of Stat3 in gp130 mutant mice promotes gastric hyperproliferation and desensitizes TGF-beta signaling. *Nat Med 11:845.*
16. Ulloa, L., J. Doody, and J. Massague. 1999. Inhibition of transforming growth factor-beta/SMAD signalling by the interferon-gamma/STAT pathway. *Nature 397:710.*
17. Bitzer, M., G. von Gersdorff, D. Liang, A. Dominguez-Rosales, A. A. Beg, M. Rojkind, and E. P. Bottinger. 2000. A mechanism of suppression of TGF-beta/SMAD signaling by NF-kappa B/RelA. *Genes Dev 14:187.*
18. Hori, S., T. Nomura, and S. Sakaguchi. 2003. Control of regulatory T cell development by the transcription factor Foxp3. *Science 299:1057.*
19. Bennett, C. L., J. Christie, F. Ramsdell, M. E. Brunkow, P. J. Ferguson, L. Whitesell, T. E. Kelly, F. T. Saulsbury, P. F. Chance, and H. D. Ochs. 2001. The immune dysregulation, polyendocrinopathy, enteropathy, X-linked syndrome (IPEX) is caused by mutations of FOXP3. *Nat Genet 27:20.*
20. Wildin, R. S., F. Ramsdell, J. Peake, F. Faravelli, J. L. Casanova, N. Buist, E. Levy-Lahad, M. Mazzella, O. Goulet, L. Perroni, F. D. Bricarelli, G. Byrne, M. McEuen, S. Proll,

M. Appleby, and M. E. Brunkow. 2001. X-linked neonatal diabetes mellitus, enteropathy and endocrinopathy syndrome is the human equivalent of mouse scurfy. *Nat Genet 27:18*.
21. Fontenot, J. D., M. A. Gavin, and A. Y. Rudensky. 2003. Foxp3 programs the development and function of CD4+CD25+ regulatory T cells. *Nat Immunol 4:330*.
22. Asano, M., M. Toda, N. Sakaguchi, and S. Sakaguchi. 1996. Autoimmune disease as a consequence of developmental abnormality of a T cell subpopulation. *J Exp Med 184:387*.
23. Suri-Payer, E., A. Z. Amar, A. M. Thornton, and E. M. Shevach. 1998. CD4+CD25+ T cells inhibit both the induction and effector function of autoreactive T cells and represent a unique lineage of immunoregulatory cells. *J Immunol 160:1212*.
24. Powrie, F., M. W. Leach, S. Mauze, L. B. Caddle, and R. L. Coffman. 1993. Phenotypically distinct subsets of CD4+ T cells induce or protect from chronic intestinal inflammation in C. B-17 scid mice. *Int Immunol 5:1461*.
25. Aranda, R., B. C. Sydora, P. L. McAllister, S. W. Binder, H. Y. Yang, S. R. Targan, and M. Kronenberg. 1997. Analysis of intestinal lymphocytes in mouse colitis mediated by transfer of CD4+, CD45RBhigh T cells to SCID recipients. *J Immunol 158:3464*.
26. McHugh, R. S., and E. M. Shevach. 2002. Cutting edge: depletion of CD4+CD25+ regulatory T cells is necessary, but not sufficient, for induction of organ-specific autoimmune disease. *J Immunol 168:5979*.
27. Huber, S., C. Schramm, H. A. Lehr, A. Mann, S. Schmitt, C. Becker, M. Protschka, P. R. Galle, M. F. Neurath, and M. Blessing. 2004. Cutting edge: TGF-beta signaling is required for the in vivo expansion and immunosuppressive capacity of regulatory CD4+CD25+ T cells. *J Immunol 173:6526*.
28. Maloy, K. J., L. Salaun, R. Cahill, G. Dougan, N. J. Saunders, and F. Powrie. 2003. CD4+CD25+ T(R) cells suppress innate immune pathology through cytokine-dependent mechanisms. *J Exp Med 197:111*.
29. Kelsen, J., J. Agnholt, H. J. Hoffmann, J. L. Romer, C. L. Hvas, and J. F. Dahlerup. 2005. FoxP3(+)CD4(+)CD25(+) T cells with regulatory properties can be cultured from colonic mucosa of patients with Crohn's disease. *Clin Exp Immunol 141:549*.
30. Maul, J., C. Loddenkemper, P. Mundt, E. Berg, T. Giese, A. Stallmach, M. Zeitz, and R. Duchmann. 2005. Peripheral and intestinal regulatory CD4+ CD25(high) T cells in inflammatory bowel disease. *Gastroenterology 128:1868*.
31. Chen, W., W. Jin, N. Hardegen, K. J. Lei, L. Li, N. Marinos, G. McGrady, and S. M. Wahl. 2003. Conversion of peripheral CD4+CD25- naive T cells to CD4+CD25+ regulatory T cells by TGF-beta induction of transcription factor Foxp3. *J Exp Med 198:1875*.
32. Fantini, M. C., C. Becker, G. Monteleone, F. Pallone, P. R. Galle, and M. F. Neurath. 2004. Cutting edge: TGF-beta induces a regulatory phenotype in CD4+CD25- T cells through Foxp3 induction and down-regulation of Smad7. *J Immunol 172:5149*.
33. Fu, S., N. Zhang, A. C. Yopp, D. Chen, M. Mao, H. Zhang, Y. Ding, and J. S. Bromberg. 2004. TGF-beta induces Foxp3 + T-regulatory cells from CD4 + CD25 - precursors. *Am J Transplant 4:1614*.
34. Wan, Y. Y., and R. A. Flavell. 2005. Identifying Foxp3-expressing suppressor T cells with a bicistronic reporter. *Proc Natl Acad Sci USA 102:5126*.
35. Dominitzki, S., M. C. Fantini, C. Neufert, A. Nikolaev, P. R. Galle, J. Scheller, G. Monteleone, S. Rose-John, M. F. Neurath, and C. Becker. 2007. Cutting edge: transsignaling via the soluble IL-6R abrogates the induction of FoxP3 in naive CD4+CD25 T Cells. *J Immunol 179:2041*.
36. Bettelli, E., Y. Carrier, W. Gao, T. Korn, T. B. Strom, M. Oukka, H. L. Weiner, and V. K. Kuchroo. 2006. Reciprocal developmental pathways for the generation of pathogenic effector TH17 and regulatory T cells. *Nature 441:235*.
37. Mangan, P. R., L. E. Harrington, D. B. O'Quinn, W. S. Helms, D. C. Bullard, C. O. Elson, R. D. Hatton, S. M. Wahl, T. R. Schoeb, and C. T. Weaver. 2006. Transforming growth factor-beta induces development of the T(H)17 lineage. *Nature 441:231*.

38. Nurieva, R., X. O. Yang, G. Martinez, Y. Zhang, A. D. Panopoulos, L. Ma, K. Schluns, Q. Tian, S. S. Watowich, A. M. Jetten, and C. Dong. 2007. Essential autocrine regulation by IL-21 in the generation of inflammatory T cells. *Nature 448:480.*
39. Korn, T., E. Bettelli, W. Gao, A. Awasthi, A. Jager, T. B. Strom, M. Oukka, and V. K. Kuchroo. 2007. IL-21 initiates an alternative pathway to induce proinflammatory T(H)17 cells. *Nature 448:484.*
40. Fantini, M. C., C. Becker, I. Tubbe, A. Nikolaev, H. A. Lehr, P. Galle, and M. F. Neurath. 2006. Transforming growth factor beta induced FoxP3+ regulatory T cells suppress Th1 mediated experimental colitis. *Gut 55:671.*
41. Walker, M. R., D. J. Kasprowicz, V. H. Gersuk, A. Benard, M. Van Landeghen, J. H. Buckner, and S. F. Ziegler. 2003. Induction of FoxP3 and acquisition of T regulatory activity by stimulated human CD4+CD25- T cells. *J Clin Invest 112:1437.*
42. Tran, D. Q., H. Ramsey, and E. M. Shevach. 2007. Induction of FOXP3 expression in naive human CD4+FOXP3- T cells by T cell receptor stimulation is TGF{beta}-dependent but does not confer a regulatory phenotype. *Blood 110:2983.*
43. Nakamura, K., A. Kitani, I. Fuss, A. Pedersen, N. Harada, H. Nawata, and W. Strober. 2004. TGF-beta 1 plays an important role in the mechanism of CD4+CD25+ regulatory T cell activity in both humans and mice. *J Immunol 172:834.*
44. Nakamura, K., A. Kitani, and W. Strober. 2001. Cell contact-dependent immunosuppression by CD4(+)CD25(+) regulatory T cells is mediated by cell surface-bound transforming growth factor beta. *J Exp Med 194:629.*
45. Piccirillo, C. A., J. J. Letterio, A. M. Thornton, R. S. McHugh, M. Mamura, H. Mizuhara, and E. M. Shevach. 2002. CD4(+)CD25(+) regulatory T cells can mediate suppressor function in the absence of transforming growth factor beta1 production and responsiveness. *J Exp Med 196:237.*
46. Fahlen, L., S. Read, L. Gorelik, S. D. Hurst, R. L. Coffman, R. A. Flavell, and F. Powrie. 2005. T cells that cannot respond to TGF-beta escape control by CD4(+)CD25(+) regulatory T cells. *J Exp Med 201:737.*
47. Boirivant, M., F. Pallone, C. Di Giacinto, D. Fina, I. Monteleone, M. Marinaro, R. Caruso, A. Colantoni, G. Palmieri, M. Sanchez, W. Strober, T. T. MacDonald, and G. Monteleone. 2006. Inhibition of Smad7 with a specific antisense oligonucleotide facilitates TGF-beta1-mediated suppression of colitis. *Gastroenterology 131:1786.*

Chapter 15
Induction of Adaptive CD4+CD25+Foxp3+ Regulatory T Cell Response in Autoimmune Disease

Jian Hong, Sheri Skinner, and Jingwu Zhang

Abstract CD4+CD25+Foxp3+ regulatory T cells (Tregs) represent one of the most critical regulatory components of the immune system. They are vitally important for homeostasis of the immune system and adequate immune responses. Tregs contain heterogeneous populations that share common markers and regulatory function but differ in the origin of differentiation or conversion. Deficiencies either in the number of Tregs or the expression of Foxp3 lead to various autoimmune pathologies in both humans and experimental animals. Here the role of adaptive Tregs that are converted from CD4+CD25− T cells in the periphery is reviewed in relationship to their induction by cytokines, such as gamma-interferon, and therapeutic modalities, including T cell vaccination and Copolymer-I currently being used or tested in multiple sclerosis. In particular, important issues related to the potential therapeutic application of adaptive Tregs in autoimmune disease are discussed.

Introduction

If one defines CD4+CD25+Foxp3+ Tregs based upon their origins, there are two types of CD4+CD25+Foxp3+ Tregs. Naturally occurring CD4+CD25+Foxp3+ regulatory T cells (nTregs) are committed regulatory T cell lineage differentiated and produced in thymus, while adaptive CD4+CD25+Foxp3+ regulatory T cells (aTregs) are induced from naïve T cells as a consequence of a particular mode of antigen exposure or cytokine milieu in the periphery. The need to maintain immune homeostasis in the face of dysfunctions such as autoimmune disease would seem to necessitate a level of redundancy in Treg mechanisms such as these subpopulations would suggest. In this regard, much attention has been paid to the delineation of Treg populations on the bases of molecular and cellular markers, cytokine dependency, immune suppressive and control mechanisms and cellular developmental pathways [1].

J. Zhang
GlaxoSmithKline R&D China, Shanghai 201203, China
e-mail: jingwu.z.zang@gsk.com

Naturally occurring CD4+CD25+Foxp3+ regulatory T cells (nTregs) have been most intensively analyzed in humans and rodents. The consensus among investigators appears to establish the presence of three molecules as primarily essential for nTreg development, activity and ultimate survival. These are the transcription factor known as Foxp3, the cytokine IL-2 and the receptor CD25. Extensive study has determined that defects in or experimental elimination of these molecules do indeed predispose to autoimmune disease in both humans and animal models. Attempts at delineating Treg cell populations have utilized the specificity of the intracellular Foxp3 and the cell surface marker CD25 as well as other markers such as GITR (glucocorticoid-induced TNF-receptor family-related gene or its resultant protein). Such approaches allow the study of nTregs, their discrimination from other T cell populations, and their complicity in autoimmune pathogenesis [2].

Efforts have also been made to study the function and development of antigen and cytokine induced adaptive Tregs (aTregs), which appear to be another important component of the regulatory T cell machinery controlling the host immune responses to foreign or self-antigens. Adaptive Tregs are induced from naïve CD4+CD25− T cells in vivo and in vitro under conditions in which they are exposed to certain antigens; in particular, when the appropriate cytokine milieu exists. A number of cytokines, such as TGF-β and IL-2, are apparently required for the de novo development of Tregs from naïve T cells and their maintenance in the periphery. The similarity of aTregs to the nTreg population lie in their expression of Foxp3 as well as their suppressive effects upon dysfunctional or excessive immune responses to self or non-self antigenic stimulation. Further, it has been seen that many aspects of Treg activity, including expansion, activation and expression, can be fine-tuned through the cytokine milieu as well as through Treg expressed accessory molecules such as CD28 and CTLA-4 [1].

However, it is not clear how aTregs are induced in vivo in the periphery and what condition or mechanism is required for their induction. Are aTregs committed regulatory cell lineage or transient stages of CD4+ T cells? Most importantly, how are aTregs relevant to the development of autoimmune disease? Data available so far provide some indications, but not proof, toward these questions. For example, our recent work indicates that aTreg is an uncommitted regulatory cell lineage. Once induced by TCR stimulation in the presence of TGF-b, they appear to lose stable expression of Foxp3 and regulatory function after withdrawal of TGF-b [unpublished results]. Furthermore, our EAE study has revealed that there is an active process of peripheral conversion of CD4+CD25− T cells to aTregs expressing CD25 and Foxp3 and aquring regulatory activity toward autoreactive T cell responses critical for EAE pathology [3]. While it is important to understand that the development of autoimmune inflammation is often accompanied by progressive conversion of aTregs to expand the regulatory T cell pool, the molecular and cellular requirements involved in this process become essential as they may have therapeutic potential (Fig. 15.1). In this chapter, recent research progress in this regard is reviewed to address some of the important issues described above.

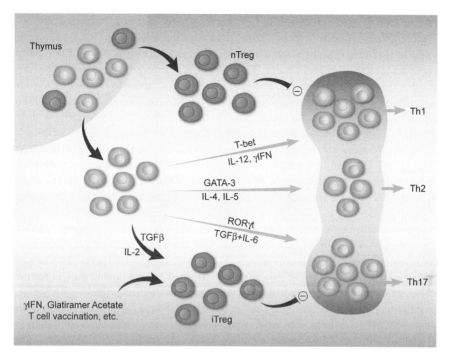

Fig. 15.1 Unlike naturally occurring Tregs (nTreg) that are differentiated in and migrate out of thymus, inducible Tregs are induced or converted from peripheral CD4+CD25− T cell pool by TGF-beta, gamma-interferon and other conditions that are reviewed in this chapter. Tregs of both types work in concert to keep various effectors (Th1, Th2 and Th17) in check *(See also Color Insert)*

Role of Gamma-IFN in the Induction of Adaptive Tregs

In addition to TGF-beta, other cytokines may also contribute to the peripheral conversion of CD4+CD25− T cells to CD4+CD25+Foxp3+ Tregs and the induction of adaptive Tregs. We recently described that heightened susceptibility of IFN-γ gene knockout mice to EAE is attributed to uncontrolled and altered encephalitogenic T cell responses, which directly result from impaired conversion of CD4+CD25− T cells to CD4+ regulatory T cells in the absence of IFN-γ in the disease state [3]. The study has provided new evidence demonstrating that lack of γ–IFN is associated with marked impairment of conversion of $CD4^+CD25^-$ T cells to $CD4^+$ regulatory T cells, leading to heightened susceptibility to EAE in γ–IFN deficient mice. The conclusion is made based upon the following observations. (1) There is reduced frequency and function of CD4+CD25+ regulatory T cells accompanied by decreased Foxp3 expression as a result of impaired conversion of CD4+CD25− T cells to $CD4^+$ regulatory T cells during acute EAE in IFN-γ gene knockout mice. (2) There is direct evidence that in vitro treatment of CD4+CD25− T cells with IFN-γ in the presence or absence of TCR co-stimulation leads to conversion of CD4+CD25− T cells to CD4+ regulatory T cells as shown by increased regulatory function and Foxp3 expression in both mouse and human experimental systems. (3) CD4+CD25– T cells converted through in vitro treatment with IFN-γ acquire the ability to suppress EAE by adoptive transfer. There is a proportion of CD4+CD25– T cells undergoing conversion to CD4+ regulatory T cells during acute EAE, a process that requires the function of IFN-γ. In GKO mice, failure to produce IFN-γ and the resulting deficit in CD4+CD25+ regulatory T cell function during acute inflammation has significant immunological and pathological consequences. The findings provide convincing evidence indicating that IFN-γ plays an essential role in the self-regulatory mechanisms of the immune system in response to acute inflammation through the induction of transcription factor Foxp3. The observations satisfactorily explained the paradoxical role of IFN-γ in the development of EAE and provided a detailed account for the added severity in the encephalitogenic responses, altered cytokine profile and clinical course of EAE in IFN-γ gene knockout mice. It is in accordance with the underlying mechanism for heightened susceptibility of IFN-γ gene knockout mice to EAE and experimental autoimmune uveitis (EAU). There was a 10- to 16-fold increase of activated T cells (CD4+, CD44hi) accumulated in the central nervous system at the onset of EAE in IFN-γ deficient mice compared to that in wild-type mice and these T cells proliferated extensively with antigen stimulation, which response could be inhibited by IFN-γ. The IFN-γ might limit the extent of EAE by suppressing expansion of activated CD4+ T cells. CD4+CD25– T cells from IFN-γ deficient mice tend to be more susceptible to IFN-γ in the induction of Foxp3 expression when compared to their wild-type counterparts.

The induction of Foxp3 in CD4+CD25– T cells and subsequent conversion to CD4+ T cells of regulatory potential do not necessarily require co-stimulation with the T cell receptors or other cytokines. A similar property of TGF-α in the conversion of CD4+CD25– T cells was also demonstrated. However, co-stimulation of CD4+CD25- T cells with T cell receptors is required to achieve conversion to CD4+CD25+ regulatory T cells through the induction of Foxp3 expression. Furthermore, the role of IFN-γ in the conversion of CD4+CD25– T cells to CD4+ regulatory T cells is consistent with the discovery that STAT1, a signaling molecule involved in the IFN-γ signaling pathway, is critical to the induction of CD4+CD25+ regulatory T cells. STAT1-deficient mice expressing a transgenic T cell receptor against myelin basic protein spontaneously developed EAE, a response attributable to a functional impairment of CD4+CD25+ regulatory T cells in STAT1-deficient mice. Further, the heightened susceptibility of IFN-γ receptor knockout mice to collagen-induced arthritis is associated with impaired function of CD4+CD25+ regulatory T cells.

The complex role of IFN-γ in the immune response and immune regulation has direct relevance to various pathological conditions, including autoimmune disease, tumor and infectious pathology. IFN-γ may act as a double-edged sword in immune responses and inflammatory processes. On one hand, it promotes a Th1 response and T cell migration to the site of inflammation and amplifies an inflammatory cascade through its ability to mediate a variety of signaling events leading to the production of inflammatory molecules. On the other hand, the intensity of Th1-induced inflammation as quantifiable by the production of IFN-γ potentially triggers the immune system to invoke a controlling mechanism. In such a scenario, IFN-γ produced at a high concentration either locally or systemically may be seen as a danger signal to the immune system that is programmed to activate various cellular and molecular events in converting CD4+CD25− T cells to a CD4+ regulatory T cell response for the purpose of regulating overt inflammation. The observed role of IFN-γ in immune regulation may represent, in part, the tremendous ability of the immune system to regulate itself to prevent overheated immune responses in various pathological conditions.

The role of IFN-γ in the induction of adaptive Tregs has particular relevance to autoimmune conditions, such as multiple sclerosis (MS). It was demonstrated that there is a functional deficit in CD4+CD25+ regulatory T cells in patients with MS. There is a significant decrease in the expression of Foxp3 in the CD4+CD25+ T cell pool in MS compared to that of healthy individuals. A large proportion of the $CD4^+CD25^+$ T cell population seen in blood of MS patients is likely to represent activated autoreactive or inflammatory T cells, rather than regulatory T cells. High amounts of IFN-γ are likely to accumulate locally in inflammatory lesions, which may trigger the induction of Foxp3 expression in T cells. Hence, it is tempting to speculate that the characteristic self-limiting feature of relapsing-remitting MS, the most inflammatory form of MS, may somehow relate to this mechanism. Furthermore, the possibility of genetic and functional defects potentially involved in the IFN-γ signaling events of transcription factor Foxp3 in MS may affect this particular function of IFN-γ.

Induction of Adaptive Tregs by Copolymer-I

Copolymer-I (COP-I), also called glatiramer acetate (GA), is a random polymer of four amino acids (Glutamic acid, Lysine, Alanine and Tyrosine) enriched in myelin basic protein (MBP) and has proven treatment efficacy for multiple sclerosis (MS) [4–6]. Although COP-I has been approved as a treatment option for MS since 1995, the mechanism of action of COP-I in relationship to its treatment efficacy is unclear [7]. COP-I has been found to activate T cells in both human and animal experimental systems [8–10]. The relevance of this unique property of COP-I to its efficacy in MS is thought to partially involve a biased induction of Th2 immunity. There is evidence suggesting that human T cell lines generated by COP-I initially secrete both Th1 (IL-2 and IFN-γ) and Th2 (IL-4, IL-6, and IL-10) cytokines in response to COP-I [11]. However, repeated in vitro stimulation of these T cell lines progressively shifts cytokine production toward the Th2 response [12, 13]. Similarly, repeated COP-I injections may lead to deviation from Th1 to Th2 response in patients with MS [14, 15]. Studies reported by other investigators, including our own, however, indicate that the effect of COP-I on the induction of T cell activation is not entirely selective for Th2 cells and that it consistently activates the production of both Th1 and Th2 cytokines in MS [16]. The exact mechanism of action of COP-I has always been elusive.

Our recent studies indicate, however, that COP-I exerts its regulatory activity, at least in part, through the induction of adaptive Tregs [17]. COP-I was found to induce in vitro conversion of unprimed human CD4+CD25– T cells to CD4+CD25+ T cells that acquired high expression of Foxp3 and the inhibitory functions. The resulting CD4+CD25+ regulatory T cells appeared to expand in response to COP-I from the CD4+ T cell pool of both CD45RA and CD45RO phenotypes. Moreover, COP-I specific T cell lines generated by repeated stimulation cycles exhibited very high expression levels of Foxp3 and high inhibition rate when compared to single stimulation of CD4+CD25– T cells with COP-I and control T cell lines. It has been confirmed that the expression of Foxp3 correlates with the inhibitory function in CD4+ T cells. Furthermore, standard treatment with COP-I in multiple sclerosis (MS) patients and administration of COP-I in mice resulted in significant increase of Foxp3 expression in T cells, providing the compelling in vivo evidence for the role of COP-I in the induction of Foxp3. Consistent with what is discussed in the preceding section is the finding that the unique property of COP-I in the induction of CD4+CD25+ regulatory T cells is mediated through its ability to induce the production of IFN-γ. The role of IFN-γ in the induction of Foxp3 expression and CD4+CD25+ regulatory T cell response is supported by the following experimental evidence: (1) Blocking of IFN-γ (but not other cytokines also induced by COP-I) by specific antibody resulted in significant inhibition of the effect of COP-I. (2) Direct induction of Foxp3 expression by recombinant IFN-γ in CD4+ T cells. (3) COP-I failed to induce Foxp3 expression in T cells of IFN-γ knockout mice in both in vitro and in vivo settings. The role of IFN-γ appears different from that of TGF-β1 in the induction of Foxp3 and conversion of CD4+CD25– T cells to CD4+CD25+ regulatory T cells. In contrast to IFN-γ, when used alone TGF-β1 is insufficient to directly induce Foxp3 expression in T cells. TGF-β1 requires co-stimulation of T cells with an anti-CD3 antibody to induce Foxp3 expression. The observation that the induction of Foxp3 expression by COP-I is partially blocked by antibody to TGF-β1 supports the possibility that the involvement of TGF-β1 in the induction of Foxp3 expression requires a stimulatory signal provided by COP-I.

The unique property of COP-I in the induction of adaptive Tregs is highly significant in the understanding of the mechanism of action of COP-I in relation to its treatment efficacy in MS. COP-I may act through its ability to induce CD4+CD25+ regulatory T cell response to compensate for a functional deficit in this important regulatory mechanism in MS. This unique property of COP-I may contribute, at least in part, to the treatment efficacy of COP-I in MS. It is also likely to offer a reasonable explanation for or reconcile with some previously described regulatory functions of COP-I of unknown mechanism. That includes so-called "bystander" inhibitory effect of COP-I on T cell activation, which is a frequently reported phenomenon. The antigen non-specific inhibitory effect of aTregs induced by COP-I is consistent with the spectrum of inhibition induced by COP-I that is often not limited to one antigen (i.e. MBP) and includes various other myelin antigens. Furthermore, if the induction of CD4+CD25+ regulatory T cells is closely associated with the treatment effect of COP-I in MS, one potentially important aspect of the clinical significance of the study may involve the role of Foxp3 as a surrogate biomarker in measurement of treatment efficacy. Currently, the treatment efficacy of COP-I can only be measured roughly nine months after the treatment using the standard clinical and magnetic resonance imaging techniques.

Induction of Adaptive Tregs by T Cell Vaccination

Another important aspect of the induction of adaptive Tregs is related to the role of T cell vaccination. Immunization with inactivated autoreactive T cells (T cell vaccination) has been demonstrated to induce regulatory immune responses in autoimmune conditions of both human and experimental animal models, which are attributed to the suppression of disease activities [18–20]. At least two types of regulatory T cell responses have been identified, which include anti-idiotypic and so-called anti-ergotypic T cell responses [19–24]. Recent clinical trials in patients with multiple sclerosis (MS) have shown that anti-idiotypic T cells induced by T cell vaccination are represented by CD8+ cytolytic T cells capable of killing target T cells used for immunization through interaction with target T cell receptor components in the context of MHC class I molecules [21, 23, 25]. CD8+ anti-idiotypic T cell response is thought to contribute directly to depletion of circulating autoreactive T cells [21, 22, 26]. It was also evident in a number of studies that in addition to CD8+ anti-idiotypic T cells, T cell vaccination induces CD4+ regulatory T cell responses in both EAE and human MS [24, 27]. CD4+ regulatory T cell responses may represent an important immune regulatory component relevant to clinical effects of T cell vaccination [24, 26, 28]. However, the nature and functional properties of these CD4+ regulatory T cells are poorly understood. It has been suggested that CD4+ regulatory T cell responses may be induced through interaction with certain T cell markers loosely and collectively called "ergotopes" [24, 28, 29]. There have been some indications that T cell activation molecules such as IL-2 receptor and heat-shock protein (HSP) 60 may be among candidate ergotopes [28–31]. It has been a topic of great interest to characterize the nature and functional properties of human CD4+ regulatory T cell responses induced by T cell vaccination in clinical trials as the findings would have direct implications in the understanding of the potential therapeutic role of T cell vaccination in human autoimmune conditions.

We recently concluded, in a clinical trial with MS patients, that CD4+ regulatory T cells induced by T cell vaccination are largely comprised of T cell populations of two distinct patterns distinguishable by the expression of Foxp3 and cytokine profile. The majority of CD4+ regulatory T cells obtained, which represents the induced adaptive Tregs, expresses high levels of Foxp3 and produces both IL-10 and interferon-γ when compared to those characterized by low expression of Foxp3 and selective IL-10 production. Although CD4+ regulatory T cells described here all share the same functional feature in the inhibition of activated T cells, they may exert inhibitory function through different mechanisms. Regulatory T cells of the CD4+CD25+Foxp3+ pattern bear a remarkable resemblance to CD4+CD25+Foxp3+ regulatory T cells and may directly expand from the existing

CD4+CD25+ regulatory T cell pool or be induced from CD4+CD25− T cells by T cell vaccination. This is supported by the evidence that there was significantly elevated expression of Foxp3 and inhibitory function in CD4+CD25+ T cells derived from post-immunization PBMC compared to baseline specimens. The observations suggest that T cell vaccination induces up-regulation and expansion of the CD4+CD25+ regulatory T cell pool that is found deficient in MS patients [32]. Similar findings were seen in another T cell vaccination trial in rheumatoid arthritis patients in which selected synovial T cells were irradiated and used for immunization [33].

It was found that CD4+ regulatory T cells induced by T cell vaccination, mainly the CD4+CD25+Foxp3+ population, preferentially recognize and are activated by an epitope corresponding to residues 61–73 of the IL-2 receptor α-chain. This result is consistent with the observed reactivity of anti-ergotypic T cells with IL-2 receptor α-chain in rodents. Furthermore, this unique property of peptide 61–73 becomes more evident as it is able to elicit CD4+ regulatory T cell responses in post-immunization specimens. It is conceivable that CD4+ regulatory T cells recognizing an epitope in peptide 61–73 are sensitized by repeated immunization with autologous activated MBP-reactive T cells. This epitope may play an instrumental role in the induction of CD4+ regulatory T cell responses in T cell vaccination. Similar to CD4+ regulatory T cells induced by T cell vaccination, T cells induced in vitro by stimulation with peptide 61–73 of the IL-2 receptor α-chain largely share the same features of the CD4+CD25+Foxp3+ T cell population. Interestingly, a similar CD4+ regulatory T cell response can be induced by TCR peptides in humans and rodents. The relevance of the identified peptide of the IL-2 receptor α-chain and its effectiveness in the induction of CD4+ regulatory T cell responses, when used alone or in combination with selected TCR peptides, warrant further investigation to explore its potential therapeutic implications.

The detailed characterization of CD4+ regulatory T cell responses provides direct relevance to the understanding of their role in the immunological and perhaps therapeutic effects of T cell vaccination. It is likely that the CD8+ cytotoxic anti-idiotypic T cell response is mainly responsible for the marked and rather rapid depletion of circulating autoreactive T cells as seen in several independent MS clinical trials. The mechanism of this CD8+ anti-idiotypic T cell regulation is relatively well characterized and is found to involve components of target T cell receptor in the context of MHC class I molecules. However, it remains to be determined whether depletion of circulating autoreactive T cells alone would be sufficient to induce or account for clinical improvements seen in some MS patients treated with T cell vaccination. Further, it is shown in EAE experiments that only activated but not resting encephalitogenic T cells could induce full protection through T cell vaccination. This observation suggests that CD4+ regulatory T cells induced by ergotopes expressed on activated T cells play an important role in the induction of other regulatory functions. These regulatory functions may not directly result in rapid depletion of circulating autoreactive T cells but are critical to maintaining homeostasis of the immune system through cytokine regulation and CD4+CD25+ regulatory T cells. This relatively broad spectrum of regulation irrespective of antigen specificity, when combined with specific depletion of autoreactive T cells by CD8+ cytotoxic anti-idiotypic T cells, becomes highly effective. The combined regulatory mechanisms confer the full protection seen in EAE and may produce treatment effects for autoimmune conditions such as MS. The combined regulatory mechanisms are particularly important in MS in which aberrant autoimmunity is not simply limited to myelin autoreactive T cells but also involves a cytokine milieu seemingly skewed to a pro-inflammatory characteristic and a functional defect of CD4+CD25+ regulatory T cell response that normally keeps autoreactive and inflammatory T cells in check.

Therapeutic Considerations

The therapeutic potential of CD4+ regulatory T cells has long been a subject of discussion. While it has been known for some years that adoptive transfer of Tregs into autoimmune-prone animals can profoundly effect disease incidence and progression [35], Sakaguchi and colleagues [34] more recently demonstrated the induction of multiorgan autoimmunity in recipient animals after adoptive transfer of a T cell population depleted of CD4+CD25+ cells. Results of such studies have prompted the current discussions of the manner in which these regulatory T cells might be of therapeutic benefit in an array of human autoimmune diseases. However, there are a number of issues that must be considered and resolved before such a therapy may be adavanced to clinical settings.

Ideally, sufficient numbers of antigen specific Tregs will provide the most potent protection against autoimmune response in vivo. In the NOD model of autoimmune diabetes, islet antigen-specific Tregs were far more effective at preventing diabetes than polyclonal Tregs, as islet antigen-specific Tregs are at least 50-fold more potent than polyclonal Tregs [36]. The number of Tregs needed to achieve therapeutic effect in vivo is likely to vary depending upon the disease in question. Many dose titration experiments described in the literature relied on the use of adoptive transfer systems in which the number of pathogenic T cells is known. In most of these experiments, a large number of Tregs, up to 3:1 Tregs to T-effector cell ratio, were needed, unless antigen-specific Tregs were used. In intact animals and patients, it is nearly impossible to estimate the total numbers of pathogenic T cells in LN and tissues; thus, the therapeutic dosage must be determined empirically. While it has been reported that in the NOD mouse model, large numbers of Tregs ($>20 \times 10^6$/mouse) provided only partial protection from diabetes development, the degree of homeostatic proliferation of Tregs or even of the T effector population, is not clear. The inexact nature of such cellular determinations complicates any attempts at estimating therapeutic dosages. Indeed, it is doubtful that it would be possible to purify sufficient numbers of nTregs from one individal for transfer in a human system. For example, expansion of nTregs 100-fold would necessitate the sorting of cells from 200–250 ml of blood. Of additional concern is the anergic nature of nTregs. All these considerations place limits on the effective degree of expansion possible for nTregs in vitro [34].

In considering the disadvantages of naturally occurring Tregs in therapeutics, one appreciates that the adaptive Treg population offers unique advantages in preparation of sufficient numbers of Tregs for use in the management of autoimmune diseases such as multiple sclerosis and rheumatoid arthritis. In addition, antigen specificity may be considered when CD4+CD25− T cells are used as a source of cells for preparation of aTregs. After T cell vaccination was proved clinically effective in clinical trials on multiple sclerosis, Tregs induced by T cell vaccination were shown to recognize a T cell activation marker which is expressed abundantly on pathogenic T cells in autoimmune diseases. Such antigen specific aTregs will preferentially recognize and suppress the activated T cells involved in the immune response. However, T cell vaccination-induced Tregs are still lacking in fine specificity since they can also inhibit other immune response beyond those of the autoimmune response as long as activated T cells are presented. On the other hand, aTregs can overcome the difficulty in preparing a large therapeutic dose of Tregs. Since large numbers of CD4+CD25− naïve T cells are usually much easier to obtain than nTregs, one large scale conversion of CD4+CD25− T cells to aTregs would be expected to make sufficient numbers of Tregs to be used in vivo. For example, TGF-beta can stimulate the conversion of CD4+CD25− T cells to aTregs at a minimum 30% conversion rate. In order to prepare 20×10^6 aTregs, approximately 200 ml of blood will be needed considering the approximate 40% proportion of CD4+CD25− T cells present in PBMC. Cytokine driven conversion of CD4+CD25− T cells to aTregs offers another important approach for preparation of sufficient numbers of Tregs for use in therapy. As compared to T cell vaccination, this would be a rather simple way to make polyclonal or even antigen specific adaptive Tregs.

It is of great advantage that adaptive Tregs exhibit a similar phenotype to that of freshly isolated naturally occurring Tregs. This suggests the possibility that the efficacy of aTregs might not be lower than that of nTregs when used in vivo. Expanded Tregs remain Foxp3+ and CD25hi when compared with similarly expanded Th cells. In addition, other molecules implicated in Treg function, such as CTLA-4, remain highly expressed on the cell surface of expanded Tregs. In some settings, but not all, enhanced production of TGF-β and IL-10 was observed. However, it is also important to consider that not all CD4+CD25− T cells will be converted to adaptive Tregs in any attempted induction assay. The remaining portion of unconverted cells could be activated to become effective T cells which actively participate in the immune response. It is possible to purify induced aTregs with a combination of cell surface markers since the phenotype of both aTregs and nTregs are similar. Thus, it is imperative to sort out the adaptive T cells for in vivo use from the heterogeneous population in the standard aTreg preparatory protocol.

Conclusion

In summary, adaptive Tregs are capable of induction from CD4+CD25− naïve T cells when TCR signaling and a suitable cytokine milieu are present. They show enormous similarity to naturally occurring Tregs, thus allowing an alternative approach for in vitro preparation of Tregs. Most importantly, induced aTregs retain their suppressive function indistinguishable from that of nTregs. It was recently reported that, in addition to TGF-β, IFN-γ, COP-I and T cell vaccination were able to induce CD4+CD25+Foxp3+ aTregs from peripheral CD4+CD25– naïve T cells. As it is technically feasible to induce adaptive Tregs for therapeutic use in vivo, we suggest that this approach to autoimmune intervention may be a good starting point for clinical testing of Treg-based cell therapy.

References

1. Sakaguchi, S. Regulatory T cells: Meden Agan. Immunol Rev, 2006 Aug, 212, 5–7.
2. Sakaguchi, S; Ono, M; Setoguchi, R; Yagi, H; Hori, S; Fehervari, Z; Shimizu, J; Takahashi, T; Nomura, T. Foxp3+ CD25+ CD4+ natural regulatory T cells in dominant self-tolerance and autoimmune disease. Immunol Rev, 2006 Aug, 212, 8–27.
3. Wang, Z; Hong, J; Sun, W; Xu, G; Li, N; Chen, X; Liu, A; Xu, L; Sun, B; Zhang, JZ. Role of IFN-gamma in induction of Foxp3 and conversion of CD4+ CD25- T cells to CD4+ Tregs. J Clin Invest, 2006 Sep, 116(9), 2434–41.
4. Bornstein, MB; Miller, A; Slagle, S; Weitzman, M; Crystal, H; Drexler, E; Keilson, M; Merriam, A; Wassertheil-Smoller, S; Spada, V. A pilot trial of Cop 1 in exacerbating-remitting multiple sclerosis. N Engl J Med, 1987 Aug 13, 317(7), 408–14.
5. Gran, B; Tranquill, LR; Chen, M; Bielekova, B; Zhou, W; Dhib-Jalbut, S; Martin, R. Mechanisms of immunomodulation by glatiramer acetate. Neurology, 2000 Dec 12, 55(11), 1704–14.
6. Teitelbaum, D; Milo, R; Arnon, R; Sela, M. Synthetic copolymer 1 inhibits human T-cell lines specific for myelin basic protein. Proc Natl Acad Sci USA, 1992 Jan 1, 89(1), 137–41.
7. Zhang, J; Hutton, G. Role of magnetic resonance imaging and immunotherapy in treating multiple sclerosis. Annu Rev Med, 2005, 56, 237–302.
8. Miller, A; Shapiro, S; Gershtein, R; Kinarty, A; Rawashdeh, H; Honigman, S; Lahat, N. Treatment of multiple sclerosis with copolymer-1 (Copaxone): implicating mechanisms of Th1 to Th2/Th3 immune-deviation. J Neuroimmunol, 1998 Dec 1, 92(1–2), 113–21.
9. Dhib-Jalbut, S; Chen, M; Said, A; Zhan, M; Johnson, KP; Martin, R. Glatiramer acetate-reactive peripheral blood mononuclear cells respond to multiple myelin antigens with a Th2-biased phenotype. J Neuroimmunol, 2003 Jul, 140(1–2), 163–71.
10. Chen, M; Gran, B; Costello, K; Johnson, K; Martin, R; Dhib-Jalbut, S. Glatiramer acetate induces a Th2-biased response and crossreactivity with myelin basic protein in patients with MS. Mult Scler, 2001 Aug, 7(4), 209–19.
11. Aharoni, R; Teitelbaum, D; Sela, M; Arnon, R. Bystander suppression of experimental autoimmune encephalomyelitis by T cell lines and clones of the Th2 type induced by copolymer 1. J Neuroimmunol, 1998 Nov 2, 91(1–2), 135–46.
12. Duda, PW; Schmied, MC; Cook, SL; Krieger, JI; Hafler, DA. Glatiramer acetate (Copaxone) induces degenerate, Th2-polarized immune responses in patients with multiple sclerosis. J Clin Invest, 2000 Apr, 105(7), 967–76.
13. Neuhaus, O; Farina, C; Yassouridis, A; Wiendl, H; Then Bergh, F; Dose, T; Wekerle, H; Hohlfeld, R. Multiple sclerosis: comparison of copolymer-1- reactive T cell lines from treated and untreated subjects reveals cytokine shift from T helper 1 to T helper 2 cells. Proc Natl Acad Sci USA, 2000 Jun 20, 97(13), 7452–7.
14. Musette, P; Benveniste, O; Lim, A; Bequet, D; Kourilsky, P; Dormont, D; Gachelin, G. The pattern of production of cytokine mRNAs is markedly altered at the onset of multiple sclerosis. Res Immunol, 1996 Sep, 147(7), 435–41.

15. Brod, SA; Nelson, LD; Khan, M; Wolinsky, JS. Increased in vitro induced CD4+ and CD8+ T cell IFN-gamma and CD4+ T cell IL-10 production in stable relapsing multiple sclerosis. Int J Neurosci, 1997 Aug, 90(3–4), 187–202.
16. Ziemssen, T; Kumpfel, T; Klinkert, WE; Neuhaus, O; Hohlfeld, R. Glatiramer acetate-specific T-helper 1- and 2-type cell lines produce BDNF: implications for multiple sclerosis therapy. Brain-derived neurotrophic factor. Brain, 2002 Nov, 125(Pt 11), 2381–91.
17. Hong, J; Li, N; Zhang, X; Zheng, B; Zhang, JZ. Induction of CD4+CD25+ regulatory T cells by copolymer-I through activation of transcription factor Foxp3. Proc Natl Acad Sci U S A, 2005 May 3, 102(18), 6449–54.
18. Ben-Nun, A; Wekerle, H; Cohen, IR. Vaccination against autoimmune encephalomyelitis with T-lymphocyte line cells reactive against myelin basic protein. Nature, 1981 Jul 2, 292(5818), 60–1.
19. Lider, O; Beraud, E; Reshef, T; Friedman, A; Cohen, IR. Vaccination against experimental autoimmune encephalomyelitis using a subencephalitogenic dose of autoimmune effector T cells. (2). Induction of a protective anti-idiotypic response. J Autoimmun, 1989 Feb, 2(1), 87–99.
20. Zhang, J; Medaer, R; Stinissen, P; Hafler, D; Raus, J. MHC-restricted depletion of human myelin basic protein-reactive T cells by T cell vaccination. Science, 1993 Sep 10, 261(5127), 1451–4.
21. Zhang, J; Vandevyver, C; Stinissen, P; Raus, J. In vivo clonotypic regulation of human myelin basic protein-reactive T cells by T cell vaccination. J Immunol, 1995 Dec 15, 155(12), 5868–77.
22. Correale, J; Lund, B; McMillan, M; Ko, DY; McCarthy, K; Weiner, LP. cell vaccination in secondary progressive multiple sclerosis. J Neuroimmunol, 2000 Jul 24, 107(2), 130–9.
23. Zang, YC; Hong, J; Rivera, VM; Killian, J; Zhang, JZ. Preferential recognition of TCR hypervariable regions by human anti-idiotypic T cells induced by T cell vaccination. J Immunol, 2000 Apr 15, 164(8), 4011–7.
24. Lohse, AW; Mor, F; Karin, N; Cohen, IR. Control of experimental autoimmune encephalomyelitis by T cells responding to activated T cells. Science, 1989 May 19, 244(4906), 820–2.
25. Zang, YC; Hong, J; Rivera, VM; Killian, J; Zhang, JZ. Human anti-idiotypic T cells induced by TCR peptides corresponding to a common CDR3 sequence motif in myelin basic protein-reactive T cells. Int Immunol, 2003 Sep, 15(9), 1073–80.
26. Zhang, J. T-cell vaccination for autoimmune diseases: immunologic lessons and clinical experience in multiple sclerosis. Expert Rev Vaccines, 2002 Oct, 1(3), 285–92.
27. Zang, YC; Hong, J; Tejada-Simon, MV; Li, S; Rivera, VM; Killian, JM; Zhang, JZ. Th2 immune regulation induced by T cell vaccination in patients with multiple sclerosis. Eur J Immunol, 2000 Mar, 30(3), 908–13.
28. Mimran, A; Mor, F; Carmi, P; Quintana, FJ; Rotter, V; Cohen, IR. DNA vaccination with CD25 protects rats from adjuvant arthritis and induces an antiergotypic response. J Clin Invest, 2004 Mar, 113(6), 924–32.
29. Cohen, IR; Quintana, FJ; Mimran, A. Tregs in T cell vaccination: exploring the regulation of regulation. J Clin Invest, 2004 Nov, 114(9), 1227–32.
30. Mor, F; Reizis, B; Cohen, IR; Steinman, L. IL-2 and TNF receptors as targets of regulatory T-T interactions: isolation and characterization of cytokine receptor-reactive T cell lines in the Lewis rat. J Immunol, 1996 Dec 1, 157(11), 4855–61.
31. Quintana, FJ; Carmi, P; Mor, F; Cohen, IR. DNA fragments of the human 60-kDa heat shock protein (HSP60) vaccinate against adjuvant arthritis: identification of a regulatory HSP60 peptide. J Immunol, 2003 Oct 1, 171(7), 3533–41.
32. Viglietta, V; Baecher-Allan, C; Weiner, HL; Hafler, DA. Loss of functional suppression by CD4+CD25+ regulatory T cells in patients with multiple sclerosis. J Exp Med, 2004 Apr 5, 199(7), 971–9.
33. Chen, G; Li, N; Zang, YC; Zhang, D; He, D; Feng, G; Ni, L; Xu, R; Wang, L; Shen, B; Zhang, JZ. Vaccination with selected synovial T cells in rheumatoid arthritis. Arthritis Rheum, 2007 Feb, 56(2), 453–63.

34. Tang, Q; Bluestone, JA. Regulatory T-cell physiology and application to treat autoimmunity. Immunol Rev, 2006 Aug, 212, 217–37.
35. Mason, D; Powrie, F. Control of immune pathology by regulatory T cells. Curr Opin Immunol, 1998 Dec, 10(6), 649–55.
36. Turley, S; Poirot, L; Hattori, M; Benoist, C; Mathis, D. Physiological beta cell death triggers priming of self-reactive T cells by dendritic cells in a type-1 diabetes model. J Exp Med, 2003 Nov 17, 198(10), 1527–37.

Chapter 16
Regulatory T Cells in Transplantation

Kathryn J Wood, Andrew Bushell, Manuela Carvalho-Gaspar, Gang Feng, Ross Francis, Nick Jones, Elaine Long, Shiqiao Luo, Ian Lyons, Satish Nadig, Birgit Sawitzki, Gregor Warnecke, Bin Wei, and Joanna Więckiewicz

Abstract Clinical success in treating transplant rejection to date has been achieved primarily through the development increasingly potent immunosuppressive drugs to inhibit immune responses. These approaches require life-long treatment and suppress the entire immune system non-specifically exposing the patient to increased risks of cancer and infection. Even then long-term graft survival is not guaranteed. Evidence that regulatory T cells can control rejection and facilitate the development of specific unresponsiveness to alloantigens in vivo has been accumulating over many years. Understanding how the immune system is controlled by regulatory T cells when it responds to alloantigen and promoting the development of cells that can regulate immune responses holds the key to the development of more selective therapies that target only destructive immune responses while leaving the beneficial protective functions of the immune system intact.

Introduction

More selective therapies that target only pathogenic/destructive immune responses without adversely affecting the beneficial protective functions of the immune system that facilitate immunological unresponsiveness to defined sets of alloantigens remain a major unmet need in clinical transplantation.

Transplantation of an organ, tissue or cells triggers a dialogue between the innate and adaptive immune systems resulting in a cascade of mechanisms that can both initiate and direct tissue destruction, but that can also set in place mechanisms that can regulate immune responses that would be devastating if they remained

---

K.J. Wood

Transplantation Research Immunology Group, Nuffield Department of Surgery, John Radcliffe Hospital, University of Oxford, Oxford OX3 9DU, UK
e-mail: kathryn.wood@nds.ox.ac.uk

Financial Support: The work from the authors' own laboratory described in this review was supported by grants from The Wellcome Trust, BBSRC, British Heart Foundation, Kidney Research UK, Garfield Weston Trust and European Union through the RISET Integrated Project.

uncontrolled. This dialogue is ongoing throughout the post-transplant period. T cells play a pivotal role in the destructive, memory and regulatory/suppressor arms of the response. Indeed, both regulatory and memory populations of T cells must be maintained to ensure that the immune system is controlled throughout an individual's lifetime even when the system is perturbed such as by transplantation and treatment with immunosuppressive drugs. Insights into the molecular events that enable T cells with regulatory or suppressor activity (Treg) to develop and function hold the key to developing new approaches for manipulating or controlling the aggressive unwanted immune responses in vivo.

Regulatory T Cells in Transplantation

T cell mediated immunoregulation or suppression is one of the key mechanisms responsible for maintaining specific immunological unresponsiveness or tolerance in vivo and for controlling T cell homeostasis. This mechanism can be used to control not only immune responses to self antigens thereby preventing autoimmune disease, but also responses to non-self molecules that are introduced into the host. It is interesting to note that the existence of leukocytes able to suppress alloantigen-specific immune responses was first described by Billingham, Brent and Medawar over 50 years ago [1,2]. Thus the concept of suppression or active regulation of the immune response in the setting of transplantation is by no means a new idea. What is new is the ability to characterise the populations responsible for this phenomenon in detail at both a cellular and molecular level. This has given new impetus to studying immunoregulation in both experimental models as well as importantly in defined clinical settings such as cell and organ transplantation.

Characterisation of Regulatory T Cells that can Control Rejection

$CD25^+CD4^+foxp3^+$Treg are one of the key populations responsible. Thymus derived or naturally occurring, $CD25^+CD4^+foxp3^+$Treg (nTreg) are generated as a distinct population in the thymus [3–6]. Importantly for therapeutic strategies, $CD25^+CD4^+foxp3^+$Treg that are phenotypically and functionally similar can also be generated under certain conditions after antigen exposure in vivo (iTreg) [7–11] and ex vivo ([12]; Feng, Bushell and Wood, unpublished data). It is likely that the ability of extracellular signals to drive the stable expression of foxp3, and differentiation of $CD4^+$ cells towards a Treg phenotype following alloantigen exposure, may vary between individuals depending on recipient factors such as age, sex, genetic background and epigenetic modification [13]. Additionally, recent studies have shown that expression of foxp3 in mature Treg is required to maintain the transcriptional program established during their commitment [14] and that foxp3 regulates T cell commitment and function in a dose-dependent manner [15], suggesting that a sustained minimum level of expression would be required. Moreover, epigenetic modification of foxp3 in mice has been shown to be required for stable expression and thus stable functional activity of regulatory T cells [16]. In the setting of transplantation, the immune system is constantly exposed to donor alloantigens while graft function is maintained, this may enable mRNA expression of foxp3 to be stabilised and the functional activity of Treg to be sustained in vivo. However, strategies designed to generate T cells with regulatory activity ex vivo may need to take stabilisation of foxp3 mRNA expression into consideration as in nTreg in mice, a conserved region upstream of exon 1 has been found to control transpcriptional activity. Complete demethylation of the CpG motifs in this reguion as well as histone modifications were found in nTreg but not in Treg generated ex vivo in the presence of TGFβ where the demethylation was found to be incomplete [16].

Alloantigen induced Treg have been shown to be able to prevent acute as well as delayed graft rejection. In mouse models our own laboratory has demonstrated that Treg induced in response to alloantigen in vivo can prevent the rejection of heart allografts in naïve mice with an intact immune repertoire [17] as well as in more subtle adoptive transfer models using immunodificient hosts [7,9]. We have also demonstrated recently that these same populations of alloantigen induced Treg can prevent the development of transplant arteriosclerosis [18]. Transplant arteriosclerosis seems to be a feature of allograft rejection especially hard to control by experimental tolerance induction strategies [19] and therefore it is interesting that in an adoptive transfer system we were able to show the potential of Treg to prevent the development of transplant arteriosclerosis suggesting that in vivo cellular therapy may be a useful approach to explore.

Inflammation and Treg Function In Vivo

The ability and effectiveness of Treg to control an immune response in vivo is linked to the conditions that prevail at the time they are generated and/or function, such as the presence or absence of inflammation [20], their number and potency relative to that of potential effector cells, naïve as well as memory [21–23], and their ability to migrate and function in the relevant microenvironment in vivo [24]. Importantly, Treg generation has been shown to be prevented by inflammation, a situation that occurs commonly in vivo after transplantation of cells and organs, leading instead to the generation of $T_H 17$ cells that have been shown in other settings to mediate tissue damage [25,26]. Once established Treg infused into transplant recipients at the time of transplantation can migrate and function to prevent graft rejection as many studies have shown (e.g. [7,9]). However, the impact of inflammation on Treg generation has not been studied in detail in vivo in transplantation models. The degree of inflammation that develops at the time of transplantation may vary considerably depending on both donor and recipient factors. Organs from deceased brain dead donors are very different in molecular characteristics to organs from living donors [27,28]. Similarly organs from older donors may also trigger a different cascade of responses after transplantation compared to organs from younger donors [29]. One can speculate that the inflammation induced by the transplant procedure may delay the generation of alloantigen induced Treg [30], however clearly their induction is not prevented altogether as evidenced by the development of Treg in recipients with long term surviving allografts both in experimental rodent models and humans [31–33].

Location of Treg In Vivo

The draining lymphoid tissue is the primary, initial site of interaction between Treg and naïve T cells following transplantation ([24,34,35]; Carvalho-Gaspar et al., unpublished data). However, there are also data suggesting that regulation can also manifest itself within the allograft itself. In the setting of clinical transplantation, in both liver and kidney transplant recipients, evidence has been obtained that T cells with the phenotypic characteristics of regulatory cells are present in both the peripheral blood and the graft [32,36].

To explore whether the site of regulation changes with time after transplantaiton, we have investigated the impact of Treg generated by exposure to alloantigen in vivo on naïve donor-reactive $CD8^+$ T cells as they respond to alloantigen after transplantation. One of the mechanisms identified by which alloantigen reactive Treg initially suppress rejection in this model was through the attenuation of the generation of heavily divided $CD8^+$ effector T cells in the draining lymph tissue, confirming the published findings (Carvalho-Gaspar et al., unpublished data). Whilst one possibility to explain these data is that Treg kill activated $CD8^+$ T cells we favor the hypothesis that Treg suppress the activation/priming of $CD8^+$ T cells. Certainly this would be consistent with many other studies that have reported that Treg prevent T cell priming by acting on the APC either by direct killing [37] or by the modification of their activity by down-regulation of co-stimulatory molecules such as CD80 and CD86 [38,39]. Real-time imaging in a non-obese diabetic mouse model revealed that Treg are able to stably interact and modify antigen-bearing DC before antigen-specific T cell activation can occur [24]. Treg were shown to home to the T cell area in the dLN where they accumulated in the T-B cell zone, in a similar way to that previously described for "effector" T cells, but that in this situation, the Treg interacted with antigen-bearing DCs before they had the chance to interact with antigen-specific naive T cells thereby preventing T cell priming [24]. A similar suppressive effect of Treg has also been described in a transgenic mouse model of experimental autoimmune encephalomyelitis [40].

Interestingly, the inhibition of $CD8^+$ T cell priming that we observed in this transplant model was found to be transient, and despite the continued presence of alloantigen reactive Treg in the dLN, by 15 days after transplantation $CD8^+$ T cells had begun to proliferate, gained the ability to migrate to the graft and induced an intragraft Tc1 type inflammatory response. However, despite this evidence for activation of T cells with the ability to initiate rejection, the skin grafts were not rejected. Thus, in addition to these data suggest that other regulatory mechanisms in addition to suppression of $CD8^+$ T cell priming in the dLN exist. It is also important to note that it appears that both inhibition of T cell priming in the lymphoid tissue as well as the suppression of rejection locally in the graft are required for long-term skin graft acceptance; the transfer of Treg 10 or 15 days post transplantation was unable to prevent BM3 T cell mediated rejection.

There are a number of reports that have suggested that Treg may control CD8$^+$ effector cell function following allo [41] or tumor antigen challenge by via the production of TGF-β [35]. Even though we have shown a similar pro-inflammatory gene expression profile induced in the graft by BM3 T cells (primed in the presence or absence of Treg), we cannot fully exclude that in part Treg prevent graft rejection via the inhibition of effector function.

Interestingly, before destruction of the graft is complete, we found that a cohort of graft infiltrating CD8$^+$ effector T cells traffic out of the graft and form the peripheral memory T cell pool, resulting in a significant number of CD44$^+$ CD8$^+$ T cells being present in all lymphoid organs by 25 days after transplantation [42]. Importantly, the presence of Treg was able to prevent the generation of a peripheral memory CD8$^+$ T cell pool, despite the fact that CD8$^+$ T cells were primed and able to migrate to the graft. This finding suggests that intra-graft deletion of CD8$^+$ effector T cells mediated by intra-graft Treg played a role in the maintenance of graft survival in this model, a conclusion consistent with other reports suggesting that both mouse and human Treg are able to kill target cells (either CD4 and CD8 T cells, monocytes, DCs and B cells) through a perforin/granzyme dependent pathway [43,44].

The location of Treg in vivo is likely to be exceptionally important for the effective inhibition and control of aggressive antigen-reactive cells. Treg have been found not only in the lymphoid tissue of transplanted mice, but also at later time-points in the graft site itself, suggesting that T cell suppression of graft rejection operates beyond secondary lymphoid tissue. What remains unclear is why the location of regulation changes with time following transplantation resulting in a switch from control of T cell priming to control of effector mediated skin graft rejection within the graft itself. One possibility is that in the lymphoid tissue draining the graft site the effector T cell response becomes more potent with time and simply overrides the suppression imposed by peripheral Treg. Alternatively, it may be that T cells initially recognize alloantigen on migrating donor APCs but that with time they are primed as a result of alloantigen recognition on vascular endothelium [45]. As mouse vascular endothelial cells are not thought to express MHC class II [46] Treg may be no longer able to suppress T cell priming resulting in T cell proliferation and infiltration of the skin graft. Clearly if this were the case then the site and mechanisms of regulation may be different between control of alloreactive CD8$^+$ T cells responding to alloantigen by the direct pathway of allorecognition and indirect and direct CD4$^+$ T cell responses where recognition of alloantigen may be limited to recognition of alloantigen on APC. However, our working hypothesis is that Treg may become activated in the dLN prior to gaining the ability to home to the allograft, leaving subsequent priming of CD8$^+$ T cells to continue in the periphery. Thereafter, CD8$^+$ T cell priming in the dLN is initiated following the migration of Treg to the graft whereupon the control of CD8$^+$ T cell responses switches to within the graft.

Mechanisms of Suppression Used by Treg to Control Rejection

Treg can control immune responses in vitro and in vivo through bystander mechanisms after reactivation [47,48]. Our experimental data show that Treg capable of preventing rejection by bystander regulation can be generated using model protein antigens completely unrelated to the transplant itself in vivo [48] and ex vivo (Feng, Wood and Bushell, unpublished data) and suggest the possibility of developing clinical protocols in which bystander Treg are generated using defined, quality controlled non-cellular antigens without the pathogen risks and donor identity complications of elective alloantigen pre-treatment. One potential safety concern of bystander regulation in vivo is the potential to cause generalised immunosuppression. Interestingly, in humans, antigen induced $CD25^+CD4^+Foxp3^+$Treg generated from rapidly dividing, highly differentiated $CD4^+$ memory T cells (T_M) given the appropriate exogenous cues [49] are short-lived. To harness the full potential of Treg to control immune responses in vivo understanding where, when, how and for how long Treg specific for a known antigen interact with other leukocytes and mediate bystander regulation is critically important.

Following reactivation in vivo Treg produce IFNγ rapidly and transiently [50] and upregulate expression of a receptor for hyaluronic acid (HA) mediated migration (RHAMM) which like CD44 can bind to the extracellular matrix protein HA and α-mannosidase, suggesting that their ability to migrate and interact with other leukocytes is modified [51]. Moreover, in vivo, we have preliminary evidence for increased expression of indoleamine 2, 3 dioxygenase (IDO); an enzyme that catalyses the initial and rate limiting step of the kyneurinine pathway of tryptophan catabolism known to be subject to transcriptional regulation by IFNγ [52,53] dependent on STAT-1 signalling [54]. IDO competent DCs may control T cells locally through tryptophan depletion and over expression of IDO inhibits alloimmune responses [e.g. 55–57].

The role of IFN-γ in cellular immunity is somewhat paradoxical in that although it is usually considered to be a pro-inflammatory effector cytokine, increasing evidence suggests that it plays a non-redundant immunoregulatory role, including our own data discussed above. Experimental autoimmune encephalomyelitis (EAE) and collagen-induced arthritis (CIA) have been historically associated with IFN-γ producing Th1 dominant responses [20], however, accumulating evidence suggests that IFN-γ plays a protective role [58]. For example, mice deficient in IFN-γ or IFN-γ receptor develop EAE at an accelerated rate [59] and similarly, deficiency in IFN-γ or IFN-γ receptor results in more severe collagen or antigen induced arthritis in susceptible mouse strains and the development of arthritis in otherwise non-susceptible strains [60].

A paradoxical role for IFN-γ in also seen in organ transplantation [61]. On the one hand, allograft rejection is a process frequently associated with a dominant Th1 IFN-γ response whereas the absence of intra-graft IFN-γ often correlates with long term graft survival [62,63]. This association of IFN-γ and rejection is partially explained by the fact that IFN-γ can induce macrophage activation, mononuclear infiltration, oedema, and tissue necrosis. Furthermore, IFN-γ promotes cytotoxic T lymphocyte (CTL) and delayed type hypersensitivity (DTH) responses which are considered to be the principle effector mechanisms of allograft rejection. On the other hand, IFN-γ appears not to be essential for acute cellular rejection as both IFN-γ deficient and wild type mice reject cardiac allografts with similar kinetics [64] and at least one study has demonstrated that IFN-γ$^{-/-}$ recipients reject skin allografts more rapidly than their wild type littermates [65]. In fact, IFN-γ may be required for successful engraftment [62,65].

Although the classical view of IFN-γ is that it favours Th1 cell development, often dependent on IL-12 [66], it is becoming apparent that IFN-γ can also play an important regulatory role. For example, IFN-γ can limit the number of Th1 effector cells via inducible nitric oxide synthase (iNOS) [50], can inhibit the proliferation of IL-4 producing Th2 cells [67] and can suppress the development of the recently identified Th17 effector cell sub-set now known to play an important role in many autoimmune models [30,60]. In addition, IFN-γ also plays an important role in the maintenance of T cell homeostasis by inducing apoptosis-dependent activation-induced cell death to limit T cell expansion following antigen encounter [68].

CD152 (CTLA-4) has powerful immunomodulatory effects and can act as a negative regulator of T cell activation [69]. CTLA-4 is expressed by Treg and critical for their function in vivo (e.g. [9,70–74]). Engagement of CD80/86 on DC can induce IDO expression [75] and is a potent activator of LFA-1 clustering and adhesion [76]. CTLA-4 can modulate T cell motility and the threshold for T cell activation [77]. Regulator for cell adhesion and polarisation type 1 (Rap-1), an allosteric regulatory element that switches between inactive GDP-bound and active GTP-bound conformations, is transiently activated during TCR ligation and regulates integrin mediated leukocyte adhesion. Expression of an active GTP-bound mutant Rap-1 resulted in defective T cell priming in vivo due to the presence of an increased fraction of Treg, despite increased expression of LFA-1 [78]. In vivo, Treg have been shown to have prolonged interactions with dendritic cells in the draining lymph nodes (dLN) that preceded the inhibition of T cell activation [24] and reduced rates of killing by CTL [35]. The full range of functions mediated by CTLA-4 in relation to the function of Treg in vivo require further investigation.

Identifying Gene Expression Patterns Related to Long Term Graft Outcome and Treg Function

Identifying molecular markers whose expression either within the graft itself, fluids draining from the graft, such as urine from a kidney transplant or lavage fluid from a lung transplant, or ideally the peripheral blood correlates with either rejection or long term graft function may be one way of monitoring the success or failure of an immunosuppression minimisation strategy or tolerance induction therapy. One would predict that the profile of gene expression found in the peripheral blood of a transplant recipient would not be identical to that found at the same time in the graft itself or fluid draining the graft. Nevertheless, one might expect that the gene expression profile found at each site would correlate with the clinical status of the graft enabling for example, the onset of rejection to be predicted in advance of the graft sustaining any damage.

Many investigators have attempted to identify markers of immune function to help diagnose allograft injury, acute rejection or the onset of chronic rejection and to distinguish them from other disorders [79–84]. Genes expressed by cytotoxic T cells such as perforin and granzyme B have been shown to be up-regulated within the graft during rejection episodes [85–88] as well as in urine samples during acute and chronic rejection [89]. In the peripheral blood, perforin mRNA expression was found to predict acute rejection episodes [90]. More recently, microarray technology has been used to identify genes whose expression is increased in the graft during acute rejection of mouse hearts [91] and human renal allografts [82,92,93], and to predict the development of chronic renal allograft rejection [94]. Furthermore, Deng et al. have identified a set of genes whose expression analysis enables the non-invasive discrimination of rejection in cardiac allograft recipients [95]. Hoffmann et al. have studied the gene expression in biopsies from normal kidneys, transplanted kidney with stable graft function, allografts with subclinical rejection and clinical rejection. They identified T-bet, FasL and CD152 as good markers to distinguish between transplant recipients in the different categories [96]. Whether the markers identified by studies of this type to date can be used to monitor induction and maintenance of tolerance is as yet unknown as more work s required in different patient populations to validate the utility of the genes identified to predict the onset of rejection or a transplant that is functioning well and not at risk of rejection.

As part of an ongoing collaboration between Berlin and Oxford, we developed a strategy to enable us to define and validate reliable gene markers whose expression in the graft and in the peripheral blood had the potential be used to monitor the success or failure of novel tolerance inducing strategies or conventional immunosuppressive therapies. The expression profiles of the 2 genes was identified and confirmed as being associated with either rejection and/or permanent acceptance, tolerance associated gene -1 (TOAG-1) and alph-1,2-mannosidase, in 3 different transplant models in 2 species by performing a kinetic expression analysis using qRT-PCR [51].

TOAG-1 is identical with the filed sequence "Rattus norvegicus similar to hypothetical protein DKFZp313N0621, Acc.No. XM_34508". Although highly conserved among different species, the sequence shows no homology to any known gene and the function of the corresponding protein is therefore unknown. Preliminary data indicate that TOAG-1 is a mitochondrial protein regulating T cell apoptosis (Gube et al. unpublished observation). The second gene, alpha-1,2-mannosidase, is important for the N-glycosylation of membrane bound and secreted proteins. Inhibition of alpha-1,2-mannosidase during ConA and anti-CD3 mAb mediated T cell activation resulted in increased IL-2 production [97]. Our results demonstrate that the magnitude of an alloresponse is higher if alpha-1,2-mannosidase activity is inhibited in T cells. Surprisingly, T cells of beta1,6 N-acetylglucosaminyltransferase V (Mgat5) deficient mice, an enzyme which is also important for the N-glycosylation of proteins, display a reduced activation threshold due to an enhanced TCR clustering [98]. These mice develop kidney autoimmune diseases, show enhanced delayed type hypersensitivity responses and an increased susceptibility to experimental autoimmune encephalomyelitis [99]. Furthermore, Morgan et al. could demonstrate that N-acetylglucosaminyltransferase V (Mgat5)-mediated N-glycosylation negatively regulates Th1 cytokine production by T cells [100]. Thus N-glycosylation of T cell surface proteins may be important for the negative regulation of T cell activation and explain the high expression of alpha-1,2-mannosidase in long term surviving grafts. Thus N-glycosylation of T cell surface proteins appears to be important for the negative regulation of T cell activation. Therefore the high expression of alpha-1,2-mannosidase in graft infiltrating T of long term surviving grafts may be an important mechanism for the attenuation of alloreactive T cell responses.

Expression of TOAG-1 and alpha-1,2-mannosidase was reduced during rejection and high in long term surviving grafts in both mice and rats. In long term surviving kidney grafts no significant transient down-regulation of TOAG-1 and alpha-1,2-mannosidase expression could be detected. In contrast, in long term surviving heart allografts an early transient down-regulation of TOAG-1 and alpha-1,2-Mannosidase expression was observed. This early dramatically reduced TOAG-1 and alpha-1,2-mannosidase expression was associated with sustained transcription of CD3 and the T cell activation marker CD69 within the graft and with histological signs of graft vasculopathy. Furthermore, TOAG-1 and alpha-1,2-mannosidase expression correlated with graft function of long term surviving heart grafts. Furthermore, expression of alpha-1,2-Mannosidase and especially TOAG-1 was also regulated in an experimental autoimmune model. These results further support the importance of both genes for negative regulation of T cell activation.

Interestingly, expression of both TOAG-1 and alpha-1,2-mannosidase was down-regulated in the peripheral blood 3 to 5 days before acute rejection of mouse heart and rat kidney allografts. No significant decrease of TOAG-1 and alpha-1,2-Mannosidase transcription in peripheral blood samples of syngeneic recipients was observed. Thus, expression analysis of these two markers might have potential for detecting acute rejection episodes before clinical signs are apparent.

Recently, several other investigators have also studied the gene expression pattern associated with tolerance induction in different transplant models. A set of genes (TGF-ß2, ppENK, GM2a, GITR, IL-1R2) were identified by SAGE screening that are specifically up-regulated in regulatory T cells associated with tolerance of skin allografts [101,102]. Matsui et al identified two genes (H2-Ea, Frzb), which are highly expressed in long term surviving heart allografts [103]. In this model long term survival was induced by costimulatory blockade. Similar studies were performed by 2 other groups [104,105]. Unfortunately, none of the identified genes overlap with those identified here. In contrast to the studies mentioned above, the expression pattern of TOAG-1 and alpha-1,2-mannosidase identified in our own work was validated in heart, liver and kidney transplant models in mice and rats and in peripheral blood samples. Using different transplant models and tolerance induction protocols may explain the missing overlap. To investigate this further a European Research Network – RISET www.risetfp6.org – is undertaking a collaborative study to assess the expression of genes whose expression has been identified as associated with long term graft survival and tolerance induction in a variety of different transplant models and cell types, including regulatory T cells.

Treg, Homeostatic Proliferation and Immunological Memory

Normally, the overall size and composition of the naïve and memory T cell pool is normally tightly regulated by cytokine dependent homeostatic mechanisms [106], and is influenced by the number of Treg and age [107]. Inflammatory mediators produced at the time of transplantation and by different therapeutic interventions, most notably leukocyte depletion, can markedly influence these processes. Memory T cells reactive with alloantigens present in transplant recipients are spared [108,109] or generated despite treatment with leukocyte depleting agents and/or immunosuppressive drugs can have a negative impact on graft survival [42] and tolerance induction in experimental models [22,110].

Memory T cells that are reactive with alloantigens can either be generated following previous exposure to donor alloantigens or can develop following immune responses to a series of environmental antigens that results in the generation of a memory population that can cross-react with donor alloantigens. Memory T cells have been reported to prevent the induction of transplantation tolerance [22]. To compare the capacity of Treg to prevent rejection initiated either by naïve or memory T cells of identical TCR specificity, we have used 2 TCR transgenic mouse models to enable naïve and memory T cells with identical specificity for alloantigen to be generated. While this type of model cannot fully replicate the complexity of a polyclonal allogeneic response, it does allow for precise identification of graft reactive effector T cells thus greatly enabling the differentiation between qualitative and quantitative effects.

When the ability of Treg to prevent allograft rejection initiated by either TCR transgenic $CD4^+$ or $CD8^+$ memory T cells was investigated, Treg were found to be incapable of controlling rejection initiated by memory T cells when they were used at the same cell dose that was capable of preventing rejection initiated by the identical population of naïve T cells [23]. Thus one of the ways in which alloreactive memory T cells may form a barrier to tolerance induction is through a reduced susceptibility to suppression by alloreactive Treg. In this same study, we also found that very high "doses" of Tregs with the same specificity as the $CD4^+$ memory T cells were able to inhibit the proliferation of memory T cell proliferation and cytokine production. This raises the possibility that differences seen between naïve and memory T cells are quantitative and not qualitative. Thus extremely high numbers of Treg might be able to modulate memory T cell mediated graft rejection, but such numbers would not be possible to achieve easily either in experimental systems or physiologically.

Conclusions

New insights into the potential of regulatory T cells that can control the immune response to a transplant may enable the development of tools for monitoring the development of specific unresponsiveness to donor alloantigens as well as more selective therapies that target only pathogenic/destructive immune responses without adversely affecting the beneficial protective functions of the immune system.

References

1. Billingham, R.E., L. Brent, and P.B. Medawar. 1953. Actively acquired tolerance of foreign cells. *Nature* 172:603–606.
2. Billingham, R.E., L. Brent, and P.B. Medawar. 1956. Quantitative studies on tissue transplantation immunity. III. Actively acquired tolerance. *Philos trans R Soc Lond B* 239:357–412.
3. Sakaguchi, S., N. Sakaguchi, M. Asano, M. Itoh, and M. Toda. 1995. Immunologic self tolerance maintained by activated T cells expressing IL-2 receptor alpha chains (CD25). Breakdown of a single mechanism of self tolerance causes various autoimmune diseases. *J Immunol* 155:1151–1164.
4. Suri-Payer, E., A. Amar, A. Thornton, and E. Shevach. 1998. $CD4^+CD25^+$ T cells inhibit both the induction and effector function of autoreactive T cells and represent a unique lineage of immunoregulatory cells. *J Immunol* 160:1212–1218.
5. Baecher-Allan, C., J. Brown, G. Freeman, and D. Hafler. 2001. $CD4^+$ $CD25^+$ high regulatory cells in human peripheral blood. *J Immunol* 167:1245–1253.
6. Stephens, L., C. Mottet, D. Mason, and F. Powrie. 2001. Human CD4+CD25+ thymocytes and peripheral T cells have immune suppressive activity in vitro. *Eur J Immunol* 31:1247–1254.
7. Hara, M., C. Kingsley, M. Niimi, S. Read, S. Turvey, A. Bushell, P. Morris, F. Powrie, and K. Wood. 2001. IL-10 is required for regulatory T cells to mediate tolerance to alloantigens in vivo. *J Immunol* 166:3789–3796.
8. Thorstenson, K., and A. Khoruts. 2001. Generation of anergic and potentially immunoregulatory CD25+CD4+ T cells in vivo after induction of peripheral tolerance with intravenous oral antigen. *J Immunol* 167:188–195.
9. Kingsley, C.I., M. Karim, A.R. Bushell, and K.J. Wood. 2002. CD25+CD4+ regulatory T cells prevent graft rejection: CTLA-4- and IL-10-dependent immunoregulation of alloresponses. *J Immunol* 168:1080–1086.
10. Apostolou, I., and H. Von Boehmer. 2004. In vivo instruction of suppressor commitment in naive T cells. *J Exp Med* 199:1401–1408.
11. Cobbold, S.P., R. Castejon, E. Adams, D. Zelenika, L. Graca, S. Humm, and H. Waldmann. 2004. Induction of foxP3+ regulatory T cells in the periphery of T cell receptor transgenic mice tolerized to transplants. *J Immunol* 172:6003–6010.
12. Marangoni, F., S. Trifari, S. Scaramuzza, C. Panaroni, S. Martino, L.D. Notarangelo, Z. Baz, A. Metin, F. Cattaneo, A. Villa, A. Aiuti, M. Battaglia, M.-G. Roncarolo, and L. Dupre. 2007. WASP regulates suppressor activity of human and murine CD4+CD25+FOXP3+ natural regulatory T cells. *J Exp Med* 204:369–380.
13. Long, E., and K.J. Wood. 2007. Understanding FOXP3: progress towards achieving transplantation tolerance. *Transplantation* 84:459–461.
14. Williams, L., and A. Rudensky. 2007. Maintenance of the Foxp3-dependent developmental program in mature regulatory T cells requires continued expression of Foxp3. *Nat Immunol* 8:277–284.
15. Wan, Y., and R. Flavell. 2007. Regulatory T-cell functions are subverted and converted owing to attenuated Foxp3 expression. *Nature* 445:766–770.

16. Floess, S., J. Freyer, C. Siewert, U. Baron, S. Olek, J.K. Polansky, K. Schlawe, H.-D. Chang, T. Bopp, E. Schmitt, S. Klein-Hessling, E. Serfling, A. Hamann, and J. Huehn. 2007. Epigenetic control of the foxp3 locus in regulatory T cells. *PLOS Biol* 5:169–178
17. Bushell, A., P. Morris, and K. Wood. 1995. Transplantation tolerance induced by antigen pretreatment and depleting anti-CD4 antibody depends on CD4$^+$ T cell regulation during the induction phase of the response. *Eur J Immunol* 25:2643–2649.
18. Warnecke, G., A. Bushell, S. Nadig, and K.J. Wood. 2007. Regulation of transplant arteriosclerosis by CD25+CD4+ T cells generated to alloantigen in vivo. *Transplantation* 83:1459–1465.
19. Koshiba, T., T. Kitade, and B. Van Damme. 2003. Regulatory cell-mediated tolerance does not protect against chronic rejection. *Transplantation* 76:588–596.
20. Weaver, C., L. Harrington, P. Mangan, M. Gavrieli, and K. Murphy. 2006. Th17: an effector CD4 T cell lineage with regulatory T cell ties. *Immunity* 24:677–688.
21. Wells, A., X. Li, Y. Li, M. Walsh, X. Zheng, Z. Wu, G. Nunez, A. Tang, M. Sayegh, W. Hancock, T. Strom, and L. Turka. 1999. Requirement for T cell apoptosis in the induction of peripheral transplantation tolerance. *Nat Med* 5:1303–1307.
22. Adams, A., M.A. Williams, T. Jones, N. Shiasugi, M.M. Durham, S. Kacch, E. Wherry, T. Onami, J. Lanier, K. Kokko, T. Pearson, R. Ahmed, and C.P. Larsen. 2003. Heterologous immunity provides a potent barrier to transplantation tolerance. *J Clin Invest* 111:1887–1895.
23. Yang, J., M. Brook, M. Carvalho-Gaspar, J. Zhang, H. Ramon, M. Sayegh, K. Wood, L. Turka, and N. Jones. 2007. Allograft rejection mediated by memory T cells is resistant to regulation. *Proc Natl Acad Sci USA* 104:19954–19959.
24. Tang, Q., J. Adams, A. Tooley, M. Bi, B. Fife, P. Serra, P. Santamaria, R. Locksley, M. Krummel, and J. Bluestone. 2006. Visualizing regulatory T cell control of autoimmune responses in nonobese diabetic mice. *Nat Immunol* 7:83–92.
25. Veldhoen, M., R. Hocking, C. Atkins, R. Locksley, and B. Stockinger. 2006. TGFβ in the context of an inflammatory cytokine milieu supports de novo differentiation of IL-17-producing T cells. *Immunity* 24:179–189.
26. Bettelli, E., Y. Carrier, W. Gao, T. Korn, T. Strom, M. Oukka, H. Weiner, and V. Kuchroo. 2006. Reciprocal developmental pathways for the generation of pathogenic effector TH17 and regulatory T cells. *Nature* 441:235–238.
27. Pratschke, J., M.J. Wilhelm, I. Laskowski, M. Kusaka, F. Beato, S.G. Tullius, P. Neuhaus, W.W. Hancock, and N.L. Tilney. 2001. Influence of donor brain death on chronic rejection of renal transplants in rats. *J Am Soc Nephrol* 12:2474–2481.
28. Weiss, S., K. Kotsch, M. Francuski, A. Reutzel-Selke, M. Mantouvalou, R. Klemz, O. Kuecuek, S. Jonas, C. Wesslau, F. Ulrich, A. Pascher, H.D. Volk, S.G. Tullius, P. Neuhaus, and J. Pratschke. 2007. Brain death activates donor organs and is associated with a worse I/R injury after liver transplantation. *Am J Transplant* 7:1584–1593.
29. Pratschke, J., D. Paz, M. Wilhelm, I. Laskowski, G. Kofla, A. Vergopoulos, H. MacKenzie, S. Tullius, P. Neuhaus, W.W. Hancock, H.-D. Volk, and N. Tilney. 2004. Donor hypertension increases graft immunogenicity and intensifies chronic changes in long-surviving renal allografts. *Transplantation* 77:43–48.
30. Chen, Y., and K.J. Wood. 2007. Interleukin-23 and TH17 cells in transplantation immunity: does 23+17 equal rejection? *Transplantation* 84:1071–1074.
31. van Maurik, A., K.J. Wood, and N. Jones. 2002. Cutting edge:CD4(+)CD25(+) alloantigen-specific immunoregulatory cells that can prevent CD8(+) T cell-mediated graft rejection: implications for anti-CD154 immunotherapy. *J Immunol* 169:5401–5404.
32. Li, Y., T. Koshiba, A. Yoshizawa, Y. Yonekawa, K. Masuda, A. Ito, M. Ueda, T. Mori, H. Kawamoto, Y. Tanaka, S. Sakaguchi, N. Minato, K.J. Wood, and K. Tanaka. 2004. Analyses of peripheral blood mononuclear cells in operational tolerance after pediatric living donor liver transplantation. *Am J Transplant* 4:2118–2125.
33. Martinez-Llordella, M., I. Puig-Pey, G. Orlando, M. Ramoni, G. Tisone, A. Rimola, J. Lerut, D. Latinne, C. Margarit, I. Bilbao, S. Brouard, M. Hernandez-Fuentes, J.P. Soulillou, and

A. Sanchez-Fueyo. 2007. Multiparameter immune profiling of operational tolerance in liver transplantation. *Am J Transplant* 7:309–319.
34. Graca, L., S.P. Cobbold, and H. Waldmann. 2002. Identification of Regulatory T Cells in Tolerated Allografts. *J Exp Med* 195:1641–1646.
35. Mempel, T.R., M. Pittet, K. Khazaie, W. Weninger, R. Weissleder, H. Von Boehmer, and U.H. von Andrian. 2006. Regulatory T cells reversibly suppress cytotoxic T cell function independent of effector differentiation. *Immunity* 25:129–141.
36. Akl, A., N.D. Jones, N. Rogers, M.A. Bakr, A. Mostafa, E.L.M. El Shehawy, M.A. Ghoneim, and K.J. Wood. 2008. An investigation to assess the potential of CD25highCD4+ T cells to regulate responses to donor alloantigens in clinically stable renal transplant recipients. *Transpl Int* 21:65–73.
37. Frasca, L., C. Scotta, G. Lombardi, and E. Piccolella. 2002. Human anergic CD4+ T cells can act as suppressor cells by affecting autologous dendritic cell conditioning and survival. *J Immunol* 168:1060–1068.
38. Vendetti, S., J.-G. Chai, J. Dyson, E. Simpson, G. Lombardi, and R. Lechler. 2000. Anergic T cell sinhibit the antigen presenting funciton of dendritic cells. *J Immunol* 165:1175–1181.
39. Cederbom, L., H. Hall, and F. Ivars. 2000. CD4+CD25+ regulatory T cells down-regulate costimulatory molecules on antigen presenting cells. *Eur J Immunol* 30:1538–1543.
40. Tadokoro, C.E., G. Shakhar, S. Shen, Y. Ding, A.C. Lino, A. Maraver, J.J. Lafaille, and M.L. Dustin. 2006. Regulatory T cells inhibit stable contacts between CD4+ T cells and dendritic cells in vivo. *J Exp Med* 203:505–511.
41. Lin, C.-Y., L. Graca, S.P. Cobbold, and H. Waldmann. 2002. Dominant transplantation tolerance impairs CD8[+] T cell function but not expansion. *Nat Immunol* 3:1208–1213.
42. Jones, N.D., M. Carvalho-Gaspar, S. Luo, M.O. Brook, L. Martin, and K.J. Wood. 2006. Effector and memory CD8+ T cells can be generated in response to alloantigen independently of CD4+ T cell help. *J Immunol* 176:2316–2323.
43. Grossman, W., J. Verbsky, W. Barchet, M. Colonna, J. Atkinson, and T. Ley. 2004. Human T regulatory cells can use the perforin pathway to cause autologous target cell death. *Immunity* 21:589–601.
44. Gondek, D.C., L.-F. Lu, S.A. Quezada, S. Sakaguchi, and R.J. Noelle. 2005. Cutting edge: Contact-mediated suppression by CD4+CD25+ regulatory cells involves a granzyme B-dependent, perforin-independent mechanism. *J Immunol* 174:1783–1786.
45. Kreisel, D., A. Krupnick, A. Gelman, F. Engels, S. Popma, A. Krasinskas, K. Balsara, W. Szeto, L. Turka, and B. Rosengard. 2002. Non-hematopoietic allograft cells directly activate CD8+ T cells and trigger acute rejection: an alternative mechanism of allorecognition. *Nat Med* 8:233–239.
46. Choo, J., J. Seebach, V. Nickeleit, A. Shimizu, H. Lei, D. Sachs, and J. Madsen. 1997. Species differences in the expression of major histocompatibility complex class II antigens on coronary artery endothelium. *Transplantation* 64:1315–1322.
47. Thornton, A., and E.M. Shevach. 2000. Suppressor effector function of CD4+CD25+ immunoregulatory T cells is antigen non-specific. *J Immunol* 164:183–190.
48. Karim, M., G. Feng, K.J. Wood, and A. Bushell. 2005. CD25+CD4+ regulatory T cells generated by exposure to a model protein antigen prevent allograft rejection: antigen specific reactivation in vivo is critical for bystander regulation. *Blood* 105:4871–4877.
49. Vukmanovic-Stejic, M., Y. Zhang, J. Cook, J. Fletcher, A. McQuaid, J. Maters, M.H. Rustin, L. Taams, P. Beverley, D. Macallan, and A.N. Akbar. 2006. Human CD4+ CD25hi Foxp3+ regulatory T cells are derived by rapid turnover of memory populations in vivo. *J Clin Invest* 116:2423–2433.
50. Sawitzki, B., C.I. Kingsley, V. Oliveira, M. Karim, M. Herber, and K.J. Wood. 2005. Interferon gamma production by alloantigen reactive CD25[+]CD4[+] regulatory T cells is important for their regulatory function in vivo. *J Exp Med* 201:1925–1935.
51. Sawitzki, B., A. Bushell, U. Steger, N. Jones, K. Risch, A. Siepert, M. Lehmann, I. Schmitt-Knosalla, K. Vogt, I. Gebuhr, K. Wood, and H.D. Volk. 2007. Identification of gene

markers for the prediction of allograft rejection or permanent acceptance. *Am J Transplant* 7: 1091–1102.
52. Hassanain, H.C., SY, and S. Gupta. 1993. Differential regulation of human indoleamine 2,3-dioxygenase gene expression by interferons-γ and α:analysis of the regulatory region of the gene and identification of an interferon-γ inducible DNA binding factor. *J Biol Chem* 268:5077–5084.
53. Orabona, C., P. Puccetti, C. Vacca, S. Bicciato, A. Luchini, F. Fallorino, R. Bianchi, E. Velardi, K. Perrucio, A. Verlardi, V. Bronte, M. Fioretti, and U. Grohmann. 2006. Towards the identification of a tolerogenic signature of IDO competent dendritic cells. *Blood* 107:2846–2854.
54. Grohmann, U., C. Orabona, F. Fallorino, C. Vacca, F. Calcinaro, A. Falorni, P. Candeloro, M. Belladonna, R. Bianchi, M. Fioretti, and P. Puccetti. 2002. CTLA-4Ig regulates tryptophan catabolism in vivo. *Nat Immunol* 3:1097–1101.
55. Mellor, A., D. Keskin, T. Johnson, P. Chandler, and D. Munn. 2002. Cells expressing 2,3-dioxygenase inhibit T cell responses. *J Immunol* 168:3771–3776.
56. Terness, P., T. Bauer, L. Rose, C. Dufter, A. Watzlik, H. Simon, and G. Opelz. 2002. Inhibition of allogeneic T cell proliferation by indoleamine 2,3-dioxygenase-expressing dendritic cells: mediation of suppression by tryptophan metabolites. *J Exp Med* 196: 447–457.
57. Beutelspacher, S., R. Pillai, M. Watson, P. Tan, J. Tsang, M. McClure, A. George, and D. Larkin. 2006. Function of indole-amine 2,3-dioxygensae in corneal allograft rejection and prolongation of allograft survival by over-expression. *Eur J Immunol* 36:690–700.
58. Billiau, A., H. Heremans, F. Vandekerckhove, R. Dijkmans, H. Sobis, E. Meulepas, and H. Carton. 1988. Enhancement of experimental allergic encephalomyelitis in mice by antibodies against IFN-gamma. *J Immunol* 140:1506–1510.
59. Kelchtermans, H., B. De Klerck, T. Mitera, M. Van Balen, D. Bullens, A. Billiau, G. Leclercq, and P. Matthys. 2005. Defective CD4+CD25+ regulatory T cell functioning in collagen-induced arthritis: an important factor in pathogenesis, counter-regulated by endogenous IFN-gamma. *Arthritis Res Ther* 7:R402–R415.
60. Irmler, I.M., M. Gajda, and R. Brauer. 2007. Exacerbation of antigen-induced arthritis in IFN-{gamma}-deficient mice as a result of unrestricted IL-17 response. *J Immunol* 179:6228–6236.
61. Wood, K.J., and B. Sawitzki. 2006. Interferon γ: a crucial role in the function of induced regulatory T cells *in vivo*. *Trends in Immunology* 27:183–187.
62. Konieczny, B., Z. Dai, E. Elwood, S. Saleem, P. Linsley, F. Baddoura, C. Larsen, T. Pearson, and F. Lakkis. 1998. IFN-g is critical for long-term allograft survival indcued by blocking the CD28 and CD40 ligand T cell costimulation pathways. *J Immunol* 160:2059–2064.
63. Yu, X.-Z., M.H. Albert, P.J. Martin, and C. Anasetti. 2004. CD28 ligation induces transplantation tolerance by IFN-{gamma}'dependent depletion of T cells that recognize alloantigens. *J Clin Invest* 113:1624–1630.
64. Saleem, S., B. Konieczny, R. Lowry, F. Baddoura, and F. Lakkis. 1996. Acute rejection of vascularized heart allografts in the absence of IFNγ. *Transplantation* 62:1908–1911.
65. Markees, T., N. Phillips, E. Gordon, R. Noelle, L. Shults, J. MOrdes, D. Greiner, and A. Rossini. 1998. Long term survival of skin allografts induced by donor splenocytes and anti-CD154 antibody in thymectomised mice requires CD4[+] T cells, interferon-γ and CTLA-4. *J Clin Invest* 101:2446–2455.
66. O'Garra, A. 1998. Cytokines induce the development of functionally heterogenous T helper cell subsets. *Immunity* 8:275–283.
67. Gajewski, T.F., and F.W. Fitch. 1988. Anti-proliferative effect of IFN-gamma in immune regulation. I. IFN- gamma inhibits the proliferation of Th2 but not Th1 murine helper T lymphocyte clones. *J Immunol* 140:4245–4252.
68. Tewari, K., Y. Nakayama, and M. Suresh. 2007. Role of Direct Effects of IFN-{gamma} on T Cells in the Regulation of CD8 T Cell Homeostasis. *J Immunol* 179:2115–2125.

69. Thompson, C., and J. Allison. 1997. The emerging role of CTLA-4 as an immune attenuator. *Immunity* 7:445–450.
70. Walunas, T., D. Lenschow, C. Bakker, P. Linsley, G. Freeman, J. Green, C. Thompson, and J. Bluestone. 1994. CTLA-4 can function as a negative regulator of T cell activation. *Immunity* 1:405–413.
71. Waterhouse, P., J. Penninger, E. Timms, A. Wakeham, A. Shahinian, K. Lee, C. Thompson, H. Griesser, and T. Mak. 1995. Lymphoproliferative disorders with early lethality in mice deficient in CTLA-4. *Science* 270:985–988.
72. Takahashi, T., T. Tagami, S. Yamazaki, T. Uede, J. Shimuzu, N. Sakaguchi, T. Mak, and S. Sakaguchi. 2000. Immunologic self tolerance is maintained by CD25+CD4+ regulatory T cells constitutively expressing cytotoxic T lymphocyte associated antigen 4. *J Exp Med* 192:303–310.
73. Read, S., V. Malmstrom, and F. Powrie. 2000. Cytotoxic T lymphocyte associated antigen 4 plays an essential role in the function of CD25+CD4+ regulatory cells that control intestinal inflammation. *J Exp Med* 192:295–302.
74. Azimzadeh, A.M., S. Pfeiffer, G. Wu, C. Schroder, G.L. Zorn, 3rd, S.S. Kelishadi, E. Ozkaynak, M. Kehry, J.B. Atkinson, G.G. Miller, and R.N. Pierson, 3rd. 2006. Alloimmunity in primate heart recipients with CD154 blockade: evidence for alternative costimulation mechanisms. *Transplantation* 81:255–264.
75. Fallarino, F., U. Grohmann, K. Hwang, C. Orabona, C. Vacca, R. Bianchi, M. Belladonna, M. Fioretti, M. Alegre, and P. Puccetti. 2003. Modulation of tryptophan catabolism by regulatory T cells. *Nat Immunol* 4:1206–12012.
76. Schneider, H., E. Valk, S. da Rocha, B. Wei, and C. Rudd. 2005. CTLA-4 upregulation of lymphocyte function-associated antigen 1 adhesion and clustering as an alternative basis for co-receptor function. *Proc Natl Acad Sci USA* 102:12861–12866.
77. Schneider, H., J. Downey, A. Smith, B.H. Zinselmeyer, C. Rush, J.M. Brewer, B. Wei, N. Hogg, P. Garside, and C.E. Rudd. 2006. Reversal of the TCR Stop Signal by CTLA-4. *Science* 313:1972–1975.
78. Li, L., R.J. Greenwald, E.M. Lafuente, D. Tzachanis, A. Berezovskaya, G.J. Freeman, A.H. Sharpe, and V.A. Boussiotis. 2005. Rap1-GTP is a negative regulator of Th cell function and promotes the generation of CD4+CD103+ regulatory T cells in vivo. *J Immunol* 175:3133–3139.
79. Hernandez-Fuentes, M., A. Warrens, and R. Lechler. 2003. Immunological monitoring. *Immunol Rev* 196:247.
80. Soulillou, J.P. 2001. Immune monitoring for rejection of kidney transplants. *N Engl J Med* 344:1006.
81. Stordeur, P., L. Zhou, and M. Goldman. 2002. Analysis of spontaneous mRNA cytokine production in peripheral blood. *J Immunol Methods* 261:195.
82. Sarwal, M., M. Chua, and N. Kambham. 2003. Molecular heterogeneity in acute renal allograft rejection identified by DNA microarray profiling. *N Engl J Med* 349:125.
83. Brouard, S., E. Mansfield, C. Braud, L. Li, M. Giral, S.-c. Hsieh, D. Baeten, M. Zhang, J. Ashton-Chess, C. Braudeau, F. Hsieh, A. Dupont, A. Pallier, A. Moreau, S. Louis, C. Ruiz, O. Salvatierra, J.-P. Soulillou, and M. Sarwal. 2007. Identification of a peripheral blood transcriptional biomarker panel associated with operational renal allograft tolerance. *Proc Natl Acad Sci* 104:15448–15453.
84. Mueller, T.F., J. Reeve, G.S. Jhangri, M. Mengel, Z. Jacaj, L. Cairo, M. Obeidat, G. Todd, R. Moore, K.S. Famulski, J. Cruz, D. Wishart, C. Meng, B. Sis, K. Solez, B. Kaplan, and P.F. Halloran. 2008. The transcriptome of the implant biopsy identifies donor kidneys at increased risk of delayed graft function. *Am J Transplant* 8:78–85.
85. Nickel, P., J. Lacha, and S. Ode-Hakim. 2001. Cytotoxic efector molecule gene expression in acute renal allograft rejection: correlation with clinical outcome; histopathology and function of the allograft. *Transplantation* 72:1158.

86. Sabek, O., M. Dorak, M. Kotb, A.O. Gaber, and L. Gaber. 2002. Quantitative detection of T-cell activation markers by real-time PCR in renal transplant rejection and correlation with histopathologic evaluation. *Transplantation* 74:701.
87. Shulzhenko, N., A. Morgun, and X. Zheng. 2001. Intragraft activation of genes encoding cytotoxic T lymphocyte effector molecules precedes the histological evidence of rejection in human cardiac transplantation. *Transplantation* 72:1705.
88. Strehlau, J., M. Pavlakis, M. Lipman, W. Maslinski, M. Shaprio, and T. Strom. 1996. The intragraft gene activation of markers reflecting T-cell-activation and -cytotoxicity analyzed by quantitative RT-PCR in renal transplantation. *Clin Nephrol* 46:30.
89. Li, B., C. Hartono, and R. Ding. 2001. Noninvasive diagnosis of renal-allograft rejection by measurement of messenger RNA for perforin and granzyme B in urine. *N Engl J Med* 344:947.
90. Simon, T., G. Opelz, M. Wiesel, R. Ott, and C. Susal. 2003. Serial peripheral blood perforin and granzyme B gene expression measurements for prediction of acute rejection in kidney graft recipients. *Am J Transplant* 3:1121.
91. Saiura, A., C. Mataki, and T. Murakami. 2001. A comparison of gene expression in murine cardiac allografts and isografts by means DNA microarray analysis. *Transplantation* 72:320.
92. Akalin, E., R. Hendrix, and R. Polavarapu. 2001. Gene expression analysis in human renal allograft biopsy samples using high-density oligoarray technology. *Transplantation* 72:948.
93. Flechner, S., S. Kurian, and S. Head. 2004. Kidney transplant rejection and tissue injury by gene profiling of biopsies and peripheral blood lymphocytes. *Am J Transplant* 4:1475.
94. Scherer, A., A. Krause, J. Walker, A. Korn, D. Niese, and F. Raulf. 2003. Early prognosis of the development of renal chronic allograft rejection by gene expression profiling of human protocol biopsies. *Transplantation* 75:1323.
95. Deng, M., H.J. Eisen, and M. Mehra. 2006. Noninvasive discrimination of rejection in cardiac allograft recipients using gene expression profiling. *Am J Transplant* 6:150.
96. Hoffmann, S., D.A. Hale, and D. Kleiner. 2005. Functionally significant renal allograft rejection is defined by transcriptional criteria. *Am J Transplant* 5:573.
97. Kosuge, T., T. Tamura, H. Nariuchi, and S. Toyoshima. 2000. Effect of inhibitors of glycoprotein processing on cytokine secretion and production in anti-CD3 stimulated T cells. *Biol Pharm Bull* 23:1–5.
98. Demetriou, M., M. Granovsky, S. Quaggin, and J. Dennis. 2001. Negative regulation of T cell activation and autoimmunity by Mgat5 N-glycosylation. *Nature* 409:733–739.
99. Dennis, J., J. Pawling, P. Cheung, E. Partridge, and M. Demetriou. 2002. UDP-N-acetylglucosamine:alpha-6-D-mannoside beta1,6 N-acetylglucosaminyltransferase V (Mgat5) deficient mice. *Biochem Biophys Acta* 1573:414.
100. Morgan, R., G. Gao, J. Pawling, J. Dennis, M. Demetriou, and B. Li. 2004. N-acetylglucosaminyltransferase V (Mgat5)-mediated N-glycosylation negatively regulates Th1 cytokine production by T cells. *J Immunol* 173:7200.
101. Cobbold, S.P., K. Nolan, and L. Graca. 2003. Regulatory T cells and dendritic cells in transplantation tolerance: molecular markers and mechanisms. *Immunol Rev* 196:109.
102. Zelenika, D., E. Adams, S. Humm, L. Graca, S. Thompson, S.P. Cobbold, and H. Waldmann. 2002. Regulatory T cells overexpress a subset of Th2 gene transpcripts. *J Immunol* 168:1069–1079.
103. Matsui, Y., A. Saiura, and Y. Sugawara. 2003. Identification of gene expression profile in tolerizing murine cardiac allograft by costimulatory blockade. *Physiol Genomics* 15:199.
104. Louvet, C., E. Chiffoleau, and M. Heslan. 2005. Identification of a new member of the CD20/FcepsilonRIbeta family overexpressed in tolerated allografts. *Am J Transplant* 5:2143.
105. Metcalfe, S., and D. SMPA. 2005. Transplantation tolerance: gene expression profiles comparing allotolerance vs. allorejection. *Int Immunopharmacol* 5:33.

106. Tan, J.T., B. Ernst, W.C. Kieper, E. LeRoy, J. Sprent, and C.D. Surh. 2002. Interleukin (IL)-15 and IL-7 jointly regulate homeostatic proliferation of memory phenotype CD8+ cells but are not required for memory phenotype CD4+ cells. *J Exp Med* 195:1523–1532.
107. Murakami, M., A. Sakamoto, J. Bender, J. Kappler, and P. Marrack. 2002. CD25+CD4+ T cells contribute to the control of memory CD8+ T cells. *Proc Natl Acad Sci* 99:8832–8837.
108. Pearl, J.P., J. Parris, D.A. Hale, S.C. Hoffmann, W.B. Bernstein, K.L. McCoy, S.J. Swanson, R.B. Mannon, M. Roederer, and A.D. Kirk. 2005. Immunocompetent T-cells with a memory-like phenotype are the dominant cell type following antibody-mediated T-cell depletion. *Am J Transplant* 5:465–474.
109. Trzonkowski, P., M. Zilvetti, P. Friend, and K.J. Wood. 2006. Recipient memory-like lymphocytes remain unresponsive to graft antigens after CAMPATH-1H induction with reduced maintenance immunosuppression. *Transplantation* 82:1342–1351.
110. Valujskikh, A., B. Pantenburg, and P.S. Heeger. 2002. Primed allospecific T cells prevent the effects of costimulatory blockade on prolonged cardiac allograft survival in mice. *Am J Transplant* 2:501–509.

Chapter 17
Regulatory T-cells in Therapeutic Transplantation Tolerance

Herman Waldmann, Elizabeth Adams, Paul Fairchild, and Stephen Cobbold

Abstract Rodent studies have shown that transplantation tolerance can be achieved with short courses of therapeutic intervention with monoclonal antibodies, without the need for hemopoietic chimerism. This was first established by us using non-lytic antibody treatment directed to the co-receptors CD4 and CD8, but was later also demonstrated using antibodies to directed to co-stimulatory molecules such as CD154. Tolerance induced by these methods need not delete nor inactivate all alloreactive cells, but involves the induction, recruitment and expansion of CD4 regulatory T-cells (Treg). This includes both natural and "induced" CD4+CD25+ FoxP3+ Treg, but other forms of regulatory T-cell may also participate. The induction of therapeutic tolerance requires TGFβ, which not only raises the threshold for T-cell responses, but is also required for the conversion of naive T-cells to Treg. Treg can be found in tolerated grafts as can T-cells competent to reject. Elimination of Treg from the graft can precipitate rejection, suggesting that graft-resident T-reg confer some form of privilege onto the tissue. This is consistent with our earlier interpretation of *linked-suppression* where we hypothesised that antigen-bearing APC bring Treg into microenvironments that prevent damaging immune responses. Persisting antigen from accepted grafts is processed as if "self" antigen by quiescent host APC, and this antigen-source is sufficient to maintain T-cell unresponsiveness and recruit further Treg into the process. In this way, cohorts of new regulatory T-cells are generated through the life of the graft, through *"infectious tolerance"*. There is compelling data that Treg act not only in the induction phase, but also at later stages of the rejection response.

Introduction

Transplanted organs are currently spared from rejection through use of long-term multi-drug immunosuppression. This carries risks of various drug side-effects, infection and cancer. Minimisation of immunosuppression has become a major goal

---

H. Waldmann
Sir William Dunn School of Pathology, University of Oxford, Oxford OX1 3RE, UK
e-mail: herman.waldmann@path.ox.ac.uk

in transplantation, and exploitation of tolerance processes a means to achieve that end. The discovery of *"infectious tolerance"* and *regulation* in transplantation tolerance, has provided new directions in therapy through exploitation of regulatory T-cells (Treg) as key players.

The emphasis in strategies for achieving therapeutic tolerance has undergone significant change in the past decade. From a time when mixed haematopoietic chimerism seemed the only way forward, we now have an opportunity to identify new drugs and negative vaccination strategies to harness regulation as a dominant physiological response of the immune system to the graft. Moreover the "ideal" of therapeutic mixed chimerism requires significant and invasive therapies targeting the host haematopoietic and immune systems. However desirable that may be, the practicalities of execution may be hard to achieve [1].

Tolerance Induction by Co-Receptor Blockade

In 1986 we demonstrated for the first time that short-term co-receptor blockade with CD4 antibodies could induce tolerance to foreign proteins [2, 3]. Tolerance could be sustained with regular antigen exposure, and expired in the absence of antigen. This raised the interesting question of how the immune system would deal with a tolerated transplant, where antigen would persist indefinitely.

To answer this we investigated whether co-receptor blockade could enable tolerance to transplants in rodent models. In 1989 we did indeed generate "classical transplantation tolerance" in the adult mouse, where donor marrow (as a source of stem cells) had been infused under an umbrella of the co-receptor antibody blockade [4]. Skin grafts from marrow donors were accepted indefinitely.

At first sight this might have been explained through traditional deletional mechanisms of tolerance. However, data emerged which provided compelling arguments for tolerated antigen driving some form of regulation. First, tolerance to skin grafts could be obtained despite very low-level blood chimerism. Second, tolerance could not be broken by infusions of naïve lymphocytes (*resistance*). By tracking expression of allo-antigen-specific T-cell receptors we showed that tolerance did not depend on bulk deletion of antigen-specific cells, but instead, the antigen-specific T-cells had become refractory to antigen stimulation. Yet, how though could such anergy be compatible with the observed *resistance*?

We came up with Civil Service Model as an explanation [5]. We proposed that *anergic* T-cells themselves were the *regulators*. The subsequent discovery of "*natural*" CD25+ Treg, the demonstration that these behave as if anergic *in vitro* [6,7], and the finding that Foxp3 transduced T-cells are rendered anergic [8], are all findings consistent with this Civil Service Model.

Given that high-level chimerism proved unnecessary, we reasoned that we should be able to achieve tolerance to the transplanted tissues without the need for any chimerism at all, as proved to be the case [3,9]. It emerged that the peripheral immune system had the full capacity to permit this form of dominant therapeutic tolerance, as *tolerance* and *resistance* could be generated within an adult-thymectomized mouse.

Direct evidence that T-cells were responsible for the regulation came from adoptive transfer studies which implicated CD4 T-cells [10–12]. What has emerged is that CD4+ cells are perpetually recruited into regulation throughout the life of the graft, a process we have coined *infectious tolerance* [10]. This process can be interrupted if T-cells are removed from their source of reinforcing antigen, and must therefore be antigen driven. As the antigens derived from a tolerated graft are, by their nature, processed in the same way as self-antigens, then it must be this form of danger-free antigen that continues to recruit new Treg through the life of the tolerated graft.

The principles underlying therapeutic reprogramming with CD4 antibodies, as described here, have subsequently been shown to determine tolerance induction through co-stimulation blockade [13,14] as well as anti-TCR blockade with CD3 antibodies [15]. This demonstrates that the potential to establish regulatory pathways pre-exists in any tolerisable recipient, and that tolerising regimens are able to recruit them. The functional and phenotypic overlaps with regulatory cells found in the different transplant-tolerance models, with those found in control of autoimmunity and immunopathology, suggest that regulation in all these systems is mediated by broadly similar types of CD4+ T-cells and mechanisms of action.

Antigen Specificity to Regulation?

Traditionally, the demonstration of regulation or suppression has often been validated through transfer of a mix of tolerant T-cells and naive T-cells into lymphopenic hosts. Using such methods, quantitative titrations of tolerant CD4+ T-cells transferred together with fixed numbers of naïve T-cells into lymphopenic recipients, demonstrated unequivocally that Treg can suppress graft rejection by either CD4+ or CD8+ T-cells, and by primed as well as naïve T-cells [11,12]. However, lymphopenic readouts are not ideal, as they encourage unusually potent expansion of small populations of lymphocytes which can, of course, create experimental artefacts [16,17]. When introduced into a lymphopenic environment, Treg are able to inhibit "homeostatic" expansion of other T-cells, and seem to do so with seemingly limited specificity for the test transplantation antigens [18]. In other words Treg of the **A-type** rendered tolerant of **B-type**, will, in such readouts, prevent aggression towards third party **C-type** grafts. Although creating difficulties for those who wish to follow the antigen-specificity of physiological regulation, this feature of Treg does offer some practical clinical applications, as in the control of Graft versus Host disease, exploiting the fact that Treg can act in a promiscuous but protective way [19,20].

That caveat aside, adoptive transfer studies into lymphocyte-replete hosts have established unequivocally that Treg can behave as if antigen-specific in readouts of *resistance*, *adoptive transfer of suppression* and *infectious tolerance*.

Linked Suppression

How might regulation work? We still remain ignorant of the major mechanisms. A critical clue came from the discovery of *linked-suppression* [21–23]. **A-type** recipients holding **B-type** grafts were able to accept grafts from **(BxC)F1-type** donors. Eventually, many of the mice holding **(BxC)F1**-type grafts became tolerant of **C-type** tissue. It appears as if **B-type** and **C-type** antigens need to be within in the same graft, as **C-type** grafts were simply rejected when placed next to, and in the same graft bed, as the tolerated **B-type** graft. This highlights a key feature of Treg, where once induced to one set of alloantigens, they can extend tolerance to other antigens within the same tissue.

Linked suppression requires that Treg operate within local microenvironments using antigen to bring naive T-cells within range of their suppressive influence. Antigen-presenting cells would seem the most likely cells capable of this focussing role, be this in lymphoid tissues, in the graft itself, or in both sites.

Sustained Regulation Exploits the Indirect Pathway of Antigen Presentation

As stated previously tolerance tends to be lost in the absence of antigen [9,24]. What then might be would be the requirements for antigen-presentation to maintain it? By selective choice of particular recipient-donor combinations, we were able to show that antigens processed by the *indirect pathway* could maintain *dominant tolerance* as well as *linked suppression* [25]. As APC presenting through the *direct* pathway would have a limited life-span, *indirect presentation* offers the likliest route for long-term antigen-dependent regulation. Also, as tolerance is sustained in immunological quiescent conditions, it would be the resting rather than activated host APC, that will presenting donor antigens.

A Unifying Mechanism to Explain Tolerance Without the Need for Chimerism

It is hard to imagine that anti-receptor, co-receptor or co-stimulation blockade evoke mechanisms that are not a normal part of the physiology of the immune system? Why are certain therapeutic antibodies better at achieving tolerance than others? We suggest that the "successful" antibody strategies are those which efficiently target molecules critical to immune function. In this way short-term treatment with such antibodies creates a complete ceasefire free of "sniper" activity. In the absence of immune-mediated inflammation, the grafts have a chance to heal. Once healed, they continue to release antigens for reprocessing by host APC. These quiescent APC are specialised (we argue) to induce and maintain tolerance, and as part of that role they induce and maintain regulation. That is why many different therapeutic agents can give the same outcome of dominant tolerance, and why therapy is only needed for a short period so as to tip the system towards tolerance induction and its maintenance.

What is Going on in the Grafts?

The finding of *linked-suppression* suggested a role regulation within the graft itself. By transfer of tolerated grafts to lymphocyte-deficient hosts, we demonstrated the existence of graft-derived host- colonising T-cells that could confer *resistance* on the host. Naïve T-cells that were later transfused into such "colonised" recipients could not reject either the original donor graft nor fresh donor skin [26]. This demonstrated that functional Treg had been harboured within the original tolerated grafts, and indicated that they might have been be regulating within the tolerated tissue. Ablation of CD25+ T-cells (presumed regulators) at the time of placing the tolerated graft onto a lymphopenic host was shown to unleash a strong rejection response [27]. Such rejection could only have been mediated by the remaining (non-regulatory) graft resident T-cells.

The finding that graft resident CD25+ T-cells could promote graft survival was accompanied by a demonstration of high levels of newly generated FoxP3 mRNA in tolerated grafts [28], and also by data showing that tolerance through co-stimulation blockade could also be prevented by prior depletion of CD25+ T-cells [29].

How might Treg be attracted to their grafts? A role for chemokines has been suggested through the observation that both CCL22 and its receptor CCR4 were upregulated in tolerated grafts, and that CCR4 gene knock-out mice could not be tolerised [29].

What then, are regulatory T cells doing in the tissues? To understand this, we can no longer think of transplanted tissues merely as passive targets awaiting their own destruction, but rather as adaptable organs with some capacity to engage in their own defence. We have suggested that Treg promote such tissue defence by helping the tissue develop a state of *acquired immunological privilege* [27,30,31]. Through *linked-suppression* they decommission local microenvironments to create pockets of acquired privilege. If the number of such pockets is large, then whole tissue might protected by coalescence of all the pockets.

The Pathway to Regulation

Co-receptor blockade does not confer immediate tolerance on T-cells. The process seems to take some weeks as demonstrated in adoptive transfer studies [9]. This is consistent with the therapeutic antibodies firstly blocking function in a reversible way, and following a period of blockade, inducing and recruiting Treg into a policing role. This process has been clarified in a T-cell receptor-transgenic mouse (A1.Rag-/-) where the only T-cells were CD4+ and B-cells were absent [28,32]. Females of this strain carry a TCR directed to a male specific transplantation antigen

Dby, and have no "natural" CD4+CD25+ T-cells, nor Foxp3 in their lymphoid tissues. Normally they reject male grafts, yet are exquisitely tolerisable to these grafts by CD4 co-receptor antibody blockade. When tolerant they exhibit *resistance*, and accumulate newly induced FoxP3+ CD4$^+$CD25$^+$ T-cells in the tolerated grafts and to some extent, the spleen [28]. TGFβ was found to be necessary for therapeutic tolerance and Treg induction [28,33]. Remarkably, CD4 antibody blockade *in-vitro* could also induce Foxp3 mRNA expression in T-cells exposed to dendritic cells bearing Dby peptide. This too required TGFβ although the cellular source was not established [28]. This is quite consistent with a previous demonstration of a need for both TGFβ and TCR to induce anergy and Foxp3 expression in naïve CD4+CD25- T-cells [34].

Although we know that TGFβ is required for the induction of therapeutic dominant tolerance, it is less clear what part it plays in the maintenance phase. We have proposed that antigen presented in a manner that does not allow full activation may ensure that the induced Treg are maintained in an active state, and that such antigen is able to convert further naïve T-cells to regulatory function (*infectious tolerance*). In the first example, repeated administration of Dby peptide in danger-free conditions in an oil/water emulsion given subcutaneously [30] was able to produce transplantation tolerance. Tolerance was accompanied by *resistance* to the infusion of naïve T-cells. Splenic T-cells from treated animals were found to be anergic, and just as for tolerance with co-receptor blockade, FoxP3 mRNA and CD4+CD25+ T-cells were found in spleens and tolerated grafts.

The second example involved use of an altered peptide ligand (APL) of the Dby male peptide [35]. This form of modified male peptide stimulates T cells poorly *in vitro*, leading us to classify it as a partial agonist. Multiple spaced doses of APL given to female recipients rendered them tolerant to male grafts. Tolerance was, yet again, associated with resistance, and induction of CD4+CD25+ FoxP3+ T-cells in the tolerated grafts and spleens.

The third example, involved a Graft versus Host Disease-model where female TCR$^+$ anti-Dby T-cells were infused into male RAG-1$^{-/-}$ hosts [35]. Donor T-cells did not produce GVHD, but became profoundly *anergic* and *suppressive* as measured *in-vitro*. In this case, however, we could demonstrate induction neither of CD4+CD25+ T-cells nor Foxp3 mRNA, indicating that a different form of regulation had been established.

Fourth, we have recently demonstrated that immature antigen-bearing myeloid dendritic cells (DC) as well as DC pharmacologically modified to prevent full activation potential, are all competent to induce transplantation tolerance and the *de novo* induction of FoxP3+ T-cells [36].

Putting all these experiments together, we propose that the key requirement for the induction of Treg is a continued exposure to a form of antigen that cannot signal T-cells fully. The quality of the Treg induced may depend on other local factors (e.g. local cytokine milieu), so explaining why sometimes Tr1, other times FoxP3+Treg, and other times neither Tr1 nor FoxP3+. Recent data from von Boehmer and colleagues providing danger free peptides through a continuous infusion pump, also argue for this notion of prolonged exposure of a poorly immunogenic form of antigen [37].

Antigen Drives the Creation of Tolerising Microenvironments

In summary then, we believe that T cells exposed to persisting antigen where full activation is not possible, will be directed towards *anergy* and *regulatory* function, the exact outcome dependent on particular aspects of the local micro-environment. The presence of TGFβ may enhance conversion towards FoxP3$^+$CD25$^+$ T cells, while IL-10 may promote Tr1-like cells [38,39].

Therapeutic agents which enable dominant tolerance would therefore be expected to permit a level of TCR signalling [40]. A sustained supply of processed donor antigens (from the accepted transplant) would promote and maintain the first cohort of Treg, while also converting further naive T cells into regulatory function.

To explain *linked-suppression* we propose that induced or natural Treg, would be drawn into a cluster with antigen-bearing APC and naive T-cells, and would consequently exert an inhibitory or decommissioning effect on those APC. As a result, not only would these APC fail to deliver necessary activating signals to naive T-cells, but might also might generate suppressive molecules. The upshot would be the tolerisation of any new thymic émigrés, and conversion of further new cohorts of naive T-cells to regulatory function (*infectious tolerance*).

Once dominant tolerance is in place, donor antigens become the "vaccines" that sustain and promote it. This explains why so many different therapeutic approaches can induce *dominant tolerance,* which once established utilises a pathway (i.e. antigen-driven) common to all. *Linked-suppression* and *infectious tolerance* are part of the same general biological process determined by creation of these tolerising microenvironments, where antigens draw participant (regulator and regulatee) cells together. As such microenvironments promote tolerance, they can reasonably be referred to as "*privileged*".

Inevitably, a process such as this has its limits, and the protective microenvironment could be perturbed if sufficient pro-inflammatory activity were permitted. It may therefore be necessary to provide some low level maintenance therapy to prevent any inflammatory flares that might override dominant tolerance.

The choice of such maintenance drugs may not be straight forward. One would need drugs that would spare the conversion of naive T-cells to Treg activity, as well as any decomissioning effects mediated by such T-cells. This is a task for the future-the identification of anti-inflammatory agents that spare regulation.

References

1. Fudaba Y, Spitzer TR, Shaffer J, Kawai T, Fehr T, Delmonico F, Preffer F, Tolkoff-Rubin N, Dey BR, Saidman SL, et al.: Myeloma responses and tolerance following combined kidney and nonmyeloablative marrow transplantation: in vivo and in vitro analyses. *Am J Transplant* 2006, **6**:2121–2133.
2. Benjamin RJ, Waldmann H: Induction of tolerance by monoclonal antibody therapy. *Nature* 1986, **320**:449–451.
3. Qin SX, Wise M, Cobbold SP, Leong L, Kong YC, Parnes JR, Waldmann H: Induction of tolerance in peripheral T cells with monoclonal antibodies. *Eur J Immunol* 1990, **20**:2737–2745.

4. Qin SX, Cobbold S, Benjamin R, Waldmann H: Induction of classical transplantation tolerance in the adult. *J Exp Med* 1989, **169**:779–794.
5. Waldmann H, Qin S, Cobbold S: Monoclonal antibodies as agents to reinduce tolerance in autoimmunity. *J Autoimmun* 1992, **5 Suppl A**:93–102.
6. Sakaguchi S, Sakaguchi N, Asano M, Itoh M, Toda M: Immunologic self-tolerance maintained by activated T cells expressing IL-2 receptor alpha-chains (CD25). Breakdown of a single mechanism of self-tolerance causes various autoimmune diseases. *J Immunol* 1995, **155**:1151–1164.
7. Thornton AM, Shevach EM: CD4+CD25+ immunoregulatory T cells suppress polyclonal T cell activation in vitro by inhibiting interleukin 2 production. *J Exp Med* 1998, **188**:287–296.
8. Hori S, Nomura T, Sakaguchi S: Control of regulatory T cell development by the transcription factor foxp3. *Science* 2003, **299**:1057–1061.
9. Scully R, Qin S, Cobbold S, Waldmann H: Mechanisms in CD4 antibody-mediated transplantation tolerance: kinetics of induction, antigen dependency and role of regulatory T cells. *Eur J Immunol* 1994, **24**:2383–2392.
10. Qin S, Cobbold SP, Pope H, Elliott J, Kioussis D, Davies J, Waldmann H: "Infectious" transplantation tolerance. *Science* 1993, **259**:974–977.
11. Davies JD, Martin G, Phillips J, Marshall SE, Cobbold SP, Waldmann H: T cell regulation in adult transplantation tolerance. *J Immunol* 1996, **157**:529–533.
12. Marshall SE, Cobbold SP, Davies JD, Martin GM, Phillips JM, Waldmann H: Tolerance and suppression in a primed immune system. *Transplantation* 1996, **62**:1614–1621.
13. Honey K, Cobbold SP, Waldmann H: CD40 ligand blockade induces CD4+ T cell tolerance and linked suppression. *J Immunol* 1999, **163**:4805–4810.
14. Graca L, Honey K, Adams E, Cobbold SP, Waldmann H: Cutting edge: anti-CD154 therapeutic antibodies induce infectious transplantation tolerance. *J Immunol* 2000, **165**:4783–4786.
15. Belghith M, Bluestone JA, Barriot S, Megret J, Bach JF, Chatenoud L: TGF-beta-dependent mechanisms mediate restoration of self-tolerance induced by antibodies to CD3 in overt autoimmune diabetes. *Nat Med* 2003, **9**:1202–1208.
16. Barthlott T, Kassiotis G, Stockinger B: T Cell Regulation as a Side Effect of Homeostasis and Competition. *J Exp Med* 2003, **197**:451–460.
17. Stockinger B, Kassiotis G, Bourgeois C: Homeostasis and T cell regulation. *Curr Opin Immunol* 2004, **16**:775–779.
18. Graca L, Le Moine A, Lin CY, Fairchild PJ, Cobbold SP, Waldmann H: Donor-specific transplantation tolerance: the paradoxical behavior of CD4+CD25+ T cells. *Proc Natl Acad Sci USA* 2004, **101**:10122–10126.
19. Steiner D, Brunicki N, Blazar BR, Bachar-Lustig E, Reisner Y: Tolerance induction by third-party " off-the-shelf" CD4+CD25+ Treg cells. *Exp Hematol* 2006, **34**:66–71.
20. Porter SB, Liu B, Rogosheske J, Levine BL, June CH, Kohl VK, Wagner JE, Miller JS, Blazar BR: Suppressor function of umbilical cord blood-derived CD4+CD25+ T-regulatory cells exposed to graft-versus-host disease drugs. *Transplantation* 2006, **82**:23–29.
21. Davies JD, Leong LY, Mellor A, Cobbold SP, Waldmann H: T cell suppression in transplantation tolerance through linked recognition. *J Immunol* 1996, **156**:3602–3607.
22. Bemelman F, Honey K, Adams E, Cobbold S, Waldmann H: Bone marrow transplantation induces either clonal deletion or infectious tolerance depending on the dose. *J Immunol* 1998, **160**:2645–2648.
23. Chen ZK, Cobbold SP, Waldmann H, Metcalfe S: Amplification of natural regulatory immune mechanisms for transplantation tolerance. *Transplantation* 1996, **62**:1200–1206.
24. Cobbold SP, Adams E, Marshall SE, Davies JD, Waldmann H: Mechanisms of peripheral tolerance and suppression induced by monoclonal antibodies to CD4 and CD8. *Immunol Rev* 1996, **149**:5–33.
25. Wise MP, Bemelman F, Cobbold SP, Waldmann H: Linked suppression of skin graft rejection can operate through indirect recognition. *J Immunol* 1998, **161**:5813–5816.

26. Graca L, Cobbold SP, Waldmann H: Identification of regulatory T cells in tolerated allografts. *J Exp Med* 2002, **195**:1641–1646.
27. Cobbold SP, Adams E, Graca L, Daley S, Yates S, Paterson A, Robertson NJ, Nolan KF, Fairchild PJ, Waldmann H: Immune privilege induced by regulatory T cells in transplantation tolerance. *Immunol Rev* 2006, **213**:239–255.
28. Cobbold SP, Castejon R, Adams E, Zelenika D, Graca L, Humm S, Waldmann H: Induction of foxP3+ regulatory T cells in the periphery of T cell receptor transgenic mice tolerized to transplants. *J Immunol* 2004, **172**:6003–6010.
29. Lee I, Wang L, Wells AD, Dorf ME, Ozkaynak E, Hancock WW: Recruitment of Foxp3+ T regulatory cells mediating allograft tolerance depends on the CCR4 chemokine receptor. *J Exp Med* 2005, **201**:1037–1044.
30. Waldmann H, Chen TC, Graca L, Adams E, Daley S, Cobbold S, Fairchild PJ: Regulatory T cells in transplantation. *Semin Immunol* 2006, **18**:111–119.
31. Waldmann H, Adams E, Fairchild P, Cobbold S: Infectious tolerance and the long-term acceptance of transplanted tissue. *Immunol Rev* 2006, **212**:301–313.
32. Zelenika D, Adams E, Mellor A, Simpson E, Chandler P, Stockinger B, Waldmann H, Cobbold SP: Rejection of H-Y disparate skin grafts by monospecific CD4+ Th1 and Th2 cells: no requirement for CD8+ T cells or B cells. *J Immunol* 1998, **161**:1868–1874.
33. Daley SR, Ma J, Adams E, Cobbold SP, Waldmann H: A key role for TGF-beta signaling to T cells in the long-term acceptance of allografts. *J Immunol* 2007, **179**:3648–3654.
34. Chen W, Jin W, Hardegen N, Lei KJ, Li L, Marinos N, McGrady G, Wahl SM: Conversion of peripheral CD4+CD25- naive T cells to CD4+CD25+ regulatory T cells by TGF-beta induction of transcription factor Foxp3. *J Exp Med* 2003, **198**:1875–1886.
35. Chen TC, Waldmann H, Fairchild PJ: Induction of dominant transplantation tolerance by an altered peptide ligand of the male antigen Dby. *J Clin Invest* 2004, **113**:1754–1762.
36. Yates SF, Paterson AM, Nolan KF, Cobbold SP, Saunders NJ, Waldmann H, Fairchild PJ: Induction of regulatory T cells and dominant tolerance by dendritic cells incapable of full activation. *J Immunol* 2007, **179**:967–976.
37. Apostolou I, von Boehmer H: In vivo instruction of suppressor commitment in naive T cells. *J Exp Med* 2004, **199**:1401–1408.
38. Groux H, O'Garra A, Bigler M, Rouleau M, Antonenko S, de Vries JE, Roncarolo MG: A CD4+ T-cell subset inhibits antigen-specific T-cell responses and prevents colitis. *Nature* 1997, **389**:737–742.
39. Roncarolo MG, Levings MK: The role of different subsets of T regulatory cells in controlling autoimmunity. *Curr Opin Immunol* 2000, **12**:676–683.
40. Coenen JJ, Koenen HJ, van Rijssen E, Hilbrands LB, Joosten I: Rapamycin, and not cyclosporin A, preserves the highly suppressive CD27+ subset of human CD4+CD25+ regulatory T cells. *Blood* 2006, **107**:1018–1023.

Chapter 18
CD4⁺CD25⁺ Regulatory T Cell Therapy for the Induction of Clinical Transplantation Tolerance

David S. Game, Robert I. Lechler, and Shuiping Jiang

Abstract The pursuit of transplantation tolerance is still in progress some 53 years after Medawar and colleagues' first description. It has been established beyond doubt that regulatory T cells can confer donor-specific tolerance in mouse models of transplantation. However, this is crucially dependent on the strain combination, the organ transplanted and most importantly, the ratio of Tregs to alloreactive effector T cells. The ex-vivo expansion of Tregs is one solution to increase the number of alloantigen specific cells capable of suppressing the alloresponse. This technique has been used to demonstrate long term graft survival in mouse models, where ex-vivo expanded, alloantigen specific T cells are shown to preferentially migrate to, and proliferate in, the graft and draining lymph node. When such models are selected to test the role of the different allorecognition pathways for Treg induced graft survival, it appears that only a modest direct pathway alloresponse is sufficient to abrogate tolerance in immunocompetent mice. This remains the case when Tregs are expanded with both direct and indirect pathway allospecificity. Therefore, in human transplantation it is likely that depletion of the majority of direct pathway alloreactive T cells will be required to tip the balance in favour of regulation. Ex-vivo expansion of alloantigen specific, indirect pathway human Tregs, which can cross regulate the residual direct pathway has been established. Rapid expansion of these cells is possible, while they retain antigen specificity, suppressive properties and favourable homing markers. Furthermore, considerable progress has been made in the last few years to define which immunosuppressive drugs favour the expansion and function of Tregs. It is proposed that a trial of Treg therapy in combination with depletion of alloreactive T cells and short term immunosuppression is on the near horizon for human transplantation.

S. Jiang
Department of Nephrology and Transplantation, Guy's Hospital, King's College London, London SE1 9RT, UK
e-mail: sj4774@hotmail.com

Introduction

Improvements in surgical technique, post-operative management and refinement of immunosuppressive medication have led to impressive rates of survival for both the graft and the recipient in solid organ transplantation. For renal transplantation, most centres report one year graft and patient survival at over 95%. In achieving this target, however, patients are exposed to the numerous undesirable side effects of non-specific immunosuppression, which range from gingival hyperplasia and hypertrichosis to life-threatening infection, malignancy and cardiovascular disease. Furthermore, having accepted these risks, most patients' grafts will eventually fail due to the insidious process termed chronic rejection, which has a major immunological component; only modest gains in long-term graft survival have been made over the last ten years.

Often described as the "holy grail" of transplant immunology, transplantation tolerance can be defined as *the absence of a damaging (allo)antigen-specific immune response without the need for exogenous immunosuppression and with an intact response to other antigens*. If achieved, this would revolutionise the medical treatment and quality of life for patients with end stage organ failure.

It is clear from other chapters in this book that a number of different Tregs could potentially suppress alloresponses [1]. It should also be clear that it is the $CD4^+CD25^+$ Treg population that has been most extensively studied across a wide spectrum of immune responses. Indeed, the clinical use of $CD4^+CD25^+$ Tregs in humans to prevent graft versus host disease and proposals for their use in the treatment of type I diabetes are at an advanced stage [2]: both are discussed elsewhere in this book. As discussed below, the use of Tregs for the induction of transplantation tolerance in humans has additional challenges that merit separate consideration.

That cells could confer tolerance to a naïve host by adoptive transfer (transferable tolerance) was first shown by Gershon in 1970 [3]. A series of in vivo experiments by Hall et al. in the 1980s [4] identified the $CD4^+$ population as being responsible for antigen specific transferable tolerance to allografts. In fact, the first description of $CD4^+CD25^+$ suppressor cells belongs to Hall in 1990 [5] in a transplantation model. He described $CD4^+$ suppressor cells that were $OX39^+OX22^+OX17^+$ ($CD25^+CD45RO^+MHC$ class II^+) from rats tolerised to cardiac allografts with cyclosporine. Depletion of this population abrogated the transfer of tolerance to naïve graft recipients: the importance of the finding was clearly not appreciated at the time and the modern literature on $CD4^+CD25^+$ Tregs stems from Sakaguchi's work in autoimmunity [6].

In order to optimise regulators of alloreactive T cells it is imperative to understand the way in which both the effector and the regulator see alloantigen [7]. Briefly, allorecognition can occur by two distinct but not mutually exclusive pathways, direct and indirect. The direct pathway occurs when donor MHC is recognised, intact, on the surface of an antigen presenting cell (APC). This is a uniquely vigorous reaction because of the high precursor frequency of alloreactive T cells, which may be up to 7% of the entire $CD4^+$ repertoire [8]. Once the donor "professional" APCs are depleted from the graft and finally die, subsequent direct pathway presentation is likely to be costimulation deficient and this will promote anergy or apoptosis of the direct pathway alloreactive T cells. Indirect pathway allorecognition occurs when donor molecules (usually MHC) are internalised, processed and presented in the context of self-MHC on the surface of the host APC [9].

Therefore the direct pathway would be predicted to diminish with time whereas the indirect pathway would remain intact although it may be subject to regulation. That alloantigen is identified as foreign is probably due to cross-reactivity with viral or bacterial epitopes. Since the indirect pathway mirrors the way the immune system normally sees antigen, the alloreactive T cell precursor frequency is much less than the direct pathway and of the order of 1 in 10000 [10].

In human renal transplantation, we have shown in a series of experiments that the frequency of direct pathway alloresponsive T cells is reduced in the majority of patients with long-term grafts, including those with chronic rejection [10]. This hyporesponsiveness was particularly apparent in the memory subset of CD4$^+$ cells, implying that homing to the graft is important in diminishing the response [11]. In vitro studies showed that the addition of IL2 could restore the frequencies of direct pathway alloreactive T cells in some patients [12]. This suggested that deletion of alloreactive T cells could not entirely account for the downregulation of the direct pathway and that regulation or anergy could also contribute.

Having shown that CD4$^+$CD25$^+$ Tregs are present in the circulation of healthy humans at about 5–10% of the CD4$^+$ fraction [13], we further examined their potential role in human alloresponses. We did not find increased numbers of circulating Tregs in patients with stable graft function. Furthermore, when CD25$^+$ Tregs were depleted from direct pathway in vitro assays, alloreactive T cell frequencies were not increased, nor did we demonstrate enhanced kinetics of regulation. This suggested that T cell anergy was the predominant non-deletional mechanism of direct pathway hyporesponsiveness [14]. In contrast, we and others have shown that increased frequencies of indirect pathway alloresponsive T cells correlate with chronic rejection [10,15,16]. That indirect pathway alloresponses may be held in check by Tregs was first shown in renal transplant recipients with stable graft function. Indirect pathway alloresponses were revealed in vitro by depleting CD25$^+$ Tregs [17].

Sakaguchi, Shevach and others have shown that Tregs need to be activated through their T cell receptor in order to suppress [18,19]. Therefore, as discussed in detail below (**Antigen Specificity for Transplantation Tolerance**), in the regulation of alloresponses, the antigen specificity of the Treg will be important: possibilities are shown in Fig. 18.1.

It seems likely that just as the unfractionated T cell pool has both direct and indirect pathway-specific populations, then so will the Treg subset. It could be argued from a teleological view, that as the "role" of naturally occurring Tregs is the prevention of autoimmunity, and that they are selected from the thymus with high avidity to self-MHC, then there may be a skew within the Treg population towards indirect pathway-specificity. Turning to Fig. 18.1, if self-specific Tregs can regulate direct pathway alloresponses, then one might anticipate that regulation would be strongest in cases of HLA matching. Indeed, our mixed lymphocyte reactions (MLR) between healthy volunteers of known tissue type suggest this may be the case. However,

(a) Direct pathway regulation on complete MHC mismatch – cross reactivity

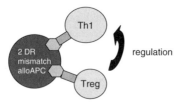

(b) Direct pathway regulation on partial mismatch – self-peptide (polyclonal) specific Treg and/or cross-reactive Treg

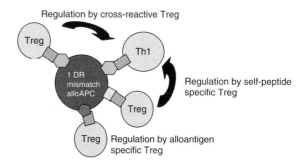

(c) Indirect pathway regulation – self-peptide (polyclonal) and/or alloantigen-specific Treg

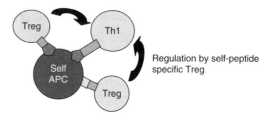

Fig. 18.1 Pathways of allorecognition and regulation. (**a**) In the case of complete mismatched direct pathway alloresponse, only those Tregs with cross-reactivity for intact alloantigen ⊙ can regulate. There are likely to be few of these because Tregs represent less than 10% of circulating CD4+ cells and Tregs are selected for high avidity to self-peptides. (**b**) With partial matching of the direct pathway, Tregs with any combination of specificity can regulate: those specific to self peptides will be able to be ligated by the matched MHC:peptide complex ◇; indirect pathway, alloantigen-specific Tregs can be ligated by the matched MHC in combination with processed allopeptide ●; cross-reactive direct pathway Tregs can suppress as before. (**c**) With self-APC, only those Tregs with self-peptide or allopeptide specificity can suppress. As discussed, most Tregs will fall into these two categories *(See also Color Insert)*

examination of the data raise an alternative explanation, namely that the degree of suppression is inversely proportional to the strength of the response [20]. That Treg suppression is superior when the reaction to be regulated is weaker is consistent with findings in different models of suppression [19,21]. That naturally occurring

Tregs suppress responses to antigens presented on autologous APCs better than a MLR has been shown for human cells in vitro [22].

Given the vigour of the direct pathway alloresponse, and the high precursor frequency of alloreactive T cells, it is generally accepted that deletion of a substantial proportion of direct pathway alloreactive T cells will be required in order to "tip the balance" from reactivity to regulation [23,24]. As discussed in **antigen specificity for transplantation tolerance**, below, very recent data concur with this hypothesis.

In order to suppress the surviving alloreactive T cells by regulation one needs sufficient numbers of effective Tregs in the right place, at the right time in an environment that favours regulation. How to achieve this and how it relates to transplantation tolerance forms the basis of the remainder of this chapter.

Adoptive Transfer of Tregs for Transplantation Tolerance in Animals

As mentioned above, it was during adoptive transfer experiments for transplantation tolerance that $CD4^+CD25^+$ Tregs were first identified [4]. In the modern era, there have been a large number of studies where Tregs have been isolated from animals which have been tolerised to alloantigen by a wide variety of techniques. The Tregs are purified or enriched and then transferred to naïve animals to induce antigen-specific tolerance of a subsequent graft. These are discussed in detail elsewhere in this book, but there are some important points pertinent to Treg therapy which will be raised here. Wood's group in Oxford were among the first to show transferable tolerance by $CD4^+CD25^+$ Tregs to skin grafts in mice [25,26]. In this case suppression was dependent on IL10 and was mediated by indirect pathway (or self-peptide) specific Tregs. In common with the majority of such studies, the Tregs were transferred to an "empty" (rag–/– or T cell depleted) host together with effector cells and then transplanted. The ratios used were often 1:1 or even 2:1 in favour of Tregs over effectors: the minimum number of Tregs required for tolerance induction in vivo was not reported. Transferable tolerance to "full" (non-depleted) mice was demonstrated in a series of experiments by Waldmann's group. These experiments also refined the concept of "infectious tolerance", where transferred Tregs could induce host T cells to themselves become Tregs and in turn be able to transfer tolerance to another host [27]. Also developed by Waldmann, and others, was the concept of linked suppression, when induction of tolerance to antigen "A" can induce suppression of responses to third party antigen "B" when "A" and "B" are presented on the same APC [28]. Effective linked suppression and infectious tolerance are likely to be very important mechanisms for Treg therapy: the former in order to widen the range of alloreactive T cells that can be suppressed and the latter to increase the Treg pool. Importantly, Tregs with indirect allospecificity could be found inside the tolerised allograft [29]. Of note, in Waldmann's experiments, tolerance was induced in "full" mice that were tolerised across minor histocompatibility mismatches thereby confining the alloresponse to the indirect pathway.

Adoptive transfer experiments in rodents therefore reinforce the notion that tolerance requires "tipping the balance" between reactivity and regulation. If one has a sufficiently high Tregs:Teffector ratio then one can tolerise against major MHC mismatches, presented by the indirect pathway, but not if this ratio is compromised.

For tolerance induction by regulation, it is therefore essential to increase the number of Tregs that can suppress alloresponses. This can be relatively easily achieved in some rodent models by a number of different induction protocols at transplantation such as costimulation blockade [30]. However, tolerance induction by these methods is crucially dependent on rodent strain combination and the nature of the graft and therefore unlikely, on its own to be translated into humans. An alternative strategy to increase the number of Tregs that can suppress alloresponses is to expand them ex-vivo.

Ex-Vivo Expansion of Tregs

Ex-vivo polyclonal expansion of Tregs that maintained suppressive properties for the prevention of graft versus host disease lethality in mice was described in 2002 [31]. We were among the first to show that human Tregs could be isolated from the circulation of healthy volunteers and expanded ex-vivo [32]. $CD4^+CD25^+$ cells were purified from an HLA-A2 negative, HLA-DR0101 positive volunteer and then cultured with autologous monocyte-derived immature DCs, HLA-A2 peptide and IL2. The cultured cells shared a similar phenotype to naturally occurring Tregs ($CD25^{hi}$, $CTLA4^+$, etc) and retained anergic and suppressive properties. The cultured Tregs suppressed antigen-driven responses of $CD4^+CD25^-$ cells and also showed linked suppression of the direct pathway by suppressing responses of semi-allogeneic DCs in the presence of HLA-A2 peptide. Linked suppression of T cell lines specific for bacterial and viral antigens restricted by HLA-DR1 was also demonstrated in the presence of HLA-A2 peptide.

Having established this "proof of principle", we then sought to refine the ex-vivo expansion. One of the limiting factors for clinical translation of Treg expansion is the time needed to expand sufficient numbers, whilst retaining specificity. We therefore purified and cultured the $CD4^+CD25^+$ as before and then massively expanded the Treg pool using beads coated in CD3/CD28 and high doses of IL2 such that 4×10^6 cells could be recovered from 1×10^5 cells within 7 days [33]. Tetramer (HLADRB1*0101:A2) staining was positive for 9% of the cell line starting from $CD4^+CD25^+$ cells and 68% of the cell line starting from $CD4^+CD25^-$ cells; tetramer staining for freshly isolated cells was negative. This has several implications. Firstly, additional purification is required to isolate a high percentage of antigen specific Tregs. Secondly, this supports the notion, above, that naturally occurring Tregs are specific for a range of autologous peptides and subsequent exposure to autologous DCs with foreign peptide leads to expansion of autoreactive clones, some of which will be cross reactive to foreign peptide. $CD4^+CD25^-$ cells in contrast, whilst self-MHC restricted, are not autoreactive so it is easier to expand clones reactive to foreign peptide. This is further evidence that without precise manipulation, alloreactive T cells can outnumber their regulatory counterparts.

Therefore in order to increase the proportion of antigen-specific Tregs in culture, we isolated Tregs by flow cytometric sorting of HLADRB1*0101:A2 tetramer stained cells. Cells could be sorted to >90% purity and the sorted cells showed more potent antigen-specific suppression than polyclonal Tregs. Importantly, these cells could also be expanded with CD3/CD28 beads and IL2 at a similar rate to that shown previously [34].

In parallel with the expansion of human Tregs, we also developed a similar line of investigation in murine models.

Tregs were purified from CBA mice and expanded with autologous bone-marrow derived DCs and K^b peptide with IL2 and IL7. In common with the human studies, these cell lines maintained the suppressive phenotype and suppressive functions in vitro [35]. The cell line expanded 5 fold weekly and could be maintained in culture for over 1 year. Upon examination of the TcR-Vβ repertoire, freshly isolated Tregs displayed heterogenous usage whereas the Treg cell line was more limited, consistent with the notion above, that once stimulated the polyclonal Tregs become more oligoclonal for foreign antigen.

Treg function in vivo was examined by grafting CBK ($H2^k+K^b$) skin onto T cell depleted/ thymectomised CBA ($H2^k$) recipients. CFSE staining of the Tregs showed that, despite being anergic in vitro, the Tregs proliferated vigorously in vivo post-transplant, consistent with other reports [36]. The Tregs homed to the graft and draining lymph node although they could also be detected at lower frequencies at other sites such as the spleen. GFP-labelled Tregs confirmed the long-lasting homing of the Tregs to the graft and draining lymph nodes. Crucially, pre-transplant adoptive transfer of the Treg cell line and effector cells at a ratio of 2:1 resulted in indefinite graft survival without any additional immunosuppression whereas freshly isolated Tregs could not protect the graft. Third party grafts were rejected at 40 days whereas 3 out of 9 semi-allogeneic F_1 (CBA/Ca × C57BL/6) grafts showed indefinite survival (MST 80 days) demonstrating some effect of linked suppression in vivo.

We then explored the potential of the expanded Tregs to protect a "full" CBA mouse from a CBK skin graft. Given the single MHC class I mismatch, rejection would be driven by $CD8^+$ direct pathway alloreactive T cells and $CD4^+$ indirect pathway alloreactive T cells. Without any exogenous immunosuppression, 2×10^6 Tregs were able to extend graft survival from 15 to 23 days. Therefore, consistent with previous data from adoptive transfer experiments in "full" versus "empty"/reconstituted mice , it appears that in the absence of immunosuppression, even modest direct pathway alloreactivity is sufficient to overwhelm the suppressive effects of Tregs. This leads to interesting questions about the required antigen specificity of Tregs for tolerance induction.

Antigen Specificity for Transplantation Tolerance

Tregs need to be activated by ligation of their TcR in order to suppress [37]. As previously discussed and shown in Fig. 18.1, Tregs could potentially regulate by specificity to intact allogeneic MHC molecules (cross reactive direct pathway), allogeneic molecules in the context of self-MHC molecules (cross reactive indirect pathway) or by self-specificity and linked suppression. In the CBK to full CBA murine model above, the CBK APCs would present direct pathway allogeneic K^b and also self peptides via $H2$-K^k. Therefore, if the Tregs could recognise self-peptide then they should be able to regulate the direct pathway by linked suppression, and indeed this may be the mechanism behind the prolonged graft survival; regulation of the indirect pathway by $H2^k$:K^b specific Tregs is likely to play a role.

Cross regulation of the direct pathway by indirect pathway Tregs by linked suppression has been shown in a number of other systems. For example, in a rat model using tolerogenic DCs with kidney allografts, Mirenda et al. demonstrated that indirect pathway alloreactive $CD4^+$ cells were able to regulate both the direct pathway and the indirect pathway in vitro. The regulatory cells also showed "linked suppression" in vitro; regulatory properties were abrogated by depletion of $CD25^+$ cells [38]. However, the Tregs were unable to prevent direct pathway driven rejection in vivo, which was achieved using a short course of CysA.

We next explored whether a Treg line with specificity for both the direct and indirect pathway could offer improved graft survival. In this model, the Treg lines were generated from BL/6 ($H2^b$) $CD4^+CD25^+$ cells stimulated by fully mismatched BALB/c ($H2^d$) DCs [39]. This should enrich for direct pathway specificity, which must be polyclonal. Indirect specificity was conferred by TcR gene transduction of the Tregs with TcR specificity for K^d presented by $H2$-A^b. Consistent with our previous experiments, the transduced Tregs maintained their regulatory phenotype and in vitro suppressive properties. In vitro suppression of T cell lines corresponding to alloantigen-specific direct and indirect pathways was confirmed. In vivo, the Tregs homed and proliferated to the graft and draining lymph node in response to specific antigen. In a model analogous to our previous experiments skin grafts were mismatched for a single class I antigen (K^d) and placed onto immunocompetent BL/6 recipients. In this case however, the Tregs would have all 3 specificities shown in Fig. 18.1: direct via intact K^d, indirect via A^b:K^d and self-peptides via any $H2^b$. The range of self-peptide specificity will be less than Tregs expanded by indirect pathway stimulation but should still be substantial. Non-transduced (direct pathway/self) Tregs could not prolong graft survival suggesting that indirect pathway regulation was essential. Somewhat disappointingly, transduced Tregs could only prolong skin graft survival from 15 to 27 days. Thus even with multiple alloantigen specificities, Tregs were still unable to significantly regulate a modest direct pathway alloresponse in an immunocompetent host without any immunosuppression.

Returning to the hypothesis that a deletional strategy is required to reduce the direct pathway alloresponse, we tested the same Tregs in a cardiac transplant model after depleting $CD8^+$ cells. With a haplotype mismatch, the non-transduced (direct pathway/self) Treg line was able to prolong graft survival from 27 to 43 days whereas the transduced Tregs conferred indefinite graft survival without any exogenous immunosuppression. Importantly, the $CD8^+$ cell count returned at 100 days without adverse effect on the graft. It had been shown previously that high numbers of Tregs could not prevent rejection of $CD4^+$ driven rejection in BALB/c to BL/6 cardiac allografts [40]. We therefore used a short course of rapamycin aiming to suppress the direct pathway (see section Treg Therapy for Transplantation Tolerance in Humans below) in this challenging strain combination. The use of only rapamycin and depleting anti-CD8 antibody prolonged graft survival from 24 to 59 days, whereas combination of these agents and Treg therapy with the transduced Tregs resulted in indefinite graft survival.

Taken together, these experiments support the view that deletion of direct pathway alloreactive T cells is required for transplantation tolerance in challenging strain combinations and therefore humans. These experiments also strongly imply that Tregs with indirect pathway specificity are required for the induction of transplantation tolerance, consistent with earlier reports [25,41]. The allospecific Tregs preferentially home to the graft and draining lymph node and then divide. It is likely that infectious tolerance is conferred on recipient T cells during this process. Indeed in a model of Tregs preventing autoimmune diabetes in NOD mice, the origin of Tregs in the draining lymph node shifted from donor to host within 10 days [42]. In addition, once in-situ and activated by their alloantigen, the Tregs have the potential to suppress in an antigen-independent manner, probably via cytokines such as IL10 or TGFβ [37]. The likely reason that Tregs with direct pathway or self-specificity (or any polyclonal population) are not able to confer tolerance is that these Tregs are diluted throughout the reticulo-endothelial system and are simply of insufficient number at the graft and draining lymph node to effectively regulate the alloresponse. Consistent with this, in a model of primary immune responses to CFA, Tregs divided and accumulated in the draining lymph preferentially to the effector T cells: the magnitude of the immune response resulted from their inverse quantitative relationship [43]. Returning to transplantation, if $CD25^-$ cells outnumber Tregs in a semi-allogeneic setting, then linked suppression can become "linked recognition" and the breaking of self-tolerance [44].

The advantage of the transduction of Tregs was that it conferred indirect alloantigen specificity much more efficiently than autologous DC culture with the human Treg cell lines and therefore increased the number of Tregs in the graft and draining lymph node. Additionally, this provides a "proof of principle" that Tregs can be genetically transduced without compromising their intrinsic suppressive properties.

Ex-Vivo Manipulation of Tregs for Transplantation Tolerance

There are a number of proteins that could potentially further improve Treg suppression in vivo if their genes are transduced into Treg lines or Tregs selected for these markers are expanded.

When described by Sakaguchi, Tregs were described by the presence of CD4 and CD25 on the cell surface [6]. Since then a large number of markers within the Treg pool have been associated with improved suppression: a selection of these is shown in Table 18.1. Some of these markers, such as CD62L are upregulated on the surface of ex-vivo expanded Tregs [35,39]. This almost certainly contributes to

Table 18.1 Markers of suppression by tregs

Marker	Mouse/human	Function	Reference
FoxP3	Both	Transcription factor for development and function of Tregs	[76]
GITR	Both	Possible role in the generation of Tregs	[77]
CD45RO	Both	Marker of memory phenotype	[22]
CTLA4	Both	Negative costimulator	[78]
CD62L	Both	Binding to high endothelial venules	[79]
CD27	human	? Intrinsic marker of suppression or binding to lamina propria	[45]
CD30	mice	Signalling to Treg by APC	[80]
CD101	mice	Signalling to Treg	[46]
CD103	mice	Binding to E-cadherin – homing to gut associated lymphoid tissue	[81]
CCR4+8	human	Homing to inflammatory CCLs	[82]
CCR5	mice	Homing to inflammatory CCLs	[83]
CCR6	both	Homing to central nervous system	[84]
Galectin1	both	Regulator of T cell homeostasis	[85]

the enhanced homing characteristics of the expanded Treg population. Indeed in the model of expanded Tregs protecting NOD mice from diabetes, those Tregs that were CD62L negative were unable to prevent diabetes and far fewer Tregs were seen in the draining lymph node [42]. From Table 18.1, it can be seen that different cell markers could alter the tissue homing specificity of Tregs. Apart from specificity to alloantigen, it is currently uncertain whether improved tissue specificity beyond high endothelial venule trafficking (CD62L) is advantageous in inducing tolerance to vascularised allografts.

Other markers are associated with improved suppression that cannot be explained by differences in homing and are likely to represent an intrinsic quality of suppression. Many of these are used to define the Treg population such as FoxP3, GITR and CTLA4, although none are absolutely specific for Tregs. Others, however, represent a subpopulation of more efficient Tregs such CD27 [45] or CD101 [46]. It is likely that expression of these "super Treg" markers will reduce the number of expanded Tregs required to promote transplantation tolerance, but this has not yet been reported.

In summary, there are a number of potential candidate genes for transduction into Tregs that could improve their capability to orchestrate transplantation tolerance, but the only one tested is the TcR gene experiment described above. There remains considerable concern about the use of cells transfected with retrovirus in humans and alternative transduction methods may need to be tested. In addition, the requirement for good manufacturing practice (GMP) conditions for cell therapy in humans will have a strong influence on the ex-vivo manipulation and selection of the best Tregs: flow-based cell sorting techniques do not currently have GMP approval [47]. In the meantime we must optimise selection and expansion methods of Tregs within GMP to maximise expression of the favourable markers and compare efficiency between populations.

Tregs and Opportunistic Infection

One concern of Treg therapy is that if, once activated, Tregs suppress in an antigen non-specific way, they may be no better than the non-specific immunosuppressive drugs in terms of increased risk of infection and/or malignancy. To specifically address this question, Bushell et al. measured anti-viral CTL activity and clearance of influenza in mice that had been tolerised with Tregs operating as bystander suppressors [48]. They found that they were no different to the antiviral responses in untreated mice. The most likely explanation for this is consistent with the conclusions of **antigen specificity for transplantation tolerance**, namely that localisation of the Tregs to graft and draining lymph nodes offers the required specificity to prevent systemic infection.

In addition, Treg therapy is unlikely to produce the plethora of non-immune side effects attributed to currently available immunosuppressive drugs.

Treg Therapy for Transplantation Tolerance in Humans

There is no doubt that tolerance induction in humans and non-human primates is more challenging than in rodents. This is partly due to heterologous immunity: exposure to viral or bacterial antigens can abrogate tolerance in some murine transplantation models [49]. Humans are clearly exposed to significantly more antigens than these tolerance-resistant mice and therefore the precursor frequency of direct pathway cross-reactive allospecific T cells may even exceed the 7% demonstrated in mice by Turka's group [8].

From the preceding discussion it is clear that Treg therapy in humans will require deletion of direct pathway alloreactive T cells and ex-vivo expansion of Tregs with indirect allospecificity. Induction therapy with T cell depleting agents has been long-established in human transplantation with the use of anti-lymphocyte globulin and more recently with Campath-1H. Apart from the removal of direct pathway alloreactive T cells, the combination of T cell depletion with Treg therapy has an additional potential advantage. Recovery from lymphopenia is influenced by homeostatic proliferation of the lymphocyte compartment. In a BALB/c to CBA cardiac transplant model, homeostatic proliferation favoured expansion of alloreactive memory T cells and transplant rejection [50]. The infusion of as few as 5×10^4 Tregs was able to promote long-term graft function, suggesting that the infused Tregs proliferated vigorously in the lymphopenic environment. In humans, without Treg therapy, currently used protocols show that despite the impressive degree of lymphodepletion achieved, acute rejection can still occur whilst taking immunosuppressive medication implying the survival and/or homeostatic expansion of alloreactive T cells, predominantly of the memory phenotype [51,52]. This, together with murine data, suggests that a period of immunosuppressive drug therapy will be required after transplantation until the Tregs have established themselves. Clearly, the key is to create an environment which favours the establishment of tolerance: sometimes called the tolerogenic milieu.

The Tolerogenic Milieu

There are several data to suggest that calcineurin inhibitors can be detrimental to tolerance induction. This may be because of the mode of action: calcineurin inhibition blocks intracellular pathways of T cell activation. In vitro and in vivo studies have shown that decreased activation of alloreactive T cells by cyclosporine can be anti-anergic [53,54], possibly because of decreased apoptosis [55,56]. In addition there are accumulating data that calcineurin inhibitors are detrimental to Tregs. This may be because of reduced Treg induction, expansion [57] and function [58].

Conversely, rapamycin appears to be tolerance permissive by selectively inducing apoptosis or necrosis of alloreactive cells whilst promoting Treg induction [59], expansion [60] and function [61]. This suggests that rapamycin is the ideal candidate for short term therapy post-depletion in humans. However, rapamycin monotherapy post-depletion is associated with a high risk of acute rejection [62]. It is unclear whether the concomitant therapy with Tregs would be sufficient to prevent this or whether further immunosuppression, for example with mycophenolate mofetil [63] will be required in the short term.

It may be possible to use some of the emerging immunosuppressive drugs for Treg therapy, once their routine clinical use is established. FTY720 is a sphingosine 1-phosphate agonist that is immunosuppressive by sequestering T cells from blood and spleen into secondary lymphoid organs and preventing their migration to sites of inflammation. If trafficking of Tregs is compromised then, it is clear from the discussion above, that capacity for tolerance induction would be reduced. In fact, it appears that not only is Treg sequestration differentially affected, but Treg activity is enhanced with the addition of FTY720 in vitro [64].

Alternative strategies to improve the tolerogenic milieu include the manipulation of cytokine signalling such as the use of cytokine fragment crystallisable (Fc) fusion proteins as developed by Strom's group. They have shown that fusion proteins of pro-IL2 and blocking-IL15 together with rapamycin can promote tolerance [65,66]. The rationale was that blockade of IL15 was required to terminate T cell proliferative responses, and IL2 signalling was required for pro-apoptotic signals in effector cells and for the survival/ expansion of Tregs. More recent work from Strom, with Kuchroo, looking at the differentiation of Th17 cells has shown that TGFβ with IL6 or IL21 are pivotal cytokines in determining whether Tregs or effector Th17 cells are generated from $CD4^+$ precursors [67]. This raises the potential use of blocking-IL6 and IL21 fusion proteins to help promote a tolerogenic milieu. To date there are no reports combining this technology with ex-vivo expanded Tregs.

How Many Cells and How Often?

This is unknown. The most optimistic extrapolation from the murine studies would be that a single pre-transplant infusion of Tregs will be sufficient. It should be clear from the above discussion that this is unlikely to be the case, and that multiple infusions will be required in transplant patients.

Living Related and Deceased Donor Transplantation

Ex vivo expansion of indirect pathway alloreactive T cells should be relatively straightforward for living related (or unrelated) donation, when there is usually plenty of time to define the best immunodominant peptide for Treg antigen specificity, to maximise the supply of Tregs prior to transplantation and to optimise the tolerance induction protocol depending on immunological risk. Deceased donor transplantation is more problematic because transplantation needs to occur as soon as possible once the donor and recipient pair is identified, and certainly within 24 hours. There are several possible solutions. Firstly, the most obvious choice would be to expand Tregs as soon as the required specificity in known and the patient would only receive Treg therapy once sufficient Tregs have expanded. In the meantime they would receive tolerance/ Treg permissive immunosuppressive drugs (as discussed above). Secondly, each patient on the waiting list could have Tregs expanded to a number of different alloantigens and sufficient cells to enable Treg therapy for a week or two are frozen at the transplant centre. Once an organ becomes available, the frozen Tregs are thawed and used as therapy until enough fresh Tregs are expanded in the usual way. One could imagine that this would be easy for a patient on dialysis who is waiting for a kidney transplant because they would be attending hospital frequently. The obvious difficulty would be predicting to which alloantigens the Tregs should be raised. An alternative would be to expand and store polyclonal Tregs until the alloantigen-specific cells are available. Finally, an interesting paper by Karim et al. suggests that antigen-specific Tregs raised against one alloantigen could protect a third party graft by bystander suppression in a mouse model [68]. As discussed in **antigen specificity for transplantation tolerance**, it is uncertain whether these latter possibilities would result in sufficient Tregs at the graft and draining lymph nodes to be tolerogenic in the demanding transplant environment of humans.

Immune Monitoring

If a tolerance induction strategy is employed then there needs to be a method of assessing tolerance. The easiest way is simply to measure graft function but, of course, one would like to predict and prevent impending graft injury, perhaps with a "booster" infusion of Tregs. The whole topic of immune monitoring is beyond the scope of this chapter, and there is currently no single assay that can accurately predict graft rejection [69]. In addition, tolerance can only be said to exist retrospectively – the only real proof of tolerance is that it has already occurred. Therefore tolerance assays will be a surrogate such that a given result is associated with a very low chance of rejection. Furthermore, whichever parameter is measured, the cut-off for rejection risk is likely to change with time after transplantation or with intercurrent illness. There are a large number of candidates for a tolerance assay and their use in Treg therapy is speculative: a few will be discussed here.

Perhaps the most obvious method of testing for tolerance in Treg therapy would be to measure Treg number and/or function post-transplant. This should test for Treg engraftment, expansion and infectious tolerance. Measurement of urinary FoxP3 mRNA has been shown to predict renal allograft rejection in some patients [70]. It is possible that after Treg therapy, with the much higher number of FoxP3$^+$ cells, the predictive power of this completely non-invasive technique will improve and may also be useful in non-renal transplantation. Conversely, the recently reported association of high soluble CD30 levels post-transplant with graft failure [71] may be confounded by the presence of large numbers of Tregs, which themselves may express high levels of this marker. The Immuknow (Cylex Inc, Columbia) assay measures the amount of ATP released by CD4+ cells stimulated by phytohaemaglutinin and is therefore a marker of "global immunity". High levels of ATP release correlated with rejection, low levels with infection and median levels with stable graft function [72]. This sort of readout might be useful in trying to find the right amount of non-specific immunosuppression by drugs, but is probably insufficiently sensitive to monitor antigen-specific suppression. It might be possible to modify the assay, perhaps testing ATP release to the alloantigens to which the Tregs have been expanded ex vivo, although whether the Tregs themselves could confound such a test is uncertain.

We have been involved in an international collaboration trying to define the "fingerprint of transplantation tolerance". The data from this collaboration are not yet published, but a number of the tests we employed could be of use in Treg therapy including gene expression signatures [73]. The trans-vivo delayed-type hypersensitivity (DTH) assay has been shown to detect antigen-specific immune responses by transfer of peripheral blood mononuclear cells and donor alloantigen to the footpad or ear pinnae of immunodeficient mice [74]. It has recently been shown that regulation by indirect pathway Tregs can be demonstrated using this technique [75] and therefore this technique could prove useful in Treg therapy.

Conclusions

In order for Tregs to suppress immune responses, they need to be at the anatomical site of the immune response. In transplantation, this is in the graft and draining lymph node. Once there, they need to be of sufficient quantity in comparison to the number of alloreactive effector cells. After activation through their T cell receptor, Tregs can suppress in an antigen-independent way and so polyclonally activated Tregs could conceptually regulate alloresponses but might also suppress desirable immune responses. In animal models ex-vivo expanded, alloantigen specific T cells are shown to preferentially migrate to, and proliferate in, the graft and draining lymph node [35]. When alloantigen-specific Tregs are used to confer tolerance, the immune response to bacterial or viral antigens remains intact, probably because of anatomical localisation of the Tregs. "Proof-of-principle" studies have shown that Tregs can be transduced with genes that confer indirect pathway antigen specificity and it is possible that genes for other proteins that improve the suppressive ability of Tregs can be transduced to engineer super-efficient Tregs ex-vivo [39].

Examination of the role of the different allorecognition pathways for Treg induced graft survival, shows that only a modest direct pathway alloresponse is sufficient to abrogate tolerance in immunocompetent mice. This remains the case when Tregs are expanded with both direct and indirect pathway allospecificity.

In human transplantation it is likely that depletion of the majority of direct pathway alloreactive T cells will be required to tip the balance in favour of regulation. Ex-vivo expansion of alloantigen specific, indirect pathway human Tregs, which can cross regulate the residual direct pathway has been established [33]. Rapid expansion of these cells is possible, while they retain antigen specificity, suppressive properties and favourable homing markers [34]. Furthermore, considerable progress has been made to define which immunosuppressive drugs favour the expansion and function of Tregs [61]. Current data suggest that rapamycin and not calcineurin inhibitors are tolerogenic and the role of newer drugs and cytokine fusion proteins are under evaluation.

A number of different techniques to monitor the immune response in human transplant recipients have been described. These will need to be tested in the domain of Treg therapy as some will prove more useful than others: the quantification of FoxP3 in the urine or blood is one attractive tool in this context.

It is proposed that a trial of Treg therapy in combination with depletion of alloreactive T cells and short term immunosuppression is on the near horizon for human transplantation.

References

1. Jiang S, Lechler RI. Regulatory T cells in the control of transplantation tolerance and autoimmunity. Am J Trans. 2003; 3:516–24.
2. Bluestone JA. Regulatory T-cell therapy: Is it ready for the clinic? Nat Rev Immunol. 2005 Apr; 5(4):343–9.
3. Gershon RK, Kondo K. Cell interactions in the induction of tolerance: The role of thymic lymphocytes. Immunology. 1970; 18:723–37.
4. Hall BM, Jelbart ME, Dorsch SE. Suppressor T cells in rats with prolonged cardiac allograft survival after treatment with cyclosporine. Transplantation. 1984 Jun; 37(6):595–600.
5. Hall BM, Pearce NW, Gurley KE, Dorsch SE. Specific unresponsiveness in rats with prolonged cardiac allograft survival after treatment with cyclosporine. III. Further characterization of the CD4+ suppressor cell and its mechanisms of action. J Exp Med. 1990 Jan 1; 171(1):141–57.
6. Sakaguchi S, Sakaguchi N, Asano M, Itoh M, Toda M. Immunologic self-tolerance maintained by activated T cells expressing IL-2 receptor alpha-chains (CD25). Breakdown of a single mechanism of self-tolerance causes various autoimmune diseases. J Immunol. 1995 Aug 1; 155(3):1151–64.
7. Game DS, Lechler RI. Pathways of allorecognition: Implications for transplantation tolerance. Trans Immunol. 2002; 10(2–3):101–8.
8. Suchin EJ, Langmuir PB, Palmer E, Sayegh MH, Wells AD, Turka LA. Quantifying the frequency of alloreactive T cells in vivo: New answers to an old question. J Immunol. 2001; 166:973–81.

9. Lechler RI, Batchelor JR. Immunogenicity of retransplanted rat kidney allografts. Effect of inducing chimerism in the first recipient and quantitative studies on immunosuppression of the second recipient. J Exp Med. 1982a; 156(6):1835–41.
10. Baker RJ, Hernandez-Fuentes MP, Brookes PA, Chaudhry AN, Cook HT, Lechler RI. Assessment of the immunological contribution to chronic allograft nephropathy. J Immunol. 2001a; 167:7199–206.
11. Baker RJ, Hernandez-Fuentes MP, Brookes PA, Chaudhry AN, Lechler RI. The role of the allograft in the induction of donor-specific T cell hyporesponsiveness. Transplantation. 2001b; 72(3):480–5.
12. Ng WF, Hernandez-Fuentes MP, Baker RJ, Chaudhry AN, Lechler RI. Reversibility with interleukin 2 suggests that T cell anergy contribute to donor-specific hyporesponsiveness in renal transplant patients. J Am Soc Nephrol. 2002; 13(12):2983–9.
13. Ng WF, Duggan PJ, Ponchel F, Matarese G, Lombardi G, Edwards AD, et al. Human CD4+CD25+ T cells: A naturally occuring population of regulatory T cells. Blood. 2001; 98(9):2736–44.
14. Game DS, Hernandez-Fuentes MP, Chaudhry AN, Lechler RI. CD4+CD25+ regulatory T cells do not significantly contribute to direct pathway hyporesponsiveness in stable renal transplant patients. J Am Soc Nephrol. 2003; 14(6):1652–61.
15. Suciu-Foca N, Ciubbotaria R, Itescu S, Rose EA, Cortesini R. Indirect allorecognition of donor HLA-DR peptides in chronic rejection of heart allografts. Transplant proc. 1998; 30:3999–4000.
16. Hornick P, Mason P, Baker R, Hernandez-Fuentes M, Frasca L, Lombardi G, et al. Significant frequencies of T cells with indirect specificity in heart graft recipients with chronic rejection. Circulation. 2000; 101:2405–16.
17. Salama AD, Najafian N, Clarkson MR, Harmon WE, Sayegh MH. Regulatory CD25+ T cells in human kidney transplant recipients. J Am Soc Nephrol. 2003; 14(6):1643–51.
18. Itoh M, Takahashi T, Sakaguchi N, Kuniyasu Y, Shimizu J, Otsuka F, et al. Thymus and autommunity: Production of CD25+CD4+ naturally anergic and suppressive T cells as a key function of the thymus in maintaining immunological self-tolerance. J Immunol. 1999; 162:5317–26.
19. Thornton A, Shevach EM. CD4+CD25+ Immunoregulatory T cells suppress polyclonal T cell activation in vitro by inhibiting interleukin 2 production. J Exp Med. 1998; 188:287–96.
20. Game DS, Lechler RI. HLA-DR matching and immune regulation. Am J Trans. 2004; 4(10):1725–6.
21. Baecher-Allan C, Viglietta V, Hafler DA. Inhibition of human CD4(+)CD25(+high) regulatory T cell function. J Immunol. 2002 Dec 1; 169(11):6210–7.
22. Taams LS, Vukmanovic-Stejic M, Smith J, Dunne PJ, Fletcher JM, Plunkett FJ, et al. Antigen-specific T cell suppression by human CD4+CD25+ regulatory T cells. Eur J Immunol. 2002 Jun; 32(6):1621–30.
23. Lechler RI, Garden OA, Turka LA. The complementary roles of deletion and regulation in transplantation tolerance. Nat Rev Immunol. 2003; 3:147–58.
24. Li XC, Strom TB, Turka LA, Wells AD. T cell death and transplantation tolerance. Immunity. 2001b; 14:407–16.
25. Hara M, Kingsley CI, Niimi M, Read S, Turvey SE, Bushell AR, et al. IL-10 is required for regulatory T cells to mediate tolerance to alloantigens in vivo. J Immunol. 2001 Mar 15; 166(6):3789–96.
26. Kingsley CI, Karim M, Bushell AR, Wood KJ. CD25+CD4+ regulatory T cells prevent graft rejection: CTLA-4 and IL10 dependent immunoregulation of alloresponses. J Immunol. 2002; 168:1080–6.
27. Qin S, Cobbold SP, Pope H, Elliot J, Kioussis D, Davies J, et al. Infectious transplantation tolerance. Science. 1993; 259:974.
28. Davies JD, Leong LY, Mellor A, Cobbold SP, Waldmann H. T cell suppression in transplantation tolerance through linked recognition. J Immunol. 1996; 156(10):3602–7.

29. Graca L, Cobbold SP, Waldmann H. Identification of regulatory T cells in tolerated allografts. J Exp Med. 2002a Jun 17; 195(12):1641–6.
30. Taylor PA, Noelle RJ, Blazar BR. CD4+CD25+ Immune regulatory cells are required for induction of tolerance to alloantigen via costimulatory blockade. J Exp Med. 2001; 193(11):1311–7.
31. Taylor PA, Lees CJ, Blazar BR. The infusion of ex vivo activated and expanded CD4(+)CD25(+) immune regulatory cells inhibits graft-versus-host disease lethality. Blood. 2002a May 15; 99(10):3493–9.
32. Jiang S, Camara NOS, Lombardi G, Lechler RI. Induction of allopeptide-specific human CD4+CD25+ regulatory T cells ex-vivo. Blood. 2003; 102(6):2180–6.
33. Jiang S, Tsang J, Game DS, Stevenson S, Lombardi G, Lechler RI. Generation and expansion of human CD4+ CD25+ regulatory T cells with indirect allospecificity: Potential reagents to promote donor-specific transplantation tolerance. Transplantation. 2006 Dec 27; 82(12):1738–43.
34. Jiang S, Tsang J, Tanriver Y, Leung E, Lechler R. In-vitro generated human CD4+CD25high regulatory T cells with indirect allospecificity as potential patient-specific reagents to promote donor-specific transplantation tolerance. Am J Transplant. 2007; 7(s2):173.
35. Golshayan D, Jiang S, Tsang J, Garin MI, Mottet C, Lechler RI. In vitro-expanded donor alloantigen-specific CD4+CD25+ regulatory T cells promote experimental transplantation tolerance. Blood. 2007 Jan 15; 109(2):827–35.
36. Fisson S, Darrasse-Jeze G, Litvinova E, Septier F, Klatzmann D, Liblau R, et al. Continuous activation of autoreactive CD4+ CD25+ regulatory T cells in the steady state. J Exp Med. 2003 Sep 1; 198(5):737–46.
37. Thornton AM, Shevach EM. Suppressor effector function of CD4+CD25+ immunoregulatory T Cells is antigen nonspecific. J Immunol. 2000; 164:183–90.
38. Mirenda V, Berton I, Read J, Cook T, Smith J, Dorling A, et al. Modified dendritic cells coexpressing self and allogeneic major histocompatability complex molecules: an efficient way to induce indirect pathway regulation. J Am Soc Nephrol. 2004 Apr; 15(4):987–97.
39. Tsang J, Tanriver Y, Jiang S, Xue S, Stauss HJ, Bucy RP, et al. Conferring indirect allospecificity on CD4+CD25+ regulatory T cells by T cell receptor gene transfer: a novel strategy for the induction of transplanation tolerance. J Clin Invest. 2008, in press.
40. Xia G, He J, Zhang Z, Leventhal JR. Targeting acute allograft rejection by immunotherapy with ex vivo-expanded natural CD4+ CD25+ regulatory T cells. Transplantation. 2006 Dec 27; 82(12):1749–55.
41. Yamada A, Chandrakar A, Laufer TM, Gerth AJ, Sayegh MH, Auchincloss H. Recipient MHC class II expression is required to achieve long term survival of murine cardiac allografts after costimulatory blockade. J Immunol. 2001; 167:5522–6.
42. Tarbell KV, Petit L, Zuo X, Toy P, Luo X, Mqadmi A, et al. Dendritic cell-expanded, islet-specific CD4+ CD25+ CD62L+ regulatory T cells restore normoglycemia in diabetic NOD mice. J Exp Med. 2007 Jan 22; 204(1):191–201.
43. Haribhai D, Lin W, Relland LM, Truong N, Williams CB, Chatila TA. Regulatory T cells dynamically control the primary immune response to foreign antigen. J Immunol. 2007 Mar 1; 178(5):2961–72.
44. Fucs R, Jesus JT, Souza Junior PH, Franco L, Vericimo M, Bellio M, et al. Frequency of natural regulatory CD4+CD25+ T lymphocytes determines the outcome of tolerance across fully mismatched MHC barrier through linked recognition of self and allogeneic stimuli. J Immunol. 2006 Feb 15; 176(4):2324–9.
45. Koenen HJ, Fasse E, Joosten I. CD27/CFSE-based ex vivo selection of highly suppressive alloantigen-specific human regulatory T cells. J Immunol. 2005 Jun 15; 174(12):7573–83.
46. Fernandez I, Zeiser R, Karsunky H, Kambham N, Beilhack A, Soderstrom K, et al. CD101 Surface expression discriminates potency among murine FoxP3+ regulatory T cells. J Immunol. 2007 Sep 1; 179(5):2808–14.

47. Bluestone JA, Thomson AW, Shevach EM, Weiner HL. What does the future hold for cell-based tolerogenic therapy? Nat Rev Immunol. 2007 Aug; 7(8):650–4.
48. Bushell A, Jones E, Gallimore A, Wood K. The generation of CD25+ CD4+ regulatory T cells that prevent allograft rejection does not compromise immunity to a viral pathogen. J Immunol. 2005 Mar 15; 174(6):3290–7.
49. Adams AB, Pearson TC, Larsen CP. Heterologous immunity: an overlooked barrier to tolerance. Immunol Rev. 2003 Dec; 196:147–60.
50. Neujahr DC, Chen C, Huang X, Markmann JF, Cobbold S, Waldmann H, et al. Accelerated memory cell homeostasis during T cell depletion and approaches to overcome it. J Immunol. 2006 Apr 15; 176(8):4632–9.
51. Pearl JP, Parris J, Hale DA, Hoffmann SC, Bernstein WB, McCoy KL, et al. Immunocompetent T-cells with a memory-like phenotype are the dominant cell type following antibody-mediated T-cell depletion. Am J Transplant. 2005 Mar; 5(3):465–74.
52. Trzonkowski P, Zilvetti M, Friend P, Wood KJ. Recipient memory-like lymphocytes remain unresponsive to graft antigens after CAMPATH-1H induction with reduced maintenance immunosuppression. Transplantation. 2006 Nov 27; 82(10):1342–51.
53. Schwartz RH. Models of T cell anergy: is there a common molecular mechanism? J Exp Med. 1996 Jul 1; 184(1):1–8.
54. Macian F, Garcia-Cozar F, Im S-H, Horton HF, Byrne MC, Rao A. Transcriptional mechanisms underlying lymphocyte tolerance. Cell. 2002; 109:719–31.
55. Wells AD, Li XC, Li Y, Walsh MC, Zheng XX, Wu Z, et al. Requirement for T-cell apoptosis in the induction of peripheral transplantation tolerance. Nat Med. 1999 Nov; 5(11):1303–7.
56. Wekerle T, Kurtz J, Sayegh M, Ito H, Wells A, Bensinger S, et al. Peripheral deletion after bone marrow transplantation with costimulatory blockade has features of both activation-induced cell death and passive cell death. J Immunol. 2001 Feb 15; 166(4):2311–6.
57. Noris M, Casiraghi F, Todeschini M, Cravedi P, Cugini D, Monteferrante G, et al. Regulatory T cells and T cell depletion: Role of immunosuppressive drugs. J Am Soc Nephrol. 2007 Mar; 18(3):1007–18.
58. Baan CC, van der Mast BJ, Klepper M, Mol WM, Peeters AM, Korevaar SS, et al. Differential effect of calcineurin inhibitors, anti-CD25 antibodies and rapamycin on the induction of FOXP3 in human T cells. Transplantation. 2005 Jul 15; 80(1):110–7.
59. Tian L, Lu L, Yuan Z, Lamb JR, Tam PK. Acceleration of apoptosis in CD4+CD8+ thymocytes by rapamycin accompanied by increased CD4+CD25+ T cells in the periphery. Transplantation. 2004 Jan 27; 77(2):183–9.
60. Gao W, Lu Y, El Essawy B, Oukka M, Kuchroo VK, Strom TB. Contrasting effects of cyclosporine and rapamycin in de novo generation of alloantigen-specific regulatory T cells. Am J Transplant. 2007 Jul; 7(7):1722–32.
61. Game DS, Hernandes-Fuentes MP, Lechler RI. Everolimus and basiliximab permit suppression by human CD4CD25 Cells in vitro. Am J Transplant. 2005; 5(3):454–64.
62. Knechtle SJ, Pirsch JD, H. Fechner J J, Becker BN, Friedl A, Colvin RB, et al. Campath-1H induction plus rapamycin monotherapy for renal transplantation: Results of a pilot study. Am J Transplant. 2003 Jun; 3(6):722–30.
63. Flechner SM, Goldfarb D, Solez K, Modlin CS, Mastroianni B, Savas K, et al. Kidney transplantation with sirolimus and mycophenolate mofetil-based immunosuppression: 5-year results of a randomized prospective trial compared to calcineurin inhibitor drugs. Transplantation. 2007 Apr 15; 83(7):883–92.
64. Sawicka E, Dubois G, Jarai G, Edwards M, Thomas M, Nicholls A, et al. The sphingosine 1-phosphate receptor agonist FTY720 differentially affects the sequestration of CD4+/CD25+ T-regulatory cells and enhances their functional activity. J Immunol. 2005 Dec 15; 175(12):7973–80.
65. Zheng XX, Sanchez-Fueyo A, Sho M, Domenig C, Sayegh MH, Strom TB. Favorably tipping the balance between cytopathic and regulatory T cells to create transplantation tolerance. Immunity. 2003a Oct; 19(4):503–14.

66. Koulmanda M, Budo E, Bonner-Weir S, Qipo A, Putheti P, Degauque N, et al. Modification of adverse inflammation is required to cure new-onset type 1 diabetic hosts. Proc Natl Acad Sci USA. 2007 Aug 7; 104(32):13074–9.
67. Korn T, Bettelli E, Gao W, Awasthi A, Jager A, Strom TB, et al. IL-21 initiates an alternative pathway to induce proinflammatory T(H)17 cells. Nature. 2007 Jul 26; 448(7152): 484–7.
68. Karim M, Feng G, Wood KJ, Bushell AR. CD25+CD4+ regulatory T cells generated by exposure to a model protein antigen prevent allograft rejection: Antigen-specific reactivation in vivo is critical for bystander regulation. Blood. 2005 Jun 15; 105(12):4871–7.
69. Newell KA, Larsen CP. Tolerance assays: Measuring the unknown. Transplantation. 2006 Jun 15; 81(11):1503–9.
70. Muthukumar T, Dadhania D, Ding R, Snopkowski C, Naqvi R, Lee JB, et al. Messenger RNA for FOXP3 in the urine of renal-allograft recipients. N Engl J Med. 2005 Dec 1; 353(22): 2342–51.
71. Langan LL, Park LP, Hughes TL, Irish A, Luxton G, Witt CS, et al. Post-transplant HLA class II antibodies and high soluble CD30 levels are independently associated with poor kidney graft survival. Am J Transplant. 2007 Apr; 7(4):847–56.
72. Kowalski RJ, Post DR, Mannon RB, Sebastian A, Wright HI, Sigle G, et al. Assessing relative risks of infection and rejection: A meta-analysis using an immune function assay. Transplantation. 2006 Sep 15; 82(5):663–8.
73. Martinez-Llordella M, Puig-Pey I, Orlando G, Ramoni M, Tisone G, Rimola A, et al. Multiparameter immune profiling of operational tolerance in liver transplantation. Am J Transplant. 2007 Feb; 7(2):309–19.
74. VanBuskirk AM, Burlingham WJ, Jankowska-Gan E, Chin T, Kusaka S, Geissler F, et al. Human allograft acceptance is associated with immune regulation. J Clin Invest. 2000 Jul; 106(1):145–55.
75. Warnecke G, Chapman SJ, Bushell A, Hernandez-Fuentes M, Wood KJ. Dependency of the trans vivo delayed type hypersensitivity response on the action of regulatory T cells: Implications for monitoring transplant tolerance. Transplantation. 2007 Aug 15; 84(3):392–9.
76. Fontenot JD, Gavin MA, Rudensky AY. Foxp3 programs the development and function of CD4+CD25+ regulatory T cells. Nat Immunol. 2003 Apr; 4(4):330–6.
77. Shimizu J, Yamazaki S, Takahashi T, Yasumasa I, Sakaguchi S. Stimulation of CD25+CD4+ regulatory T cells through GITR breaks immunological self-tolerance. Nat Immunol. 2002; 3(2):135–42.
78. Takahashi T, Tagami T, Yamazaki S, Uede T, Shimizu J, Sakaguchi N, et al. Immunologic self-tolerance maintained by CD25(+)CD4(+) regulatory T cells constitutively expressing cytotoxic T lymphocyte-associated antigen 4. J Exp Med. 2000 Jul 17; 192(2):303–10.
79. Fu S, Yopp AC, Mao X, Chen D, Zhang N, Mao M, et al. CD4+ CD25+ CD62+ T-regulatory cell subset has optimal suppressive and proliferative potential. Am J Transplant. 2004 Jan; 4(1):65–78.
80. Zeiser R, Nguyen VH, Hou JZ, Beilhack A, Zambricki E, Buess M, et al. Early CD30 signaling is critical for adoptively transferred CD4+CD25+ regulatory T cells in prevention of acute graft-versus-host disease. Blood. 2007 Mar 1; 109(5):2225–33.
81. Banz A, Peixoto A, Pontoux C, Cordier C, Rocha B, Papiernik M. A unique subpopulation of CD4+ regulatory T cells controls wasting disease, IL-10 secretion and T cell homeostasis. Eur J Immunol. 2003 Sep; 33(9):2419–28.
82. Iellem A, Mariani M, Lang R, Recalde H, Panina-Bordignon P, Sinigaglia F, et al. Unique chemotactic response profile and specific expression of chemokine receptors CCR4 and CCR8 by CD4(+)CD25(+) regulatory T cells. J Exp Med. 2001 Sep 17; 194(6):847–53.
83. Wysocki CA, Jiang Q, Panoskaltsis-Mortari A, Taylor PA, McKinnon KP, Su L, et al. Critical role for CCR5 in the function of donor CD4+CD25+ regulatory T cells during acute graft-versus-host disease. Blood. 2005 Nov 1; 106(9):3300–7.

84. Kleinewietfeld M, Puentes F, Borsellino G, Battistini L, Rotzschke O, Falk K. CCR6 expression defines regulatory effector/memory-like cells within the CD25(+)CD4+ T-cell subset. Blood. 2005 Apr 1; 105(7):2877–86.
85. Garin MI, Chu CC, Golshayan D, Cernuda-Morollon E, Wait R, Lechler RI. Galectin-1: a key effector of regulation mediated by CD4+CD25+ T cells. Blood. 2007 Mar 1; 109(5):2058–65.

Chapter 19
Regulatory T Cells in Allergic Disease

Catherine Hawrylowicz

Abstract Allergic diseases, including asthma, are associated with the development of allergen-specific Th2 and IgE responses, which regulate the early and late phase allergic reactions. Naturally occurring T regulatory cells (Treg), which are present in all healthy individuals together with antigen-induced Treg that secrete inhibitory cytokines such as IL-10 and/or TGFβ, in the periphery. The balance between allergen-specific disease-promoting T helper 2 cells (Th2) and various Treg populations appears to be decisive in the development of a disease promoting allergic versus a non-disease promoting or tolerogenic immune response respectively. Treg specific for common environmental allergens represent the dominant subset in non-atopic individuals implying a state of natural or active tolerance to allergen in health. In contrast, there is a high frequency of allergen-specific Th2 cells in allergic individuals. The function of both naturally occurring and adaptive or inducible Treg appears to be impaired in active allergic disease. Therapies associated with amelioration of disease symptoms, including allergen immunotherapy and glucocorticoids, have been shown to modulate favourably Treg function. Strategies to improve current therapeutic regimens are increasingly focusing on manipulation of Treg for patient benefit.

Allergic Disease, Including Asthma

Allergy can be defined as a clinically evident reaction to ubiquitous environmental, allergens. Clinical symptoms vary greatly, which is in part dependent on how the allergen enters the body. For example aeroallergens, such as pollen, house dust mite or cat dander, cause allergic rhinitis, conjunctivitis and, in some individuals, asthma. Allergic responses to food allergens such as peanuts or shellfish, can be very severe and even fatal, and are associated with vomiting, diarrhea as well as in some cases atopic dermatitis. Reactions to bee and wasp venom can also be very severe and are

---

C. Hawrylowicz
Department of Asthma, Allergy and Respiratory Science, 5th Floor Thomas Guy House, Guy's Hospital, King's College London, London SE1 9RT, UK
e-mail: catherine.hawrylowicz@kcl.ac.uk

characterized by responses distal from the site of initial entry. Despite these diverse clinical reactions a common immunological pathway and mechanisms have been proposed to account for allergic symptoms [1,2].

Upon initial exposure to allergen susceptible individuals mount allergen-specific T helper 2 (Th2) responses associated with the cytokines IL-4, IL-5, and IL-13 as well as an increase in IgE synthesis, which together drive disease pathology [2]. The entry of allergens via the gut or respiratory route is more frequently associated with allergic sensitization than other routes of antigen administration. This likely reflects in part the nature of the antigen presenting cells present at these sites, environmental signals and the physical properties and concentration, generally low, of the allergen itself. During allergic sensitization the Th2 cytokines IL-4 and IL-13 promote allergen specific IgE synthesis. The IgE binds to high affinity IgE receptors (FcεRI) expressed on the surface of mast cells present at mucosal sites reflecting common sites of allergen entry into the body. Subsequent allergen challenge of a sensitized individual leads to the early or acute allergic response (EAR) and the rapid activation of mast cells via allergen cross-linking of surface bound IgE. Activation leads to the release of pre-formed mediators present in mast cell granules, and newly synthesized mediators. These include histamines, leukotrienes, and cytokines, which promote vascular permeability, smooth muscle contraction and mucus production. The EAR occurs within minutes of exposure to allergen. Chemokines released by mast cells and other cells lead to the late phase allergic response (LAR) and the recruitment and activation of cells such as Th2 cells, and eosinophils, which release cytokines leukotrienes and basic proteins. The LAR generally occurs several hours following allergen challenge and can be severe and prolonged [1,2].

There is a clear genetic component leading to a strong pre-disposition to develop allergic disease within families [3,4]. However there has also been a sharp rise in the incidence of allergic conditions, including asthma, in the developed world since the middle of the twentieth century, indicating an additional role for environmental changes during that period [5]. Around a third of individuals in developed countries can suffer from an allergic reaction at sometime in their life, although the incidence varies enormously between different countries (e.g. 16% in Albacete, Spain versus 46% in Christchurch, New Zealand) [6]. In contrast, despite exposure to helminthes and other infections that predispose to the development of strong Th2 responses, a much lower incidence of allergic conditions is seen in the developing world (reviewed in [7]). The reasons for this are of considerable interest immunologically, however the evidence that the incidence of Th1-mediated immune pathologies are also increasing in developed countries has led to the prevailing concept that inadequate exposure to environmental microorganisms in early life leads to the failure to mount appropriate immunoregulatory controls (reviewed in [8,9,10,11]).

Current Therapies and Their Limitations

Common treatments for allergic symptoms are predominantly non-specific in nature and do not provide long-term relief [2,12]. These include the prophylactic use of anti-histamines and the topical, inhaled or systemic use of the anti-inflammatory mediators, glucocorticoids. Glucocorticoids represent the cornerstone of asthma therapy and are safe and effective in the majority of patients, however they fail to provide long-term relief or cure from disease symptoms and subsequent exacerbations require further treatment [12]. Furthermore, around 10–20% of asthma patients are unable to adequately control their symptoms with glucocorticoids and are termed glucocorticoid insensitive. They represent those individuals most at risk of hospitalization and death, from their asthma. Providing more effective, specific and longer lasting therapies is therefore highly desirable.

Allergy is unique among immunological conditions in that the disease-promoting antigen has generally been identified, as well as the relevant immunodominant epitopes. Undoubtedly related to this is the fact that a disease modifying antigen-specific therapy is currently available for the treatment of allergic disease. Allergen desensitisation immunotherapy (IT) was first described around a hundred years ago by von Pirquet and involves the injection of increasing amounts, starting with very low concentrations, of allergen to which the patient is sensitive [13,14]. It requires prolonged treatment over several years for maximal efficacy. This type of therapy is not effective in all patients, and most effective in those with high IgE levels and in IgE mediated diseases such as allergic rhinitis, asthma and venom anaphylaxis. Unsurprisingly, the treatment itself is associated with a significant risk of adverse effects, and even occasionally death, and for this reason it is not commonly used for the treatment of allergic asthma in the UK. There remains a need for improved and novel therapies for the control of allergic diseases, including asthma, and current interest in this area has focussed on immunoregulatory pathways and especially T regulatory cells.

Prevention of the Development of Allergic Responses

The role of thymic selection for the control of deleterious immune responses to self antigens is imperfect. In allergic disease the importance of this process to prevent deleterious immune responses to allergen is even less clear since allergens themselves are unlikely to be present at the time of thymic selection, although negative selection on unknown cross-reactive determinants may occur. Additional controls are therefore in place in the periphery to prevent disease promoting immune responses to allergen. Allergens themselves are often not very immunogenic compounds. They frequently enter the body via the respiratory tract or via the gut, which unlike other routes of immunization, often result in the induction of tolerance [15]. Additionally, specialized T cell subsets, broadly termed regulatory T cells (Treg), play an essential and protective role in immune suppression in the periphery.

Regulatory T Cells

Two broad subsets of CD3+CD4+ suppressive or regulatory T (Treg) cells have been described namely constitutive or naturally occurring Treg versus adaptive or inducible Treg [16,17,18,19,20,21]. However other T regulatory populations clearly exist, although much less is known concerning their role in the control of allergic disease. This includes CD8+ Treg cells with the capacity to both inhibit or exacerbate allergic airway disease in mice [22,23], double negative (CD4–CD8–) TCRαβ+ Treg cells that mediate tolerance in several experimental autoimmune diseases [24] and TCRγδ Treg cells which are suggested to play a role in the inhibition of immune responses to tumors [25,26,27,28]. Many other cell types are also capable of playing an immunoregulatory role. This potentially includes any cell capable of secreting IL-10 and/or transforming growth factor beta (TGFβ), including macrophages, B cells and dendritic cells, (DC) [29,30,31]. In addition natural killer (NK) cells, epithelial cells, macrophages, glial cells and even conventional effector T cells populations such as Th1 and Th2 synthesise suppressor cytokines such as IL-10 and TGFβ [32,33,34], which may contribute either directly to disease control or indirectly through promoting the induction of a variety of Treg populations. The localization, and time at which these different cell types secrete these mediators is likely to influence their role, although as yet we know little regarding their capacity and importance in modulating the allergic response. Nevertheless it seems highly likely that some of these cells efficiently contribute to the generation and maintenance of a regulatory/suppressor type immune response.

CD4+CD25+Foxp3+ Tregs and Allergy

Specific pathways for the selection of naturally occurring Treg from the thymus have been described [35,36]. These cells are characterized by expression of the intracellular forkhead winged transcription factor Foxp3 (forkhead box P3), which appears central to lineage development and function. A number of studies suggest that thymic differentiation accounts for many Treg cells that are specific for self-peptides, however it is less clear whether Treg specific for environmental antigens are generated by this pathway.

More recently pathways for the induction of CD4+CD25+Foxp3+Treg from CD4+CD25− naïve T cells in the periphery have been described. They may be more relevant for the generation of environmental antigen-specific T cells for which an undesired immune response results in pathology. Experimental protocols for the conversion of naive Foxp3 negative T cells into Foxp3-expressing regulatory T cells that are described as indistinguishable from "natural" Treg involve the delivery of subimmunogenic peptide by osmotic minipumps or by peptide containing DEC 205 antibodies [37,38,39]. In addition conversion of peripheral CD4+CD25− naive T cells to Foxp3+CD4+CD25+ regulatory T cells by TGFβ has also been reported [40,34]. In a murine asthma model, TGFβ-induced Treg prevented house dust mite-induced allergic pathogenesis in lungs [40]. Overall, expression of Foxp3 by murine T cells appears to more closely correlate with regulatory activity. In humans there is evidence that the activation of CD4+ T cells results in the transient upregulation of FoxP3, but these cell populations do not always exhibit regulatory function.

In vitro studies suggest that CD4+CD25+Treg inhibit T cell function via cell contact dependent pathways. The majority of studies suggest that in vitro

CD4+CD25+Treg cells do not secrete significant quantities of inhibitory cytokines e.g. IL-10, TGFβ when compared to either inducible Treg or other populations such as macrophages. Furthermore, human thymus-derived CD25+ Treg are less efficient in the inhibition of Th2 responses in comparison to suppression of Th1 responses [41]. An independent study reported that CD25+Treg cells inhibited Th2 differentiation, but were less effective in the suppression of cytokine production and proliferation of established Th2 cells. Pre-activation of the CD4+CD25+ T cells in vitro induced stronger inhibition of Th2 responses [42].

Mutations affecting the gene encoding Foxp3 have been described in mice and humans and lead to the loss of the naturally occurring Treg compartment. In humans this has been termed the X-linked autoimmune and allergic dysregulation syndrome (XLAAD) or immune dysregulation polyendocrinopathy enteropathy-X-linked syndrome (IPEX). It affects boys at a very early age, who suffer from autoimmunity and severe atopy, including eczema, food allergy, and eosinophilic inflammation [43,44,45]. Marked skewing of T lymphocyte responses towards the Th2 phenotype was observed following analysis of two kindreds with XLAAD. In the scurfy mouse and more recently studies of a targeted loss-of-function mutation in the murine Foxp3 gene, revealed an intense multiorgan inflammatory response in the mice, which was associated with allergic airway inflammation, a striking hyper-immunoglobulinemia E, eosinophilia, and dysregulated T(H)1 and T(H)2 cytokine production. However an absence of overt T(H)2 skewing was reported in the mice compared to the human studies [46]. These data may reflect the absence of Foxp3+Treg control at the time of initial allergen exposure and could therefore support a role for the naturally occurring Treg compartment in prevention of sensitisation to allergen.

Studies of the allergen responsiveness of T cells from healthy non-atopic human donors further suggest that active control of the allergic response occurs by CD4+CD25+ Treg, rather than simply a lack of T cells with allergic potential. This is demonstrated by experiments looking at the allergen-specific responsiveness of CD4+ T cells from non-atopic subjects, where poor proliferative and Th2 cytokine responses to allergen in culture are observed in comparison to responses seen with cells from atopic patients. If, however, peripheral blood mononuclear cells from non-atopic donors are depleted of CD4+CD25+ cells prior to stimulation with allergen, increased proliferative and Th2 cytokine responses are observed [47]. Importantly this has been observed for a range of allergens including grass pollen, nickel and milk allergens e.g. [48, 49, 47].

Animal Models of CD4+CD25+Treg Activity

The mechanism of suppression by CD4+CD25+Treg in vivo differs dependent on the experimental system being studied. This is likely to reflect differences in the tissue, the type of inflammation and animal model under study, as highlighted by studies of infection and autoimmune disease. Emerging data suggest diverse, but not necessarily mutually exclusive mechanisms of CD4+CD25+Treg activity in allergic airway disease. Most, but not all, early studies demonstrated the capacity of CD4+ CD25+Treg to inhibit allergic airway disease in mice (reviewed and fully referenced in [50,51]). More recently several mechanistic studies of CD4+CD25+ Treg cell suppression of allergic airway disease have been reported and suggest a role for inhibitory cytokines. In one such study, transfer of ovalbumin (OVA) peptide-specific CD4+CD25+ T cells to OVA-sensitized mice reduced airway hyperreactivity (AHR), recruitment of eosinophils, and T helper type 2 (Th2) cytokine expression in the lung after allergen challenge. The suppression by the OVA peptide-specific CD4+CD25+ T cells was inhibited by an anti-IL-10 receptor antibody. However, the adoptive transfer of CD4+CD25+Treg from IL-10 knockout mice still effectively inhibited this response and the CD4+CD25+Treg were shown to induce expression of IL-10 by resident lung CD4+ T cells [52]. A second study also suggested that the capacity of CD4+CD25+ T cells present in the lung to regulate airway allergic responses was dependent on IL-10, but that this was dependent on the prior induction of TGFβ [53]. A further study concluded that inhibition was mediated by CD4+CD25+ Treg cell suppression of DC activation and that the absence of this regulatory pathway contributed to disease susceptibility [54]. The maintenance of protective Treg activity appears to depend on continuing allergen stimulation [55]. Whilst most studies to date have indicated at least some capacity to inhibit allergic airway inflammation, several of these more recent studies have also highlighted that some inhibition of airway hyper responsiveness occurred. Reports on the capacity to inhibit structural remodeling in the airways are awaited with interest.

Loss of Suppression During Acute Inflammatory Responses

Pro-inflammatory signals appear to abrogate the regulatory function of CD4+CD25+Treg function and this may occur directly or indirectly via effects on effector T cells and/or APC. For example the activation of dendritic cells (DC) through Toll like receptors (TLR), which facilitate the recognition of structurally conserved components of pathogens by cells of the innate immune system, was shown to lead to the production of signals, including both soluble IL-6 and undefined membrane bound ones, which blocked the suppressive effect of CD4+CD25+Treg [56]. In a mouse model of allergic airway disease IL-6 has been proposed to act via two mechanisms to promote disease, the direct enhancement of Th2 responses as well as by overcoming the suppressive function of CD4+CD25+Treg [57]. However it IL-6 may not be as effective in dampening human CD4+CD25+Treg function as in the mouse [58]. TNF [59,60] as well as IL-7 and IL-15 [61] have also been proposed to overcome regulatory activity in human autoimmune disease. In a mouse model of allergic airway disease a TLR2 agonist ameliorated established airway inflammation but this was not via effects on regulatory T cells, but by promoting a Th1 response [62]. However recent evidence also suggests that certain T cell, both T effector and regulatory, populations also express TLR (reviewed by [63]). Inflammatory stimuli delivered directly to

TLR expressing T cells may dampen down Treg function either directly or through rendering effector T cells refractory to suppression. These data have been used to support a model in which during early inflammatory responses, Treg are stimulated to expand, but their regulatory activity is impaired allowing effector T cell populations to resolve the inflammatory insult. Upon resolution of the inflammation, an expanded Treg population would regain its functional activity in order to maintain immune homeostasis and prevent host damage and autoimmunity that might arise from activated effector cells. Thus TLR ligation may directly, and/or indirectly inhibit Treg activity [63]. For example TLR-induced signals from dendritic cells and/or by enhanced IL-2 secretion by effector cells could make effector cells more refractory to suppression and/or impair Treg function. The evidence that TLR5 stimulated CD4+ effector T cells are refractory to suppression [64], and that TLR8 ligation abrogates CD25+Treg function [65], suggest that the direct actions of other TLRs on both effector and regulatory populations may exist with the potential to impair regulatory readouts. The importance and complexities of TLR on Treg function in allergic disease has yet to be fully elucidated. However it is highly topical since the therapeutic potential of TLR agonists either alone or conjugated with allergen are of considerable research interest for the treatment of allergic disease [66]. For example TLR9 ligand, CpG conjugated to allergen has been used in clinical trials of allergy with reported success [67]. The proposed mechanism is via promotion of Th2 to Th1 deviation, but further studies on Treg function in patients are awaited with interest [68].

Several studies report that in humans exhibiting strong responses to allergen as well as in patients' assessed during the hayfever season, regulation by peripheral blood-derived CD4+CD25+ Treg is impaired [69,47,70]. The mechanism for this is unclear and a number of different explanations are possible. Since Treg selection was performed in many cases by MACS bead selection and these individuals were undergoing immune activation due to allergen exposure, an increased frequency of CD4+CD25+ activated effector T cells may have been isolated and studied rather than Treg. Allergen-specific Treg may have exited the blood and migrated to sites of allergen exposure [53], allergen-activated effector T cells and/or APC might be more refractory to suppression or the Treg function might be directly impaired. One study attempting to address these issues has suggested that only a proportion of allergic patients exhibit defective CD4+CD25+Treg activity because the function of Treg is dependent on the concentration and the type of the respective allergen with different thresholds for individual allergens and patients [71]. A recent study also importantly demonstrated quantitative and functional impairment of pulmonary CD4+CD25^{high} Treg, implicating Treg absence at the active site of disease, in paediatric asthma patients as compared to children with cough or control children. Encouragingly, inhaled glucocorticoids increased the percentage of CD4+CD25high T cells in peripheral blood and bronchoalveolar fluid in the same study [72]. A further study reported impaired CD4+CD25+Foxp3+ T cell infiltration in the skin in atopic dermatitis [73]. In conclusion, CD4+CD25+ Treg clearly have the capacity to control disease promoting Th2 responses to allergen in both animal models and in human cultures. Whilst it is desirable to fully restore the activity of this Treg compartment in allergic patients, it is as yet unclear whether this would be sufficient for long-term resolution of allergic disease.

Pro-inflammatory signals can directly and/or indirectly overcome CD25+Treg activity. Therefore reducing the pro-inflammatory environment is predicted to restore CD25+Treg function and studies with anti-inflammatory agents have attempted to address this. For example, glucocorticoids, represent the cornerstone of asthma therapy, with well-established anti-inflammatory properties. Studies of the effects of glucocorticoids on CD4+CD25+Treg activity in vitro and in vivo have been reported, with all implying beneficial effects on regulatory function [74,75,76,72]. Another example is of the immunosuppressive agent rapamycin that reportedly expands human CD25+Treg in vitro as well as in type I diabetic patients in whom a defect in these cells has been reported [77].

Therapeutic Application of Naturally Occurring Treg

The capacity to either expand existing CD25+Treg populations or convert CD4+CD25– T cells to a regulatory phenotype offers the potential for adoptive transfer therapy for various immune mediated pathologies [78,38,79]. One major limitation may be the capacity to generation sufficient numbers of allergen-specific T cells [80]. However it is debatable whether such therapy will be appropriate for the treatment of allergic disease, which in many cases whilst severely impacting on the quality of life is not life-threatening, based on both safety considerations and cost.

Inducible T Regulatory Cells

Naïve peripheral T cells are able to acquire regulatory characteristics as a consequence of peripheral induction after exposure to antigen (induced or adaptive Treg) [18,50]. These cells are characterised by the expression of the anti-inflammatory cytokines IL-10 (Tr1 or IL-10-Treg) and/or TGFβ (Th3), and in many cases also express CD25 and a number of other markers associated with natural Treg. Factors influencing development of adaptive Treg include soluble antigen or peptide in the absence of adjuvant, immature or cytokine activated dendritic cell populations, and certain soluble mediators in the micro-environment in which antigen is encountered. Additional evidence is also emerging that previously activated T cells may be able to deviate towards a regulatory phenotype, possibly reflecting end stage differentiation.

Transforming Growth Factor Beta (TGFβ) and Th3

TGFβ is produced by a wide range of cell types, including T lymphocytes (reviewed in [33,34]). It plays a central role in the maintenance of tolerance within the immune system, particularly oral tolerance (reviewed and fully referenced in [33,81]). TGFβ inhibits both B and T lymphocyte proliferation, differentiation and survival. It inhibits most immunoglobulin (Ig) isotype switching, but promotes the differentiation of IgA secreting plasma cells, associated with defence to microbial infection in the mucosa. TGFβ promotes the differentiation of Langerhan cells and DC with an immature phenotype and acts on monocytes and macrophages, promoting a pro-inflammatory phenotype in the former, but a largely inhibitory phenotype in macrophages. In mast cells TGFβ promotes mast cell chemotaxis, but may inhibit FcϵRI expression. Eosinophils can produce significant quantities of TGFβ, which has complex effects on their survival and activation. TGFβ is also associated with the resolution of immune responses, and the induction of regulatory T cell populations, as discussed above.

Given these broad ranging functional properties, the effects of TGFβ in allergic disease are complex, with evidence of both disease inhibitory and promoting effects. TGFβ has a demonstrated capacity to inhibit human Th2 responses in vitro [82]. In a murine model of OVA-induced allergic airway asthma, over-expression of TGFβ-1 in ovalbumin (OVA)-specific CD4+ T cells abolished airway hyperresponsiveness (AHS) and airway inflammation induced by OVA-specific Th2 cells [83]. A recent study of TGFβ-1 heterozygous mice, producing smaller quantities of TGFβ-1 in comparison to wild-type mice, exhibited exacerbated allergic airway immunopathology [84], suggesting that endogenous TGFβ-1 can suppress airway disease in mice. T cells appear to be an important target of TGFβ-1 action, since antigen-induced airway inflammation and airway reactivity were enhanced in Smad7 transgenic mice, with a blockade in TGFβ signalling. Both OVA-induced airway hypersensitivity and airway inflammation was enhanced in the transgenic mice and was associated with high production of Th2 cytokines [85]. Involvement of TGFβ in the regulation of allergic airway disease by naturally occurring Treg has also been reported. In a study by Ostroukhova et al, tolerance induced by repeated exposure to low doses of inhaled allergen involved CD4+ T cells expressing both membrane bound TGFβ and Foxp3 [86], whilst Joetham et al suggested lung CD4+CD25+Treg controlled airway disease through local induction of TGFβ production [53]. These studies suggest that endogenous TGFβ acts to minimize airway sensitization and inflammation and all use mouse studies that represent acute models of airway inflammation. They do not address the potential of TGFβ for inducing structural remodeling in the lung, where clear associations of TGFβ with chronic human lung disease have been documented [87]. In a mouse model using prolonged exposure to allergen that exhibited many properties of chronic asthma, blockade of TGFβ even after the onset of established eosinophilic airway inflammation, significantly reduced peribronchiolar extracellular matrix deposition, airway smooth muscle cell proliferation, and mucus production in the lung without affecting established airway inflammation and Th2 cytokine production [88] leading the authors to suggest it may be possible to uncouple airway inflammation and remodeling during prolonged allergen exposure.

The contribution of TGFβ in the regulation of human asthma remains to be fully elucidated. TGFβ-1 levels are reportedly higher in the airways of atopic asthmatics compared to normal controls [89] and these increase further following segmental bronchoprovocation with allergen. These authors conclude their findings are consistent with the hypothesis that TGFβ-1 is implicated in airway wall remodeling in asthma. However an overview of all the published data in patients and mice, of TGFβ-1 suggest it has the potential to (i) provide a negative feedback mechanism to control airway inflammation; (ii) may be involved in the repair of asthmatic airways; (iii) induce fibrosis to exaggerate disease development [90]. The highly diverse nature of these data make it difficult to conclude about the likely therapeutic potential of TGFβ in allergy, and particularly allergic airway disease.

Interleukin 10 and IL-10 Secreting Regulatory T Cells

IL-10 inhibits many effector cells and functions associated with allergic disease. It is inversely correlated with disease incidence and severity (reviewed and fully referenced in [32,91,50,92]). IL-10 is synthesized by a wide range of cell types, which includes APC and T cells. It inhibits pro-inflammatory cytokine production, as well as Th1 and Th2 cell activation, which is likely to be largely attributable to effects on APC, although direct effects on T cell function have also been noted. IL-10 impairs mast cell and eosinophil activation, effector cells associated with early and late phase allergic responses. It promotes IgG_4 synthesis and the induction of IL-10 following allergen immunotherapy has been suggested to account for the favourable change in IgG_4:IgE ratios associated with successful allergen immunotherapy (discussed below). Studies in both murine models and in humans have proposed that IL-10 plays a role in maintaining immune homeostasis in the lung in health [50,92].

An inverse correlation exists between the presence of IL-10 and the incidence and severity of asthmatic disease in the lung [93,94]. For example, the bronchoalveolar lavage of asthmatic patients contains lower levels of IL-10 than healthy individuals and T cells from asthmatic children produce less IL-10 than those from healthy ones [50,93]. Additionally an inverse correlation also exists with the levels of IL-10 and skin prick test reactivity to allergen [95]. IL-10 positive T cells consistently represent the dominant subset specific for common environmental allergens in the peripheral blood of healthy individuals; in contrast, there is a high frequency of allergen-specific IL-4-secreting T cells in allergic individuals [96]. In addition bee keepers who are naturally tolerant to bee venom allergen, due to multiple bee stings, demonstrate a similar high IL-10 response [97]. IL-10 production by T cell clones derived from the blood of children with cow's milk allergy produced IL-4 and IL-13, whilst those from control non-allergic children were characterized by the production of IL-10 and, to a lesser extent, IFNγ [98]. All these studies suggest that increased levels of IL-10 synthesis, including allergen specific production, are associated with the lack of allergic symptoms.

A large body of published data describes the capacity of IL-10 to control allergic airway disease in mice. For example, in studies where recombinant murine IL-10 was instilled intranasally at the same time as allergen challenge, or of IL-10 gene delivery to the airways demonstrated suppression of cellular recruitment and airway inflammation [99,100]. Similarly, CD4 T-helper cells engineered to produce IL-10 prevented allergen-induced airway hyperreactivity and inflammation [101], whilst adoptive transfer of IL-10 secreting cells (Tr1) inhibited a Th2-specific response in vivo [102]. Studies describing the induction of tolerance in the airways highlight the association of IL-10 with protection [29,103,104]. Suppression of airway eosinophilia by a killed *Mycobacterium vaccae* preparation induced allergen-specific regulatory T-cells and protection was associated with the synthesis of both IL-10 and TGFβ [105]. Finally as discussed above, protection by adoptive transfer of naturally occurring CD4+CD25^{high} Treg was dependent on the induction of IL-10 in some studies [52,53].

Generation and Therapeutic Application of Inducible Treg Cells

A wide variety of protocols have been developed to generate induced Treg cells from naive cells either in vitro, using antigen, cytokines, and dendritic cells, or in vivo by antigen administration (protocols of tolerance induction) (extensively reviewed and referenced in [17,18,50]). Other inducible Treg cells have been described as deriving from naive T cell populations after, for example, in vitro induction in the presence of IL-10 alone, IL-10 together with other cytokines e.g. IL-4 or IFNα, or after repetitive stimulation in the presence of immature dendritic cells. Similarly, a wide range of antigen administration protocols have been used to induce Treg cells not only in vitro but in vivo as well [18]. A common feature of many, but not all of these induced Treg populations is the production not only of IL-10 and/or TGFβ, but also Th1 and Th2 associated cytokines, and therefore they have the potential to exacerbate inflammatory reactions. In all cases the stability of the profile of cytokine production over time in vivo is unclear. Additionally, since many exhibit a limited potential to expand, they may have a restricted clinical application.

A strategy designed to generate Treg secreting IL-10, in the absence of Th1 and Th2-associated cytokines, employed in vitro stimulation of CD4+ cells in the presence of glucocorticoids, either alone or more effectively, together with the active form of vitamin D, 1α25-dihydroxyvitamin D3. These cells were able to regulate autoimmunity in vivo (experimental allergic encephalomyelitis or EAE) in an IL-10 dependent manner [106,107]. Interestingly in mice, these drug-induced IL-10-Treg cells were shown to have similar functional suppressive properties to the naturally occurring Treg cells in vitro and were able to also inhibit naïve T cells via cell contact dependent pathways, even though they did not express Foxp3 [108]. In humans these drug-induced IL-10-Treg inhibited the proliferation and cytokine responses of naïve T cells, as well as established Th1 and Th2 cells, including allergen–specific Th2 cell lines [109]. It may be possible to use other drug combinations to effectively generate IL-10-secreting Treg populations which act via effects on T cell and/or the antigen presenting cell compartment [50,110,111].

Therapies Associated with the Induction of Treg and the Production of IL-10 and /or TGFβ

A number of examples exist of treatments for allergic diseases that are associated with the induction of IL-10, as well as in some cases TGFβ. This includes both antigen specific and non antigen-specific therapies.

Immunotherapy

Allergen-specific immunotherapy (IT) is most effective in individuals with IgE-mediated diseases such as rhinitis, conjunctivitis, and venom hypersensitivity [112,113,114,115]. Importantly it has also been shown to prevent onset of new sensitisations in children [116] and reduce development of asthma in patients with rhinitis caused by inhalant allergens [117,118]. Immunotherapy can improve asthma, but concern exists over the risk of adverse events [119]. In addition, the duration of treatment involving an initial course of treatment over several months and then less frequent maintenance therapy over several years requires considerable commitment by the patient to undergo a full course of treatment. Thus there remains considerable scope for improving the safety, delivery regimen and efficacy of allergen immunotherapy.

Understanding the mechanism(s) by which allergen IT works is an important route to improving efficacy (Table 19.1). Activated T cells and their products play a major role in the pathogenesis of allergic diseases and allergen-specific T cells are considered the primary target of IT [1,2,114,120,121,97,122,123,124,125,126,127]. A host of early studies suggested the association of successful IT with a

Table 19.1 The effects of allergen-specific immunotherapy on immunological parameters related to immune tolerance

	Inhibition	Enhancement
T cells	Decreased allergen-induced proliferation	Induction of Treg cells
	Suppression of Th2 cells and cytokines	Increased secretion of IL-10 and TGF-beta
	Decreased T cell numbers in late phase response	
B cells	Decreased specific IgE production	Increased specific IgG4 production
	Suppressed IgE-facilitated antigen presentation	Increased specific IgA production
Dendritic cells	Suppressed IgE-facilitated antigen presentation	
Eosinophils	Reduction of tissue numbers	
	Decrease in mediator release	
Mast cells	Reduction of tissue numbers	
	Decrease in mediator release	
	Decrease in pro-inflammatory cytokine production	
Basophils	Decrease in mediator release	
	Decrease in pro-inflammatory cytokine production	

decrease in IL-4 and IL-5 production by CD4+ Th2 cells, and a shift towards increased IFN-γ production. However, more recent studies have suggested that increased IL-10 secretion is most closely associated with successful IT. This was originally shown in IT for venom allergies, but has now also been demonstrated in studies of IT for aeroallergens. The cellular origin of IL-10 was identified as the antigen-specific T cell population, which represented CD4+CD25+ T cells, as well as monocytes and B cells [97]. In a study of bee venom IT, neutralization of IL-10 was shown to restore proliferative and effector cytokine responses to allergen in culture following IT, suggesting active control or suppression of disease promoting Th2 responses, rather than their deletion. It is interesting to note that a natural state of tolerance, induced in bee keepers following occupational exposure to allergen, is also associated with increased IL-10 production. In healthy individuals IL-10-secreting Tr1 or IL-10-Treg cells represent the dominant subset responsive to common environmental allergens, whereas a high frequency of allergen-specific IL-4 secreting T cells (Th2-like) is found in allergic individuals [96]. It would appear that some attempt to restore the "norm" is occurring in successful allergen IT.

A study of immune responses to the major house-dust mite (HDM) and birch pollen allergens, Dermatophagoides pteroynyssinus (Der p)1 and Bet v 1 respectively, demonstrated not only much lower proliferative T cell, Th1 (IFN-gamma) and Th2 (IL-5, IL-13) cytokine responses, but increased IL-10 and also TGF-beta secretion by allergen-specific T cells in healthy as compared to allergic donors [125]. Neutralization of cytokine activity indicated that T cell suppression by IL-10 and TGF-beta occurred both in normal "healthy" immunity and following IT to the mucosal allergens. Unlike with these mucosal allergens, no increases in TGF-beta production during IT were observed in venom allergy. Differences in the control mechanisms that regulate immune responses to venoms versus aeroallergens are as yet unclear, but different routes of allergen exposure are likely to be of relevance.

Inevitably the majority of studies that report changes in T cell responses to allergen in health, allergic disease and post IT, use cultures of PBMC and not from the active site of disease. However T cell responses have been examined in nasal mucosal and skin tissue after grass pollen IT, where increased IL-10 mRNA-expressing cells following IT was demonstrated during the grass pollen season. Notably, IL-10 was not increased in non-atopic subjects exposed during the pollen season, unlike studies with PBMC although increased Th1 activity was demonstrated both in the skin and nasal mucosa [128,129,130,131,132]. Studies of local tissue responses are scarce, but are highly relevant to our understanding of disease pathogenesis and pertinent mechanisms associated with health, natural tolerance and disease resolution. To date the findings in tissues do not necessarily reflect induction of peripheral tolerance and/or Treg induction, but may represent a number of other mechanisms including cell apoptosis, migration, homing and survival signals, which are likely to be dependent upon natural allergen exposure and environmental factors [133]. The effects of allergen IT specifically on naturally occurring Treg is less well documented, but published studies suggest little effect [134].

A number of pre-clinical and clinical initiatives to improve both the safety and efficacy of allergen IT are under investigation [135]. For example, modified allergens, with lower allergenicity and less likely to cause adverse events, are being developed for IT [136]). The use of allergen derived peptide fragments have also been tested, which based on their failure to bind IgE were thought unlikely to cause EAR. Despite isolated LAR at high peptide doses through as yet undefined mechanisms, cat allergen peptide IT was demonstrated to have some efficacy and to promote peptide specific, IL-10 secreting T cells [137]. Often allergen dose for optimal efficacy of IT is limited by risk of adverse advents and pretreatment for 9 weeks with the monoclonal anti-IgE antibody omalizumab reduced systemic reactions during rush immunotherapy 5-fold and allowed further build-up at weekly intervals without systemic reactions [138].

Novel adjuvants to use in combination with allergen tolerising regimens are also of interest to improve both safety and efficacy. Ongoing initiatives include the use of bacterial products likely to either promote Treg or deviate immune responses away from the disease promoting Th2 response. For example TLR agonists are of interest, either alone or directly conjugated to allergen (reviewed in [66]). The TLR9 ligand, an immunostimulatory DNA sequence – cytosine phosphorothionate guanosine (CpG) DNA – conjugated to the major allergen of ragweed resulted in impressive improvement in symptoms during the first pollen season that persisted during the second pollen season without any further administration of the conjugate. The conjugate was reportedly less immunogenic and limited mechanistic studies demonstrated deviation of disease promoting Th2 responses towards Th1 responses [68,67]. A role for Treg in such therapy has not yet been reported. A number of other microbial derived components are of interest as adjuvants to promote either T cell deviation or induction of Treg populations e.g. probiotics, Mycobacterial derived derivatives [105,139,140].

Non antigen-specific therapies, such as glucocorticoids, also appear to promote Treg function. For example glucocorticoids induce the synthesis of IL-10 both in patients [141,142] and in vitro [106]. The likely relevance to clinical efficacy of steroid action is illustrated by the significantly reduced capacity of T cells from asthma patients who are clinically insensitive to steroid treatment, to respond to dexamethasone in culture by the synthesis of IL-10. This is in marked contrast to T cells from either healthy donors or steroid-sensitive asthma patients of comparable age and disease severity [143]. These steroid insensitive patients represent a challenge to the respiratory physician, utilize a disproportionate percent of the asthma health care budgets and represent those most at risk of hospitalization or death from their asthma. The active form of vitamin D, $1\alpha 25$-dihydroxyvitamin D3 enhances steroid induced IL-10 production and an IL-10-Treg phenotype with the capacity to inhibit naïve T cells as well as Th1 and Th2 effector cells [107]. Notably, combining dexamethasone and $1\alpha 25$-dihydroxyvitamin D3 in cultures of T cells from steroid-insensitive asthma patients greatly enhanced IL-10 synthesis, and these were now comparable to levels observed in steroid–sensitive patients stimulated in the presence of steroid alone [109]. In a small study ingestion of $1\alpha 25$-dihydroxyvitamin D3 (calcitriol/rolcatrol), by 3 steroid insensitive asthma patients for just one week, was sufficient to restore their responsiveness to steroids for the induction of IL-10 synthesis [109]. The dose of calcitriol used was the standard dose recommended by the British National Formulary for steroid-induced osteoporosis. These studies open up the exciting possibility of a potential clinical steroid-sparing effect of 1α25dihydroxyvitamin D3 in asthma. This could be applicable not only to those patients who are completely refractory to steroid treatment, but also to those who demonstrate high dose long-term dependency for steroids. Limited mechanistic studies suggest 1α25dihydroxyvitamin D3 may act to rescue the glucocorticoid receptor from ligand-dependent down regulation [109].

Conclusions

A fine balance between allergen-specific Th2 cells and Treg, both naturally occurring and antigen inducible inhibitory-cytokine secreting Treg, is likely to be crucial in determining the development of allergic diseases versus the maintenance of health. The importance of the naturally occurring Treg compartment is highlighted by studies in IPEX/XLAAD patients who lack this population and suffer from severe allergic manifestations from an early age. Additionally naturally occurring Treg function appears to be overcome by inflammatory stimuli, including cytokines, associated with inflammation, and active or severe allergic disease. The potential of inducible cytokine secreting Treg, with the capacity to inhibit pro-inflammatory cytokines, to restore the naturally occurring Treg compartment and immune balance thus exists.

Existing successful therapies for allergic disease, notably allergen immunotherapy, appear to act by favourably altering this effector to regulatory T cell balance, and promoting inhibitory cytokine secreting Treg. Nevertheless this treatment is only successful in selected patient groups and is associated with significant risk of adverse events. A need for novel and improved therapies remains. These include the use of adjuvants or modified allergen preparations, to increase safety and efficacy. Potential adjuvants being considered include small-molecular weight compounds, pathogen-derived molecules and non-specific agents such as glucocorticoids that promote the generation of Treg cells or increase their suppressive properties.

Allergen-specific Treg cell populations are difficult to grow, expand and clone in vitro, and adoptive transfer of these cells in other immune-mediated conditions such as transplantation is being considered. However it is debatable whether these will ever be considered a safe, cost effective and practical treatment option in allergic and asthmatic disease. An attractive alternative therapeutic approach is to induce de novo, boost or restore endogenous Treg numbers and/or function in patients in vivo. Allergen immunotherapy appears to work via the induction of Treg populations, however these regimens require further refining for improved efficacy and safety.

In addition to the treatment of established allergy, it is essential to consider prophylactic approaches before initial sensitization has taken place, for example in young children or even in utero. Preventive vaccines that induce Treg responses could be developed, and allergen-specific Treg cells, causing them to become predominant may in turn dampen both the Th1 and Th2 cells and their cytokines, ensuring a well-balanced immune response. This might be of great value, but particularly challenging to develop, for children at high familial risk of the development of allergic disease.

In conclusion, the application of recent knowledge in peripheral tolerance mechanisms and Treg biology, offers a more rational and safer approach for the treatment, prevention and cure of allergic diseases.

Acknowledgments The author gratefully acknowledges support from the European Union (EURO-Thymaide), the Medical Research Council of Great Britain, and Asthma UK.

References

1. Kay, A.B. 2001. Allergy and allergic diseases. First of two parts. *N Engl J Med* 344:30–37.
2. Kay, A.B. 2001. Allergy and allergic diseases. Second of two parts. *N Engl J Med* 344: 109–113.
3. Cookson, W.O.C.M., and M.F. Moffatt. 1997. Asthma: An epidemic in the absence of infection? *Science* 275:41–42.
4. Jones, G. 2007. Susceptibility to asthma and eczema from mucosal and epidermal expression of distinctive genes. *Curr Allergy Asthma Rep* 7:11–17.
5. Anderson, H.R., R. Gupta, D.P. Strachan, and E.S. Limb. 2007. 50 years of asthma: UK trends from 1955 to 2004. *Thorax* 62:85–90.
6. Burney, P., E. Malmberg, S. Chinn, D. Jarvis, C. Luczynska, and E. Lai. 1997. The distribution of total and specific serum IgE in the European community respiratory health survey. *J Allergy Clin Immunol* 99:314–322.
7. Yazdanbakhsh, M., P.G. Kremsner, and R. van Ree. 2002. Allergy, parasites, and the hygiene hypothesis. *Science* 296:490–494.
8. Strachan, D.P. 1989. Hay fever, hygiene, and household size. *BMJ Clin Res Ed* 299: 1259–1260.
9. Strachan, D. 1996. Socioeconomic factors and the development of allergy. *Toxicol lett* 86:199–203.
10. Wills-Karp, M., J. Santeliz, and C.L. Karp. 2001. The germless theory of allergic disease: Revisiting the hygiene hypothesis. *Nat Rev Immunol* 1:69–75.
11. Bach, J.F. 2002. The effect of infections on susceptibility to autoimmune and allergic diseases. *N Engl J Med* 347:911–920.
12. Barnes, P.J. 2006. Drugs for asthma. *Br J Pharmacol* 147 Suppl 1:S297–303.
13. Larche, M., C.A. Akdis, and R. Valenta. 2006. Immunological mechanisms of allergen-specific immunotherapy. *Nat Rev Immunol* 6:761–771.
14. Passalacqua, G., and S.R. Durham. 2007. Allergic rhinitis and its impact on asthma update: Allergen immunotherapy. *J Allergy Clin Immunol* 119:881–891.
15. Macaubas, C., R.H. DeKruyff, and D.T. Umetsu. 2003. Respiratory tolerance in the protection against asthma. *Curr Drug Targets Inflamm Allergy* 2:175–186.
16. Maloy, K.J., and F. Powrie. 2001. Regulatory T cells in the control of immune pathology. *Nat Immunol* 2:816–822.
17. Bluestone, J.A., and A.K. Abbas. 2003. Natural versus adaptive regulatory T cells. *Nat Rev Immunol* 3:253–257.
18. O'Garra, A., and P. Vieira. 2004. Regulatory T cells and mechanisms of immune system control. *Nat Med* 10:801–805.
19. Shevach, E. 2004. CD4+CD25+ suppressor T cells: More questions than answers. *Nat Rev Immunol* 2:389–400.

20. Kronenberg, M., and A. Rudensky. 2005. Regulation of immunity by self-reactive T cells. *Nature* 435:598–604.
21. Sakaguchi, S., M. Ono, R. Setoguchi, H. Yagi, S. Hori, Z. Fehervari, J. Shimizu, T. Takahashi, and T. Nomura. 2006. Foxp3+ CD25+ CD4+ natural regulatory T cells in dominant self-tolerance and autoimmune disease. *Immunol Rev* 212:8–27.
22. Stock, P., T. Kallinich, O. Akbari, D. Quarcoo, K. Gerhold, U. Wahn, D.T. Umetsu, and E. Hamelmann. 2004. CD8(+) T cells regulate immune responses in a murine model of allergen-induced sensitization and airway inflammation. *Eur J Immunol* 34:1817–1827.
23. Noble, A., A. Giorgini, and J.A. Leggat. 2006. Cytokine-induced IL-10-secreting CD8 T cells represent a phenotypically distinct suppressor T-cell lineage. *Blood* 107:4475–4483.
24. Strober, S., L. Cheng, D. Zeng, R. Palathumpat, S. Dejbakhsh-Jones, P. Huie, and R. Sibley. 1996. Double negative (CD4–CD8– alpha beta+) T cells which promote tolerance induction and regulate autoimmunity. *Immunol Rev* 149:217–230.
25. Seo, N., Y. Tokura, M. Takigawa, and K. Egawa. 1999. Depletion of IL-10- and TGF-beta-producing regulatory gamma delta T cells by administering a daunomycin-conjugated specific monoclonal antibody in early tumor lesions augments the activity of CTLs and NK cells. *J Immunol* 163:242–249.
26. Hayday, A., and R. Tigelaar. 2003. Immunoregulation in the tissues by gammadelta T cells. *Nat Rev Immunol* 3:233–242.
27. Jiang, S., R.I. Lechler, X.S. He, and J.F. Huang. 2006. Regulatory T cells and transplantation tolerance. *Hum Immunol* 67:765–776.
28. Thomson, C.W., B.P. Lee, and L. Zhang. 2006. Double-negative regulatory T cells: nonconventional regulators. *Immunol Res* 35:163–178.
29. Akbari, O., R.H. DeKruyff, and D.T. Umetsu. 2001. Pulmonary dendritic cells producing IL-10 mediate tolerance induced by respiratory exposure to antigen. *Nat Immunol* 2:725–731.
30. Steinbrink, K., M. Wolfl, H. Jonuleit, J. Knop, and A. Enk. 1997. Induction of tolerance by IL-10-treated dendritic cells. *J Immunol* 159:4772–4780.
31. Mauri, C., D. Gray, N. Mushtaq, and M. Londei. 2003. Prevention of arthritis by interleukin 10-producing B cells. *J Exp Med* 197:489–501.
32. Moore, K.W., A. O'Garra, R. de Waal Malefyt, P. Vieira, and T.R. Mosmann. 1993. Interleukin-10. *Annu Rev Immunol* 11:165–190.
33. Li, M.O., Y.Y. Wan, S. Sanjabi, A.K. Robertson, and R.A. Flavell. 2006. Transforming growth factor-beta regulation of immune responses. *Annu Rev Immunol* 24:99–146.
34. Rubtsov, Y.P., and A.Y. Rudensky. 2007. TGFbeta signalling in control of T-cell-mediated self-reactivity. *Nat Rev Immunol* 7:443–453.
35. Jordan, M.S., A. Boesteanu, A.J. Reed, A.L. Petrone, A.E. Holenbeck, M.A. Lerman, A. Naji, and A.J. Caton. 2001. Thymic selection of CD4+CD25+ regulatory T cells induced by an agonist self-peptide. *Nat Immunol* 2:301–306.
36. Liston, A., and A.Y. Rudensky. 2007. Thymic development and peripheral homeostasis of regulatory T cells. *Curr Opin Immunol* 19:176–185.
37. Apostolou, I., and H. von Boehmer. 2004. In vivo instruction of suppressor commitment in naive T cells. *J Exp Med* 199:1401–1408.
38. Kretschmer, K., I. Apostolou, D. Hawiger, K. Khazaie, M.C. Nussenzweig, and H. von Boehmer. 2005. Inducing and expanding regulatory T cell populations by foreign antigen. *Nat Immunol* 6:1219–1227.
39. von Boehmer, H. 2005. Peptide-based instruction of suppressor commitment in naive T cells and dynamics of immunosuppression in vivo. *Scand J Immunol* 62(Suppl 1):49–54.
40. Chen, W., W. Jin, N. Hardegen, K.J. Lei, L. Li, N. Marinos, G. McGrady, and S.M. Wahl. 2003. Conversion of peripheral CD4+CD25– naive T cells to CD4+CD25+ regulatory T cells by TGF-beta induction of transcription factor Foxp3. *J Exp Med* 198:1875–1886.
41. Cosmi, L., F. Liotta, R. Angeli, B. Mazzinghi, V. Santarlasci, R. Manetti, L. Lasagni, V. Vanini, P. Romagnani, E. Maggi, F. Annunziato, and S. Romagnani. 2004. Th2 cells are

less susceptible than Th1 cells to the suppressive activity of CD25+ regulatory thymocytes because of their responsiveness to different cytokines. *Blood* 103:3117–3121.
42. Stassen, M., H. Jonuleit, C. Muller, M. Klein, C. Richter, T. Bopp, S. Schmitt, and E. Schmitt. 2004. Differential regulatory capacity of CD25+ T regulatory cells and pre-activated CD25+ T regulatory cells on development, functional activation, and proliferation of Th2 cells. *J Immunol* 173:267–274.
43. Bennett, C.L., J. Christie, F. Ramsdell, M.E. Brunkow, P.J. Ferguson, L. Whitesell, T.E. Kelly, F.T. Saulsbury, P.F. Chance, and H.D. Ochs. 2001. The immune dysregulation, polyendocrinopathy, enteropathy, X-linked syndrome (IPEX) is caused by mutations of FOXP3. *Nat Genet* 27:20–21.
44. Chatila, T.A., F. Blaeser, N. Ho, H.M. Lederman, C. Voulgaropoulos, C. Helms, and A.M. Bowcock. 2000. JM2, encoding a fork head-related protein, is mutated in X-linked autoimmunity-allergic disregulation syndrome. *J Clin Invest* 106:R75–81.
45. Wildin, R.S., F. Ramsdell, J. Peake, F. Faravelli, J.L. Casanova, N. Buist, E. Levy-Lahad, M. Mazzella, O. Goulet, L. Perroni, F.D. Bricarelli, G. Byrne, M. McEuen, S. Proll, M. Appleby, and M.E. Brunkow. 2001. X-linked neonatal diabetes mellitus, enteropathy and endocrinopathy syndrome is the human equivalent of mouse scurfy. *Nat Genet* 27: 18–20.
46. Lin, W., N. Truong, W.J. Grossman, D. Haribhai, C.B. Williams, J. Wang, M.G. Martin, and T.A. Chatila. 2005. Allergic dysregulation and hyperimmunoglobulinemia E in Foxp3 mutant mice. *J Allergy Clin Immunol* 116:1106–1115.
47. Ling, E.M., T. Smith, X.D. Nguyen, C. Pridgeon, M. Dallman, J. Arbery, V.A. Carr, and D.S. Robinson. 2004. Relation of CD4+CD25+ regulatory T-cell suppression of allergen-driven T-cell activation to atopic status and expression of allergic disease. *Lancet* 363: 608–615.
48. Cavani, A., F. Nasorri, C. Ottaviani, S. Sebastiani, O. De Pita, and G. Girolomoni. 2003. Human CD25+ regulatory T cells maintain immune tolerance to nickel in healthy, nonallergic individuals. *J Immunol* 171:5760–5768.
49. Karlsson, M.R., J. Rugtveit, and P. Brandtzaeg. 2004. Allergen-responsive CD4+CD25+ regulatory T cells in children who have outgrown cow's milk allergy. *J Exp Med* 199: 1679–1688.
50. Hawrylowicz, C.M., and A. O'Garra. 2005. Potential role of interleukin-10-secreting regulatory T cells in allergy and asthma. *Nat Rev Immunol* 5:271–283.
51. Hawrylowicz, C.M. 2005. Regulatory T cells and IL-10 in allergic inflammation. *J Exp Med* 202:1459–1463.
52. Kearley, J., J.E. Barker, D.S. Robinson, and C.M. Lloyd. 2005. Resolution of airway inflammation and hyperreactivity after in vivo transfer of CD4+CD25+ regulatory T cells is interleukin 10 dependent. *J Exp Med* 202:1539–1547.
53. Joetham, A., K. Takada, C. Taube, N. Miyahara, S. Matsubara, T. Koya, Y.H. Rha, A. Dakhama, and E.W. Gelfand. 2007. Naturally occurring lung CD4+CD25+ T cell regulation of airway allergic responses depends on IL-10 induction of TGF-beta. *J Immunol* 178: 1433–1442.
54. Lewkowich, I.P., N.S. Herman, K.W. Schleifer, M.P. Dance, B.L. Chen, K.M. Dienger, A.A. Sproles, J.S. Shah, J. Kohl, Y. Belkaid, and M. Wills-Karp. 2005. CD4+CD25+ T cells protect against experimentally induced asthma and alter pulmonary dendritic cell phenotype and function. *J Exp Med* 202:1549–1561.
55. Strickland, D.H., P.A. Stumbles, G.R. Zosky, L.S. Subrata, J.A. Thomas, D.J. Turner, P.D. Sly, and P.G. Holt. 2006. Reversal of airway hyperresponsiveness by induction of airway mucosal CD4+CD25+ regulatory T cells. *J Exp Med* 203:2649–2660.
56. Pasare, C., and R. Medzhitov. 2003. Toll pathway-dependent blockade of CD4+CD25+ T cell-mediated suppression by dendritic cells. *Science* 299:1030–1031.
57. Doganci, A., T. Eigenbrod, N. Krug, G.T. De Sanctis, M. Hausding, V.J. Erpenbeck, B. Haddad el, H.A. Lehr, E. Schmitt, T. Bopp, K.J. Kallen, U. Herz, S. Schmitt, C. Luft,

O. Hecht, J.M. Hohlfeld, H. Ito, N. Nishimoto, K. Yoshizaki, T. Kishimoto, S. Rose-John, H. Renz, M.F. Neurath, P.R. Galle, and S. Finotto. 2005. The IL-6R alpha chain controls lung CD4+CD25+ Treg development and function during allergic airway inflammation in vivo. *J Clin Invest* 115:313–325.
58. van Amelsfort, J.M., J.A. van Roon, M. Noordegraaf, K.M. Jacobs, J.W. Bijlsma, F.P. Lafeber, and L.S. Taams. 2007. Proinflammatory mediator-induced reversal of CD4+,CD25+ regulatory T cell-mediated suppression in rheumatoid arthritis. *Arthritis Rheum* 56:732–742.
59. Nadkarni, S., C. Mauri, and M.R. Ehrenstein. 2007. Anti-TNF-alpha therapy induces a distinct regulatory T cell population in patients with rheumatoid arthritis via TGF-beta. *J Exp Med* 204:33–39.
60. Valencia, X., G. Stephens, R. Goldbach-Mansky, M. Wilson, E.M. Shevach, and P.E. Lipsky. 2006. TNF downmodulates the function of human CD4+CD25hi T-regulatory cells. *Blood* 108:253–261.
61. Ruprecht, C.R., M. Gattorno, F. Ferlito, A. Gregorio, A. Martini, A. Lanzavecchia, and F. Sallusto. 2005. Coexpression of CD25 and CD27 identifies FoxP3+ regulatory T cells in inflamed synovia. *J Exp Med* 201:1793–1803.
62. Patel, M., D. Xu, P. Kewin, B. Choo-Kang, C. McSharry, N.C. Thomson, and F.Y. Liew. 2005. TLR2 agonist ameliorates established allergic airway inflammation by promoting Th1 response and not via regulatory T cells. *J Immunol* 174:7558–7563.
63. Sutmuller, R.P., M.E. Morgan, M.G. Netea, O. Grauer, and G.J. Adema. 2006. Toll-like receptors on regulatory T cells: expanding immune regulation. *Trends immunol* 27:387–393.
64. Crellin, N.K., R.V. Garcia, O. Hadisfar, S.E. Allan, T.S. Steiner, and M.K. Levings. 2005. Human CD4+ T cells express TLR5 and its ligand flagellin enhances the suppressive capacity and expression of FOXP3 in CD4+CD25+ T regulatory cells. *J Immunol* 175:8051–8059.
65. Peng, G., Z. Guo, Y. Kiniwa, K.S. Voo, W. Peng, T. Fu, D.Y. Wang, Y. Li, H.Y. Wang, and R.F. Wang. 2005. Toll-like receptor 8-mediated reversal of CD4+ regulatory T cell function. *Science* 309:1380–1384.
66. Goldman, M. 2007. Translational mini-review series on Toll-like receptors: Toll-like receptor ligands as novel pharmaceuticals for allergic disorders. *Clin Exp Immunol* 147:208–216.
67. Creticos, P.S., J.T. Schroeder, R.G. Hamilton, S.L. Balcer-Whaley, A.P. Khattignavong, R. Lindblad, H. Li, R. Coffman, V. Seyfert, J.J. Eiden, and D. Broide. 2006. Immunotherapy with a ragweed-toll-like receptor 9 agonist vaccine for allergic rhinitis. *N Engl J Med* 355:1445–1455.
68. Simons, F.E., Y. Shikishima, G. Van Nest, J.J. Eiden, and K.T. HayGlass. 2004. Selective immune redirection in humans with ragweed allergy by injecting Amb a 1 linked to immunostimulatory DNA. *J Allergy Clin Immunol* 113:1144–1151.
69. Bellinghausen, I., B. Klostermann, J. Knop, and J. Saloga. 2003. Human CD4+CD25+ T cells derived from the majority of atopic donors are able to suppress TH1 and TH2 cytokine production. *J Allergy Clin Immunol* 111:862–868.
70. Grindebacke, H., K. Wing, A.C. Andersson, E. Suri-Payer, S. Rak, and A. Rudin. 2004. Defective suppression of Th2 cytokines by CD4CD25 regulatory T cells in birch allergics during birch pollen season. *Clin Exp Allergy* 34:1364–1372.
71. Bellinghausen, I., B. Konig, I. Bottcher, J. Knop, and J. Saloga. 2005. Regulatory activity of human CD4 CD25 T cells depends on allergen concentration, type of allergen and atopy status of the donor. *Immunology* 116:103–111.
72. Hartl, D., B. Koller, A.T. Mehlhorn, D. Reinhardt, T. Nicolai, D.J. Schendel, M. Griese, and S. Krauss-Etschmann. 2007. Quantitative and functional impairment of pulmonary CD4+CD25hi regulatory T cells in pediatric asthma. *J Allergy Clin Immunol* 119: 1258–1266.
73. Verhagen, J., M. Akdis, C. Traidl-Hoffmann, P. Schmid-Grendelmeier, D. Hijnen, E.F. Knol, H. Behrendt, K. Blaser, and C.A. Akdis. 2006. Absence of T-regulatory cell expression and function in atopic dermatitis skin. *J Allergy Clin Immunol* 117:176–183.

74. Karagiannidis, C., M. Akdis, P. Holopainen, N.J. Woolley, G. Hense, B. Ruckert, P.Y. Mantel, G. Menz, C.A. Akdis, K. Blaser, and C.B. Schmidt-Weber. 2004. Glucocorticoids upregulate FOXP3 expression and regulatory T cells in asthma. *J Allergy Clin Immunol* 114:1425–1433.
75. Chen, X., J.J. Oppenheim, R.T. Winkler-Pickett, J.R. Ortaldo, and O.M. Howard. 2006. Glucocorticoid amplifies IL-2-dependent expansion of functional FoxP3(+)CD4(+)CD25(+) T regulatory cells in vivo and enhances their capacity to suppress EAE. *Eur J Immunol* 36:2139–2149.
76. Dao Nguyen, X., and D.S. Robinson. 2004. Fluticasone propionate increases CD4CD25 T regulatory cell suppression of allergen-stimulated CD4CD25 T cells by an IL-10-dependent mechanism. *J Allergy Clin Immunol* 114:296–301.
77. Battaglia, M., A. Stabilini, B. Migliavacca, J. Horejs-Hoeck, T. Kaupper, and M.G. Roncarolo. 2006. Rapamycin promotes expansion of functional CD4+CD25+FOXP3+ regulatory T cells of both healthy subjects and type 1 diabetic patients. *J Immunol* 177: 8338–8347.
78. Levings, M.K., R. Sangregorio, and M.G. Roncarolo. 2001. Human CD25(+)CD4(+) t regulatory cells suppress naive and memory T cell proliferation and can be expanded in vitro without loss of function. *J. Exp. Med.* 193:1295–1302.
79. Bluestone, J.A. 2005. Regulatory T-cell therapy: is it ready for the clinic? *Nat Rev Immunol* 5:343–349.
80. Kuball, J., M.L. Dossett, M. Wolfl, W.Y. Ho, R.H. Voss, C. Fowler, and P.D. Greenberg. 2007. Facilitating matched pairing and expression of TCR-chains introduced into human T-cells. *Blood Mar* 15; 109:2331–2338.
81. Faria, A.M., and H.L. Weiner. 2006. Oral tolerance and TGF-beta-producing cells. *Inflamm Allergy Drug Targets* 5:179–190.
82. Kunzmann, S., P.-Y. Mantel, J.G. Wohlfahrt, M. Akdis, K. Blaser, and C.B. Schmidt-Weber. 2003. Histamine enhances TGF-beta1-mediated suppression of Th2 responses. *Faseb J.* 17:1089–1095.
83. Hansen, G., J.J. McIntire, V.P. Yeung, G. Berry, G.J. Thorbecke, L. Chen, R.H. DeKruyff, and D.T. Umetsu. 2000. CD4(+) T helper cells engineered to produce latent TGF-beta1 reverse allergen-induced airway hyperreactivity and inflammation. *J. Clin. Invest.* 105: 61–70.
84. Scherf, W., S. Burdach, and G. Hansen. 2005. Reduced expression of transforming growth factor beta 1 exacerbates pathology in an experimental asthma model. *Eur J Immunol* 35:198–206.
85. Nakao, A., S. Miike, M. Hatano, K. Okumura, T. Tokuhisa, C. Ra, and I. Iwamoto. 2000. Blockade of transforming growth factor beta/Smad signaling in T cells by overexpression of Smad7 enhances antigen-induced airway inflammation and airway reactivity. *J Exp Med* 192:151–158.
86. Ostroukhova, M., C. Seguin-Devaux, T.B. Oriss, B. Dixon-McCarthy, L. Yang, B.T. Ameredes, T.E. Corcoran, and A. Ray. 2004. Tolerance induced by inhaled antigen involves CD4(+) T cells expressing membrane-bound TGF-beta and FOXP3. *J Clin Invest* 114:28–38.
87. Gauldie, J., M. Kolb, K. Ask, G. Martin, P. Bonniaud, and D. Warburton. 2006. Smad3 signaling involved in pulmonary fibrosis and emphysema. *Proc Am Thorac Soc* 3:696–702.
88. McMillan, S.J., G. Xanthou, and C.M. Lloyd. 2005. Manipulation of Allergen-Induced Airway Remodeling by Treatment with Anti-TGF-{beta} Antibody: Effect on the Smad Signaling Pathway. *J Immunol* 174:5774–5780.
89. Redington, A.E., J. Madden, A.J. Frew, R. Djukanovic, W.R. Roche, S.T. Holgate, and P.H. Howarth. 1997. Transforming growth factor-beta 1 in asthma. Measurement in bronchoalveolar lavage fluid. *Am J Respir Crit Care Med* 156:642–647.
90. Branton, M.H., and J.B. Kopp. 1999. TGF-beta and fibrosis. *Microbes Infect* 1:1349–1365.
91. Asadullah, K., W. Sterry, and H.D. Volk. 2003. Interleukin-10 therapy–review of a new approach. *Pharmacol Rev* 55:241–269.

92. Urry, Z., E. Xystrakis, and C.M. Hawrylowicz. 2006. Interleukin-10-secreting regulatory T cells in allergy and asthma. *Curr Allergy Asthma Rep* 6:363–371.
93. Borish, L., A. Aarons, J. Rumbyrt, P. Cvietusa, J. Negri, and S. Wenzel. 1996. Interleukin-10 regulation in normal subjects and patients with asthma. *J .Allergy Clin. Immunol.* 97: 1288–1296.
94. Lim, S., E. Crawley, P. Woo, and P.J. Barnes. 1998. Haplotype associated with low interleukin-10 production in patients with severe asthma. *Lancet* 352:113.
95. Heaton, T., J. Rowe, S. Turner, R.C. Aalberse, N. de Klerk, D. Suriyaarachchi, M. Serralha, B.J. Holt, E. Hollams, S. Yerkovich, K. Holt, P.D. Sly, J. Goldblatt, P. Le Souef, and P.G. Holt. 2005. An immunoepidemiological approach to asthma: identification of in-vitro T-cell response patterns associated with different wheezing phenotypes in children. *Lancet* 365:142–149.
96. Akdis, M., J. Verhagen, A. Taylor, F. Karamloo, C. Karagiannidis, R. Crameri, S. Thunberg, G. Deniz, R. Valenta, H. Fiebig, C. Kegel, R. Disch, C.B. Schmidt-Weber, K. Blaser, and C.A. Akdis. 2004. Immune responses in healthy and allergic individuals are characterized by a fine balance between allergen-specific T regulatory 1 and T helper 2 cells. *J Exp Med* 199:1567–1575.
97. Akdis, C.A., T. Blesken, M. Akdis, B. Wüthrich, and K. Blaser. 1998. Role of IL-10 in specific immunotherapy. *J. Clin. Invest.* 102:98–106.
98. Tiemessen, M.M., A.G. Van Ieperen-Van Dijk, C.A. Bruijnzeel-Koomen, J. Garssen, E.F. Knol, and E. Van Hoffen. 2004. Cow's milk-specific T-cell reactivity of children with and without persistent cow's milk allergy: key role for IL-10. *J Allergy Clin Immunol* 113: 932–939.
99. Stampfli, M.R., M. Cwiartka, B.U. Gajewska, D. Alvarez, S.A. Ritz, M.D. Inman, Z. Xing, and M. Jordana. 1999. Interleukin-10 gene transfer to the airway regulates allergic mucosal sensitization in mice. *Am J Respir Cell Mol Biol* 21:586–596.
100. Nakagome, K., M. Dohi, K. Okunishi, Y. Komagata, K. Nagatani, R. Tanaka, J. Miyazaki, and K. Yamamoto. 2005. In vivo IL-10 gene delivery suppresses airway eosinophilia and hyperreactivity by down-regulating APC functions and migration without impairing the antigen-specific systemic immune response in a mouse model of allergic airway inflammation. *J Immunol* 174:6955–6966.
101. Oh, J.W., C.M. Seroogy, E.H. Meyer, O. Akbari, G. Berry, C.G. Fathman, R.H. Dekruyff, and D.T. Umetsu. 2002. CD4 T-helper cells engineered to produce IL-10 prevent allergen-induced airway hyperreactivity and inflammation. *Allergy Clin Immunol* 110:460–468.
102. Cottrez, F., S.D. Hurst, R.L. Coffman, and H. Groux. 2000. T regulatory cells 1 inhibit a Th2-specific response in vivo. *J Immunol* 165:4848–4853.
103. Akbari, O., G.J. Freeman, E.H. Meyer, E.A. Greenfield, T.T. Chang, A.H. Sharpe, G. Berry, R.H. DeKruyff, and D.T. Umetsu. 2002. Antigen-specific regulatory T cells develop via the ICOS-ICOS-ligand pathway and inhibit allergen-induced airway hyperreactivity. *Nat Med* 8:1024–1032.
104. Stock, P., O. Akbari, G. Berry, G.J. Freeman, R.H. Dekruyff, and D.T. Umetsu. 2004. Induction of T helper type 1-like regulatory cells that express Foxp3 and protect against airway hyper-reactivity. *Nat Immunol* 5:1149–1156.
105. Zuany-Amorim, C., E. Sawicka, C. Manlius, A. Le Moine, L.R. Brunet, D.M. Kemeny, G. Bowen, G. Rook, and C. Walker. 2002. Suppression of airway eosinophilia by killed Mycobacterium vaccae-induced allergen-specific regulatory T-cells. *Nat. Med.* 8: 625–629.
106. Richards, D.F., M. Fernandez, J. Caulfield, and C.M. Hawrylowicz. 2000. Glucocorticoids drive human CD8(+) T cell differentiation towards a phenotype with high IL-10 and reduced IL-4, IL-5 and IL-13 production. *Eur J Immunol* 30:2344–2354.
107. Barrat, F.J., D.J. Cua, A. Boonstra, D.F. Richards, C. Crain, H.F. Savelkoul, R. de Waal-Malefyt, R.L. Coffman, C.M. Hawrylowicz, and A. O'Garra. 2002. In vitro generation of interleukin 10-producing regulatory CD4(+) T cells is induced by immunosuppressive

drugs and inhibited by T helper type 1 (Th1)- and Th2-inducing cytokines. *J Exp Med* 195: 603–616.
108. Vieira, P.L., J.R. Christensen, S. Minaee, E.J. O'Neill, F.J. Barrat, A. Boonstra, T. Barthlott, B. Stockinger, D.C. Wraith, and A. O'Garra. 2004. IL-10-secreting regulatory T cells do not express Foxp3 but have comparable regulatory function to naturally occurring CD4+CD25+ regulatory T cells. *J Immunol* 172:5986–5993.
109. Xystrakis, E., S. Kusumakar, S. Boswell, E. Peek, Z. Urry, D.F. Richards, T. Adikibi, C. Pridgeon, M. Dallman, T.K. Loke, D.S. Robinson, F.J. Barrat, A. O'Garra, P. Lavender, T.H. Lee, C. Corrigan, and C.M. Hawrylowicz. 2006. Reversing the defective induction of IL-10-secreting regulatory T cells in glucocorticoid-resistant asthma patients. *J Clin Invest* 116:146–155.
110. Adorini, L. 2002. Immunomodulatory effects of vitamin D receptor ligands in autoimmune diseases. *Int Immunopharmacol* 2:1017–1028.
111. Peek, E.J., D.F. Richards, A. Faith, P. Lavender, T.H. Lee, C.J. Corrigan, and C.M. Hawrylowicz. 2005. Interleukin-10-secreting "regulatory" T cells induced by glucocorticoids and beta2-agonists. *Am J Respir Cell Mol Biol* 33:105–111.
112. Kussebi, F., F. Karamloo, M. Akdis, K. Blaser, and C.A. Akdis. 2003. Advances in immunological treatment of allergy. *Curr Med Chem* 2:297–308.
113. Bousquet, J., R. Lockey, H.J. Malling, E. Alvarez-Cuesta, G.W. Canonica, M.D. Chapman, P.J. Creticos, J.M. Dayer, S.R. Durham, P. Demoly, R.J. Goldstein, T. Ishikawa, K. Ito, D. Kraft, P.H. Lambert, H. Lowenstein, U. Muller, P.S. Norman, R.E. Reisman, R. Valenta, E. Valovirta, and H. Yssel. 1998. Allergen immunotherapy: therapeutic vaccines for allergic diseases. World Health Organization. American academy of Allergy, Asthma and Immunology. *Ann Allergy Asthma Immunol* 81:401–405.
114. Durham, S.R., S.M. Walker, E.-V. Varga, M.R. Jacobson, F. O'Brien, W. Noble, S.J. Till, Q.A. Hamid, and K.T. Nouri-Aria. 1999. Long-term clinical efficacy of grass-pollen immunotherapy. *N Engl J Med* 341:468–475.
115. Bonifazi, F., M. Jutel, B.M. Bilo, J. Birnbaum, and U. Muller. 2005. Prevention and treatment of hymenoptera venom allergy: guidelines for clinical practice. *Allergy* 60: 1459–1470.
116. Pajno, G.B., G. Barberio, F. De Luca, L. Morabito, and S. Parmiani. 2001. Prevention of new sensitizations in asthmatic children monosensitized to house dust mite by specific immunotherapy. A six-year follow-up study. *Clin Exp Allergy* 31:1392–1397.
117. Moller, C., S. Dreborg, H.A. Ferdousi, S. Halken, A. Host, L. Jacobsen, A. Koivikko, D.Y. Koller, B. Niggemann, L.A. Norberg, R. Urbanek, E. Valovirta, and U. Wahn. 2002. Pollen immunotherapy reduces the development of asthma in children with seasonal rhinoconjunctivitis (the PAT-study). *J Allergy Clin Immunol* 109:251–256.
118. Eng, P.A., M. Reinhold, and H.P. Gnehm. 2002. Long-term efficacy of preseasonal grass pollen immunotherapy in children. *Allergy* 57:306–312.
119. Walker, S.M., G.B. Pajno, M.T. Lima, D.R. Wilson, and S.R. Durham. 2001. Grass pollen immunotherapy for seasonal rhinitis and asthma: a randomized, controlled trial. *J Allergy Clin Immunol* 107:87–93.
120. Akdis, C.A., M. Akdis, T. Blesken, D. Wymann, S.S. Alkan, U. Müller, and K. Blaser. 1996. Epitope specific T cell tolerance to phospholipase A_2 in bee venom immunotherapy and recovery by IL-2 and IL-15 in vitro. *J Clin Invest* 98:1676–1683.
121. Akdis, C.A., and K. Blaser. 1999. IL-10 induced anergy in peripheral T cell and reactivation by microenvironmental cytokines: two key steps in specific immunotherapy. *Faseb J* 13: 603–609.
122. Müller, U.R., C.A. Akdis, M. Fricker, M. Akdis, F. Bettens, T. Blesken, and K. Blaser. 1998. Successful immunotherapy with T cell epitope peptides of bee venom phospholipase A_2 induces specific T cell anergy in bee sting allergic patients. *J Allergy Clin Immunol* 101: 747–754.
123. Bellinghausen, I., G. Metz, A.H. Enk, S. Christmann, J. Knop, and J. Saloga. 1997. Insect venom immunotherapy induces interleukin-10 production and a Th2-to-Th1 shift,

and changes surface marker expression in venom-allergic subjects. *Eur J Immunol* 27: 1131–1139.
124. Marcotte, G.V., C.M. Braun, P.S. Norman, C.F. Nicodemus, A. Kagey-Sobotka, L.M. Lichtenstein, and D.M. Essayan. 1998. Effects of peptide therapy on ex vivo T-cell responses. *J Allergy Clin Immunol* 101:506–513.
125. Jutel, M., M. Akdis, F. Budak, C. Aebischer-Casaulta, M. Wrzyszcz, K. Blaser, and A.C. Akdis. 2003. IL-10 and TGF-β cooperate in regulatory T cell response to mucosal allergens in normal immunity and specific immunotherapy. *Eur J Immunol* 33: 1205–1214.
126. Jutel, M., W.J. Pichler, D. Skrbic, A. Urwyler, C. Dahinden, and U.R. Müller. 1995. Bee venom immunotherapy results in decrease of IL-4 and IL-5 and increase of IFN-g secretion in specific allergen stimulated T cell cultures. *J Immunol* 154:4178–4194.
127. Durham, S.R., and S.J. Till. 1998. Immunologic changes associated with allergen immunotherapy. *J Allergy Clin Immunol* 102:157–164.
128. Varney, V.A., Q.A. Hamid, M. Gaga, S. Ying, M. Jacobson, A.J. Frew, A.B. Kay, and S.R. Durham. 1993. Influence of grass pollen immunotherapy on cellular infiltration and cytokine mRNA expression during allergen-induced late-phase cutaneous responses. *J Clin Invest* 92:644–651.
129. Varga, E.M., P. Wachholz, K.T. Nouri-Aria, A. Verhoef, C.J. Corrigan, S.J. Till, and S.R. Durham. 2000. T cells from human allergen-induced late asthmatic responses express IL-12 receptor beta 2 subunit mRNA and respond to IL-12 in vitro. *J Immunol* 165: 2877–2885.
130. Hamid, Q.A., E. Schotman, M.R. Jacobson, S.M. Walker, and S.R. Durham. 1997. Increases in IL-12 messenger RNA+ cells accompany inhibition of allergen-induced late skin responses after successful grass pollen immunotherapy. *J Allergy Clin Immunol* 99: 254–260.
131. Durham, S.R., S. Ying, V.A. Varney, M.R. Jacobson, R.M. Sudderick, I.S. Mackay, A.B. Kay, and Q.A. Hamid. 1996. Grass pollen immunotherapy inhibits allergen-induced infiltration of CD4+ T lymphocytes and eosinophils in the nasal mucosa and increases the number of cells expressing messenger RNA for interferon-gamma. *J Allergy Clin Immunol* 97:1356–1365.
132. Wachholz, P.A., and S.R. Durham. 2004. Mechanisms of immunotherapy: IgG revisited. *Curr Opin Allergy Clin Immunol* 4:313–318.
133. Akdis, C.A., K. Blaser, and M. Akdis. 2004. Apoptosis in tissue inflammation and allergic disease. *Curr Opin Immunol* 16:717–723.
134. Smith, T.R., C. Alexander, A.B. Kay, M. Larche, and D.S. Robinson. 2004. Cat allergen peptide immunotherapy reduces CD4(+) T cell responses to cat allergen but does not alter suppression by CD4(+) CD25(+) T cells: a double-blind placebo-controlled study. *Allergy* 59:1097–1101.
135. Nelson, H.S. 2007. Allergen immunotherapy: where is it now? *J Allergy Clin Immunol* 119:769–779.
136. Valenta, R., and V. Niederberger. 2007. Recombinant allergens for immunotherapy. *J Allergy Clin Immunol* 119:826–830.
137. Larche, M. 2007. Update on the current status of peptide immunotherapy. *J Allergy Clin Immunol* 119:906–909.
138. Casale, T.B., W.W. Busse, J.N. Kline, Z.K. Ballas, M.H. Moss, R.G. Townley, M. Mokhtarani, V. Seyfert-Margolis, A. Asare, K. Bateman, and Y. Deniz. 2006. Omalizumab pretreatment decreases acute reactions after rush immunotherapy for ragweed-induced seasonal allergic rhinitis. *J Allergy Clin Immunol* 117:134–140.
139. Ulisse, S., P. Gionchetti, S. D'Alo, F.P. Russo, I. Pesce, G. Ricci, F. Rizzello, U. Helwig, M.G. Cifone, M. Campieri, and C. De Simone. 2001. Expression of cytokines, inducible nitric oxide synthase, and matrix metalloproteinases in pouchitis: effects of probiotic treatment. *Am J gastroenterol* 96:2691–2699.

140. Guarner, F., R. Bourdet-Sicard, P. Brandtzaeg, H.S. Gill, P. McGuirk, W. van Eden, J. Versalovic, J.V. Weinstock, and G.A. Rook. 2006. Mechanisms of disease: the hygiene hypothesis revisited. *Nat Clin Pract* 3:275–284.
141. Wan, S., J.L. LeClerc, and J.L. Vincent. 1997. Cytokine responses to cardiopulmonary bypass: lessons learned from cardiac transplantation. *Ann Thorac Surg* 63:269–276.
142. John, M., Lim, S., Seybold, J., Jose, P., Robichaud, A., O'Connor, B., Barnes, P.J., and Chung, K.F. 1998. Inhaled corticosteroids increase interleukin-10 but reduce macrophage inflammatory protein-1a, granulocyte-macrophage stimulating factor and Interferon-g release from alveolar macrophages in asthma. *Am J Respir Crit Care Med* 157:256–262.
143. Hawrylowicz, C., D. Richards, T.K. Loke, C. Corrigan, and T. Lee. 2002. A defect in corticosteroid-induced IL-10 production in T lymphocytes from corticosteroid-resistant asthmatic patients. *J Allergy Clin Immunol* 109:369–370.

Chapter 20
Regulatory T Cells and Tumour Immunotherapy

Ilona Kryczek and Weiping Zou

Abstract Immune responses influence the development and progression of a malignancy. However, the tumor can manipulate the immune system on its own end, often resulting in an ineffective tumor immunity and immune suppression, ablating the clinical efficacy of tumor therapy. An appreciation of the complexity of the interaction between tumor and host immune system is important for the development of effective cancer therapies. The chapter will emphasize regulatory T cells as a prominent mechanism whereby tumors escape tumor immunity and highlight the newly therapeutic strategies by targeting Treg cells in patients with cancer.

Introduction

In patients with cancer surgical cytoreduction, chemotherapy and radioactive therapy remain the most effective modalities to significantly reduce tumor mass, maximally minimize metastasized tumor and residual disease. However, patients often die at tumor metastasis and residual disease. Why, then, can't the immune system control tumor metastasis and residual disease? Recent evidence has pointed out that immune tolerance established during tumor development may prevent and suppress tumor immunity and disable clinical efficacy of tumor therapy.

It was suggested 40 years ago that tolerance is at least partially regulated by a subpopulation of T cells with suppressive activity [1]. However, this concept was met with skepticism in the scientific field during the last 25 years. The interest in suppressor T cells, or more commonly called regulatory T cells (Treg cells) has been initiated and enforced by Drs. Sakaguchi [2] and Shevach [3,4]. Dr. Sakaguchi demonstrated that CD25 could be a useful marker for Treg cells [2] (Fig. 20.1). Dr. Shevach established a functional assay to define Treg cells [4]. Further,

---

W. Zou
Department of Surgery, University of Michigan School of Medicine, C560B MSRB II, 1150 W. Medical Center Dr. Ann Arbor MI 48109-0669, USA
e-mail: wzou@umich.edu

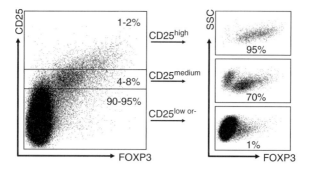

Fig. 20.1 CD25 and Foxp3 expression on human CD3$^+$CD4$^+$ cells PBMCs were isolated from healthy donor blood and stained for CD3, CD4, CD25 and Foxp3 antigens as described [88]. *Left* dot-plot shows co-staining of CD25 and Foxp3 antigens in gated CD3$^+$CD4$^+$ T cells. *Right* dot-plots show Foxp3 expression in CD25^{high}, CD25^{medium} and CD25$^{low\ or\ -}$ T cells subpopulations

Drs. Sakaguchi, Shevach and other investigators have demonstrated the immunosuppressive effects of Treg cells in multiple mouse autoimmune disease models.

Treg Cells in Patients with Cancer

In the early eighties, Drs. North, Bursuker and Berendt showed that CD4$^+$ T cells isolated from tumor bearing mice inhibit tumor rejection [5–7]. It is now well known that CD4$^+$CD25$^+$ Treg cells are responsible for this suppression [8,9]. In the last 10 years, high levels of Treg cells were found in many human tumors including ovarian cancer [10,11], lung cancer [11,12], gastric cancer [13], breast cancer [14], pancreas [14] and gastric carcinomas [13], melanoma [15], Hodgkin's [16] and non-Hodgkin's [17] lymphomas [15,18–20]. Further, tumor Treg cells are inversely associated with patient survival [10,13]. Given the suppressive nature of Treg cells in vitro and ex vivo in human studies as well as the in vivo suppressive capacity in multiple mouse models, it is reasonable to presume that Treg cells suppress tumor immunity in vivo in patients with cancer. Manipulation of Treg cells is a novel strategy to treat patients with cancer (reviewed in [21,22]) (Fig. 20.2).

Depletion of CD25$^+$ Cells in Patients with Cancer

CD25 is a useful phenotypic marker for Treg cells [2]. Anti-CD25 antibody has been used to deplete Treg cells in mice in vivo. Depletion of CD25$^+$ cells reduced tumor growth, induced tumor regression, increased survival and improved tumor immunity in mice [8,23–26]. The data lead to the human clinical trials to deplete Treg cells by administrating Ontak (denileukin diftitox). Ontak has been approved by the Food and Drug Administration (FDA) to treat CD25$^+$ cutaneous T-cell leukemia/lymphoma. In Ontak, the interleukin (IL)-2 gene is genetically fused to the enzymatically active and translocating domains of diphtheria toxin [27,28]. Ontak,

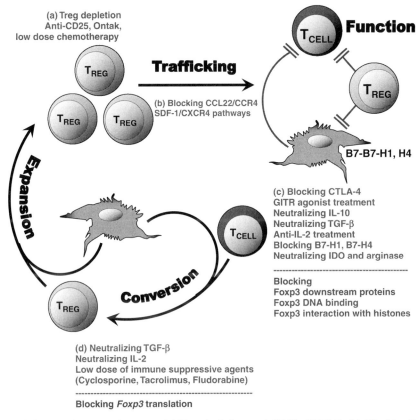

Fig. 20.2 Targeting Tregs in cancer (**a**) Treg depletion: anti-CD25, ONTAK. (**b**) Blocking Treg tumour trafficking. (**c**) Reducing Treg function. (**d**) Blocking Treg expansion and conversion *(See also Color Insert)*

after bounding by IL-2 receptor, is translocated into the cytosol, where inhibition by diphtheria toxin protein synthesis results in cell apoptosis.

In vitro experiments showed that Ontak (Table 20.1) eliminated Treg cells without affecting $CD25^{-/intermediate}$ lymphocytes in peripheral blood mononuclear cells in patients with renal cell carcinoma [29]. A single dose of Ontak (12 µg/kg) reduced phenotypic and functional peripheral $CD4^+CD25^+$ Treg cells in lung, ovarian and breast cancer patients [30]. Although an objective clinical response was observed in one ovarian cancer [30], there are insufficient data to evaluate the clinical efficacy of Ontak treatment. One study examined the combination of Ontak with dendritic cell (DC) vaccination. In this study, the administration of single dose of denileukin diftitox into renal cells carcinoma patients resulted in reduction of Treg cells in blood and in abrogation of Treg-mediated suppressive activity. In contrast, Ontak was reported to have no significant effects on Treg cells in patients with metastatic melanoma [31]. Variable effects of Ontak treatment could be related to Ontak dose, administration regimen, type of cancer and probably other factors.

Table 20.1 In vivo and in vitro effect of ONTAK on human Treg cells

Cancer type	Decreased CD4+CD25+ cells		Decreased Foxp3+ cells		Suppressive activity		Ref
	invitro	in vivo	in vitro	in vivo	in vitro	in vivo	
Renal cell carcinoma	+	+	+	+	+	+	[29]
Melanoma or Renal cell carcinoma	+	+/–*	–	+/–*	–	+/–*	[31]
Lung, ovarian or breast cancer	ND	+	ND	+	ND	+	[30]

Nonetheless, Ontak administration has to be optimized for maximal efficiency of reducing Treg cells, but not other cell depletion. For example, it was reported that more than four weekly infusions of Ontak resulted in a depletion of effector T cells [30] (Table 20.1). The study in patients with metastatic melanoma also showed that Ontak could reduce the number of CD8$^+$CD25$^+$ T cells and that this ability is strictly related to the Ontak dose [31]. Notably, the potential risk of depleting CD25 effector T cells may be higher in human than in mice. Mouse CD25$^+$ T cells are mostly Foxp3$^+$ Treg cells. It is well known that human CD4$^+$CD25^{high} cells are predominately Foxp3$^+$ cells. However, human CD25$^{low/medium}$ T cells include both Foxp3$^+$ and Foxp3$^-$ cells [32] (Fig. 20.1). It is possible that some CD25$^{low/medium}$ T cells are effector T cells and become a target of Ontak.

On the other hand, targeting Treg cells may not be as effective in patients with advanced disease as in patients with early stage diseases. In mouse tumor models, depletion of Treg cells is significantly more efficient to induce tumor regression before tumor inoculation or at the early stage of cancer than in well-established tumors [8,23,25]. It is possible that Ontak treatment in advance human cancers have to be combined with other modalities to boost TAA-specific immunity. Indeed Ontak was combined with transfected by tumor RNA-transduced DC vaccination [29]. The authors observed that tumor-specific T cell response was enhanced in patients treated with Ontak and DC vaccination compared to those receiving the vaccine alone. Tumor specific T cell immunity induced by CpG vaccination is suppressed by activated Treg in tumor tissue [15]. It is suggested that immune stimulation or vaccination after Ontak treatment could maximize the protective effects of tumor effector T cells.

Modification of Treg Cell Function

Targeting of CTLA-4$^+$ Cells

CTLA-4 binds the CD80 (B7-1) and CD86 (B7-2) ligands on APCs. The affinity of CTLA4 for binding both CD80 and CD86 is 20–100 fold higher than that of CD28 [33]. In opposition to the stimulatory role of CD28 in T cell activation, CTLA4 transmits the inhibitory signals and plays an essential role in preventing the development of immune pathology. Cross-linking of CTLA-4 downregulates T cell proliferation and cytokine production, whereas blockade of CTLA-4 enhances T cell activation [34,35]. Loss of CTLA-4 in CTLA-4-deficient mice leads to lymphoproliferative disease resulting in multiorgan and tissue destruction [36–38]. Human CTLA-4 polymorphism is associated with autoimmune disorders [39]. These observations indicate a critical role for CTLA4 in the negative regulation of the Tcell compartment.

Interestingly, CTLA-4 is constitutively expressed by Treg cells, but not naïve T cells [40–42]. Activation of Tcells induces CTLA-4 expression. However, the level of CTLA-4 at the cell surface remains limited due to the rapid endocytosis of surface CTLA-4. In mouse models, anti-CTLA-4 antibody administered before tumor inoculation or in an early stage resulted in the tumor rejection; meanwhile, when it was administered to the established tumor, tumor immunity was improved and tumor regression was observed (reviewed in [21,22]). However, the treatment appeared to result in an increase of autoimmune disease [43], especially in gastrointestinal disturbances [44,45].

Anti-CTLA-4 antibody has been used in clinical trials in treating patients with certain cancers. Anti-CTLA-4 induces durable objective clinical responses (Table 20.2) associated with the induction of autoimmune adverse effects [46]. Anti-CTLA-4 therapy induces much higher incidences of autoimmune diseases than anti-CD25. It suggests that anti-CTLA-4 and anti-CD25 may act differently including at different targets. Further, anti-CTLA-4 did not reduce the absolute numbers of $CD25^+$ Treg cells in mice. Multiple doses of anti-CTLA-4 did not change CD25 and *Foxp3* mRNA expression [47], although a slight decrease in the percentage of $CD4^+CD25^+$ was observed (Table 20.2). Anti-CTLA-4 failed to alter the suppressive activity of human $CD4^+CD25^+$ Treg cells [48–51].

Anti-CTLA4 antibody may abrogate mouse Treg cell-mediated suppression in vitro and in vivo [41,42,52–54]. However, the mechanism by which CTLA-4 influences the suppressive effect of Treg cells remains unclear in mice and humans. CTLA-4 ligation can induce TGF-β production [55]. TGF-β may in turn support Treg cell function and expansion. One possibility is that Treg cells may enable suppressor activity to APCs through CTLA-4 by triggering indolamine 2,3 dioxygenase (IDO) expression. Nonetheless, although anti-CTLA-4 may predominantly target non-Treg cells in humans, blocking CTLA4 signals could be an option for treating cancer patients.

Anti-GITR Antibody

Treg cells, but not fresh $CD25^-$ T cells express this member of the TNFR family [56,57]. But GITR is not specific for Treg cells. Expression of GITR is rapidly induced in $CD25^-$ T cells by TCR engagement [57,58] and in APCs by activation

Table 20.2 *In vivo* effect of anti-CTLA-4 antibody treatment

Cancer	Treatment anti-CTLA-4 mAb	Combination	Case (n)	In vivo effects on Treg cells	objective clinical response	Ref
melanoma	10 or 15 mg/kg every 3 months	—	12	Decreased $CD4^+CD25^+$ (2.2%)	17%	[108]
melanoma	intrapatient dose-escalation (3 to 9 mg/kg)	—	46	Decreased $CD4^+CD25^+$ (0.5 to 6.8%)	11%	[46]
melanoma	intrapatient dose-escalation (3 to 9 mg/kg)	—	4	No change in $CD4^+CD25^+$, Increased $Foxp3^+$ suppression	—	[47]
melanoma	3 mg/kg, single dose	IL-2	36	—	21%	[109]
melanoma	3 mg/kg every 3 weeks	peptide vaccine	29	Decreased $CD4^+CD25^+$ (2.6%)	14%	[45]
melanoma	Dose reduced from 3 to 1 mg/kg every 3 weeks.	peptide vaccine	27		11%	[45]
melanoma	3 mg/kg every 3 weeks	No known	113	—	13%	[110]
melanoma	3 mg/kg every 3 weeks	peptide vaccine	14	—	21%	[43]
ovarian carcinoma	3 mg/kg, single dose	tumor vaccine	2	—	—	[111]
melanoma	3 mg/kg, single dose	tumor vaccine	3	—	—	[111]
melanoma	3 mg/kg, single dose	melanosomal antigens	4	—	—	[111]

—, not tested.

stimuli (e.g. lipopolysaccharide) [57,59]. In a later phase of T cell response, GITR expression is reduced in both T cells and APC but constitutively expressed in Treg cells.

Agonistic anti-GITR monoclonal antibody, DTA1, can reduce the suppressor function of Treg cells [56,57] and co-stimulate effector function of T cells [58–61]. Administration of DTA1 protected animals against secondary melanoma [62] and stimulated immunity against MethA-induced fibrosarcoma [25]. Moreover anti-GITR treatment synergized with CTLA-4 neutralization, but not with Treg cell depletion [25]. The contribution of modification of Treg cell function to tumor regression remains controversial. It has been shown that DTA1 can stimulate Treg cell proliferation in the presence of IL-2 without antigen stimulation [59,63]. Treg cells obtained from GITR deficient mice and wild type were equally able to exert suppressor activity, suggesting that GITR is not essential for Treg cell development and function [63]. Addition of anti-GITR antibodies to cocultures of Treg cells and $CD25^-$ T cells from wild-type and $GITR^{-/-}$ mice demonstrated that the target for the Ab-mediated reversal of suppression was the $CD25^-$ T cell, but not Treg cells [61]. Nonetheless, it is possible that DTA1 treatment has certain effects on Treg cells but predominantly stimulates T cells activation. The activated T cells would escape from Treg cell mediated suppression. The potential clinical usage of anti-GITR needs to be carefully examined.

Modulation of FOXP3 Molecular Signals

The Foxp3-dependent genetic program has remained largely unknown. Recently, Wu et al. proposed that a single transcription factor, NFAT (nuclear factor of activated T cells), can direct T cell activation if it is bound to AP-1, or T cell tolerance if it interacts with FOXP3 [64]. NFAT and Foxp3 complex binds to the forkhead element in the ARRE-2 region of the *IL-2* promoter and represses *IL-2* transcription through active deacetylation of histone H3 [64,65]. Interaction of Foxp3 with the NFAT also blocked IL-4 expression by T helper in the presence of Foxp3 [66]. NFAT DNA binding activity is dependent on Ca^{2+} and controlled by calcineurin [67]. Cyclosporin A (CsA), tacrolimus (FK 506) and other calcineurin inhibitors suppress NFAT [68]. In fact, low doses of CSA induce Treg cell apoptosis and result in loss of suppressive activity of Treg cells [62,69–71]. Treatment with calcineurin inhibitors including tacrolimus [72,73] decreased Treg cells in periphery. Acting on NFAT in Treg cells may be one of the possible mechanisms of calcineurin inhibitors.

Immunosuppressants such as alkylating agents, corticosteroids and fluderabine achieve their suppressive activities through targeting the STAT1 pathway [74]. STAT1 and Jak2 expression are significantly increased after TCR stimulation in Treg cells, but not in other cells [56]. $STAT1^{-/-}$ Treg cells are functionally impaired [75]. Those results suggest that STAT1 signaling might be required for the development or/and functional activity of Treg cells. In support of this, the levels and function of Treg cells were reduced in a majority of patients with CLL treated with fludarabine-containing therapy regimens [76].

Thus, an attractive strategy to break Treg cell-mediated tolerance is to develop small molecules that block NFAT and Foxp3 interaction without interfering with the interaction of NFAT and AP-1 or that specifically interfere *FOXP3* translation activator complex and the *FOXP3* repressors.

Modification of Treg Cell Trafficking

It has been reported that a majority of human peripheral blood CD4$^+$CD25$^+$ Treg cells express lymphoid homing molecules CD62L and CCR7 [77,78]. However, Treg cells cells also express variable levels of peripheral tissue-homing molecules such as CLA, CCR4, CCR8 and α4β7 and αEβ7 integrins [10,79–85]. One hypothesis is that Treg cells may traffic into effector sites in an organ or tissue specific manner. Specific trafficking molecules may control Treg cell migration [21,22, 86,87]. It is possible that blockade of tumor-specific Treg cell trafficking may lead to potent tumor immunity and tumor regression (Fig. 20.2).

Indeed, human CCR4$^+$ Treg cells migrate toward CCL22 produced in the tumor microenvironment and blockade of CCL22 and CCR4 pathway in vivo reduces Treg cell tumor trafficking [10]. Treg cells can trigger APC IL-10 production, which in turn stimulates B7-H4 expression and renders APCs suppressive through B7-H4 [88]. It suggests that specific blockade of Treg cell trafficking into the tumor environment could elicit potent tumor immunity by promoting APC function.

Modulation of Treg Cell Expansion

Treg cell expansion can be controlled by cytokine milieu. Schramm et al. [89] and Chen et al. [90] demonstrated that TGF-β and TCR stimulation convert naive mouse CD4$^+$CD25$^-$ T cells into CD4$^+$CD25$^+$ Treg cells. These studies were confirmed by subsequent experiments in both mouse and human systems [90–99]. Further, it appears that a tumor selectively promotes the proliferation of Treg cells via TGF-β produced by a subset of DCs exhibiting a myeloid immature phenotype [100,101]. TGF-β not only impacts Treg cell expansion, but also on their natural suppressive activity [102,103]. The data suggest that blocking TGFβ production or signal pathway may be therapeutically meaningful. In support of this concept, treatment with anti-TGF-β antibody decreases the conversion of T cells into Treg cells in mice [104]. In addition to Treg cells, TGF-β can be a growth factor for multiple epithelial tumors (review in [105]). Thus, blocking the TGF-β signal pathway may function as one stone capable of killing two birds, Treg cells and tumor cells.

Several TGFβ signaling antagonists are under active investigation. NovaRx is conducting Phase I/II clinical trials on a TGFβ$_2$ antisense-modified allogeneic tumor-cell vaccine. In a preclinical study, rats bearing intracranial gliomas were immunized with gliosarcoma cells expressing TGFβ antisense plasmid. All the treated animals survived for the 12-week study period compared with only 13%

of the control animals [106]. Optimistically, 2 out of 6 patients treated with autolougus tumor cells expressing TGFβ antisense (AP 12009) showed objective clinical response, and 2 out of 6 patients exhibited stable disease [107]. In April 2005, an international open-label, active-controlled, randomized parallel-group phase IIb trial in adult patients with recurrent high-grade glioma was completed with 145 patients enrolled [105]. Patients are currently either on treatment or in follow-up. No clinical data have been published yet.

Concluding Remarks

Treg cells play a key role in induction of tumor related immune suppression. In order to engender potent tumor immunity and lead to tumor regression, Treg cells could be targeted in multiple ways (Fig. 20.2). The direct approach is to specifically deplete Treg cells. However, the lack of a specific surface marker makes this strategy less attractive. There are several other options including controlling of Treg cell expansion, their trafficking into tumors or tumor-draining lymph nodes and inhibiting their function. The increasing knowledge about molecular mechanisms controlling Foxp3 expression and action may lead to develop specific inhibitors to precisely inhibit Treg cells development and function.

References

1. Gershon, R.K. and K. Kondo, Cell interactions in the induction of tolerance: the role of thymic lymphocytes. *Immunology*, 1970, **18**(5): 723–37.
2. Sakaguchi, S., et al., Immunologic self-tolerance maintained by activated T cells expressing IL-2 receptor alpha-chains (CD25). Breakdown of a single mechanism of self-tolerance causes various autoimmune diseases. *J Immunol*, 1995, **155**(3): 1151–64.
3. Suri-Payer, E., et al., CD4+CD25+ T cells inhibit both the induction and effector function of autoreactive T cells and represent a unique lineage of immunoregulatory cells. *J Immunol*, 1998, **160**(3): 1212–8.
4. Thornton, A.M. and E.M. Shevach, CD4+CD25+ immunoregulatory T cells suppress polyclonal T cell activation in vitro by inhibiting interleukin 2 production. *J Exp Med*, 1998, **188**(2): 287–96.
5. Berendt, M.J. and R.J. North, T-cell-mediated suppression of anti-tumor immunity. An explanation for progressive growth of an immunogenic tumor. *J Exp Med*, 1980, **151**(1): 69–80.
6. Bursuker, I. and R.J. North, Generation and decay of the immune response to a progressive fibrosarcoma. II. Failure to demonstrate postexcision immunity after the onset of T cell-mediated suppression of immunity. *J Exp Med*, 1984, **159**(5): 1312–21.
7. North, R.J. and I. Bursuker, Generation and decay of the immune response to a progressive fibrosarcoma. I. Ly-1+2- suppressor T cells down-regulate the generation of Ly-1-2+ effector T cells. *J Exp Med*, 1984, **159**(5): 1295–311.
8. Onizuka, S., et al., Tumor rejection by in vivo administration of anti-CD25 (interleukin-2 receptor alpha) monoclonal antibody. *Cancer Res*, 1999, **59**(13): 3128–33.
9. Sakaguchi, S., Naturally arising Foxp3-expressing CD25+CD4+ regulatory T cells in immunological tolerance to self and non-self. *Nat Immunol*, 2005, **6**(4): 345–52.

10. Curiel, T.J., et al., Specific recruitment of regulatory T cells in ovarian carcinoma fosters immune privilege and predicts reduced survival. *Nat Med*, 2004, **10**(9): 942–9.
11. Woo, E.Y., et al., Regulatory CD4(+)CD25(+) T cells in tumors from patients with early-stage non-small cell lung cancer and late-stage ovarian cancer. *Cancer Res*, 2001, **61**(12): 4766–72.
12. Ishibashi, Y., et al., Expression of Foxp3 in non-small cell lung cancer patients is significantly higher in tumor tissues than in normal tissues, especially in tumors smaller than 30 mm. *Oncol Rep*, 2006, **15**(5): 1315–9.
13. Sasada, T., et al., CD4+CD25+ regulatory T cells in patients with gastrointestinal malignancies: possible involvement of regulatory T cells in disease progression. *Cancer*, 2003, **98**(5): 1089–99.
14. Liyanage, U.K., et al., Prevalence of regulatory T cells is increased in peripheral blood and tumor microenvironment of patients with pancreas or breast adenocarcinoma. *J Immunol*, 2002, **169**(5): 2756–61.
15. Appay, V., et al., New generation vaccine induces effective melanoma-specific CD8+ T cells in the circulation but not in the tumor site. *J Immunol*, 2006, **177**(3): 1670–8.
16. Ishida, T., et al., Specific recruitment of CC chemokine receptor 4-positive regulatory T cells in Hodgkin lymphoma fosters immune privilege. *Cancer Res*, 2006, **66**(11): 5716–22.
17. Yang, Z.Z., et al., Intratumoral CD4+CD25+ regulatory T-cell-mediated suppression of infiltrating CD4+ T cells in B-cell non-Hodgkin lymphoma. *Blood*, 2006, **107**(9): 3639–46.
18. Meloni, F., et al., Foxp3 expressing CD4+ CD25+ and CD8+CD28- T regulatory cells in the peripheral blood of patients with lung cancer and pleural mesothelioma. *Hum Immunol*, 2006, **67**(1–2): 1–12.
19. Ormandy, L.A., et al., Increased populations of regulatory T cells in peripheral blood of patients with hepatocellular carcinoma. *Cancer Res*, 2005, **65**(6): 2457–64.
20. Lau, K.M., et al., Increase in circulating Foxp3+CD4+CD25(high) regulatory T cells in nasopharyngeal carcinoma patients. *Br J Cancer*, 2007, **96**(4): 617–22.
21. Zou, W., Immunosuppressive networks in the tumour environment and their therapeutic relevance. *Nat Rev Cancer*, 2005, **5**(4): 263–74.
22. Zou, W., Regulatory T cells, tumour immunity and immunotherapy. *Nat Rev Immunol*, 2006, **6**(4): 295–307.
23. Shimizu, J., S. Yamazaki, and S. Sakaguchi, Induction of tumor immunity by removing CD25+CD4+ T cells: a common basis between tumor immunity and autoimmunity. *J Immunol*, 1999, **163**(10): 5211–8.
24. Jones, E., et al., Depletion of CD25+ regulatory cells results in suppression of melanoma growth and induction of autoreactivity in mice. *Cancer Immun*, 2002, **2**: 1.
25. Ko, K., et al., Treatment of advanced tumors with agonistic anti-GITR mAb and its effects on tumor-infiltrating Foxp3+CD25+CD4+ regulatory T cells. *J Exp Med*, 2005, **202**(7): 885–91.
26. Casares, N., et al., CD4+/CD25+ regulatory cells inhibit activation of tumor-primed CD4+ T cells with IFN-gamma-dependent antiangiogenic activity, as well as long-lasting tumor immunity elicited by peptide vaccination. *J Immunol*, 2003, **171**(11): 5931–9.
27. Foss, F.M., DAB(389)IL-2 (denileukin diftitox, ONTAK): a new fusion protein technology. *Clin Lymphoma*, 2000, **1**(Suppl 1): S27–31.
28. Foss, F.M., DAB(389)IL-2 (ONTAK): a novel fusion toxin therapy for lymphoma. *Clin Lymphoma*, 2000, **1**(2): 110–6; discussion 117.
29. Dannull, J., et al., Enhancement of vaccine-mediated antitumor immunity in cancer patients after depletion of regulatory T cells. *J Clin Invest*, 2005, **115**(12): 3623–33.
30. Barnett, B., et al., Regulatory T cells in ovarian cancer: biology and therapeutic potential. *Am J Reprod Immunol*, 2005, **54**(6): 369–77.
31. Attia, P., et al., Inability of a fusion protein of IL-2 and diphtheria toxin (Denileukin Diftitox, DAB389IL-2, ONTAK) to eliminate regulatory T lymphocytes in patients with melanoma. *J Immunother*, 2005, **28**(6): 582–92.

32. Liu, W., et al., CD127 expression inversely correlates with FoxP3 and suppressive function of human CD4+ T reg cells. *J Exp Med*, 2006, **203**(7): 1701–11.
33. Collins, A.V., et al., The interaction properties of costimulatory molecules revisited. *Immunity*, 2002, **17**(2): 201–10.
34. Walunas, T.L., et al., CTLA-4 can function as a negative regulator of T cell activation. *Immunity*, 1994, **1**(5): 405–13.
35. Krummel, M.F. and J.P. Allison, CD28 and CTLA-4 have opposing effects on the response of T cells to stimulation. *J Exp Med*, 1995, **182**(2): 459–65.
36. Waterhouse, P., et al., Lymphoproliferative disorders with early lethality in mice deficient in Ctla-4. *Science*, 1995, **270**(5238): 985–8.
37. Kulkarni, A.B., et al., Transforming growth factor beta 1 null mutation in mice causes excessive inflammatory response and early death. *Proc Natl Acad Sci USA*, 1993, **90**(2): 770–4.
38. Tivol, E.A., et al., Loss of CTLA-4 leads to massive lymphoproliferation and fatal multiorgan tissue destruction, revealing a critical negative regulatory role of CTLA-4. *Immunity*, 1995, **3**(5): 541–7.
39. Ueda, H., et al., Association of the T-cell regulatory gene CTLA4 with susceptibility to autoimmune disease. *Nature*, 2003, **423**(6939): 506–11.
40. Sansom, D.M. and L.S. Walker, The role of CD28 and cytotoxic T-lymphocyte antigen-4 (CTLA-4) in regulatory T-cell biology. *Immunol Rev*, 2006, **212**: 131–48.
41. Read, S., V. Malmstrom, and F. Powrie, Cytotoxic T lymphocyte-associated antigen 4 plays an essential role in the function of CD25(+)CD4(+) regulatory cells that control intestinal inflammation. *J Exp Med*, 2000, **192**(2): 295–302.
42. Takahashi, T., et al., Immunologic self-tolerance maintained by CD25(+)CD4(+) regulatory T cells constitutively expressing cytotoxic T lymphocyte-associated antigen 4. *J Exp Med*, 2000, **192**(2): 303–10.
43. Phan, G.Q., et al., Cancer regression and autoimmunity induced by cytotoxic T lymphocyte-associated antigen 4 blockade in patients with metastatic melanoma. *Proc Natl Acad Sci USA*, 2003, **100**(14): 8372–7.
44. Sanderson, K., et al., Autoimmunity in a phase I trial of a fully human anti-cytotoxic T-lymphocyte antigen-4 monoclonal antibody with multiple melanoma peptides and Montanide ISA 51 for patients with resected stages III and IV melanoma. *J Clin Oncol*, 2005, **23**(4): 741–50.
45. Attia, P., et al., Autoimmunity correlates with tumor regression in patients with metastatic melanoma treated with anti-cytotoxic T-lymphocyte antigen-4. *J Clin Oncol*, 2005, **23**(25): 6043–53.
46. Maker, A.V., et al., Intrapatient dose escalation of anti-CTLA-4 antibody in patients with metastatic melanoma. *J Immunother*, 2006, **29**(4): 455–63.
47. Maker, A.V., P. Attia, and S.A. Rosenberg, Analysis of the cellular mechanism of antitumor responses and autoimmunity in patients treated with CTLA-4 blockade. *J Immunol*, 2005, **175**(11): 7746–54.
48. Levings, M.K., R. Sangregorio, and M.G. Roncarolo, Human cd25(+)cd4(+) t regulatory cells suppress naive and memory T cell proliferation and can be expanded in vitro without loss of function. *J Exp Med*, 2001, **193**(11): 1295–302.
49. Jonuleit, H., et al., Identification and functional characterization of human CD4(+)CD25(+) T cells with regulatory properties isolated from peripheral blood. *J Exp Med*, 2001, **193**(11): 1285–94.
50. Baecher-Allan, C., et al., CD4+CD25+ regulatory cells from human peripheral blood express very high levels of CD25 ex vivo. *Novartis Found Symp*, 2003, **252**: 67–88; discussion 88–91, 106–14.
51. Ng, W.F., et al., Human CD4(+)CD25(+) cells: a naturally occurring population of regulatory T cells. *Blood*, 2001, **98**(9): 2736–44.
52. Liu, H., et al., CD4+CD25+ regulatory T cells cure murine colitis: the role of IL-10, TGF-beta, and CTLA4. *J Immunol*, 2003, **171**(10): 5012–7.

53. Salomon, B., et al., B7/CD28 costimulation is essential for the homeostasis of the CD4+CD25+ immunoregulatory T cells that control autoimmune diabetes. *Immunity*, 2000, **12**(4): 431–40.
54. Read, S., et al., Blockade of CTLA-4 on CD4+CD25+ regulatory T cells abrogates their function in vivo. *J Immunol*, 2006, **177**(7): 4376–83.
55. Chen, W., W. Jin, and S.M. Wahl, Engagement of cytotoxic T lymphocyte-associated antigen 4 (CTLA-4) induces transforming growth factor beta (TGF-beta) production by murine CD4(+) T cells. *J Exp Med*, 1998, **188**(10): 1849–57.
56. McHugh, R.S., et al., CD4(+)CD25(+) immunoregulatory T cells: gene expression analysis reveals a functional role for the glucocorticoid-induced TNF receptor. *Immunity*, 2002, **16**(2): 311–23.
57. Shimizu, J., et al., Stimulation of CD25(+)CD4(+) regulatory T cells through GITR breaks immunological self-tolerance. *Nat Immunol*, 2002, **3**(2): 135–42.
58. Kanamaru, F., et al., Costimulation via glucocorticoid-induced TNF receptor in both conventional and CD25+ regulatory CD4+ T cells. *J Immunol*, 2004, **172**(12): 7306–14.
59. Shevach, E.M. and G.L. Stephens, The GITR-GITRL interaction: co-stimulation or contrasuppression of regulatory activity? *Nat Rev Immunol*, 2006, **6**(8): 613–8.
60. Kohm, A.P., J.S. Williams, and S.D. Miller, Cutting edge: ligation of the glucocorticoid-induced TNF receptor enhances autoreactive CD4+ T cell activation and experimental autoimmune encephalomyelitis. *J Immunol*, 2004, **172**(8): 4686–90.
61. Stephens, G.L., et al., Engagement of glucocorticoid-induced TNFR family-related receptor on effector T cells by its ligand mediates resistance to suppression by CD4+CD25+ T cells. *J Immunol*, 2004, **173**(8): 5008–20.
62. Turk, M.J., et al., Concomitant tumor immunity to a poorly immunogenic melanoma is prevented by regulatory T cells. *J Exp Med*, 2004, **200**(6): 771–82.
63. Ronchetti, S., et al., GITR, a member of the TNF receptor superfamily, is costimulatory to mouse T lymphocyte subpopulations. *Eur J Immunol*, 2004, **34**(3): 613–22.
64. Wu, Y., et al., FOXP3 controls regulatory T cell function through cooperation with NFAT. *Cell*, 2006, **126**(2): 375–87.
65. Chen, C., et al., Transcriptional regulation by Foxp3 is associated with direct promoter occupancy and modulation of histone acetylation. *J Biol Chem*, 2006, **281**(48): 36828–34.
66. Bettelli, E., M. Dastrange, and M. Oukka, Foxp3 interacts with nuclear factor of activated T cells and NF-kappa B to repress cytokine gene expression and effector functions of T helper cells. *Proc Natl Acad Sci USA*, 2005, **102**(14): 5138–43.
67. Loh, C., et al., Calcineurin binds the transcription factor NFAT1 and reversibly regulates its activity. *J Biol Chem*, 1996, **271**(18): 10884–91.
68. Ho, S., et al., The mechanism of action of cyclosporin A and FK506. *Clin Immunol Immunopathol*, 1996, **80**(3 Pt 2): S40–5.
69. Awwad, M. and R.J. North, Cyclophosphamide (Cy)-facilitated adoptive immunotherapy of a Cy-resistant tumour. Evidence that Cy permits the expression of adoptive T-cell mediated immunity by removing suppressor T cells rather than by reducing tumour burden. *Immunology*, 1988, **65**(1): 87–92.
70. Ghiringhelli, F., et al., CD4+CD25+ regulatory T cells suppress tumor immunity but are sensitive to cyclophosphamide which allows immunotherapy of established tumors to be curative. *Eur J Immunol*, 2004, **34**(2): 336–44.
71. Lutsiak, M.E., et al., Inhibition of CD4(+)25+ T regulatory cell function implicated in enhanced immune response by low-dose cyclophosphamide. *Blood*, 2005, **105**(7): 2862–8.
72. Caproni, M., et al., The effects of tacrolimus ointment on regulatory T lymphocytes in atopic dermatitis. *J Clin Immunol*, 2006, **26**(4): 370–5.
73. San Segundo, D., et al., Calcineurin inhibitors affect circulating regulatory T cells in stable renal transplant recipients. *Transplant Proc*, 2006, **38**(8): 2391–3.
74. Frank, D.A., S. Mahajan, and J. Ritz, Fludarabine-induced immunosuppression is associated with inhibition of STAT1 signaling. *Nat Med*, 1999, **5**(4): 444–7.

75. Nishibori, T., et al., Impaired development of CD4+ CD25+ regulatory T cells in the absence of STAT1: increased susceptibility to autoimmune disease. *J Exp Med*, 2004, **199**(1): 25–34.
76. Beyer, M., et al., Reduced frequencies and suppressive function of CD4+CD25hi regulatory T cells in patients with chronic lymphocytic leukemia after therapy with fludarabine. *Blood*, 2005, **106**(6): 2018–25.
77. Baecher-Allan, C., et al., CD4+CD25high regulatory cells in human peripheral blood. *J Immunol*, 2001, **167**(3): 1245–53.
78. Hoffmann, P., et al., Large-scale in vitro expansion of polyclonal human CD4(+)CD25high regulatory T cells. *Blood*, 2004, **104**(3): 895–903.
79. Cavani, A., et al., Human CD25+ regulatory T cells maintain immune tolerance to nickel in healthy, nonallergic individuals. *J Immunol*, 2003, **171**(11): 5760–8.
80. Iellem, A., et al., Unique chemotactic response profile and specific expression of chemokine receptors CCR4 and CCR8 by CD4(+)CD25(+) regulatory T cells. *J Exp Med*, 2001, **194**(6): 847–53.
81. Iellem, A., L. Colantonio, and D. D'Ambrosio, Skin-versus gut-skewed homing receptor expression and intrinsic CCR4 expression on human peripheral blood CD4+CD25+ suppressor T cells. *Eur J Immunol*, 2003, **33**(6): 1488–96.
82. Stassen, M., et al., Human CD25+ regulatory T cells: two subsets defined by the integrins alpha 4 beta 7 or alpha 4 beta 1 confer distinct suppressive properties upon CD4+ T helper cells. *Eur J Immunol*, 2004, **34**(5): 1303–11.
83. Allakhverdi, Z., et al., Expression of CD103 identifies human regulatory T-cell subsets. *J Allergy Clin Immunol*, 2006, **118**(6): 1342–9.
84. Chen, X., et al., Pertussis toxin as an adjuvant suppresses the number and function of CD4+CD25+ T regulatory cells. *Eur J Immunol*, 2006, **36**(3): 671–80.
85. Hultkrantz, S., S. Ostman, and E. Telemo, Induction of antigen-specific regulatory T cells in the liver-draining celiac lymph node following oral antigen administration. *Immunology*, 2005, **116**(3): 362–72.
86. Wei, S., I. Kryczek, and W. Zou, Regulatory T-cell compartmentalization and trafficking. *Blood*, 2006, **108**(2): 426–31.
87. Zou, L., et al., Bone marrow is a reservoir for CD4+CD25+ regulatory T cells that traffic through CXCL12/CXCR4 signals. *Cancer Res*, 2004, **64**(22): 8451–5.
88. Kryczek, I., et al., Cutting edge: induction of B7-H4 on APCs through IL-10: novel suppressive mode for regulatory T cells. *J Immunol*, 2006, **177**(1): 40–4.
89. Schramm, C., et al., TGFbeta regulates the CD4+CD25+ T-cell pool and the expression of Foxp3 in vivo. *Int Immunol*, 2004, **16**(9): 1241–9.
90. Chen, W., et al., Conversion of peripheral CD4+CD25- naive T cells to CD4+CD25+ regulatory T cells by TGF-beta induction of transcription factor Foxp3. *J Exp Med*, 2003, **198**(12): 1875–86.
91. Walker, M.R., et al., Induction of FoxP3 and acquisition of T regulatory activity by stimulated human CD4+CD25- T cells. *J Clin Invest*, 2003, **112**(9): 1437–43.
92. Walker, M.R., et al., De novo generation of antigen-specific CD4+CD25+ regulatory T cells from human CD4+CD25- cells. *Proc Natl Acad Sci USA*, 2005, **102**(11): 4103–8.
93. Morgan, M.E., et al., Expression of FOXP3 mRNA is not confined to CD4+CD25+ T regulatory cells in humans. *Hum Immunol*, 2005, **66**(1): 13–20.
94. Curotto de Lafaille, M.A., et al., CD25- T cells generate CD25+Foxp3+ regulatory T cells by peripheral expansion. *J Immunol*, 2004, **173**(12): 7259–68.
95. Liang, S., et al., Conversion of CD4+ CD25- cells into CD4+ CD25+ regulatory T cells in vivo requires B7 costimulation, but not the thymus. *J Exp Med*, 2005, **201**(1): 127–37.
96. Apostolou, I. and H. von Boehmer, In vivo instruction of suppressor commitment in naive T cells. *J Exp Med*, 2004, **199**(10): 1401–8.
97. von Boehmer, H., Peptide-based instruction of suppressor commitment in naive T cells and dynamics of immunosuppression in vivo. *Scand J Immunol*, 2005, **62**(Suppl 1): 49–54.

98. Knoechel, B., et al., Sequential development of interleukin 2-dependent effector and regulatory T cells in response to endogenous systemic antigen. *J Exp Med*, 2005, **202**(10): 1375–86.
99. Kretschmer, K., et al., Inducing and expanding regulatory T cell populations by foreign antigen. *Nat Immunol*, 2005, **6**(12): 1219–27.
100. Ghiringhelli, F., et al., Tumor cells convert immature myeloid dendritic cells into TGF-beta-secreting cells inducing CD4+CD25+ regulatory T cell proliferation. *J Exp Med*, 2005, **202**(7): 919–29.
101. Zhou, G. and H.I. Levitsky, Natural regulatory T cells and de novo-induced regulatory T cells contribute independently to tumor-specific tolerance. *J Immunol*, 2007, **178**(4): 2155–62.
102. Nakamura, K., et al., TGF-beta 1 plays an important role in the mechanism of CD4+CD25+ regulatory T cell activity in both humans and mice. *J Immunol*, 2004, **172**(2): 834–42.
103. Ghiringhelli, F., et al., CD4+CD25+ regulatory T cells inhibit natural killer cell functions in a transforming growth factor-beta-dependent manner. *J Exp Med*, 2005, **202**(8): 1075–85.
104. Liu, V.C., et al., Tumor evasion of the immune system by converting CD4+CD25- T cells into CD4+CD25+ T regulatory cells: role of tumor-derived TGF-beta. *J Immunol*, 2007, **178**(5): 2883–92.
105. Schlingensiepen, K.H., et al., Targeted tumor therapy with the TGF-beta2 antisense compound AP 12009. *Cytokine Growth Factor Rev*, 2006, **17**(1–2): 129–39.
106. Fakhrai, H., et al., Eradication of established intracranial rat gliomas by transforming growth factor beta antisense gene therapy. *Proc Natl Acad Sci USA*, 1996, **93**(7): 2909–14.
107. Fakhrai, H., et al., Phase I clinical trial of a TGF-beta antisense-modified tumor cell vaccine in patients with advanced glioma. *Cancer Gene Ther*, 2006, **13**(12): 1052–60.
108. Reuben, J.M., et al., Biologic and immunomodulatory events after CTLA-4 blockade with ticilimumab in patients with advanced malignant melanoma. *Cancer*, 2006, **106**(11): 2437–44.
109. Maker, A.V., et al., Tumor regression and autoimmunity in patients treated with cytotoxic T lymphocyte-associated antigen 4 blockade and interleukin 2: a phase I/II study. *Ann Surg Oncol*, 2005, **12**(12): 1005–16.
110. Blansfield, J.A., et al., Cytotoxic T-lymphocyte-associated antigen-4 blockage can induce autoimmune hypophysitis in patients with metastatic melanoma and renal cancer. *J Immunother*, 2005, **28**(6): 593–8.
111. Hodi, F.S., et al., Biologic activity of cytotoxic T lymphocyte-associated antigen 4 antibody blockade in previously vaccinated metastatic melanoma and ovarian carcinoma patients. *Proc Natl Acad Sci USA*, 2003, **100**(8): 4712–7.

Chapter 21
Regulatory T Cells in Hepatitis and Hepatocellular Carcinoma

Fu-Sheng Wang and George F. Gao

Abstract Persistent viral infection in liver is still a global threat to human health. It has been suggested that compromised immune responses play a critical role for chronic viral infection. It is, at least in part, shown that impaired innate and adaptive immunity as well as liver tolerance are responsible for the compromised immune responses, but the underlying cellular and molecular mechanisms remain largely undefined. Recently, CD4+CD25+FoxP3 regulatory T cells (Treg) have been implicated in playing an important role during viral infection. This chapter not only outlines what we have learned about Treg, but also highlights the challenges we are facing regarding the origin of inducible antigen-specific Treg and how do they functionally suppress target effector immune cells such as CD4+ and CD8+ T cells, dendritic cell subsets in viral hepatitis and liver cancers in humans. Finally, we will summarize current progresses, future directions and unanswered questions regarding the potential application of Treg that could serve as a potential immunotherapeutic target for hepatitis and hepatocellular carcinoma in clinic.

Abbreviations: AHB, acute hepatitis B; CHB, chronic hepatitis B; HBV, hepatitis B virus; HCV, Hepatitis C virus; DC, dendritic cell; mDC, myeloid DC; pDC, plasmacytoid DC; HCC, hepatocellular carcinoma; LC, liver cirrhosis; Treg, T regulatory cells.

Introduction

Liver diseases such as viral hepatitis, autoimmune hepatitis and hepatocellular carcinoma (HCC) remain great challenges to human health worldwide. Patients with chronic hepatitis B virus (HBV) and/or hepatitis C virus (HCV) infection may progress to liver cirrhosis (LC) and HCC over years of infection [1,2]. HCC is

---

F.-S. Wang
Research Center for Biological Therapy, Beijing Institute of Infectious Diseases, Beijing 302 Hospital, Beijing 100039, China
e-mail: fswang@public.bta.net.cn

the fifth most common cancer worldwide with a poor prognosis. Autoimmune hepatitis is a generally progressive disease [3], for example, primary biliary cirrhosis is characterized by portal inflammation and immune-mediated destruction of the intrahepatic bile ducts, resulting in further hepatic damage, fibrosis, cirrhosis, and eventually, liver failure [4].

As for viral hepatitis, the infectant virologic and host immunological factors that serve as two major factors drive the disease progression. An increasing body of evidence has shown that host antiviral immune responses play an important role in determining the outcome of HBV infection. For example, around 90–95% of adult but only 5–10% pediatric patients who spontaneously recover from acute HBV infection typically can mount vigorous multiepitope-specific CD4+ and CD8+ T-cell responses, while 5–10% of adult and 80–90% pediatric patients who develop into chronic hepatitis B tend to have limited T-cell responses [2,5], exhibiting an anergy, exhaustion (functional inactivation) or loss of effector T cells. However, the underlying mechanisms responsible for host immune responses remain poorly understood. In addition, autoimmune liver diseases are induced by autoreactive T cells, which are activated by responding to peripheral tissue antigens. However, HCV infection can readily establish viral persistence in 75–85% individuals even in immunocompetent hosts. Recently, HBV or HCV persistence has been correlated with the increased regulatory T cells (Treg) that impair the antiviral immune responses of effector T cells, however the underlying mechanisms remain poorly understood. This raises the possibility that the interaction between antiviral effector and regulatory T cells may influence the outcome of HBV/HCV infection.

Recently, a growing body of studies has demonstrated that Treg, expressing CD4, CD25, and FoxP3 and representing 2–5% of peripheral CD4+ T cells in humans [6], play a critical role in maintaining the delicate balance of antiviral immune responses between liver immunopathology and viral resolution. However, constant inflammatory/necrotic stimulation may induce Treg to impair antiviral T-cell immunity, which favors viruses to establish and maintain a persistent infection. Many studies have shown that liver diseases, including autoimmune liver disease, viral hepatitis and liver cancer are associated with impaired Treg function [7–9]. In this chapter, we will summarize our current understanding of the prevalence and biological functions of CD4+CD25+ Treg in human liver diseases including HBV/HCV infection, hepatocellualr carcinoma and autoimmune disease, and discuss the potential of CD4+CD25+ Treg as novel immunointervention strategies for liver diseases.

CD4+ regulatory T cells are divided into two subtypes: natural (naturally occurring or innate) Treg and adaptive (acquired or induced) Treg. The best-characterized induced CD4+ Treg are Th3 and Tr1 cells. Their suppressive capacity is through a cell-to-cell contact-independent, and immunosuppressive cytokines such as IL-10 [10] (for Tr1 cells) and TGF-β (for Th3 cells) dependent mechanism [11]. Adaptive CD4+ Treg can be induced upon encountering an antigen in the absence of co-stimulation, and their induction from CD4+CD25– T cells require either a cell-to-cell contact-dependent interaction with natural CD4+CD25+ Treg or immunosuppressive cytokine IL-10 or TGF-β.

Since induced Treg can be similar to natural CD4+CD25+ Treg in phenotype, we here defined these cells that also express CD25 as adaptive CD4+CD25+ FoxP3 Treg. Surface molecule CTLA-4, a negative co-stimulation regulator, is constitutively expressed on CD4+CD25+ Treg [12]. In addition, GITR is expressed preferentially at high levels on CD4+CD25+ Treg and mediates their suppressive functions [13].

The main feature of CD4+CD25+ Treg is their ability to down-regulate both innate and adaptive immune responses. However, the underlying mechanisms have not been fully elucidated. It has been shown that the direct contact with target cells is required for the inhibitory effect of CD4+CD25+ Treg in HCV-infected individuals. However, it is, to some contents, puzzled that Treg in patients with chronic HCV infection produce IL-10 and TGF-β upon re-stimulation in-vitro, and the suppressive effect of the Treg can be neutralized by the addition of antibody against TGF-β [14]. It is likely that both activation and expansion of natural Treg and the induction of antigen-specific adaptive Treg contribute to effector T-cell hyporesponsiveness in the established viral infection, but the degree of contribution of these two subtypes of Treg remains unclear.

Treg in Immune Suppression in Chronic HBV/HCV Infection

Clinical observation and clinic studies have demonstrated that patients with chronic hepatitis manifest impaired innate and adaptive immunity as well as liver immune tolerance. However, the underlying molecular and cellular mechanisms remain elusive. Recently, a growing body of reports has showed that Treg play a role in viral infectious diseases, contributing to compromised immunity and persistent infections. These cells play a critical role in maintaining the delicate balance between preventing liver immunopathology and allowing the immune response for viral resolution.

The mechanisms responsible for the T cells to tolerate chronic HBV and HCV infection are not completely understood. It is possible that negative selection, immunological ignorance, dysregulation of lymphocytokine production, induction of anergy by high antigen doses, viral mutation escape, and deficient in antigen presentation could all contribute to the hyporesponsiveness in hosts who are continuously exposed to viral antigens [2,15]. It has been established that Treg are associated with the pathogenesis and the disease progression of HBV and HCV infection, and are correlated with HBV replication.

Treg in Chronic Hepatitis B

Three independent groups have characterized the prevalence, functions, and clinical significance of Treg in HBV-infected patients [16] and shown that Treg are associated with the chronicity of the disease in patients [17,18]. Though whether the frequency of circulating CD4+CD25+ Treg is increased in chronic HBV patients is controversial. It is prominent that the average frequency of circulating Treg was higher in HBeAg+ patients than in HBeAg- patients, indicating that increased frequency of Treg correlates with viral replication. More importantly, it was demonstrated that there was a significant increase of Treg and inflammatory cell infiltration in the liver in chronic hepatitis B, in particular, in liver cirrhosis and chronic severe hepatitis B. Franzese et al. [17] observed that depletion of CD4+CD25+ Treg increased the frequency of originally detectable HBV-specific CD8+ T cells, which could be reversed by CD4+CD25+ Treg reconstitution. Stoop et al. [18] also reported that CD4+CD25+ T cells from peripheral blood mononuclear cells (PBMC) could inhibit HBV-specific proliferation and interferon-γ production. In addition, CD4+CD25+ Treg were capable of suppressing the proliferation of autologous PBMC stimulated by HBV antigens, which implies the generation of HBV antigen-specific Treg in circulation and in the liver of HBV-infected patients. Taken together, these findings suggest that CD4+CD25+ Treg have an immunosuppressive effect on HBV-specific helper T cells, and the presence of HBV associated Treg could contribute to an inadequate immune response against the virus, leading to chronic infection. In line with the data observed in HBV-infected humans, several animal studies have shown that depletion of CD4+CD25+ Treg could enhance the HBV-specific CD8+ T-cell response primed by DNA immunization in mice [19], and increase Th1-type cytokine IFN-γ and IL-2 produced by CD4+ T cells in HBcAg immunized mice [20].

Interestingly, a longitudinal study showed that the frequency of Treg in patients with acute self-limited HBV infection was relative low at the acute phase, and significantly increased at the convalescent phase, and was restored to a normal level at the resolved phase (Fig. 21.1a,b), suggesting that Treg may limit immune-mediated

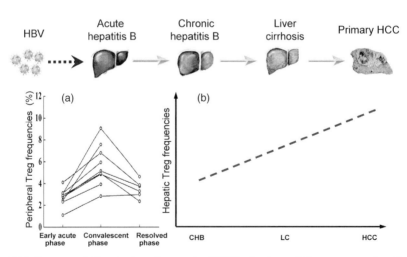

Fig. 21.1 Upper panel: Treg in the pathogenesis of HBV infection and disease progression. Down panel: (**a**) Longitudinal Treg alteration throughout the resolution of acute HBV infection; (**b**) Schematic alteration of the increased trend of hepatic Treg infiltration from CHB, to LC and finally to primary HCC *(See also Color Insert)*

liver damage by suppressing the host immune response after HBV is cleared. In addition, it was also found that there was no significant difference of phenotype in circulating Treg between patients and healthy donors, as well as among different groups of patients.

Taken together, these findings suggest that CD4+CD25+ Treg play an active role not only in modulating effectors of immune response to HBV infection, but also in influencing the disease prognosis in patients with hepatitis B [16]. In other words, this also implies that Treg-mediated suppression appears to be responsible for allowing certain viruses to establish and maintain a persistent state.

Treg in Chronic Hepatitis C

Several groups have addressed whether Treg would affect on the outcome of HCV infection by studying chronic hepatitis C infection in humans. It has been shown that the number and suppressive activity of Treg in chronic HCV patients were significant higher when compared with the controls, who have an acute resolved HCV infection. These Treg could suppress the HCV-specific reactivity of CD8+ T cells ex-vivo [21].

Cabrera et al. [14] reported that HCV-specific IFN-γ production was enhanced in PBMC after depletion of CD4+CD25+ Treg, and the production was also suppressed in PBMC enriched with CD4+CD25+ Treg via a cell-to-cell contact-dependent mechanism. There is a positive correlation of the frequency of Treg with the serum HCV RNA load, but an inverse correlation with liver inflammatory activity. Other studies have also shown that the activities of CD4+CD25+ Treg could influence the protective aspects of T-cell immunity during chronic HCV infection, providing an indirect evidence to support the notion that CD4+CD25+ Treg could influence the outcome of infection [22]. In addition, Treg have been isolated from peripheral blood and in one instance, from the liver itself, and these cells were shown to modulate peptide-specific proliferation of effector T cells have been depleted Treg in-vitro as well as the maturation of effectors of HCV-specific CD8+ T cells. The suppression was cell-to-cell contact-dependent other than via immunosuppressive cytokine IL-10 or TGF-β. Increased HCV-specific CD4+CD25+ Treg were associated with reduced HCV-specific CD4+ T-cell response in HCV-infected patients with normal versus abnormal alanine aminotransferase levels, suggesting double-edged roles of the Treg in modulating antiviral responses and controlling hepatic damage in chronic HCV carriers [23]. In addition, a recent study [24] showed that FoxP3 expression and Treg suppression were not different during early stage of infection among patients who subsequently developed persistence or resolved HCV infection spontaneously. After 6 months of infection, the resolution of disease was associated with a relative loss of functional suppression of Treg, indicating that a possible Treg suppressive function shift occurs in patients with chronic HCV infection versus the individuals with resolved HCV infection.

In accordance with these observations in human studies, CD4+CD25+ Treg also suppressed IFN-γ production, expansion, and activation-induced cell death of epitope-specific T cells in HCV-infected chimpanzees. When compared with HCV-infected human, however, contradictory data were observed that the frequency of Treg and the extent of suppression were as high in spontaneously recovered HCV-infected animals as in persistently HCV-infected animals [25], suggesting that HCV infection may influence the population of CD4+CD25+ Treg. There was also a significant difference in Treg responsiveness to IL-2 and in their TREC copy number between naïve and HCV-infected chimpanzees.

Viral Antigen-Specific CD4+CD25+ Treg and their Properties

Does viral specific Treg exist in-vivo? In other words, CD4+CD25+ Treg could preferentially inhibit HBV- or HCV-specific T-cell responses? The antigen-specific regulation of T cell immunity encompasses three stages: antigen presentation, Treg activation, and recognition of target cells by Treg [26]. Though the suppressive effect of CD4+CD25+ Treg is generally antigen nonspecific, i.e. in a bystander manner, the effect of specific inhibition of antigen-specific T-cell responses has been reported in some studies. In HIV-infected individuals, CD4+CD25+ Treg exerted their suppression in an antigen-specific manner and produced IL-10 in response to viral antigen [27]. It was shown that natural CD4+CD25+ Treg exhibited specificity for viral antigens both in their generation and function [28]. In human HBV infection, depletion of CD4+CD25+ Treg enhanced effector T-cell proliferation stimulated by HbcAg, but not by irrelevant antigens such as tetanus toxin [18]. Xu et al. [16] observed that in HBV-infected patients the depletion of CD4+CD25+ Treg led to a significant increase of IFN-γ production and cellular proliferation of PBMC stimulated by HBV antigens than by anti-CD3, and co-culture of CD4+CD25+ Treg with effector T cells led to significant suppression of HBV antigen-stimulated than anti-CD3 stimulated IFN-γ production and cellular proliferation, suggesting that suppressive effect of CD4+CD25+ Treg in HBV infection is both antigen-nonspecific and antigen-specific which are mainly endowed by activated natural CD4+CD25+ Treg and adaptive CD4+CD25+ Treg, respectively. In HCV infection, however, Treg in chronic HCV-infected patients functionally suppressed peptide-specific CD8+ T-cell responses against other viruses in addition to HCV-specific T-cell responses [22]. However, the evidence that CD4+CD25+ Treg produced IL-10 in response to viral antigen suggests that the suppressive function of Treg is viral antigen-specific [14].

In the case of epitope specificity of Treg during chronic viral infection, HBV and HCV associated antigen specific Treg were identified respectively. Using MHC class II tetramers, it was observed that the frequency of HBcAg-specific Treg declined during acute exacerbations of chronic hepatitis B. In the contrary, the frequencies of HBcAg peptide-specific cytotoxic T lymphocyte increased [29]. These results have pointed to the conclusion that HBcAg specific Treg not only exist, but also functional during HBV infection. Furthermore, Li et al. [30] identified the HCV specific epitopes of Treg. By incubation of peripheral blood mononuclear cells of HCV-positive patients with HCV protein-derived peptides, CD25high IFN-γ – Foxp3+ Treg can be rapidly induced. During the culture, HCV specific Treg failed to proliferate by stimulation with HCV-derived peptides. These Treg also did not secrete IFN-γ. The epitopes of Treg varied between patients, but within any given subject only a small number of the epitope peptides were able to stimulate Treg. These results demonstrate the occurrence of dominant Treg epitopes, suggesting that natural Treg may be implicated in host immune tolerance during HCV infection.

Double-Edged Role of Treg in Liver Diseases

The generation of Treg may be a normal physiological process that occurs to prevent immunopathological damage via suppression of cellular immune responses, while certain viruses have evolved the abilities to directly or indirectly subvert the immunosuppressive properties of Treg to help them to evade immunological destruction. On one hand, in many instances where Treg function during viral infections, they seem to impair the efficacy of protective immunity, which in turn contributes to viral persistence and perhaps to chronic disease. In such case, the Treg response would be considered as harmful. We and others have shown that the increase of prevalence of CD4+CD25+ Treg could be positively correlated with HBV or HCV viral load [14,16]. In addition, as the effector function of activated Treg can be antigen non-specific, the presence of such cells during viral infection could have bystander immunosuppressive effects against co-infecting pathogens. Furthermore, there is evidence that the efficacy of vaccines could also be hampered by the presence of CD4+CD25+ Treg.

Treg express constitutively cytotoxic T-lymphocyte antigen 4 (CTLA-4), a negative co-stimulator of immune activation. Genetic polymorphisms of CTLA-4 alleles could impair the Treg function, leading to the development of type 1 autoimmune hepatitis in individuals because of the defect of suppression by Treg [31,32]. Accordingly, liver infiltration by Treg in chronic viral hepatitis and liver cancer in situ was associated with local immune suppression leading to viral persistence and liver cancer progression [16,33]. Though Treg can be initially recruited to limit inflammation provoked by liver tissue damage, they were also required to maintain the stability of chronic inflammation. Chronic antigen exposure from tissue damage facilitated the recruitment of Treg and the establishment of the hepatic microenvironment where IL-10 and IL-4 balanced the pro-inflammatory cytokines IL-2 and IFN-γ, resulting in the stable chronic inflammation rather than fulminant hepatitis [34].

Thus, the outcome of inflammatory response in the liver is determined by the complex of local interactions between several cell types including Treg and effector cells. On the other hand, it is also evident in studies of several viral infections that Treg response may benefit the host by minimizing the tissue-damaging effects caused by the immune response against the virus. A typical example is that in HIV infection, the decline of Treg activity could hasten the progression to AIDS

[35]. In the case of persistent viral infection with HCV, evidence also supports the hypothesis that Treg play a beneficial role by inhibiting chronic inflammatory reaction to persistent replicating virus [36]. We have observed that the increase of intrahepatic FoxP3+ Treg is positively associated with the increase of CD4+ and CD8+ T cells in livers of patients with chronic hepatitis B, suggesting the enrichment of Treg is a protective response against inflammatory response in liver [16]. However, it is not clear whether Treg involved in controlling inflammation are adaptive Treg that are induced by in situ soluble inhibitory cytokines, or are natural Treg.

It seems that in the acute phase of infection, CD4+ helper T cells play a dominant role and contribute to the induction and maintenance of a functional CD8+ T-cell response, while in the chronic phase, CD4+ Treg play a dominant role and contribute to suppress virus-specific CD8+ T-cell response and thereby help the virus to persist. After the initial inflammation resolves, the suppressive ability of the CD4+CD25+ Treg would return to a normal level. We observed that there was a transient increase of circulating CD4+CD25+ Treg in patients with acute resolved hepatitis B. It is interesting that the increase of Treg was not in the early acute phase but in the convalescent phase [16]. We assume that this pattern of Treg prevalence would favor the balance of antivirus responses and tissue inflammation in acute resolved hepatitis B patients. This assumption is reinforced by the suggestion that the Treg-mediated impairment of HCV-specific CD8+ T-cell proliferation and IFN-γ secretion is relatively specific at the chronic phase of infection [16]. Further study is required to clarify this issue.

In many infections in both mice and humans, removal of CD4+CD25+ Treg has resulted in enhanced effector immune responses. Targeting the molecules involved in regulatory cell activity in-vivo, such as CTLA-4, TGF-β or IL-10, alone or in combination, has often proven effective in controlling many chronic infections [8]. In human HBV or HCV infection, however, caution should be taken when considering such strategies because removal of Treg or Treg-related factors may also enhance considerable liver damage through hyperinflammation. Perhaps the best way is to achieve selective abolishment of the Treg activity that preferentially inhibits virus-specific T-cell responses. Further exploiting our knowledge of the role and mechanisms of CD4+CD25+ Treg in immunity and tissue damage is required. Greater attention should be given to investigate the Treg properties throughout the process of HBV and HCV infection, especially in the pathogen-affected site, the liver. How host immune system delicately recalibrate the balance of quantity and quality of Treg in favor of the viral resolution needs to be further studied.

Treg promote the Progression of Primary Hepatocellular Carcinoma

Hepatocellular carcinoma (HCC) ranks the fifth most common cancer worldwide [37] with a poor prognosis despite improved diagnostic and treatment strategies. Compared with healthy controls and patients with liver cirrhosis, the frequency of circulating CD4+ CD25+ FoxP3+ Treg is increased significantly in peripheral blood

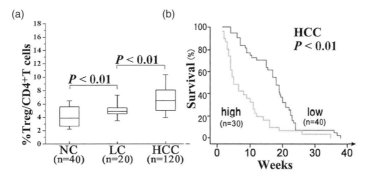

Fig. 21.2 There is an increased Treg frequency in peripheral blood, which is associated with the disease progression of primary HCC. (**a**) Treg frequency is significantly increased compared with NC and LC patients. (**b**) HCC patients with higher Treg frequency showed shorter survival time. Actual overall survival rates were analyzed by the Kaplan–Meier method and survival was measured in weeks from diagnosis to death. The log-rank test was applied to compare between the groups. Multivariate analysis of prognostic factors for overall survival was performed using the Cox proportional hazards model (Modified from Fu et al. Gastroenterology, 2007, 132: 2328–2339) *(See also Color Insert)*

[33] and associated with disease progression [7,33,38] in HCC patients (Fig. 21.2a). Furthermore, increased quantity of circulating CD4+CD25+ Treg is associated with high mortality and reduced survival time of HCC patients (Fig. 21.2b) [33]. As Treg exhibit significantly decreased levels of TRECs [39], it concluded that increased frequency of Treg in the peripheral blood of HCC patients was due to the active proliferation.

Interestingly, Fu et al. and other labs observed that abundant accumulation of Treg within liver tumor is concurrent with a significant decrease of CD8+ T cell infiltration. Immunochemical double staining analysis showed that contact of Treg and CD8+ T cells was identified in liver tumor in situ and in peritumor regions. Notably, Fu et al. have recently demonstrated that circulating Treg isolated from HCC patients potently suppressed cytokine secretion and proliferation of CD4+CD25- T cells as well as the proliferation, activation, degranulation, and production of granzymes, perforin and cytotoxic molecules (TNF-α and IFN-γ) by CD8+ T cells [7,33,38], which results in a compromised anti-tumour immunity in patients with primary HCC.

Treg in the Pathogenesis of Autoimmune Liver Diseases

Autoimmune liver disease (AILD) includes autoimmune hepatitis (AIH) and primary biliary cirrhosis (PBC) [3]. Study showed that CD4+CD25+ Treg are associated with AIH development. They can suppress the proliferation and IFN-γ production of CD8+ T cells in a direct cell-to-cell contact manner while increasing the secretion of IL-4, IL-10 and TGF-β regulatory cytokines [31]. In addition, it was observed that CD4+CD25+ Treg still retained the suppressor function to inhibit IFN-γ production by CD4+CD25- T cells despite of decreased frequency and impaired expansion ability in AIH patients [40].

Similarly, CD4+CD25+ Treg are deficient in frequency [41] and function in PBC patients [42]. These data suggest that impaired Treg function may favour the initiation and even intensify the development of AILD. In addition, immunohistochemical analyses showed that intrahepatic Treg were significantly infiltrated in patients with liver diseases [43]. There were significantly fewer intrahepatic Treg in the AIH patients than in the PBC patients. In addition, the frequency of Treg decreased in the liver of PBC patients as the pathological stage of the disease advanced. We found significantly less infiltration of CD4+ T cells in AIH than in HBV/HCV infection and PBC patients.

Treg as a Therapeutic Target for Liver Diseases

In the light of the crucial factor of Treg in the progression of liver diseases, much attention has been attracted to utilizing Treg as novel therapeutic targets. The key to the success would be to establish the correct balance between Treg and effector cells [44]. On one hand, it is possible to increase or decrease the quantity of Treg to weaken or enhance immune response. On the other hand, targeting the function of Treg per se or effecting molecules of Treg would be advisable.

In many infections in both mice and humans, removal of CD4+CD25+ Treg has resulted in enhanced effector immune responses. Depletion of Treg with anti-CD25 antibody can elicit a protective immune response that leads to the regression of tumors [45]. Moreover, combination of CTLA-4 blockade and depletion of CD4+CD25+ Treg has a synergism in the terms of boosting anti-tumor immunity [46].

On the contrary, in order to inhibit excessive immune response contributing to autoimmune liver diseases, it is necessary to enhance the suppressive function of Treg. The expansion of CD4+CD25+ Treg for adoptive transfer can be achieved by repeated stimulation of CD4+CD25+ Treg with antigen combined with APCs or with CD3/CD8 monoclonal antibodies in the presence of a high dose of IL-2 [47]. Moreover, experimental evidence has shown that antigen-specific Treg can be generated and expanded in sufficient quantities in-vitro without loss of their characteristic phenotype and regulatory properties [48].

Similarly, targeting the molecules involved in Treg activity in-vivo, such as CTLA-4, TGF-β, IL-10 or IL-35, alone or in combination, has often proven effective in controlling Treg-mediated immune balance [8].

However, strategies mentioned above are antigen non-specific, could result in the imbalance of immune system [49]. In human HBV or HCV infection, removal of Treg or Treg-related factors may also enhance hepatocyte impairment by hyperinflammation. Perhaps the best way is to achieve selective abolishment of the Treg activity that preferentially inhibits virus-specific T-cell responses. Further exploiting our knowledge of the role and mechanism of CD4+CD25+ Treg in immunity and tissue damage is important. Greater attention should be given to investigate

Treg properties in the whole process of HBV and HCV infection, especially in the pathogen-affected site: the liver.

Conclusion and Perspectives

In summary, Treg play an import role in the pathogenesis and disease progression of liver diseases including viral hepatitis, autoimmune hepatitis and hepatocellular carcinoma. Manipulation of Treg would be potential therapeutic strategies to cure chronic liver failure, but cautions should be considered. As difference in methods, reagents and samples used in studies might account for the discrepancies and controversies between different labs. In addition, un-unified statuses and ethnic populations of enrolled patients as well as the definition of Treg may also contribute to the problem. Therefore, it is necessary not only to use large cohort sample size and enough sample amounts from each individual, but also to unify the enrollment requirement as well as the measurement protocols in future studies. In addition, some useful suggestions should be considered:

(1) Viral antigen-specific Treg in humans with hepatitis C or B, or HCC tumor antigen specific Treg need to be further studied. In particular, it is important to unravel the origin of adaptive antigen-specific Treg in these patients. (2) Current progresses in the field of Treg would lead to a reliable surface marker to identify Treg, thus giving better evaluation of prognosis and survival of HCC patients. (3) In the future, it would be possible to block the Treg suppressive function by using therapeutic agents, which would lead to restore the immune impairment in patients with chronic hepatitis C or B, by clearing the virus in-vivo and promoting better outcome for patients. (4) Finally, further studies are required to understand the mechanisms responsible for the suppression of Treg on innate and adaptive immunities in-vivo. A better understanding of Treg will not only lead to elucidate the pathogenesis of liver disease including viral hepatitis, autoimmune hepatitis and hepatocellular carcinoma, but also have potential in diagnosis and treatment of these diseases.

References

1. Ganem, D. and A.M. Prince, Hepatitis B virus infection–natural history and clinical consequences. N Engl J Med, 2004, 350(11):1118–29.
2. Rehermann, B. and M. Nascimbeni, Immunology of hepatitis B virus and hepatitis C virus infection. Nat Rev Immunol, 2005, 5(3):215–29.
3. Krawitt, E.L., Autoimmune hepatitis. N Engl J Med, 2006, 354(1):54–66.
4. Kaplan, M.M. and M.E. Gershwin, Primary biliary cirrhosis. N Engl J Med, 2005, 353(12):1261–73.
5. Bertoletti, A. and A.J. Gehring, The immune response during hepatitis B virus infection. J Gen Virol, 2006, 87(Pt 6):1439–49.
6. Roncador, G., et al., Analysis of FOXP3 protein expression in human CD4+CD25+ regulatory T cells at the single-cell level-12. Eur J Immunol, 2005, 35(6):1681–91.
7. Unitt, E., et al., Compromised lymphocytes infiltrate hepatocellular carcinoma: the role of T-regulatory cells. Hepatology, 2005, 41(4):722–30.
8. Belkaid, Y. and B.T. Rouse, Natural regulatory T cells in infectious disease. Nat Immunol, 2005, 6(4):353–60.

9. Shevach, E.M., Regulatory T cells in autoimmmunity*. Annu Rev Immunol, 2000, 18:423–49.
10. Hawrylowicz, C.M. and A. O'Garra, Potential role of interleukin-10-secreting regulatory T cells in allergy and asthma. Nat Rev Immunol, 2005, 5(4):271–83.
11. Nakamura, K., A. Kitani, and W. Strober, Cell contact-dependent immunosuppression by CD4(+)CD25(+) regulatory T cells is mediated by cell surface-bound transforming growth factor beta. J Exp Med, 2001, 194(5):629–44.
12. Takahashi, T., et al., Immunologic self-tolerance maintained by CD25(+)CD4(+) regulatory T cells constitutively expressing cytotoxic T lymphocyte-associated antigen 4. J Exp Med, 2000, 192(2):303–10.
13. Kanamaru, F., et al., Costimulation via glucocorticoid-induced TNF receptor in both conventional and CD25+ regulatory CD4+ T cells. J Immunol, 2004, 172(12):7306–14.
14. Cabrera, R., et al., An immunomodulatory role for CD4(+)CD25(+) regulatory T lymphocytes in hepatitis C virus infection. Hepatology, 2004, 40(5):1062–71.
15. Kanto, T. and N. Hayashi, Immunopathogenesis of hepatitis C virus infection: multifaceted strategies subverting innate and adaptive immunity. Intern Med, 2006, 45(4):183–91.
16. Xu, D., et al., Circulating and liver resident CD4+CD25+ regulatory T cells actively influence the antiviral immune response and disease progression in patients with hepatitis B. J Immunol, 2006, 177(1):739–47.
17. Franzese, O., et al., Modulation of the CD8+-T-cell response by CD4+ CD25+ regulatory T cells in patients with hepatitis B virus infection. J Virol, 2005, 79(6):3322–8.
18. Stoop, J.N., et al., Regulatory T cells contribute to the impaired immune response in patients with chronic hepatitis B virus infection. Hepatology, 2005, 41(4):771–8.
19. Furuichi, Y., et al., Depletion of CD25+CD4+T cells (Tregs) enhances the HBV-specific CD8+ T cell response primed by DNA immunization. World J Gastroenterol, 2005, 11(24):3772–7.
20. Chichester, J.A., M.A. Feitelson, and C.E. Calkins, Transient inhibition of Th1-type cytokine production by CD4 T cells in hepatitis B core antigen immunized mice is mediated by regulatory T cells. Immunology, 2006, 118(4):438–48.
21. Sugimoto, K., et al., Suppression of HCV-specific T cells without differential hierarchy demonstrated ex vivo in persistent HCV infection. Hepatology, 2003, 38(6):1437–48.
22. Boettler, T., et al., T cells with a CD4+CD25+ regulatory phenotype suppress in vitro proliferation of virus-specific CD8+ T cells during chronic hepatitis C virus infection. J Virol, 2005, 79(12):7860–7.
23. Bolacchi, F., et al., Increased hepatitis C virus (HCV)-specific CD4+CD25+ regulatory T lymphocytes and reduced HCV-specific CD4+ T cell response in HCV-infected patients with normal versus abnormal alanine aminotransferase levels. Clin Exp Immunol, 2006, 144(2): 188–96.
24. Smyk-Pearson, S., et al., Functional Suppression by FoxP3(+)CD4(+)CD25(high) Regulatory T Cells during Acute Hepatitis C Virus Infection. J Infect Dis, 2008, 197(1):46–57.
25. Manigold, T., et al., Foxp3+CD4+CD25+ T cells control virus-specific memory T cells in chimpanzees that recovered from hepatitis C. Blood, 2006, 107(11):4424–32.
26. Grossman, W.J., et al., Human T regulatory cells can use the perforin pathway to cause autologous target cell death. Immunity, 2004, 21(4):589–601.
27. Kinter, A.L., et al., CD25(+)CD4(+) regulatory T cells from the peripheral blood of asymptomatic HIV-infected individuals regulate CD4(+) and CD8(+) HIV-specific T cell immune responses in vitro and are associated with favorable clinical markers of disease status. J Exp Med, 2004, 200(3):331–43.
28. Weiss, L., et al., Human immunodeficiency virus-driven expansion of CD4+CD25+ regulatory T cells, which suppress HIV-specific CD4 T-cell responses in HIV-infected patients. Blood, 2004, 104(10):3249–56.
29. Feng, I.C., et al., HBcAg-specific CD4+CD25+ regulatory T cells modulate immune tolerance and acute exacerbation on the natural history of chronic hepatitis B virus infection. J Biomed Sci, 2007, 14(1):43–57.

30. Li, S., et al., Defining target antigens for CD25+ FOXP3 + IFN-gamma- regulatory T cells in chronic hepatitis C virus infection. Immunol Cell Biol, 2007, 85(3):197–204.
31. Longhi, M.S., et al., Functional study of CD4+CD25+ regulatory T cells in health and autoimmune hepatitis. J Immunol, 2006, 176(7):4484–91.
32. Agarwal, K., et al., Cytotoxic T lymphocyte antigen-4 (CTLA-4) gene polymorphisms and susceptibility to type 1 autoimmune hepatitis. Hepatology, 2000, 31(1):49–53.
33. Fu, J., et al., Increased regulatory T cells correlate with CD8 T-cell impairment and poor survival in hepatocellular carcinoma patients. Gastroenterology, 2007, 132(7):2328–39.
34. Westendorf, A.M., et al., CD4+ T cell mediated intestinal immunity: chronic inflammation versus immune regulation. Gut, 2005, 54(1):60–9.
35. MacDonald, A.J., et al., CD4 T helper type 1 and regulatory T cells induced against the same epitopes on the core protein in hepatitis C virus-infected persons. J Infect Dis, 2002, 185(6):720–7.
36. Suffia, I.J., et al., Infected site-restricted Foxp3+ natural regulatory T cells are specific for microbial antigens. J Exp Med, 2006, 203(3):777–88.
37. Llovet, J.M., A. Burroughs, and J. Bruix, Hepatocellular carcinoma. Lancet, 2003, 362(9399):1907–17.
38. Ormandy, L.A., et al., Increased populations of regulatory T cells in peripheral blood of patients with hepatocellular carcinoma. Cancer Res, 2005, 65(6):2457–64.
39. Wolf, D., et al., Telomere length of in vivo expanded CD4(+)CD25 (+) regulatory T-cells is preserved in cancer patients. Cancer Immunol Immunother, 2006, 55(10):1198–208.
40. Longhi, M.S., et al., Impairment of CD4(+)CD25(+) regulatory T-cells in autoimmune liver disease. J Hepatol, 2004, 41(1):31–7.
41. Aoki, C.A., et al., IL-2 receptor alpha deficiency and features of primary biliary cirrhosis. J Autoimmun, 2006, 27(1):50–3.
42. Lan, R.Y., et al., Liver-targeted and peripheral blood alterations of regulatory T cells in primary biliary cirrhosis. Hepatology, 2006, 43(4):729–37.
43. Sakaki, M., et al., Intrahepatic status of regulatory T cells in autoimmune liver diseases and chronic viral hepatitis. Hepatol Res, 2008, 38(4):354–61.
44. Mottet, C. and D. Golshayan, CD4+CD25+Foxp3+ regulatory T cells: from basic research to potential therapeutic use. Swiss Med Wkly, 2007, 137(45–46):625–34.
45. Onizuka, S., et al., Tumor rejection by in vivo administration of anti-CD25 (interleukin-2 receptor alpha) monoclonal antibody. Cancer Res, 1999, 59(13):3128–33.
46. Sutmuller, R.P., et al., Synergism of cytotoxic T lymphocyte-associated antigen 4 blockade and depletion of CD25(+) regulatory T cells in antitumor therapy reveals alternative pathways for suppression of autoreactive cytotoxic T lymphocyte responses. J Exp Med, 2001, 194(6):823–32.
47. Hoffmann, P., et al., Large-scale in vitro expansion of polyclonal human CD4(+)CD25high regulatory T cells. Blood, 2004, 104(3):895–903.
48. Golshayan, D., et al., In vitro-expanded donor alloantigen-specific CD4+CD25+ regulatory T cells promote experimental transplantation tolerance. Blood, 2007, 109(2):827–35.
49. Yamaguchi, T., et al., Control of immune responses by antigen-specific regulatory T cells expressing the folate receptor. Immunity, 2007, 27(1):145–59.

Chapter 22
CD4⁺CD25⁺ Regulatory T Cells in Viral Infections

Wayne A. Tompkins, Mary B. Tompkins, Angela M. Mexas, and Jonathan E. Fogle

Abstract CD4$^+$CD25$^+$ Tregulatory (Treg) cells are a thymus-derived distinct lineage of T cells that recognize and suppress the expansion and function of potential self-reactive T cell clones, thus maintaining peripheral self-tolerance. It is now established that Treg cells activated in the peripheral immune compartment also modulate immune responses to pathogens. Data suggest that pathogen activated Treg cells in lymph nodes (LN) down-regulate T and B cells responding to the same pathogen, thus minimizing the immunopathology associated with primary immune responses. While Treg cells normally return to a resting state after elimination of the pathogen, in those infections that are not resolved by an anti-viral immune response, they remain chronically activated and immunosuppressive, thereby contributing to a persistent viremia. It is well-established that chronic Hepatitis B and C infections, as opposed to resolved infections, are associated with an increased number or increased activation state of Treg cells that suppress anti-viral CD4$^+$ and CD8$^+$ T cells and contribute to the long-term viremia. In murine models of HSV-1 infection, in vivo depletion of CD25$^+$ Treg cells in mice results in enhanced anti-HSV CD8$^+$ immune responses and more rapid clearance of virus. Similarly, the CD4$^+$ and CD8$^+$ immune deficiency observed in chronic AIDS lentivirus infections have been attributed to immunosuppressive Treg cells capable of suppressing virus-specific CD4$^+$ and CD8$^+$ cytokine and proliferation responses. Whether these activated Treg cells contribute to the host's failure to eliminate these lentiviruses is not known. However, it has been demonstrated that virus-specific Tcells are anergic, cannot produce IL2, and cannot expand in response to virus peptide stimulation. Activated CD4$^+$CD25$^+$ Treg cells mediate T cell immunosuppression by contact-dependent mechanisms that transduce a signal for transcriptional down-regulation of cytokine genes, including IL2, and induction of T cell anergy. Recent data suggest that Treg-induced T cell anergy is mediated through the TGF-β/TGF-βR signaling pathway. It is clear from the studies described herein that

W.A. Tompkins
Immunology Program, North Carolina State University, 4700 Hillsborough St, Raleigh, NC 27606, USA
e-mail: Wayne_Tompkins@ncsu.edu

chronic viral infections are characterized by an early and sustained activation of immunosuppressive CD4$^+$CD25$^+$ Treg cells capable of inhibiting anti-viral CD4$^+$ and CD8$^+$ immune responses, which allows for the establishment of long-term infection.

Introduction

Convincing evidence has accumulated for the existence of heterogeneous populations of CD4$^+$ regulatory T cells that maintain peripheral self tolerance and prevent autoimmunity [1]. One regulatory CD4$^+$ T cell subset that has been well characterized is distinguished by constitutive expression of cell surface CD25, the IL-2R α chain [2]. These CD4$^+$CD25$^+$ T regulatory cells, comprising 5–10% of peripheral CD4$^+$ T cells, are generated by self-antigen presentation by the thymic epithelium and are functionally mature when exported from the thymus into the periphery. When activated by engagement of their TCR by their cognate antigen, they acquire the potential of suppressing IL2 production and proliferation of self-reactivate CD4$^+$ Th and CD8$^+$ effector cells [3,4]. It is now well established that, in addition to maintaining peripheral self-tolerance, CD4$^+$CD25$^+$ Treg cells are responsive to pathogens in the peripheral lymphoid tissues and play a major role in modulating immune responses to infectious agents [5,6]. Pathogen-induced CD4$^+$CD25$^+$ Treg cells can be derived from either CD4$^+$CD25$^+$ or CD4$^+$CD25$^-$ T cells in the periphery, and are indistinguishable from the natural Treg cells phenotypically in that they up-regulate CTLA4, GITR, certain Toll-like receptors, CD62-L, surface TGF-β, and most importantly express Foxp3, which is required for their homeostasis and suppressor function [7–9]. Similarly, pathogen-induced Treg cells are anergic to antigen and mitogen stimulation and suppress proliferation of other activated CD4$^+$ and CD8$^+$ T cells in a contact-dependent manner by down-regulating IL2 [6,10].

While it is evident that pathogen-induced Treg cells play a central role in maintaining the balance between immunity and immunopathology to infectious agents, their specificity and mechanism of immune suppression remains controversial. Some studies have suggested that pathogen-induced CD4$^+$CD25$^+$ Treg cells are antigen specific [11–13], whereas other studies have indicated that they may be activated in response to pathogen associated molecular patterns (PAMP's, e.g. LPS, flagellin) and cytokines (e.g. IL2 and TGF-β) in the infection microenvironment [10,14]. In this regard, murine [14] and feline [10] Treg cells express TLR-4 and can be activated with LPS. In addition, murine [14] and feline [10] CD4$^+$CD25$^+$ Treg cells can be induced with IL2 to proliferate and express potent immunosuppressor function. Although the question of activation specificity of pathogen-induced Treg cells cannot be answered with the current information, it is evident that, at least in vitro, the antigen specificity, if it exists, can be overridden with strong mitogenic stimulation, such as IL2 or by ligands for specific signaling pathways, such as TLR-4 and TGF-βR [7,14].

Role of CD4$^+$CD25$^+$ Treg Cells in Viral Immunity and Immunopathology

Data suggest that failure to eliminate certain pathogens and the resulting chronic antigenemia may be due to the repressive immunomodulatory effects of activated CD4$^+$CD25$^+$ Treg cells on the protective CD4$^+$ and CD8$^+$ T cell immune responses [10,15,16]. Further, evidence suggests that CD4$^+$CD25$^+$ Treg cells are a normal, immunological entity linking the innate and acquired immune responses and are necessary for an ordered protective immune response with a minimum of collateral immunopathology. This is a reasonable scenario for most pathogens that are usually eliminated after inducing a strong protective primary immune response that is also associated with development of long-lasting memory. However, it is less certain that Treg cell control over the relative magnitude of immunity and immunopathology is as stringent in the case of some viral pathogens that directly infect cells of the immune system or develop latency and establish long-term persistence. For example, in the case of viral infections that establish latency or a persistent antigenemia, are CD4$^+$CD25$^+$ Treg cells activated concurrent with the innate and/or acquired T cell immune responses, and if so, would they hinder or promote disease progression? In addition, what effect does the chronic productive viral infection of Treg cells that occurs with AIDS lentiviruses [17,18] have on their function and numbers? The remainder of this chapter will address these questions and discuss what is known of the role of CD4$^+$CD25$^+$ Treg cells in maintaining a balance between immunity and immunopathology in selected viral infections.

Herpesviruses

Recent data indicate that both primary and memory immune responses to chronic viral infections such as herpes simplex virus type 1 (HSV-1) and HSV-2 are regulated by CD4$^+$CD25$^+$ Treg cells. Suvas et al. [5] reported that prior in vivo antibody depletion of CD25$^+$ T cells increased acute and memory CD8$^+$ cytotoxic T cell responses to HSV-1 challenge. Further, these authors observed that IFN-γ responses of splenic CD8$^+$ T cells stimulated by an immunodominant HSV-1 peptide are markedly increased in CD25$^+$-depleted mice as measured by flow cytometry intracellular staining and ELISpot assays. Enhanced anti-HSV1 CD8$^+$ immune responses in CD25$^+$-depleted mice correlated with a more rapid clearance of virus. Collectively, these observations suggest that immunosuppressive Treg cells are activated during acute HSV-1 infection, and that elimination of Treg cells during the acute state of infection could result in a more robust protective CD8$^+$ immune response and more rapid clearance of the virus. In a follow-up study, Suvas et al. [19] employed a murine HSV-induced stromal keratitis model to demonstrate that depletion of CD25$^+$ T cells prior to challenge with HSV-1 increased the severity of the ocular inflammatory lesions that correlated with an increase in HSV-specific CD4$^+$ T cell responses. Treg cells were also shown to modulate the keratitis induced by adoptive transfer of pathogenic CD4$^+$ T cells in HSV-1-infected SCID mice.

In another study, the potential impact of CD4$^+$CD25$^+$ Treg cells on vaccine efficacy was suggested by the observation that depletion of CD25$^+$ T cells enhanced primary and memory CD8$^+$ responses to a HSV-1 subunit vaccine [20]. In challenge experiments, memory CD8$^+$ T cells generated with plasmid DNA in mice depleted of CD25$^+$ cells cleared the virus more effectively than control mice.

Others have targeted GITR in attempts to modulate Treg responses and influence HSV-1 pathogenesis. GITR (gluccorticoid-induced TNF family receptor), a member of the TNF receptor super family, is preferentially expressed on CD4$^+$CD25$^+$ Treg cells, and engagement of GITR by its ligand abrogates suppressor function of Treg cells [21]. Suvas et al. [22] demonstrated in the aforementioned murine HSV-1-induced keratitis model that virus specific T cell responses in LN and spleen were enhanced in mice treated with anti-GITR antibodies. However, in contrast to the expected increase in the T cell-mediated inflammatory lesions, anti-GITR treatment resulted in significantly diminished T cell-mediated ocular lesions. La et al. [23] also reported that a single injection of anti-GITR antibody immediately after HSV-1 infection of mice significantly increased the number of CD4$^+$ and CD8$^+$ T cells secreting IFN-γ, suggesting that ligation of GITR on Treg cells diminished their ability to suppress CD4$^+$ and CD8$^+$ T cell responses to HSV-1 antigens. These authors did not determine the effect of anti-GITR antibody treatment on the pathogenesis and outcome of HSV-1 infection. The reason for the contradictory results of anti-GITR Treg inactivation leading to enhanced T cell immune responses, but diminished cell-mediated ocular inflammation in the keratitis model is not immediately apparent. However, the studies do point out the sometimes unpredictable immunological consequences of Treg manipulation in different animal models, as has so often been observed in models of T cell-mediated autoimmunity [24]. While the HSV-1 keratitis model may be an exception, the important issue that arises from these studies is that depletion of Treg cells allows for a more robust protective T cell response to pathogens, yet the ensuing more aggressive CD4$^+$ and CD8$^+$ anti-viral immune response may aggravate the immunopathology in the target organ.

Treg cells have also been suspected of altering T cell immune responses to EBV, HSV-2 and CMV infections in humans. EBV infection induces IL10 secreting Treg cells that inhibit Th1 responses to EBV antigens and facilitate viral persistence [25]. In addition, Diaz and Koelle [26] demonstrated by CD25$^+$ depletion and reconstitution experiments that Treg cells from healthy HSV-2-infected individuals suppress CD4$^+$ T cell responses to HSV-2 antigens and influence their antiviral and inflammatory function. In the same study, these authors demonstrated that CD25$^+$ depletion significantly increased the CMV-specific proliferative response of T cells in a subset of CMV seropositive patients. Others have reported that CD4$^+$CD25$^+$ Treg cells negatively regulate memory CD4$^+$ responses to CMV antigens [15,27]. As with studies in the murine HSV-1 model, these data demonstrate that CD4$^+$CD25$^+$ Treg cells generally have contrasting effects on virus clearance and T cell-mediated inflammatory changes in target tissues. Enhanced viral clearance and increased inflammation in most HSV models following CD25$^+$-depletion is consistent with the concept that CD4$^+$CD25$^+$ Treg cells are activated during acute and chronic HSV infection and are fully armed for potent T cell suppressor function.

Hepatitis Viruses

Hepatitis B Virus (HBV) a member of the family of DNA *Hepadniviridae* and Hepatitis C virus (HCV) a member of RNA *Flaviviridae* family are hepatotropic viruses capable of causing both acute and chronic inflammation of the liver. Recent data suggest that the CD4$^+$CD25$^+$ Treg response to these infections may determine if the host develops a strong T cell immune response and resolves the infection or has a diminished immune response and establishes a persistent viremia. HBV-specific CD8$^+$ T cells are necessary for elimination of the virus but are defective in patients with persistent HBV infection compared with those who resolve the infection [28]. Xu et al. [29] reported that in chronic severe hepatitis B patients, the frequency of CD4$^+$CD25$^+$ Treg cells in both PBMC and liver was significantly increased, which corresponded to a marked increase in Foxp3 positive T cells infiltrating the liver compared to controls. In contrast, in acute HBV infections, circulating CD4$^+$CD25$^+$ T cells initially increased in number and then returned to normal levels upon resolution of the infection. These data suggest the chronic HBV infection correlates with persistent high levels of Foxp3 positive CD4$^+$CD25$^+$ Treg cells, which could be responsible for the CD8$^+$ T cell immune deficiency observed in these patients.

Franzese et al. [30] reported that CD4$^+$CD25$^+$ T cells from HBV patients were capable of suppressing IFN-γ production by HBV antigen-stimulated CD8$^+$ T cells; however these data did not demonstrate a functional difference between the circulating CD4$^+$CD25$^+$ Treg cells of chronic and resolved HBV patients. These authors were also unable to show a difference in numbers of CD4$^+$CD25$^+$ Treg cells in chronic and resolved HBV patients, suggesting that increased numbers or functional activation of Treg cells are not necessarily associated with establishment of chronic HBV infection. There is no immediate explanation of the apparent contrasting results of these two studies. As the latter study was limited to PBMC, a more detailed study of other lymphoid tissues as well as the liver might resolve these differences. In the case of HIV-infected patients, Eggena et al. [31] reported that CD4$^+$CD25$^+$ T cells accumulate in or preferentially expand in lymphoid tissue with high virus burden when compared to the blood. In addition, Vahlenkamp et al. [10] reported that CD4$^+$CD25$^+$ Treg cell numbers in the blood of control and chronic FIV-infected cats did not differ, but the Treg cells in FIV-infected cats were phenotypically and functionally activated as compared to control cats.

In contrast to HBV infection, there is more convincing evidence that CD4$^+$CD25$^+$ Treg cells may modulate immune responses to HCV and determine the outcome of the infection. Rushbrook et al. [27] reported that depletion of CD4$^+$CD25$^+$ Treg cells from PBMC of patients with chronic HCV infection resulted in increased virus-specific CD8$^+$ T cell proliferation and IFN-γ production, suggesting that Treg cells are functionally activated in chronic HCV infection. In addition, these authors demonstrated that CD4$^+$CD25$^+$ depletion resulted in increased CD8$^+$ T cell immune responses to EBV peptides, suggesting that Treg cells, once activated, are not antigen specific in their suppressor function, which appears to be the case with all pathogen-induced Treg cells, as well as natural Treg cells [32].

Boettler et al. [16] recently reported on studies revealing a direct link between $CD4^+CD25^+$ Treg cells and impaired HCV-specific immune responses in chronic HCV infection. In vitro depletion and reconstitution experiments revealed that peptide-induced proliferation, as well as IFN-γ production by HCV-specific $CD8^+$ T cells was inhibited by $CD4^+CD25^+$ T cells. The inhibition was contact dependent and was independent of IL10 and TGF-β. As reported by others [15], and consistent with the non-specific suppressor function of activated Treg cells, these Treg cells from chronically HCV-infected patients also inhibited influenza-specific $CD8^+$ immune responses. Importantly, $CD4^+CD25^+$ Treg cells from patients recovered from HCV infections, as well as normal healthy blood donors, exhibited significantly less suppressor activity. These data, taken together with the observation of increased frequency of $CD4^+CD25^+$ Treg cells in chronically infected HCV patients, suggest that in vivo activated Treg cells directly contribute to HCV persistence and could be a target for immunotherapy.

In addition to virus-specific $CD8^+$ T cell responses, patients who spontaneously clear the virus and recover from HCV infections mount a vigorous $CD4^+$ T cell response. Yang et al. [33] reported that patients with chronic HCV infection had increased numbers and increased suppressor function of $CD4^+CD25^+$ Treg cells. $CD4^+CD25^+$ Treg cells from chronic HCV infected patients expressed high levels of Foxp3 and IL10 and were strongly suppressive for $CD4^+$ T cells proliferation and IFN-γ production in response to HCV antigen stimulation.

Bolacchi et al. [34] reported an inverse correlation between HCV-specific TGF-β production by $CD4^+CD25^+$ Treg cells and liver inflammation in chronic HCV carriers with normal alanine aminotransference levels (ALT) as compared to patients with elevated ALT levels. Depletion of $CD4^+CD25^+$ Treg cells from PBMC of patients with normal ALT levels resulted in an increase of both IFN-γ production and proliferation of HCV-specific $CD4^+$ T cells. These studies suggest that $CD4^+CD25^+$ Treg cells may not only contribute to persistent HCV infection by suppressing protective $CD4^+$ T cell responses, but paradoxically, in some patients, control chronic inflammation and hepatic damage by secreting high levels of the anti-inflammatory cytokine TGF-β.

In contrast to studies in humans, studies in HCV-infected chimpanzees, the sole animal model for HCV infection, demonstrated that the frequency of $Foxp3^+$ $CD4^+CD25^+$ Treg cells and their suppressor function was as high in spontaneously recovered animals as in persistently HCV-infected animals [35]. These authors speculated that Treg cells control HCV-specific T cells not only in persistent infection but also after recovery when they may regulate memory T cell responses by controlling their activation and preventing AICD (apoptosis).

Retroviruses

Retroviruses persist for the lifetime of their hosts and can cause T cell tumors and/or a T cell immunodeficiency. In both cases, there is evidence that $CD4^+CD25^+$ Treg cells can, to some extent, control the outcome of infection through their regulatory influence on virus-activated $CD4^+$ and $CD8^+$ T cells.

Gammaretroviruses

While the suppressive effects of Treg cells has been reported in several murine retrovirus models, including the murine leukemia virus model of murine AIDS [36], most studies have been done with the Friend murine retrovirus. Persistent infection of mice with Friend retrovirus is associated with a decreased ability to develop anti-tumor immune responses [37]. It has been shown that in these chronically infected mice, IL10 producing Treg cells expand in number and suppress the protective $CD8^+$ immune response. In vivo depletions of Treg cells results in restoration of $CD8^+$ anti-tumor function and a marked reduction in virus load [37].

Dittmer et al. [38] also reported that Treg cells facilitated Friend virus persistence by inhibiting $CD8^+$ T cell responses. Interestingly, unlike the characteristic anti-proliferation function of classical $CD4^+CD25^+$ Treg cells [10], these suppressor cells did not diminish $CD8^+$ proliferation or activation but rather inhibited effector function. In a more recent study, Robertson et al. [39] confirmed that $CD4^+CD25^+$ Treg cells isolated from chronically Friend virus-infected mice suppressed the development of effector function in naïve $CD8^+$ T cells without affecting their ability to proliferate or up-regulate activation markers. Suppression was mediated by direct cell-to-cell contact and was not antigen specific. These authors argued that the ability of $CD4^+CD25^+$ Treg cells to suppress effector function of activated CTL's is important to limiting immunopathology of $CD8^+$ T cells but in so doing also limits the anti-viral immune response, allowing the virus to establish and maintain a chronic infection. This model is an excellent example of the paradoxical nature of $CD4^+CD25^+$ Treg activation, whereby they play an important role in controlling the immunopathology associated with virus activated inflammatory T cells during acute infection, but in so doing dampen the protective T cell immune response, resulting in failure to eliminate the virus.

Deltaretroviruses

Human T cell lymphotropic virus (HTLV-1) infects $CD4^+$ T cells, and in symptomatic patients the major reservoir for the virus is $CD4^+CD25^+$ Treg cells [40]. However, these HTLV-1-infected Treg cells are not suppressive but rather are stimulatory for the HTLV-1 tax-specific proliferation of $CD8^+$ T cells. Interestingly, transfection of the tax gene into $CD4^+CD25^+$ cells resulted in inhibition of Foxp3 expression [41]. This is consistent with the finding that Foxp3 mRNA and protein expression in $CD4^+CD25^+$ Treg cells of symptomatic patients is lower than in healthy donors. Therefore, the lack of Treg activity in HTLV-1 infected patients may contribute to the HTLV-1 associated disorders with multiorgan lymphocytic infiltrates.

Lentiviruses

Perhaps, the most intriguing questions regarding the origin, specificity, mechanisms of activation and the potential immunopathological consequences of pathogen-induced

Treg cells have been addressed in AIDS lentiviruses infections. The relationship between Treg cells and AIDS lentiviruses is unusual, if not unique, from most other viruses in two major ways. Studies in our laboratory demonstrated that $CD4^+CD25^+$ Treg cells provide an important reservoir of productive FIV replication in infected cats. Susceptibility of $CD4^+CD25^+$ Treg cells to productive FIV infection correlated with over-expression of the FIV co-receptor CXCR4 and constitutive transactivation of transcription factors such as AP1 and ATF that bind to the FIV promoter [42,43]. Oswald-Richter et al. [18] also reported that Treg cells from healthy donors express the HIV co-receptor CCR5 and are highly susceptible to HIV infection and replication. Productive infection of $CD4^+CD25^+$ Treg cells could markedly compromise the delicate balance between Treg activation and development of protective T cell immune responses. While data suggest that FIV infection of Treg cells does not compromise their viability or suppressor function [10,17], Oswald-Richter et al. [18] reported that HIV infection of Treg cells is cytolytic. Thus, the fraction of Treg cells infected at any given time could determine the numbers and function of this suppressor population. A second confounding characteristic of lentivirus infection is the fact that Treg cells in these infections are constitutively activated in vivo [10,12,13]. $CD4^+CD25^+$ Treg cells recovered from acute asymptomatic FIV-infected cats express cell surface TGF-β and are fully armed for T cell suppressor function, suggesting that they have escaped the normal in vivo regulatory controls and are potentially capable of chronic T cell immunosuppression. As we will discuss below, the chronic productive infection and activation phenotype of Treg cells in lentivirus infections could have major implications for the immunopathology and, more specifically, the T cell immunodeficiency caused by these viruses.

We were the first to identify $CD4^+CD25^+$ Treg cells in the cat and describe their phenotypic and functional activation in the FIV-feline model for human AIDS [10,17,42]. Vahlenkamp et al. [10] reported that, similar to human and mouse Treg cells, $CD4^+CD25^+$ cells isolated from LN and blood of the domestic cat are anergic (arrested in the G0/G1 stage of cell cycle) and fail to produce IL2 and proliferate in response to immune stimulation. Importantly, $CD4^+CD25^+$ T cells from asymptomatic FIV-infected cats support a productive FIV infection [17]. These Treg cells are constitutively activated in vivo and suppress IL2 production and proliferation of ConA-stimulated autologous $CD4^+CD25^-$ T cells in a dose-dependent manner [10]. We speculated that the chronic productive virus infection or persistent antigenemia in FIV-infected cats may maintain long-term Treg cell activation in vivo, which could contribute to the progressive loss of $CD4^+$ and $CD8^+$ T cell immune function (anergy) and promote development of AIDS.

In the case of HIV infection, Kinter et al. [12] reported that $CD4^+CD25^+$ T cells in the majority of healthy HIV-infected patients significantly suppressed cellular proliferation and cytokine production by $CD4^+$ and $CD8^+$ T cells stimulated with HIV peptides in vitro. Immune suppression by T cells from HIV-infected patients was shown to be cell contact-dependent and IL10- and TGF-β-independent. Interestingly, the level of $CD4^+CD25^+$ Treg suppressor function correlated inversely with plasma viremia, suggesting that Treg cells may reduce the number of available "activated" $CD4^+$ targets for productive HIV infection. This speculation is consistent with the recent report by Eggena et al. [31] that a decline in CD4+CD25+ Treg cells in late stage HIV infection is associated with increased T cell immune hyperactivation and higher viremia. In another study, Weiss et al. [13] reported a significant expansion of CD4+CD25+ T cells in the blood of HIV-infected patients on HAART. These CD4+CD25+ T cells possessed the characteristics of Treg cells, in that they constitutively expressed Foxp3 mRNA, were anergic to anti-CD3/anti-CD28 co-stimulation, and suppressed the proliferative response of CD4+CD25− T cells stimulated with tuberculin. These CD4+CD25+ Treg cells, or at least a fraction of them, appeared to be specific for HIV antigen, as p24 stimulation resulted in up-regulation of TGF-β and IL10. However, immunosuppressive activity was not dependent on TGF-β or IL10. Andersson et al. [44] also reported that Foxp3 expressing T cells in the circulation were decreased in untreated viremic HIV-infected patients as compared to patients on HAART. Interestingly, these authors reported increased numbers of Foxp3 expressing cells in the tonsils of untreated patients compared with HAART patients, sugessting that Treg cells may accumulate or preferentially expand in lymphoid tissue sites of HIV replication.

In related studies, Aandahl et al. [15] reported that depletion of CD25+ T cells from PBMC of HIV-infected patients not on HAART enhanced the frequency of IFN-γ and TNF-α expressing T cells in response to stimulation with HIV and CMV antigens, indicating that HIV-induced Treg cells suppress T cell responses to specific antigens, as well as unrelated antigens. Eggena et al. [31] similarly reported that in vitro depletion of Treg cells from HIV patients' PBMC increased gag-specific CD8+ responses, as measured by IFN-γ ELISpot. While not all of these studies addressed the question of what factors are responsible for inducing immunosuppressive CD4+CD25+ Treg cells in lentivirus infection, at least a fraction of Treg cells in HIV-infected patients are responsive to HIV antigens. These observations of chronic in vivo activation of Treg cells in FIV-infected cats and HIV-infected individuals beg the question of what role these cells could play in the CD4+ and CD8+ immunodeficiency (anergy) associated with these virus infections.

CD4+CD25+ Treg Cells Play a Role in Lentivirus-Induced AIDS

The above-referenced studies collectively suggest that CD4+CD25+ Treg cells are fully armed for suppressor function in vivo in FIV/HIV infection, can potentially abort (anergize) protective CD4+ and CD8+ T cell immune responses to viral antigens, and may to the development of AIDS. If Treg cells are activated during the early acute stage of FIV/HIV infection, it might be predicted that protective T cell immune responses could be suppressed before epitope specific CD4+ Th and CD8+ T effector cells could clonally expand and develop an effective immune response. While there is no direct evidence that Treg-mediated immune suppression occurs during the acute FIV/HIV infection, data suggest that it is possible.

In both HIV and FIV infections, there is an early loss in CD4+ and CD8+ T cell immune function. Data suggest that in untreated primary HIV infection or in patients where HAART is delayed, CD4+ Th cell responses to HIV antigens develop early, as measured by lymphocyte proliferative responses, but are prematurely lost during the acute stage infection. This loss of proliferative response to HIV-specific epitopes is long lasting, as evidenced by similar HIV-epitope specific immune deficiency in long-term progressor patients with established infections [45–47]. Also, this early, acute-stage CD4$^+$ T cell immunodeficiency is specific to HIV antigens, as CD4$^+$ T cells from these patients proliferate normally to other antigens such as PPD, TT and CMV [45–47]. Similar to the CD4$^+$ T cell immune deficiency, CD8$^+$ T cells from chronically HIV-infected patients display defective proliferative responses to DC-pulsed HIV antigens but responded normally to other antigens, such as FLU, CMV and EBV [48]. However, these HIV-specific T cell clones appear not to be deleted, as CD4$^+$ and CD8$^+$ T cells from immunodeficient patients will produce IFN-γ in response to HIV peptide stimulation, despite the fact that there is no proliferative response to peptide-APC stimulation. A role for CD4$^+$CD25$^+$ Treg cells on this CD4$^+$ Th and CD8$^+$ anergy to HIV peptide stimulation is implied by the observation that depletion of CD4$^+$CD25$^+$ T cells from PBMC from HIV-infected patients increases a cytokine response to HIV antigen stimulation in vitro [15,31]. Collectively these data suggest that the mechanism(s) responsible for perturbation of HIV specific T cell responses during acute stage infection may involve epitope specific clonal anergy rather than clonal deletion, and this anergy may be mediated by activated Treg cells. Thus it is possible that in addition to induction of antigen-specific CD4$^+$ Th and CD8$^+$ effector cell responses during the acute stage infection, CD4$^+$CD25$^+$ Treg cells are also activated by viral antigens or the inflammatory microenvironment of the LN. Because of their ability to interact with and induce anergy in other activated T cells, Treg cells would restrict expansion of newly activated virus-specific CD4$^+$ Th and CD8$^+$ effector cell clones, and prematurely, and perhaps permanently, abort a protective T cell immune response. Thus, the relative balance between CD4$^+$ Th and CD4$^+$CD25$^+$ Treg immune activation during the acute stage HIV/FIV infection may determine the ultimate virus set-point and the long-term ability to control viremia, and in effect predict the development of T cell immune deficiency and disease progression.

Although there is no direct evidence in HIV infection that CD4$^+$CD25$^+$-mediated T cell immune suppression occurs during HIV infection, there is such evidence in the SIV and FIV animal models for human AIDS. Kornfield et al. [49] reported an early increase in CD4$^+$CD25$^+$ T cells following SIV infection of African green monkeys that correlated with an early TGF-β and IL10 response. In the more pathogenic SIV-Macaque infection, there was no early CD4$^+$CD25$^+$ or TGF-β responses that correlated with greater inflammation and disease, leading to the speculation that the greater pathogenicity in the SIV-infected Macaque is due to failure to control CD4$^+$ and CD8$^+$ T cell activation and the associated acute inflammation. In contrast to Kornfield et al. [49], Estes et al. [50] reported that SIV infection of Macques induced an early immunosuppressive response that correlated with a marked increase in frequency of CD4$^+$CD25$^+$Foxp3$^+$ Treg cells and TGF-β and IL10 positive T cells, suggesting that Treg cells and/or anti-inflammatory cytokines may contribute to viral persistence by prematurely limiting the antiviral immune response.

While natural CD4+CD25+ Treg cells secrete TGF-β and express cell surface TGF-β (mTGF-β) when activated for suppressor function [51–53], the involvement of TGF-β in their immunoregulatory function is controversial. Nakamura et al. [52] demonstrated that anti-CD3/APC stimulated Treg cells express mTGF-β, and treatment with anti-TGF-β neutralizing antibodies abrogated their contact-dependent suppressor function, suggesting that mTGF-β mediated the suppression. In a follow up study, Nakamura et al. [54] demonstrated that another TGF-β blocking molecule, recombinant latency-associate peptide of TGF-β1 (rLAP) reverses suppression by mouse and human CD4+CD25+ T cells. Others have demonstrated that suppressor function of both mouse and human CD4+CD25+ thymocytes are at least partially inhibited by neutralization of TGF-β [51,55,56]. In contrast, others have reported that antibody neutralization of TGF-β failed to reverse CD4+CD25+ suppressor activity [4,57]. In addition, CD4+CD25+ Treg cells from TGF-β1 deficient mice could suppress CD4+CD25− T cell proliferation, suggesting that TGF-β is not necessary for suppressor function in vitro [58].

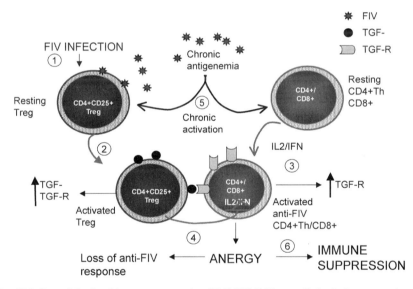

Fig. 22.1 Potential role of immunosuppressive CD4+CD25+ T reg cells in the immunopathogenesis of FIV: A model for persistent viral replication and immunosuppression FIV productively infects Treg cells (1), resulting in Treg cell activation characterized by expression of membrane TGF-β (mTGF-β) (2). Infection activates virus specific CD4+ Th and CD8+ T cells characterized by expression of TGF-βRII and secretion of IL2 and IFN-γ (3). TGF-β on the activated Treg cell is able to bind TGF-β RII on the virus activated CD4+ and CD8+ cells, transducing a signal for down-regulation of IL2 and IFN-γ gene transcription, resulting in anergy and loss of anti-FIV immune responses (4). This loss of the antiviral immune response contributes to continued virus replication, chronic antigenemia, and chronic activation of Treg cells (5). These activated Treg cells are then capable of suppressing not only anti-viral immune responses, but immune responses to other antigens, thus contributing to the global immune suppression associated with FIV infection (6) *(See also Color Insert)*

As noted above, the role of TGF-β in mediating Treg suppressor function is controversial. However, most studies have evaluated soluble TGF-β and few studies have addressed the role of cell surface TGF-β (mTGF-β) in mediating Treg suppressor function. We have recently demonstrated that CD4$^+$CD25$^+$ Treg cells from FIV-infected cats but not control cats express mTGF-β and that pretreatment with anti-TGF-β mAb or pretreatment of ConA-stimulated CD4$^+$CD25$^-$ target cells with anti-TGF-βRII antibodies abrogates suppressor function [59]. Similarly, we have demonstrated that ConA/TGF-β stimulation of CD4$^+$CD25$^-$ T cells converts them into CD4$^+$CD25$^+$Foxp3$^+$mTGF-β$^+$ Treg cells whose suppressor function can also be inhibited by anti-TGF-β (Treg) or anti-TGF-βRII (target) treatment [59]. Thus, at least in the feline AIDS lentivirus model, data suggest that TGF-β/TGF-βR signaling regulates both homeostasis and suppressor function of Treg cells. Figure 22.1 illustrates the proposed mechanism of CD4$^+$CD25$^+$-mediated CD4$^+$/CD8$^+$ immune suppression in acutely FIV-infected cats and the potential long-term immunopathological consequences.

Conclusion

Naturally occurring CD4$^+$CD25$^+$ Treg cells were first identified as a distinct lineage of thymus-derived T cells that maintained peripheral self-tolerance by their ability to recognize and suppress the function and proliferation of self-reactive T cells. Recent studies have revealed that CD4$^+$CD25$^+$ Treg cells also play a major role in modulation immune responses to microbial pathogens, including bacteria, fungi, viruses, and intracellular parasites [5,6]. Evidence suggests that CD4$^+$CD25$^+$ Treg cells are a normal immunological entity linking the innate and acquired immune responses to pathogens and provide the necessary control for an ordered protective immune response with a minimum of collateral immunopathology. Data reviewed in this chapter suggest that in the case of chronic viral infections, activation of Treg cells may have long-term immunological consequences that are not seen in infections that are resolved by an acute protective immune response. The chronic antigenemia with unresolved Hepatitis B and C infection result in an apparent unregulated Treg activation response and sustained immunosuppression of CD4$^+$ and CD8$^+$ T cells. This chronic activation of immunosuppressive Treg cells may be responsible for the inability of the host to eliminate the virus and may contribute to disease progression. In contrast, Treg cells may have a positive influence on virus-induced immune-mediated inflammatory conditions. For example, in the case of the murine HSV-induced stromal keratitis model, in vivo depletion of CD25$^+$ cells increases the severity of the ocular inflammatory lesions that correlates with an increase in HSV-specific CD4$^+$ T cell responses.

Similar to chronic hepatitis and HSV infections, Treg cells appear to be persistently activated throughout the course of AIDS lentivirus infections. FIV infection of cats activates Treg cells early in the acute stage of infection and they remain activated and immunosuppressive for CD4$^+$ and CD8$^+$ T cells throughout the long-term asymptomatic stage of infection. At this time, it is not known what actual role immunosuppressive Treg cells have on the T cell immune dysfunction that is the hallmark of AIDS lentivirus infection. However, it is clear from in vitro $CD25^+$ depletion studies that Treg cells from acute, as well as asymptomatic infections suppress antigen-induced cytokine production and proliferation of $CD4^+$ and $CD8^+$ T cells. It is also conceivable that activated Treg cells may have a positive influence in AIDS lentivirus infections by mitigating the pathology associated with T cell activation. Studies with SIV-infected primates suggest that Treg cells are activated during acute-stage infection, and they may play a positive role in down-regulating the immunopathology associated with activated $CD4^+$ and $CD8^+$ T cells.

These studies collectively show that in the case of chronic viral infections, activated Treg cells generally have contrasting effects on virus clearance and T cell-mediated inflammation in target tissue. Thus, virus activated Treg cells may play an important role in reducing the immunopathology associated with T cell activation during the acute stage infection, but in so doing they limit the protective immune response, which fails to eliminate the virus, resulting in chronic infection and persistent activation of immunosuppressive Treg cells.

References

1. Bluestone, J.A.; Abbas, A.K. Natural versus adaptive regulatory T cells. Nat Rev Immunol, 2003, 3: 253–7.
2. Sakaguchi, S. Naturally arising CD4+ regulatory t cells for immunologic self-tolerance and negative control of immune responses. Annu Rev Immunol, 2004, 22: 531–62.
3. Suri-Payer, E.; Amar, A.Z.; Thornton, A.M.; Shevach, E.M. CD4+CD25+ T cells inhibit both the induction and effector function of autoreactive T cells and represent a unique lineage of immunoregulatory cells. J Immunol, 1998, 160: 1212–8.
4. Thornton, A.M.; Shevach, E.M. CD4+CD25+ immunoregulatory T cells suppress polyclonal T cell activation in vitro by inhibiting interleukin 2 production. J Exp Med, 1998, 188: 287–96.
5. Suvas, S.; Kumaraguru, U.; Pack, C.D.; Lee, S.; Rouse, B.T. CD4+CD25+ T Cells Regulate Virus-specific Primary and Memory CD8+ T Cell Responses. J Exp Med, 2003, 198: 889–901.
6. Belkaid, Y.; Rouse, B.T. Natural regulatory T cells in infectious disease. Nat Immunol, 2005, 6: 353–60.
7. Chen, W.; Jin, W.; Hardegen, N.; Lei, K.J.; Li, L.; Marinos, N.; McGrady, G.; Wahl, S.M. Conversion of peripheral CD4+CD25- naive T cells to CD4+CD25+ regulatory T cells by TGF-beta induction of transcription factor Foxp3. J Exp Med, 2003, 198: 1875–86.
8. Fantini, M.C.; Becker, C.; Monteleone, G.; Pallone, F.; Galle, P.R.; Neurath, M.F. Cutting edge: TGF-beta induces a regulatory phenotype in CD4+CD25- T cells through Foxp3 induction and down-regulation of Smad7. J Immunol, 2004, 172: 5149–53.
9. Walker, M.R.; Kasprowicz, D.J.; Gersuk, V.H.; Benard, A.; Van Landeghen, M.; Buckner, J.H.; Ziegler, S.F. Induction of FoxP3 and acquisition of T regulatory activity by stimulated human CD4+CD25- T cells. J Clin Invest, 2003, 112: 1437–43.
10. Vahlenkamp, T.; Tompkins, M.; Tompkins, W. Feline immunodeficiency virus (FIV) infection phenotypically and functionally activates immunosuppressive CD4+CD25+ T regulatory (Treg) cells. J Immunol, 2004, 172: 4752–61.
11. Belkaid, Y.; Piccirillo, C.A.; Mendez, S.; Shevach, E.M.; Sacks, D.L. CD4+CD25+ regulatory T cells control leishmania major persistence and immunity. Nature, 2002, 420: 502–7.
12. Kinter, A.L.; Hennessey, M.; Bell, A.; Kern, S.; Lin, Y.; Daucher, M.; Planta, M.; McGlaughlin, M.; Jackson, R.; Ziegler, S.F.; Fauci, A.S. CD25+CD4+ regulatory T cells from

the peripheral blood of asymptomatic HIV-infected individuals regulate CD4+ and CD8+ HIV-specific T cell immune responses in vitro and are associated with favorable clinical markers of disease status. J Exp Med, 2004, 200: 331–43.
13. Weiss, L.; Donkova-Petrini, V.; Caccavelli, L.; Balbo, M.; Carbonneil, C.; Levy, Y. Human immunodeficiency virus-driven expansion of CD4+CD25+ regulatory T cells which suppress HIV-specific CD4 T-cell responses in HIV-infected patients. Blood, 2004, 104:3249–56.
14. Caramalho, I.; Lopes-Carvalho, T.; Ostler, D.; Zelenay, S.; Haury, M.; Demengeot, J. Regulatory T cells selectively express toll-like receptors and are activated by lipopolysaccharide. J Exp Med, 2003, 197: 403–11.
15. Aandahl, E.M.; Michaelsson, J.; Moretto, W.J.; Hecht, F.M.; Nixon, D.F. Human CD4+ CD25+ regulatory T cells control T-cell responses to human immunodeficiency virus and cytomegalovirus antigens. J Virol, 2004, 78: 2454–9.
16. Boettler, T.; Spangenberg, H.C.; Neumann-Haefelin, C.; Panther, E.; Urbani, S.; Ferrari, C.; Blum, H.E.; von Weizsacker, F.; Thimme, R. T cells with a CD4+CD25+ regulatory phenotype suppress in vitro proliferation of virus-specific CD8+ T cells during chronic hepatitis C virus infection. J Virol, 2005, 79: 7860–7.
17. Joshi, A.; Garg, H.; Tompkins, M.B.; Tompkins, W.A. Preferential feline immunodeficiency virus (FIV) infection of CD4+ CD25+ T-regulatory cells correlates both with surface expression of CXCR4 and activation of FIV long terminal repeat binding cellular transcriptional factors. J Virol, 2005, 79: 4965–76.
18. Oswald-Richter, K.; Grill, S.M.; Shariat, N.; Leelawong, M.; Sundrud, M.S.; Haas, D.W.; Unutmaz, D. HIV infection of naturally occurring and genetically reprogrammed human regulatory T-cells. PLoS Biol, 2004, 2: E198.
19. Suvas, S.; Azkur, A.K.; Kim, B.S.; Kumaraguru, U.; Rouse, B.T. CD4+CD25+ regulatory T cells control the severity of viral immunoinflammatory lesions. J Immunol, 2004, 172: 4123–32.
20. Toka, F.N.; Suvas, S.; Rouse, B.T. CD4+ CD25+ T cells regulate vaccine-generated primary and memory CD8+ T-cell responses against herpes simplex virus type 1. J Virol, 2004, 78: 13082–9.
21. McHugh, R.S.; Whitters, M.J.; Piccirillo, C.A.; Young, D.A.; Shevach, E.M.; Collins, M.; Byrne, M.C. CD4(+)CD25(+) immunoregulatory T cells: gene expression analysis reveals a functional role for the glucocorticoid-induced TNF receptor. Immunity, 2002, 16: 311–23.
22. Suvas, S.; Kim, B.; Sarangi, P.P.; Tone, M.; Waldmann, H.; Rouse, B.T. In vivo kinetics of GITR and GITR ligand expression and their functional significance in regulating viral immunopathology. J Virol, 2005, 79: 11935–42.
23. La, S.; Kim, E.; Kwon, B. In vivo ligation of glucocorticoid-induced TNF receptor enhances the T-cell immunity to herpes simplex virus type 1. Exp Mol Med, 2005, 37: 193–8.
24. Maloy, K.J.; Powrie, F. Regulatory T cells in the control of immune pathology. Nat Immunol, 2001, 2: 816–22.
25. Marshall, N.A.; Vickers, M.A.; Barker, R.N. Regulatory T cells secreting IL-10 dominate the immune response to EBV latent membrane protein 1. J Immunol, 2003, 170: 6183–9.
26. Diaz, G.A.; Koelle, D.M. Human CD4+ CD25 high cells suppress proliferative memory lymphocyte responses to herpes simplex virus type 2. J Virol, 2006, 80: 8271–3.
27. Rushbrook, S.M.; Ward, S.M.; Unitt, E.; Vowler, S.L.; Lucas, M.; Klenerman, P.; Alexander, G.J. Regulatory T cells suppress in vitro proliferation of virus-specific CD8+ T cells during persistent hepatitis C virus infection. J Virol, 2005, 79: 7852–9.
28. Chisari, F.V. Cytotoxic T cells and viral hepatitis. J Clin Invest, 1997, 99: 1472–7.
29. Xu, D.; Fu, J.; Jin, L.; Zhang, H.; Zhou, C.; Zou, Z.; Zhao, J.M.; Zhang, B.; Shi, M.; Ding, X.; Tang, Z.; Fu, Y.X.; Wang, F.S. Circulating and liver resident CD4+CD25+ regulatory T cells actively influence the antiviral immune response and disease progression in patients with hepatitis B. J Immunol, 2006, 177: 739–47.

30. Franzese, O.; Kennedy, P.T.; Gehring, A.J.; Gotto, J.; Williams, R.; Maini, M.K.; Bertoletti, A. Modulation of the CD8+-T-cell response by CD4+ CD25+ regulatory T cells in patients with hepatitis B virus infection. J Virol, 2005, 79: 3322–8.
31. Eggena, M.P.; Barugahare, B.; Jones, N.; Okello, M.; Mutalya, S.; Kityo, C.; Mugyenyi, P.; Cao, H. Depletion of regulatory T cells in HIV infection is associated with immune activation. J Immunol, 2005, 174: 4407–14.
32. Thornton, A.M.; Shevach, E.M. Suppressor effector function of CD4+CD25+ immunoregulatory T cells is antigen nonspecific. J Immunol, 2000, 164: 183–90.
33. Yang, J.H.; Zhang, Y.X.; Yu, R.B.; Su, C.; Sun, N.X. [CD4+ CD25+ regulatory T cells suppress CD4+ T cell responses in patients with persistent hepatitis C virus infection]. Zhonghua Nei Ke Za Zhi, 2006, 45: 29–33.
34. Bolacchi, F.; Sinistro, A.; Ciaprini, C.; Demin, F.; Capozzi, M.; Carducci, F.C.; Drapeau, C.M.; Rocchi, G.; Bergamini, A. Increased hepatitis C virus (HCV)-specific CD4+CD25+ regulatory T lymphocytes and reduced HCV-specific CD4+ T cell response in HCV-infected patients with normal versus abnormal alanine aminotransferase levels. Clin Exp Immunol, 2006, 144: 188–96.
35. Manigold, T.; Shin, E.C.; Mizukoshi, E.; Mihalik, K.; Murthy, K.K.; Rice, C.M.; Piccirillo, C.A.; Rehermann, B. Foxp3+CD4+CD25+ T cells control virus-specific memory T cells in chimpanzees that recovered from hepatitis C. Blood, 2006, 107: 4424–32.
36. Beilharz, M.W.; Sammels, L.M.; Paun, A.; Shaw, K.; van Eeden, P.; Watson, M.W.; Ashdown, M.L. Timed ablation of regulatory CD4+ T cells can prevent murine AIDS progression. J Immunol, 2004, 172: 4917–25.
37. Iwashiro, M.; Messer, R.J.; Peterson, K.E.; Stromnes, I.M.; Sugie, T.; Hasenkrug, K.J. Immunosuppression by CD4+ regulatory T cells induced by chronic retroviral infection. Proc Natl Acad Sci USA, 2001, 98: 9226–30.
38. Dittmer, U.; He, H.; Messer, R.J.; Schimmer, S.; Olbrich, A.R.; Ohlen, C.; Greenberg, P.D.; Stromnes, I.M.; Iwashiro, M.; Sakaguchi, S.; Evans, L.H.; Peterson, K.E.; Yang, G.; Hasenkrug, K.J. Functional impairment of CD8(+) T cells by regulatory T cells during persistent retroviral infection. Immunity, 2004, 20: 293–303.
39. Robertson, S.J.; Messer, R.J.; Carmody, A.B.; Hasenkrug, K.J. In vitro suppression of CD8+ T cell function by Friend virus-induced regulatory T cells. J Immunol, 2006, 176: 3342–9.
40. Yamano, Y.; Cohen, C.J.; Takenouchi, N.; Yao, K.; Tomaru, U.; Li, H.C.; Reiter, Y.; Jacobson, S. Increased expression of human T lymphocyte virus type I (HTLV-I) Tax11-19 peptide-human histocompatibility leukocyte antigen A*201 complexes on CD4+ CD25+ T Cells detected by peptide-specific, major histocompatibility complex-restricted antibodies in patients with HTLV-I-associated neurologic disease. J Exp Med, 2004, 199: 1367–77.
41. Yamano, Y.; Takenouchi, N.; Li, H.C.; Tomaru, U.; Yao, K.; Grant, C.W.; Maric, D.A.; Jacobson, S. Virus-induced dysfunction of CD4+CD25+ T cells in patients with HTLV-I-associated neuroimmunological disease. J Clin Invest, 2005, 115: 1361–8.
42. Joshi, A.; Vahlenkamp, T.W.; Garg, H.; Tompkins, W.A.; Tompkins, M.B. Preferential replication of FIV in activated CD4(+)CD25(+)T cells independent of cellular proliferation. Virology, 2004, 321: 307–22.
43. Joshi, A.; Garg, H.; Tompkins, M.B.; Tompkins, W.A. Different thresholds of T cell activation regulate FIV infection of CD4(+)CD25(+) and CD4(+)CD25(-) cells. Virology, 2005, 335: 212–21.
44. Andersson, J.; Boasso, A.; Nilsson, J.; Zhang, R.; Shire, N.J.; Lindback, S.; Shearer, G.M.; Chougnet, C.A. The prevalence of regulatory T cells in lymphoid tissue is correlated with viral load in HIV-infected patients. J Immunol, 2005, 174: 3143–7.
45. Kelker, H.C.; Seidlin, M.; Vogler, M.; Valentine, F.T. Lymphocytes from some long-term seronegative heterosexual partners of HIV-infected individuals proliferate in response to HIV antigens. AIDS Res Hum Retroviruses, 1992, 8: 1355–9.

46. Reddy, M.M.; Englard, A.; Brown, D.; Buimovici-Klien, E.; Grieco, M.H. Lymphoproliferative responses to human immunodeficiency virus antigen in asymptomatic intravenous drug abusers and in patients with lymphadenopathy or AIDS. J Infect Dis, 1987, 156: 374–6.
47. Wahren, B.; Morfeldt-Mansson, L.; Biberfeld, G.; Moberg, L.; Sonnerborg, A.; Ljungman, P.; Werner, A.; Kurth, R.; Gallo, R.; Bolognesi, D. Characteristics of the specific cell-mediated immune response in human immunodeficiency virus infection. J Virol, 1987, 61: 2017–23.
48. Arrode, G.; Finke, J.S.; Zebroski, H.; Siegal, F.P.; Steinman, R.M. CD8+ T cells from most HIV-1-infected patients, even when challenged with mature dendritic cells, lack functional recall memory to HIV gag but not other viruses. Eur J Immunol, 2005, 35: 159–70.
49. Kornfeld, C.; Ploquin, M.J.; Pandrea, I.; Faye, A.; Onanga, R.; Apetrei, C.; Poaty-Mavoungou, V.; Rouquet, P.; Estaquier, J.; Mortara, L.; Desoutter, J.F.; Butor, C.; Le Grand, R.; Roques, P.; Simon, F.; Barre-Sinoussi, F.; Diop, O.M.; Muller-Trutwin, M.C. Antiinflammatory profiles during primary SIV infection in African green monkeys are associated with protection against AIDS. J Clin Invest, 2005, 115: 1082–91.
50. Estes, J.D.; Li, Q.; Reynolds, M.R.; Wietgrefe, S.; Duan, L.; Schacker, T.; Picker, L.J.; Watkins, D.I.; Lifson, J.D.; Reilly, C.; Carlis, J.; Haase, A.T. Premature induction of an immunosuppressive regulatory T cell response during acute simian immunodeficiency virus infection. J Infect Dis, 2006, 193: 703–12.
51. Annunziato, F.; Cosmi, L.; Liotta, F.; Lazzeri, E.; Manetti, R.; Vanini, V.; Romagnani, P.; Maggi, E.; Romagnani, S. Phenotype, localization, and mechanism of suppression of CD4(+)CD25(+) human thymocytes. J Exp Med, 2002, 196: 379–87.
52. Nakamura, K.; Kitani, A.; Strober, W. Cell contact-dependent immunosuppression by CD4(+)CD25(+) regulatory T cells is mediated by cell surface-bound transforming growth factor beta. J Exp Med, 2001, 194: 629–44.
53. Oida, T.; Zhang, X.; Goto, M.; Hachimura, S.; Totsuka, M.; Kaminogawa, S.; Weiner, H.L. CD4+CD25− T cells that express latency-associated peptide on the surface suppress CD4+CD45RBhigh-induced colitis by a TGF-beta-dependent mechanism. J Immunol, 2003, 170: 2516–22.
54. Nakamura, K.; Kitani, A.; Fuss, I.; Pedersen, A.; Harada, N.; Nawata, H.; Strober, W. TGF-beta 1 plays an important role in the mechanism of CD4+CD25+ regulatory T cell activity in both humans and mice. J Immunol, 2004, 172: 834–42.
55. Zhang, X.; Reddy, J.; Ochi, H.; Frenkel, D.; Kuchroo, V.K.; Weiner, H.L. Recovery from experimental allergic encephalomyelitis is TGF-beta dependent and associated with increases in CD4+LAP+ and CD4+CD25+ T cells. Int Immunol, 2006, 18: 495–503.
56. Zhang, X.; Izikson, L.; Liu, L.; Weiner, H.L. Activation of CD25(+)CD4(+) regulatory T cells by oral antigen administration. J Immunol, 2001, 167: 4245–53.
57. Takahashi, T.; Kuniyasu, Y.; Toda, M.; Sakaguchi, N.; Itoh, M.; Iwata, M.; Shimizu, J.; Sakaguchi, S. Immunologic self-tolerance maintained by CD25+CD4+ naturally anergic and suppressive T cells: induction of autoimmune disease by breaking their anergic/suppressive state. Int Immunol, 1998, 10: 1969–80.
58. Piccirillo, C.A.; Letterio, J.J.; Thornton, A.M.; McHugh, R.S.; Mamura, M.; Mizuhara, H.; Shevach, E.M. CD4(+)CD25(+) regulatory T cells can mediate suppressor function in the absence of transforming growth factor beta1 production and responsiveness. J Exp Med, 2002, 196: 237–46.
59. Petty, C.S., Feline lentivirus enhanced CD4+CD25+ T regulatory conversion of CD4+CD25− T cells to phenotypic and functional T reg cells via TGF-beta/TGF-betaRII signaling pathway, in Immunology. 2006, North Carolina State University: Raleigh NC.

Chapter 23
IL-10 and TGF-β-Producing Regulatory T Cells in Infection

P.J. Dunne, A.G. Rowan, J.M. Fletcher, and Kingston H.G. Mills

Abstract Protective immunity against viruses, bacteria, parasites and fungi is mediated by innate and adaptive immune responses of the hosts. However, pathogens have developed strategies for subverting these responses and thereby establishing persistent or chronic infections. Furthermore, anti-pathogen effector immune responses must be tightly regulated to prevent collateral damage to host tissues. Regulatory T (Treg) cells that secrete the immunosuppressive cytokines IL-10 and /or TGF-β have been implicated in both processes, assisting the host to prevent immunopathology and helping the pathogen to evade protective immunity. This chapter discusses the induction of antigen-specific IL-10 and TGF-β-producing regulatory T cells during infection and gives some examples of their different roles in immunity to parasites, bacteria, viruses and fungi.

Introduction

Regulatory T cells (Treg cells) were originally identified through their function in controlling autoimmunity, where they mediate immunological tolerance to self antigens [1]. However, it later became apparent that Treg cells also control immune responses to foreign antigens on pathogens [2]. The function of Treg cells during infection appears to be primarily to prevent collateral damage from unrestrained immune responses to the pathogen, with a number of studies demonstrating enhanced immunopathology with defective or depleted Treg cells. However, it also appears that the induction of Treg cells during infection is an immune subversion strategy evolved by certain pathogens in order to prolong their survival in the host [3]. The first demonstration of pathogen-induced Treg cells showed that interleukin- (IL-) 10 and or transforming growth factor-beta (TGF-β) -producing T cells are induced during infection with bacteria [4], viruses [5] or parasites [6]. These inducible Treg cells, termed Tr1 cells, for predominantly IL-10-secreting Treg

---

K.H.G. Mills
School of Biochemistry and Immunology, Trinity College Dublin 2, Dubin, Ireland
e-mail: kingston.mills@tcd.ie

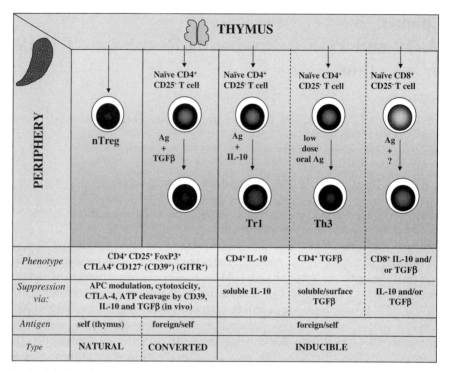

Fig. 23.1 Schematic representation of Treg subsets in humans and mice *(See also Color Insert)*

cells [7] or Th3 cells for predominantly TGF-β-secreting Treg cells [8], can be generated from naïve T cells during infection and appear to be a distinct population from the natural Treg (nTreg) cells that emerge from the thymus (Fig. 23.1). More recently it has been demonstrated that nTreg cells can suppress protective immune responses to pathogens and that removal of nTreg cells can enhance pathogen-specific effector T cell responses and promote pathogen elimination [9].

Natural Treg Cells

Fully differentiated natural Treg (nTreg) cells are derived from the thymus, when CD4 T cells with higher than normal affinity for self antigens are positively selected [10]. However it is now thought that nTreg cells with as yet indistinguishable phenotypic and functional characteristics can also be generated in the periphery [11–13]. In mice naïve CD4$^+$ T cells stimulated by antigen in the presence of TGF-β and IL-2 (Fig. 23.1) acquire FoxP3 expression and suppressive capacity [14]. In humans, however, there is still some controversy as to whether TGF-β converted Treg cells are suppressive [15,16]. Other mechanisms have been suggested for the extrathymic generation of nTreg cells including antigen exposure [17,18] and anergy induction [19,20].

nTreg cells, initially identified only by their surface expression of CD4 and CD25, have more recently have been shown to express the transcription factor FoxP3 which is essential for directing Treg function [21,22]. Lack of FoxP3 expression in humans with IPEX syndrome or in scurfy mice results in uncontrollable autoimmune disease due to lack of regulatory function [23]. Therefore, FoxP3 has become the nTreg marker of choice, although it cannot be used for cell sorting experiments as it is expressed intracellularly. Furthermore in humans, like CD25, FoxP3 is upregulated upon activation of non-Treg cells limiting its' usefulness as a Treg marker, particularly in disease settings [24]. More recently, FoxP3 expression has also been correlated with cell surface expression of CD39 [25,26] and inversely correlated with CD127 [27]. In combination with CD25, these surface markers may facilitate sorting of Treg cells, although since they are also modulated by activation, the problem of distinguishing Treg cells from recently activated effector cells remains.

nTreg function is usually measured by means of a standard suppression assay, where effector cells are stimulated in the presence of various ratios of nTreg and the amount of suppression measured by proliferation. The suppressive function of nTreg cells in vitro requires cell contact with responder cells, although no single mechanism described to date can account for all cases of Treg suppression. Mechanisms described thus far include modulation of APC function, cytotoxic activity of Treg cells and the induction of tryprophan catabolism via CTLA-4 [28]. The ectonuclease CD39 is of interest since it is the first cell surface Treg marker described that has an obvious suppressive function. Extracellular ATP, which has multiple pro-inflammatory effects, is cleaved by CD39 to form adenosine [25,26]. CD39 is expressed on all FoxP3$^+$ T cells in mice, however the situation in humans is less clear with only a subset of FoxP3$^+$ cells co-expressing CD39 [26]. While no role for soluble mediators can be demonstrated in in vitro suppression experiments, IL-10 and to a lesser extent TGF-β have been shown to contribute to Treg cell function in vivo [28].

Inducible Treg Cells

Inducible Treg cells, including Tr1, Th3 and various subsets of CD8$^+$ regulatory cells are derived in the periphery from uncommitted naïve T cells stimulated by antigen under specific conditions. Tr1 cells exert their suppressive function primarily via IL-10 secretion, and can be generated in vitro by stimulating naïve CD4$^+$ T cells in the presence of IL-10 [29]. In vivo, conditions that favour IL-10 secretion by antigen presenting cells are likely to contribute to Tr1 generation [30]. Th3 cells secrete primarily TGF-β and have been associated with the induction of oral tolerance. The induction of Th3 cells from naive CD4$^+$ T cells is favoured by the presence of TGF-β in vitro and also in vivo, where administration of low dose oral antigen leads to antigen presentation within the TGF-β-rich gut milieu [31].

Since thymically derived nTreg cells are specific for self antigens, they are thought to play a role mostly in preventing autoimmunity, where as inducible Treg cells are more likely to be generated in response to foreign antigens during infection. However the lines between natural and inducible Treg cells are somewhat blurred by the fact that peripheral conversion of nTreg cells can also occur, and it is not clear how these converted cells differ in terms of phenotype, function or antigen specificity from 'classical' thymically derived nTreg cells. Furthermore, although suppression involving IL-10 and TGF-β are the hallmark of inducible Treg cells, the suppressive function of nTreg cells in vivo is often cytokine dependent. Therefore the lack of markers, apart from IL-10 or TGF-β secretion, to identify Tr1 or Th3 cells respectively can hamper the interpretation of experiments.

There is no doubt that the various types of Treg cells play a crucial role in maintaining a healthy balance in the immune system. Treg cells have been shown to have a protective role in autoimmunity, allergy and transplantation, while in cancer Treg cells may prevent appropriate immune response to the tumour. During infection by viruses, bacteria or parasites, Treg cells can act to the advantage of either the host or the pathogen or sometimes both [3]. Certain pathogens have specific mechanisms to induce Treg cells, thereby subverting the immune response to favour the infection. Alternatively, Treg induction may favour the host by limiting an excessive immune response to the infection, thereby avoiding destructive immunopathology.

IL-10 and TGF-β-Producing Treg Cells in Parasite Infection

Malaria

Malaria is a major health problem that affects 300–500 million people worldwide, but particularly those living in sub-Saharan Africa. It can cause significant morbidity and mortality especially in young children. It is transmitted by mosquito bites from infected female mosquitoes. After transmission, the parasite develops and multiplies, causing periodic bouts of flu-like symptoms, including fever, headache, and chills. The developing parasites destroy red blood cells, which may cause death by severe anaemia as well as by blocking capillaries that supply the brain or other vital organs with blood. The most virulent of the four species of the parasite is *Plasmodium falciparum*, a species most likely to be transmitted by the mosquito *Anopheles gambiae*. Recent research has shed new light on the involvement of Treg cells, particularly those that secrete IL-10 and TGF-β, in suppressing immune responses in infected humans and mice (Table 23.1).

In *P. falciparum* infection of both unvaccinated and vaccinated (not protected) humans the peak plasma TGF-β concentration has been shown to correlate with elevated parasitaemia 7 days post sporozoite-initiated infection and this TGF-β caused a delay in the secretion of both IFN-γ and IL-6 [32]. Furthermore, the frequency of $CD3^+CD4^+$ cells expressing CD25 and FoxP3 peaked 7 days post infection. TGF-β, produced mainly by monocytes in infected humans could be responsible

23 IL-10 and TGF-β-Producing Regulatory T Cells in Infection

Table 23.1 Treg cells in parasitic infections

Parasite	Species	Phenotype	Inducible Treg cells	Mechanism of Suppression	Antigen-specific	References
P. falciparum	Human (clearance)	CD4$^+$ CD25$^+$ FoxP3$^+$	TGF-β induced	TGF-β secreted by monocytes		[32]
P. falciparum	Human neonates (reduced immunity)			IL-10 and CTLA-4		[34]
P. yoelii	Susceptible BALB/c mouse	CD4$^+$ CD25$^+$		Low IL-10 produced by CD4$^+$ CD25$^+$ cells		[33]
P. yoelii	Resistant DAB/2 mouse			2 early bursts of IL-10		[33]
L. viannia braziliensis	Human self healing cutaneous lesions	CD4$^+$ CD25$^+$ FoxP3$^+$	Yes	IL-10 and TGF-β	Yes	[35]
L. major	C57BL/6 mouse self healing cutaneous lesions	CD4$^+$ CD25$^+$	Yes	IL-10-independent	Yes	[37]
L. amazonensis	C57BL/6 mouse non-healing cutaneous lesions	CD4$^+$ CD25$^+$ FoxP3$^+$		IL-10 and signalling via CD86		[39]
S. mansoni	Mouse	CD4$^+$ CD25$^+$ FoxP3$^+$		IL-10-independent	Yes	[44]
S. mansoni	Mouse	CD4$^+$ CD25$^+$ FoxP3$^+$		IL-10-independent	Yes	[45]
S. mansoni	Mouse	CD4$^+$ CD25$^+$ FoxP3$^+$		Low IL-10; principle source Th2 cell	Yes	[47]
H. polygyrus	Mouse	CD4$^+$ CD25$^+$ FoxP3$^-$	Yes	IL-10	Yes	[48]
L. sigmodontis	Mouse	CD4$^+$ CD25$^+$ FoxP3$^-$	Yes	IL-10-independent		[50]

for the conversion of $CD4^+CD25^-$ naïve T cells into $CD25^+FoxP3^+$ Treg cells at the later stages of infection with *P. Falciparum* [14,33]. Whether the cells involved are nTreg cells that are thymically-derived or whether they are generated from effector T cells in response to TGF-β remains to be seen. What is apparent is that these cells restrain the immune response and prevent extensive tissue damage in the host.

Research on mothers infected with *P. falciparum* has revealed that freshly isolated cord blood contains a significant proportion of $CD4^+CD25^+IL-10^+CTLA-4^+$ T cells when compared with non-infected controls [34]. These Treg cells were shown to suppress IFN-γ secreted by Th1 cells and $CD8^+$ T cells and decrease MHC class I and class II expression on monocytes [34]. Although the foetus developed a protective Th1 response, Treg cells acquired from the mother secreted IL-10, which suppressed any significant activation. Children born to mothers with *P. Falciparum* infection are more susceptible to infection than other children, and the increase in the frequency of IL-10-secreting Treg cells might be the cause [34]. Clearly, a balance must be found between the promotion of Treg cell expansion, which prevents immune-mediated tissue destruction, and the generation of a protective *P. Falciparum*-specific Th1 response. It is possible that Treg cells induced by TGF-β late in the immune response in humans might be more beneficial than IL-10-secreting Treg cells soon after the onset of infection.

An examination of susceptible (BALB/c) and resistant (DBA/2) mice to a lethal strain of malaria *P. yoelii* (17XL) [33] has yielded interesting results on Treg cells that secrete regulatory cytokines in malaria infection. In these studies IFNγ was detected early in infection; however, the concentration of IFNγ produced in resistant DBA/2 mice was significantly greater than in susceptible BALB/c mice. In addition, the frequency of $CD4^+CD25^+$ cells peaked soon after infection in BALB/c mice and was steadily reduced thereafter, whereas in resistant mice, the peak in the frequency of $CD4^+CD25^+$ occurred later and was maintained throughout infection. Spleen cells isolated from infected BALB/c mice produced very little IL-10, whereas spleen cells from resistant DAB/2 mice were found to secrete two significant bursts of the regulatory cytokine 3 and 6 days post infection with *P. Yoelii* [33].

The delayed peak in the frequency of $CD4^+CD25^+$ regulatory cells post infection in DAB/2 mice could be responsible for preventing significant tissue damage and promoting host survival. It is also possible that the source of IL-10 post infection in DAB/2 mice was the $CD4^+CD25^+$ nTreg cell [33]. The delayed appearance of Treg cells (around day 6 post infection in both mice and humans) prevented tissue damage and promoted host survival. The obvious discrepancy between studies carried out on humans and mice is that in human infection TGF-β, secreted by monocytes might convert $CD4^+$ T cells into a regulatory phenotype which suppress the immune response whereas in the murine model it is proposed that natural Treg cells inhibit the Th1 response by IL-10 particularly in infection-resistant DAB/2 mice [33]. However, a very recent study has shown that naïve $CD4^+$ T cells activated in the presence of TGF-β are not capable of suppressing effector T cell responses in humans [15].

Leishmania

Leishmaniasis, caused by infection with protozoan parasites, is believed to have infected 12 million people worldwide and the number of cases of Leishmania/HIV co-infection is expected to rise significantly in the coming years. The infection in humans causes self-healing skin lesions, diffuse cutaneous and mucocutaneous disease and potentially lethal visceral lesions. The symptoms result from parasite replication in dermal macrophages, nasopharyngeal mucosa and other phagocytes.

$CD4^+CD25^+FoxP3^+$ cells that secrete IL-10 and TGF-β have been isolated from the cutaneous lesions of patients with *Leishmania viannia braziliensis* [35]. In this report these cells suppressed PHA-activated peripheral blood mononuclear cells (PBMC) at a Treg to T effector ratio of 1:10, similar to natural Treg cells [35]. Therefore, nTreg cells were capable of curtailing immunity to this strain of parasite in humans, presumably by secretion of IL-10 and TGF-β [35]. $CD4^+CD25^+$ cells that secrete TGF-β and inhibit the proliferation of $CD4^+CD25^-$ T cells in an antigen-specific manner were also detected in PBMC (isolated from healthy human subjects) exposed to live *L. guyanensis* [36]. Interestingly, these cells inhibited IL-10 and IL-2 production by $CD4^+CD25^-$ T cells, indicating that TGF-β, and not IL-10, plays a role in suppressing both Th1 and Th2 responses in humans exposed to *L. guyanensis* for the first time [36].

$CD4^+CD25^+$ T cells have also been detected in cutaneous lesions in C57BL/6 mice infected with *L. major*. These cells were shown to suppress anti-CD3-induced activation of $CD4^+$ $CD25^-$ T cells in vitro [37]. Up until recently it has been accepted that IL-10 plays a significant role in suppressing the adaptive immune response to *L. major* in mice and humans [38]. However, blocking studies with anti-IL-10 receptor antibodies have demonstrated that IL-10 does not play a role, at least in in vitro suppression [37]. Experiments with $IL-10^{-/-}$ (IL-10-deficient) mice have shown that the Treg-mediated promotion of parasite survival is not affected, which might indicate a level of redundancy, with other cytokines such as TGF-β, substituting for IL-10 secreted by the $CD4^+CD25^+$ T cells in the cutaneous lesions. Despite this, a significant amount of IL-10 mRNA can be detected in the $CD4^+CD25^+$ cells isolated from cutaneous lesions in mice. Moreover, $CD4^+CD25^+$ Treg cells do produce IL-10 in response to *L. major*-infected skin dendritic cells (DC), but not to anti-CD40 stimulation or with DC infected with *Toxplasma gondii*, indicating that the Treg cells accumulating within these skin lesions are Leishmania-specific [37].

Treg cells that accumulate early in infection with *L. major* might allow the development of chronic infection and the potential for disease reactivation. This benefit to the parasite may also confer an advantage to the host, whereby life-long immunity to the parasite is maintained. This would explain why those individuals infected with *L. major* develop lesions that heal and are resistant to subsequent re-infection. IL-10 might have a role to play in this durable immunity in infected humans and mice since IL-10 knockout mice that cleared infection displayed poor immunity to re-infection with *L. major*.

As discussed above, *L. major* is generally a contained cutaneous infection and is usually self healing. On the other hand, *L. Amazonensis* is neither self healing nor contained within the skin and can lead to the development of visceral lesions [39]. In *L. amazonensis* infection severe combined immunodeficient mice (SCID), MHC II$^{-/-}$ and RAG$^{-/-}$ mice were found to clear the parasite, supporting other studies which show that T cells and Treg cells in particular play a role in promoting the chronic infection of the host with this parasite [39]. A significant increase in the absolute numbers of CD4$^+$CD25$^+$CD86$^+$FoxP3$^+$ T cells and considerable quantities of IL-10 receptor mRNA were detected in the cells isolated from the lymph nodes and spleens of infected mice at 1 and 3 weeks post infection [39]. In addition, CD4$^+$CD25$^+$ T cells isolated from the lesions of these mice suppressed proliferation of CD4$^+$CD25$^-$ cells in vitro. In this experimental system IL-10 secreted by the CD4$^+$CD25$^+$ cells was responsible for the suppression observed [39].

CD4$^+$CD25$^+$ Treg cells which secrete IL-10 accumulate in the local lymph nodes soon after infection with *L. amazonensis* [39]. At this juncture, these cells serve to protect the host from immune-induced tissue damage, and to inhibit IFNγ secreted by Th1 cells that promotes survival of the parasite at the onset of infection. However, it has also been demonstrated that the beneficial effect of the Treg cell is short-lived and the disease ultimately progresses to the development of visceral lesions [39].

Different strains of mice are more or less susceptible to infection with Leishmania and disease progression is influenced by the strain of the parasite, a fact that makes results yielded from studying infection in these models difficult to interpret. What is clear however is that Treg cells have a role to play in Leishmania infection in both humans [35,40,41] and mice [37,39]. Both IL-10 and TGF-β may play significant roles in inhibiting the immune response early in infection in humans. However, the situation in mice is less clear. Experiments with blocking antibodies and in IL-10$^{-/-}$ mice have suggested Treg cells in Leishmania-infected mice act in an IL-10-independent fashion. However, in wild type mice Treg cells do produce IL-10 during infection. It is possible that TGF-β can act in a redundant manner in the absence of IL-10. In addition, the early appearance of Treg cells that secrete IL-10 at the site of infection would seem to permit a balance between parasitic survival and concomitant infection which in turn confers life long immunity to the host. Once again, the balance between Treg cells that produce regulatory cytokines and effector T helper responses is crucial to disease progression and survival of the host.

It is interesting to note that the Treg cells isolated from animals and humans infected with *L. viannia braziliensis* and *L. major* are specific for parasitic antigens. Since nTreg cells have been described as self-antigen-specific [42], it could be speculated that the Treg cells described in Leishmania infection started out as parasite-specific effectors that some how differentiated into Treg cells. There is evidence to suggest that inducible Treg cells are in the main pathogen-specific [43] unlike natural Treg cells which are generally self antigen-specific.

In contrast to malaria, where TGF-β seems to play a significant role in infected humans, IL-10 appears to have a greater role in Leishmania infection (the exception being first time exposure of human PBMC to *L. guyanensis* where TGF-β plays a significant role [36]). This of course might not be particularly significant given the redundant nature of these cytokines. It is apparent that a slight delay in the infiltration and accumulation of Treg cells that secrete either IL-10 or TGF-β has a beneficial effect for the host (limited tissue damage and enhanced immunity). The exception being children born to mothers infected with *P. falciparum* who seem to have reduced immunity to the parasite, presumably as a direct result of Treg-mediated suppression of Th1 responses [34].

Schistosoma

Schistosomiasis is a chronic disease that affects greater than 200 million people worldwide and is caused by a helminth parasite infection of the genus *Schistosoma*. The pathology and clinical symptoms observed result from the immune response to parasite eggs that become trapped in the liver, intestine and urogenital tract. The eggs evoke a brief initial Th1 response followed by a strongly polarised Th2 response, which fails to eradicate the parasite and leads to the formation of granulomas, in which the eggs survive for prolonged periods. One of the complications of the chronic disease is severe fibrosis of the liver, which can be fatal to the host.

$CD4^+CD25^+FoxP3^+$ cells from *S. mansoni*-infected mice were shown to suppress T cell proliferation even more so than Treg cells from non-infected animals, indicating that these Treg cells are antigen-specific [44]. Adoptive transfer of $CD25^+$ T cells diminished the Th1 response as defined by IFN-γ secretion even in $IL-10^{-/-}$ animals, indicating that these cells can act in an IL-10-independent manner [44]. Nevertheless, the Treg cells in *S. mansoni* infection in wild-type mice still produced IL-10. These Treg cells most likely prevent an aggressive Th1 response against the parasite, as is the case in $IL-10^{-/-}$ animals, which develop severe disease, where the parasite survives within granulomas [44].

A significant expansion of IL-10-secreting *S. mansoni*-specific $FoxP3^+$ Treg cells was observed 7 days post infection of mice with *S. mansoni* eggs [45]. These Treg cells suppressed both Th1 and Th2 responses in an IL-10-independent manner, (the Treg cells isolated from $IL-10^{-/-}$ mice still suppressed CD4 T cell activation in vitro) [45]. TGF-β is unlikely to play a role in the absence of IL-10, as it has been demonstrated that TGF-β is not required for the development of IL-13-induced fibrosis in mice [46]. Nevertheless it is possible that IL-10 and TGF-β act in a redundant manor in *Schistosoma* infection in mice.

Conversely it has been demonstrated that the ratio of $CD4^+FoxP3^+$ nTreg cells to effector T cells is not altered in mice infected with *S. mansoni* when absolute numbers and not percentages of $CD4^+FoxP3^+$ are taken into account [47]. Experiments involving helminth-infected 4get mice, which have a bicistronic reporter that permits the co-expression of IL-4 with the fluorescent molecule GFP (green fluorescent protein) have yielded interesting results [47]. $CD4^+$ T cells that secrete IL-4 in this system co-express GFP and can subsequently be detected by flow cytometry. Using these mice it has been demonstrated that the IL-10 secreted during infection emanates from Th2 cells and not from $CD4^+CD25^+FoxP3^+$ Treg cells [47].

Furthermore, depletion of CD25$^+$ cells in infected animals resulted in increased IL-10 production by Th2 cells and IFNγ by Th1 cells [47]. This finding supports the idea that Treg cells in *S. mansoni* infection act in an IL-10-independent manner.

In conclusion, expanded antigen-specific Treg cells that secrete IL-10 play a significant role in inhibiting the Th1 and Th2 responses against the parasite, which in turn contributes to the establishment of a chronic parasite infection of the host. However, the role of IL-10 secreted by these cells is limited, as experiments with IL-10$^{-/-}$ mice and antibody-depletion have shown that these cells most likely inhibit both Th1 and IL-10-secreting Th2 arms of the immune response in a cell-contact mediated manner.

Heligmosomoides Polygyrus

H. polygyrus is a gastrointestinal nematode parasite of mice, which follows a direct transmission cycle. Orally ingested larvae invade the intestinal mucosa and emerge as adult worms 9–11 days later. In most strains of mice they survive as chronic luminal dwelling infections that can persist for up to 300 days [48]. T cells isolated from the mesenteric lymph nodes (MLN) of infected mice were shown express intracellular IL-10 and surface TGF-β. CD4$^+$CD25$^+$ cells outgrew CD4$^+$CD25$^-$ cells from day 7 onwards and were maintained at significant frequencies throughout infection. CD25 expression on these cells is not simply an indicator of activation since these cells suppressed CD4$^+$CD25$^-$ cells in vitro in an antigen-specific manner [48].

Interestingly FoxP3 expression did not increase significantly over time, indicating that inducible or FoxP3$^-$ Treg cells might play a role in *H. polygyrus* infection [48]. CTLA-4 and GITR, surface markers associated with Treg, were found almost exclusively on CD4$^+$CD25$^-$ early in infection (day 20), coupled with a rise in surface TGF-β. CD103 (α4β7 integrin found on activated CD4$^+$ Treg cells) was closely associated with FoxP3$^+$ cells, while TGF-β was detected mostly on FoxP3$^-$ cells [48].

The T cells found in *H. polygyrus* infection could be generated from naïve precursors that upon activation produced both IL-10 and TGF-β, but primarily the latter. It has been demonstrated that TGF-β plays a very significant role in dampening the immune response to this parasite in the intestines of infected mice [49]. The antigen-specific nature of these cells supports this conclusion. Furthermore, it is interesting to note that Treg cells isolated from *S. mansoni*-infected mice are also suppressive in an antigen-specific manner [44].

Litomosoides Sigmodontis

L. sigmodontis is a close relative of Brugia and Wuchereria species which are responsible for in excess of 200 million cases of lymphatic filariasis in humans worldwide. Research on this parasite has shown that with the exception of a short burst in the expansion of CD4$^+$CD25$^+$FoxP3$^+$ T cells at the beginning of infection, the ratio of FoxP3$^+$ Treg cells to FoxP3$^-$ effector T cells does not change significantly in infected mice [50]. Phenotyping of CD4$^+$ T cells isolated from the draining lymph nodes of infected mice revealed that the majority of cells coexpressed the regulatory markers, GITR and CTLA-4 [50]. While the most intense expression of these surface markers was found on CD4$^+$FoxP3$^+$ T cells, the majority of GITR$^+$CTLA-4$^+$ cells were FoxP3$^-$ and therefore assumed not to be nTreg cells [50].

The role of Treg cells in filariasis was examined using an anti-CD25 depleting antibody and an agonistic anti-GITR antibody [50]. Treatment with a combination of antibodies resulted in complete clearance of the worms or encapsulation by immune cells in infected mice. Eradication was associated with increased concentrations of IL-5 [50]. Intriguingly, the concentration of IL-10 actually increased in antibody-treated mice, indicating that the Treg cells which suppress immunity against the worm inhibit the Th1 and Th2 arms of the immune response in an IL-10-independent fashion [50].

The CD4$^+$CD25$^-$FoxP3$^-$GITR$^+$CTLA4$^+$ T cell population observed in mice infected with *L sigmodontis* might be 'conditioned' effector cells that remain unresponsive to specific antigen in vitro. While the Treg cells described suppress in an IL-10-independent manner, some IL-10 is produced during infection. Although a significant amount of latent TGF-β has been detected in infected mice, it remains unclear as to how the CD4$^+$CD25$^+$FoxP3$^+$ T cells suppress immunity during the first 20 days of infection [50].

IL-10 and TGF-β-Producing Treg Cells in Bacterial Infection

Lactobacillus

Probiotic bacteria are defined as live microbial food ingredients that are beneficial to health in general and are also commensal bacteria of healthy human intestinal micro flora [51]. One of the most important genera is *Lactobacillus* which has been shown to have beneficial effects in the treatment of asthma [52] and gastrointestinal disease [53]. The beneficial effects of *Lactobacillus* have been ascribed to the generation of Treg cells that produce IL-10 and TGF-β at the gut mucosa.

Investigations into the immunosuppressive properties of *Lactobacillus* revealed that *L. paracasei* was able to induce Treg cells that secrete TGF-β and IL-10, which subsequently inhibited the growth of effector CD4$^+$ T cells in vitro [54]. Furthermore, *L. reuteri* and *L. casei*-treated DC could convert effector CD4$^+$ T cells into Treg cells, which in turn inhibited proliferation of conventional CD4$^+$ T cells [51]. These cells acted in an IL-10-dependent manner, since neutralisation of IL-10 abrogated the suppressive effect [51]. In appears that both *L. reuteri* and *L. casei* prime DC in the gut mucosa to promote the development of Treg cells, which inhibit the

Table 23.2 Treg cells in bacterial and fungal infections

Pathogen	Species	Phenotype	Inducible Treg cells	Mechanism of Suppression	Antigen-specific	References
Bacteria						
L. reuteri, L. casei, L. paracasei	Human	CD4+ CD25+	Yes	IL-10 and TGF-β		[38,52,53]
M. tuberculosis	Human	CD4+ CD25+ FoxP3+		Surface-bound TGF-β	Yes	[58,59]
H. pylori	Mice	CD4+ CD25+ FoxP3+	Yes	IL-10 and TGF-β	Yes	[18,62]
B. pertussis	Mice	CD4+ CD25+	Yes	IL-10	Yes	[4,64]
Fungi						
P. braziliensis	Human	CD4+ CD25+		Surface-bound TGF-β		[65]
A. fumigatus	Mice	CD4+ CD25+		IL-10 and TGF-β		[66]
C. albicans	Mice	CD4+ CD25+		IL-10		[71]

activation and proliferation of adjacent effector CD4+ T cells in an IL-10-dependent fashion. It is thought that these particular strains of *Lactobacilli* modulate DC by binding to the C-type lectin dendritic cell-specific intercellular adhesion molecule 3-grabbing non-integrin (DC-SIGN) [51]. Lactobacilli appear to play a role in maintaining a suppressive environment within the intestinal mucosa and they accomplish this by indirectly generating Treg cells, that secret IL-10 and TGF-β (Table 23.2)

Mycobacterium Tuberculosis

Infection with *M. tuberculosis* is one of the leading causes of mortality in the world today as a result of a single infectious agent and is attributable to approximately 2 million deaths and 8 million new cases annually [55]. TGF-β and IL-10 secreted by T cells have been implicated in the persistence of infection with *M. tuberculosis* and in muting host immune effector functions [56,57]. However, recent studies have questioned the role of both these immunomodulatory cytokines in suppressing immune responses. Blocking TGF-β and IL-10 with neutralising antibodies did not enhance IFNγ secretion by PBMC from infected patients in response to the *M. tuberculosis* antigen, heparin-binding haemagglutinin (HBHA) [58]. However, depletion of CD4+CD25+ cells from the same PBMC did result in increased IFNγ production and proliferation of CD4+ T cells. Treg cells from *M. tuberculosis* infected patients were found to suppress effector responses in an IL-10 and TGF-β-independent manner [58].

Intriguingly, depletion of CD4+CD25+ cells from PBMC isolated from patients with pulmonary tuberculosis (PTB) resulted in increased concentrations of IFNγ and IL-10 [59]. The CD4+CD25+ cells from infected patients could only suppress

Bacille Calmette Guérin (BCG) and early-secreted antigen target-6 (ESAT-6) antigen-specific activation and not responses to irrelevant antigen. However, there is no overall difference in the total FoxP3 expression relative to the number of CD4$^+$ T cells in from patients with PTB and healthy controls. It is likely that antigen-specific CD4$^+$CD25$^+$FoxP3$^+$ T cells are induced in response to *M. tuberculosis* infection and these cells suppress both Th1-and Th2-associated cytokines. Although IL-10 is widely associated with immunosuppression, it can also stimulate B cells and T cells, particularly CD8$^+$ T cells [60]. Therefore, while Treg cells have a major role to play in promoting the survival of *M. tuberculosis* within its host, it is likely that neither TGF-β nor IL-10 are essential for the immune suppression. However, the possibility of surface-bound TGF-β having a part to play in the suppression of immunity to *M. tuberculosis* can not be ruled out.

Helicobacter Pylori

H. pylori is a bacterium that causes chronic gastritis in humans and is the most common cause of ulcers worldwide [61]. Infected individuals usually carry the infection indefinitely unless they are treated. Complications associated with chronic infection include stomach cancer, mucosa-associated lymphoid tissue (MALT) lymphoma as well as ulcers of the stomach and duodenum [61].

In mice, a strong pro-inflammatory response involving IL-12p40 and TNFα has been observed soon after infection [62]. This was followed by a steady increase in immunomodulatory cytokines, such as IL-10 and TGF-β as well as expression of FoxP3 (based on mRNA expression in the gastric mucosa) [62]. Depletion of CD4$^+$CD25$^+$ T cells from mice infected with *H. pylori* resulted in marked gastric inflammation and restructuring of the gastric mucosa within four weeks. In addition, a strong correlation was found between FoxP3 and IL-10 expression over time in the gastric mucosa of *H. pylori*-infected animals [62]. IL-10 is important in preventing destruction of the gastric mucosa, since progressive atrophy of the mucosal wall and eradication of the bacteria was observed in IL-10$^{-/-}$ mice infected with *H. pylori*. IL-10 produced after the initial phase of infection might induce the development of CD4$^+$ Tr1 cells in the gastric mucosa. This is consistent with the observation that the Treg cells from *H. pylori*-infected mice suppressed in an antigen-specific manner only and not in the presence of irrelevant antigen or mitogen such as phytohaemagglutin (PHA) [62]. These observations are further supported by the fact that pathogen-specific, inducible CD4$^+$CD25$^+$FoxP3$^+$ Treg cells can also be generated at the site of inflammation in humans [18].

In *H. pylori* infection, the initial Th1 response is followed by an increase in either nTreg cells or CD4$^+$CD25$^-$ cells converted to a Treg phenotype by the high local concentration of IL-10. This accumulation of Treg cells in the gastric mucosa results in survival of the bacteria in high density clusters and allows the host to avoid gastric pathology.

Bordetella Pertussis

Respiratory infection with the gram negative bacterium *B. pertussis* causes whooping cough in humans, a severe debilitating disease which still affects up to 40 million young children worldwide [3]. Complications of infection include convulsions, encephalopathy, encephalitis, permanent brain damage and death [63]. Immunity against *B. pertussis* develops after natural infection and confers long-lived protection against subsequent infection.

B. pertussis infection in mice is asymptomatic, but provides a useful model for studying the immunomodulatory properties of this microbe in vivo. Filamentous haemagglutinin (FHA), a virulence factor produced by *B. pertussis*, has been shown to interact directly with DC to produce IL-10 [4], which in turn inhibited LPS-induced IL-12 and inflammatory cytokine induction. The bacteria stimulated innate IL-10 and promoted the induction of Tr1 cells from naïve $CD4^+$ T cells, which could subsequently inhibit *B. pertussis*-specific IFNγ secretion and proliferation of effector $CD4^+$ T cells in an IL-10-dependent fashion [4].

In addition to virulence factors such as FHA, LPS from *B. pertussis* can stimulate DC to mature and produce IL-10 as well as proinflammatory cytokines via ligation of surface TLR4 [64]. It has been demonstrated that the secretion of innate IL-10 results in the induction of Tr1 cells that produce IL-10 upon activation with pathogen-specific antigen [64]. TLR4-defective mice have enhanced antigen-specific IFNγ secretion and decreased antigen-specific IL-10 secretion, as well as enhanced immunopathology in response to infection [64]. It appears that virulence factors and TLR4 agonists, such as LPS, play a significant role in limiting inflammation and tissue destruction in response to infection with *B. pertussis*, thereby promoting the survival of the bacterium and conferring life-long immunity to the host.

IL-10 and TGF-β-Producing Treg Cells in Fungal Infection

Paracoccodioides Braziliensis

Paracoccidioidomycosis (PCM) is a deep granulomatous mycosis caused by a dimorphic fungus, *P. braziliensis* [65]. Infection usually involves the skin, mouth, throat, and lymph nodes, although it sometimes appears in the lungs, liver, or spleen. Chronically-infected individuals have persistent antigenic stimulation and develop effective control of fungal growth and dissemination [65]. The acute form of the disease is very severe and often mimics lymphoproliferative disease in children and occurs in adults as result of the re-activation of quiescent foci [65].

$CD4^+CD25^+$ cells isolated from patients with PCM expressed intracellular FoxP3 and surface latent associated protein-1 (LAP-1), which is noncovalently linked to TGF-β in its inactive form [65]. These Treg cells were significantly more potent at suppressing autologous conventional $CD4^+$ T cells than

CD4+CD25+ cells isolated from healthy donors [65] (Table 23.2). The absolute numbers of CD4+CD25+ cells within cutaneous lesions expressing intracellular IFNγ and IL-10 were similar to non-infected controls. However, the total number of CD4+CD25+TGF-β+ cells was significantly higher in PCM-infected lesions [65]. Interestingly, the percentage of CD4+CD25+IFNγ+IL-10+TGF-β+ cells was found to be higher in infected biopsies than in controls. Therefore, although TGF-β plays a significant part in suppressing the CD4+ T cell-mediated response in PCM, a role for IL-10 cannot be excluded.

Aspergillus Fumigatus

A. fumigatus is a thermotolerant saprophyte associated with a wide expression of disease from severe infection to allergy [66]. The conidia are normally eliminated in the immunocompetent host, however, this infection can cause fatal invasive aspergilliosis, in immunocompromised patients, in particular those who have acquired immunodeficiency syndrome (AIDS) [67]. Aspergilliosis is the most common mould infection worldwide.

Investigation of CD4+CD25+ Treg cells isolated from mice early in infection (day 3) revealed that these cells produce high concentrations of IL-10, whereas during late infection Treg cells secreted mainly TGF-β [66]. Infiltrating Treg at the start of infection dampened the Th1 response against the fungus and were vital to the long term survival of the pathogen since early elimination of Treg cells from infected animals, using cyclophosphamide, enhanced pathology and lead to a marked reduction in fungal growth [66]. Treg cells arriving later in infection (day 10) were much more potent at inhibiting CD4+ T cell effector proliferation than the CD4+CD25+ Treg cells found at the onset of fungal infection [66].

Treg cells isolated from *A. fumigatus* infected mice at the start of infection were found to be very effective at suppressing the innate immune response. These cells inhibited TNFα secretion from monocytes and IFNγ by polymorphonuclear phagocytes (PMN) and suppressed phagocytosis, oxidant production and fungicidal activity [66]. These cells suppressed both the innate and adaptive phase of the immune response via surface CTLA-4/B7.1 interaction, which in turn induced indoleamine 2,3-dioxygenase (IDO) within DC and macrophages [66]. IDO catalyses oxidative degradation of the amino acid tryptophan and these decreased levels have an inhibitory effect on proliferation of T cells, directly or indirectly via activation of Treg cells [68]. In contrast, Treg cells isolated from infected mice 10 days post infection (late) inhibited the Th2-associated cytokine IL-5, the total number of eosinophils in bronchiolar lavage samples and total serum IgE [66]. This suppression was largely attributed to TGF-β secreted by Treg cells and to a lesser extent IL-10 [66].

Candida Albicans

Candidiasis is the fourth most common cause of nosocomial infectious disease in the world and can have serious consequences for the immunocompromised host [69].

Disseminated candidiasis as a result of infection with *C. albicans* can originate in the mucosa of the gastrointestinal tract when the pathogen enters epithelial microvilli. In both cases, organisms enter the vasculature for dissemination into tissues, such as the kidney. A vigorous host response occurs at this site, involving both mononuclear and PMN cells [69].

Recently, TLR2 that binds PAMPs such as Pam3CSK4, have been described on Treg cells [70] and TLR2 is one of the receptors engaged by *C. albicans* [71]. TLR2$^{-/-}$ mice injected with a lethal dose of *C. albicans* survive longer than control animals [71]. In murine candidiasis the kidneys are the principle target for dissemination and TLR-2$^{-/-}$ mice have been shown to have 100-fold decreased load of the fungus when compared with infected wildtype mice [71]. While TLR4$^{-/-}$ mice posses normal absolute numbers of CD4$^+$CD25$^+$ T cells, TLR2$^{-/-}$ mice have a 50% reduction in this T cell subset [71]. Moreover, in vitro studies showed that the concentration of IL-10 produced by activated spleen cells isolated from TLR2$^{-/-}$ infected mice was significantly reduced. Mice depleted of CD25$^+$ cells showed a 10-fold decrease of fungal outgrowth in the kidneys 7 days post infection [71]. Taken together, these findings suggest that CD4$^+$CD25$^+$ T cells suppress immunity against *C. albicans* infection, particularly at the later stage of infection and more than likely via IL-10. The organism can enhance the suppressive capacity of infiltrating natural Treg cells by binding directly to TLR-2 on its surface and subsequently prolong its survival within the host.

IL-10 and TGF-β-Producing Treg Cells in Viral Infection

The character of a protective antiviral immune response varies according to the life cycle of each individual virus; this is reflected in the heterogeneity of virus specific Treg cells described in infection (Table 23.3). Treg cells are most commonly described in persistent virus infections, this may reflect specific induction of regulatory response by these viruses, or a host mechanism to reduce pathology in the face of persistent viral replication.

Friend Virus

Friend virus (FV) is a murine retrovirus which can cause erythroleukaemia in susceptible mice. FV consists of a retroviral complex composed of nonpathogenic, replication competent friend murine leukemia virus, and the pathogenic but replication defective spleen focus forming virus. Although strong virus-specific CTL responses resolve acute FV infection, viral replication usually persists in splenic B cells for the lifetime of the host, which have a reduced capacity to reject experimentally induced tumours [72]. Depletion of CD8$^+$ T cells during the persistent phase of disease had no effect on virus levels, indicating a diminished protective role for virus-specific CTL at this stage of infection [73]. This reduced antiviral CD8$^+$ cell response was not due to viral epitope mutation, clonal deletion or a reduction in

23 IL-10 and TGF-β-Producing Regulatory T Cells in Infection 439

Table 23.3 Treg cells in viral infections

Virus	Species	Phenotype	Inducible Treg cells	Mechanism of Suppression	Antigen-specific	References
FV	Mouse	CD4+CD25±	Yes	IL-10-independent		[74]
HTLV	Human	CD4+ CD8+		IL-10		[79]
SIV	Macaque	CD4+ CD25+ FoxP3+	T reg frequency increased	TGF-β		[83]
SIV	Macaque	CD3+	T reg frequency increased	IL-10		[83]
HIV	Human	CD4+ CD25+	Yes	IL-10 and TGF-β -independent	Yes	[87]
HIV	Human	CD8+	Yes	TGF-β	Yes	[88]
EBV	Human	CD8+	Yes	IL-10	Yes	[89,90]
EBV	Human	CD4+CD25+FoxP3± GITR±	Yes		Yes	[92]
EBV	Human	CD4+	Yes	IL-10	Yes	[93]
HBV	Human	CD4+	Yes	IL-10	Yes	[95]
HCV	Human	CD4+	Yes	IL-10	Yes	[5,96]
HCV	Human	CD8+	Yes	IL-10	Yes	[99]
HCV	Human	CD8+	Yes	IL-10 and TGF-β	Yes	[101]
HCV	Human	CD4+CD25+	Yes	IL-10 and TGF-β	Yes	[97,98]
HSV	Mouse	CD4+CD25+		IL-10		[103]

antigen presentation by infected cells [74]. In the natural course of FV infection, the number of virus specific $CD8^+$ T cells, detected by tetramer staining and flow cytometric analysis, peaked two weeks after exposure to the virus. Even at this early stage, CTL responses were functionally impaired, as shown by suboptimal expression of effector molecules granzyme B and CD107 by activated ($CD43^+$) $CD8^+$ T cells [75,76].

Immune dysfunction in acute FV infection can be exacerbated by adoptive transfer of $CD4^+$ T cells from persistently infected mice. $CD4^+$ T cells isolated from persistently infected mice expressed IL-10, and induced IL-10 production by host $CD4^+$ T cells when transferred into acutely infected mice. Both $CD25^+$ and $CD25^-$ T cells from persistently infected animals were capable of suppressing acute responses in vivo. Similarly, fully competent virus specific $CD8^+$ T cells from acutely infected mice transferred into persistently infected mice could not eliminate residual virus, and the transferred cells lost their ability to make IFN-γ. Blockade of IL-10R signalling in vivo did not prevent the loss of $CD8^+$ IFN-γ^+ T cells [74], suggesting that IL-10 did not mediate the suppression.

Induction of functional impairment has been shown to be temporally linked to increased numbers of $CD4^+CD25^+$, $CD4^+GITR^+$ and $CD4^+CD103^+$ T cells, a proportion of which co-express FoxP3. Depletion of total $CD4^+$ T cells in the second week of infection resulted in decreased viral load, indicating enhanced clearance by virus specific CTLs, while removal prior to day 10, the stage at which the Treg cells emerge, had no effect. Also, transfer of $CD4^+CD25^+$ and $CD4^+CD25^-CD103^+$ to acutely infected mice caused a reduction in the total number of $CD8^+$granzyme A^+ T cells [75].

Examination of $CD4^+CD25^+$ T cells isolated from chronically infected mice revealed that these cells expressed enhanced levels of GITR, LAG-3, CD69 and CD103. Unstimulated $CD4^+CD25^+$ cells from infected mice, but not naive mice, inhibited IFN-γ and granzyme B production of antigen stimulated FV TCR transgenic and OVA TCR transgenic $CD8^+$ T cells in vitro, while proliferative responses were not inhibited. Similarly, in vivo primed FV TCR transgenic CTLs were suppressed in a contact-dependent manner, but were not affected by neutralisation of IL-10 or TGF-β [77].

In another study it has been demonstrated that treatment of acutely infected mice with agonistic anti-GITR antibody resulted in earlier induction of protective Th1 responses, reduced viral load and pathology [78]. Anti-GITR administration alone was not sufficient to cure persistent infection, but could reduce viral loads when combined with transfer of $CD8^+$ T cells from animals with acute infection [74]. These studies do not identify the signals driving Treg cell induction, or demonstrate virus specific Treg cell generation. They do, however provide a fascinating insight, and robust evidence that inducible Treg cells can block virus-specific CTL responses *in vivo*.

Human T Cell Lymphotropic Virus

Human T cell lymphotropic virus-1 (HTLV-1) belongs to the cancer causing oncovirus family. 1–5% of chronically infected patients develop adult T cell leukaemia (ATLL) 20–30 years after infection. HTLV-1 is a retrovirus which infects T cells, and is associated with a range of chronic inflammatory disorders, including infectious dermatitis, uveitis, arthropathy and HTLV-1 associated myelopathy or topical spastic parapresis (HAM/TSP). Chronically infected patients can range from asymptomatic, oligosymptomatic or HAM/TSP according to the severity of their symptoms. Studies carried out on unstimulated monocytes from all three groups have shown that these cells spontaneously produce proinflammatory TNF-α. Significant numbers of IL-10-producing $CD4^+$ and $CD8^+$ T cells could also be detected in asymptomatic, but not oligosymptomatic or HAM/TSP patients. The number of IL-10 producing cells negatively correlated with the amount of TNF-α produced by monocytes, suggesting a possible role for Tr1 in preventing immune pathology. In symptomatic patients there was no such correlation, instead, increased numbers of IFN-γ^+ T cells were detected [79], furthermore, $CD4^+$ T cells isolated from HAM/TSP patients had reduced expression of FoxP3 [80].

Human Immunodeficiency Virus

Human immunodeficiency virus (HIV) is a retrovirus that predominantly infects $CD4^+$ T cells. Over the course of infection the number of circulating $CD4^+$ T cells drops dramatically, resulting in a progressive generalised immunodeficiency known as AIDS. AIDS patients have increased susceptibility to opportunistic infections and unusual cancers. HIV is a human pathogen, but the course of disease may be closely mimicked by simian immunodeficiency virus infection (SIV) of macaques. Dysregulation of virus specific $CD4^+$ and $CD8^+$ T cell responses is evident from the earliest stages of infection, and is compounded by constant viral mutation and chronic exposure to viral antigens.

$CD4^+CD25^+$ cells are expanded in the peripheral blood of HIV infected individuals [81,82]. Disease progression in SIV and HIV infection is associated with increased numbers of $FoxP3^+$ and $CTLA-4^+$ cells, and increased expression of TGF-β, but not IL-10, mRNA in lymphoid tissue [83,84]. There is an accumulation of $CD4^+CD25^+FoxP3^+TGF-\beta^+$ and $CD3^+IL-10^+$ T cells during acute experimental infection of macaques. This increase may be driven by antiviral immune responses, as Treg cell frequency correlates strongly with the magnitude of immune activation [83]. However, a high perforin/FoxP3 ratio correlates with lack of disease progression [84], suggesting that the efficiency of immune effector functions determines disease outcome. Depletion of $CD4^+CD25^+$ cells results in increased expression of virus-specific IFN-γ and TNF-α production by $CD4^+$ and $CD8^+$ T cells in both SIV and HIV infection [85,86].

$CD4^+CD25^+$ T cells isolated from HIV positive patients were found to produce IFN-γ, IL-2, IL-10 and TGF-β in response to stimulation with anti-CD3 and anti-CD28. Stimulation of $CD4^+CD25^+$ cells with HIV p24, however, resulted in production of IL-10 and TGF-β, with no IFN-γ or IL-2 [87], revealing a population of antigen-specific Treg cells within the $CD4^+CD25^+$ subset. $CD4^+CD25^+$ cells did not proliferate in response to p24, and could suppress proliferation of $CD4^+CD25^-$

cells stimulated with p24 or an unrelated recall antigen. Neutralisation of TGF-β and IL-10 did not significantly restore proliferation [87].

CD8$^+$ T cells with regulatory activity were first described in HIV infection by Garba et al., who observed an inverse relationship between IFN-γ and TGF-β production by PBMC from HIV positive patients in response to a panel of HIV-derived peptides [88]. CD8$^+$ T cells were the source of the TGF-β, and neutralisation of TGF-β resulted in increased proliferation and IFN-γ production by effector CD4$^+$ T cells. HIV-specific CD8$^+$TGF-β$^+$ T cells were also capable of suppressing recall responses to other viruses [88]. Similarly, comparing effector CD8$^+$ responses in HIV positive patients revealed that an increased frequency of gag-specific IL-10 secreting CD8$^+$ T cells correlated with reduced expression of gag-specific CD107 and IFN-γ by other CD8$^+$ T cells [89]. Induction of antigen specific CD8$^+$IL-10$^+$ cells is also associated with loss of CD4$^+$ T cells and increased viral load. During successful antiviral therapy, the frequency of these cells declines [89]. Depletion of gag-specific IL-10 secreting CD8$^+$ T cells results in enhanced expression of CD107 and IL-2, but not IFN-γ, in response to gag, and other unrelated recall antigens. Direct cell-cell contact was required for maximal suppression of CTL effector responses [90].

A possible mechanism by which HIV may enhance the frequency of Treg cells has been identified. Exposure of a mixed population of CD4$^+$ T cells to inactivated HIV virions enhanced survival and increased expansion of FoxP3 expressing T cells [84]. This was unlikely to be due to peripheral conversion of CD4$^+$CD25$^-$ T cells to a regulatory phenotype, as TGF-β expression was not detected. Soluble anti-CD4 inhibited virion induced expansion of FoxP3$^+$ cells. Heat-treated virions, which selectively loose gp120 from the viral envelope were unable to expand FoxP3$^+$ T cells. Gp120 directly binds to CD4 during viral entry, and an antibody directed against the same epitope on CD4 can also direct the FoxP3$^+$ T cell expansion [84]. Also, HIV infected immature monocytes-derived dendritic cells from HIV infected patients, but not normal individuals primed autologous T cells to produce a mixture of IFN-γ and IL-10 [91]. These T cells suppressed proliferation of allogenic T cells in an IL-10 dependant manner, and the amount of IL-10 produced correlated with the degree of suppression observed [91].

There is conflicting evidence as to whether Treg activity is beneficial or detrimental in HIV infection. Although there is evidence that they effectively suppress potentially protective immune responses, they may also serve to protect against chronic immune activation, which is thought to lead to exhaustion of virus specific T cells and immunopathology.

Epstein Barr Virus

Epstein-Barr virus (EBV) is a B cell tropic human pathogen, of the γ-herpesvirus family. Over 90% of the adult population are persistently infected with EBV. Although predominantly asymptomatic, clinical manifestations of EBV infection include infectious mononucleosis (IM), caused by hyperproliferation of infected B cells, and chronic active EBV (CAEBV), in which T and NK cells are infected. Hodgkin's lymphoma (HL) and nasopharyngeal carcinoma (NPC) are associated with EBV infection, and viral protein expression may be detected in malignant tissue in some of these patients.

CTL responses directed against viral proteins expressed in the initial phase of infection have been shown to control the acute infection, but invariably, the virus proceeds to enter a persistent latent phase in memory B cells, one of the many mechanisms by which EBV evades immune responses. During latency, a limited selection of genes, including latent membrane protein 1 (LMP-1) and epstein-barr nuclear antigen 1 (EBNA1) are expressed.

Suppressive EBNA-specific CD4$^+$ T cell lines and clones have been generated from PBMC of EBV seropositive donors [92], and these were found to inhibit proliferation and IL-2 production by anti-CD3 stimulated CD4$^+$ T cells. A subset of EBNA-specific Treg cells express CD25, GITR and FoxP3, but not TGF-β or IL-10, suggesting that more than one phenotype of Treg cell is induced in response to EBNA. In agreement with this, a study examining EBNA specific suppression revealed that inhibition could be either contact-dependant or mediated by unidentified soluble factors [92].

In comparison with EBNA-1, there is a distinct lack of LMP-1-specific CTL in infected individuals. Studies have shown that stimulation of EBV seropositive PBMC with LMP-1 protein or LMP-1 derived peptides resulted in IL-10, but not IFN-γ production or proliferation by CD4$^+$ T cells [93]. LMP-1 stimulation also suppressed IFN-γ production and proliferation in response to unrelated recall antigens in an IL-10 dependant manner. This effect was not observed in PBMC from EBV seronegative donors. LMP-1 specific Treg cells could also reprogram responses to recall antigens. Prolonged antigenic stimulation in the presence of LMP-1 activated Treg cells caused a switch from IFN-γ producing Th1 to an IL-10 producing Tr1 phenotype [93]. Transfection of B cells with DNA encoding LMP-1 normally enhances their allogeneic T cell stimulatory capacity, however B cells transfected with LMP-1 isolated from a NPC patient were shown to produce significant amounts of IL-10. These cells are also less efficient at stimulating allospecific CD4$^+$ T cell responses than B cells expressing LMP-1 isolated from a latently infected B cell. Neutralisation of IL-10 restored allo-specific proliferation and IFN-γ production from effector CD4$^+$ T cells [94]. These studies suggest that Tr1 cells play an important role in the persistence of EBV infection.

Hepatitis B Virus

Hepatitis B virus (HBV) is a non-cytopathic DNA virus which infects the liver. Infected adults normally undergo acute self limiting hepatitis, which proceeds to chronicity in approximately 10% of infected individuals. Long term infection with HBV is a risk factor for the development of liver cirrhosis and hepatocellular carcinoma. Control of the disease is associated with vigorous multispecific CD4$^+$ and CD8$^+$ T cell responses, whereas HBV-specific responses are weak or undetectable in chronically infected individuals. CD4+ T cells isolated from patients with self limiting HBV infection have been shown to produce IL-2 and IFN-γ in response to stimulation with HBV core antigen. In contrast, chronically infected patients displayed a Tr1-like response, and produced IL-10, in the absence of Th1 cytokines [95]. Suppression of Th1 responses was found to be IL-10 dependant, as IL-10 neutralisation enhanced antigen specific IL-2 and IFN-γ production.

Hepatitis C Virus

Hepatitis C virus (HCV) is a blood borne RNA virus which primarily replicates in the liver. Up to 80% of infected patients fail to clear the virus, which, although frequently asymptomatic, carries a significantly increased risk of developing chronic hepatitis, cirrhosis, fibrosis and hepatocellular carcinoma. Clearance of the virus is associated with robust multispecific CD4+ and CD8+ T cell responses, and viral persistence with weak and sometimes undetectable responses against a limited range of viral epitopes.

Research in our laboratory led to the first definitive identification of classical Tr1 cells in HCV infection [5]. PBMC isolated from HCV infected patients produced IL-10 and IFN-γ when stimulated with a panel of peptides derived from HCV core protein. To clarify the cellular source of these cytokines, a series of HCV core-specific CD4+ T cell lines and clones were generated. Several $IL-10^+IL-4^-IFN-\gamma^{low}$ Tr1-type T cell clones were identified, including those that recognised the same HCV core epitopes as Th1 cells. A subsequent study showed that neutralisation of IL-10 enhanced HCV specific IFN-γ production by PBMC from chronically infected patients [96]. HCV non-structural protein 4 (NS4) may contribute to the generation of HCV specific Tr1 cells, as it inhibited the production of Th1 polarising cytokine, IL-12, and enhanced the production of Tr1 inducing IL-10 by monocytes both from HCV infected patients and normal individuals [96].

CD4+CD25+ T cell frequency is increased in patients chronically infected with HCV, when compared with those that resolve infection or with healthy controls [97]. Depletion of CD4+CD25+ T cells increased proliferation and IFN-γ production by PBMC stimulated with HCV antigens and decreased antigen-specific IL-10 and TGF-β production. Supplementation of PBMC with CD4+CD25+ T cells had the opposite effect. Suppression of proliferation was found to be contact dependant, but inhibition of IFN-γ production was TGF-β-dependant. The effect of neutralising IL-10 was not tested; despite the fact that a subset of antigen stimulated CD4+CD25+ cells were shown to produce IL-10 [97].

Total CD4+ T cells from chronically infected patients with elevated serum levels of alanine aminotransferase (ALT), an indirect marker for liver injury, produce more IFN-γ in response to HCV proteins than total CD4+ T cells from patients with normal ALT. CD4+CD25+ T cells isolated from chronically infected patients with normal ALT produced TGF-β in response to stimulation with HCV antigens [98], and there was an inverse relationship between CD4+CD25+ T cell TGF-β production and liver pathology. CD4+CD25+ from patients with elevated ALT produced less TGF-β. Depletion experiments revealed that CD4+CD25+ Treg cells could suppress IFN-γ and proliferative responses to HCV more efficiently in patients with normal ALT than patients with elevated ALT, despite equivalent frequencies of these cells in both groups. This suggests a defect in controlling proinflammatory HCV-specific responses in patients which go on to develop liver disease. This could be due to loss of antigen specific CD4+CD25+ or loss of function of these cells. Another explanation could be that the antigen specific CD25− T cells are less susceptible to suppression in patients who go on to develop liver pathology.

HCV-specific CTL which produce IL-10 but not IFN-γ have been detected in liver infiltrating lymphocytes (LIL) from HCV infected patients stimulated with HCV tetramers [99]. Low frequency of CD8+tetramer+IL-10+ T cells in liver biopsies prohibited their purification for use in suppressor assays. Nevertheless, proliferation of anti-CD3 and anti-CD28 stimulated PBMC could be reproducibly inhibited in the presence of CD8+ T cells isolated from LIL. Suppression of proliferation was IL-10 dependant in cultures which produced IL-10, however, CD8+ LIL from some patients did not produce IL-10, but were still capable of suppressing proliferation of PBMC. Tetramer-stimulated CD8+ LIL could also inhibit HCV-specific proliferation by autologous PBMC in an IL-10 dependant manner [99]. Liver inflammation was found to correlate directly with frequency of CD8+tetramer+IFN-γ+ T cells, but inversely with the frequency of CD8+tetramer+IL-10+ T cells [99]. A related study, using confocal microscopy, showed that CD8+tetramer+IL-10+ T cells are located in regions with low fibrosis, whereas CD8+ T cells in regions exhibiting high levels of apoptosis, and high laminin expression (a marker for fibrosis) were tetramer− and tended to produce more IFN-γ than IL-10 [100]. No CD8+tetramer+IFN-γ+ cells were identified in this study. HCV-specific TGF-β producing CD8+ T cells have also been detected in the peripheral blood of chronically infected patients [101]. IFN-γ responses to the same epitopes were enhanced by neutralisation of TGF-β, which also had the effect of enhancing IL-10 production. Similarly, depletion of CD8+ T cells enhanced IFN-γ production by antigen specific CD4+ T cells [101].

Herpes simplex Virus

Herpes simplex virus (HSV-1) is a ubiquitously distributed pathogen which can infect both humans and mice. Infection is initially established in the oral mucosa or corneal epithelium, causing vesicular lesions. It rapidly spreads to proximal sensory nerve endings, where it is transported to peripheral ganglions, and enters a period of latency where it remains dormant until neural perturbation reactivates the virus. Ocular infection with HSV can result in the formation of immunoinflammatory lesions in the corneal stroma (stromal keratitis), which is the most frequent cause of infectious blindness in the western hemisphere [102]. It has been shown that Th1 cells of unknown specificity are responsible for pathology in these lesions. CD4+CD25+ Treg cells were also detected in lesions, however they were found to have a protective role, as CD4+CD25+ Treg cell depletion increased lesion severity despite enhancing virus-specific immune responses [103]. CD4+CD25+ cells isolated from lesions produced IL-10 in response to polyclonal TCR stimulation, and were capable of suppressing proliferation of pathogenic lesion-derived CD4+CD25+ T cells.

Lymphocytic Choriomeningitis Virus

Although predominantly a murine pathogen, lymphocytic choriomeningitis virus (LCMV) is a useful model virus, and has provided much information about viral interactions with the immune system. LCMV is tropic for the nervous system, and infection results in febrile disease with occasional CNS involvement followed by viral clearance, or, in the case of immunosuppressed hosts, long term persistance of the virus. Clearance of the virus requires virus specific CD4+ and CD8+ T cell responses, but is less reliant on antiviral humoral responses.

Virus-specific CD8+ T cell responses in persistantly infected mice are notably defective, exhibiting reduced IFN-γ production and killing of target cells, when compared with CD8+ cells from mice which resolve infection [104]. The outcome of infection is strain dependant, with mice infected with the LCMV armstrong strain efficiently clearing the virus within 8 days, and mice infected with a macrophage tropic mutant strain, clone 13, become persistently infected. Prolonged elevation of serum IL-10 is a key feature of persistant infection with clone 13, and is not observed in mice infected with the armstrong strain [105]. This is accompanied by enhanced constitutive production of IL-10 by spleen cells [106]. Antigen-specific IL-10 production was detected in splenic CD4+ T cells up to day 5 after infection with clone 13 [106]. Lack of detection of IL-10-producing T cells in the spleen after this stage, however, may simply reflect migration to the site of infection. After day 5, splenic DC, and later in infection macrophages and B cells, were the predominant source of IL-10. Infection with clone 13 resulted in deletion of the CD8α+ DC subset, and an increase in the relative proportion of CD8α− DC, although IL-10 production was not detected by DC ex vivo [107]. CD8α− DC isolated from mice infected with clone 13, but not the armstrong strain, enhanced IL-10 production by LCMV-specific TCR transgenic T cells, providing a potential mechanism of virus specific Treg cell induction.

Administration of anti-IL-10R antibody, both prophylactically and therapeutically, to mice infected with clone 13 dramatically reduced viral titres and restored virus specific IFN-γ production by CD8+ T cells [106,107], demonstrating a crucial role for IL-10 in viral persistance. Similarly, IL-10$^{-/-}$ mice infected with clone 13 had enhanced virus-specific CD4+ and CD8+ T cell responses early in infection [105,106]. There is evidence that other factors also contribute to virus induced anergy, as antigen-specific T cells were still lost in some IL-10$^{-/-}$ mice after prolonged infection [105]. Although these studies do not directly attribute suppression to Treg cells, they demonstrate the powerful influence of IL-10, regardless of its source, on antiviral T cell responses.

Conclusions

There is growing evidence of a role for Treg cells induced or expanded during infection and of a role for these cells in controlling effector immune responses that function to eliminate pathogens (Fig. 23.2). Indeed it had been recognized for some time that IL-10 and TGF-β were produced by the host during a range of different infections, where they functioned to suppress immune responses against the pathogen. However, it was only in the last 5–6 years that it was recognized that Treg cells are a major source of these immunosuppressive and anti-inflammatory cytokines. Most of the studies to date on pathogen-induced Treg cells have focused on chronic infection where it is well established that Treg cells contribute to chronicity by suppressing protective immune responses. However, there is growing evidence that Treg cells are also generated during acute infections and indeed it will probably emerge that these cells play a role in regulating immune responses in all infectious diseases. It may appear counterintuitive that the induction of suppressive T cells could benefit the host in immunity to infection where the objective is to generate potent pathogen-specific effector T cells and antibody responses. However, the morbidity and indeed the mortality in certain infections is mediated not by the pathogen itself but by the innate and adaptive immune responses generated against the pathogen and these responses must be tightly regulated in order to prevent excessive immunopathology. However, over-regulation of effector responses can lead to pathogen persistence.

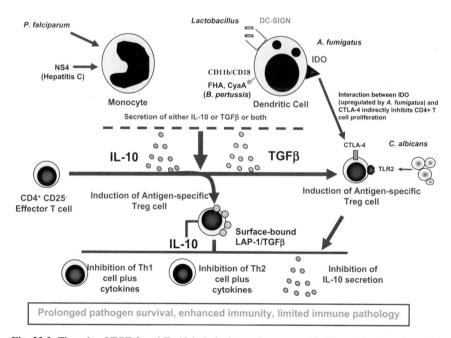

Fig. 23.2 The role of TGF-β and IL-10 in inducing pathogen-specific Treg cells (*See also Color Insert*)

Therefore, the balance between regulatory and effector T cells is critical to the outcome of an infection and regulating this balance may hold the key to development of new therapies against infectious diseases.

References

1. Sakaguchi S, Sakaguchi N, Asano M, Itoh M and Toda M. Immunologic self-tolerance maintained by activated T cells expressing IL-2 receptor alpha-chains (CD25). Breakdown of a single mechanism of self-tolerance causes various autoimmune diseases. J Immunol 1995; 155:1151–64
2. McGuirk P, Mills KH. Pathogen-specific regulatory T cells provoke a shift in the Th1/Th2 paradigm in immunity to infectious diseases. Trends Immunol 2002; 23:450–5.
3. Mills KH. Regulatory T cells: friend or foe in immunity to infection? Nat Rev Immunol 2004; 4:841–55
4. McGuirk P, McCann C and Mills KH. Pathogen-specific T regulatory 1 cells induced in the respiratory tract by a bacterial molecule that stimulates interleukin 10 production by dendritic cells: a novel strategy for evasion of protective T helper type 1 responses by Bordetella pertussis. J Exp Med 2002; 195:221–31
5. MacDonald AJ, Duffy M, Brady MT, et al. CD4 T helper type 1 and regulatory T cells induced against the same epitopes on the core protein in hepatitis C virus-infected persons. J Infect Dis 2002; 185:720–7
6. Doetze A, Satoguina J, Burchard G, et al. Antigen-specific cellular hyporesponsiveness in a chronic human helminth infection is mediated by T(h)3/T(r)1-type cytokines IL-10 and transforming growth factor-beta but not by a T(h)1 to T(h)2 shift. Int Immunol 2000; 12: 623–30
7. Groux H, O'Garra A, Bigler M, et al. A CD4+ T-cell subset inhibits antigen-specific T-cell responses and prevents colitis. Nature 1997; 389:737–42
8. Chen Y, Kuchroo VK, Inobe J, Hafler DA and Weiner HL. Regulatory T cell clones induced by oral tolerance: suppression of autoimmune encephalomyelitis. Science 1994; 265: 1237–40
9. Hisaeda H, Maekawa Y, Iwakawa D, et al. Escape of malaria parasites from host immunity requires CD4+CD25+ regulatory T cells. 2004; 10:29–30
10. Liston A, Rudensky AY. Thymic development and peripheral homeostasis of regulatory T cells. Curr Opin Immunol 2007; 19:176–85
11. Akbar AN, Vukmanovic-Stejic M, Taams LS and Macallan DC. The dynamic co-evolution of memory and regulatory CD4+ T cells in the periphery. Nat Rev Immunol 2007; 7:231–7
12. Lohr J, Knoechel B and Abbas AK. Regulatory T cells in the periphery. Immunol Rev 2006; 212:149–62
13. Sakaguchi S. The origin of FOXP3-expressing CD4+ regulatory T cells: thymus or periphery. J Clin Invest 2003; 112:1310–2
14. Chen W, Jin W, Hardegen N, et al. Conversion of peripheral CD4+CD25- naive T cells to CD4+CD25+ regulatory T cells by TGF-beta induction of transcription factor Foxp3. J Exp Med 2003; 198:1875–86
15. Tran DQ, Ramsey H and Shevach EM. Induction of FOXP3 expression in naive human CD4+FOXP3- T cells by T cell receptor stimulation is TGF{beta}-dependent but does not confer a regulatory phenotype. Blood 2007; 110:2983–90.
16. Fantini MC, Becker C, Monteleone G, Pallone F, Galle PR and Neurath MF. Cutting edge: TGF-beta induces a regulatory phenotype in CD4+CD25- T cells through Foxp3 induction and down-regulation of Smad7. J Immunol 2004; 172:5149–53
17. Kretschmer K, Apostolou I, Hawiger D, Khazaie K, Nussenzweig MC and von Boehmer H. Inducing and expanding regulatory T cell populations by foreign antigen. Nat Immunol 2005; 6:1219–27

18. Walker MR, Carson BD, Nepom GT, Ziegler SF and Buckner JH. De novo generation of antigen-specific CD4+CD25+ regulatory T cells from human CD4+CD25- cells. Proc Natl Acad Sci USA 2005; 102:4103–8
19. Chai JG, Bartok I, Chandler P, et al. Anergic T cells act as suppressor cells in vitro and in vivo. Eur J Immunol 1999; 29:686–92
20. Taams LS, van Rensen AJ, Poelen MC, et al. Anergic T cells actively suppress T cell responses via the antigen-presenting cell. Eur J Immunol 1998; 28:2902–12
21. Hori S, Sakaguchi S. Foxp3: a critical regulator of the development and function of regulatory T cells. Microbes Infect 2004; 6:745–51
22. Fontenot JD, Gavin MA and Rudensky AY. Foxp3 programs the development and function of CD4+CD25+ regulatory T cells. Nat Immunol 2003; 4:330–6
23. Wildin RS, Smyk-Pearson S and Filipovich AH. Clinical and molecular features of the immunodysregulation, polyendocrinopathy, enteropathy, X linked (IPEX) syndrome. J Med Genet 2002; 39:537–45
24. Roncador G, Brown PJ, Maestre L, et al. Analysis of FOXP3 protein expression in human CD4+CD25+ regulatory T cells at the single-cell level. Eur J Immunol 2005; 35: 1681–91
25. Deaglio S, Dwyer KM, Gao W, et al. Adenosine generation catalyzed by CD39 and CD73 expressed on regulatory T cells mediates immune suppression. J Exp Med 2007; 204: 1257–65
26. Borsellino G, Kleinewietfeld M, Di Mitri D, et al. Expression of ectonucleotidase CD39 by Foxp3+ Treg cells: hydrolysis of extracellular ATP and immune suppression. Blood 2007; 110:1225–32
27. Liu W, Putnam AL, Xu-Yu Z, et al. CD127 expression inversely correlates with FoxP3 and suppressive function of human CD4+ T reg cells. J Exp Med 2006; 203:1701–11
28. Miyara M, Sakaguchi S. Natural regulatory T cells: mechanisms of suppression. Trends Mol Med 2007; 13:108–16
29. Roncarolo MG, Gregori S, Battaglia M, Bacchetta R, Fleischhauer K and Levings MK. Interleukin-10-secreting type 1 regulatory T cells in rodents and humans. Immunol Rev 2006; 212:28–50
30. Mills KH, McGuirk P. Antigen-specific regulatory T cells–their induction and role in infection. Semin Immunol 2004; 16:107–17
31. Weiner HL. Induction and mechanism of action of transforming growth factor-beta-secreting Th3 regulatory cells. Immunol Rev 2001; 182:207–14
32. Walther M, Tongren JE, Andrews L, et al. Upregulation of TGF-[beta], FOXP3, and CD4+CD25+ Regulatory T Cells Correlates with More Rapid Parasite Growth in Human Malaria Infection. Immunity 2005; 23:287–96
33. Wu Y, Wang Q-H, Zheng L, et al. Plasmodium yoelii: Distinct CD4+CD25+ regulatory T cell responses during the early stages of infection in susceptible and resistant mice. Exp Parasitol 2007; 115:301–4
34. Brustoski K, Möller U, Kramer M, et al. Reduced cord blood immune effector-cell responsiveness mediated by CD4[+] cells induced in utero as a consequence of placental plasmodium falciparum infection. J Infect Dis 2006; 193:146–154
35. Campanelli AP, Roselino AM, Cavassani KA, et al. CD4+CD25+ T cells in skin lesions of patients with cutaneous leishmaniasis exhibit phenotypic and functional characteristics of natural regulatory T cells. J Infect Dis 2006; 193:1313–22
36. Kariminia A, Bourreau E, Pascalis H, et al. Transforming growth factor beta 1 production by CD4+ CD25+ regulatory T cells in peripheral blood mononuclear cells from healthy subjects stimulated with Leishmania guyanensis. Infect Immun 2005; 73:5908–14
37. Belkaid Y, Piccirillo CA, Mendez S, Shevach EM and Sacks DL. CD4+CD25+ regulatory T cells control Leishmania major persistence and immunity. 2002; 420:502–7
38. Smits HH, Engering A, van der Kleij D, et al. Selective probiotic bacteria induce IL-10-producing regulatory T cells in vitro by modulating dendritic cell function through

dendritic cell-specific intercellular adhesion molecule 3-grabbing nonintegrin. J Allergy Clin Immunol 2005; 115:1260–7

39. Ji J, Masterson J, Sun J and Soong L. CD4+CD25+ Regulatory T Cells Restrain Pathogenic Responses during Leishmania amazonensis Infection. J Immunol 2005; 174:7147–53
40. Mendez S, Reckling SK, Piccirillo CA, Sacks D and Belkaid Y. Role for CD4+ CD25+ Regulatory T Cells in Reactivation of Persistent Leishmaniasis and Control of Concomitant Immunity 10.1084/jem.20040298. J Exp Med 2004; 200:201–10
41. Suffia* I, Reckling* SK, Salay* G and Belkaid* Y. A Role for CD103 in the Retention of CD4+CD25+ Treg and Control of Leishmania major Infection. J Immunol 2005; 174: 5444–55
42. Sakaguchi S. Naturally arising CD4+ regulatory t cells for immunologic self-tolerance and negative control of immune responses. Annu Rev Immunol 2004; 22:531–62
43. Mills KHG, McGuirk P. Antigen-specific regulatory T cells–their induction and role in infection. Seminars in Immunology Regulatory T Cells 2004; 16:107–17
44. McKee AS, Pearce EJ. CD25+CD4+ Cells Contribute to Th2 Polarization during Helminth Infection by Suppressing Th1 Response Development. J Immunol 2004; 173: 1224–31
45. Taylor JJ, Mohrs M and Pearce EJ. Regulatory T Cell Responses Develop in Parallel to Th Responses and Control the Magnitude and Phenotype of the Th Effector Populatio. J Immunol 2006; 176:5839–47
46. Kaviratne M, Hesse M, Leusink M, et al. IL-13 Activates a Mechanism of Tissue Fibrosis That Is Completely TGF-{beta} Independent. J Immunol 2004; 173:4020–29
47. Baumgart M, Tompkins F, Leng J and Hesse M. Naturally Occurring CD4+Foxp3+ Regulatory T Cells Are an Essential, IL-10-Independent Part of the Immunoregulatory Network in Schistosoma mansoni Egg-Induced Inflammation. J Immunol 2006; 176:5374–87
48. Finney CA, Taylor MD, Wilson MS and Maizels RM. Expansion and activation of CD4(+)CD25(+) regulatory T cells in Heligmosomoides polygyrus infection. Eur J Immunol 2007; 37:1874–86
49. Doligalska M, Rzepecka J, Drela N, Donskow K and Gerwel-Wronka M. The role of TGF-beta in mice infected with Heligmosomoides polygyrus. Parasite Immunol 2006; 28:387–95
50. Taylor MD, LeGoff L, Harris A, Malone E, Allen JE and Maizels RM. Removal of regulatory T cell activity reverses Hyporesponsiveness and leads to filarial parasite clearance in vivo. J Immunol 2005; 174:4924–33
51. Smits HH, Engering A, van der Kleij D, et al. Selective probiotic bacteria induce IL-10-producing regulatory T cells in vitro by modulating dendritic cell function through dendritic cell-specific intercellular adhesion molecule 3-grabbing nonintegrin. J Allergy Clin Immunol 2005; 115:1260–67
52. Laiho K, Ouwehand A, Salminen S and Isolauri E. Inventing probiotic functional foods for patients with allergic disease. Ann Allergy Asthma Immunol 2002; 89:75–82
53. Isolauri E, Kirjavainen PV and Salminen S. Probiotics: a role in the treatment of intestinal infection and inflammation? Gut 2002; 50 Suppl 3:54–9
54. von der Weid T, Bulliard C and Schiffrin EJ. Induction by a lactic acid bacterium of a population of CD4(+) T cells with low proliferative capacity that produce transforming growth factor beta and interleukin-10. Clin Diagn Lab Immunol 2001; 8:695–701
55. Dye C, Scheele S, Dolin P, Pathania V, Raviglione MC and for the WHO Global Surveillance and Monitoring Project. Global burden of tuberculosis: estimated incidence, prevalence, and mortality by country 10.1001/jama.282.7.677. JAMA 1999; 282:677–86
56. Hirsch CS, Hussain R, Toossi Z, Dawood G, Shahid F and Ellner JJ. Cross-modulation by transforming growth factor beta in human tuberculosis: Suppression of antigen-driven blastogenesis and interferon gamma production 10.1073/pnas.93.8.3193. PNAS 1996; 93: 3193–98
57. Boussiotis VA, Tsai EY, Yunis EJ, et al. IL-10-producing T cells suppress immune responses in anergic tuberculosis patients. J. Clin. Invest. 2000; 105:1317–25

58. Hougardy JM, Place S, Hildebrand M, et al. Regulatory T cells depress immune responses to protective antigens in active tuberculosis. Am J Respir Crit Care Med 2007; 176:409–16.
59. Chen X, Zhou B, Li M, et al. CD4(+)CD25(+)FoxP3(+) regulatory T cells suppress Mycobacterium tuberculosis immunity in patients with active disease. Clin Immunol 2007; 123:50–9
60. Groux H, Bigler M, de Vries JE and Roncarolo MG. Inhibitory and stimulatory effects of IL-10 on human CD8+ T cells. J Immunol 1998; 160:3188–93
61. Logan RPH, Walker MM. ABC of the upper gastrointestinal tract: Epidemiology and diagnosis of Helicobacter pylori infection 10.1136/bmj.323.7318.920. BMJ 2001; 323:920–22
62. Rad R, Brenner L, Bauer S, et al. CD25+/Foxp3+ T cells regulate gastric inflammation and Helicobacter pylori colonization in vivo. Gastroenterology 2006; 131:525–37
63. Mills KH. Immunity to Bordetella pertussis. Microbes Infect 2001; 3:655–77
64. Higgins SC, Lavelle EC, McCann C, et al. Toll-like receptor 4-mediated innate IL-10 activates antigen-specific regulatory T cells and confers resistance to Bordetella pertussis by inhibiting inflammatory pathology. J Immunol 2003; 171:3119–27
65. Cavassani KA, Campanelli AP, Moreira AP, et al. Systemic and local characterization of regulatory T cells in a chronic fungal infection in humans. J Immunol 2006; 177:5811–18
66. Montagnoli C, Fallarino F, Gaziano R, et al. Immunity and tolerance to aspergillus involve functionally distinct regulatory T cells and tryptophan catabolism. J Immunol 2006; 176:1712–23
67. Latge JP. Aspergillus fumigatus and aspergillosis. Clin Microbiol Rev 1999; 12:310–50
68. Raitala A, Pertovaara M, Karjalainen J, Oja SS and Hurme M. Association of Interferon-gamma +874(T/A) single nucleotide polymorphism with the rate of tryptophan catabolism in healthy individuals doi:10.1111/j.1365-3083.2005.01586.x. Scand J Immunol 2005; 61:387–90
69. Chauhan N, Latge J-P and Calderone R. Signalling and oxidant adaptation in Candida albicans and Aspergillus fumigatus. Nat Rev Microbiol 2006; 4:435–44
70. Liu H, Komai-Koma M, Xu D and Liew FY. Toll-like receptor 2 signaling modulates the functions of CD4+ CD25+ regulatory T cells. Proc Natl Acad Sci USA 2006; 103:7048–53
71. Netea MG, Sutmuller R, Hermann C, et al. Toll-Like receptor 2 suppresses immunity against candida albicans through induction of IL-10 and regulatory T cells. J Immunol 2004; 172:3712–18
72. Iwashiro M, Messer RJ, Peterson KE, Stromnes IM, Sugie T and Hasenkrug KJ. Immunosuppression by CD4+ regulatory T cells induced by chronic retroviral infection. Proc Natl Acad Sci USA 2001; 98:9226–30.
73. Hasenkrug KJ, Brooks DM and Dittmer U. Critical role for CD4(+) T cells in controlling retrovirus replication and spread in persistently infected mice. J Virol 1998; 72: 6559–64
74. Dittmer U, He H, Messer RJ, et al. Functional impairment of CD8(+) T cells by regulatory T cells during persistent retroviral infection. Immunity 2004; 20:293–303
75. Zelinskyy G, Kraft AR, Schimmer S, Arndt T and Dittmer U. Kinetics of CD8+ effector T cell responses and induced CD4+ regulatory T cell responses during Friend retrovirus infection. Eur J Immunol 2006; 36:2658–70
76. Zelinskyy G, Robertson SJ, Schimmer S, Messer RJ, Hasenkrug KJ and Dittmer U. CD8+ T-cell dysfunction due to cytolytic granule deficiency in persistent Friend retrovirus infection. J Virol 2005; 79:10619–26
77. Robertson SJ, Messer RJ, Carmody AB and Hasenkrug KJ. In vitro suppression of CD8+ T cell function by Friend virus-induced regulatory T cells. J Immunol 2006; 176:3342–9
78. He H, Messer RJ, Sakaguchi S, Yang G, Robertson SJ and Hasenkrug KJ. Reduction of retrovirus-induced immunosuppression by in vivo modulation of T cells during acute infection. J Virol 2004; 78:11641–7
79. Brito-Melo GE, Peruhype-Magalhaes V, Teixeira-Carvalho A, et al. IL-10 produced by CD4+ and CD8+ T cells emerge as a putative immunoregulatory mechanism to counterbal-

ance the monocyte-derived TNF-alpha and guarantee asymptomatic clinical status during chronic HTLV-I infection. Clin Exp Immunol 2007; 147:35–44
80. Oh U, Grant C, Griffith C, Fugo K, Takenouchi N and Jacobson S. Reduced Foxp3 protein expression is associated with inflammatory disease during human t lymphotropic virus type 1 Infection. J Infect Dis 2006; 193:1557–66
81. Levy Y, Durier C, Krzysiek R, et al. Effects of interleukin-2 therapy combined with highly active antiretroviral therapy on immune restoration in HIV-1 infection: a randomized controlled trial. Aids 2003; 17:343–51
82. Miedema F, Petit AJ, Terpstra FG, et al. Immunological abnormalities in human immunodeficiency virus (HIV)-infected asymptomatic homosexual men. HIV affects the immune system before CD4+ T helper cell depletion occurs. J Clin Invest 1988; 82:1908–14
83. Estes JD, Li Q, Reynolds MR, et al. Premature induction of an immunosuppressive regulatory T cell response during acute simian immunodeficiency virus infection. J Infect Dis 2006; 193:703–12
84. Nilsson J, Boasso A, Velilla PA, et al. HIV-1-driven regulatory T-cell accumulation in lymphoid tissues is associated with disease progression in HIV/AIDS. Blood 2006; 108: 3808–17
85. Hryniewicz A, Boasso A, Edghill-Smith Y, et al. CTLA-4 blockade decreases TGF-beta, IDO, and viral RNA expression in tissues of SIVmac251-infected macaques. Blood 2006; 108:3834–42
86. Aandahl EM, Michaelsson J, Moretto WJ, Hecht FM and Nixon DF. Human CD4+ CD25+ regulatory T cells control T-cell responses to human immunodeficiency virus and cytomegalovirus antigens. J Virol 2004; 78:2454–9
87. Weiss L, Donkova-Petrini V, Caccavelli L, Balbo M, Carbonneil C and Levy Y. Human immunodeficiency virus-driven expansion of CD4+CD25+ regulatory T cells, which suppress HIV-specific CD4 T-cell responses in HIV-infected patients. Blood 2004; 104:3249–56
88. Garba ML, Pilcher CD, Bingham AL, Eron J and Frelinger JA. HIV antigens can induce TGF-beta(1)-producing immunoregulatory CD8+ T cells. J Immunol 2002; 168:2247–54
89. Elrefaei M, Barugahare B, Ssali F, Mugyenyi P and Cao H. HIV-specific IL-10-positive CD8+ T cells are increased in advanced disease and are associated with decreased HIV-specific cytolysis. J Immunol 2006; 176:1274–80
90. Elrefaei M, Ventura FL, Baker CA, Clark R, Bangsberg DR and Cao H. HIV-specific IL-10-positive CD8+ T cells suppress cytolysis and IL-2 production by CD8+ T cells. J Immunol 2007; 178:3265–71
91. Granelli-Piperno A, Golebiowska A, Trumpfheller C, Siegal F and Steinman R. HIV-1-infected monocyte-derived dendritic cells do not undergo maturation but can elicit IL-10 production and T cell regulation. Proc Natl Acad Sci USA 2004; 101:7669–74
92. Voo KS, Peng G, Guo Z, et al. Functional characterization of EBV-encoded nuclear antigen 1-specific CD4+ helper and regulatory T cells elicited by in vitro peptide stimulation. Cancer Res 2005; 65:1577–86
93. Marshall NA, Vickers MA and Barker RN. Regulatory T cells secreting IL-10 dominate the immune response to EBV latent membrane protein 1. J Immunol 2003; 170:6183–9
94. Pai S, O'Sullivan B, Abdul-Jabbar I, et al. Nasopharyngeal carcinoma-associated Epstein-Barr virus-encoded oncogene latent membrane protein 1 potentiates regulatory T-cell function. Immunol Cell Biol 2007; 85:370–7
95. Szkaradkiewicz A, Jopek A, Wysocki J, Grzymislawski M, Malecka I and Wozniak A. HBcAg-specific cytokine production by CD4 T lymphocytes of children with acute and chronic hepatitis B. Virus Res 2003; 97:127–33
96. Brady MT, MacDonald AJ, Rowan AG and Mills KH. Hepatitis C virus non-structural protein 4 suppresses Th1 responses by stimulating IL-10 production from monocytes. Eur J Immunol 2003; 33:3448–57
97. Cabrera R, Tu Z, Xu Y, et al. An immunomodulatory role for CD4(+)CD25(+) regulatory T lymphocytes in hepatitis C virus infection. Hepatology 2004; 40:1062–71

98. Bolacchi F, Sinistro A, Ciaprini C, et al. Increased hepatitis C virus (HCV)-specific CD4+CD25+ regulatory T lymphocytes and reduced HCV-specific CD4+ T cell response in HCV-infected patients with normal versus abnormal alanine aminotransferase levels. Clin Exp Immunol 2006; 144:188–96
99. Accapezzato D, Francavilla V, Paroli M, et al. Hepatic expansion of a virus-specific regulatory CD8(+) T cell population in chronic hepatitis C virus infection. J Clin Invest 2004; 113:963–72
100. Abel M, Sene D, Pol S, et al. Intrahepatic virus-specific IL-10-producing CD8 T cells prevent liver damage during chronic hepatitis C virus infection. Hepatology 2006; 44:1607–16
101. Alatrakchi N, Graham CS, van der Vliet HJ, Sherman KE, Exley MA and Koziel MJ. Hepatitis C virus (HCV)-specific CD8(+) cells produce transforming growth factor beta that can suppress HCV-specific T-cell responses. J Virol 2007; 81:5882–92
102. Streilein JW, Dana MR and Ksander BR. Immunity causing blindness: five differen paths to herpes stromal keratitis. Immunol Today 1997; 18:433
103. Suvas S, Azkur AK, Kim BS, Kumaraguru U and Rouse BT. CD4+CD25+ regulatory T cells control the severity of viral immunoinflammatory lesions. J Immunol 2004; 172: 4123–32
104. Zajac AJ, Blattman JN, Murali-Krishna K, et al. Viral immune evasion due to persistence of activated T cells without effector function. J Exp Med 1998; 188:2205–13
105. Maris CH, Chappell CP and Jacob J. Interleukin-10 plays an early role in generating virus-specific T cell anergy. BMC Immunol 2007; 8:8
106. Brooks DG, Trifilo MJ, Edelmann KH, Teyton L, McGavern DB and Oldstone MB. Interleukin-10 determines viral clearance or persistence in vivo. Nat Med 2006; 12:1301–9
107. Ejrnaes M, Filippi CM, Martinic MM, et al. Resolution of a chronic viral infection after interleukin-10 receptor blockade. J Exp Med 2006; 203:2461–72

Chapter 24
Human Type 1 T Regulatory Cells

Manuela Battaglia, Silvia Gregori, Rosa Bacchetta, and Maria Grazia Roncarolo

Abstract IL-10 producing T regulatory type 1 (Tr1) cells were first identified in 1994 in severe combined immunodeficient patients successfully transplanted with allogeneic hemtaopoietic stem cells. Interestingly, the presence of Tr1 cells in vivo correlated with long term allograft tolerance without the need of immunosuppressive treatment. Since then, Tr1 cells were further characterized and efficient ways for their in vitro generation were developed. It is now evident that Tr1 cells represent a unique T cell subset, distinct from Th1, Th2, and Th17 cells and also from other T regulatory cells. Tr1 cells arise from naïve precursors and are characterized by the production of high levels of IL-10 in the absence of IL-4. They are anergic cells which suppress proliferation of T cells via an IL-10-dependent mechanism. Human Tr1 cells have been clearly shown to maintain tolerance to self and non-self non harmful antigens, and to allo-antigens but also to infectious agents and tumor-bearing cells. In some clinical settings, transfer of ex vivo generated Tr1 cells can be envisaged as an efficient and safe approach to promote or restore immunological tolerance. One clinical trial with transfer of Tr1 cells in patients transplanted with hematopoietic stem cells is already ongoing at our institute and it will pave the road for future clinical Tr1 cell applications.

Introduction

The immune system discriminates between self and non-self antigens to establish and preserve immunological tolerance. Tolerance is first maintained by central thymic clonal deletion of self-reactive T and B cells exposed to self-antigens at immature stages of their development. Central deletion mechanisms are, however, not 100% efficient and there is ample evidence that potentially hazardous

---

M.G. Roncarolo
San Raffaele Telethon Institute for Gene Therapy (HSR-TIGET); Università Vita-Salute
San Raffaele, Via Olgettina 58, Milan 20132, Italy
e-mail: m.roncarolo@hsr.it

self-reactive lymphocytes are present in the periphery of normal individuals. Self-reactive T cells escaping central deletion are kept under control by mechanisms of peripheral tolerance, which include: deletion, anergy, and/or active immunoregulation mediated by T regulatory cells (Tregs).

The concept of Tregs was proposed more than 30 years ago as a dominant form of immunological tolerance involving a specialized population of "suppressor" T cells that act to terminate conventional immune responses and to prevent autoimmune pathology [1]. Early studies hypothesized a suppressor cell cascade involving multiple suppressor factors, anti-idiotypic T cell networks, "suppressor inducer", and "contra-suppressor" cells [2]. In the last few years modern technologies and new experimental approaches consented a rebirth of suppressor cells (now called regulatory T cells, Tregs), which are considered at present, as one of the central players in peripheral immune regulation. Among all the subsets with regulatory activity that have been described, $CD4^+$ type 1 T regulatory (Tr1) cells and $CD4^+CD25^+$ $FOXP3^+$ Tregs are two of the most studied. The main difference between these two Treg populations resides in their ontogeny: the $CD4^+CD25^+$ $FOXP3^+$ Tregs develop in the thymus as a separate T cell lineage, while Tr1 cells are generated in the periphery under tolerogenic conditions.

In this chapter, the $CD4^+$ Tr1 cells, originally described and extensively characterized by our group, will be reviewed from discovery until the most recent findings. We will focus on human Tr1 cells. Nevertheless, some instrumental data generated in the mouse will also be described.

Defining Tr1 Cells

Main Features

$CD4^+$ Tr1 cells represent a T cell subset distinct from Th1, Th2, or Th17 cells, but, like the latter, are distinguished by their cytokine production profile. Upon activation via the T cell receptor (TCR), $CD4^+$ Tr1 cells produce high amounts of IL-10 in the absence of IL-4 and very low levels of IL-2 [3]. IL-10 secreted by Tr1 cells is detectable as soon as 4 hours after activation. Thereafter, the levels of IL-10 increase rapidly and the highest concentration is reached 12–24 hours after activation [4]. Production of TGF-ß and IL-5 has also been ascribed to $CD4^+$ Tr1 cells [3] but not in all experimental conditions. Human Tr1 cells produce also IFN-γ, although at levels that are at least 1 log lower than those produced by Th1 cells [5]. Depending on the experimental approach the specific cytokines produced by $CD4^+$ Tr1 cells might slightly change. Nevertheless, $CD4^+$ T cells producing high levels of IL-10 and undetectable levels of IL-4 are deemed as Tr1 cells.

$CD4^+$ Tr1 cells are anergic since they have a very low proliferative capacity upon TCR activation, which is at least partially due to the autocrine antiproliferative effect of IL-10 [5]. Blockade of IL-10 activity can indeed partially restore Tr1 cell proliferation. Importantly, they can be expanded in vitro in the presence of cytokines such IL-2 and IL-15 [5]. Despite their low in vitro proliferative capacity, Tr1 cell clones express normal levels of activation markers such as CD25, CD28, CD40L, CD69, human leucocyte antigen-DR (HLA-DR) and cytotoxic T-lymphocyte antigen-4 (CTLA4) following TCR-mediated stimulation, whereas they constitutively express high levels of the IL-2Rß and γ chains independently on their activation status [5].

The main effector function of Tr1 cells is their ability to suppress immune responses through secretion of the immunosuppressive cytokines IL-10 and TGF-ß, although some cell-cell contact suppressive mechanisms cannot be excluded (Fig. 24.1).

IL-10 producing $CD8^+$ Tr1-like cells have also been described [6–8]. These cells are anergic and suppress proliferative responses via IL-10. However, their characterization is still very limited.

The Origins

Tr1 cells were originally identified by our group in 1994 in a severe combined immunodeficient (SCID) patient successfully transplanted with HLA-mismatched hematopoietic stem cells. $CD4^+$ host-reactive T cell clones isolated from this patient produced very high levels of IL-10 and IL-5 in the absence of IL-4 and IL-2 after antigen-specific stimulation in vitro. Their in vivo presence correlated with the absence of graft versus host disease (GvHD) and with long-term graft tolerance without the need of immunosupression [4]. Follow up experiments demonstrated

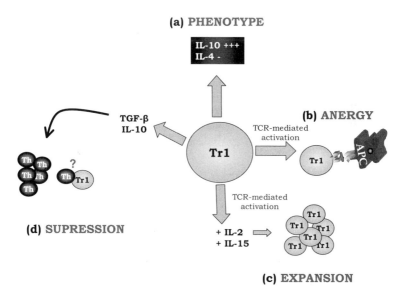

Fig. 24.1 Main features of human $CD4^+$ Tr1 cells $CD4^+$ Tr1 cells are distinguished by their ability to produce IL-10 in the absence of IL-4 (**a**). Autocrine production of IL-10 render these cells highly anergic upon TCR-mediated activation (**b**). but addition of exogenous cytokines, such as IL-2 and IL-15, can revert this anergic phenotype allowing their in vitro expansion (**c**). Upon TCR engagement, $CD4^+$ Tr1 cells exert their suppressive function in a non antigen-specific manner by secreting TGF-β and IL-10, although some cell-cell contact mechanisms cannot be excluded (**d**) *(See also Color Insert)*

Table 24.1 Ten years of Tr1 cells

1992	IL-10 inhibits allogeneic proliferative and cytotoxic T cell responses [67]
1994	IL-10 producing CD4$^+$ host reactive T cell clones were identified in tolerant SCID patients [4]
1996	Antigen specific anergy was induced in vitro by exogenous IL-10 added during mixed lymphocyte cultures [9]
1997	IL-10 producing cells were named Tr1 cells. Human and mouse CD4$^+$ Tr1 cells were generated in vitro [3]
2001	Improved generation of human Tr1 cells in vitro [11]
2001	First clinical trial using Tr1 cells induced in vitro with IL-10, in cancer patients undergoing hematopoietic stem cell transplantation (Bacchetta et al. unpublished).
2002	IL-15 and IL-2 were identified as the inducers of Tr1 cell proliferation in vitro [5]

that CD4$^+$ T cells, activated by allogeneic monocytes in a primary mixed lymphocyte reaction (MLR) in the presence of exogenous IL-10 became anergic and specifically failed to proliferate after re-stimulation with the same alloantigens [9]. These data pointed to an important tolerogenic role for IL-10. Subsequently, in 1997 another study performed by our group demonstrated that in vitro activation of human or mouse CD4$^+$ T cells in the presence of high doses of exogenous IL-10 resulted in the generation of T-cell clones with a cytokine production profile distinct from that of Th1 or Th2 cells but superimposable to that of host-reactive T-cell clones isolated from the SCID patients [3]. Those T-cell clones suppressed proliferation of CD4$^+$ T cells in response to antigens in vitro, and prevented colitis induced in SCID mice immune-reconstituted with pathogenic CD4$^+$CD45RB^{high} splenic T cells [3]. This study was fundamental for the definitive description and characterization of those T cells found in vivo in SCID patients [4] or generated in vitro in the presence of IL-10 [9]. These cells were termed Tr1 cells [3] based on the fact the they have regulatory activity and they were the first type of Tregs isolated and characterized at the single cell level. Follow up breakthrough studies identified better ways for human Tr1 cell induction [10,11] and expansion [5], and led to the first clinical trial using ex vivo donor IL-10 anergized T cells for the prevention of GvHD (Table 24.1).

Distinguishing Tr1 Cells

As a separate T cell subset, Tr1 cells have distinguishing characteristics that lead to unique activities. Some of these features are solved while others are still undefined. For example, on the contrary to Th1 and Th2 cells, nothing is known about the downstream signaling events that lead to Tr1 cell generation. It is still unresolved the mechanisms that guide to increased IL-10 production with concomitant "switch off" of IL-4 production.

Markers Other than the Cytokine Production Profile

While the cytokine production profile is a unique hallmark for Tr1 cells, this biological feature is technically difficult to be used as marker for their specific in vivo detection and in vitro selection. Therefore, in the past years, we and others performed extensive studies with the aim to identify specific cell marker/s to distinguish Tr1 cells from other CD4$^+$ T cell subsets.

The selective expression of chemokine receptors on CD4$^+$ Tr1 cells has been explored. In the resting phase, human Tr1 cells express both Th1-associated (CXCR3 and CCR5) and Th2-associated (CCR3, CCR4 and CCR8) chemokine receptors [12]. Interestingly, CCR8 is expressed on Tr1 cells at higher levels compared to Th2 cells and, upon activation, human Tr1 cells migrate preferentially in response to I-309, a ligand for CCR8 [12]. CCR8 seems also to be relevant in vivo, since it has been demonstrated that in a mouse model of helminth infection, CCR8 is expressed on IL-10 producing CD4$^+$ T cells, which resemble Tr1 cells [13]. On the other hand, CD4$^+$ T cells with a Tr1 cell phenotype isolated from normal donors have been found to express CCR9, suggesting that these cells home to the gut [14]. It is therefore possible that Tr1 cells do not constitutively and uniformly express one or a group of chemokine receptors but depending on the patho-physiological condition and their migratory requirements, they might express different chemokine receptor/s.

FOXP3 is a transcription factor that is constitutively expressed at high levels by CD4$^+$CD25$^+$ Tregs and controls their differentiation [15]. Since its discovery [16], FOXP3 was thought to be a major player in all cells with regulatory activity. However, we and others clearly showed that Tr1 cells do not constitutively express FOXP3, which in turn can be up-regulated upon activation, similarly to what occurs in CD4$^+$CD25$^-$ T cells [17,18].

In conclusion, up to date Tr1 cells are still orphan of a specific marker that could significantly help in their isolation, and further characterization. Therefore, a genomic approach is ongoing in our group to identify such marker/s. However, as it turned out for Th1 and Th2 cells, the cytokine production profile might represent the solely distinguishing factor for Tr1 cells over other CD4$^+$ T cells.

Suppressive Functions

The suppressive mechanisms of Tr1 cells represent another distinguishing feature that render this cell subset unique. Tr1 cells regulate immune responses through secretion of the immunosuppressive cytokines IL-10 and TGF-β, and suppress proliferation of both naïve and memory T cells in vivo and in vitro [3,4,11,19]. Moreover, human CD4$^+$ Tr1 cell clones have been shown to suppress immunoglobulin production by B cells [20]. These functions are accomplished since IL-10 inhibits cytokine production by T cells and monocytes/macrophages, and induces long-lasting antigen-specific anergy in both CD4$^+$ and CD8$^+$ T cells [3,21,22]. Furthermore, IL-10 down-regulates the expression of MHC class II, co-stimulatory and adhesion molecules [23–25], inhibits the release of inflammatory cytokines, and modulates the stimulatory capacity of dendritic cells (DC) and other antigen presenting cells (APC) [25]. Similarly, TGF-β inhibits proliferation and cytokine production by T cells and down-regulates the APC functions (reviewed in [26]). The suppressive effects of Tr1 cells are either partially or fully reversed by addition of anti-IL10 and anti-TGF-β neutralizing mAb [3,4,10]. Nevertheless, a mechanism implicating direct cell-cell contact with the target cells has also been postulated [27]. Apart from IL-10 and TGF-β, a granzyme B /perforin-dependent suppressive mechanism has also been ascribed to the suppressive function of $CD4^+$ Tr1 cells differentiated by CD3/CD46 cross-linking [28,29].

Importantly, $CD4^+$ Tr1 cells are inducible, antigen-specific, and need to be activated via their TCR in order to exert their suppressive functions. Although Tr1 cells must encounter the antigen toward which they are specific to exert their suppressive functions, once activated they suppress in an antigen non-specific manner. Presumably, this bystander suppression is due to the release of IL-10 and TGF-ß [3].

Generating Tr1 Cells

Tr1 cells are inducible cells and for this reason, similar to Th1, Th2 cells, and Th17 cells, they arise from naive precursors and can be differentiated both in vitro and in vivo. IL-10 is considered the driving force for Tr1 cell generation, as shown by experiments in which antigen-specific Tr1 cells can be induced in vitro by repeated TCR stimulation in the presence of high doses of IL-10 [3]. Similarly, T cells stimulated with anti-CD3 plus anti-CD28 mAbs in the presence of vitamin D3, dexamethasone, induce Tr1 cell differentiation via autocrine IL-10 production [19]. IL-10 is therefore not only responsible for the regulatory function of Tr1 cells but it is also fundamental for their differentiation.

In Vitro Differentiation

Additional experiments following the first description of Tr1 cell differentiation in vitro with exogenous IL-10 [3] demonstrated that, in many experimental settings IL-10 is necessary but probably not sufficient for human Tr1 cell differentiation. In a system where artificial APC (i.e. a murine fibroblast cell line) expressing high levels of hCD32 (Fc-γRII-Ly17), hCD58 (CD2 ligand), and hCD80 (CD28 ligand) were used in combination with anti-CD3 mAb to activate naïve human $CD4^+$ T cells, addition of exogenous IL-10 results in a relatively small increase in IL-10–producing T cells. IFN-α, a crucial cytokine for clearing viral infections and increasing IL-10 production by T cells, synergized with IL-10 in this in vitro system leading to differentiation of human $CD4^+$ Tr1 cells [11]. Interestingly, costimulation via CD2 alone, in the absence of costimulations through CD28– or LFA-1, induces T cell anergy in an IL-10-independent pathway along with the differentiation of Ag-specific Tr1 cells [30]. Differently, Atkinson and colleagues identified co-signaling via CD46 as another physiological inducer of Tr1 cells [31]. However, it is still unclear whether these CD3/CD46-stimulated T cells are bona fide Tr1 cells or if they represent a distinct inducible Treg subset. Finally, we showed that a new monoclonal antibody, which recognizes the RO and the RB isoforms of CD45 on human cells, induces anergic antigen-specific CD4$^+$ Tr1 cells and CD8$^+$ Tr1-like cells with suppressive activity [7].

Alternatively to the use of specific compounds acting directly on the T cells and leading to elevated IL-10 production, indirect generation of Tr1 cells through APC has been extensively described. DC are professional APC that commonly initiate Ag-specific immune responses upon antigen encounter. This process involves the terminal maturation of DC (mDC) which efficiently present the processed antigens to effector T cells that initiate an active immune response. In contrast, in steady states, DC remain immature (iDC) and growing evidence indicate that iDC can induce tolerance mainly through deletion of antigen-specific T cells or induction of Tregs [32]. For example, repetitive stimulation of naïve cord blood CD4$^+$ T cells with allogeneic iDC results in the differentiation of IL-10-producing Tregs [33]. Furthermore, we reported that repeated stimulation of naïve peripheral blood CD4$^+$ T cells with allogeneic iDC induces the differentiation of human Tr1 cells in vitro [10]. In this system, T cells become increasingly hypo-responsive to re-activation with mDC and after three rounds of stimulation with iDC, they are profoundly anergic and acquire regulatory function.

Specialized subsets of DC can also promote Tr1 cell differentiation, independently on their maturation stage. Tolerogenic DC can be induced by either biological or pharmacological agents. For example, we recently demonstrated that DC generated in the presence of exogenous IL-10 (DC-10) secrete significantly higher levels of IL-10 compared to iDC, whereas the amounts of IL-12 are low and comparable to those produced by iDC. Importantly, this ratio of cytokine production by DCTEN is maintained upon activation with LPS and IFN-γ. DC-10 are more powerful than iDC in inducing Tr1 cells, since they reproducibly induce anergic T cell lines with strong suppressive activity after only one round of stimulation (Gregori S. et al, manuscript submitted). IL-10 modulated myeloid iDC can also induce CD8$^+$ Tr1 cells, which are allo-antigen specific and anergic [8]. Finally, myeloid DC activated via CD40L in the presence of the anti-human CD45RO/RB mAb display a phenotype super-imposable to that of untreated mDC but induce the differentiation of anergic-suppressor T cells that display a cytokine production profile similar to that of Tr1 cells (Gregori S. personal communication).

Not only myeloid DC but also other subsets of DC, including plasmacytoid DC (pDC), can induce the differentiation of Tregs. pDC polarize T cells towards IL-10 production and regulatory activity in vitro [6,34]. Furthermore, activated pDC prime naïve CD8$^+$ T cells to become CD8$^+$ Tr1-like cells, which display poor secondary proliferative and cytolytic responses, and mediate suppression through IL-10 [6].

Based on these data, one can hypothesize that, in vivo, in steady state conditions Tr1 cells are generated and expanded by iDC, whereas during an active immune response, specialized subsets of DC are required for their generation.

In Vivo Differentiation

In animal models several treatments have been shown to be potent in vivo Tr1 cell inducers. However, very few information on in vivo induction of Tr1 cells in humans is available. Treatment of mice with a killed *Mycobacterium vaccae* suspension gives rise to allergen-specific Tr1 cells that confer protection against airway inflammation [35]. In addition, treatment with filamentous hemagglutinin from *Bordetella pertussis* enhances IL-10 production from macrophages and DC, which in turn promotes the induction of Tr1 cells [36]. Moreover, in mice immunised with cholera toxin in the presence of antigen antigen-specific Tr1 cells are generated [37]. Interestinlgy, we showed that 30 days of an in vivo pharmacological treatment with rapamycin+IL-10 are able to prevent allograft rejection in a mouse model of islet transplantation. This treatment not only prevents acute allograft rejection but also leads to active long-term tolerance via induction of antigen-specific Tr1 cells [38]. The same treatment given to pre-diabetic NOD mice (strain susceptible of spontaneous autoimmune diabetes development) for longer period (i.e. 20 weeks) blocks disease development. Long term tolerance is achieved through the generation of Tr1 cells, induced by IL-10, and expansion of $CD4^+CD25^+FoxP3^+$ Tregs, mediated by rapamycin. These two Treg subsets act in concert to block the autoaggressive disease and re-establish self tolerance [39]. It remains to be defined whether this therapy is efficacious also in humans for the in vivo differentiation of antigen-specific Tr1 cells. Interestingly, in both mouse models [38,39], in vivo administration of IL-10 alone does not protect from allograft rejection or diabetes development. Although IL-10 monotherapy generates Tr1 cells in vivo in both experimental models, this is not sufficient to counteract the expansion and function of effector T cells. Addition of rapamycin was required to achieve long-term tolerance. Similarly, a peri-transplant treatment of diabetic non-human primates (NHP) with anti-CD3 immunotoxin and deoxyspergualin induces stable rejection-free tolerance to allogeneic pancreatic islet transplants, which is associated with sustained elevation in serum IL-10 levels [40]. In this in vivo regimen, anti-CD3 immunotoxin depletes effector T cells, whereas deoxyspergualin arrests the production of proinflammatory cytokines and the maturation of DC. Interestingly, frequencies of Tr1 cells are significantly increased in PBMC of long-term tolerant NHP recipients as compared to controls [40]. Similar to our own findings, these results suggest that induction of Ag-specific long-term tolerance requires the synergic effect of drugs that down-modulate inflammation, block effector T cells and differentiate Tr1 cells.

In humans, new onset type 1 diabetic patients were treated in a phase I trial with a humanized anti-CD3 mAb, which led to the improvement in insulin production in 75% of the patients (as assessed by the C peptide) [41,42]. This treatment resulted in the induction of IL-10 producing $CD4^+$ T cells resembling Tr1 cells [43], suggesting that it might promote the in vivo induction of Tr1 cells. However, further data are necessary to confirm this observation.

In summary, these data indicate that selected immunomodulatory compounds combined with standard immunosuppressive drugs can induce Tr1 cells in vivo which mediate long-term immunological tolerance. Importantly, the pharmacological agents used for suppression of undesired immune responses should be tested and selected based on their capacity to promote Treg function.

Tolerance and Tr1 Cells

In steady states, human Tr1 cells have been shown to promote tolerance to self and non-self innocuous antigens. Tolerance interruption due to absence or defects in Tr1 cell function, might lead to autoimmune diseases, food intolerance, and allergies. Tr1 cells play also an important tolerogenic role in non-physiological conditions such as in transplants (Fig. 24.2). Conversely, an increase in number or function of Tr1 cells can exert a negative role impeding host protective immunity leading to infections and/or tumors.

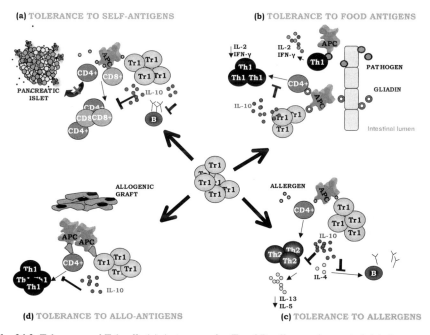

Fig. 24.2 Tolerance and Tr1 cells (**a**) Auto-agressive T and B cells escaping central deletion may get activated by self antigens (such as pancreas-related antigens) and develop an auto-aggressive immune response that leads to self-tissues destruction (such as destruction of insulin-producing cells, in type 1 diabetes) and autoimmunity. IL-10 produced by Tr1 cells can efficiently keep under control the auto-aggressive reaction and prevent disease development. (**b**) Tr1 cells present in the intestinal lumen can control undesired immune responses to non-self non harmful antigens such as gliadin, through IL-10 production. Despite this activity, Tr1 cells permit active immune responses versus pathogens, which are very abundant in the mucosal system. (**c**) The Th2-mediated immune response to allergens and the consequent IgE production can be actively controlled by Tr1 cells. (**d**) Tr1 cells avoid the immunological reactions activated by allo-antigens introduced by transplantation. Importantly, in all conditions depicted in the figure, Tr1 cells need to be first activated by their specific antigens in order to exert their suppressive function *(See also Color Insert)*

Tolerance to Self Antigens

The first proof that human Tr1 cells are fundamental to prevent the development of immune responses to self antigens came from the study of Kitani and colleagues who isolated self-MHC-reactive Tr1 cell clones from peripheral blood of healthy individuals. These Tr1 cells inhibit proliferation of primary CD4$^+$ T cells and tetanus toxoid-specific T-cell clones via IL-10 and TGF-β [20]. Tr1 cells specific for desmoglein 3 (Dsg3), which is the autoantigens of pemphigus vulgaris (PV), were isolated from 80% of healthy carriers of PV-associated HLA class II alleles but only in 17% of PV patients [44]. Similarly, T cells from non-diabetic individuals carrying HLA class II molecules associated with type 1 diabetes produce IL-10 in response to islet peptides while T cells from diabetic subjects secrete predominantly IFN-γ [45]. Furthermore, CD4$^+$ T cells producing IL-10 but not IL-2 and IL-4, were isolated from peripheral blood and synovial tissue of rheumatoid arthritis (RA) patients, but their frequency was significantly lower compared to that detected in control patients with non-autoimmune mediated joints inflammation. Indeed, it was reported that the frequency of IL-10-producing CD4$^+$ Tr1 cells in the synovia of RA patients inversely correlates with the frequency of Th1 cells [46]. IFN-γ– or IL-10–producing CD4$^+$ T cells specific for the major red blood cell autoantigen, the RhD protein, were concomitantly found in the peripheral blood of patients with autoimmune haemolytic anemia (AHA) [47], indicating that, although present, Tr1 cells cannot fully inhibit the development of this autoimmune disease. Overall these data demonstrate that self-Ag-specific Tr1 cells are present in men, play a key role in maintaining tolerance to self-constituents and hence prevent autoimmunity.

Tolerance to Non self Non Harmful Antigens

The mucosal system has to precisely discriminate between pathogens entering the organism through the intestine, and a large variety of harmless antigens such as food antigens. An immune response against nonpathogenic antigens can lead to hypersensitivity responses against dietary antigens [48]. Tr1 cells have been shown to play central roles in maintaining this delicate balance. For example, celiac disease is a common disorder of small intestine due to permanent intolerance to dietary wheat gluten. An abnormal T-cell mediated immune response to gliadin, together with the absence of specific immune regulation, plays a crucial role in inducing the celiac enteropathy [49]. In accordance with this observation, we found that gliadin-specific T cell activation was suppressed in celiac intestinal mucosa cultured 24 hours ex-vivo with gliadin and IL-10. Interestingly, these gliadin-specific T-cell lines generated in the presence of IL-10 were anergic in response to gliadin [50]. Moreover, we recently isolated gliadin-specific Tr1 cell clones from intestinal mucosa of a treated celiac patient. These Tr1 cell clones were anergic, produced IL-10 and TGF-β, and had a strong inhibitory capacity on gliadin-specific T cell response [51]. Based on these findings we hypothesize that Tr1 cells, although present in the inflamed intestinal mucosa of celiac patients, are not sufficient to down-modulate the massive anti-gliadin immune response. Addition of exogenous

IL-10 might have an anti-inflammatory effect on gliadin presentation by local APC and induce gliadin specific Tr1 cells leading to down-modulation of adverse gliadin specific immune response (Fig. 24.2).

Similarly to food antigens, allergens are proteins recognized as "dangerous" only by a subgroup of allergic individuals. Allergic diseases are inflammatory disorders in which aberrant Th2 immune responses lead to IgE production by B cells, eosinophil development, and allergic tissues homing of Th2 cells (reviewed in [52]). Immunological studies demonstrated the presence of allergen specific antibodies in non allergic subjects indicating that they are not "ignorant" toward the allergens but rather have intrinsic mechanisms of active tolerance. Indeed, it has been demonstrated that Tr1 cells are important in down-regulating immune responses towards allergens such as nickel [53], insect venom [54] and cat allergens [55]. Furthermore, it is now evident that during the early phase of allergen-specific immunotherapy, which efficiently reduces allergic reactions in patients, Tr1 cells are generated in vivo (reviewed in [56]).

These data clearly indicate the involvement of Tr1 cells in suppressing immune responses towards innocuous environmental substances, such as food antigens and allergens, to which our organism is daily exposed. Therefore, Tr1 cells can be clinically exploited to control chronic diseases such as celiac diseases or allergy and asthma.

Tolerance to Allo-Antigens

Tr1 cells are clearly involved in tolerance induction after transplantation. Precisely, it was in a model of transplantation where the existence of an IL-10 producing T regulatory cell subset was elucidated [4]. Studies in severe combined immunodeficiency (SCID) patients successfully transplanted with HLA-mismatched allogenic stem cells showed that, despite the HLA disparity, some of these patients did not develop a GvHD. Long-term tolerance to the allograft correlated with a higher concentration of IL-10 in the peripheral blood of these patients. In addition, the presence of the IL-10 promoter polymorphism associated with high transcription levels of IL-10 has been shown to be an independent protective factor for severe acute GvHD [57]. A correlation between high spontaneously produced levels of IL-10 in the PBMCs of patients before allogenic transplantation and lower incidence of GvHD and transplant related mortality has been also reported [58,59]. Finally, spontaneous development of tolerance to kidney and liver allograft in transplanted patients is associated with the presence of $CD4^+$ T cells that suppress T cell responses via production of IL-10 and TGF-β [60]. In conclusion these data indicate that Tr1 cells can regulate allograft immune responses and induce tolerance in human after bone marrow and solid organ transplantation.

Tolerance to Infectious Agents and Tumors

In addition to the protective role, Tr1 cells can also interfere with host protective immunity. Tr1 cells specific for infectious agents or tumor antigens can indeed impede host's immune response and thus be detrimental. Presumably, it is advantageous for pathogens to evolve strategies to enhance the differentiation of Tr1 cells, which would then limit the protective immunity and allow long-term infection of the host [36]. In chronic helminthes infections, where patients have relatively little sign of dermatitis despite the presence of small worms in the skin, antigen-specific Tr1 cells were isolated. These cells were able to inhibit proliferation of effector T cell clones [61]. Moreover, Leishmania major, hepatitis C, filarial nematodes and Helicobacter hepaticus have been shown to induce IL-10 secretion by APC as well as differentiation of Tr1 cells in vivo [62].

Similarly to infectious agents, several studies report the existence of Tr1 cells specific for tumor antigens. It has been shown that myeloma cells prime DC towards a state that favors the generation of T cells with a Tr1 rather than an effector phenotype [63]. Exposure to cyclooxygenase-2–overexpressing glioma induces mature DC to over-express IL-10 and decreased IL-12p70 production. These DC induce a Tr1-cell response, which is characterized by robust secretion of IL-10 and TGF-β with negligible IL-4 secretion by $CD4^+$ T cells, and an inhibitory effect on responder T cells [64]. In addition, Hodgkin lymphoma infiltrating $CD4^+$ lymphocytes (HLILs), unlike their PBMC counterpart, have been found to be anergic, and suppress cell proliferation in vitro. Furthermore, HLILs contain large populations of IL-10–secreting Tr1 cells. Thus, HLILs are highly enriched in Tregs, which create a profoundly immunosuppressive environment that provides an ineffective immune clearance of cancer cells [65]. It has been also described the presence of $CD8^+$ Tr1 cells in tumors. In an ovarian carcinoma environment, pDC induce tumor-associated antigen-specific $CD8^+$ IL-10 producing T cells. These cells are able to induce tolerance through IL-10 production and contribute to the immunosuppressive tumor environment [66].

It is now evident that the balance between effector T cells and Tr1 cells need to be precisely tuned in order to provide optimal control of immune responses and to prevent suppression in settings in which it can be detrimental.

Cellular Therapy with Tr1 Cells

It is now firmly established that Tr1 cells are an indispensable component of the immune system, which can actively control immune responses. Therefore, adoptive trasfer of ex vivo generated Tr1 cells might be highly advantageous in several T-cell mediated diseases. HLA-haploidentical transplantation offers a valuable source of hematopoietic stem cells (HSC) to most of the patients in the need of a bone marrow transplant when matched donors are unavailable. A megadose of highly purified $CD34^+$ HSC depleted of mature T cells is crucial for promoting engraftment without GvHD. However, T-cell–depleted transplants are at high risk of recurrent life-threatening infections, and of disease relapse due to the absence of graft versus leukemia (GvL) effect. To overcome these limitations, we are currently performing in our institute a clinical trial in which donor cells anergized ex vivo in the presence of IL-10 are used as post-transplant cellular therapy in hematological cancer patients undergoing HLA-haploidentical HSC transplantation. Donor PBMC cultured in vitro with irradiated host PBMC in the presence of IL-10 for 10 days become anergic towards the host Ags, while they preserve the ability to proliferate in response to third party and nominal Ags. Importantly, these IL-10 anergized T cells are highly enriched for alloAg-specific Tr1 cell precursors (Bacchetta et al. unpublished data). The administered cells should therefore include alloantigen specific T cells with the ability to differentiate in fully competent Tr1 cells but also T cells able to respond to infectious agents and presumably to provide a GvL effect. This approach, which is currently being tested in the context of HLA-haploidentical HSC transplantation, has the potential to be applied to unrelated bone marrow transplants and to solid organ transplantations.

Conclusions

Following their initial description in 1994 [4] Tr1 cells have been extensively characterized and are considered today a professional subset of Tregs able to prevent immune-mediated diseases and to maintain immunological tolerance. Important questions that were set forward more than 10 years ago have been successfully addressed, whereas others still remain unresolved. It is still largely unknown, for example, the Tr1 cell migratory ability in vivo, their homeostasis, and the molecular and cellular basis of their development. Results from the ongoing proof of concepts trials and the proposed phase I/II clinical trials with Tr1 cell based immunotherapy will help us in understanding the biology of these cells and will be fundamental for their future exploitation in the clinic.

References

1. Gershon, R. K., and K. Kondo. 1971. Infectious immunological tolerance. *Immunology* 21:903–914.
2. Dorf, M. E., and B. Benacerraf. 1984. Suppressor cells and immunoregulation. *Annu Rev Immunol* 2:127–157.
3. Groux, H., A. O'Garra, M. Bigler, M. Rouleau, S. Antonenko, J. E. de Vries, and M. G. Roncarolo. 1997. A CD4+ T-cell subset inhibits antigen-specific T-cell responses and prevents colitis. *Nature* 389:737–742.
4. Bacchetta, R., M. Bigler, J. L. Touraine, R. Parkman, P. A. Tovo, J. Abrams, R. de Waal Malefyt, J. E. de Vries, and M. G. Roncarolo. 1994. High levels of interleukin 10 production in vivo are associated with tolerance in SCID patients transplanted with HLA mismatched hematopoietic stem cells. *J Exp Med* 179:493–502.
5. Bacchetta, R., C. Sartirana, M. K. Levings, C. Bordignon, S. Narula, and M. G. Roncarolo. 2002. Growth and expansion of human T regulatory type 1 cells are independent from TCR activation but require exogenous cytokines. *Eur J Immunol* 32:2237–2245.
6. Gilliet, M., and Y. J. Liu. 2002. Generation of human CD8 T regulatory cells by CD40 ligand-activated plasmacytoid dendritic cells. *J Exp Med* 195:695–704.
7. Gregori, S., P. Mangia, R. Bacchetta, E. Tresoldi, F. Kolbinger, C. Traversari, J. M. Carballido, J. E. de Vries, U. Korthauer, and M. G. Roncarolo. 2005. An anti-CD45RO/RB monoclonal

antibody modulates T cell responses via induction of apoptosis and generation of regulatory T cells. *J Exp Med* 201:1293–1305.
8. Steinbrink, K., H. Jonuleit, G. Muller, G. Schuler, J. Knop, and A. H. Enk. 1999. Interleukin-10-treated human dendritic cells induce a melanoma-antigen-specific anergy in CD8(+) T cells resulting in a failure to lyse tumor cells. *Blood* 93:1634–1642.
9. Groux, H., M. Bigler, J. E. de Vries, and M. G. Roncarolo. 1996. Interleukin-10 induces a long-term antigen-specific anergic state in human CD4+ T cells. *J Exp Med* 184: 19–29.
10. Levings, M. K., S. Gregori, E. Tresoldi, S. Cazzaniga, C. Bonini, and M. G. Roncarolo. 2005. Differentiation of Tr1 cells by immature dendritic cells requires IL-10 but not CD25+CD4+ Tr cells. *Blood* 105:1162–1169.
11. Levings, M. K., R. Sangregorio, F. Galbiati, S. Squadrone, R. de Waal Malefyt, and M. G. Roncarolo. 2001. IFN-alpha and IL-10 induce the differentiation of human type 1 T regulatory cells. *J Immunol* 166:5530–5539.
12. Sebastiani, S., P. Allavena, C. Albanesi, F. Nasorri, G. Bianchi, C. Traidl, S. Sozzani, G. Girolomoni, and A. Cavani. 2001. Chemokine receptor expression and function in CD4+ T lymphocytes with regulatory activity. *J Immunol* 166:996–1002.
13. Freeman, C. M., B. C. Chiu, V. R. Stolberg, J. Hu, K. Zeibecoglou, N. W. Lukacs, S. A. Lira, S. L. Kunkel, and S. W. Chensue. 2005. CCR8 is expressed by antigen-elicited, IL-10-producing CD4+CD25+ T cells, which regulate Th2-mediated granuloma formation in mice. *J Immunol* 174:1962–1970.
14. Papadakis, K. A., C. Landers, J. Prehn, E. A. Kouroumalis, S. T. Moreno, J. C. Gutierrez-Ramos, M. R. Hodge, and S. R. Targan. 2003. CC chemokine receptor 9 expression defines a subset of peripheral blood lymphocytes with mucosal T cell phenotype and Th1 or T-regulatory 1 cytokine profile. *J Immunol* 171:159–165.
15. Fontenot, J. D., and A. Y. Rudensky. 2005. A well adapted regulatory contrivance: regulatory T cell development and the forkhead family transcription factor Foxp3. *Nat Immunol* 6: 331–337.
16. Hori, S., T. Nomura, and S. Sakaguchi. 2003. Control of regulatory T cell development by the transcription factor Foxp3. *Science* 299:1057–1061.
17. Allan, S. E., L. Passerini, R. Bacchetta, N. Crellin, M. Dai, P. C. Orban, S. F. Ziegler, M. G. Roncarolo, and M. K. Levings. 2005. The role of 2 FOXP3 isoforms in the generation of human CD4+ Tregs. *J Clin Invest* 115:3276–3284.
18. Walker, M. R., D. J. Kasprowicz, V. H. Gersuk, A. Benard, M. Van Landeghen, J. H. Buckner, and S. F. Ziegler. 2003. Induction of FoxP3 and acquisition of T regulatory activity by stimulated human CD4+CD25- T cells. *J Clin Invest* 112:1437–1443.
19. Barrat, F. J., D. J. Cua, A. Boonstra, D. F. Richards, C. Crain, H. F. Savelkoul, R. de Waal-Malefyt, R. L. Coffman, C. M. Hawrylowicz, and A. O'Garra. 2002. In vitro generation of interleukin 10-producing regulatory CD4(+) T cells is induced by immunosuppressive drugs and inhibited by T helper type 1 (Th1)- and Th2-inducing cytokines. *J Exp Med* 195: 603–616.
20. Kitani, A., K. Chua, K. Nakamura, and W. Strober. 2000. Activated self-MHC-reactive T cells have the cytokine phenotype of Th3/T regulatory cell 1 T cells. *J Immunol* 165:691–702.
21. Steinbrink, K., M. Wolfl, H. Jonuleit, J. Knop, and A. H. Enk. 1997. Induction of tolerance by IL-10-treated dendritic cells. *J Immunol* 159:4772–4780.
22. Steinbrink, K., E. Graulich, S. Kubsch, J. Knop, and A. H. Enk. 2002. CD4(+) and CD8(+) anergic T cells induced by interleukin-10-treated human dendritic cells display antigen-specific suppressor activity. *Blood* 99:2468–2476.
23. Willems, F., A. Marchant, J. P. Delville, C. Gerard, A. Delvaux, T. Velu, M. de Boer, and M. Goldman. 1994. Interleukin-10 inhibits B7 and intercellular adhesion molecule-1 expression on human monocytes. *Eur J Immunol* 24:1007–1009.
24. Fiorentino, D. F., A. Zlotnik, T. R. Mosmann, M. Howard, and A. O'Garra. 1991. IL-10 inhibits cytokine production by activated macrophages. *J Immunol* 147:3815–3822.

25. Allavena, P., L. Piemonti, D. Longoni, S. Bernasconi, A. Stoppacciaro, L. Ruco, and A. Mantovani. 1998. IL-10 prevents the differentiation of monocytes to dendritic cells but promotes their maturation to macrophages. *Eur J Immunol* 28:359–369.
26. Li, M. O., Y. Y. Wan, S. Sanjabi, A. K. Robertson, and R. A. Flavell. 2006. Transforming growth factor-beta regulation of immune responses. *Annu Rev Immunol* 24:99–146.
27. Hawrylowicz, C. M., and A. O'Garra. 2005. Potential role of interleukin-10-secreting regulatory T cells in allergy and asthma. *Nat Rev Immunol* 5:271–283.
28. Grossman, W. J., J. W. Verbsky, W. Barchet, M. Colonna, J. P. Atkinson, and T. J. Ley. 2004. Human T regulatory cells can use the perforin pathway to cause autologous target cell death. *Immunity* 21:589–601.
29. Grossman, W. J., J. W. Verbsky, B. L. Tollefsen, C. Kemper, J. P. Atkinson, and T. J. Ley. 2004. Differential expression of granzymes A and B in human cytotoxic lymphocyte subsets and T regulatory cells. *Blood* 104:2840–2848.
30. Cottrez, F., and H. Groux. 2001. Regulation of TGF-beta response during T cell activation is modulated by IL-10. *J Immunol* 167:773–778.
31. Kemper, C., A. C. Chan, J. M. Green, K. A. Brett, K. M. Murphy, and J. P. Atkinson. 2003. Activation of human CD4+ cells with CD3 and CD46 induces a T-regulatory cell 1 phenotype. *Nature* 421:388–392.
32. Steinman, R. M., D. Hawiger, and M. C. Nussenzweig. 2003. Tolerogenic dendritic cells. *Annu Rev Immunol* 21:685–711.
33. Jonuleit, H., E. Schmitt, G. Schuler, J. Knop, and A. H. Enk. 2000. Induction of interleukin 10-producing, nonproliferating CD4(+) T cells with regulatory properties by repetitive stimulation with allogeneic immature human dendritic cells. *J Exp Med* 192:1213–1222.
34. Kuwana, M., J. Kaburaki, T. M. Wright, Y. Kawakami, and Y. Ikeda. 2001. Induction of antigen-specific human CD4(+) T cell anergy by peripheral blood DC2 precursors. *Eur J Immunol* 31:2547–2557.
35. Zuany-Amorim, C., E. Sawicka, C. Manlius, A. Le Moine, L. R. Brunet, D. M. Kemeny, G. Bowen, G. Rook, and C. Walker. 2002. Suppression of airway eosinophilia by killed Mycobacterium vaccae-induced allergen-specific regulatory T-cells. *Nat Med* 8:625–629.
36. McGuirk, P., C. McCann, and K. H. Mills. 2002. Pathogen-specific T regulatory 1 cells induced in the respiratory tract by a bacterial molecule that stimulates interleukin 10 production by dendritic cells: a novel strategy for evasion of protective T helper type 1 responses by Bordetella pertussis. *J Exp Med* 195:221–231.
37. Lavelle, E. C., E. McNeela, M. E. Armstrong, O. Leavy, S. C. Higgins, and K. H. Mills. 2003. Cholera toxin promotes the induction of regulatory T cells specific for bystander antigens by modulating dendritic cell activation. *J Immunol* 171:2384–2392.
38. Battaglia, M., A. Stabilini, E. Draghici, S. Gregori, C. Mocchetti, E. Bonifacio, and M. G. Roncarolo. 2006. Rapamycin and interleukin-10 treatment induces T regulatory type 1 cells that mediate antigen-specific transplantation tolerance. *Diabetes* 55:40–49.
39. Battaglia, M., A. Stabilini, E. Draghici, B. Migliavacca, S. Gregori, E. Bonifacio, and M. G. Roncarolo. 2006. Induction of Tolerance in Type 1 Diabetes via Both CD4+CD25+ T Regulatory Cells and T Regulatory Type 1 Cells. *Diabetes* 55:1571–1580.
40. Asiedu, C. K., K. J. Goodwin, G. Balgansuren, S. M. Jenkins, S. Le Bas-Bernardet, U. Jargal, D. M. Neville, Jr., and J. M. Thomas. 2005. Elevated T regulatory cells in long-term stable transplant tolerance in rhesus macaques induced by anti-CD3 immunotoxin and deoxyspergualin. *J Immunol* 175:8060–8068.
41. Herold, K. C., J. B. Burton, F. Francois, E. Poumian-Ruiz, M. Glandt, and J. A. Bluestone. 2003. Activation of human T cells by FcR nonbinding anti-CD3 mAb, hOKT3gamma1(Ala-Ala). *J Clin Invest* 111:409–418.
42. Chatenoud, L. 2003. CD3-specific antibody-induced active tolerance: from bench to bedside. *Nat Rev Immunol* 3:123–132.

43. Herold, K. C., W. Hagopian, J. A. Auger, E. Poumian-Ruiz, L. Taylor, D. Donaldson, S. E. Gitelman, D. M. Harlan, D. Xu, R. A. Zivin, and J. A. Bluestone. 2002. Anti-CD3 monoclonal antibody in new-onset type 1 diabetes mellitus. *N Engl J Med* 346: 1692–1698.
44. Veldman, C., A. Hohne, D. Dieckmann, G. Schuler, and M. Hertl. 2004. Type I regulatory T cells specific for desmoglein 3 are more frequently detected in healthy individuals than in patients with pemphigus vulgaris. *J Immunol* 172:6468–6475.
45. Arif, S., T. I. Tree, T. P. Astill, J. M. Tremble, A. J. Bishop, C. M. Dayan, B. O. Roep, and M. Peakman. 2004. Autoreactive T cell responses show proinflammatory polarization in diabetes but a regulatory phenotype in health. *J Clin Invest* 113:451–463.
46. Yudoh, K., H. Matsuno, F. Nakazawa, T. Yonezawa, and T. Kimura. 2000. Reduced expression of the regulatory CD4+ T cell subset is related to Th1/Th2 balance and disease severity in rheumatoid arthritis. *Arthritis Rheum* 43:617–627.
47. Hall, A. M., F. J. Ward, M. A. Vickers, L. M. Stott, S. J. Urbaniak, and R. N. Barker. 2002. Interleukin-10-mediated regulatory T-cell responses to epitopes on a human red blood cell autoantigen. *Blood* 100:4529–4536.
48. Mowat, A. M. 2003. Anatomical basis of tolerance and immunity to intestinal antigens. *Nat Rev Immunol* 3:331–341.
49. Troncone, R., C. Gianfrani, G. Mazzarella, L. Greco, J. Guardiola, S. Auricchio, and P. De Berardinis. 1998. Majority of gliadin-specific T-cell clones from celiac small intestinal mucosa produce interferon-gamma and interleukin-4. *Dig Dis Sci* 43:156–161.
50. Salvati, V. M., G. Mazzarella, C. Gianfrani, M. K. Levings, R. Stefanile, B. De Giulio, G. Iaquinto, N. Giardullo, S. Auricchio, M. G. Roncarolo, and R. Troncone. 2005. Recombinant human interleukin 10 suppresses gliadin dependent T cell activation in ex vivo cultured coeliac intestinal mucosa. *Gut* 54:46–53.
51. Gianfrani, C., M. K. Levings, C. Sartirana, G. Mazzarella, G. Barba, D. Zanzi, A. Camarca, G. Iaquinto, N. Giardullo, S. Auricchio, R. Troncone, and M. G. Roncarolo. 2006. Gliadin-specific type 1 regulatory T cells from the intestinal mucosa of treated celiac patients inhibit pathogenic T cells. *J Immunol* 177:4178–4186.
52. Umetsu, D. T., and R. H. DeKruyff. 2006. The regulation of allergy and asthma. *Immunol Rev* 212:238–255.
53. Cavani, A., F. Nasorri, C. Prezzi, S. Sebastiani, C. Albanesi, and G. Girolomoni. 2000. Human CD4+ T lymphocytes with remarkable regulatory functions on dendritic cells and nickel-specific Th1 immune responses. *J Invest Dermatol* 114:295–302.
54. Saloga, J., I. Bellinghausen, and J. Knop. 1999. Do Tr1 cells play a role in immunotherapy? *Int Arch Allergy Immunol* 118:210–211.
55. Reefer, A. J., R. M. Carneiro, N. J. Custis, T. A. Platts-Mills, S. S. Sung, J. Hammer, and J. A. Woodfolk. 2004. A role for IL-10-mediated HLA-DR7-restricted T cell-dependent events in development of the modified Th2 response to cat allergen. *J Immunol* 172: 2763–2772.
56. Akdis, M. 2006. Healthy immune response to allergens: T regulatory cells and more. *Curr Opin Immunol* 18:738–744.
57. Weston, L., A. Geczy, and H. Briscoe. 2005. Production od IL-10 by alloreactive sibling donor cells and its influence on teh development of acute GvHD. *Bone Marrow Transplant* 37:207–212.
58. Holler, E., M. G. Roncarolo, R. Hintermeier-Knabe, G. Eissner, B. Ertl, U. Schulz, H. Knabe, H. J. Kolb, R. Andreesen, and W. Wilmanns. 2000. Prognostic significance of increased IL-10 production in patients prior to allogeneic bone marrow transplantation. *Bone Marrow Transplant* 25:237–241.
59. Baker, K. S., M. G. Roncarolo, C. Peters, M. Bigler, T. DeFor, and B. R. Blazar. 1999. High spontaneous IL-10 production in unrelated bone marrow transplant recipients is associated with fewer transplant-related complications and early deaths. *Bone Marrow Transplant* 23:1123–1129.

60. VanBuskirk, A. M., W. J. Burlingham, E. Jankowska-Gan, T. Chin, S. Kusaka, F. Geissler, R. P. Pelletier, and C. G. Orosz. 2000. Human allograft acceptance is associated with immune regulation. *J Clin Invest* 106:145–155.
61. Satoguina, J., M. Mempel, J. Larbi, M. Badusche, C. Loliger, O. Adjei, G. Gachelin, B. Fleischer, and A. Hoerauf. 2002. Antigen-specific T regulatory-1 cells are associated with immunosuppression in a chronic helminth infection (onchocerciasis). *Microbes Infect* 4: 1291–1300.
62. Kullberg, M. C., A. G. Rothfuchs, D. Jankovic, P. Caspar, T. A. Wynn, P. L. Gorelick, A. W. Cheever, and A. Sher. 2001. Helicobacter hepaticus-induced colitis in interleukin-10-deficient mice: cytokine requirements for the induction and maintenance of intestinal inflammation. *Infect Immun* 69:4232–4241.
63. Fiore, F., B. Nuschak, S. Peola, S. Mariani, M. Muraro, M. Foglietta, M. Coscia, B. Bruno, M. Boccadoro, and M. Massaia. 2005. Exposure to myeloma cell lysates affects the immune competence of dendritic cells and favors the induction of Tr1-like regulatory T cells. *Eur J Immunol* 35:1155–1163.
64. Akasaki, Y., G. Liu, N. H. Chung, M. Ehtesham, K. L. Black, and J. S. Yu. 2004. Induction of a CD4+ T regulatory type 1 response by cyclooxygenase-2-overexpressing glioma. *J Immunol* 173:4352–4359.
65. Marshall, N. A., L. E. Christie, L. R. Munro, D. J. Culligan, P. W. Johnston, R. N. Barker, and M. A. Vickers. 2004. Immunosuppressive regulatory T cells are abundant in the reactive lymphocytes of Hodgkin lymphoma. *Blood* 103:1755–1762.
66. Wei, S., I. Kryczek, L. Zou, B. Daniel, P. Cheng, P. Mottram, T. Curiel, A. Lange, and W. Zou. 2005. Plasmacytoid dendritic cells induce CD8+ regulatory T cells in human ovarian carcinoma. *Cancer Res* 65:5020–5026.
67. Bejarano, M. T., R. de Waal Malefyt, J. S. Abrams, M. Bigler, R. Bacchetta, J. E. de Vries, and M. G. Roncarolo. 1992. Interleukin 10 inhibits allogeneic proliferative and cytotoxic T cell responses generated in primary mixed lymphocyte cultures. *Int Immunol* 4:1389–1397.

Chapter 25
CD8$^+$ T Regulatory Cells in Eye Derive Tolerance

Joan Stein-Streilein and Hiroshi Keino

Abstract Regulatory T cells or Tregs are critical to the development of self-tolerance and a natural way to terminate an immune response. While most studies are directed toward understanding thymic-derived natural afferent CD4+ CD25+ Tregs that regulate induction of immune responses, our studies focus on the antigen specific efferent CD8+ Treg cells that develop in the periphery and limit immune responses during their effector stage. The development of CD8+ Treg cells is one of the mechanisms that contribute to the eye being an immune privileged site. Immune privilege is a term that is associated with sites and tissue that enjoy long-term survival of tissue grafts of foreign derivation. Immune privilege is a dynamic process that allows for immune responses that lack inflammation thus permitting immune protection in the absence of tissue damage. A model to study immune privilege is called anterior chamber associated immune deviation or ACAID. Through the study of ACAID, cellular and molecular mechanisms were observed that show that the development of CD8+ Treg cells post intracameral inoculation of antigen is dependent on specialized antigen presenting cells (F4/80+ APC), NKT cells, T cells and, marginal zone derived B cells that meet not in the T cell areas of the secondary lymphoid tissue but in the marginal zone of the spleen. These studies were expanded by recent investigations that explored the mechanisms used by CD8+ Treg cells to regulate immune CD4+ T cell effector function. Understanding how to generate efferent Treg cells that limit ongoing immune inflammation may lead to novel therapy for immune inflammatory diseases in the eye and the periphery.

Introduction

Eye derived tolerance has been historically studied by using an animal model called anterior chamber associated immune deviation or ACAID. The understanding of eye induced antigen specific peripheral tolerance contributes, in part, to a broader

---

J. Stein-Streilein
Schepens Eye Research Institute, Department of Ophthalmology, Harvard Medical School, 20 Staniford Street, Boston, MA 02114, USA.
e-mail: Joan.stein@schepens.harvard.edu

understanding of the immune privileged mechanisms of the eye. Immune privilege is a transplantation term originally defining tissue sites that allowed for the extended survival of foreign grafts [1,2]. Tissues may also be immune privileged, and if so they are easily transplanted to foreign hosts and survive indefinitely. Thus, the term immune privilege is an operational term. As we now know, immune privilege is promoted by a complex interaction of mechanisms that allow for host defenses against infections and tumor invasion in the absence of tissue damaging inflammation. Lessons learned from the study of immune privilege are applicable to promotion of graft survival, and treatment of autoimmune diseases. Since tumors often survive by setting up ad hoc immune privilege environments, understanding how tolerance or immune privilege can be broken may lead to novel therapies for cancer treatment.

It is generally believed that although the mechanisms of immune regulation presented here are encountered in immune privileged sites like the eye, reproductive tract, or the brain, we propose that the same tolerance mechanisms are used, in part, by many non-immune privileged tissues in the body. It is apparent that immune regulation is customized for each tissue and organ and the biological functions of the organs have a great influence on both the immune functions and their regulation. Thus by understanding T regulatory (Treg) cells induced by an immune privileged site like the eye information is gathered about T regulatory cells in general and thus may lead to novel therapeutic approaches in autoimmune and immune inflammatory diseases.

Immune Privilege

The immune response is designed to protect the body from danger, infection and malignant cells [3]. Medawar actually coined the term, immune privileged, for organs and tissues that accepted foreign grafts for an extended period of time [1,2]. When the field of immunology exploded in the 60 s and 70 s, models of immune privilege were developed. A variety of immune privileged sites exist in the body including the female reproductive tract [4], testis, [5–7] and the brain [8–10]. The immune privilege of the eye is best studied with an animal model called ACAID [11,12]. Some of the mechanisms that contribute to the differentiation of $CD8^+$ Treg cells in the ACAID model are true for other immune privileged sites as well [5,9]. Inoculation of antigen into the anterior chamber (a.c.) induces an active immune response that eventuates into the production of peripheral $CD8^+$ Treg cell that suppress both Th1 and Th2 effector cells [13,14]. The divergence of the eye immune response from the expected immune response appears to be along two lines of immune cellular behavior since it includes immune defense without inflammation and a strong Treg cell response to control the immune responses that do occur.

Anterior Chamber Associated Immune Deviation (ACAID)

There are many reviews on mechanisms involved in ACAID [12,15–19]. Antigen specific peripheral tolerance or ACAID is induced when antigen is placed in the anterior chamber of the eye (Fig. 25.1).

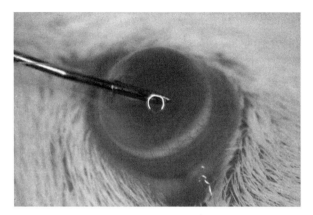

Fig. 25.1 Anterior chamber inoculation After 2 μl of aqueous humor is removed, the antigen is delivered into the chamber through a custom-made glass needle inserted in the hole made by the extraction needle. Injected eyes are followed for three days for signs of inflammation and used only if the cornea and eye are clear *(See also Color Insert)*

The indigenous APC then carries the antigen through the trabecular meshwork into the blood, ending up in the marginal zone [20] of the spleen either directly from the eye or after a detour through the thymus [21]. It seems that the eye must remain in tact for three days since enucleation of the eye removes the tolerance outcome [22]. To date it is unknown what happens in the eye during this time.

Early studies by Wilbanks reported that both an afferent $CD4^+$ Treg and efferent $CD8^+$ Treg developed in response to antigen inoculated in the anterior chamber [12,23]. Both of these regulatory T cell responses are expressed in the periphery and, although never tested, presumed to function in the local eye environment. Because the assay used to test for suppression involves the suppression of the T cell effector response or delayed hypersensitivity (DH) responses in vivo, most of the information generated about Tregs in ACAID relates to the efferent $CD8^+$ Treg cells [4,22]. However, Hiroshi Keino recently reported new insights into both the afferent $CD4^+$ Treg and the efferent $CD8^+$ Treg cells using transgenic T cells in vitro ACAID assays [24,25].

The cell that is central to the development of both the $CD4^+$ and $CD8^+$ Treg in ACAID is the $F4/80^+$ antigen-presenting cell (APC) that leaves the eye, and transports the antigen to the marginal zone (MZ) of the spleen. The putative eye-derived APC secretes chemokines (MIP-2) along the way that recruit a $CD4^+$ invariant iNKT cell to the zone to [20,26] interact with $CD1d^+$ MZ B cells [27], and T cells. The F4/80 protein is required for the successful interaction of the cell aggregates to generate the efferent $CD8^+$ Treg since F4/80 null mice are unable to suppress DH to ovalbumin (OVA) following a.c. injection of OVA [28]. The requirement for F4/80 protein expression is also relevant for the generation of the $CD8^+$ Treg cell in low dose oral tolerance, another peripheral tolerance model that requires iNKT cells [28, 29].

Unique Traffic Patterns for Induction of CD8⁺ Treg Cells in Eye Induced Tolerance (Fig. 25.2).

One would think intuitively that antigens inoculated into the eye induce intravenous (i.v.) tolerance since about 98% of the antigen goes directly into the blood stream post a.c. inoculation [30]. However, it has clearly been shown that the injection of antigen directly into the blood will not induce an ACAID like tolerance. Furthermore, efferent CD8$^+$ Treg cells are not induced when antigen is injected i.v. [31].

Certain aspects of ACAID (discussed below), like a requirement for NKT cells [28,32] and the F4/80 molecules, are not involved in the development of i.v. induced tolerance although, as mentioned before, are involved in the induction of low dose oral tolerance [28,29]. Interestingly the ACAID efferent CD8$^+$ Treg cells do not require a traditional CD4$^+$ T cell for its development since CD8$^+$ Tregs can be generated in Class II KO mice following antigen inoculation into the a.c. [33]. (Fig. 25.2)

The iNKT cell required for ACAID induction expresses CD4 because removal of the NKT cell with antibodies to CD4 abrogate the ability of the Class II KO

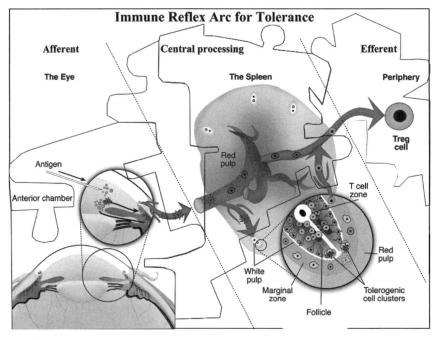

Fig. 25.2 Immune reflex arc for tolerance Illustration of the traffic of eye-derived F4/80+ cells (*red cells*) showing it path through the blood to the marginal zone of the T cell areas in the spleen. Five to seven days post a.c. inoculation, antigen specific Treg cells are in the peripheral tissue and can be assessed from by testing the cells in the spleen for its ability to suppress Th1 and Th2 responses *(See also Color Insert)*

mice to respond to the a.c. inoculation with the development of CD8$^+$ Treg and the suppression of DTH in local adoptive transfer assays [33]. In addition, Keino et al. demonstrated that ACAID to antigen inoculated into the anterior chamber was induced in mice depleted of CD25$^+$ T cells [24]. This implies that ACAID induction is independent of, and perhaps even irrelevant to, the presence of natural CD4$^+$CD25$^+$ Tregs. Despite these strong data, others report that CD4$^+$ cells are needed for ACAID [34,35]. Since systems are not the same, it is difficult to reconcile these contrasting reports.

It is becoming more apparent that the type of APC that presents an antigen is critical to the immune outcome. In keeping with this notion, the APC indigenous to the eye that induces the CD8$^+$ Treg cell post a.c. inoculation are naturally bathed in immunosuppressive cytokines. We and others have made surrogate ACAID tolerogenic APC that induced T regulatory cells in vitro by exposing a variety of APC types to TGFβ [18,24,31,36]. It is important that the APC that picks up the eye inoculated antigen be the indigenous F4/80$^+$ APC [37]. When the F4/80$^+$ cell is observed in the tissue, its morphology is similar to a macrophage and indeed until Lin and colleagues showed that the F4/80 protein was needed for the development of tolerance the F4/80 protein was widely accepted only as a marker of tissue macrophages [28]. Whether the F4/80$^+$ cell remains a "macrophage" as it travels to the spleen is debatable, since in vivo and in vitro-generated tolerogenic APC have characteristics of semi-mature dendritic cells [14]. Cone and colleagues reported that the APC transporting the eye-derived antigen travels through the thymus, since thymectomy prevented the development of ACAID [38]. In contrast, others report that neonatal thymectomy does not seem to prevent the development of eye induced tolerance [39].

A few years after Wilbanks' report, Faunce et al., showed that the F480$^+$ cell responsible for eye-induced tolerance localized in the marginal zone of the spleen rather than the T cell areas where inflammatory immune responses are induced [20]. The eye-derived F4/80$^+$ APC associated with more F4/80$^+$ cells recruited from the periphery as well as invariant NKT cells [32], and T cells. An important function of the F4/80$^+$ APC is to secrete MIP-2 chemokine that is responsible for the recruitment of the NKT cell [40]. D'Orazio and Niederkorn showed that the B cell was needed for the differentiation of the CD8$^+$ Treg cell [41]. In this report, they suggest that the F4/80$^+$ APC transfer their antigen to the B cells and through a Qa-2 dependent process present the antigen to the T cell. Later, Sonoda and colleagues showed that the CD1d expressing marginal zone B cells and not the splenic follicular cells were required for the differentiation of the CD8$^+$ Treg cell [42].

NKT Cells and ACAID.

NKT cells make up a relatively rare population of cells that were fairly well characterized before a function was ascribed to them. One of the first biological functions described for this cell type was its role in eye-derived tolerance and the development of CD8$^+$ Treg cells [32]. This cell is absolutely required for the generation of the CD8$^+$ Treg cell that ultimately down regulates CD4$^+$ T cell mediated effector cell responses during the expression of ACAID. A major population of NKT cells express the invariant TCR Vα14 jα18. An analogous cell population (Vα 24 Jα invariant NKT cells) exists in humans. Moreover, the role of NKT cells in self-tolerance in implied by the many autoimmune diseases that correlate with either reduced or functionally impaired NKT cells [43]. The invariant TCR NKT cells in mice and humans respond to and are dependent on CD1d for their development. Thus, the mechanisms that are revealed by studies of tolerance in the mouse are directly translatable to human immune regulation.

Ocular Mechanisms that Lead to the Development of Treg Cells

Immune privilege is achieved within the eye by at least two overlapping and related mechanisms: (1) soluble immunosuppressive and anti-inflammatory factors within the microenvironment (TGF-β2 and neuropeptides) [44,45] and (2) surface molecules expressed on ocular parenchymal cells especially the pigment epithelium [46–49], the corneal epithelium [50] and the corneal endothelium [51].

Aqueous Humor or its Components Generate Tregs.

Cells that surround the anterior chamber are bathed in AqH and are not only protected by the immunosuppressive qualities but are modified by it (Fig. 25.3).

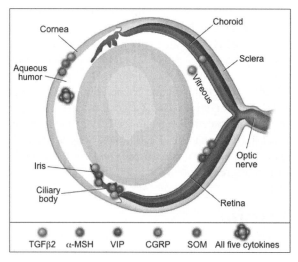

Figure 25.3 Illustration of a Mouse Eye The diagram shows the anatomy of the eye with symbols representing each of the various immunosuppressive factors that collaborate to establish an immunosuppressive environment *(See also Color Insert)*

TGF-β2 is known to be a crucial immunomodulatory factor within the eye [44,45,52]. However, the distinction between TGF-β2 and TGF-β1 becomes blurred since bone marrow cells that are exposed to TGF-β2 produce TGF-β1. TGFβ is a pleiotropic cytokine/growth factor that also has the capacity to suppress aspects of immunity [53–55]. Besides TGF-β2, there are other major factors found in the eye that can directly influence the behavior of T cells [56] (Fig. 25.1). The presence of such factors was first demonstrated by the ability of healthy aqueous humor to suppress IFN-γ production by Th1 cells [55,57].

In addition, other immunosuppressive substances believed to be in AqH have been used to induce T reg cells. Taylor and Yee report that somatostatin is capable of inducing $CD4^+$ $CD25^+$ T regs [24]. Alpha melanocortin stimulating hormone (□MSH) exhibits a variety of suppressive functions in vitro including suppression of IFNγ production and promotion of TGFβ production by T cells [16,58].

Complement regulatory proteins are also available in the Aq H and contribute the eye's ability to tightly regulate complement mediated inflammation [50,59,60].

Membrane Bound Molecules Contribute to Suppression and Generation of Treg Cells

In addition to the soluble factors in the fluids of the eye, the cells in the eye express surface molecules that aid in preventing immune inflammation. Many cells throughout the eye express Fas ligand (FasL, CD95L), a molecule that when engaged by FAS (that is induced on activated cells) leads to programmed cell death in the FAS + cell [51]. FasL is constitutively expressed on cells immune privileged tissues and organs. Another feature of cells in the eye is their poor expression of Major Histocompatibility Complex (MHC) molecules, MHC Class I and II, [19]. Cells within the eye do express surface molecules associated with suppression [61] class I B molecules like Qa-2 in the mouse and HLA G in humans [12,14].

The posterior surfaces of the iris, the ciliary body, and neural retina are lined with pigment epithelium (PE) that partially surrounds the privileged sites [12] of the anterior chamber [62], the vitreous cavity [63], and the subretinal space [64,65] respectively. Both cultured and freshly prepared PE cells from iris, ciliary body and retina are capable of suppressing TCR-dependent activation and promote the generation of T regulatory cells [48] of naïve and primed T cells in culture [46,47,66]. Sugita and colleagues suggest that the ability of ocular PE to suppress T effector cell activity and to convert activated T cells into regulators is an immune privilege strategy to limit immunogenic inflammation in the eye.

Iris PE cells (IPE) are unique from other PE cells in the eye because unlike ciliary body PE (CBPE) and retina PE (RPE), they use a cell surface contact-dependent mechanism exclusively to suppress T cell activation in vitro [46]. Both fresh and cultured IPE constitutively express B7-1 and B7-2 on their surface and are necessary to convert naïve T cells to T regulatory cells. T cells that are converted are $CD8^+$ and express CTLA4. Soluble factors are also involved in PE suppression, since neutralizing TGF-β antibody allowed for anti CD3 activation of naïve cells even in the presence of IPE in the cultures. A role for membrane bound TGF-β [67] was tested in a subsequent paper [68]. In that report they show that IPE through their co-stimulatory molecules, B7-1 and B7-2, make contact with CTLA-4 on a subpopulation of CD8+ T cells and when engaged IPE membrane associated, active TGF-β is delivered precisely to the T cells. The targeted CD8+ T cells (IPE Treg) in turn up-regulated expression of their B7 molecules and TGF-β1/TGF-β2. The membrane associated TGF-β may be as important as the soluble form in maintaining the privilege of the eye. Thus, IPE, CBPE, and RPE all secrete soluble active TGF-β but, in the case of IPE, some of the active molecules are functioning on the cell surface. Cells that expressed a dominant/negative TGF-β RII were incapable of being suppressed when exposed to IPE. Histological analysis showed that the surface TGF-β on IPE was uneven and localized to punctate areas on the adjacent T cells where TGF-β RII co-localized.

There are many reports that show that Tregs express surface TGF-β and secrete a soluble form [67,69–74]. But it has also been reported that CD4+ CD25+ Tregs suppress in the absence of TGF-β [75]. Since active soluble TGF-β may be deleterious as well as suppressive and contributes to scarring, a mechanisms whereby the TGF-β may be activated but controlled within punctuate areas on the cells may be a more secure way of allowing suppression on a cell to cell basis [68].

Peripheral CD8+ T Regulatory Cells

Little is known about how the efferent CD8+ Treg cell interferes with T cell effector responses. Unlike the naturally occurring CD4+ CD25+ Treg cells that can be harvested from naïve mice and make up 5–10 % of the lymphocyte population in the spleen of a naïve mouse [76], the antigen specific CD8+ Tregs in the mice with ACAID are far less. For this reason, Keino and colleagues used OTI T cells (OVA TCR transgene for OVA presented in the context of K^b) and generated CD8+ Treg by exposing them to tolerogenic APC [25]. After 72 hours, newly expressed genes in the CD8+ Treg cells were grouped as (1) genes related to synthesis, secretion, activation and receptor binding of TGFβ☐ (2) genes associated with inhibition or loss of NK or CD8+ T effector function; (3) genes that promote preferential localization to non-lymphoid sites or antigen deposition; and (4) genes associated with a resistance to TCR-ligation induced apoptosis. Like natural CD4+ CD25+ Tregs, antigen specific efferent CD8+ Tregs express CD103. Moreover, CD103 null mice were unable to develop CD8+ Treg cells post a.c. inoculation of antigen [25]. In addition, CD8+ Tregs from spleens of mice with ACAID proved to be unable to transfer suppression of delayed hypersensitivity (DH) to OVA-primed mice if the splenic T cells were first depleted of CD103-bearing cells [25]. Furthermore, CD103-depleted ACAID T regulators fail to suppress DH expression in the local adoptive transfer (LAT) assay [25].

There are at least two ways in which CD103 interactions with its E-cadherin ligand could lead to the development of Treg cells. ACAID-mediating APC up-regulate their expression of E-cadherin [25]; thus, the E-cadherin expressed by APC may facilitate interaction with specific T cell [77,78]. Another possibility is that CD103 interactions with E-cadherin may be important for the Treg cell's activity with vascular endothelial cells [58]. Recent papers have shown that CD103 functions to promote the entry or retention of CD8$^+$ cytotoxic T cells into the epithelial compartment of allograft [79,80]. Together these findings lead us to propose the possibility that up-regulation of CD103 on CD8$^+$ Tregs generates a novel functional program that promotes migration of efferent T regulators to the non-lymphoid sites where antigen is deposited, i.e. ear pinnae where OVA is injected to elicit DH. CD103 interactions with E-cadherin may be important for the Treg cell's activity with vascular endothelial cells [58]. Another possibility is that the F4/80 tolerogenic APC up-regulate their expression of E-cadherin [25] so that the it is able to facilitate interaction with specific CD103$^+$ T cell [77,78].

Besides CD103, MMP9 and retinoic acid receptor (RAR) were upregulated 20 fold or more in the eye-induced CD8$^+$ Treg cell. Recently it was reported that vitamin A metabolite, retinoic acid plays a key role in regaling TGF b dependent immune responses. This pathway is capable of inhibiting the IL-6-driven induction of proinflammatory Th17 cells and promoting anti-inflammatory T reg cell differentiation [81,82].

Very little is known about MMP9 in lymphocytes. MMP9 in monocytes and macrophages has been well studied where its function is to allow leukocytes to move through the extracellular matrix to arrive in the tissues and sites of inflammation [83]. Thus the obvious explanation for MMP9's function is that is helps the CD8$^+$ Treg cell to move to the site of immune inflammation to regulation the Th1 or Th2 response. MMP9 also has the capacity to convert latent form to active form of TGFβ [84,85]. Moreover, Retinoic acid induces the activation of latent TGFβ [82]. In this context, MMP9 and RAR could assist in the activation of the latent TGFβ in the Treg cell or intimate vicinity and participate in the down regulation of the CD4$^+$ immune mediated response.

Recent unpublished work in our laboratory (Joan Stein-Streilein) supports the postulate that MMP9 KO mice do not get ACAID. Moreover, CD8 null mice do not generate CD8$^+$ Treg cells unless reconstituted with WT CD8 T cells but not with CD8$^+$ T cells from MMP9 KO mice. These observations support the notion that CD8 Treg cells must express the MMP9. The details of this malfunction are not clear and it remains possible that the CD8$^+$ T cells are generated but not functional since adoptive transfer of the cells to a transferred DH response in the ear of a naïve mice were unable to suppress the immune inflammation. While this could be an inability to activate TGFβ, one cannot rule out that the cells are also not able to migrate through the tissue, an activity well established as a role for MMP9.

It is known that the CD8$^+$ Treg cell produced IL-10 and TGFβ, however blocking antibody studies were unable to interfere with its suppressor function. The CD8$^+$ Treg cell may function through membrane bound TGFβ. Also related to regulatory function is unpublished data from H. Keino's fellowship work and our laboratory which shows that the CD8$^+$ Treg cells do not express CD28 but do express

FOXP3. These observations fit with other descriptions of the $CD8^+$ Treg cell in other models [86].

TGF-β2 is known to be a crucial immunomodulatory factor within the eye [44,45,52]. However, the distinction between TGF-β2 and TGF-β1 becomes blurred since bone marrow cells that are exposed to TGF-β2 produce TGF-β1. TGFβ is a pleiotropic cytokine/growth factor that also has the capacity to suppress aspects of immunity [53–55]. Besides TGF-β2, there are other major factors found in the eye that can directly influence the behavior of T cells [56] (Fig. 25.1). The presence of such factors was first demonstrated by the ability of healthy aqueous humor to suppress IFN-γ production by Th1 cells [55,57].

The posterior surfaces of the iris, the ciliary body, and neural retina are lined with pigment epithelium (PE) that partially surrounds the privileged sites [12] of the anterior chamber [62], the vitreous cavity [63], and the subretinal space [64,65] respectively. Both cultured and freshly prepared PE cells from iris, ciliary body and retina are capable of suppressing TCR-dependent activation and promote the generation of T regulatory cells [48] of naïve and primed T cells in culture [46,47,66]. Sugita and colleagues suggest that the ability of ocular PE to suppress T effector cell activity and to convert activated T cells into regulators is an immune privilege strategy to limit immunogenic inflammation in the eye.

Iris PE cells (IPE) are unique from other PE cells in the eye because unlike ciliary body PE (CBPE) and retina PE (RPE), they use a cell surface contact-dependent mechanism exclusively to suppress T cell activation in vitro [46]. Both fresh and cultured IPE constitutively express B7-1 and B7-2 on their surface and are necessary to convert naïve T cells to T regulatory cells. T cells that are converted are $CD8^+$ and express CTLA4. Soluble factors are also involved in PE suppression, since neutralizing TGF-β antibody allowed for anti CD3 activation of naïve cells even in the presence of IPE in the cultures. A role for membrane bound TGF-β [67] was tested in a subsequent paper [68]. In that report they show that IPE through their co-stimulatory molecules, B7-1 and B7-2, make contact with CTLA-4 on a subpopulation of $CD8^+$ T cells and when engaged IPE membrane associated, active TGF-β is delivered precisely to the T cells. The targeted $CD8^+$ T cells (IPE Treg) in turn up-regulated expression of their B7 molecules and TGF-β1/TGF-β2. The membrane associated TGF-β may be as important as the soluble form in maintaining the privilege of the eye. Thus, IPE, CBPE, and RPE all secrete soluble active TGF-β but, in the case of IPE, some of the active molecules are functioning on the cell surface. Cells that expressed a dominant/negative TGF-β RII were incapable of being suppressed when exposed to IPE. Histological analysis showed that the surface TGF-β on IPE was uneven and localized to punctate areas on the adjacent T cells where TGF-β RII co-localized.

There are many reports that show that Tregs express surface TGF-β and secrete a soluble form [67,69–74]. But it has also been reported that $CD4^+$ $CD25^+$ Tregs suppress in the absence of TGF-β [75]. Since active soluble TGF-β may be deleterious as well as suppressive and contributes to scarring, a mechanisms whereby the TGF-β may be activated but controlled within punctuate areas on the cells may be a more secure way of allowing suppression on a cell to cell basis [68].

Clinical Applications for the Induction of ACAID

It has been reported that ACAID can reverse ongoing immune responses [12,13,87,88]. Faunce, (EAE), and later Zhang-Hoover, (OVA induced hyperreactivity), showed that transfer of in vitro generated tolerogenic APC ameliorated immune mediated disease [14,89]. Transfer of ACAID mechanisms with tolerogenic APC also modulated pulmonary interstitial fibrosis in the mouse [90]. In these studies, it was shown that CD8$^+$ Tregs were generated and could transfer the tolerance to another group of mice. Thus, by transferring the mechanisms that induce tolerance with an ACAID-like APC, the recipient generates its own antigen specific CD8$^+$ Treg cells and the immune inflammatory response is altered. It takes only a few tolerogenic APC to induce the process [31].

While it is generally thought that the antigen specific CD8$^+$ Treg cell in ACAID regulates CD4$^+$ effector cells, a few studies have shown that CD8$^+$ Treg cells negatively regulated antigen presenting cells [91]. Since a major hallmark of the ACAID response is that both the APC and the CD8$^+$ Treg cells are able to suppress ongoing immune responses, eye induced tolerance mechanisms are candidates for cell based regulatory therapy. Importantly, the mechanisms of tolerance induced via the eye leads to antigen specific T cell regulation with essentially no bystander effects.

Acknowledgments The authors thank Ms. Amelia Margolis for her assistance with the preparation of this review. We also thank Peter Mallen for the graphics prepared for this manuscript.
Funding: Research supported by grants from the NIH, EY11983 and EY016476.
Joan Stein-Streilein Patent Pending (#10/468,944)

References

1. Medawar, P. B. 1948. Immunity to homologous grafted skin. III. The fate of skin homografts transplanted to the brain, to subcutaneous tissue and to the anterior chamber of the eye. *Br J Exp Pathol* 29:58–69.
2. Medawar, P. B. 1945. A second study of the behavior and fate of skin homografts in rabbits. (A report to the War Wounds Committee of the Medical Research Council). *J Anat* 79: 157–188.
3. Matzinger, P. 1994. Tolerance, danger, and the extended family. *Annu Rev Immunol* 12: 991–1045.
4. Niederkorn, J. Y. 2006. See no evil, hear no evil, do no evil: the lessons of immune privilege. *Nat Immunol* 7:354–359.
5. Schuppe, H. C., and A. Meinhardt. 2005. Immune privilege and inflammation of the testis. *Chem Immunol Allergy* 88:1–14.
6. Nasr, I. W., Y. Wang, G. Gao, S. Deng, L. Diggs, D. M. Rothstein, G. Tellides, F. G. Lakkis, and Z. Dai. 2005. Testicular immune privilege promotes transplantation tolerance by altering the balance between memory and regulatory T cells. *J Immunol* 174:6161–6168.
7. Dai, Z., I. W. Nasr, M. Reel, S. Deng, L. Diggs, C. P. Larsen, D. M. Rothstein, and F. G. Lakkis. 2005. Impaired recall of CD8 memory T cells in immunologically privileged tissue. *J Immunol* 174:1165–1170.
8. Suter, T., G. Biolaz, D. Gatto, L. Bernasconi, T. Herren, W. Reith, and A. Fontana. 2003. The brain as an immune privileged site: dendritic cells of the central nervous system inhibit T cell activation. *Eur J Immunol* 33:2998–3006.

9. Wenkel, H., J. W. Streilein, and M. J. Young. 2000. Systemic immune deviation in the brain that does not depend on the integrity of the blood-brain barrier. *J Immunol* 164:5125–5131.
10. Ksander, B. R., and J. W. Streilein. 1994. Regulation of the immune response within privileged sites. *Chem Immunol* 58:117–145.
11. Kaplan, H. J., J. W. Streilein, and T. R. Stevens. 1975. Transplantation immunology of the anterior chamber of the eye. II. Immmune response to allogeneic cells. *J Immunol* 118: 809–814.
12. Streilein, J. W. 2003. Ocular immune privilege: therapeutic opportunities from an experiment of nature. *Nat Rev Immunol* 3:878–889.
13. Katagiri, K., J. Zhang-Hoover, J. S. Mo, J. Stein-Streilein, and J. W. Streilein. 2002. Using tolerance induced via the anterior chamber of the eye to inhibit Th2-dependent pulmonary pathology. *J Immunol* 169:84–89.
14. Zhang-Hoover, J., P. Finn, and J. Stein-Streilein. 2005. Modulation of ovalbumin-induced airway inflammation and hyperreactivity by tolerogenic APC. *J Immunol* 175:7117–7124.
15. Stein-Streilein, J. 2005. A privileged view of NKT cells and peripheral tolerance through the eye. *Ocul Immunol Inflamm* 13:111–117.
16. Stein-Streilein, J., and A. W. Taylor. 2007. An eye's view of T regulatory cells. *J Leukoc Biol* 81:593–598.
17. Stein-Streilein, J., and C. Watte. 2007. Cross Talk among Cells Promoting Anterior Chamber-Associated Immune Deviation. *Chem Immunol Allergy* 92:115–130.
18. Zhang-Hoover, J., and J. Stein-Streilein. 2007. Therapies based on principles of ocular immune privilege. *Chem Immunol Allergy* 92:317–327.
19. Niederkorn, J. Y. 2002. Immune privilege in the anterior chamber of the eye. *Crit Rev Immunol* 22:13–46.
20. Faunce, D. E., K.-H. Sonoda, and J. Stein-Streilein. 2001. See reference 6678. *J Immunol* 166: 313–321.
21. Wang, Y., I. Goldschneider, J. O'Rourke, and R. E. Cone. 2001. Blood mononuclear cells induce regulatory NK T thymocytes in anterior chamber-associated immune deviation. *J Leukoc Biol* 69:741–746.
22. Streilein, J. W. 2003. Ocular immune privilege: the eye takes a dim but practical view of immunity and inflammation. *J Leukoc Biol* 74:179–185.
23. Wilbanks, G. A., and J. W. Streilein. 1990. Characterization of suppressor cells in anterior chamber-associated immune deviation (ACAID) induced by soluble antigen. Evidence of two functionally and phenotypically distinct T-suppressor cell populations. *Immunology* 71: 383–389.
24. Keino, H., M. Takeuchi, T. Kezuka, M. Usui, O. Taguchi, J. W. Streilein, and J. Stein-Streilein. 2006. Induction of eye-derived tolerance does not depend on naturally occurring CD4+CD25+ T regulatory cells. *Invest Ophthalmol Vis Sci* 47:1047–1055.
25. Keino, H., S. Masli, S. Sasaki, J. W. Streilein, and J. Stein-Streilein. 2006. CD8* T regulatory cells use a novel genetic program that includes CD103 to suppress Th1 immunity in eye-derived tolerance. *Invest Ohthalmol Vis Sci* 47:1533–1543.
26. Faunce, D. E., and J. Stein-Streilein. 2002. NKT cell-derived RANTES recruits APCs and CD8 + T cells to the spleen during the generation of regulatory T cells in tolerance. *J Immunol* 169:31–38.
27. Sonoda, K.-H., and J. Stein-Streilein. 2002. CD1d on antigen-transporting APC and splenic marginal zone B cells promotes NKT cell-dependent tolerance. *Eur J Immunol* 32:848–857.
28. Lin, H. H., D. E. Faunce, M. Stacey, A. Terajewicz, T. Nakamura, J. Zhang-Hoover, M. Kerley, M. L. Mucenski, S. Gordon, and J. Stein-Streilein. 2005. The macrophage F4/80 receptor is required for the induction of antigen-specific efferent regulatory T cells in peripheral tolerance. *J Exp Med* 201:1615–1625.
29. Roelofs-Haarhuis, K., X. Wu, and E. Gleichmann. 2004. Oral tolerance to nickel requires CD4(+) invariant NKT cells for the infectious spread of tolerance and the induction of specific regulatory T cells. *J Immunol* 173:1043–1050.

30. Dullforce, P. A., K. L. Garman, G. W. Seitz, R. J. Fleischmann, S. M. Crespo, S. R. Planck, D. C. Parker, and J. T. Rosenbaum. 2004. APCs in the anterior uveal tract do not migrate to draining lymph nodes. *J Immunol* 172:6701–6708.
31. Hara, Y., R. R. Caspi, B. Wiggert, M. Dorf, and J. W. Streilein. 1992. Analysis of an in vitro-generated signal that induces systemic immune deviation similar to that elicited by antigen injected into the anterior chamber of the eye. *J Immunol* 149:1531–1538.
32. Sonoda, K.-H., M. Exley, S. Snapper, S. Balk, and J. Stein-Streilein. 1999. CD1 reactive NKT cells are required for development of systemic tolerance through an immune privileged site. *J Exp Med* 190:1215–1225.
33. Nakamura, T., K. H. Sonoda, D. E. Faunce, J. Gumperz, T. Yamamura, S. Miyake, and J. Stein-Streilein. 2003. $CD4^+$ NKT cells, but not conventional $CD4^+$ T cells, are required to generate efferent $CD8^+$ T regulatory cells following antigen inoculation in an immune privileged site. *J Immunol* 171:1266–1271.
34. Skelsey, M. E., E. Mayhew, and J. Y. Niederkorn. 2003. $CD25^+$, interleukin-10-producing $CD4^+$ T cells are required for suppressor cell production and immune privilege in the anterior chamber of the eye. *Immunol* 110:18–29.
35. Meng, Q., P. Yang, B. Li, H. Zhou, X. Huang, L. Zhu, Y. Ren, and A. Kijlstra. 2006. CD4+PD-1+ T cells acting as regulatory cells during the induction of anterior chamber-associated immune deviation. *Invest Ophthalmol Vis Sci* 47:4444–4452.
36. D'Orazio, T. J., and J. Y. Niederkorn. 1996. A novel role for TGF β and IL-10 in the induction of immune privilege. *J Immunol* 160:2089–2098.
37. Wilbanks, G. A., and J. W. Streilein. 1991. Studies on the induction of anterior chamber-associated immune deviation (ACAID). I. Evidence that an antigen-specific, ACAID-inducing, cell-associated signal exists in the peripheral blood. *J Immunol* 146:2610–2617.
38. Wang, Y., I. Goldschneider, D. Foss, D. Y. Wu, J. O'Rourke, and R. E. Cone. 1997. Direct thymic involvement in anterior chamber-associated immune deviation: Evidence for a non-deletional mechanism of centrally induced tolerance to extrathymic antigens in adult mice. *J Immunol* 158:2150–2155.
39. Kosiewicz, M. M., and P. Alard. 2004. Tolerogenic antigen-presenting cells: regulation of the immune response by TGF-beta-treated antigen-presenting cells. *Immunol Res* 30:155–170.
40. Faunce, D. E., K. H. Sonoda, and J. Stein-Streilein. 2001. MIP-2 recruits NKT cells to the spleen during tolerance induction. *J Immunol* 166:313–321.
41. D'Orazio, T. J., and J. Y. Niederkorn. 1998. Splenic B cells are required for tolerogenic antigen presentation in the induction of anterior chamber-associated immune deviation (ACAID). *Immunology* 95:47–55.
42. Sonoda, K. H., and J. Stein-Streilein. 2002. CD1d on antigen-transporting APC and splenic marginal zone B cells promotes NKT cell-dependent tolerance. *Eur J Immunol* 32:848–857.
43. Nowak, M., and J. Stein-Streilein. 2007. Invariant NKT cells and tolerance. *Int Rev Immunol* 26:95–119.
44. Cousins, S. W., M. M. McCabe, D. Danielpour, and J. W. Streilein. 1991. Identification of transforming growth factor-beta as an immunosuppressive factor in aqueous humor. *Invest. Ophthalmol. Vis. Sci* 32:2201–2211.
45. Taylor, A. W. 1999. Ocular immunosuppressive microenvironment. *Chem Immunol* 73:72–89.
46. Sugita, S., and J. W. Streilein. 2003. Iris pigment epithelium expressing CD86 (B7-2) directly suppresses T cell activation in vitro via binding to cytotoxic T lymphocyte-associated antigen 4. *J Exp Med* 198:161–171.
47. Sugita, S., T. F. Ng, J. Schwartzkopff, and J. W. Streilein. 2004. CTLA-4+CD8+ T cells that encounter B7-2+ iris pigment epithelial cells express their own B7-2 to achieve global suppression of T cell activation. *J Immunol* 172:4184–4194.
48. Yoshida, M., T. Kezuka, and J. W. Streilein. 2000. Participation of pigment epithelium of iris and ciliary body in ocular immune privilege. 2. Generation of TGF-beta-producing regulatory T cells. *Invest Ophthalmol Vis Sci* 41:3862–3870.

49. Yoshida, M., M. Takeuchi, and J. W. Streilein. 2000. Participation of pigment epithelium of iris and ciliary body in ocular immune privilege. 1. Inhibition of T-cell activation in vitro by direct cell-to-cell contact. *Invest Ophthalmol Vis Sci* 41:811–821.
50. Bora, N. S., C. L. Gobleman, J. P. Atkinson, J. S. Pepose, and H. J. Kaplan. 1993. Differential expression of the complement regulatory proteins in the human eye. *Invest Ophthalmol Vis Sci* 34:3579–3584.
51. Griffith, T. S., T. Brunner, S. M. Fletcher, D. R. Green, and T. A. Ferguson. 1995. Fas ligand-induced apoptosis as a mechanism of immune privilege. *Science* 270:1189–1192.
52. Streilein, J. W. 1999. Immunoregulatory mechanisms of the eye. *Prog Retin Eye Res* 18:357–370.
53. Kingsley, D. M. 1994. The TGF-beta superfamily: new members, new receptors, and new genetic tests of function in different organisms. *Genes Dev* 8:133–146.
54. Lowrance, J. H., F. X. O'Sullivan, T. E. Caver, W. Waegell, and H. D. Gresham. 1994. Spontaneous elaboration of transforming growth factor beta suppresses host defense against bacterial infection in autoimmune MRL/lpr mice. *J Exp Med* 180:1693–1703.
55. Cousins, S. W., W. B. Trattler, and J. W. Streilein. 1991. Immune privilege and suppression of immunogenic inflammation in the anterior chamber of the eye. *Curr Eye Res* 10: 287 297.
56. Taylor, A. 2003. A review of the influence of aqueous humor on immunity. *Ocul Immunol Inflamm* 11:231–241.
57. Taylor, A. W., J. W. Streilein, and S. W. Cousins. 1992. Identification of alpha-melanocyte stimulating hormone as a potential immunosuppressive factor in aqueous humor. *Current Eye Research* 11:1199–1206.
58. Huehn, J., K. Siegmund, J. C. Lehmann, C. Siewert, U. Haubold, M. Feuerer, G. F. Debes, J. Lauber, O. Frey, G. K. Przybylski, U. Niesner, M. de la Rosa, C. A. Schmidt, R. Brauer, J. Buer, A. Scheffold, and A. Hamann. 2004. Developmental stage, phenotype, and migration distinguish naive- and effector/memory-like CD4+ regulatory T cells. *J Exp Med* 199: 303–313.
59. Goslings, W. R. O., Prodeus, A.P., Streilein, J. W., Carroll, M. C., Jager, M. S., and Taylor, A. W. 1998. A small molecular weight factor in aqueous humor acts on C1q to prevent antibody-dependent complement activation. *Invest Ophthalmol Vis Sci* 39:989–995.
60. Lass, L. H., E. I. Walter, T. E. Burris, H. E. Grossniklaus, M. I. Roat, D. l. Skelnik, L. Needham, M. Singer, and M. E. Medof. 1990. Expression of two molecular forms of the complement decay-accelerating factor in the eye and lacrimal gland. *Invest Opthahalmol Vis Sci* 31:1136–1148.
61. Wilbanks, G. A., and J. W. Streilein. 1990. Characterization of suppressor cells in anterior chamber-associated immune deviation (ACAID) induced by soluble antigen. Evidence of two functionally and phenotypically distinct T-suppressor cell populations. *Immunology* 71: 383–389.
62. Stein-Streilein, J., and J. W. Streilein. 2002. Anterior chamber associated immune deviation (ACAID); regulation, biological relevance, and implications for therapy. *Int Rev Immunol* 21:123–152.
63. Sonoda, K. H., T. Sakamoto, H. Qiao, T. Hisatomi, T. Oshima, C. Tsutsumi-Miyahara, M. Exley, S. P. Balk, M. Taniguchi, and T. Ishibashi. 2005. The analysis of systemic tolerance elicited by antigen inoculation into the vitreous cavity: vitreous cavity-associated immune deviation. *Immunology* 116:390–399.
64. Wenkel, H., Chen, P.W., Ksander, B.R., and Streilein, J. W. 1999. Immune privilge is extended, then withdrawn, from allogeneic tumor cell grafts place in the subretinal space. *Invest Opthahalmol Vis Sci* 40:3203–3208.
65. Streilein, J. W., N. Ma, H. Wenkel, T. F. Ng, and P. Zamiri. 2002. Immunobiology and privilege of neuronal retina and pigment epithelium transplants. *Vision Res* 42:487–495.
66. Ishida, K., N. Panjwani, Z. Cao, and J. W. Streilein. 2003. Participation of pigment epithelium in ocular immune privilege. 3. Epithelia cultured from iris, ciliary body, and retina suppress T-cell activation by partially non-overlapping mechanisms. *Ocul Immunol Inflamm* 11:91–105.

67. Nakamura, K., A. Kitani, and W. Strober. 2001. Cell contact-dependent immunosuppression by CD4+CD25+ regulatory T cells is mediated by cell surface-bound transforming growth factor β. *J Exp Med* 194:629–644.
68. Sugita, S., T. F. Ng, P. J. Lucas, R. E. Gress, and J. W. Streilein. 2006. B7+ iris pigment epithelium induce CD8+ T regulatory cells; both suppress CTLA-4+ T cells. *J Immunol* 176:118–127.
69. Chen, W., and S. M. Wahl. 2003. TGF-beta: the missing link in CD4+CD25+ regulatory T cell-mediated immunosuppression. *Cytokine Growth Factor Rev* 14:85–89.
70. Levings, M. K., R. Sangregorio, C. Sartirana, A. L. Moschin, M. Battaglia, P. C. Orban, and M. G. Roncarolo. 2002. Human CD25+CD4+ T suppressor cell clones produce transforming growth factor beta, but not interleukin 10, and are distinct from type 1 T regulatory cells. *J Exp Med* 196:1335–1346.
71. Nakamura, K., A. Kitani, I. Fuss, A. Pedersen, N. Harada, H. Nawata, and W. Strober. 2004. TGF-beta 1 plays an important role in the mechanism of CD4+CD25+ regulatory T cell activity in both humans and mice. *J Immunol* 172:834–842.
72. Zheng, S. G., J. D. Gray, K. Ohtsuka, S. Yamagiwa, and D. A. Horwitz. 2002. Generation ex vivo of TGF-beta-producing regulatory T cells from CD4+CD25– precursors. *J Immunol* 169:4183–4189.
73. Cosmi, L., F. Liotta, E. Lazzeri, M. Francalanci, R. Angeli, B. Mazzinghi, V. Santarlasci, R. Manetti, V. Vanini, P. Romagnani, E. Maggi, S. Romagnani, and F. Annunziato. 2003. Human CD8+CD25+ thymocytes share phenotypic and functional features with CD4+CD25+ regulatory thymocytes. *Blood* 102:4107–4114.
74. Chen, W., W. Jin, N. Hardegen, K. J. Lei, L. Li, N. Marinos, G. McGrady, and S. M. Wahl. 2003. Conversion of peripheral CD4+CD25– naive T cells to CD4+CD25+ regulatory T cells by TGF-beta induction of transcription factor Foxp3. *J Exp Med* 198:1875–1886.
75. Piccirillo, C. A., J. J. Letterio, A. M. Thornton, R. S. McHugh, M. Mamura, H. Mizuhara, and E. M. Shevach. 2002. CD4(+)CD25(+) regulatory T cells can mediate suppressor function in the absence of transforming growth factor beta1 production and responsiveness. *J Exp Med* 196:237–246.
76. Shevach, E. M. 2002. CD4+CD25+ suppressor T cells: More questions than answers. *Nature Rev Immunol* 2:389–400.
77. Riedl, E., J. Stockl, O. Majdic, C. Scheinecker, K. Rappersberger, W. Knapp, and H. Strobl. 2000. Functional involvement of E-cadherin in TGF-beta 1-induced cell cluster formation of in vitro developing human Langerhans-type dendritic cells. *J Immunol* 165:1381–1386.
78. Higgins, J. M., D. A. Mandlebrot, S. K. Shaw, G. J. Russell, E. A. Murphy, Y. T. Chen, W. J. Nelson, C. M. Parker, and M. B. Brenner. 1998. Direct and regulated interaction of integrin alphaEbeta7 with E-cadherin. *J Cell Biol* 140:197–210.
79. Feng, Y., D. Wang, R. Yuan, C. M. Parker, D. L. Farber, and G. A. Hadley. 2002. CD103 expression is required for destruction of pancreatic islet allografts by CD8(+) T cells. *J Exp Med* 196:877–886.
80. Wang, D., R. Yuan, Y. Feng, R. El-Asady, D. L. Farber, R. E. Gress, P. J. Lucas, and G. A. Hadley. 2004. Regulation of CD103 expression by CD8+ T cells responding to renal allografts. *J Immunol* 172:214–221.
81. Mucida, D., Y. Park, G. Kim, O. Turovskaya, I. Scott, M. Kronenberg, and H. Cheroutre. 2007. Reciprocal TH17 and regulatory T cell differentiation mediated by retinoic acid. *Science* 317:256–260.
82. Imai, S., M. Okuno, H. Moriwaki, Y. Muto, K. Murakami, K. Shudo, Y. Suzuki, and S. Kojima. 1997. 9,13-di-cis-Retinoic acid induces the production of tPA and activation of latent TGF-beta via RAR alpha in a human liver stellate cell line, LI90. *FEBS Lett* 411:102–106.
83. Webster, N. L., and S. M. Crowe. 2006. Matrix metalloproteinases, their production by monocytes and macrophages and their potential role in HIV-related diseases. *J Leukoc Biol* 80:1052–1066.
84. Dallas, S. L., J. L. Rosser, G. R. Mundy, and L. F. Bonewald. 2002. Proteolysis of latent transforming growth factor-beta (TGF-beta)-binding protein-1 by osteoclasts. A cellular mechanism for release of TGF-beta from bone matrix. *J Biol Chem* 277:21352–21360.

85. Yu, Q., and I. Stamenkovic. 2000. Cell surface-localized matrix metalloproteinase-9 proteolytically activates TGF-beta and promotes tumor invasion and angiogenesis. *Genes Dev* 14:163–176.
86. Jiang, L., P. Yang, H. He, B. Li, X. Lin, S. Hou, H. Zhou, X. Huang, and K. Aize. 2007. Increased expression of Foxp3 in splenic CD8+ T cells from mice with anterior chamber-associated immune deviation. *Mol Vis* 13:968–974.
87. Kosiewicz, M. M., S. Okamoto, S. Miki, B. R. Ksander, T. Shimizu, and J. W. Streilein. 1994. Imposing deviant immunity on the presensitized state. *J Immunol* 153:2962–2973.
88. Okamoto, S., M. M. Kosiewicz, R. R. Caspi, and J. W. Streilein. 1994. ACAID as a potential therapy for establishmental autoimmune uveitis. In *Advances in Ocular Immunology*. R. B. Nussenblatt, S. M. Whitcup, R. R. Caspi, and I. Gery, eds. Elsevier Science, Amsterdam. 195–198.
89. Faunce, D. E., A. Terajewicz, and J. Stein-Streilein. 2004. Cutting edge: In vitro-generated tolerogenic APC induce CD8$^+$ T regulatory cells that can suppress ongoing experimental autoimmune encephalomyelitis. *J Immunol* 172:1991–1995.
90. Zhang-Hoover, J., and J. Stein-Streilein. 2004. Tolerogenic APC generate CD8$^+$ T regulatory cells that modulate pulmonary interstitial fibrosis. *J Immunol* 172:178–185.
91. Yokoi, H., and J. W. Streilein. 2004. Antigen-presenting cells are targets of regulatory T cells similar to those that mediate anterior chamber-associated immune deviation. *Ocul Immunol Inflamm* 12:101–114.

Chapter 26
Immune Suppression by a Novel Population of CD8αα+TCRαβ+ Regulatory T cells

Trevor R.F. Smith and Vipin Kumar

Abstract Peripheral tolerance mechanisms involve CD4+CD25+FoxP3+ as well as CD8+ Treg populations, each operating in a distinct manner. Here we have focused on immune suppression mediated by a novel population of CD8αα+TCRαβ+ Treg that target only activated T cells. These Treg are reactive to a peptide derived from the conserved region of the beta chain of the T cell receptor (TCR) utilized by pathogenic T cells and are restricted by a class Ib MHC molecule, Qa-1. Upregulation of Qa-1 on activated T cells for a limited time provides a window of opportunity for the suppression to occur. Furthermore expansion of CD8αα+TCRαβ+ Treg is dependent upon the provision of IL-2 externally and upon activation of dendritic cells. This brief review summarizes some of the important features of CD8αα+TCRαβ+ Treg and highlights their importance in the negative feedback regulation of the immune response.

Introduction

Thymic, or central, tolerance mechanisms purge the body of most T cells reactive to self-antigens, but some potentially pathogenic autoreactive T cells escape the thymus and reach the periphery. These cells are subjected to regulation through peripheral tolerance mechanisms. Activation induced cell death, anergy, and active regulation by regulatory T cells (Treg) all aid in the prevention of autoimmune diseases caused by self-reactive T cells [27,46,34,48]. The induction of active regulation by Treg – that act against pathogenic autoreactive T cells, has become the goal of many immunotherapeutic strategies aimed at treating or preventing autoimmune diseases such as MS, RA and Crohn's disease [18]. Identifying Treg populations

---

V. Kumar
Laboratory of Autoimmunity, Torrey Pines Institute for Molecular Studies, 3550 General Atomics Court, San Diego, CA 92121, USA
e-mail: vkumar@tpims.org

This work was supported by grants from the National Institutes of Health, USA, MSNRC and DNRG to VK.

and delineating their mechanism of regulation is of paramount importance for the success of such therapeutic strategies.

Both CD4+ and CD8+ Treg subsets have been identified. The CD4+ Treg cells have been covered in depth in this book and in many other recent review articles. In this chapter we will focus on a novel population CD8+ Treg that specifically suppress the pathogenic T cells mediating experimental autoimmune disease. CD8+ Treg cells of various phenotypes have been described in many systems of immune regulation [53]. The first in vivo evidence for the involvement of CD8+ T cells in regulation was detected after the immunization of animals with self-reactive attenuated T lymphoblasts, a treatment called T cell vaccination (TCV) [6]. In the graft-versus-host disease model (GVHR), Wilson and colleagues demonstrated that immunization of F1 rats with alloreactive T cells originating from the parental strain could, on transfer, induce T cell responses in the recipient specific to the MHC molecules expressed on the transferred cells. The anti-MHC responses were shown to reside within the cytotoxic T cell population and protect the recipient from GVHD [30]. Vaccination with attenuated myelin basic protein–reactive (MBP-reactive) cloned CD4+ T cell lines prevented MBP-induced experimental autoimmune encephalomyelitis (EAE) in Lewis rats, further suggesting the induction of immunity against the antigen receptors on autoimmune lymphocytes [4]. Cytotoxic CD8+ T cell lines capable of responding to T cells were induced in rats recovering from cell–mediated EAE [52]. In human studies CD8+ T cells isolated from patients vaccinated with MBP-reactive CD4+ T cells specifically lysed the immunizing CD4+ T cells in vitro, and vaccination resulted in a decrease in the frequency of MBP-reactive T cells in the peripheral blood lymphocyte pool [58], although no correlation of these cells with disease activity has been shown.

CD8+ Treg can be induced by other means than TCV. Cells of the phenotype CD8+CD28- can be induced in a mixed lymphocyte reaction (MLR). These cells have the ability to modulate the APC's stimulatory capacity rendering them tolerogenic so that alloreactive CD4+ T cells subsequently interacting with the modulated APC are anergized [11]. Naturally occurring CD8+CD25+ and CD8+CD45RC^{low} Treg cells have been described that display regulatory function akin to their CD4+ counterparts [14,57]. CD8+ Treg that are defined by their expression of PD-1 have recently been described. These cells can be induced in vivo by ICOS-B7h blockade and function to inhibit the expansion of CD4+ T cells in a PD1-dependent manner. The exact target and mechanism of suppression for these Treg has yet to be defined [25]. Here we will focus on a novel population of peripheral CD8+ T cells with regulatory function. We have identified and characterized them as TCRαβ+ CD8αα+, recognizing a peptide derived from the TCR in the context of a class Ib MHC molecule [54].

CD8αα+ Treg

The peripheral CD8αα+ Treg cells that we have characterized both as clones and short-term cell lines are restricted by the non-classical MHC class I molecule Qa-1a, and display reactivity toward a self-antigen derived from the TCR expressed

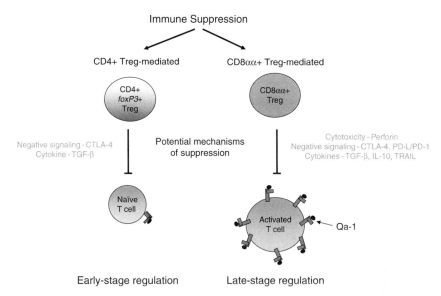

Fig. 26.1 Targets and mechanisms of CD4+CD25+FOXP3+ and CD8αα+TCRαβ+ Treg cell-mediated suppression. See text for details *(See also Color Insert)*

on disease-mediating T cells ([54,60]). These cells are similar in phenotype to the CD8αα+ intra-epithelial lymphocytes (IELs) residing primarily in the gut [12], although it is not yet established whether these two populations originate from a common lineage. In this chapter we will outline one of the important mechanisms of peripheral immune regulation mediated by Qa-1-restricted CD8αα+ Treg cells. The action of Treg can be categorized as specific or non-specific. Bystander regulation mediated by the release immunosuppressive cytokines like IL-10 or TGF-β or Th1/2 counter-inhibitory cytokine profiles is non-specific. Cell-contact dependent regulation is usually associated with some degree of specificity. The Treg we are discussing could act either on APCs to modulate their stimulatory capacity, or directly on the disease-mediating T effector cell. In our system CD8αα+ T cell regulation is specific in that the Treg cell is primed by antigenic determinants derived from the conserved V-region of the beta chain of the effector T cell's TCR. We hypothesized earlier that CD4+CD25+ Treg cells regulate the early immune response; whereas CD8αα+ Treg suppress late in the response and target only a specific activated T effector cell population (see Fig. 26.1).

CD8αα+ Treg and CD8αα+ IEL

CD8αα+ TCRαβ+ cells represent a significant population of intraepithelial lymphocytes (IELs) in mice. CD8αα+ IELs express a repertoire of TCRαβ's that are enriched for self-reactivity, and it appears that their thymic selection and development follows an unconventional pathway [35]. The CD8αα+ co-receptor binds with high affinity to the non-classical MHC class I molecule thymic leukemia

(TL) antigen [36]. Thus these cells can be distinguished from conventional CD8αβ heterodimers (weak TL binders) by the use of fluorescent TL tetramers. The restriction element associated with CD8αα+ IELs is related to a MHC class I molecule, as they are absent in β2-microglobulin-deficient mice [15]. However, in mice lacking the classical MHC class I molecules (K^b/D^b double knockouts) or CD1 molecules, CD8αα+ IEL are still present [40]. This suggests that a non-classical MHC class I restriction molecule may be associated with their development and function. It is not yet clear whether a non-classical MHC class I molecule is required for their thymic selection and expansion in the periphery.

Interestingly, it was observed that in a TCR transgenic NOD mouse, fetal thymic organ cultures produced low numbers CD8αα+ T cells in response to agonist peptides [59]. Further experiments showed that the genetic control elements for the clonal deletion rescue mechanisms of CD8αα+ and CD4+Foxp3+ Treg cells are shared [21]. Due to their connection with FoxP3+ Treg and their paucity in the NOD mouse, one might predict that CD8αα+ T cells may have a role in the downregulation of spontaneous diabetes in NOD mice. However, based upon the available data in the TCR transgenic mice it is not clear whether CD8αα+ T cells contribute to or inhibit the pathogenic response.

There are some important similarities between the peripheral CD8αα+ Treg and CD8αα+ IEL. Functionally, CD8αα+ IEL have been shown to regulate the CD4+ T cell-mediated colitis that occurs in severe combined immunodeficient (SCID) mice [41]. Here, the investigators demonstrated that prior reconstitution of SCID recipients with self-reactive intraintestinal TCRαβ+CD4-CD8α+β- T cells could prevent colitis. The mechanism of protection was IL-10–dependent. In humans it was shown that a population of lamina propria (LP) non-classical MHC-restricted regulatory CD8+ T cells can suppress antibody responses in a cell contact-dependent manner, notably there is a deficit of these cells in the LP of inflammatory bowel disease patients [9]. Recent genotyping of TCR-transgenic CD8αα+ IELs identified upregulation of genes (TGF-β3 and LAG-3) involved in immune regulation in this population [17]. It has also been reported that the Vβ repertoire of CD8αα+ IELs is nonrandom with subtypes including Vβ11 and Vβ6 over expressed [44]. Similarly, we have found preferential use of the Vβ6 TCR in CD8αα+ Treg populations inhibiting EAE in the B10.PL mouse. [55]. We are currently investigating whether CD8αα+ Treg preferentially utilize particular V-beta regions in other models of autoimmune disease.

Qa-1 and CD8αα+ Treg

The class Ib MHC molecules Qa-1 (mouse) and HLA-E (human) are considered functional equivalents based on their ability to act as a ligand for the CD94/NKG2A receptor complex on natural killer cells (or classical CD8+ T cells) and to bind class I leader sequence-derived peptides [8]. It has been almost 30 years since Qa-1 was first identified (by serological means) as a key molecule in the function of CD8+ suppressor cells [10]. However, the molecular characterization of the Qa-1 role in

CD8+ T cell-mediated immune down-regulation has only recently been revealed [23]. Here investigators demonstrated that Qa-1-restricted cytotoxic CD8+ T cells targeted Qa-1 molecules up-regulated on effector CD4+ T cells, since in the absence of Qa-1 molecules, or when Qa-1 fails to express the cognate Ag, CD4+ T cells could not be killed [39]. On the other hand, Qa-1-restricted CD8+ T cells have been reported to be important in priming the CD4+ immune responses [13]. Here mice were injected with DCs coated with a CD4+ T cell cognate listeriolysin O peptide (LLO), with or without the CD8+ T cell cognate Qa-1 peptide GroEL. Results showed that priming of the LLO-specific CD4+ T cell response was significantly enhanced in the mice receiving DCs containing both peptides. Thus, in this model Qa-1-restricted CD8+ T cells appear to be important in augmenting CD4+ T cell priming. The mechanism exploited here by the CD8+ T cell to enhance CD4+ T cell priming has yet to be determined. It is possible that the CD8+ T cells modulate the DC, secrete cytokines (e.g. IFN-γ), or act directly through a cognate molecule mechanism (e.g. Qa-1) on the CD4+ T cell.

Compared to the repertoire presented by classical MHC class Ia molecules, the number of peptides identified that bind to Qa-1 and are presented to CD8+ T cells is small. This may partly be due to the instability of the Qa-1 peptide complex [28]. The dissociation rate of Qdm with Qa-1 is at least 10-fold faster than that of OVA and H2k^b. The Qa-1 determinant modifier (Qdm) is a hydrophobic peptide derived from the class Ia leader sequence and is a strong binding peptide for Qa-1. Other Qa-1 binding molecules include peptides derived from the TCR, a heat-shock 60, insulin, and Salmonella GroEL [16,54,51,37]. Our identification of a TCR-derived antigenic determinant that binds to Qa-1 provides the first direct evidence that Qa-1 restricted Tregs target TCR regions for down-regulation [54,55]. After immunizing H2^u mice with overlapping 9-10-mer peptides derived from a conserved region of the Vβ8.2 TCR and analyzing the CD8+ T cell recall response the TCRVβ8.2-derived p42-50 peptide was identified as a target epitope. Binding of p42-50 with Qa-1a was confirmed using immunological, genetic, and biochemical assays: Qdm but not Class Ia MHC-binding molecules were able to compete with p42-50 in direct biochemical binding with Qa-1a molecules. The binding affinity of the p42-50 peptide, akin to the insulin peptide [56], was found to be lower than that for Qdm. However, p42-50 is able to compete successfully at 2 fold higher molar concentration than Qdm. Thus it is likely that high expression of Treg activating Qa-1 peptides is required to suppress immunity.

Priming of CD8αα+ Treg

Through TAP-independent mechanisms many cell types can present exogenous ligands in the context of Qa-1 [56]. We have begun to decipher the mechanism by which professional antigen presenting cells can present exogenous peptide to Qa-1-restricted CD8αα+ Treg [60]. As stated above, CD8αα+ Treg react to peptide derived from the Vβ8.2 TCR. Our experiments indicate that a Qa-1/TCR-peptide complex is expressed on the surface of activated CD4+ T cells and that this complex is the target for CD8αα+ T cell regulation. It is predicted that upon TCR ligation the CD4+ Vβ8.2+ T cells internalize their TCRs, which then are either degraded in lysosomes or enter the proteaosome pathway. TCR peptide fragments may be recycled and enter the Qa-1 presentation pathway. These cells would then become targets for effector CD8+ T cell-mediated killing by the recognition of a Qa-1-TCR peptide complex. However, T cells lack the essential levels of co-stimulatory molecules that would be necessary for efficient priming of these Treg. We now have evidence that professional antigen presenting cells provide this "help". It is widely recognized that dendritic cells can phagocytose apoptotic cells, including T and B cells, and cross-present the antigens derived from these cells in the context of MHC class I and II [1,24,45]. Ronchetti and colleagues demonstrated in vivo that CTLs reactive to an antigenic determinant associated with T cell lymphoma could be cross-primed by injected DCs loaded with apoptotic tumor cells [45]. We hypothesize that during the normal cell turnover, and at an increased rate during the EAE inflammatory response, apoptotic TCR Vβ8.2+ T cells are captured by the host's DC population, and the TCR-derived antigenic determinants are processed and presented on the cell surface of the DC. In in vitro mixing assays with apoptotic cell-loaded DCs have demonstrated their ability to prime both CD4+ and CD8αα+ TCR-reactive Treg clones. Optimal Treg priming was seen after treatment of the apoptotic cell-loaded DCs with Toll-like receptor agonists (Smith et al., in preparation).

To date, little is known concerning how exogenous cellular antigens are cross-presented in the context of Qa-1. Previous studies have shown the Qa-1 presentation of insulin as a soluble exogenous antigen was not significantly altered in TAP-deficient mice, thus implying that a cytosolic independent pathway was involved [56]. Chloroquine, however, significantly inhibits antigen presentation. Together these results indicate an endosomal pathway is functioning in the presentation of soluble insulin by Qa-1. We have demonstrated that the Qa-1 presentation of TCR peptide p42 derived from apoptotic T cells is sensitive to the effects of inhibitors of endosomal protease activity and proteasome degradation, but inhibitors of MHC egress from the endoplasmic reticulum have no effect. We therefore predict that TCR-derived antigenic determinants are not loaded onto Qa-1 molecules in the ER and possibly proceed along a novel pathway. In this pathway, after proteolytic degradation, the peptide fragments may be retro-transported back into the phagosome. It has been reported, but remains controversial, that the phagosome can acquire all the machinery for successful presentation of MHC class I peptides [20,22].

CD8αα+ Treg Mechanism of Suppression

The induction of CD8αα+ Treg-mediated suppression appears to be dependent on the presence of Th1-like cytokines at the time of priming. We have shown that under Th-2-deviating conditions the priming of anti-TCR Treg is hampered [32,7], and for apoptotic killing of Vβ8.2+TCR CD4+ T cells IFN-γ-producing CD4+ Treg cells are needed [38]. The depletion of high-avidity activated Th1 cells by the combined action of CD4+ and CD8+ Treg allows expansion of the slower reacting compartment of low-avidity, MBP-reactive Th2 cells; This results in immune deviation towards a non-pathogenic anti-MBP response [33,32,7]. We are currently investigating the mechanism of killing, as discussed below. It appears that the mechanism of regulation exploited by CD8αα+ IEL and CD8αα+ Treg may differ. Here will summarize the mechanisms of suppression that have been defined for CD8+ Treg cells.

Cytokines

Earlier data suggest that CD8αα+ IELs down-regulate the immune response in a bystander fashion, utilizing cytokines such as IL-10 and/or TGF-β [41, 17]. However, characterization of our CD8αα+ Treg has failed to detect increased IL-10 or TGF-β secretion, the cytokine profile revealed a Tc1 phenotype ([54]; unpublished data). We are also examining the involvement of another immunosuppressive cytokine, TRAIL, secreted by CD8+ T cells in some circumstances and shown to mediate the down-regulation of CD4+ T cell immunity [19].

Negative Signaling

Inhibitory cell surface molecules (B7 family members CTLA-4, PD-1) have been shown to be critical for regulating T cell activation in models of autoimmunity and transplantation [29]. CD4+CD25+ and CD8+CD25+ Treg cells constitutively express CTLA-4, and CTLA-4 blockade has been shown to inhibit Treg function in a model of intestinal inflammation [43,47,5]. PD-1 expression has recently been reported to be over-expressed on functionally impaired/exhausted CD8+ T cells [3]. As mentioned previously CD8+ Treg cells induced in vivo by ICOS-B7h blockade inhibit the expansion of CD4+ T cells in a PD1-dependent manner [25]. Additionally, PD-1 and CTLA-4 have been reported as crucial molecules for peripheral CD8+ T cell tolerance induced by resting dendritic cells [42]. Our recent data suggest that some of these members of the TNFR/TNF-related molecules are expressed on CD8αα+ Treg populations (Unpublished data).

Killing of the Target Effectors

Our data indicate that CD8αα+ Treg regulate immune response in a direct and specific manner by recognition of Qa-1/TCR peptide complexes on activated T cells, leading to their killing [54]. We and others have in vitro and in vivo data suggesting that CD8+ Treg preferentially target Th-1 cells [26,31]. Following regulation and the killing of the pathogenic Vβ8.2 TCR+ T cell population we have observed that the immune response to MBP is deviated in a type 2 direction. This suggests that the CD8αα+ Treg population is targeting only the activated Vβ8.2+ Th1 cells. Recently, we have found that the Fas/Fas Ligand pathway is dispensable for

CD8αα+ Treg-associated killing (Beeston et al. manuscript in preparation). This finding is supported by a report by Badovinac and colleagues who show that Fas or Fas Ligand knockout mice display a normal contraction phase of CD8+ T cells after activation; however, in perforin knockout mice this contraction phase is absent [2]. Additionally, familial lymphohistiocytosis, characterized by uncontrolled activation of T cells, has been associated with a perforin gene defect [50]. These observations strongly suggest that it is the perforin pathway that is involved in CD8αα+ Treg-mediated regulation.

Summary

We are currently investigating the means by which CD8αα+ Treg cells differentially target Qa-1-expressing T cells. Of particular interest are the dynamics of Qa-1 and other class I molecule expression on the surface of activated T cells [54]. Efficient CD8αα+ Treg effector function appears to be dependent on two events: first, a high level of stimulation provided by an APC and the presence of CD4+ Treg; and second, increased levels of Qa-1 expression on the target T cell. Both resting TCR+ Vβ8.2 T cells and Vβ8.2+ hybridomas express low levels of Qa-1 and are not targeted for regulation. We are also investigating whether CD8αα+ Treg only see self-peptides in the context of Qa1 or whether they might also recognize peptides presented by Class Ia molecules.

CD8αα+ Treg-mediated regulation targets the late phase of the immune response, once significant expansion of the effector T cell population has already occurred. Therefore, this is more a form of homeostatic or feedback regulation limiting the over-expansion of effector T cells, rather than complete suppression of an immune response. It should be noted that CD8αα+ Treg exist in a very low numbers in the periphery even during a regulatory episode, and not much is known about their persistence. We predict the low numbers of peripheral CD8αα+ Treg will not hinder the generation of subsequent TCR Vβ-restricted primary immune responses. This is important when considering the need for the generation of multiple protective responses driven by a TCR Vβ-restricted T cell repertoire throughout an animal's lifespan. Unlike the non-specific regulation provided by CD4+CD25+FoxP3+ Treg cells, there is only a window of time (due to transient Qa-1 expression on target CD4+ T cells) when the antigen specific CD8αα+ Treg can act to dampen the immune response (Fig. 26.1).

References

1. Albert, M. L., B. Sauter, and N. Bhardwaj. 1998. Dendritic cells acquire antigen from apoptotic cells and induce class I-restricted CTLs. *Nature* 392:86.
2. Badovinac, V. P., A. R. Tvinnereim, and J. T. Harty. 2000. Regulation of antigen-specific CD8+ T cell homeostasis by perforin and interferon-gamma. *Science* 290:1354.

3. Barber, D. L., E. J. Wherry, D. Masopust, B. Zhu, J. P. Allison, A. H. Sharpe, G. J. Freeman, and R. Ahmed. 2006. Restoring function in exhausted CD8 T cells during chronic viral infection. *Nature* 439:682.
4. Ben-Nun, A., H. Wekerle, and I. R. Cohen. 1981. Vaccination against autoimmune encephalomyelitis with T-lymphocyte line cells reactive against myelin basic protein. *Nature* 292:60.
5. Bienvenu, B., B. Martin, C. Auffray, C. Cordier, C. Becourt, and B. Lucas. 2005. Peripheral CD8+CD25+ T lymphocytes from MHC class II-deficient mice exhibit regulatory activity. *J Immunol* 175:246.
6. Binz, H., and H. Wigzell. 1976. Specific transplantation tolerance induced by autoimmunization against the individual's own, naturally occurring idiotypic, antigen-binding receptors. *J Exp Med* 144:1438.
7. Braciak, T. A., B. Pedersen, J. Chin, C. Hsiao, E. S. Ward, I. Maricic, A. Jahng, F. L. Graham, J. Gauldie, E. E. Sercarz, and V. Kumar. 2003. Protection against experimental autoimmune encephalomyelitis generated by a recombinant adenovirus vector expressing the V beta 8.2 TCR is disrupted by coadministration with vectors expressing either IL-4 or -10. *J Immunol* 170:765.
8. Braud, V. M., D. S. Allan, C. A. O'Callaghan, K. Soderstrom, A. D'Andrea, G. S. Ogg, S. Lazetic, N. T. Young, J. I. Bell, J. H. Phillips, L. L. Lanier, and A. J. McMichael. 1998. HLA-E binds to natural killer cell receptors CD94/NKG2A, B and C. *Nature* 391:795.
9. Brimnes, J., M. Allez, I. Dotan, L. Shao, A. Nakazawa, and L. Mayer. 2005. Defects in CD8+ regulatory T cells in the lamina propria of patients with inflammatory bowel disease. *J Immunol* 174:5814.
10. Cantor, H., J. Hugenberger, L. Vay-Boudreau, D. D. Eardley, J. Kemp, F. W. Shen, and R. K. Gershon. 1978. Immunoregulatory circuits among T-cell sets. Identification of a subpopulation of T-helper cells that induces feedback inhibition. *J Exp Med* 148:871.
11. Chang, C. C., R. Ciubotariu, J. S. Manavalan, J. Yuan, A. I. Colovai, F. Piazza, S. Lederman, M. Colonna, R. Cortesini, R. la-Favera, and N. Suciu-Foca. 2002. Tolerization of dendritic cells by T(S) cells: the crucial role of inhibitory receptors ILT3 and ILT4. *Nat Immunol* 3:237.
12. Cheroutre, H. 2005. IELs: enforcing law and order in the court of the intestinal epithelium. *Immunol Rev* 206:114.
13. Chow, M. T., S. Dhanji, J. Cross, P. Johnson, and H. S. Teh. 2006. H2-M3-restricted T cells participate in the priming of antigen-specific CD4+ T cells. *J Immunol* 177:5098.
14. Cosmi, L., F. Liotta, R. Angeli, B. Mazzinghi, V. Santarlasci, R. Manetti, L. Lasagni, V. Vanini, P. Romagnani, E. Maggi, F. Annunziato, and S. Romagnani. 2004. Th2 cells are less susceptible than Th1 cells to the suppressive activity of CD25+ regulatory thymocytes because of their responsiveness to different cytokines. *Blood* 103:3117.
15. Das, G., and C. A. Janeway, Jr. 1999. Development of CD8alpha/alpha and CD8alpha/beta T cells in major histocompatibility complex class I-deficient mice. *J Exp Med* 190:881.
16. Davies, A., S. Kalb, B. Liang, C. J. Aldrich, F. A. Lemonnier, H. Jiang, R. Cotter, and M. J. Soloski. 2003. A peptide from heat shock protein 60 is the dominant peptide bound to Qa-1 in the absence of the MHC class Ia leader sequence peptide Qdm. *J Immunol* 170:5027.
17. Denning, T. L., S. Granger, D. Mucida, R. Graddy, G. Leclercq, W. Zhang, K. Honey, J. P. Rasmussen, H. Cheroutre, A. Y. Rudensky, and M. Kroncnbcrg. 2007. Mouse TCRalphabeta+CD8alphaalpha intraepithelial lymphocytes express genes that down-regulate their antigen reactivity and suppress immune responses. *J Immunol* 178:4230.
18. Feldmann, M., and L. Steinman. 2005. Design of effective immunotherapy for human autoimmunity. *Nature* 435:612.
19. Griffith, T. S., H. Kazama, R. L. VanOosten, J. K. Earle, Jr., J. M. Herndon, D. R. Green, and T. A. Ferguson. 2007. Apoptotic cells induce tolerance by generating helpless CD8+ T cells that produce TRAIL. *J Immunol* 178:2679.

20. Guermonprez, P., L. Saveanu, M. Kleijmeer, J. Davoust, E. P. van, and S. Amigorena. 2003. ER-phagosome fusion defines an MHC class I cross-presentation compartment in dendritic cells. *Nature* 425:397.
21. Holler, P. D., T. Yamagata, W. Jiang, M. Feuerer, C. Benoist, and D. Mathis. 2007. The same genomic region conditions clonal deletion and clonal deviation to the CD8alphaalpha and regulatory T cell lineages in NOD versus C57BL/6 mice. *Proc Natl Acad Sci USA* 104:7187.
22. Houde, M., S. Bertholet, E. Gagnon, S. Brunet, G. Goyette, A. Laplante, M. F. Princiotta, P. Thibault, D. Sacks, and M. Desjardins. 2003. Phagosomes are competent organelles for antigen cross-presentation. *Nature* 425:402.
23. Hu, D., K. Ikizawa, L. Lu, M. E. Sanchirico, M. L. Shinohara, and H. Cantor. 2004. Analysis of regulatory CD8 T cells in Qa-1-deficient mice. *Nat Immunol* 5:516.
24. Inaba, K., S. Turley, F. Yamaide, T. Iyoda, K. Mahnke, M. Inaba, M. Pack, M. Subklewe, B. Sauter, D. Sheff, M. Albert, N. Bhardwaj, I. Mellman, and R. M. Steinman. 1998. Efficient presentation of phagocytosed cellular fragments on the major histocompatibility complex class II products of dendritic cells. *J Exp Med* 188:2163.
25. Izawa, A., K. Yamaura, M. J. Albin, M. Jurewicz, K. Tanaka, M. R. Clarkson, T. Ueno, A. Habicht, G. J. Freeman, H. Yagita, R. Abdi, T. Pearson, D. L. Greiner, M. H. Sayegh, and N. Najafian. 2007. A Novel Alloantigen-Specific CD8+PD1+ Regulatory T Cell Induced by ICOS-B7h Blockade In Vivo. *J Immunol* 179:786.
26. Jiang, H., N. S. Braunstein, B. Yu, R. Winchester, and L. Chess. 2001. CD8+ T cells control the TH phenotype of MBP-reactive CD4+ T cells in EAE mice. *Proc Natl Acad Sci USA* 98:6301.
27. Kabelitz, D., T. Pohl, and K. Pechhold. 1993. Activation-induced cell death (apoptosis) of mature peripheral T lymphocytes. *Immunol Today* 14:338.
28. Kambayashi, T., J. R. Kraft-Leavy, J. G. Dauner, B. A. Sullivan, O. Laur, and P. E. Jensen. 2004. The nonclassical MHC class I molecule Qa-1 forms unstable peptide complexes. *J Immunol* 172:1661.
29. Keir, M. E., and A. H. Sharpe. 2005. The B7/CD28 costimulatory family in autoimmunity. *Immunol Rev* 204:128.
30. Kimura, H., and D. B. Wilson. 1984. Anti-idiotypic cytotoxic T cells in rats with graft-versus-host disease. *Nature* 308:463.
31. Kumar, V., and E. Sercarz. 1998. Induction or protection from experimental autoimmune encephalomyelitis depends on the cytokine secretion profile of TCR peptide-specific regulatory CD4 T cells. *J Immunol* 161:6585.
32. Kumar, V., J. Maglione, J. Thatte, B. Pederson, E. Sercarz, and E. S. Ward. 2001. Induction of a type 1 regulatory CD4 T cell response following V beta 8.2 DNA vaccination results in immune deviation and protection from experimental autoimmune encephalomyelitis. *Int Immunol* 13:835.
33. Kumar, V., and E. Sercarz. 2001. An integrative model of regulation centered on recognition of TCR peptide/MHC complexes. *Immunol Rev* 182:113.
34. Kumar, V. 2004. Homeostatic control of immunity by TCR peptide-specific Tregs. *J Clin Invest* 114:1222.
35. Lambolez, F., M. Kronenberg, and H. Cheroutre. 2007. Thymic differentiation of TCR alpha beta(+) CD8 alpha alpha(+) IELs. *Immunol Rev* 215:178.
36. Leishman, A. J., O. V. Naidenko, A. Attinger, F. Koning, C. J. Lena, Y. Xiong, H. C. Chang, E. Reinherz, M. Kronenberg, and H. Cheroutre. 2001. T cell responses modulated through interaction between CD8alphaalpha and the nonclassical MHC class I molecule, TL. *Science* 294:1936.
37. Lo, W. F., A. S. Woods, A. DeCloux, R. J. Cotter, E. S. Metcalf, and M. J. Soloski. 2000. Molecular mimicry mediated by MHC class Ib molecules after infection with gram-negative pathogens. *Nat Med* 6:215.

38. Madakamutil, L. T., I. Maricic, E. Sercarz, and V. Kumar. 2003. Regulatory T cells control autoimmunity in vivo by inducing apoptotic depletion of activated pathogenic lymphocytes. *J Immunol* 170:2985.
39. Panoutsakopoulou, V., K. M. Huster, N. McCarty, E. Feinberg, R. Wang, K. W. Wucherpfennig, and H. Cantor. 2004. Suppression of autoimmune disease after vaccination with autoreactive T cells that express Qa-1 peptide complexes. *J Clin Invest* 113:1218.
40. Park, S. H., D. Guy-Grand, F. A. Lemonnier, C. R. Wang, A. Bendelac, and B. Jabri. 1999. Selection and expansion of CD8alpha/alpha(1) T cell receptor alpha/beta(1) intestinal intraepithelial lymphocytes in the absence of both classical major histocompatibility complex class I and nonclassical CD1 molecules. *J Exp Med* 190:885.
41. Poussier, P., T. Ning, D. Banerjee, and M. Julius. 2002. A unique subset of self-specific intraintestinal T cells maintains gut integrity. *J Exp Med* 195:1491.
42. Probst, H. C., K. McCoy, T. Okazaki, T. Honjo, and B. M. van den. 2005. Resting dendritic cells induce peripheral CD8+ T cell tolerance through PD-1 and CTLA-4. *Nat Immunol* 6:280.
43. Read, S., V. Malmstrom, and F. Powrie. 2000. Cytotoxic T lymphocyte-associated antigen 4 plays an essential role in the function of CD25(+)CD4(+) regulatory cells that control intestinal inflammation. *J Exp Med* 192:295.
44. Rocha, B., P. Vassalli, and D. Guy-Grand. 1991. The V beta repertoire of mouse gut homodimeric alpha CD8+ intraepithelial T cell receptor alpha/beta + lymphocytes reveals a major extrathymic pathway of T cell differentiation. *J Exp Med* 173:483.
45. Ronchetti, A., P. Rovere, G. Iezzi, G. Galati, S. Heltai, M. P. Protti, M. P. Garancini, A. A. Manfredi, C. Rugarli, and M. Bellone. 1999. Immunogenicity of apoptotic cells in vivo: role of antigen load, antigen-presenting cells, and cytokines. *J Immunol* 163:130.
46. Schwartz, R. H. 2003. T cell anergy. *Annu Rev Immunol* 21:305.
47. Shevach, E. M. 2002. CD4+ CD25+ suppressor T cells: more questions than answers. *Nat Rev Immunol* 2:389.
48. Smith, T. R. F., X. Tang, and V. Kumar. 2007. Priming regulatory T cells and antigen-specific suppression of autoimmune disease. *Immune Regulation and Immunotherapy in Autoimmune Disease. Springer Science; 20–25.*
49. Smith, T. R. F., and V. Kumar. 2007. A novel population of CD8αα+ regulatory T cells: Identification, characterization and induction. T cell vaccination. Eds., Drs. Zhang and Cohen, Nova Sci
50. Stepp, S. E., R. Dufourcq-Lagelouse, D. F. Le, S. Bhawan, S. Certain, P. A. Mathew, J. I. Henter, M. Bennett, A. Fischer, B. G. de Saint, and V. Kumar. 1999. Perforin gene defects in familial hemophagocytic lymphohistiocytosis. *Science* 286:1957.
51. Sullivan, B. A., P. Kraj, D. A. Weber, L. Ignatowicz, and P. E. Jensen. 2002. Positive selection of a Qa-1-restricted T cell receptor with specificity for insulin. *Immunity* 17:95.
52. Sun, D., Y. Qin, J. Chluba, J. T. Epplen, and H. Wekerle. 1988. Suppression of experimentally induced autoimmune encephalomyelitis by cytolytic T-T cell interactions. *Nature* 332:843.
53. Tang, X. L., T. R. Smith, and V. Kumar. 2005. Specific control of immunity by regulatory CD8 T cells. *Cell Mol Immunol* 2:11.
54. Tang, X., I. Maricic, N. Purohit, B. Bakamjian, L. M. Reed-Loisel, T. Beeston, P. Jensen, and V. Kumar. 2006. Regulation of immunity by a novel population of Qa-1-restricted CD8alphaalpha+TCRalphabeta+ T cells. *J Immunol* 177:7645.
55. Tang, X., I. Maricic, and V. Kumar. 2007. Anti-TCR antibody treatment activates a novel population of nonintestinal CD8 alpha alpha+ TCR alpha beta+ regulatory T cells and prevents experimental autoimmune encephalomyelitis. *J Immunol* 178:6043.
56. Tompkins, S. M., J. R. Kraft, C. T. Dao, M. J. Soloski, and P. E. Jensen. 1998. Transporters associated with antigen processing (TAP)-independent presentation of soluble insulin to alpha/beta T cells by the class Ib gene product, Qa-1(b). *J Exp Med* 188:961.
57. Xystrakis, E., A. S. Dejean, I. Bernard, P. Druet, R. Liblau, D. Gonzalez-Dunia, and A. Saoudi. 2004. Identification of a novel natural regulatory CD8 T-cell subset and analysis of its mechanism of regulation. *Blood* 104:3294.

58. Zhang, J., R. Medaer, P. Stinissen, D. Hafler, and J. Raus. 1993. MHC-restricted depletion of human myelin basic protein-reactive T cells by T cell vaccination. *Science* 261:1451.
59. Zucchelli, S., P. Holler, T. Yamagata, M. Roy, C. Benoist, and D. Mathis. 2005. Defective central tolerance induction in NOD mice: genomics and genetics. *Immunity* 22:385.
60. Smith TR, Kumar V. Revival of CD8(+) Treg-mediated suppression. Trends Immunol. 2008 Jul; 29:337.

Chapter 27
Innate Regulatory iNKT Cells

Dalam Ly and Terry L. Delovitch

Abstract The complexity of the immune system requires mechanisms, cellular and molecular, that coordinate early innate immune responses to those of late adaptive immune responses. One type of immune cell that is able to mediate this bridge between innate immunity and adaptive immunity is the invariant natural killer T (iNKT) cell. In recent years, much research has been focused on describing the role of iNKT cells in a variety of immune responses, from pathogen clearance, cancer immunity, to autoimmune regulation. In each of these immune conditions, iNKT cells have been shown to play direct or indirect roles in the coordinating immune responses leading to downstream effector activation. In this review, we highlight our current understanding of iNKT cell biology, and provide an overview of iNKT cell antigen specificities and of the role of iNKT cells in regulating immune responses.

Introduction

The innate immune system has evolved many conserved mechanisms in host defense, including the ability to recognize pathogen-associated molecular patterns (PAMPs) to initiate adaptive immunological responses towards foreign pathogens. Among the cells that encode recognition receptors are macrophages, dendritic cells (DC), mast cells, neutrophils, eosinophils, and natural killer (NK) cells, most of which become activated during an inflammatory response that can rid the host of an infection without the need of adaptive immunity. However, there are instances where the innate immune system alone is unable to deal with an infection and must activate the adaptive immune system [1]. To help orchestrate signals between innate and adaptive immune responses, various costimulatory molecules and regulatory cell responses have evolved to ensure proper immune activation that does not

---

T.L. Delovitch
Laboratory of Autoimmune Diabetes, Robarts Research Institute; Department of Microbiology and Immunology, University of Western Ontario; FOCIS Centre for Clinical Immunology and Immunotherapeutics, London, ON N6A5K8, Canada
e-mail: del@robarts.ca

lead to host autoimmunity. One of the more recently defined regulatory cells that help orchestrate signals between innate and adaptive immunity are a group of cells collectively termed natural killer T (NKT) cells.

Originally described as an $\alpha\beta$ T cell receptor (TCR)$^+$ T cell that expresses the NK1.1 marker usually associated with NK cells, NKT cells have more recently been better categorized according to our growing understanding of this T cell subset. As many commonly used mouse strains (e.g. BALB/c, CBA, and NOD) do not express NK1.1, and because CD1d- (an MHC class-I like molecule) dependent and -independent subsets of NK1.1$^+$ T cells exist, NKT cells are now classified into the Type I (classical; invariant; CD1d-dependent), Type II (non-classical; non-invariant; CD1d-dependent) and Type III NKT-like (NK1.1$^+$, CD1d-independent) cell subsets [2]. This review will focus on the most widely studied subset of NKT cells, the CD1d-restricted Type I classical NKT cell or invariant (i) NKT cell, and will summarize our current understanding of how these innate regulatory cells may help to orchestrate adaptive immune responses in health and disease.

Defining iNKT Cells and Natural Antigen Specificity

Murine iNKT cells express an invariant Vα14-Jα18 (Vα24-Jα18 in humans) TCR rearrangement that preferentially pairs with a Vβ2, Vβ7, or Vβ8.2 (Vβ11 in humans) chain. This invariant TCR rearrangement allows for the recognition of glycolipid antigen presented in the context of the conserved MHC class-I like molecule CD1d. Unlike MHC molecules which present peptide fragments, the antigen-binding site of CD1d consists of highly hydrophobic pockets that accommodate hydrocarbon chains of lipids, and allow for the polar head group of the glycolipid to be recognized by the invariant TCR [3]. The majority of murine iNKT cells are CD4$^+$ or double negative (CD4$^-$CD8$^-$), whereas in humans, some iNKT cells express CD8α. In mice, iNKT cells are found at the highest frequency in the liver (10–40% of liver lymphocytes), but are also found at lower frequencies (typically <1%) in the spleen, lymph nodes, thymus, and blood. As mentioned previously, in addition to surface TCR, iNKT cells also express a wide assortment of NK cell-associated receptors (NKRs) encoded on the NK-gene complex (NKC) on mouse chromosome 6. These include members of the Nkrp1 (NK1.1), Ly-49 (Ly-49a), and NKG2D families [4]. Furthermore, iNKT cells constitutively express markers associated with an activated/memory phenotype such as CD69, CD44, and CD122. The remarkable characteristic of iNKT cells are their ability, when activated, to release large amounts of immunomodulatory cytokines such as IL-4 and IFN-γ, which mediate their effector functions [5–9].

Classically, iNKT cells are defined by their ability to react to the glycosphingolipid α-galactosylceramide (α-GalCer). Originally isolated from the marine sponge *Agelas mauritianus*, this compound was shown to potently activate murine iNKT cells [10]. It is now clear that α-GalCer and its synthetic equivalent KRN7000 (Kirin Breweries, Japan) bind CD1d molecules from mouse and humans and stimulate iNKT cells from either species in a highly conserved manner [11]. Moreover, the use of α-GalCer loaded CD1d tetramers is the most accepted way to identify iNKT cells by flow cytometry [12,13]. The attractiveness of α-GalCer is its ability to potently activate iNKT cells, which can elicit a wide range of immunomodulatory activities that may be exploited for therapeutic purposes (see below). Interestingly, α-anomeric glycosphingolipids are absent in mammalian cells, and thus α-GalCer is probably not a physiological antigen for iNKT cells [3].

Recent advances have been made in the identification of natural iNKT cell antigens. These natural antigens are derived from either the host (endogenous) or foreign pathogens (exogenous). The search for endogenous antigens began with studies demonstrating that natural glycosylphosphatidylinositols bind to CD1d [14,15], but their purpose was not understood in the scheme of CD1d antigen presentation, as only in rare circumstances could they stimulate iNKT cells. Currently, it is believed that these endogenous lipids may function similarly to the invariant chain-derived CLIP peptide for MHC class II mediated antigen presentation. Thus, natural glyco- and phospholipids may reside in the CD1d binding pocket until these ligands are exchanged in late endocytic compartments with glycolipids that when presented on the APC cell surface can stimulate iNKT cells [16,17].

Supportive evidence for this process was revealed by the demonstration that mice deficient in the saposin family of lipid transfer proteins are also deficient in iNKT cells [18,19], suggesting that this deficiency results in the failure to load an endogenous thymic selecting antigen onto CD1d molecules for presentation. Shortly thereafter, it was shown that isoglobotrihexosylceramide (iGb3), a mammalian self glycolipid, was indeed required for positive selection of iNKT cells [20]. In mice, the enzyme β-hexosaminidase B is required to cleave N-acetylgalactosamine groups from glycolipids and is involved in the degradation of globosides and isoglobosides, specifically converting Gb4 to Gb3 and iGb4 to iGb3. Thus, mice deficient in β-hexosaminidase B (Hexb$^{-/-}$) do not express the iGb3 product and cannot positively select iNKT cells in the thymus for their differentiation. This results in a numerical and functional deficiency of iNKT cells in these mice. Interestingly, the finding that synthetic iGb3 activates iNKT cells [20] led to the observation that LPS from *Salmonella typhimurium*, though not directly recognized by iNKT cells, can activate iNKT cells indirectly possibly by the upregulation of iGb3 expression by antigen presenting cells (APCs). This was further supported by transfers of iNKT cells into Hexb$^{-/-}$ mice which did not express iGb3, in these mice iNKT cell responses to LPS were reduced [21,22]. Hence, iNKT cells may be recruited by APC to sites of inflammation via the upregulation of self-glycolipids to aid in immune responses like pathogen clearance, cancer immunity or autoimmune regulation.

Several exogenous lipids derived from bacterial species have also been shown to directly bind CD1d and activate iNKT cells. It follows that iNKT cells may have evolved not only to indirectly respond to pathogens, but also to directly recognize bacterial glycolipid products bound to CD1d. Previously, it was shown that mice injected with de-proteinized cell walls prepared from *Mycobacterium tuberculosis* developed granuloma-like lesions in which NKT cells were the predominant infiltrate [23]. It was later shown that the cell wall constituent responsible for iNKT cells activation was phosphatidylinositol mannosides (PIM), which can bind directly to CD1d and activate iNKT cells to secrete IFN-γ *in vivo* [24,25]. Another example of iNKT cell activation by bacterial glycolipids emerged from three independent groups. iNKT cells were found to directly recognize glycosphingolipids from *Sphingomonas*, a gram-negative bacteria found ubiquitously in the environment [21,26,27]. Interestingly, this gram-negative bacteria lacks LPS, and provides an example of the unique innate-type ability of iNKT cells to recognize antigen that cannot be detected by other pattern recognition receptors such as toll-like receptor (TLR) 4. Thus, the iNKT cell receptor, when presented with the proper ligand in the context of CD1d, more closely resembles pattern recognition receptors expressed by innate immune cells, rather than the diverse antigen-specific receptors of the adaptive immune system. In particular, the α-glucuronosylceramide (GSL-1) glycolipid binds CD1d and activates iNKT cells *in vivo*. More importantly, bacterial clearance is dependent on iNKT cell activity, demonstrating that iNKT cells can directly recognize microbial antigens and contribute to host defense against gram-negative, LPS negative microorganisms [21,26].

Until recently, it was questioned why such a recognition pattern was evolutionarily maintained, as *Sphingomonas* are not highly pathogenic. A recent report revealed that iNKT cells also directly recognize glycolipids from *Borrelia burgdorferi*, a gram-negative, LPS negative bacteria that is the causative agent of Lyme disease [28]. Based on the knowledge that iNKT cell deficient mice injected with *Borrelia burgdorferi* result in a higher bacterial burden and increased thickening of the tibiotarsal joint indicative of arthritis [29], it was subsequently shown that iNKT cells may be directly activated by glycolipids of *Borrelia burgdorferi*. Mice injected with live *Borrelia burgdorferi* stimulate iNKT cells in a TLR independent manner, and the diacylglycerol 1,2-diacyl-3-*O*-α-galactosyl-*sn*-glycerol (BbGL-II) from the bacteria directly bind CD1d and stimulate iNKT cells *in vivo* and *in vitro* [28]. These results demonstrate that iNKT cells can play direct roles in the recognition and effector function of host defense against bacterial pathogens.

Regulation of iNKT Cell Polarization by Synthetic Ligand Analogs

A hallmark of iNKT cells is their ability to secrete large amounts of immunomodulatory cytokines in a CD1d-dependent manner soon after activation. This rapid cytokine release in turn transactivates several leukocyte subsets, including macrophages, DC, NK cells, B cells, and T cells. The ability of iNKT cells to interact with other cell subsets is one of the most important aspects of iNKT cell function and one of the most appealing aspects for use of iNKT cells in therapeutic settings.

Nonetheless, a potential downfall of these observations is that most studies use α-GalCer as the activating stimulus. Although α-GalCer is a rather potent stimulus it is not a physiological antigen for iNKT cells, and hence α-GalCer stimulation may not reflect the normal physiological outcome of iNKT cell activation. Notwithstanding this caveat, the use of α-GalCer to activate iNKT cells has provided a valuable structural and functional resource for not only advancing our understanding of iNKT cell biology but also refining the development of therapeutic strategies in several disease models.

α-GalCer was originally identified during a screen for anti-metastatic agents, and it was not until another screen of glycolipid antigens was performed using splenocytes from Vα14+ transgenic mice that revealed it as specific ligand for iNKT cells [8,10]. These studies provided a specific antigen for the activation of iNKT cells, and together with the development of α-GalCer/CD1d tetramers provided a way to track and study these cells *in vivo* [12,13].

Numerous studies have shown that α-GalCer activation of iNKT cells results in the rapid secretion of prototypical Th1 cytokines (IFN-γ, TNF-α) important for responses against tumors and pathogens, and Th2 cytokines (IL-4, IL-10, IL-13) that promote protection against autoimmunity within hours after injection [6,7,9]. The rapid production of cytokines and upregulation of CD40L on iNKT cells results in the activation of DC, NK cells, and other cell types, an ability reminiscent of innate immune cells. A functional consequence is that iNKT cells can suppress certain immune responses usually via the release of immunosuppressive cytokines, such as IL-10. In contrast, during other types of immune responses, iNKT cells can promote cell-mediated immunity via the production of Th1 type cytokines [6,9]. Thus, activated iNKT cells can both prevent autoimmunity by downregulating autoaggressive immune responses and promote cell-mediated rejection of cancers and pathogens.

Despite being capable of Th1- and Th2-type cytokine production, the mechanism of polarization of iNKT cells towards a Th-1 or Th2-type immune response is unclear and remains a key challenge in the field of iNKT cell research. Factors that influence this polarization of iNKT cells include the timing, dosing, and context in which a glycolipid ligand is presented. For example, IL-4 is detectable in the serum of mice as early as 2 hours post-injection of α-GalCer, and IFN-γ then predominates in the serum at 6–24 hours after injection. Upon multiple injections and chronic exposure to α-GalCer, iNKT cells are polarized towards Th2 cytokine secretion [30,31], with very little IFN-γ production upon re-challenge. This may be due to the finding that α-GalCer may induce the functional inactivation or anergy of iNKT cells [32,33]. After a single dose of α-GalCer, iNKT cells remain unresponsive to this antigen upon re-challenge. Unresponsive iNKT cells continue to produce low levels of IL-4, but IFN-γ secretion is strongly blunted. As with conventional T cells, this iNKT cell anergy may be broken upon exposure to exogenous IL-2 [32,33]. The context in which α-GalCer (or other glycolipids) is presented also modulates iNKT cell responses. Thus, α-GalCer loaded DC activate iNKT cells to secrete predominantly IFN-γ rather than IL-4 [34,35]. In agreement with this finding, mice depleted of DC have reduced iNKT cell responses towards α-GalCer, particularly with regard to IFN-γ production [36]. Furthermore, while α-GalCer-loaded DC elicit a robust iNKT cell response accompanied by IFN-γ and low IL-4 secretion, α-GalCer-loaded B cells elicit reduced responses with a bias towards IL-4 cytokine production [36]. These results may explain why variable responses to α-GalCer stimulation are observed, especially following the systemic administration of α-GalCer. Accordingly, depending on the subset of APCs that present the glycolipid, different responses may ensue.

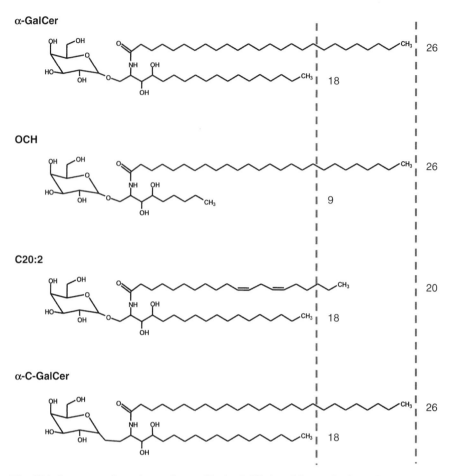

Fig. 27.1 Structure of α-galactosylceramide (α-GalCer) and its synthetic analogs Structures of α-GalCer and its analogs are discussed in the text

One approach to overcome this variability associated with α-GalCer induced responses has involved the generation of analogs (Fig. 27.1) that can "bias" an iNKT cell response towards Th2 cytokine secretion [37]. Truncation of the sphingosine chain of α-GalCer yields the synthetic analog OCH, which has proven more efficient than α-GalCer in the prevention and treatment of experimental autoimmune encephalitis (EAE). This was due to the ability of OCH to preferentially induce a Th2 biased cytokine response by iNKT cells [37]. OCH also prevents the spontaneous development of type 1 diabetes (T1D) in nonobese diabetic (NOD) mice. Although the incidence of T1D is the same after α-GalCer or OCH, OCH does reduce the severity of insulitis [38]. The half-life for iNKT cell stimulation is shorter for OCH than α-GalCer due to its less-stable association with CD1d. Accordingly, while the induction of IFN-γ production by iNKT cells requires a long lasting interaction beween glycolipid and CD1d, a short-lived glycolipid/CD1d interaction is sufficient to induce IL-4 secretion [39].

Another approach to achieving Th2-biased responses has resorted to a modification of the acyl chain in α-GalCer, in which this chain was reduced from C26 to C20 residues and two unsaturated bonds were introduced. This modification gave rise to an analog termed C20:2, which elicits such a Th2-biased response [40]. Interestingly, as the majority of the α-GalCer molecule was left intact in C20:2 compared to OCH, C20:2 and α-GalCer are essentially equivalent in their ability to activate iNKT cells as they both yield similar elevated levels of IL-2 and IL-4 secretion and a decreased level of IFN-γ secretion. In addition, C20:2 and α-GalCer possess a similar binding affinity for CD1d [40]. Nonetheless, differences between the two glycolipids are detectable downstream of iNKT cell activation, as C20:2 is less effective than α-GalCer in its capacity to either induce IFN-γ secretion by NK cells or stimulate CD69 expression by activated B cells. Because C20:2 can activate iNKT cells with similar potency to that of α-GalCer, it is believed that C20:2 may preferentially induce IL-4 secretion *in vivo* because it fails to optimally induce secondary NK cell activation and consequent IFN-γ secretion. Thus, C20:2 may function via a different mechanism of action than that of OCH. However, as C20:2 binds to CD1d with higher affinity than OCH [40], activation of iNKT cells by C20:2 may be more efficacious therapeutically as C20:2 can be recognized by a larger iNKT cell population that is not limited to only high affinity iNKT cell subsets [40].

Given that some α-GalCer analogs can polarize iNKT cells towards Th2 cytokine responses, other investigators have sought to design an analog that promotes Th1 biased responses. Substitution of the O-glycosidic linkage of α-GalCer by CH2 resulted in a compound termed α–C-galactosylceramide (α-C-GalCer), which prolongs IFN-γ and IL-12 production and decreases IL-4 production relative to α-GalCer [41]. With these alterations, α-C-GalCer acquires increased anti-malarial and anti-metastatic activity compared to α-GalCer, and these activities depend on the ability of the glycolipids to induce downstream IFN-γ secretion by NK cells [41]. The latter increase in Th1-type activity is attributable to the greater binding capacity of α-C-GalCer than α-GalCer for CD1d on DC, which leads to longer lasting iNKT cell responses *in vivo* [42].

iNKT Cells and Host Defense Against Infection

iNKT cells can regulate immune responses towards a variety of bacterial, viral, and parasitic infections [43,44] by various mechanisms. In some cases, iNKT cells can directly recognize glycolipid antigen derived from bacteria and mount specific responses against the infection. In other instances, mice that lack NKT cells are generally more susceptible to infection by certain pathogens, suggesting that iNKT cells may have a natural role in either direct recognition of antigen or indirect recognition by aiding the recruitment and activation of innate cells. Studies have also shown that α-GalCer induced activation of iNKT cells can improve disease outcome, which may not demonstrate a physiological role for iNKT cells but rather highlight their therapeutic potential.

In addition to the ability of iNKT cells to recognize and respond to glycolipid antigens from *Mycobacterium tuberculosis* [23–25], *Sphingomonas* [21,26,27], and *Borrelia burgdorferi* [28,29], iNKT cells play a critical role in the clearance of an acute infection in a murine model of *Pseudomonas aeruginosa*-induced pneumonia [45]. Mice deficient in CD1d-restricted iNKT cells show reduced eradication of *P. aeruginosa*, which is associated with lower levels of macrophage inflammatory protein-2 (MIP-2) and reduced numbers of neutrophils in bronchoalveolar lavage fluids. In the same study, the authors treated mice with α-GalCer prior to infection and found enhanced bacterial clearance associated with increased recruitment of macrophages to the lung. Although it is unknown whether iNKT cells can directly recognize *P. aeruginosa* glycolipids, these findings raise the possibility that iNKT cell activation leads to the recruitment of innate immune cells involved in clearance.

A role for iNKT cells in viral infection has also been demonstrated. CD1d deficient mice are more susceptible to herpes simplex virus type 1 (HSV-1) infection, develop larger skin lesions and are impaired in their viral clearance compared to wild-type littermates [46]. Immune evasion by HSV-1 can be achieved by inhibition of the transport of CD1d to the cell surface, thereby inhibiting the ability to activate CD1d-restricted iNKT cells and highlighting the importance of this subset in viral clearance [47,48]. In a mouse model of murine cytomegalovirus (MCMV) infection, the viral load does not differ between iNKT cell deficient Jα281$^{-/-}$ mice and wild-type mice, suggesting that iNKT cells may not be involved in the natural clearance of MCMV infection [49]. Nonetheless, α-GalCer treated mice showed increased MCMV clearance that was dependent on enhanced expression of IFN-γ by NK cells, as this α-GalCer induced effect was abrogated in IFN-γ$^{-/-}$ mice and in mice treated with an anti-asialo-GM1 antibody (depletes NK cells). These results demonstrate the therapeutic potential of α-GalCer.

Current evidence suggests that viruses may not express antigenic glycolipids recognized by iNKT cells, suggesting that activation of iNKT cells during viral infection may occur through self-glycolipids. The subset of plasmacytoid DC (pDC) that express TLR-7 and -9, but not CD1d, mediate the recognition of viral PAMPs such as CpG DNA. Upon activation via TLR-9 during viral infection, pDC rapidly produce type I interferons (IFN-α/β) that result in the maturation of conventional myeloid DC (mDC) as well as activation of NK and T cells involved in viral clearance [50,51]. Similarly, iNKT cell activation results in many of the same innate responses, e.g. mDC maturation and activation of downstream effector functions. As iNKT cells do not express TLR-9, the mechanism of how iNKT cells mediate TLR-9 induced activation is presently unclear. CpG signaling through TLR-9 can increase iNKT cell responsiveness to α-GalCer, as measured by increased IFN-γ secretion and CD69 expression [52]. Analyses of purified of human iNKT cells, pDC, and conventional mDC in co-cultures have demonstrated that CpG oligodeoxynucleotides can "license" iNKT cells through soluble factors (possibly IL-15) released by pDC [53]. This "licensing" allows iNKT cells to recognize glycolipid antigen in the context of CD1d expressed on mDC and mature them allowing for downstream activation of antiviral clearance mechanisms. Thus, iNKT cells through indirect signals from pDC can mediate crosstalk between pDC and mDC, which express mutually exclusive TLR and cytokine responses. Although the specific self-glycolipid recognized by iNKT cells was not identified, these results highlight the role iNKT cells may play in orchestrating signals between innate and adaptive immune cells to aid in the clearance of viral targets.

The role of iNKT cells during parasitic infections is less clear, and varies according to type of infection, extent of Th1 or Th2 polarization, and mouse host strain. iNKT cells promote the required Th1-type responses to combat *Leishmania major* infection, and C57BL/6 mice deficient in iNKT cells are more susceptible to *L. major* infection than their resistant wild-type counterparts [54]. The host mouse strain and iNKT cell response is also particularly important during malarial infection. Infection with *Plasmodium* parasites involves two stages. Infectious *Plasmodium* sporozoites infect the host and travel through the blood to infect hepatocytes, where they undergo asexual division to produce merozoites. Once hepatocytes are lysed, merozoites are able to invade erythrocytes and begin a new infectious cycle of asexual division in the blood stage [44]. Immune responses differ in the liver stage and blood stage, with the liver response generally being clinically silent and short lived. A strong Th1 dependency results in anti-malarial activity. Therefore, mice administered α-GalCer are protected against a *Plasmodium yoelii* liver stage infection, which is dependent on IFN-γ responses and iNKT cells [55]. Pathogenesis during *Plasmodium* infection is largely mediated during the asexual blood stages of infection - both humans and experimental animal models suffer symptoms, with the most severe complication being anaemia and cerebral malaria. Cerebral malaria results in the adherence of parasitized red blood cells to the vascular endothelium that blocks capillaries to the brain, a process amplified by pro-inflammatory cytokines (e.g TNF-α, IFN-γ) that upregulate the expression of adherence molecules such as ICAM-1 [44].

The role iNKT cells play in this stage of *Plasmodium* infection in experimental mouse models is dependent on the genetic background. The disease progression of *Plasmodium bergheri* parasitized erythrocyte infection has been tested using wild-type and $CD1d^{-/-}$ mice in disease-resistant BALB/c and disease-susceptible C57BL/6 (B6) mice [56]. iNKT cells protect BALB/c mice from infection, as $CD1d^{-/-}$ mice on this background are susceptible to disease. In contrast, iNKT cells promote disease in C57BL/6 mice susceptible mice, as $CD1d^{-/-}$ and $J\alpha281^{-/-}$ mice on the B6 background deficient in iNKT cells are protected against cerebral malaria. Thus, whereas in BALB/c mice iNKT cells may promote a Th2 response and downregulate the activity of pathogenic pro-inflammatory cytokines, in C57/BL6 mice iNKT cells promote IFN-γ responses that favor disease pathogenesis. It is possible that different molecules of the NK-gene complex (NKC) are expressed on iNKT cells in BALB/c and C57BL/6 mice. Support for this is provided by studies with NKC congenic strains, in which BALB/c.B6-Cmv1 mice that carry a B6 derived NKC are susceptible to cerebral malaria in BALB/c mice that are naturally resistant. Furthermore, these mice display increased IFN-γ secretion after challenge with α-GalCer [56].

iNKT Cells and Cancer

Much of what is understood about the role of iNKT cells in tumor surveillance and immunity comes from studies using α-GalCer. Activation of iNKT cells by α-GalCer can result in direct cytolysis of tumor cell targets [10] or indirectly activate downstream innate and adaptor effector mechanisms of tumor immunity. Early studies demonstrated that iNKT cells are necessary for IL-12-mediated tumor therapy in mice [57], and considerable recent evidence suggests that iNKT cells may not need to directly kill tumor targets *in vivo* but instead recruit or promote downstream effectors. TCR and CD28 dependent iNKT-DC interactions elicit increased expression of CD40L by iNKT cells. CD40L interacts with CD40 on DC and results in the maturation/activation of DC to produce IL-12. IL-12, IL-15, and IL-2 produced by DC can activate iNKT cells to secrete IFN-γ, which promotes NK cell activation and enhanced IFN-γ production as well as $CD8^+$ CTL responses against a broad range of tumors [58–60]. The pivotal role of IFN-γ in iNKT cell mediated tumor responses to α-GalCer has been highlighted [61,62]. Consistent with this role is the evidence that the C-glycoside analog of α-GalCer, α-C-GalCer, which can prolong IFN-γ responses, is more potent at preventing metastases of the B16 melanoma [41,42].

Central to iNKT cell activation is the ability of iNKT cells to translate innate activation responses to adaptive immune responses through interactions with DC. Activation of iNKT cells with α-GalCer can functionally mature DC capable of inducing adaptive immune responses to a neo-antigen (e.g. ovalbumin). Mice given α-GalCer concurrently with soluble ovalbumin can mount specific Th1 $CD4^+$ and $CD8^+$ T cell responses, and reject ovalbumin-expressing tumors [63,64]. These responses are dependent on the maturation of and priming by DC and are also dependent on CD40 ligation. iNKT cells from mice deficient in CD40 and CD40L can not prime T cells even though upregulation of MHC and CD80/86 results from α-GalCer treatment [63,64]. These results demonstrate that iNKT cells activated by α-GalCer or certain analogs can promote adaptive immune responses against tumor antigens in a manner dependent on DC maturation.

The physiological role of iNKT cells in tumor rejection has been directly explored in mouse models deficient in iNKT cells. $J\alpha 281^{-/-}$ mice deficient in iNKT cells are more susceptible to the formation of tumors induced by the carcinogen methylcholanthrene (MCA) [65,66]. Sarcomas form with much higher incidence in these mice than their wild-type counterparts, but the adoptive transfer of iNKT cells can block tumor development. Furthermore, iNKT cells can lyse MCA-induced tumor cell targets derived from $J\alpha 281^{-/-}$ mice, indicating that iNKT cells may play a role in natural tumor surveillance. As iNKT cells secrete IL-4 and induce Th2 responses, this suggested role in tumor surveillance is supported by the finding that specific subsets of iNKT cells possess different anti-tumor capacities [67]. Liver- but neither thymus- nor spleen-derived iNKT cells can inhibit the growth of an MCA-induced tumor in $J\alpha 281^{-/-}$ mice, and interestingly the $CD4^-$ fraction of liver-derived iNKT cells was most efficient at tumor rejection. Since the amounts of intracellular IFN-γ and IL-4 from $CD4^+$ and $CD4^-$ liver- derived iNKT cells were similar after α-GalCer stimulation, differences in effector cytokines may not account for the differences in function.

Given that α-GalCer stimulates iNKT cell-dependent tumor immunity in mice and that activated human Vα24 iNKT cells are cytotoxic against tumor targets *in vitro* [68–70], several clinical trials were performed to determine the efficacy of iNKT cell activation against tumors in patients. A phase I clinical trial conducted in patients with solid tumors administered α-GalCer intravenously demonstrated α-GalCer to be well tolerated over a wide dose range. The number of iNKT cells in the peripheral blood of these cancer patients was significantly less than in healthy controls. Although only those patients with normal iNKT cell numbers showed signs of iNKT cell activation, this did not result in clinical responses, indicating the clinical safety of α-GalCer [71]. The finding that α-GalCer-pulsed DC promotes increased IFN-γ responses and anti-tumor activity in mice [34,35] prompted several phase I trials to determine the safety and efficacy of α-GalCer pulsed autologous DC transfer [72–74]. Re-infusion (i.v.) of autologous DC loaded with α-GalCer into patients revealed this therapy to be well tolerated and effective at expanding iNKT cells *in vivo* and stimulating increased levels in serum IFN-γ. Furthermore, T cells and NK cells were transactivated as monitored by CD69 expression, and NK cell cytotoxicity against tumor targets was enhanced in 5 of 11 patients [72]. In another recent phase I trial conducted in patients with non-small lung cancer re-infused intravenously with autologous peripheral blood mononuclear cells (PBMC) previously expanded *in vitro* for several weeks with α-GalCer and IL-2, no adverse side effects were observed in any of the patients and several patients showed increases in PBMC-derived iNKT cells and serum IFN-γ [75].

iNKT Cells and Autoimmune Disease

Evidence from studies of many experimental autoimmune diseases in mice reveals that glycoplipid activated iNKT cells can either protect from or promote disease. These outcomes differ widely from the defined suppressive role of $CD4^+CD25^+$ regulatory T cells (T_{regs}) in the pathogenesis of autoimmune disease [76–78]. Thus, a more complete understanding of iNKT cell biology and function in autoimmune disease is essential for the design of improved, novel, safe and effective iNKT cell-based therapeutics.

Our understanding of the role of iNKT cells in autoimmune disease derives largely from studies of the pathogenesis of type 1 diabetes (T1D). T1D results from a T cell mediated destruction of insulin producing β-cells of the islets of Langerhans in the pancreas. Analyses of non-obese diabetic (NOD) mice, which spontaneously develop T1D with some similarities to the immunopathological profile seen in human T1D [79,80], have identified a numerical and functional deficiency in iNKT cells in NOD mice [9,81,82]. Transgenic overexpression of the Vα14-Jα18 rearrangement protects NOD mice from T1D [83]. In NOD mice, activated iNKT cells produce less IL-4 after stimulation when compared to iNKT cells in non-autoimmune prone strains [84–88]. Therapeutically, this can be overcome by repeated administration of α-GalCer or its analogs [38,84–88]. Nonetheless, our understanding of the mechanism of iNKT cell mediated protection regarding the nature of interacting cell subset(s) and the iNKT cell derived cytokines require further experimentation.

Our analyses of NOD mice deficient in IL-4 and IL-10 expression demonstrate that α-GalCer activated iNKT cells transfer protection against T1D in an IL-4-dependent and IL-10-independent manner [89]. However, other groups have not found any differences in the incidence of spontaneous T1D in α-GalCer treated NOD and NOD.IL-4$^{-/-}$ mice [90,91]. The discrepancy in these data may derive in part from the transfer vs. spontaneous models of T1D used.

The ability of α-GalCer stimulated iNKT cells to transactivate T cells, B cells, NK cells and DC, is compatible with an earlier report that α-GalCer induces the activation and recruitment of tolerogenic CD11c$^+$CD8$^-$ DC to pancreatic lymph nodes (PLN) where these DC can tolerize islet β cell autoreactive T cells [92]. More recent evidence raises the possibility that soluble factors, other than IL-4, IL-10, IFN-γ, IL-13, TGF-β, and TNF-α from iNKT cells activated upon repeated administration of α-GalCer, may induce the maturation of DC and create a tolerogenic environment capable of preventing T1D induced by the transfer of NOD islet autoantigen reactive TCR transgenic AI4 CD8$^+$ T cells [90]. AI4 T cells transferred into α-GalCer treated NOD recipients proliferate in the PLN and induce tolerance rather than disease. Alternatively, iNKT cells themselves may directly inhibit autoreactive T cells by a contact-dependent mechanism similar to that observed for CD4$^+$CD25$^+$ T$_{regs}$ in vitro. In this regard, it is noteworthy that NOD transgenic BDC2.5 islet β cell autoantigen reactive T cells did not transfer T1D to Vα14 Cα$^{-/-}$ NOD recipient mice that are enriched in iNKT cells but did transfer T1D to Cα$^{-/-}$ NOD recipients that lack both iNKT and conventional T cells [93]. The differentiation of BDC2.5 T cells into IFN-γ producing cells was blocked in the recipient mice enriched in iNKT cells. Thus, as noted for AI4 autoreactive T cells, the proliferation but not activation of BDC2.5 T cells was observed which led to tolerance rather than disease. In addition, while NOD.β2m$^{-/-}$ CD4$^+$ T cells that are devoid of iNKT cells do not prevent the transfer of T1D by BDC2.5 T cells, wildtype CD4$^+$ T cells that do contain NKT cells can prevent the transfer of T1D in an IFN-γ dependent manner [94]. When examined in vitro, transwell co-cultures of Vα14 Cα$^{-/-}$ NOD T cells enriched for iNKT cells with BDC2.5 T cells revealed that iNKT cells require direct cell contact with BDC2.5 T cells to inhibit their differentiation [91]. Furthermore, this contact-dependent mechanism appears to occur independent of IL-4, IL-10, IL-13, and TGF-β, as blocking antibodies to these cytokines did not inhibit iNKT cell mediated suppression of BDC2.5 T cells. Taken together, these studies indicate that iNKT cell activation requires contact with CD1d expressing APC, such that activation of iNKT cells may mature CD1d-expressing APC towards a tolerogenic rather than activating phenotype.

Interestingly, the mechanism of contact-dependent interaction between iNKT cells and APC is reminiscent of that suggested for CD4$^+$CD25$^+$ T$_{reg}$-DC interactions. Two-photon imaging studies illustrate that T$_{regs}$ interact directly with antigen-specific DC rather than BDC2.5 autoreactive effector T cells in vivo, and T$_{regs}$

thereby modulate the ability of the effector T cells to be activated and elicit T1D [95,96]. A similar mechanism of protection from T1D may apply to tolerogenic iNKT-DC interactions. Notwithstanding, another model of T1D induced by adoptive transfer of islet autoantigen specific CD8$^+$ T cells demonstrated that iNKT cells when present in high frequency can enhance IFN-γ –dependent CD8$^+$ T cell effector function and exacerbate rather than protect from disease [97]. Presumably, the presence of disease-activating DC rather than tolerogenic DC explains the latter findings.

Thus far, the reports on the role of iNKT cells in the control of human T1D in patients have been controversial. Patients with T1D were shown to have a reduced frequency of Vα24 iNKT cells compared to their non-diabetic identical twin, and NKT cell clones isolated from diabetic individuals secrete only IFN-γ while clones from non-diabetic individuals can secrete IFN-γ and IL-4 [98]. These data identified numerical and function deficiencies iNKT cells similar to that described for iNKT cells from NOD mice. Other studies have not found this correlation [99,100], and in fact found even higher numbers of iNKT cells in diabetic patients [99]. The discrepancy could be due to the intrinsic high variability in the number of iNKT cells between patients as well as differences in the method of detection of iNKT cells between the studies [100]. Also, the deficiencies in iNKT cells observed in NOD mice derive mainly from observations within tissue rather than peripheral blood [101]. Nevertheless, one study recently showed that iNKT cell clones from the PLN of diabetic patients do indeed have reduced IL-4 secretion, however, this study awaits independent verification [102].

In experimental autoimmune encephalomyelitis (EAE), a mouse model of multiple sclerosis (MS), induced by immunizing susceptible strains with adjuvant and myelin-derived antigens (such as MBP, or MOG), iNKT cells have been also shown to play conflicting roles. Results using α-GalCer therapy to prevent disease have had mixed results, with some studies able to prevent disease [103–105], while others demonstrating exacerbation of disease [103]. These discrepancies maybe due to the timing of doses, as activation of iNKT cells simultaneously with autoreactive T cells results in exacerbation of disease, while prior activation of iNKT cell protects [103]. As mentioned above, an approach to overcome this discrepancy can involve the use of the analog OCH, which when administered simultaneously with autoantigen protects against disease [37] unlike α-GalCer which exacerbates disease [103]. Transgenic mice that overexpress iNKT cells when given MOG and adjuvant are resistant to EAE induction, suggesting that iNKT cells may play a natural role in the prevention of EAE [106]. As with T1D, the role of cytokines in iNKT cell responses in EAE is also controversial, with studies demonstrating dependent [37,103,104] and independent [105,106] roles for IL-4, and requirements for IFN-γ [105]. Clinically, CD4$^+$ Vα24 iNKT cell lines from the peripheral blood of MS patients in remission or relapse have been compared to healthy control subjects. iNKT cells from patients in remission have increased secretion of IL-4 when stimulated with α-GalCer as compared to patients in relapse or healthy controls, which suggests a regulatory role for iNKT cells in MS [107].

Another autoimmune disease mediated by regulatory iNKT cells is systemic lupus erythematosus (SLE). SLE is characterized by autoantibodies directed against a variety of self-antigens, including double stranded DNA, which ultimately leads to multi-organ failure. Development of SLE is associated with the generation of autoreactive Th1 cells and a reduction in regulatory T cells [108]. Similar to the situation in T1D and EAE/MS, animals and human patients with SLE seem to have a reduced frequency and function of iNKT cells [109–111]. Various mouse models of the disease including spontaneously developing MRL-FAS$^{lpr/lpr}$ mice and (NZB × NZW) F1 mice develop specific organ failure that can be regulated by iNKT cells. For example, while a CD1d deficiency exacerbates inflammatory dermatitis in MRL-FAS$^{lpr/lpr}$ mice [112], treatment with repeated doses of α-GalCer protects against inflammatory dermatitis in these mice [113]. On the other hand, in (NZB × NZW) F1 mice that develop severe glomerulonephritis, iNKT cells contribute to the pathogenesis of the disease. iNKT cell activation is increased during disease progression [114] and can exacerbate lupus symptoms that result in increased anti-dsDNA antibodies [115]. Thus, iNKT cell activation can both potentiate and suppress the development of T1D, EAE and SLE in experimental animal models of these autoimmune diseases, depending on the mouse strain, autoantigen, dose and time of antigen administration.

CD4$^+$CD25$^+$FOXP3$^+$ Regulatory T Cell Control of iNKT Cells

CD4$^+$CD25$^+$Foxp3$^+$ T$_{regs}$ represent a very well studied subset of regulatory T cells. They comprise a population of CD4$^+$ T cells that constitutively express the surface antigen CD25 and transcription factor Foxp3. Like iNKT cells, a role for T$_{regs}$ has also been implicated in the outcome of various models of autoimmune disease, infectious disease and cancer [76–78]. It is commonly held that T$_{regs}$ inhibit immune responses, whether in models that suppress either autoimmunity or immunity against cancer. This property of T$_{regs}$ differs from that of iNKT cells, which seem to have an activating function, potentiating immune responses towards specific antigens and targets rather than suppressing them [78]. In the case of autoimmune disease, iNKT cells may activate DC towards a tolerogenic phenotype, resulting in the suppression of autoreactive T cells.

Recently, the role that T$_{reg}$ cells and iNKT cells play in regulating each others function has been highlighted [116]. iNKT cell mediated generation of T$_{regs}$ was initially shown to be important in nickel induced oral tolerance of iNKT cell deficient mice [117]. Oral administration of nickel to wild-type C57BL/6 mice induces the generation of T cells and APC capable of transferring tolerance to naïve recipient mice. Moreover, the adoptive transfer of tolerogenic cells induces antigen-specific T$_{regs}$ in recipients. Interestingly, when nickel was administered orally to mice deficient in iNKT cells, the induction of T$_{regs}$ was blocked in recipients of adoptively transferred APC. Thus, iNKT cells are required to induce tolerogenic APC capable of inducing antigen-specific T$_{regs}$.

In agreement with a role for iNKT cells in the induction of T_{reg} activity, α-GalCer activation of iNKT cells was found to increase the number of T_{regs} in a model of autoimmune myasthenia gravis (MG). T_{regs} are crucial for α-GalCer mediated protection against MG, as prior inactivation of T_{regs} by an anti-CD25 antibody results in the onset of MG [118]. In humans, one study suggests that the activation of iNKT cells and resultant IL-2 secretion is critical for the proliferation and activation of human T_{regs} [119]. When DC, irradiated CD4$^+$ iNKT cells and T_{regs} were co-cultured with α-GalCer, IL-2 secreted by iNKT cells increased the proliferation of T_{regs} and enabled these T_{regs} to retain their suppressive function *in vitro*. These observations suggest that activated iNKT cells can regulate the activation of T_{reg} responses, which may be crucial for the control of tolerance induction and prevention of autoimmune disease.

In a reciprocal manner, T_{regs} effectively suppress immune responses of various antigen-specific T cells, including iNKT cells. Human T_{regs} can suppress the activation of iNKT cell clones by α-GalCer *in vitro* [120]. This T_{reg} mediated suppression of iNKT cells is contact dependent, and results in reduced iNKT cell proliferation and cytokine secretion. Consistent with this finding, T_{regs} responding to serologically defined autoantigens can also suppress iNKT cell mediated tumor immunity [121]. In addition, our recent studies of the mechanism of α-GalCer mediated protection against T1D have also demonstrated that collaboration between T_{regs} and iNKT cells is required for this protection in NOD mice [122]. However, contrary to the findings with the MG model, we did not observe any increases in the frequency or activity of T_{regs} in α-GalCer treated NOD mice, suggesting that iNKT cells do not promote T_{reg} activity in these mice. Nevertheless, we found that iNKT cells require the activity of T_{regs} as treatment of NOD mice with anti-CD25 to inactivate T_{regs} prior to administration of α-GalCer abrogated iNKT protection and elicited the onset of T1D. Furthermore, we showed that T_{reg} inactivation induced an increase in iNKT cell secretion of IL-2, IL-4, IL-10 and IFN-γ. Moreover, T_{reg} inactivation increased the capacity of iNKT cells to transactivate B cells, NK cells, conventional T cells and DC. Thus, T_{regs} appear to regulate the activity of iNKT cells, either directly by downregulating the activity of iNKT cells and returning them to a state of homeostasis or indirectly by the activation of DC that in turn downregulate iNKT cell activity. Further experimentation is required to test these possible pathways.

Conclusion

iNKT cells represent a unique subset of T cells that can link innate recognition signals to adaptive immune responses in a variety of immune conditions, from pathogen clearance, cancer immunity, to the regulation of autoimmunity, according to a model of iNKT cell activation shown in Fig. 27.2. Uniquely, upon interaction with selected subsets of APCs, activated iNKT cells seems to potentiate rather than suppress immune responses. iNKT cells can mediate adaptive immunity towards specific targets, and in autoimmune prone strains can activate tolerogenic mechanisms required to prevent autoimmune disease. The natural role of iNKT cells in

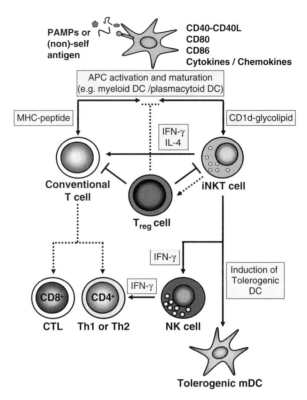

Fig. 27.2 Schematic model of iNKT cell activation and regulation iNKT cells upon activation with endogenous or exogenous glycolipid can potentiate immune responses. Upon interaction with glycolipid-CD1d complexes expressed on dendritic cells (DC), iNKT cells can mature DC to activate conventional T cells, which can be further activated with soluble signals derived from both DC and iNKT cells. During cell mediated immune responses, iNKT cells can activate NK cells and direct the activation of conventional T cells towards Th1 or Th2 immunity to antigen-specific targets. During prevention from autoimmune disease, activated iNKT cells can induce the activation and recruitment of tolerogenic DC or otherwise tolerize autoreactive pathogenic T cells through cell-cell contact dependent mechanisms. $CD4^+CD25^+Foxp3^+$ T_{reg} cells at the centre of this activation schema can regulate iNKT cell or conventional T cell responses via iNKT-DC and/or T-cell-DC interactions, or during downstream iNKT mediated responses *(See also Color Insert)*

pathogen clearance is currently better understood than their role in autoimmune disease and cancer. A central question that requires further study is what natural glycolipids activate iNKT cells in these models of immune regulation and how these glycolipids are physiologically regulated during an immune response. With a more clearly defined role for iNKT cells during physiological responses, safer therapeutic use of α-GalCer or its analogs could be devised that would alleviate the concern of whether iNKT cells exacerbate or prevent autoimmunity. Additionally, a much better understanding of how iNKT cells interact with other subsets of immune cells, particularly DC and T_{regs}, is of interest, in view of the wide array of responses that can be regulated by the collaboration of iNKT cells with DC and T_{regs}. Thus, the therapeutic use of α-GalCer and its analogs that differentially activate certain iNKT cell responses coupled with clinical studies demonstrating the non-toxic nature of α-GalCer present unique opportunities in the future to modulate iNKT cell mediated innate and adaptive immune responses.

References

1. Janeway, CA. Jr; Medzhitov, R. Innate immune recognition. Annu. Rev. Immunol. 2002, 20: 197–216.
2. Godfrey, DI; MacDonald, HR; Kronenberg, M; Smyth, MJ; Van Kaer, L. NKT cells: what's in a name? Nat. Rev. Immunol. 2004, 4: 231–237.
3. Vincent, MS; Gumperz, JE; Brenner, MB. Understanding the function of CD1-restricted T cells. Nat. Immunol. 2003, 4: 517–523.
4. Yokoyama, WM; Plougastel, BF. Immune functions encoded by the natural killer gene complex. Nat. Rev. Immunol. 2003, 3: 304–316.
5. Bendelac, A; Rivera, MN; Park, SH; Roark, JH. Mouse CD1-specific NK1 T cells: development, specificity, and function. Annu. Rev. Immunol. 1997, 15: 535–562.
6. Godfrey, DI; Kronenberg, M. Going both ways: immune regulation via CD1d-dependent NKT cells. J. Clin. Invest. 2004, 114: 1379–1388.
7. Kronenberg, M. Toward an understanding of NKT cell biology: progress and paradoxes. Annu. Rev. Immunol. 2005, 23: 877–900.
8. Taniguchi, M; Harada, M; Kojo, S; Nakayama, T; Wakao, H. The regulatory role of Valpha14 NKT cells in innate and acquired immune response. Annu. Rev. Immunol. 2003, 21: 483–513.
9. Wilson, SB; Delovitch, TL. Janus-like role of regulatory iNKT cells in autoimmune disease and tumour immunity. Nat. Rev. Immunol. 2003, 3: 211–222.
10. Kawano, T; Cui, J; Koezuka, Y; Toura, I; Kaneko, Y; Motoki, K; Ueno, H; Nakagawa, R; Sato, H; Kondo, E; Koseki, H; Taniguchi, M. CD1d-restricted and TCR-mediated activation of Valpha14 NKT cells by glycosylceramides. Science. 1997, 278: 1626–1629.
11. Brossay, L; Chioda, M; Burdin, N; Koezuka, Y; Casorati, G; Dellabona, P; Kronenberg, M. CD1d-mediated recognition of an alpha-galactosylceramide by natural killer T cells is highly conserved through mammalian evolution. J. Exp. Med. 1998, 188: 1521–1528.
12. Matsuda, JL; Naidenko, OV; Gapin, L; Nakayama, T; Taniguchi, M; Wang, CR; Koezuka, Y; Kronenberg, M. Tracking the response of natural killer T cells to a glycolipid antigen using CD1d tetramers. J. Exp. Med. 2000, 192: 741–754.
13. Sidobre, S; Kronenberg, M. CD1 tetramers: a powerful tool for the analysis of glycolipid-reactive T cells. J. Immunol. Methods 2002, 268: 107–121.
14. Joyce, S; Woods, AS; Yewdell, JW; Bennink, JR; De Silva, AD; Boesteanu, A; Balk, SP; Cotter, RJ; Brutkiewicz, RR. Natural ligand of mouse CD1d1: cellular glycosylphosphatidylinositol. Science 1998, 279: 1541–1544.
15. Giabbai, B; Sidobre, S; Crispin, MD; Sanchez-Ruiz, Y; Bachi, A; Kronenberg, M; Wilson, IA; Degano, M. Crystal structure of mouse CD1d bound to the self ligand phosphatidylcholine: a molecular basis for NKT cell activation. J. Immunol. 2005, 175: 977–984.
16. Brutkiewicz, RR. CD1d ligands: the good, the bad, and the ugly. J. Immunol. 2006, 177: 769–775.
17. Roberts, TJ; Sriram, V; Spence, PM; Gui, M; Hayakawa, K; Bacik, I; Bennink, JR; Yewdell, JW; Brutkiewicz, RR. Recycling CD1d1 molecules present endogenous antigens processed in an endocytic compartment to NKT cells. J. Immunol. 2002, 168: 5409–5414.
18. Kang, SJ; Cresswell, P. Saposins facilitate CD1d-restricted presentation of an exogenous lipid antigen to T cells. Nat. Immunol. 2004, 5: 175–181.

19. Major, AS; Joyce, S; Van Kaer, L. Lipid metabolism, atherogenesis and CD1-restricted antigen presentation. Trends Mol. Med. 2006, 12: 270–278.
20. Zhou, D; Mattner, J; Cantu, C III; Schrantz, N; Yin, N; Gao, Y; Sagiv, Y; Hudspeth, K; Wu, YP; Yamashita, T; Teneberg, S; Wang, D; Proia, RL; Levery, SB; Savage, PB; Teyton, L; Bendelac, A. Lysosomal glycosphingolipid recognition by NKT cells. Science 2004, 306: 1786–1789.
21. Mattner, J; Debord, KL; Ismail, N; Goff, RD; Cantu, C III; Zhou, D; Saint-Mezard, P; Wang, V; Gao, Y; Yin, N; Hoebe, K; Schneewind, O; Walker, D; Beutler, B; Teyton, L; Savage, P.B; Bendelac, A. Exogenous and endogenous glycolipid antigens activate NKT cells during microbial infections. Nature 2005, 434: 525–529.
22. Savage, PB; Teyton, L; Bendelac, A. Glycolipids for natural killer T cells. Chem. Soc. Rev. 2006, 35: 771–779.
23. Apostolou, I; Takahama, Y; Belmant, C; Kawano, T; Huerre, M; Marchal, G; Cui, J; Taniguchi, M; Nakauchi, H; Fournie, JJ; Kourilsky, P; Gachelin, G. Murine natural killer T(NKT) cells [correction of natural killer cells] contribute to the granulomatous reaction caused by mycobacterial cell walls. Proc. Natl. Acad. Sci. U. S. A. 1999, 96: 5141–5146.
24. Fischer, K; Scotet, E; Niemeyer, M; Koebernick, H; Zerrahn, J; Maillet, S; Hurwitz, R; Kursar, M; Bonneville, M; Kaufmann, SH; Schaible, UE. Mycobacterial phosphatidylinositol mannoside is a natural antigen for CD1d-restricted T cells. Proc. Natl. Acad. Sci. U. S. A. 2004, 101: 10685–10690.
25. Gilleron, M; Ronet, C; Mempel, M; Monsarrat, B; Gachelin, G; Puzo, G. Acylation state of the phosphatidylinositol mannosides from Mycobacterium bovis bacillus Calmette Guerin and ability to induce granuloma and recruit natural killer T cells. J. Biol. Chem. 2001, 276: 34896–34904.
26. Kinjo, Y; Wu, D; Kim, G; Xing, GW; Poles, MA; Ho, DD; Tsuji, M; Kawahara, K; Wong, CH; Kronenberg, M. Recognition of bacterial glycosphingolipids by natural killer T cells. Nature 2005, 434: 520–525.
27. Sriram, V; Du, W; Gervay-Hague, J; Brutkiewicz, RR. Cell wall glycosphingolipids of Sphingomonas paucimobilis are CD1d-specific ligands for NKT cells. Eur. J. Immunol. 2005, 35: 1692–1701.
28. Kinjo, Y; Tupin, E; Wu, D; Fujio, M; Garcia-Navarro, R; Benhnia, MR; Zajonc, DM; Ben Menachem, G; Ainge, GD; Painter, GF; Khurana, A; Hoebe, K; Behar, SM; Beutler, B; Wilson, IA; Tsuji, M; Sellati, TJ; Wong, CH; Kronenberg, M. Natural killer T cells recognize diacylglycerol antigens from pathogenic bacteria. Nat. Immunol. 2006, 7: 978–986.
29. Kumar, H; Belperron, A; Barthold, SW; Bockenstedt, LK. Cutting edge: CD1d deficiency impairs murine host defense against the spirochete, Borrelia burgdorferi. J. Immunol. 2000, 165: 4797–4801.
30. Burdin, N; Brossay, L; Kronenberg, M. Immunization with alpha-galactosylceramide polarizes CD1-reactive NK T cells towards Th2 cytokine synthesis. Eur. J. Immunol. 1999, 29: 2014–2025.
31. Singh, N; Hong, S; Scherer, DC; Serizawa, I; Burdin, N; Kronenberg, M; Koezuka, Y; Van Kaer, L. Cutting edge: activation of NK T cells by CD1d and alpha-galactosylceramide directs conventional T cells to the acquisition of a Th2 phenotype. J. Immunol. 1999, 163: 2373–2377.
32. Parekh, VV; Wilson, MT; Olivares-Villagomez, D; Singh, AK; Wu, L; Wang, CR; Joyce, S; Van Kaer, L. Glycolipid antigen induces long-term natural killer T cell anergy in mice. J. Clin. Invest. 2005, 115: 2572–2583.
33. Uldrich, AP; Crowe, NY; Kyparissoudis, K; Pellicci, DG; Zhan, Y; Lew, AM; Bouillet, P; Strasser, A; Smyth, MJ; Godfrey, DI. NKT cell stimulation with glycolipid antigen in vivo: costimulation-dependent expansion, Bim-dependent contraction, and hyporesponsiveness to further antigenic challenge. J. Immunol. 2005, 175: 3092–3101.

34. Fujii, S; Shimizu, K; Kronenberg, M; Steinman, RM. Prolonged IFN-gamma-producing NKT response induced with alpha-galactosylceramide-loaded DCs. Nat. Immunol. 2002, 3: 867–874.
35. Toura, I; Kawano, T; Akutsu, Y; Nakayama, T; Ochiai, T; Taniguchi, M. Cutting edge: inhibition of experimental tumor metastasis by dendritic cells pulsed with alpha-galactosylceramide. J. Immunol. 1999, 163: 2387–2391.
36. Bezbradica, JS; Stanic, AK; Matsuki, N; Bour-Jordan, H; Bluestone, JA; Thomas, JW; Unutmaz, D; Van Kaer, L; Joyce, S. Distinct roles of dendritic cells and B cells in Va14Ja18 natural T cell activation in vivo. J. Immunol. 2005, 174: 4696–4705.
37. Miyamoto, K; Miyake, S; Yamamura, T. A synthetic glycolipid prevents autoimmune encephalomyelitis by inducing TH2 bias of natural killer T cells. Nature 2001, 413: 531–534.
38. Mizuno, M; Masumura, M; Tomi, C; Chiba, A; Oki, S; Yamamura, T; Miyake, S. Synthetic glycolipid OCH prevents insulitis and diabetes in NOD mice. J. Autoimmun. 2004, 23: 293–300.
39. Oki, S; Chiba, A; Yamamura, T; Miyake, S. The clinical implication and molecular mechanism of preferential IL-4 production by modified glycolipid-stimulated NKT cells. J. Clin. Invest. 2004, 113: 1631–1640.
40. Yu, KO; Im, JS; Molano, A; Dutronc, Y; Illarionov, PA; Forestier, C; Fujiwara, N; Arias, I; Miyake, S; Yamamura, T; Chang, YT; Besra, GS; Porcelli, SA. Modulation of CD1d-restricted NKT cell responses by using N-acyl variants of alpha-galactosylceramides. Proc. Natl. Acad. Sci. U. S. A. 2005, 102: 3383–3388.
41. Schmieg, J; Yang, G; Franck, RW; Tsuji, M. Superior protection against malaria and melanoma metastases by a C-glycoside analogue of the natural killer T cell ligand alpha-Galactosylceramide. J. Exp. Med. 2003, 198: 1631–1641.
42. Fujii, S; Shimizu, K; Hemmi, H; Fukui, M; Bonito, AJ; Chen, G; Franck, RW; Tsuji, M; Steinman, RM. Glycolipid alpha-C-galactosylceramide is a distinct inducer of dendritic cell function during innate and adaptive immune responses of mice. Proc. Natl. Acad. Sci. U. S. A. 2006, 103: 11252–11257.
43. Skold, M; Behar, SM. Role of CD1d-restricted NKT cells in microbial immunity. Infect. Immun. 2003, 71: 5447–5455.
44. Hansen, DS; Schofield, L. Regulation of immunity and pathogenesis in infectious diseases by CD1d-restricted NKT cells. Int. J. Parasitol. 2004, 34: 15–25.
45. Nieuwenhuis, EE; Matsumoto, T; Exley, M; Schleipman, RA; Glickman, J; Bailey, DT; Corazza, N; Colgan, SP; Onderdonk, AB; Blumberg, RS. CD1d-dependent macrophage-mediated clearance of Pseudomonas aeruginosa from lung. Nat. Med. 2002, 8: 588–593.
46. Grubor-Bauk, B; Simmons, A; Mayrhofer, G; Speck, PG. Impaired clearance of herpes simplex virus type 1 from mice lacking CD1d or NKT cells expressing the semivariant V alpha 14-J alpha 281 TCR. J. Immunol. 2003, 170: 1430–1434.
47. Raftery, MJ; Winau, F; Kaufmann, SH; Schaible, UE; Schonrich, G. CD1 antigen presentation by human dendritic cells as a target for herpes simplex virus immune evasion. J. Immunol. 2006, 177: 6207–6214.
48. Yuan, W; Dasgupta, A; Cresswell, P. Herpes simplex virus evades natural killer T cell recognition by suppressing CD1d recycling. Nat. Immunol. 2006, 7: 835–842.
49. van Dommelen, SL; Tabarias, HA; Smyth, MJ; Degli-Esposti, MA. Activation of natural killer (NK) T cells during murine cytomegalovirus infection enhances the antiviral response mediated by NK cells. J. Virol. 2003, 77: 1877–1884.
50. Liu, YJ. IPC: professional type 1 interferon-producing cells and plasmacytoid dendritic cell precursors. Annu. Rev. Immunol. 2005, 23: 275–306.
51. Colonna, M; Trinchieri, G; Liu, J. Plasmacytoid dendritic cells in immunity. Nat. Immunol. 2004, 5: 1219–1226.
52. Marschner, A; Rothenfusser, S; Hornung, V; Prell, D; Krug, A; Kerkmann, M; Wellisch, D; Poeck, H; Greinacher, A; Giese, T; Endres, S; Hartmann, G. CpG ODN enhance

antigen-specific NKT cell activation via plasmacytoid dendritic cells. Eur. J. Immunol. 2005, 35: 2347–2357.
53. Montoya, CJ; Jie, HB; Al Harthi, L; Mulder, C; Patino, PJ; Rugeles, MT; Krieg, AM; Landay, AL; Wilson, SB. Activation of plasmacytoid dendritic cells with TLR9 agonists initiates invariant NKT cell-mediated cross-talk with myeloid dendritic cells. J. Immunol. 2006, 177: 1028–1039.
54. Ishikawa, H; Hisaeda, H; Taniguchi, M; Nakayama, T; Sakai, T; Maekawa, Y; Nakano, Y; Zhang, M; Zhang, T; Nishitani, M; Takashima, M; Himeno, K. CD4(+) v(alpha)14 NKT cells play a crucial role in an early stage of protective immunity against infection with Leishmania major. Int. Immunol. 2000, 12: 1267–1274.
55. Gonzalez-Aseguinolaza, G; de Oliveira, C; Tomaska, M; Hong, S; Bruna-Romero, O; Nakayama, T; Taniguchi, M; Bendelac, A; Van Kaer, L; Koezuka, Y; Tsuji, M. alpha-galactosylceramide-activated Valpha 14 natural killer T cells mediate protection against murine malaria. Proc. Natl. Acad. Sci. U. S. A. 2000, 97: 8461–8466.
56. Hansen, DS; Siomos, MA; Buckingham, L; Scalzo, AA; Schofield, L. Regulation of murine cerebral malaria pathogenesis by CD1d-restricted NKT cells and the natural killer complex. Immunity. 2003, 18: 391–402.
57. Cui, J; Shin, T; Kawano, T; Sato, H; Kondo, E; Toura, I; Kaneko, Y; Koseki, H; Kanno, M; Taniguchi, M. Requirement for Valpha14 NKT cells in IL-12-mediated rejection of tumors. Science 1997, 278: 1623–1626.
58. Munz, C; Steinman, RM; Fujii, S. Dendritic cell maturation by innate lymphocytes: coordinated stimulation of innate and adaptive immunity. J. Exp. Med. 2005, 202: 203–207.
59. Seino, K; Motohashi, S; Fujisawa, T; Nakayama, T; Taniguchi, M. Natural killer T cell-mediated antitumor immune responses and their clinical applications. Cancer Sci. 2006, 97: 807–812.
60. Smyth, MJ; Crowe, NY; Hayakawa, Y; Takeda, K; Yagita, H; Godfrey, DI. NKT cells – conductors of tumor immunity? Curr. Opin. Immunol. 2002, 14: 165–171.
61. Hayakawa, Y; Takeda, K; Yagita, H; Kakuta, S; Iwakura, Y; Van Kaer, L; Saiki, I; Okumura, K. Critical contribution of IFN-gamma and NK cells, but not perforin-mediated cytotoxicity, to anti-metastatic effect of alpha-galactosylceramide. Eur. J. Immunol. 2001, 31: 1720–1727.
62. Smyth, MJ; Crowe, NY; Pellicci, DG; Kyparissoudis, K; Kelly, JM; Takeda, K; Yagita, H; Godfrey, DI. Sequential production of interferon-gamma by NK1.1(+) T cells and natural killer cells is essential for the antimetastatic effect of alpha-galactosylceramide. Blood 2002, 99: 1259–1266.
63. Fujii, S; Shimizu, K; Smith, C; Bonifaz, L; Steinman, RM. Activation of natural killer T cells by alpha-galactosylceramide rapidly induces the full maturation of dendritic cells in vivo and thereby acts as an adjuvant for combined CD4 and CD8 T cell immunity to a coadministered protein. J. Exp. Med. 2003, 198: 267–279.
64. Fujii, S; Liu, K; Smith, C; Bonito, AJ; Steinman, RM. The linkage of innate to adaptive immunity via maturing dendritic cells in vivo requires CD40 ligation in addition to antigen presentation and CD80/86 costimulation. J. Exp. Med. 2004, 199: 1607–1618.
65. Smyth, MJ; Thia, KY; Street, SE; Cretney, E; Trapani, JA; Taniguchi, M; Kawano, T; Pelikan, SB; Crowe, NY; Godfrey, DI. Differential tumor surveillance by natural killer (NK) and NKT cells. J. Exp. Med. 2000, 191: 661–668.
66. Crowe, NY; Smyth, MJ; Godfrey, DI. A critical role for natural killer T cells in immunosurveillance of methylcholanthrene-induced sarcomas. J. Exp. Med. 2002, 196: 119–127.
67. Crowe, NY; Coquet, JM; Berzins, SP; Kyparissoudis, K; Keating, R; Pellicci, DG; Hayakawa, Y; Godfrey, DI; Smyth, MJ. Differential antitumor immunity mediated by NKT cell subsets in vivo. J. Exp. Med. 2005, 202: 1279–1288.
68. Kawano, T; Nakayama, T; Kamada, N; Kaneko, Y; Harada, M; Ogura, N; Akutsu, Y; Motohashi, S; Iizasa, T; Endo, H; Fujisawa, T; Shinkai, H; Taniguchi, M. Antitumor

cytotoxicity mediated by ligand-activated human V alpha24 NKT cells. Cancer Res. 1999, 59: 5102–5105.
69. Rogers, PR; Matsumoto, A; Naidenko, O; Kronenberg, M; Mikayama, T; Kato, S. Expansion of human Valpha24+ NKT cells by repeated stimulation with KRN7000. J. Immunol. Methods 2004, 285: 197–214.
70. van der Vliet, HJ; Nishi, N; Koezuka, Y; von Blomberg, BM; van den Eertwegh, AJ; Porcelli, SA; Pinedo, HM; Scheper, RJ; Giaccone, G. Potent expansion of human natural killer T cells using alpha-galactosylceramide (KRN7000)-loaded monocyte-derived dendritic cells, cultured in the presence of IL-7 and IL-15. J. Immunol. Methods 2001, 247: 61–72.
71. Giaccone, G; Punt, C.J; Ando, Y; Ruijter, R; Nishi, N; Peters, M; von Blomberg, BM; Scheper, RJ; van der Vliet, HJ; van den Eertwegh, AJ; Roelvink, M; Beijnen, J; Zwierzina, H; Pinedo, HM. A phase I study of the natural killer T-cell ligand alpha-galactosylceramide (KRN7000) in patients with solid tumors. Clin. Cancer Res. 2002, 8: 3702–3709.
72. Nieda, M; Okai, M; Tazbirkova, A; Lin, H; Yamaura, A; Ide, K; Abraham, R; Juji, T; Macfarlane, DJ; Nicol, AJ. Therapeutic activation of Valpha24+Vbeta11+ NKT cells in human subjects results in highly coordinated secondary activation of acquired and innate immunity. Blood 2004, 103: 383–389.
73. Chang, DH; Osman, K; Connolly, J; Kukreja, A; Krasovsky, J; Pack, M; Hutchinson, A; Geller, M; Liu, N; Annable, R; Shay, J; Kirchhoff, K; Nishi, N; Ando, Y; Hayashi, K; Hassoun, H; Steinman, RM; Dhodapkar, MV. Sustained expansion of NKT cells and antigen-specific T cells after injection of alpha-galactosyl-ceramide loaded mature dendritic cells in cancer patients. J. Exp. Med. 2005, 201: 1503–1517.
74. Ishikawa, A; Motohashi, S; Ishikawa, E; Fuchida, H; Higashino, K; Otsuji, M; Iizasa, T; Nakayama, T; Taniguchi, M; Fujisawa, T. A phase I study of alpha-galactosylceramide (KRN7000)-pulsed dendritic cells in patients with advanced and recurrent non-small cell lung cancer. Clin. Cancer Res. 2005, 11: 1910–1917.
75. Motohashi, S; Ishikawa, A; Ishikawa, E; Otsuji, M; Iizasa, T; Hanaoka, H; Shimizu, N; Horiguchi, S; Okamoto, Y; Fujii, S; Taniguchi, M; Fujisawa, T; Nakayama, T. A phase I study of in vitro expanded natural killer T cells in patients with advanced and recurrent non-small cell lung cancer. Clin. Cancer Res. 2006, 12: 6079–6086.
76. Shevach, EM. Regulatory T cells in autoimmmunity. Annu. Rev. Immunol. 2000, 18: 423–449.
77. Sakaguchi, S. Naturally arising CD4+ regulatory t cells for immunologic self-tolerance and negative control of immune responses. Annu. Rev. Immunol. 2004, 22: 531–562.
78. Kronenberg, M; Rudensky, A. Regulation of immunity by self-reactive T cells. Nature. 2005, 435: 598–604.
79. Delovitch, TL; Singh, B. The nonobese diabetic mouse as a model of autoimmune diabetes: immune dysregulation gets the NOD. Immunity. 1997, 7: 727–738.
80. Shoda, LK; Young, DL; Ramanujan, S; Whiting, CC; Atkinson, MA; Bluestone, JA; Eisenbarth, GS; Mathis, D; Rossini, AA; Campbell, SE; Kahn, R; Kreuwel, HT. A comprehensive review of interventions in the NOD mouse and implications for translation. Immunity 2005, 23: 115–126.
81. Van Kaer, L. alpha-Galactosylceramide therapy for autoimmune diseases: prospects and obstacles. Nat. Rev. Immunol. 2005, 5: 31–42.
82. Hammond, KJ; Kronenberg, M. Natural killer T cells: natural or unnatural regulators of autoimmunity? Curr. Opin. Immunol. 2003, 15: 683–689.
83. Lehuen, A; Lantz, O; Beaudoin, L; Laloux, V; Carnaud, C; Bendelac, A; Bach, JF; Monteiro, RC. Overexpression of natural killer T cells protects Valpha14- Jalpha281 transgenic nonobese diabetic mice against diabetes. J. Exp. Med. 1998, 188: 1831–1839.
84. Hammond, KJ; Pellicci, DG; Poulton, LD; Naidenko, OV; Scalzo, AA; Baxter, AG; Godfrey, DI. CD1d-restricted NKT cells: an interstrain comparison. J. Immunol. 2001, 167: 1164–1173.

85. Hong, S; Wilson, MT; Serizawa, I; Wu, L; Singh, N; Naidenko, OV; Miura, T; Haba, T; Scherer, DC; Wei, J; Kronenberg, M; Koezuka, Y; Van Kaer, L. The natural killer T-cell ligand alpha-galactosylceramide prevents autoimmune diabetes in non-obese diabetic mice. Nat. Med. 2001, 7: 1052–1056.
86. Poulton, LD; Smyth, MJ; Hawke, CG; Silveira, P; Shepherd, D; Naidenko, OV; Godfrey, DI; Baxter, A.G. Cytometric and functional analyses of NK and NKT cell deficiencies in NOD mice. Int. Immunol. 2001, 13: 887–896.
87. Sharif, S; Arreaza, GA; Zucker, P; Mi, QS; Sondhi, J; Naidenko, OV; Kronenberg, M; Koezuka, Y; Delovitch, TL; Gombert, JM; Leite-De-Moraes, M; Gouarin, C; Zhu, R; Hameg, A; Nakayama, T; Taniguchi, M; Lepault, F; Lehuen, A; Bach, JF; Herbelin, A. Activation of natural killer T cells by alpha-galactosylceramide treatment prevents the onset and recurrence of autoimmune Type 1 diabetes. Nat. Med. 2001, 7: 1057–1062.
88. Wang, B; Geng, YB; Wang, CR. CD1-restricted NK T cells protect nonobese diabetic mice from developing diabetes. J. Exp. Med. 2001, 194: 313–320.
89. Mi, QS; Ly, D; Zucker, P; McGarry, M; Delovitch, TL. Interleukin-4 but not interleukin-10 protects against spontaneous and recurrent type 1 diabetes by activated CD1d-restricted invariant natural killer T-cells. Diabetes 2004, 53: 1303–1310.
90. Chen, YG; Choisy-Rossi, CM; Holl, TM; Chapman, HD; Besra, GS; Porcelli, SA; Shaffer, DJ; Roopenian, D; Wilson, SB; Serreze, DV. Activated NKT cells inhibit autoimmune diabetes through tolerogenic recruitment of dendritic cells to pancreatic lymph nodes. J. Immunol. 2005, 174: 1196–1204.
91. Novak, J; Beaudoin, L; Griseri, T; Lehuen, A. Inhibition of T cell differentiation into effectors by NKT cells requires cell contacts. J. Immunol. 2005, 174: 1954–1961.
92. Naumov, YN; Bahjat, KS; Gausling, R; Abraham, R; Exley, MA; Koezuka, Y; Balk, SB; Strominger, JL; Clare-Salzer, M; Wilson, SB. Activation of CD1d-restricted T cells protects NOD mice from developing diabetes by regulating dendritic cell subsets. Proc. Natl. Acad. Sci. U. S. A. 2001, 98: 13838–13843.
93. Beaudoin, L; Laloux, V; Novak, J; Lucas, B; Lehuen, A. NKT cells inhibit the onset of diabetes by impairing the development of pathogenic T cells specific for pancreatic beta cells. Immunity 2002, 17: 725–736.
94. Cain, JA; Smith, JA; Ondr, JK; Wang, B; Katz, JD. NKT cells and IFN-gamma establish the regulatory environment for the control of diabetogenic T cells in the nonobese diabetic mouse. J. Immunol. 2006, 176: 1645–1654.
95. Tang, Q; Adams, JY; Tooley, AJ; Bi, M; Fife, BT; Serra, P; Santamaria, P; Locksley, RM; Krummel, MF; Bluestone, JA. Visualizing regulatory T cell control of autoimmune responses in nonobese diabetic mice. Nat. Immunol. 2006, 7: 83–92.
96. Tadokoro, CE; Shakhar, G; Shen, S; Ding, Y; Lino, AC; Maraver, A; Lafaille, JJ; Dustin, ML. Regulatory T cells inhibit stable contacts between CD4+ T cells and dendritic cells in vivo. J. Exp. Med. 2006, 203: 505–511.
97. Griseri, T; Beaudoin, L; Novak, J; Mars, LT; Lepault, F; Liblau, R; Lehuen, A. Invariant NKT cells exacerbate type 1 diabetes induced by CD8 T cells. J. Immunol. 2005, 175: 2091–2101.
98. Wilson, SB; Kent, SC; Patton, KT; Orban, T; Jackson, RA; Exley, M; Porcelli, S; Schatz, DA; Atkinson, MA; Balk, SP; Strominger, JL; Hafler, DA. Extreme Th1 bias of invariant Valpha24JalphaQ T cells in type 1 diabetes. Nature 1998, 391: 177–181.
99. Oikawa, Y; Shimada, A; Yamada, S; Motohashi, Y; Nakagawa, Y; Irie, J; Maruyama, T; Saruta, T. High frequency of valpha24(+) vbeta11(+) T-cells observed in type 1 diabetes. Diabetes Care 2002, 25: 1818–1823.
100. Lee, PT; Putnam, A; Benlagha, K; Teyton, L; Gottlieb, PA; Bendelac, A. Testing the NKT cell hypothesis of human IDDM pathogenesis. J. Clin. Invest. 2002, 110: 793–800.
101. Berzins, SP; Kyparissoudis, K; Pellicci, DG; Hammond, KJ; Sidobre, S; Baxter, A; Smyth, MJ; Kronenberg, M; Godfrey, D.I. Systemic NKT cell deficiency in NOD mice is

not detected in peripheral blood: implications for human studies. Immunol. Cell Biol. 2004, 82: 247–252.
102. Kent, SC; Chen, Y; Clemmings, SM; Viglietta, V; Kenyon, NS; Ricordi, C; Hering, B; Hafler, DA. Loss of IL-4 secretion from human type 1a diabetic pancreatic draining lymph node NKT cells. J. Immunol. 2005, 175: 4458–4464.
103. Jahng, AW; Maricic, I; Pedersen, B; Burdin, N; Naidenko, O; Kronenberg, M; Koezuka, Y; Kumar, V. Activation of natural killer T cells potentiates or prevents experimental autoimmune encephalomyelitis. J. Exp. Med. 2001, 194: 1789–1799.
104. Singh, AK; Wilson, MT; Hong, S; Olivares-Villagomez, D; Du, C; Stanic, AK; Joyce, S; Sriram, S; Koezuka, Y; Van Kaer, L. Natural killer T cell activation protects mice against experimental autoimmune encephalomyelitis. J. Exp. Med. 2001, 194: 1801–1811.
105. Furlan, R; Bergami, A; Cantarella, D; Brambilla, E; Taniguchi, M; Dellabona, P; Casorati, G; Martino, G. Activation of invariant NKT cells by alphaGalCer administration protects mice from MOG35–55-induced EAE: critical roles for administration route and IFN-gamma. Eur. J. Immunol. 2003, 33: 1830–1838.
106. Mars, LT; Laloux, V; Goude, K; Desbois, S; Saoudi, A; Van Kaer, L; Lassmann, H; Herbelin, A; Lehuen, A; Liblau, RS. Cutting edge: V alpha 14-J alpha 281 NKT cells naturally regulate experimental autoimmune encephalomyelitis in nonobese diabetic mice. J. Immunol. 2002, 168: 6007–6011.
107. Araki, M; Kondo, T; Gumperz, JE; Brenner, MB; Miyake, S; Yamamura, T. Th2 bias of CD4+ NKT cells derived from multiple sclerosis in remission. Int. Immunol. 2003, 15: 279–288.
108. Major, AS; Singh, RR; Joyce, S; Van Kaer, L. The role of invariant natural killer T cells in lupus and atherogenesis. Immunol. Res. 2006, 34: 49–66.
109. Mieza, MA; Itoh, T; Cui, JQ; Makino, Y; Kawano, T; Tsuchida, K; Koike, T; Shirai, T; Yagita, H; Matsuzawa, A; Koseki, H; Taniguchi, M. Selective reduction of V alpha 14+ NK T cells associated with disease development in autoimmune-prone mice. J. Immunol. 1996, 156: 4035–4040.
110. Oishi, Y; Sumida, T; Sakamoto, A; Kita, Y; Kurasawa, K; Nawata, Y; Takabayashi, K; Takahashi, H; Yoshida, S; Taniguchi, M; Saito, Y; Iwamoto, I. Selective reduction and recovery of invariant Valpha24JalphaQ T cell receptor T cells in correlation with disease activity in patients with systemic lupus erythematosus. J. Rheumatol. 2001, 28: 275–283.
111. Kojo, S; Adachi, Y; Keino, H; Taniguchi, M; Sumida, T. Dysfunction of T cell receptor AV24AJ18+, BV11+ double-negative regulatory natural killer T cells in autoimmune diseases. Arthritis Rheum. 2001, 44: 1127–1138.
112. Yang, JQ; Chun, T; Liu, H; Hong, S; Bui, H; Van Kaer, L; Wang, CR; Singh, RR. CD1d deficiency exacerbates inflammatory dermatitis in MRL-lpr/lpr mice. Eur. J. Immunol. 2004, 34: 1723–1732.
113. Yang, JQ; Saxena, V; Xu, H; Van Kaer, L; Wang, CR; Singh, RR. Repeated alpha-galactosylceramide administration results in expansion of NK T cells and alleviates inflammatory dermatitis in MRL-lpr/lpr mice. J. Immunol. 2003, 171: 4439–4446.
114. Forestier, C; Molano, A; Im, JS; Dutronc, Y; Diamond, B; Davidson, A; Illarionov, PA; Besra, GS; Porcelli, SA. Expansion and hyperactivity of CD1d-restricted NKT cells during the progression of systemic lupus erythematosus in (New Zealand Black x New Zealand White)F1 mice. J. Immunol. 2005, 175: 763–770.
115. Zeng, D; Liu, Y; Sidobre, S; Kronenberg, M; Strober, S. Activation of natural killer T cells in NZB/W mice induces Th1-type immune responses exacerbating lupus. J. Clin. Invest. 2003, 112: 1211–1222.
116. La Cava, A; Van Kaer, L; Fu, DS. CD4+CD25+ Tregs and NKT cells: regulators regulating regulators. Trends Immunol. 2006, 27: 322–327.
117. Roelofs-Haarhuis, K; Wu, X; Gleichmann, E. Oral tolerance to nickel requires CD4+ invariant NKT cells for the infectious spread of tolerance and the induction of specific regulatory T cells. J. Immunol. 2004, 173: 1043–1050.

118. Liu, R; La Cava, A; Bai, XF; Jee, Y; Price, M; Campagnolo, DI; Christadoss, P; Vollmer, TL; Van Kaer, L; Shi, FD. Cooperation of invariant NKT cells and CD4+CD25+ T regulatory cells in the prevention of autoimmune myasthenia. J. Immunol. 2005, 175: 7898–7904.
119. Jiang, S; Game, DS; Davies, D; Lombardi, G; Lechler, RI. Activated CD1d-restricted natural killer T cells secrete IL-2: innate help for CD4+CD25+ regulatory T cells? Eur. J. Immunol. 2005, 35: 1193–1200.
120. Azuma, T; Takahashi, T; Kunisato, A; Kitamura, T; Hirai, H. Human CD4+ CD25+ regulatory T cells suppress NKT cell functions. Cancer Res. 2003, 63: 4516–4520.
121. Nishikawa, H; Kato, T; Tanida, K; Hiasa, A; Tawara, I; Ikeda, H; Ikarashi, Y; Wakasugi, H; Kronenberg, M; Nakayama, T; Taniguchi, M; Kuribayashi, K; Old, LJ; Shiku, H. CD4+ CD25+ T cells responding to serologically defined autoantigens suppress antitumor immune responses. Proc. Natl. Acad. Sci. U. S. A. 2003, 100: 10902–10906.
122. Ly, D; Mi, QS; Hussain, S; Delovitch, TL. Protection from type 1 diabetes by invariant NK T cells requires the activity of CD4+CD25+ regulatory T cells. J. Immunol. 2006, 177: 3695–3704.

Chapter 28
Natural Killer T Cells Regulate the Development of Asthma

Muriel Pichavant, Rosemarie H. DeKruyff, and Dale T. Umetsu

Abstract NKT cells play an important role at the interface between innate and acquired immunity, because of their rapid production of Th1 as well as Th2 cytokines upon activation, and their expression of germline encoded T cell receptors. In addition, NKT cells critically regulate the development of airway hypeerreactivity (AHR), a cardinal feature of asthma, by producing IL-4 and IL-13, and by interacting with T cells and inducing Th2 cell production of IL-4 and IL-13. NKT cells thus play an important proinflammatory role in asthma, in contrast to suppressive $CD4^+$ $CD25^+$ natural and adaptive regulatory T cells, which inhibit airway inflammation and AHR.

Introduction

Asthma

Overview and Definition

Bronchial asthma is a major public-health problem in industrialized countries that has increased dramatically in prevalence over the past two decades [1–3], presumably due to rapid changes in the environment. These environmental changes include changes in the incidence of many infectious diseases, and in sanitation and air quality [1,3]. Examination of the effects of these changes on asthma is difficult to interpret however because the term "asthma" refers to a diverse collection of syndromes caused by a range of genetic and environmental factors, including allergen sensitization, infection, obesity, aspirin sensitivity, as well as exposure to air pollution [4]. In addition, asthma affects individuals of all ages, and is manifest by a range of severities, from mild intermittent to severe persistent [1,5,6]. Moreover, diverse

---

D.T. Umetsu
Karp Laboratories, Division of Immunology and Allergy, Children's Hospital, Harvard Medical School, Harvard University, Boston, MA 02115, USA
e-mail: Dale.Umetsu@childrens.harvard.edu

cellular and molecular mechanisms have been proposed to explain the pathogenesis of various forms of asthma, including allergic inflammation, Th2 cells, eosinophils, basophils [7], neutrophils [8], and oxidative stress [9], and a shared disease mechanism for all forms of asthma has not been established. Nevertheless, all forms of asthma are associated with reversible airway obstruction with wheezing, and with airway hyperreactivity (AHR), an increased sensitivity of the airways to nonspecific stimuli such as cold air or respiratory irritants, and quantitated by responsiveness to methacholine or histamine [10].

AHR is an important feature of asthma, since it correlates with the severity of reversible airway obstruction and the symptoms of asthma, and distinguishes asthma from other non-asthmatic pulmonary inflammatory states such as pneuomonia, bronchiectasis or sarcoidosis. In contrast, other measures of asthma (airway inflammation, airway eosinophilia, exhaled NO, FEV1) do not always show a relationship with the presence or severity of asthma [11], and therefore direct assessment of AHR may be the most accurate test for asthma, particularly in experimental models of asthma, although other measurements (e.g., with or sputum eosinophilia, airway NO or FEV1) may be helpful in clinically managing and dosing therapy for asthma.

The Immune System in Asthma

Allergic asthma, which is the most common form of asthma affecting 70–80% of patients, is characterized by an inflammatory process driven by eosinophils, basophils, mast cells and allergen specific adaptive $CD4^+$ Th2 cells producing IL-4, IL-5 and IL-13. $CD4^+$ Th2 cells and the cytokines that they produce are particularly important in asthma since they promote asthma by enhancing the growth, differentiation and recruitment of eosinophils, basophils, mast cells, IgE-producing B cells and Th2 cells, and directly inducing AHR [1,10,12,13]. In animal models, depletion of $CD4^+$ T cells prevents the development of AHR [14], and furthermore, $CD4^+$ cells are present in the lungs of virtually all patients with asthma [15]. Finally, the critical role of allergen-specific $CD4^+$ Th2 cells in asthma supports the Th1/Th2 paradigm, first proposed more than twenty years ago [16], and is consistent with the idea that allergic asthma is driven by allergen sensitization [17].

Although conventional $CD4^+$ Th2 cells play an important role in the pathogenesis of allergic asthma [15,18], Th2 cells are not sufficient for the development of allergic asthma, since only about one third of patients with allergen-sensitization (i.e., with allergic rhinitis) have asthma. This suggests that other elements in addition to $CD4^+$ Th2 cells may be required for the full development of asthma. For example, these additional elements may include pathways leading to airway remodelling [19], and components localized to the lower respiratory tract, such as bronchial epithelial cells. In addition, since the CD4 molecule is expressed not only by conventional MHC-II restricted $CD4^+$ T cells but also on Natural Killer T (NKT) cells, $CD4^+$ NKT cells may also provide an essential element in the pathogenesis of asthma.

NKT Cells

Characteristics of NKT Cells

NKT cells constitute a subpopulation of lymphocytes that express characteristics of both natural killer cells such as NK1.1 and CD161, and of conventional T cells such as TCR. NKT cells are present in the thymus, spleen, liver and bone marrow, and at low numbers in normal lung. NKT cells can be divided into three types based upon the specificities of their TCRs (Table 28.1).2 Type 1 NKT cells express a highly conserved invariant TCR α-chain (Vα14-Jα18 in mice and Vα24-Jα18 in humans). Invariant TCR$^+$ NKT cells (*i*NKT) are restricted by the MHC-class-I-like molecule CD1d. Type 1 *i*NKT cells are either CD4$^+$ or are double negative CD4$^-$ CD8$^-$ (DN), and a small subset of human *i*NKT cells are CD8$^+$. The CD4$^+$ *i*NKT cell subset differs from the DN *i*NKT cell subset, in that CD4$^+$ *i*NKT cells produce greater amounts of IL-4, IL-13 and GM-CSF (Th2 cytokines), while the DN *i*NKT cells exhibit greater cytotoxicity and produce only IFN-γ and TNF-α [20–23]. Importantly, both CD4$^+$ and DN *i*NKT cell subsets produce IFN-γ. Like Type 1 *i*NKT cells, Type 2 NKT cells are restricted by CD1d, but Type 2 NKT cells have a diverse TCR repertoire. Type 3 NKT cells are restricted to MHC Class I and class II molecules rather than CD1d.

Type 1 and Type 2 NKT cells recognize glycolipid antigens rather than peptides presented by CD1d [24,25], which is widely expressed by airway and intestinal epithelial cells, B cells, macrophages and dendritic cells [26]. Type 1 NKT cells are very efficiently activated by a synthetic glycolipid, originally derived from a marine sponge, known as alpha-Galactosyl ceramide (α-GalCer) [27]. This compound binds with high affinity to CD1d, and the α-GalCer CD1d complex is recognized by the invariant TCR of *i*NKT cells [28], leading to activation of both murine and human NKT cells. Recognition of CD1d-α-GalCer by *i*NKT cells is thus highly

Table 28.1 Subsets of mouse NKT cells

	Type 1 NKT	Type 2 NKT	Type 3 NKT
Repertoire	Vα14-Jα18 Vβ8.2/7/2	Semi-diverse Vα3.2-Jα9/Vα8, Vβ8	Vα diverse Vβ diverse
Co-receptor	CD4$^+$ or DN	CD4$^+$ or DN	CD8$^+$, CD4$^+$ or DN
Reactivity	α-GalCer	ND	Self-agonist (?)
Antigen-presenting molecule	CD1d	CD1d	MHC class I and others
NK receptors	DX5$^-$ Mostly NK1.1$^+$	DX5 (?) NK1.1$^{+/-}$	DX5$^{+/-}$ NK1.1$^+$
Location	Thymus, lung, liver, spleen and bone marrow	Thymus (?), lung, liver, spleen and bone marrow (?)	Liver, lung (?), spleen and bone marrow

If no J region is indicated, T-cell receptors are diverse. α-GalCer, α-galactosylceramide; DN, double negative; J, joining region; NK, natural killer; V, variable region.
From Kronenberg M et al., Nature Reviews Immunology, 2002 [30]

conserved across phylogeny, suggesting that *i*NKT cells play a very pivotal role in immunity [29,30].

Other glycolipid antigens are thought to bind to CD1d and activate *i*NKT cells, but only a handful of other glycolipids have been identified as yet. These glycolipids may be endogenous or exogenous in origin [29]. For example, endogenous glyclipids may be expressed in the thymus by thymocytes, which select and expand *i*NKT cells [24], or may be expressed by somatic cells when they are injured or damaged during infection [31]. Endogenous glycolipids include the lysosomal glycosphingolipid, isoglobotrihexosylceramide (iGb3) [32–34], but additional endogenous CD1d binding lipids may be involved in thymic selection of *i*NKT cells [35]. Furthermore, other endogenous lipids can bind to CD1d and activate NKT cells or other CD1d-restricted T cells, including the natural lipid sulfatide found in myelin. Sulfatide can bind to group 1 CD1 molecules CD1a, b, and c [36,37], as well as CD1d, and may be important in multiple sclerosis. In mice, sulfatide can be recognized by a subset of NKT cells [38], and regulate the development of experimental autoimmune encephalomyelitis (EAE). Two additional natural lipid components of the cell membrane, the disialoganglioside GD3 and PI, have been also shown to be CD1d-presented antigens capable of stimulating some NKT cell hybridomas [39, 40]. Exogenous glycolipids include those expressed by infectious pathogens including *Sphingomonas*, *Borrelia burgdorferi*, *Ehrlichia* and *Schistosoma mansoni* [32,41–43], and will be discussed below. Finally, human *i*NKT cells have been shown to recognize lipids from plant pollens [44,45], also discussed below.

Functions of NKT Cells

The rapid production of cytokines by *i*NKT cells regulates the development of antimicrobial, autoimmune, antitumor and antitransplant immune responses and atherosclerosis [31,46–51], providing either disease-causing or disease-protective effects. In particular, CD1d-restricted *i*NKT cells appear to play an important role in the intestinal tract and in the respiratory tract.

In the gastrointestinal tract, NKT cells have been proposed to provide both protective and pathogenic contributions. For example, NKT cells appear to be involved in the maintenance of mucosal homeostasis [52–54]. On the other hand, Type 1 NKT cells are required for the development of oxazalone-induced colitis in mice, and Type 2 NKT cells play a pathogenic role in human ulcerative colitis [55,56]. The specific glycolipids (exogenous or endogenous) that normally activate NKT cells in the intestines are not clear. In addition, NKT cells are also thought to play a pathological role in primary biliary cirrhosis. In this instance however, NKT cells appear to be responding to glycolipids from the bacteria *Sphingomonas*. Support for the role of *Sphingomonas* in primary biliary cirrhosis includes the observations that NKT cells and *Sphingomonas* are present in the biliary tree in these patients, and antibody against *Sphingomonas* correlates with disease [42,46,57]. *Novosphingobium* is a gram negative, LPS negative bacteria, and glycolipids from *Novosphingobium* are thought to activate NKT cells, which normally would eliminate the infection. In primary biliary cirrhosis however, activated NKT cells may be the cause of pathologic inflammation in the liver.

In addition to *Novosphingobium*, several other microbial pathogens may activate CD1d-restricted NKT cells [58], which may have evolved to respond to certain types of infectious agents. NKT cells recognize glycolipids from *Borrelia burgdorferi*, the cause of Lyme disease, and NKT cell deficient mice have difficulty in clearly infection with *Borrelia* [41,59]. *Salmonella* may activate NKT cells, although this may occur via a TLR-2 dependant mechanism [60]. Finally, in humans, Epstein-Barr virus (EBV) may activate NKT cells, although the evidence is indirect. Thus, in patients with X-linked lymphoproliferative (XLP) syndrome, due to mutations in SAP [61,62] or in X-linked inactivator of apoptosis (XIAP) [63], severe EBV infections occur, thought to be due to the absence of NKT cells.

*i*NKT Cells and Asthma

While CD4$^+$ Th2 cells play an important role in the pathophysiology of allergic asthma, there is growing evidence that CD4$^+$ IL-4 producing *i*NKT cells also play a prominent role in asthma. In mouse models of asthma, *i*NKT cells have been shown to be required for the development of allergen-induced AHR [64–66]. Thus, *i*NKT cell deficient mice (CD1d$^{-/-}$ mice, which lack the restriction element of *i*NKT cells and Jα18$^{-/-}$ mice, which lack the invariant TCRα chain) fail to develop AHR. The requirement for *i*NKT cells in the development of allergen-induced AHR is independent of Th2 responses, which develop normally in these mice [67]. The development of AHR could be restored in the Jα18$^{-/-}$ mice by adoptive transfer of *i*NKT cells from wildtype mice [64]. In addition, the production of IL-4 and IL-13 but not IFN-γ by *i*NKT cells was required for the development of AHR. These results indicate that *i*NKT cells provide an intrinsic element that permits Th2 responses to drive the full development of AHR and asthma [64], and may explain why only one third of individuals with allergic rhinitis (caused by Th2 responses in the upper respiratory tract) develop asthma.

*i*NKT Cells in Human Asthma

The importance of *i*NKT cells in murine models of asthma suggested that *i*NKT cells could play an important role in human asthma. However, since mouse models of asthma do not replicate all features of human asthma, direct assessment of *i*NKT cell function in humans with asthma was required. As such, examination of lungs in patients with moderate to severe asthma showed the presence of significant numbers of *i*NKT cells, as defined by expression of the invariant T cell receptor on cells in bronchoalveolar lavage (BAL) fluid [68]. In contrast, *i*NKT cells were absent in the lungs of normal individuals or in patients with sarcoidosis, an inflammatory lung disease, associated with large numbers of CD4$^+$ Th1 cells in the lungs. Further, stimulation of BAL fluid cells with α-GalCer rapidly induced the production of large amounts of IL-4 and IL-13 but little IFN-γ. The finding of increased numbers of *i*NKT cells present in the lungs of patients with asthma compared to normal non-asthmatic individuals has been confirmed by at least six additional groups [45,69–73]. On the other hand, two other groups concluded that *i*NKT cells were not increased in asthma, although these investigators did not examine control BAL fluid from normal individuals (in whom *i*NKT cell numbers are generally <1%) [74,75]. This omission is important, since these investigators found that up to 2% of the T cells in the lungs of patients with asthma were *i*NKT cells. Taken together, these reports strongly suggest that *i*NKT cells are indeed present in the lungs of patients with asthma, but that the specific number may vary considerably (from 1% up to 60% of the T cells), possibly correlating with disease severity.

Mechanisms of Asthma

Although the number of *i*NKT cells in the lungs of different patients with asthma appears to vary considerably, even small numbers of pulmonary *i*NKT cells may have potent effects. Thus, in mice, direct activation of even the small numbers of *i*NKT cells present in the lungs of naïve mice with α-GalCer rapidly induced the development of AHR, and this response occurred in the complete absence of conventional $CD4^+$ T cells [76]. These results indicate not only that *i*NKT cell activation alone is sufficient for the induction of AHR, but also that *i*NKT cells, like mast cells in anaphylaxis [77], are potent such that activation of even a few *i*NKT cells in the lungs can trigger AHR. On the other hand, the number of *i*NKT cells in the lungs can increase, for example after an inflammatory response is initiated, attracted by the chemokines TARC/CCL17 and MDC/CCL22, which are produced by activated bronchial epithelial cells [78]. Thus, a subset of *i*NKT cells expressing CCR4, the receptor for both TARC/CCL17 and MDC/CCL22, and producing IL-4 and IL-13, appears to be specifically involved in the development of AHR [79]. The $CCR4^+$ *i*NKT cell subset is distinct from other *i*NKT cell subsets, which express other chemokine receptors and which are specifically found in other organs such as the liver or intestines.

In allergic asthma, *i*NKT cells producing IL-4 and IL-13 that enter the lungs are thought to become activated by endogenous glycolipids expressed by dendritic cells in the inflammatory environment induced by allergen-specific Th2 cells. In other words, allergen-specific Th2 cells responding to exogenous protein antigens can establish an inflammatory milieu in the lungs, which then leads to glycolipid formation and the activation of *i*NKT cells. Therefore Th2 cells and *i*NKT cells synergize in the development of allergic asthma. While the specific endogenous glycolipids that activate the *i*NKT cells in asthma are not yet identified, *i*NKT cells may also enlarge the repertoire of exogenous antigens that can generate pulmonary inflammation. Thus, *i*NKT cells might respond to glycolipid antigens from plant pollens and induce AHR. This idea is supported by the observation that human *i*NKT cells recognize lipids from cypress tree pollen [44,45], suggesting that glycolipids from plant pollens may act as adjuvants by activating *i*NKT cells and enhancing Th2 response to pollen proteins. Finally, glycolipids from pulmonary infectious pathogens might directly activate *i*NKT cells resulting in the development of AHR, independently of Th2 cells. Clearly, much more work is needed to identify the specific glycolipids in the lungs that activate *i*NKT cells, and how *i*NKT cells interact with other inflammatory cells in the lungs.

Other Forms of Asthma

The idea that different types of glycolipids can activate *i*NKT cells to induce AHR suggests that *i*NKT cells activated by other endogenous or exogenous glycolipids may be responsible for other forms of asthma (e.g., infectious asthma, or asthma associated with exposure to aspirin or air pollution). For example, mice challenged with α-GalCer or its analog PBS57 develop primarily airway neutrophilia, in association with the presence of an NK1.1$^-$ *i*NKT cell subset producing IL-17 [80]. Airway neutrophilia is observed in some patients with asthma [81], particularly severe asthma, and may develop via pathways distinct from those involving Th2 biased immune responses, e.g., involving IL-17 and CXCL8/IL-8, rather than IL-4 and IL-13. At this point in time however, although IL-17 can be found in the induced sputum of some patients with asthma [82], the role of IL-17 in asthma is controversial, with some studies indicating a protective role for IL-17 in asthma [83]. Therefore whether IL-17 producing *i*NKT cells enhance or inhibit AHR is not yet clear. Nevertheless, different subsets of *i*NKT cells producing distinct cytokines may be involved in different forms of asthma, in which *i*NKT cells represent a common disease mechanism for the development of AHR.

*i*NKT cells however, may not be required in all forms of AHR, for example that which develops in β2-microglobulin$^{-/-}$ mice. β2-microglobulin$^{-/-}$ mice, which lack *i*NKT cells, develop AHR normally when sensitized and challenged with allergen [84,85]. However, although β2-microglobulin$^{-/-}$ mice do not express the heterodimeric CD1d molecule (consisting of CD1d and β2 microglobulin chains), they do retain a β2-microglobulin-independent form of CD1d [86], and contain Type 2 and Type 3 NKT cells. The development of AHR in these mice that lack both CD8$^+$ T cells and *i*NKT cells suggests that Th2 cells may step in under certain circumstances to act as effector cells for AHR, or alternatively that Type 2 or Type 3 NKT cells may substitute for *i*NKT cells in inducing AHR. In any case, understanding the precise mechanisms by which AHR develops in other forms of asthma may help to clarify the various cell types that are capable of inducing the development of AHR.

Treatment of Asthma

Since *i*NKT cells appear to play such a significant pathogenic role in asthma, targeting of *i*NKT cells may provide very effective therapy for asthma. Importantly, current therapies for asthma, for example with corticosteroids, appear to be ineffective against *i*NKT cells, which are thought to be corticosteroid resistant [87,88]. Corticosteroids are effective in asthma because they induce apoptosis in eosinophils and conventional T cells, and prevent the influx of inflammatory cells (including iNKT cells) into the airways. However, it is possible that once iNKT cells become established in the lungs, corticosteroids may be unable to dampen their function, resulting in corticosteroid resistant asthma, which affects 10–30% of asthmatics [89]. Therefore, the development of iNKT cell specific therapies may provide important new therapy for asthma, particularly for corticosteroid resistant forms.

There may be several methods to specifically target iNKT cells in the treatment of asthma. First, since the CD1d molecule restricts iNKT cells, anti-CD1d antibody can block the activation of iNKT cells and prevent the development of AHR [66]. Second, treatments that anergize or paralyze iNKT cells, for example with strong iNKT cell activating glycolipids, have been shown to prevent the subsequent development of AHR [90–92]. This approach may be problematic however, since activation of iNKT cells may initially exacerbate asthma symptoms, and control of symptoms may require continuous therapy. Alternative approaches might involve agents, e.g., glycolipids, which could specifically block the activation of iNKT cells or which might anergize iNKT cells without inducing excessive cytokine production from$i$NKT cells.

Regulation of Asthma by Regulatory T (T_{Reg}) Cells

While iNKT cells play a critical pro-inflammatory role in asthma, other regulatory cells are responsible for down modulating inflammation in asthma. In the past, Th1 cells were thought to counterbalance Th2 cells, but it has become apparent that Th1 cells are pro-inflammatory [93]. Recently, there has been great interest in anti-inflammatory T cells with suppressive activity, including $CD25^+$ natural T_{Reg} cells and antigen-specific adaptive T_{Reg} cells.

In allergic individuals, suppressive $CD25^+$ natural T_{Reg} cells expressing Foxp3 have been shown to be present in reduced numbers [94], or to have reduced activity [95]. In mouse models, $CD25^+$ natural T_{Reg} cells have been shown to suppress the development of AHR [96]. These results together suggest that allergic asthma might result from a reduction in the quantity or function of natural T_{Reg} cells or an enhanced development of Th2 responses. Since allergic individuals do not have an increased incidence of autoimmune disease, however, as might be predicted by a decrease in natural T_{Reg} cells, it is likely that the "immunological defect" in allergic individuals is confined to mucosal immune responses, or to allergen-specific immune responses [97,98].

Thus, allergic individuals may have a deficit in terms of the development of allergen-specific adaptive T_{Reg} cells. These adaptive T_{Reg} cells develop on exposure to specific allergen, presumably from $CD25^-$ non-regulatory precursor cells, although such T_{Reg} cells may represent a regulatory cell lineage distinct from other antigen-specific T cells. Adaptive T_{Reg} cells include Th3 cells, which develop primarily in the gastrointestinal tract [99] and Tr1 cells, which develop from antigen-specific cells exposed in vitro to IL-10 [100]. In other cases, respiratory exposure to allergen induces the development of allergen-specific T_{Reg} cells that produce IL-10, and expressing Foxp3 and GATA-3 (Th2-like T_{Reg} cells) [101–103]. Alternatively, immunization with the adjuvant heat-killed *Listeria monocytogenes* induces the development of antigen-specific T_{Reg} cells expressing IL-10, IFN-γ, T-bet and Foxp3 (Th1-like T_{Reg} cells) [104]. The range of phenotypes of T_{Reg} cells suggests that a spectrum of allergen-specific T_{Reg} cells may exist, and that several pathways may be involved in their development.

Balance Between Allergen-Specific T_{Reg} Cells, Th2 Cells and *i*NKT Cells

In situations when adequate allergen-specific T_{Reg} cells develop, the clinical features of asthma are not likely to arise. On the other hand, when increased numbers of Th2 cells or *i*NKT cells develop in the airways due to environmental exposures, the inhibitory effects of T_{Reg} cells may be overwhelmed, resulting in the development of asthma (Table 28.2). This situation appears to occur more readily in individuals over the past two decades resulting in a higher prevalence of asthma. Environmental factors that result in the activation of *i*NKT cells may be particularly important, although the specific environmental changes that are responsible are not fullyunderstood. Further study of the factors that activate and expand the number of

Table 28.2 Comparison of NKT Cells and $CD25^+CD4^+$ Tregs

	NKT cells	$CD25^+CD4^+$ Tregs
Distribution and specificity		
Prevalence (spleen and thymus)	0.5–1%	5–10%
Specificity	Glycolipids plus CD1d	Peptides plus MHC class II
Autoreactivity	Yes	Yes, in some cases
TCRs	Invariant Vα, limited Vβ	Diverse
Development		
Positive selection	CD1d-expressing thymocytes	MHC class II$^+$ thymic epithelium
Unique requirements	IL-15 pathway, lymphotoxin, fyn, Ets, AP-1, and NF-κB transcript factors. CD1d antigen presentation pathway(?): cathepsin L, AP-3, saposins	IL-2, foxp3 B7-mediated costimulation
Ontogeny	First week after birth	First week after birth
Functions		
Mode of action	Suppression or activation	Suppression
Cytokines	Many, including Th1 and Th2	IL-10, TGF-β
Function-related cell surface molecules	CD40L, FAS-L	TGF-β, CTLA-4

FAS-L, Fas ligand.

iNKT cells in the lungs will greatly enhance our understanding of the pathogenesis of asthma.

Suppressive Effects of NKT Cells

While iNKT cells are responsible for the development of AHR, in many other disease models systems, iNKT cells suppress the development of pathology. For example, in Type 1 diabetes in non-obese diabetic (NOD) mice, dysregulation of regulatory T cell development results in the activation of effector Th1 cells and diabetes. However, iNKT cell activation can prevent diabetes by inhibiting the differentiation of anti-islet T cells [105]. Thus, treatment with α-GalCer prevents the development of diabetes in wild-type but not CD1d-deficient NOD mice [106,107]. Activated iNKT cells elicit diabetes protection in NOD mice by producing a soluble factor(s) that induces DC maturation and accumulation in pancreatic lymph nodes, where they then recruit and tolerize pathogenic T cells [108]. iNKT also play a suppressive role on experimental autoimmune encephalomyelitis (EAE). Indeed, a single injection of an analogue of α-GalCer, OCH, which preferentially induces IL-4 production in iNKT cells, consistently induce a Th2 bias of autoimmune T cells leading to suppression of EAE [109]. It is not clear however, if certain subsets of iNKT cells have a direct suppressive role, or whether subsets of iNKT cells enhance the development of T_{Reg} cells, which then inhibit disease.

Conclusion and Future Directions

The importance of iNKT cells in asthma is a novel and surprising finding that greatly alters our understanding of the pathobiology of asthma. The study of iNKT cells has only recently become possible with the availability of reagents to identify iNKT cells, and therefore many more experiments and studies are needed to clarify the precise role of iNKT cells in various forms of asthma. Although a number of clinical investigators are currently focused on establishing the "correct" number of iNKT cells that are present in the lungs of patients with asthma, it is likely that the number of pulmonary iNKT cells varies considerably, possibly with disease severity, particularly given the potent and complex function of iNKT cells in asthma. Multiple iNKT subsets appear to exist, and multiple different glycolipids are likely to be recognized by the conserved, invariant TCR of iNKT cells. Thus, the role of iNKT may be extremely adaptable such that iNKT cells can serve to directly induce AHR by responding to exogenous or endogenous glycolipids, or serve to enhance Th2 sensitization to exogenous allergens. Moreover, identification of the specific glycolipid antigens that activate iNKT cells is only beginning, but these glycolipids, which include both endogenous as well as exogenous glycolipids, may differentially induce distinct iNKT cell cytokine profiles and functions. These multiple functions may explain how this novel cell type may critically regulate the development of distinct forms of asthma. Finally, if our hypothesis regarding *i*NKT cells in asthma is correct, then therapies which disrupt the activation or effector function of pulmonary *i*NKT cells may prove to be extremely effective in the treatment of multiple forms of human asthma, especially forms that are currently recalcitrant to available therapies such as in corticosteroid resistant asthma.

References

1. Busse, W.W., and R.F. Lemanske, Jr. 2001. Asthma. *The New England journal of medicine* 344:350–362.
2. Mannino, D.M., D.M. Homa, L.J. Akinbami, J.E. Moorman, C. Gwynn, and S.C. Redd. 2002. Surveillance for asthma – United States, 1980–1999. *MMWR Surveill Summ* 51:1–13.
3. Umetsu, D.T., J.J. McIntire, O. Akbari, C. Macaubas, and R.H. DeKruyff. 2002. Asthma: an epidemic of dysregulated immunity. *Nature immunology* 3:715–720.
4. Gold, D.R., and R. Wright. 2005. Population disparities in asthma. *Annual review of public health* 26:89–113.
5. Kiley, J., R. Smith, and P. Noel. 2007. Asthma phenotypes. *Current opinion in pulmonary medicine* 13:19–23.
6. Umetsu, D.T., and R.H. Dekruyff. 2006. Immune dysregulation in asthma. *Current opinion in immunology* 18:727–732.
7. Voehringer, D., T.A. Reese, X. Huang, K. Shinkai, and R.M. Locksley. 2006. Type 2 immunity is controlled by IL-4/IL-13 expression in hematopoietic non-eosinophil cells of the innate immune system. *The Journal of experimental medicine* 203:1435–1446.
8. Gibson, P.G., J.L. Simpson, and N. Saltos. 2001. Heterogeneity of airway inflammation in persistent asthma : evidence of neutrophilic inflammation and increased sputum interleukin-8. *Chest* 119:1329–1336.
9. Li, N., J. Alam, M.I. Venkatesan, A. Eiguren-Fernandez, D. Schmitz, E. Di Stefano, N. Slaughter, E. Killeen, X. Wang, A. Huang, M. Wang, A.H. Miguel, A. Cho, C. Sioutas, and A.E. Nel. 2004. Nrf2 is a key transcription factor that regulates antioxidant defense in macrophages and epithelial cells: protecting against the proinflammatory and oxidizing effects of diesel exhaust chemicals. *Journal of Immunology* 173:3467–3481.
10. Wills-Karp, M. 1999. Immunologic basis of antigen-induced airway hyperresponsiveness. *Annual review of immunology* 17:255–281.
11. Brusasco, V., E. Crimi, and R. Pellegrino. 1998. Airway hyperresponsiveness in asthma: not just a matter of airway inflammation. *Thorax* 53:992–998.
12. Grunig, G., M. Warnock, A.E. Wakil, R. Venkayya, F. Brombacher, D.M. Rennick, D. Sheppard, M. Mohrs, D.D. Donaldson, R.M. Locksley, and D.B. Corry. 1998. Requirement for IL-13 independently of IL-4 in experimental asthma. *Science (New York)* 282: 2261–2263.
13. Romagnani, S. 2004. Immunologic influences on allergy and the TH1/TH2 balance. *The Journal of allergy and clinical immunology* 113:395–400.
14. Gavett, S.H., X. Chen, F. Finkelman, and M. Wills-Karp. 1994. Depletion of murine CD4+ T lymphocytes prevents antigen-induced airway hyperreactivity and pulmonary eosinophilia. *American journal of respiratory cell and molecular biology* 10:587–593.
15. Robinson, D.S., Q. Hamid, S. Ying, A. Tsicopoulos, J. Barkans, A.M. Bentley, C. Corrigan, S.R. Durham, and A.B. Kay. 1992. Predominant TH2-like bronchoalveolar T-lymphocyte population in atopic asthma. *The New England journal of medicine* 326:298–304.
16. Mosmann, T.R., H. Cherwinski, M.W. Bond, M.A. Giedlin, and R.L. Coffman. 2005. Two types of murine helper T cell clone. I. Definition according to profiles of lymphokine activities and secreted proteins *Journal of Immunology* 175:5–14.
17. Wills-Karp, M., and S.L. Ewart. 1997. The genetics of allergen-induced airway hyperresponsiveness in mice. *American journal of respiratory and critical care medicine* 156:S89–96.

18. Cohn, L., J.A. Elias, and G.L. Chupp. 2004. Asthma: mechanisms of disease persistence and progression. *Annual review of immunology* 22:789–815.
19. Holgate, S.T. 2000. Epithelial damage and response. *Clinical and Experimental Allergy* 30(Suppl 1):37–41.
20. Crowe, N.Y., J.M. Coquet, S.P. Berzins, K. Kyparissoudis, R. Keating, D.G. Pellicci, Y. Hayakawa, D.I. Godfrey, and M.J. Smyth. 2005. Differential antitumor immunity mediated by NKT cell subsets in vivo. *The Journal of experimental medicine* 202:1279–1288.
21. Gumperz, J.E., S. Miyake, T. Yamamura, and M.B. Brenner. 2002. Functionally distinct subsets of CD1d-restricted natural killer T cells revealed by CD1d tetramer staining. *The Journal of experimental medicine* 195:625–636.
22. Kim, C.H., E.C. Butcher, and B. Johnston. 2002. Distinct subsets of human Valpha24-invariant NKT cells: cytokine responses and chemokine receptor expression. *Trends in immunology* 23:516–519.
23. Lee, P.T., K. Benlagha, L. Teyton, and A. Bendelac. 2002. Distinct functional lineages of human V(alpha)24 natural killer T cells. *The Journal of experimental medicine* 195:637–641.
24. Bendelac, A., M.N. Rivera, S.H. Park, and J.H. Roark. 1997. Mouse CD1-specific NK1 T cells: development, specificity, and function. *Annual review of immunology* 15:535–562.
25. Brossay, L., M. Chioda, N. Burdin, Y. Koezuka, G. Casorati, P. Dellabona, and M. Kronenberg. 1998. CD1d-mediated recognition of an alpha-galactosylceramide by natural killer T cells is highly conserved through mammalian evolution. *The Journal of experimental medicine* 188:1521–1528.
26. Umetsu, D.T., and R.H. DeKruyff. 2006. A role for natural killer T cells in asthma. *Nature reviews* 6:953–958.
27. Hayakawa, Y., D.I. Godfrey, and M.J. Smyth. 2004. Alpha-galactosylceramide: potential immunomodulatory activity and future application. *Current medicinal chemistry* 11:241–252.
28. Sidobre, S., O.V. Naidenko, B.C. Sim, N.R. Gascoigne, K.C. Garcia, and M. Kronenberg. 2002. The V alpha 14 NKT cell TCR exhibits high-affinity binding to a glycolipid/CD1d complex. *J Immunol* 169:1340–1348.
29. Brutkiewicz, R.R. 2006. CD1d ligands: the good, the bad, and the ugly. *Journal of Immunology* 177:769–775.
30. Kronenberg, M., and L. Gapin. 2002. The unconventional lifestyle of NKT cells. *Nature reviews* 2:557–568.
31. Kronenberg, M. 2005. Toward an understanding of NKT cell biology: progress and paradoxes. *Annual review of immunology* 23:877–900.
32. Mattner, J., N. Donhauser, G. Werner-Felmayer, and C. Bogdan. 2006. NKT cells mediate organ-specific resistance against Leishmania major infection. *Microbes and infection/Institut Pasteur* 8:354–362.
33. Zhou, D., C. Cantu, 3rd, Y. Sagiv, N. Schrantz, A.B. Kulkarni, X. Qi, D.J. Mahuran, C.R. Morales, G.A. Grabowski, K. Benlagha, P. Savage, A. Bendelac, and L. Teyton. 2004. Editing of CD1d-bound lipid antigens by endosomal lipid transfer proteins. *Science (New York)* 303:523–527.
34. Zhou, D., J. Mattner, C. Cantu, 3rd, N. Schrantz, N. Yin, Y. Gao, Y. Sagiv, K. Hudspeth, Y.P. Wu, T. Yamashita, S. Teneberg, D. Wang, R.L. Proia, S.B. Levery, P.B. Savage, L. Teyton, and A. Bendelac. 2004. Lysosomal glycosphingolipid recognition by NKT cells. *Science (New York)* 306:1786–1789.
35. Porubsky, S., A.O. Speak, B. Luckow, V. Cerundolo, F.M. Platt, and H.J. Grone. 2007. Normal development and function of invariant natural killer T cells in mice with isoglobotrihexosylceramide (iGb3) deficiency. *Proceedings of the National Academy of Sciences of the United States of America* 104:5977–5982.

36. Shamshiev, A., A. Donda, I. Carena, L. Mori, L. Kappos, and G. De Libero. 1999. Self glycolipids as T-cell autoantigens. *European Journal of Immunology* 29:1667–1675.
37. Shamshiev, A., H.J. Gober, A. Donda, Z. Mazorra, L. Mori, and G. De Libero. 2002. Presentation of the same glycolipid by different CD1 molecules. *The Journal of experimental medicine* 195:1013–1021.
38. Jahng, A., I. Maricic, C. Aguilera, S. Cardell, R.C. Halder, and V. Kumar. 2004. Prevention of autoimmunity by targeting a distinct, noninvariant CD1d-reactive T cell population reactive to sulfatide. *The Journal of experimental medicine* 199:947–957.
39. Wu, D.Y., N.H. Segal, S. Sidobre, M. Kronenberg, and P.B. Chapman. 2003. Cross-presentation of disialoganglioside GD3 to natural killer T cells. *The Journal of experimental medicine* 198:173–181.
40. Gumperz, J.E., C. Roy, A. Makowska, D. Lum, M. Sugita, T. Podrebarac, Y. Koezuka, S.A. Porcelli, S. Cardell, M.B. Brenner, and S.M. Behar. 2000. Murine CD1d-restricted T cell recognition of cellular lipids. *Immunity* 12:211–221.
41. Kinjo, Y., E. Tupin, D. Wu, M. Fujio, R. Garcia-Navarro, M.R. Benhnia, D.M. Zajonc, G. Ben-Menachem, G.D. Ainge, G.F. Painter, A. Khurana, K. Hoebe, S.M. Behar, B. Beutler, I.A. Wilson, M. Tsuji, T.J. Sellati, C.H. Wong, and M. Kronenberg. 2006. Natural killer T cells recognize diacylglycerol antigens from pathogenic bacteria. *Nature immunology* 7: 978–986.
42. Kinjo, Y., D. Wu, G. Kim, G.W. Xing, M.A. Poles, D.D. Ho, M. Tsuji, K. Kawahara, C.H. Wong, and M. Kronenberg. 2005. Recognition of bacterial glycosphingolipids by natural killer T cells. *Nature* 434:520–525.
43. Wu, D., G.W. Xing, M.A. Poles, A. Horowitz, Y. Kinjo, B. Sullivan, V. Bodmer-Narkevitch, O. Plettenburg, M. Kronenberg, M. Tsuji, D.D. Ho, and C.H. Wong. 2005. Bacterial glycolipids and analogs as antigens for CD1d-restricted NKT cells. *Proceedings of the National Academy of Sciences of the United States of America* 102:1351–1356.
44. Agea, E., A. Russano, O. Bistoni, R. Mannucci, I. Nicoletti, L. Corazzi, A.D. Postle, G. De Libero, S.A. Porcelli, and F. Spinozzi. 2005. Human CD1-restricted T cell recognition of lipids from pollens. *The Journal of experimental medicine* 202:295–308.
45. Spinozzi, F., and S.A. Porcelli. 2007. Recognition of lipids from pollens by CD1-restricted T cells. *Immunology and allergy clinics of North America* 27:79–92.
46. Bendelac, A., P.B. Savage, and L. Teyton. 2007. The biology of NKT cells. *Annual review of immunology* 25:297–336.
47. Carnaud, C., D. Lee, O. Donnars, S.H. Park, A. Beavis, Y. Koezuka, and A. Bendelac. 1999. Cutting edge: Cross-talk between cells of the innate immune system: NKT cells rapidly activate NK cells. *Journal of Immunology* 163:4647–4650.
48. Cui, J., N. Watanabe, T. Kawano, M. Yamashita, T. Kamata, C. Shimizu, M. Kimura, E. Shimizu, J. Koike, H. Koseki, Y. Tanaka, M. Taniguchi, and T. Nakayama. 1999. Inhibition of T helper cell type 2 cell differentiation and immunoglobulin E response by ligand-activated Valpha14 natural killer T cells. *The Journal of experimental medicine* 190:783–792.
49. Taniguchi, M., M. Harada, S. Kojo, T. Nakayama, and H. Wakao. 2003. The regulatory role of Valpha14 NKT cells in innate and acquired immune response. *Annual review of immunology* 21:483–513.
50. Taniguchi, M., K. Seino, and T. Nakayama. 2003. The NKT cell system: bridging innate and acquired immunity. *Nature immunology* 4:1164–1165.
51. Yoshimoto, T., and W.E. Paul. 1994. CD4pos, NK1.1pos T cells promptly produce interleukin 4 in response to in vivo challenge with anti-CD3. *The Journal of experimental medicine* 179:1285–1295.
52. Galli, G., S. Nuti, S. Tavarini, L. Galli-Stampino, C. De Lalla, G. Casorati, P. Dellabona, and S. Abrignani. 2003. CD1d-restricted help to B cells by human invariant natural killer T lymphocytes. *The Journal of experimental medicine* 197:1051–1057.
53. Ueno, Y., S. Tanaka, M. Sumii, S. Miyake, S. Tazuma, M. Taniguchi, T. Yamamura, and K. Chayama. 2005. Single dose of OCH improves mucosal T helper type 1/T helper type 2

cytokine balance and prevents experimental colitis in the presence of valpha14 natural killer T cells in mice. *Inflammatory bowel diseases* 11:35–41.
54. van Dieren, J.M., C.J. van der Woude, E.J. Kuipers, J.C. Escher, J.N. Samsom, R.S. Blumberg, and E.E. Nieuwenhuis. 2007. Roles of CD1d-restricted NKT cells in the intestine. *Inflammatory bowel diseases* 13:1146–1152.
55. Fuss, I.J., F. Heller, M. Boirivant, F. Leon, M. Yoshida, S. Fichtner-Feigl, Z. Yang, M. Exley, A. Kitani, R.S. Blumberg, P. Mannon, and W. Strober. 2004. Nonclassical CD1d-restricted NK T cells that produce IL-13 characterize an atypical Th2 response in ulcerative colitis. *The Journal of clinical investigation* 113:1490–1497.
56. Kaser, A., and R.S. Blumberg. 2004. The other way round: colitis regulates regulatory T cells. *Gastroenterology* 126:1903–1906.
57. Kita, H., O.V. Naidenko, M. Kronenberg, A.A. Ansari, P. Rogers, X.S. He, F. Koning, T. Mikayama, J. Van De Water, R.L. Coppel, M. Kaplan, and M.E. Gershwin. 2002. Quantitation and phenotypic analysis of natural killer T cells in primary biliary cirrhosis using a human CD1d tetramer. *Gastroenterology* 123:1031–1043.
58. Brigl, M., L. Bry, S.C. Kent, J.E. Gumperz, and M.B. Brenner. 2003. Mechanism of CD1d-restricted natural killer T cell activation during microbial infection. *Nature immunology* 4:1230–1237.
59. Godfrey, D.I., and S.P. Berzins. 2006. NKT cells join the war on Lyme disease. *Nature immunology* 7:904–906.
60. Shimizu, H., T. Matsuguchi, Y. Fukuda, I. Nakano, T. Hayakawa, O. Takeuchi, S. Akira, M. Umemura, T. Suda, and Y. Yoshikai. 2002. Toll-like receptor 2 contributes to liver injury by Salmonella infection through Fas ligand expression on NKT cells in mice. *Gastroenterology* 123:1265–1277.
61. Nichols, K.E., J. Hom, S.Y. Gong, A. Ganguly, C.S. Ma, J.L. Cannons, S.G. Tangye, P.L. Schwartzberg, G.A. Koretzky, and P.L. Stein. 2005. Regulation of NKT cell development by SAP, the protein defective in XLP. *Nature medicine* 11:340–345.
62. Pasquier, B., L. Yin, M.C. Fondaneche, F. Relouzat, C. Bloch-Queyrat, N. Lambert, A. Fischer, G. de Saint-Basile, and S. Latour. 2005. Defective NKT cell development in mice and humans lacking the adapter SAP, the X-linked lymphoproliferative syndrome gene product. *The Journal of experimental medicine* 201:695–701.
63. Rigaud, S., M.C. Fondaneche, N. Lambert, B. Pasquier, V. Mateo, P. Soulas, L. Galicier, F. Le Deist, F. Rieux-Laucat, P. Revy, A. Fischer, G. de Saint Basile, and S. Latour. 2006. XIAP deficiency in humans causes an X-linked lymphoproliferative syndrome. *Nature* 444:110–114.
64. Akbari, O., P. Stock, E. Meyer, M. Kronenberg, S. Sidobre, T. Nakayama, M. Taniguchi, M.J. Grusby, R.H. DeKruyff, and D.T. Umetsu. 2003. Essential role of NKT cells producing IL-4 and IL-13 in the development of allergen-induced airway hyperreactivity. *Nature medicine* 9:582–588.
65. Joetham, A., K. Takeda, C. Taube, N. Miyahara, A. Kanehiro, A. Dakhama, and E.W. Gelfand. 2005. Airway hyperresponsiveness in the absence of CD4+ T cells after primary but not secondary challenge. *American journal of respiratory cell and molecular biology* 33:89–96.
66. Lisbonne, M., S. Diem, A. de Castro Keller, J. Lefort, L.M. Araujo, P. Hachem, J.M. Fourneau, S. Sidobre, M. Kronenberg, M. Taniguchi, P. Van Endert, M. Dy, P. Askenase, M. Russo, B.B. Vargaftig, A. Herbelin, and M.C. Leite-de-Moraes. 2003. Cutting edge: invariant V alpha 14 NKT cells are required for allergen-induced airway inflammation and hyperreactivity in an experimental asthma model. *Journal of Immunology* 171: 1637–1641.
67. Smiley, S.T., M.H. Kaplan, and M.J. Grusby. 1997. Immunoglobulin E production in the absence of interleukin-4-secreting CD1-dependent cells. *Science (New York)* 275: 977–979.

68. Akbari, O., J.L. Faul, E.G. Hoyte, G.J. Berry, J. Wahlstrom, M. Kronenberg, R.H. DeKruyff, and D.T. Umetsu. 2006. CD4+ invariant T-cell-receptor+ natural killer T cells in bronchial asthma. *The New England journal of medicine* 354:1117–1129.
69. Fujiki R., Y.T., Watson R.M., Gauvreau G.M., O'Byrne P.M., Hamilton M. 2007. Natural killer T cells in sputum and peripheral blood from subjects with allergen-induced late asthmatic aesponses. Manuscript in preparation.
70. Hamzaoui, A., S.C. Rouhou, H. Grairi, H. Abid, J. Ammar, H. Chelbi, and K. Hamzaoui. 2006. NKT cells in the induced sputum of severe asthmatics. *Mediators of inflammation* 2006:71214.
71. Pham-Thi, N., J. de Blic, and M.C. Leite-de-Moraes. 2006. Invariant natural killer T cells in bronchial asthma. *The New England journal of medicine* 354:2613–2616; author reply 2613–2616.
72. Sen, Y., B. Yongyi, H. Yuling, X. Luokun, H. Li, X. Jie, D. Tao, Z. Gang, L. Junyan, H. Chunsong, X. Zhang, J. Youxin, G. Feili, J. Boquan, and T. Jinquan. 2005. V alpha 24-invariant NKT cells from patients with allergic asthma express CCR9 at high frequency and induce Th2 bias of CD3+ T cells upon CD226 engagement. *Journal of Immunology* 175:4914–4926.
73. Thomas, S.Y., A. Banerji, B.D. Medoff, C.M. Lilly, and A.D. Luster. 2007. Multiple chemokine receptors, including CCR6 and CXCR3, regulate antigen-induced T cell homing to the human asthmatic airway. *Journal of Immunology* 179:1901–1912.
74. Thomas, S.Y., C.M. Lilly, and A.D. Luster. 2006. Invariant natural killer T cells in bronchial asthma. *The New England journal of medicine* 354:2613–2616; author reply 2613–2616.
75. Vijayanand, P., G. Seumois, C. Pickard, R.M. Powell, G. Angco, D. Sammut, S.D. Gadola, P.S. Friedmann, and R. Djukanovic. 2007. Invariant natural killer T cells in asthma and chronic obstructive pulmonary disease. *The New England journal of medicine* 356:1410–1422.
76. Meyer, E.H., S. Goya, O. Akbari, G.J. Berry, P.B. Savage, M. Kronenberg, T. Nakayama, R.H. DeKruyff, and D.T. Umetsu. 2006. Glycolipid activation of invariant T cell receptor+ NK T cells is sufficient to induce airway hyperreactivity independent of conventional CD4+ T cells. *Proceedings of the National Academy of Sciences of the United States of America* 103:2782–2787.
77. Brightling, C.E., P. Bradding, F.A. Symon, S.T. Holgate, A.J. Wardlaw, and I.D. Pavord. 2002. Mast-cell infiltration of airway smooth muscle in asthma. *The New England journal of medicine* 346:1699–1705.
78. Sekiya, T., M. Miyamasu, M. Imanishi, H. Yamada, T. Nakajima, M. Yamaguchi, T. Fujisawa, R. Pawankar, Y. Sano, K. Ohta, A. Ishii, Y. Morita, K. Yamamoto, K. Matsushima, O. Yoshie, and K. Hirai. 2000. Inducible expression of a Th2-type CC chemokine thymus- and activation-regulated chemokine by human bronchial epithelial cells. *Journal of Immunology* 165:2205–2213.
79. Meyer, E., Wurbel, M.A., Staton, T.L., Pichavant, M., Kan, M., Savage, P.B., Dekruyff, R.H., Butcher, E.C., Campbell, J.J., Umetsu, D.T. 2007. A specialized subset of iNKT cells require CCR4 to localize to the airways and to induce airway hyperreactivity. *Journal of Immunology*, Oct 1; 179:4661–4671.
80. Michel, M.L., A.C. Keller, C. Paget, M. Fujio, F. Trottein, P.B. Savage, C.H. Wong, E. Schneider, M. Dy, and M.C. Leite-de-Moraes. 2007. Identification of an IL-17-producing NK1.1(neg) iNKT cell population involved in airway neutrophilia. *The Journal of experimental medicine* 204:995–1001.
81. Hellings, P.W., A. Kasran, Z. Liu, P. Vandekerckhove, A. Wuyts, L. Overbergh, C. Mathieu, and J.L. Ceuppens. 2003. Interleukin-17 orchestrates the granulocyte influx into airways after allergen inhalation in a mouse model of allergic asthma. *American journal of respiratory cell and molecular biology* 28:42–50.
82. Barczyk, A., W. Pierzchala, and E. Sozanska. 2003. Interleukin-17 in sputum correlates with airway hyperresponsiveness to methacholine. *Respiratory medicine* 97:726–733.

83. Schnyder-Candrian, S., D. Togbe, I. Couillin, I. Mercier, F. Brombacher, V. Quesniaux, F. Fossiez, B. Ryffel, and B. Schnyder. 2006. Interleukin-17 is a negative regulator of established allergic asthma. *The Journal of experimental medicine* 203:2715–2725.
84. Maeda, M., A. Shadeo, A.M. MacFadyen, and F. Takei. 2004. CD1d-independent NKT cells in beta 2-microglobulin-deficient mice have hybrid phenotype and function of NK and T cells. *Journal of Immunology* 172:6115–6122.
85. Zhang, Y., K.H. Rogers, and D.B. Lewis. 1996. Beta 2-microglobulin-dependent T cells are dispensable for allergen-induced T helper 2 responses. *The Journal of experimental medicine* 184:1507–1512.
86. Kaser, A., E.E. Nieuwenhuis, W. Strober, L. Mayer, I. Fuss, S. Colgan, and R.S. Blumberg. 2004. Natural killer T cells in mucosal homeostasis. *Annals of the New York Academy of Sciences* 1029:154–168.
87. Milner, J.D., S.C. Kent, T.A. Ashley, S.B. Wilson, J.L. Strominger, and D.A. Hafler. 1999. Differential responses of invariant V alpha 24J alpha Q T cells and MHC class II-restricted CD4+ T cells to dexamethasone. *Journal of Immunology* 163:2522–2529.
88. Tamada, K., M. Harada, K. Abe, T. Li, and K. Nomoto. 1998. IL-4-producing NK1.1+ T cells are resistant to glucocorticoid-induced apoptosis: implications for the Th1/Th2 balance. *Journal of Immunology* 161:1239–1247.
89. Ito, K., K.F. Chung, and I.M. Adcock. 2006. Update on glucocorticoid action and resistance. *The Journal of allergy and clinical immunology* 117:522–543.
90. Hachem, P., M. Lisbonne, M.L. Michel, S. Diem, S. Roongapinun, J. Lefort, G. Marchal, A. Herbelin, P.W. Askenase, M. Dy, and M.C. Leite-de-Moraes. 2005. Alpha-galactosylceramide-induced iNKT cells suppress experimental allergic asthma in sensitized mice: role of IFN-gamma. *European Journal of Immunology* 35:2793–2802.
91. Matsuda, H., T. Suda, J. Sato, T. Nagata, Y. Koide, K. Chida, and H. Nakamura. 2005. alpha-Galactosylceramide, a ligand of natural killer T cells, inhibits allergic airway inflammation. *American journal of respiratory cell and molecular biology* 33:22–31.
92. Morishima, Y., Y. Ishii, T. Kimura, A. Shibuya, K. Shibuya, A.E. Hegab, T. Iizuka, T. Kiwamoto, Y. Matsuno, T. Sakamoto, A. Nomura, M. Taniguchi, and K. Sekizawa. 2005. Suppression of eosinophilic airway inflammation by treatment with alpha-galactosylceramide. *European Journal of Immunology* 35:2803–2814.
93. Hansen, G., G. Berry, R.H. DeKruyff, and D.T. Umetsu. 1999. Allergen-specific Th1 cells fail to counterbalance Th2 cell-induced airway hyperreactivity but cause severe airway inflammation. *The Journal of clinical investigation* 103:175–183.
94. Akdis, M., K. Blaser, and C.A. Akdis. 2005. T regulatory cells in allergy: novel concepts in the pathogenesis, prevention, and treatment of allergic diseases. *The Journal of allergy and clinical immunology* 116:961–968; quiz 969.
95. Ling, E.M., T. Smith, X.D. Nguyen, C. Pridgeon, M. Dallman, J. Arbery, V.A. Carr, and D.S. Robinson. 2004. Relation of CD4+CD25+ regulatory T-cell suppression of allergen-driven T-cell activation to atopic status and expression of allergic disease. *Lancet* 363: 608–615.
96. Lewkowich, I.P., N.S. Herman, K.W. Schleifer, M.P. Dance, B.L. Chen, K.M. Dienger, A.A. Sproles, J.S. Shah, J. Kohl, Y. Belkaid, and M. Wills-Karp. 2005. CD4+CD25+ T cells protect against experimentally induced asthma and alter pulmonary dendritic cell phenotype and function. *The Journal of experimental medicine* 202:1549–1561.
97. Romagnani, S. 2006. Regulation of the T cell response. *Clinical and Experimental Allergy* 36:1357–1366.
98. Stock, P., R.H. DeKruyff, and D.T. Umetsu. 2006. Inhibition of the allergic response by regulatory T cells. *Current opinion in allergy and clinical immunology* 6:12–16.
99. Weiner, H.L. 2001. The mucosal milieu creates tolerogenic dendritic cells and T(R)1 and T(H)3 regulatory cells. *Nature immunology* 2:671–672.
100. Bellinghausen, I., B. Konig, I. Bottcher, J. Knop, and J. Saloga. 2006. Inhibition of human allergic T-helper type 2 immune responses by induced regulatory T cells requires the

combination of interleukin-10-treated dendritic cells and transforming growth factor-beta for their induction. *Clinical and Experimental Allergy* 36:1546–1555.
101. Akbari, O., R.H. DeKruyff, and D.T. Umetsu. 2001. Pulmonary dendritic cells producing IL-10 mediate tolerance induced by respiratory exposure to antigen. *Nature immunology* 2: 725–731.
102. Akbari, O., G.J. Freeman, E.H. Meyer, E.A. Greenfield, T.T. Chang, A.H. Sharpe, G. Berry, R.H. DeKruyff, and D.T. Umetsu. 2002. Antigen-specific regulatory T cells develop via the ICOS-ICOS-ligand pathway and inhibit allergen-induced airway hyperreactivity. *Nature medicine* 8:1024–1032.
103. Ostroukhova, M., C. Seguin-Devaux, T.B. Oriss, B. Dixon-McCarthy, L. Yang, B.T. Ameredes, T.E. Corcoran, and A. Ray. 2004. Tolerance induced by inhaled antigen involves CD4(+) T cells expressing membrane-bound TGF-beta and FOXP3. *The Journal of clinical investigation* 114:28–38.
104. Stock, P., O. Akbari, G. Berry, G.J. Freeman, R.H. Dekruyff, and D.T. Umetsu. 2004. Induction of T helper type 1-like regulatory cells that express Foxp3 and protect against airway hyper-reactivity. *Nature immunology* 5:1149–1156.
105. Novak, J., L. Beaudoin, S. Park, T. Griseri, L. Teyton, A. Bendelac, and A. Lehuen. 2007. Prevention of type 1 diabetes by invariant NKT cells is independent of peripheral CD1d expression. *Journal of Immunology* 178:1332–1340.
106. Hong, S., M.T. Wilson, I. Serizawa, L. Wu, N. Singh, O.V. Naidenko, T. Miura, T. Haba, D.C. Scherer, J. Wei, M. Kronenberg, Y. Koezuka, and L. Van Kaer. 2001. The natural killer T-cell ligand alpha-galactosylceramide prevents autoimmune diabetes in non-obese diabetic mice. *Nature medicine* 7:1052–1056.
107. Sharif, S., G.A. Arreaza, P. Zucker, Q.S. Mi, J. Sondhi, O.V. Naidenko, M. Kronenberg, Y. Koezuka, T.L. Delovitch, J.M. Gombert, M. Leite-De-Moraes, C. Gouarin, R. Zhu, A. Hameg, T. Nakayama, M. Taniguchi, F. Lepault, A. Lehuen, J.F. Bach, and A. Herbelin. 2001. Activation of natural killer T cells by alpha-galactosylceramide treatment prevents the onset and recurrence of autoimmune Type 1 diabetes. *Nature medicine* 7:1057–1062.
108. Chen, Y.G., C.M. Choisy-Rossi, T.M. Holl, H.D. Chapman, G.S. Besra, S.A. Porcelli, D.J. Shaffer, D. Roopenian, S.B. Wilson, and D.V. Serreze. 2005. Activated NKT cells inhibit autoimmune diabetes through tolerogenic recruitment of dendritic cells to pancreatic lymph nodes. *Journal of Immunology* 174:1196–1204.
109. Miyamoto, K., S. Miyake, and T. Yamamura. 2001. A synthetic glycolipid prevents autoimmune encephalomyelitis by inducing TH2 bias of natural killer T cells. *Nature* 413:531–534.

Chapter 29
The Development, Activation, Function and Mechanisms of Immunosuppressive Double Negative (DN) T Cells

Megan S. Ford and Li Zhang

Abstract Double negative (DN) T cells are a subset of T cells, present in the peripheral lymphatic organs and blood in very low numbers (1–2% of lymphocytes) in mice and humans. DN T cells have been shown to inhibit transplant rejection, lymphoma development and graft versus host disease. Furthermore, recent studies have suggested that DN T cells may play a role in the prevention and treatment of autoimmune diseases and bacterial and viral infections. This chapter will discuss the development, activation, functions and mechanism of DN T cells in the suppression of immune responses. The data discussed here suggest that DN T cells might be used as a novel therapy to prevent human transplant rejection, limit tumor survival, inhibit autoimmune disease development and control pathogen infection in an antigen-specific manner.

Introduction

$TCR\alpha\beta^+CD4^-CD8^-NK1.1^-$, DN T cells were first identified in mice as possessing regulatory function following donor lymphocyte infusion (DLI). DLI given prior to transplantation can induce specific tolerance to donor type tissue [1–5]. The total numbers and the percentage of DN T cells increased in the spleen and lymph nodes following one MHC class I mismatched DLI [6]. Furthermore, in 2C transgenic (Tg) mice, DN T cells, but not $CD4^+$ or $CD8^+$ T cells, could suppress the proliferation of naïve anti-donor T cells *in vitro* [6]. Moreover, adoptive transfer of DN T cells from the spleen and lymph nodes of DLI-treated mice to naïve mice prevented allogeneic and xenogeneic transplant rejection in a donor specific fashion [6,7]. These data were the first to demonstrate that DN T cells possess regulatory function in mouse models. Subsequently, human DN T cells have also been identified and shown to

---

L. Zhang
Departments of Laboratory Medicine and Pathobiology, Immunology, Toronto General Research Institute, University Health Network, University of Toronto, TMDT 2-807, 101 College Street, Toronto, Ontario, M5G, 1L7, Canada
e-mail: lzhang@uhnres.utoronto.ca

suppress the proliferation of CD8$^+$ T cells in an antigen-specific fashion [8]. These studies indicate that DN T cells may have an evolutionarily conserved function as a regulatory cell subset. Thus, learning the various functions of DN T cells, as well as their mechanism of development, activation and suppression is important in order to devise better strategies to utilize these cells to treat or prevent immune pathologies.

DN T Cell Development

Recent studies into the developmental requirements of regulatory T (Treg) cells have largely been focused on the CD4$^+$CD25$^+$ Treg cell subset. The development of naturally occurring CD4$^+$CD25$^+$ Tregs, which have been shown to be potent inhibitors of autoimmune diseases, requires thymic maturation [9–11]. Neonatal thymectomy significantly reduces the number of CD4$^+$CD25$^+$ Treg cells and promotes the development of autoimmune diseases [12]. In addition, data has shown that some CD4$^+$CD25$^+$ Treg cells undergo a positive selection process in the thymus that involves interaction with self-peptides presented in the context of MHC class II [13–15]. Further, haematopoitic chimera mice that express superantigen only on thymic epithelial cells were shown to have enhanced development of superantigen-specific CD4$^+$CD25$^+$ Treg cells [16]. A recent study also shows that high-affinity self-TCR expressing CD4$^+$CD25$^+$FoxP3$^+$ Treg cells are selected on thymic epithelial cells as early as the CD4$^+$CD8$^+$ stage of development [17]. On the other hand, studies have also shown that adult thymectomy along with depletion of CD25$^+$ cells could not prevent the development of CD4$^+$CD25$^+$ Treg from CD4$^+$CD25$^-$ precursors [18]. These adaptive Treg could prevent allotransplant rejection when adoptively transferred into naive recipient mice [18]. This suggests that some CD4$^+$CD25$^+$ Treg cells can develop outside the thymus. Collectively, these findings indicate that although the thymus may be required for the initial development of some Treg cells, the gain of Treg cell function can also happen in the absence of the thymus.

The origins and developmental pathways involved in DN T cell maturation have not been extensively studied. A recent study done by Aifantis et al. has demonstrated that DN T cells can develop in the thymus, spleen and lymph nodes of transgenic, as well as non-transgenic mice, under the control of the E delta enhancer [19]. DN phenotype cells were also shown to develop during fetal thymic organ culture [19]. Furthermore, CD1d–/– and Tcrd–/– mice have DN T cells, indicating that they are not NK or γδ lineage cells [19]. Although a significant portion of T cells in the thymus possess the DN T cell phenotype, whether any of them go on to become Tregs in the lymphatic periphery is not known. Wang et al. showed that thymic double positive cells can down-regulate CD4 and CD8 molecules and become DN T cells when stimulated with high affinity antigens in reaggregate cultures. Interestingly, these DN T cells were also able to suppress T cell proliferation *in vitro* [20]. Experiments have suggested that some DN T cell populations may develop extrathymically in the appendix [21], female genital tract [22], nasal associated lymphoid tissue (NALT) [23], or the liver [24,25]. However, the suppressive function of these DN T cells is yet to be determined. Our data has also suggested that DN T cells can develop outside of the thymus. Thymectomy did not prevent the development of DN T cells from bone marrow cells, and these DN T cells were able to suppress and kill autologous T cells, suggesting that the thymus is not required for regulatory DN T cell development [26]. However, it is probable that the DN T cell population comprises a heterogeneous group of cells that possesses varied functions and may originate and develop from diverse locations *in vivo*.

Other studies have focussed on the developmental pathways whereby DN T cells may originate. In one study, $CD8^+$ T cells treated with IL-4 and ionomycin were shown to differentiate into DN T cells that produce IL-4 and other Th2 cytokines [27]. However, other work has indicated the DN T cells do not develop from $CD8^+$ T cell precursors. DN T cells were shown to be present in the lymph nodes and spleen of $CD8\alpha^{-/-}$ mice, and these DN T cells could kill autologous $CD8^+$ T cells when activated by alloantigens [26]. Further, adoptive transfer of Tg TCR expressing $CD8^+$ cells into Tg TCR-antigen-expressing mice did not result in the development of DN T cells [26]. Taken together, this data suggests that CD8 expression is not essential for the maturation and development of regulatory DN T cells.

DN T Cell Activation

Experiments suggest that DN T cells require stimulation through their TCR in order to gain suppressive function. DLI from either alloantigen MHC class I-mismatched or concordant xenoantigen mismatched animals has been shown to increase DN T cell numbers *in vivo*, and enhances donor-specific graft survival [6,28,29]. However, untreated recipient mice can reject donor grafts within a rapid time-course [6,28,29]. This suggests that DLI may specifically activate the suppressive function of DN T cells either by increasing their numbers *in vivo*, or by enhancing the expression of suppressive molecules by DN T cells. Interestingly, *in vitro* data suggests that lipopolysaccharide (LPS) activated APC, including B cells and mDC are more effective at inducing DN T cell proliferation and suppressive function compared to naïve APC (Ford, MS, manuscript in preparation). This suggests that DN T cells may function to inhibit immune responses in the context of inflammation. Human DN T cells have also been shown to possess suppressive function following activation with antigen expressing DC. Human DN T cells that were primed with Melan A-pulsed DC could inhibit the proliferation of Melan A-specific autologous $CD8^+$ T cells *in vitro* [8].

Analysis of the cell surface and cytokine expression by DN T cells has revealed a unique phenotype. Both murine and human DN T cells lack expression of NK cell markers including NKG2D and CD56 [6,8]. Further, naïve murine and freshly isolated human DN T cells express low to no levels of CD25, CD69 and CTLA-4 [6,8]. However, activated DN T cells cloned from transplant tolerant mice express CD25 and low levels of CD30, but have little to no expression of CD28 or CD44 on their cell surface [6]. In addition, DN T cells activated by i.v. injection of TCR-specific peptides express a high level of CD69 and low levels of CD25. RNA analysis has also shown that DN T cell clones express Fas, FasL, IFN-γ, TNF-α and TGF-β, but no IL-2, IL-4, IL-10 or IL-13 [6]. Furthermore, human DN T cells were also found to express high levels of IFN-γ and undetectable amounts of IL-2, IL-4 and IL-10 [8]. Since the cell surface marker profile and the cytokine expression levels between murine and human DN T cells is remarkably similar, this suggests that the functions and mechanisms of DN T cells may have been evolutionarily conserved across these two species.

DN T cells have also been shown to increase proliferation, and suppress autologous T cell proliferation in the presence of exogenous IL-2 and IL-4 [6,8]. The specific function of these cytokines on DN T cell proliferation and function has not yet been identified. However IL-2 is required for DN T cell proliferation in the presence of alloantigen or TCR-crosslinking antibody, suggesting that similar to conventional T cells, DN T cells may require additional signals to the TCR in order to induce proliferation [30] (Ford, MS, unpublished data). Interestingly, unlike $CD4^+$ or $CD8^+$ T cells which are sensitive to activation induced cell death, DN T cells are resistant to apoptosis induction both *in vitro* and *in vivo*. For example, when the TCR is crosslinked using mAb to CD3, $CD8^+$ T cells undergo apoptosis but DN T cells do not [30,31]. However, when DN T cells are cultured without IL-4, TCR crosslinking does induce DN T cell apoptosis [31]. This suggests IL-4 promotes DN T cell survival. Indeed, DN T cells that are cultured in the presence of IL-4 have been shown to upregulate the expression of BclxL, which has been shown to protect T cells from apoptosis [31]. Further studies have shown that DN T cells are resistant to apoptosis *in vivo* as they persist for a much longer period of time compared to $CD8^+$ T cells following adoptive transfer into alloantigen expressing mice [32]. This feature may allow DN T cells to function for a prolonged period of time to regulate immune responses. Interestingly, when DN T cells are cultured *in vitro* with IL-4 and IL-10, their ability to resist apoptosis is abolished [30,33]. This sensitivity to IL-10 may represent a pathway that is used to modulate the function of DN T cells *in vivo*. It also suggests that the cytokine milieu within the tissues where DN T cells are present is important for their ability to survive and function.

DN T Cells in Transplantation

DN T cells that have been activated by single MHC class I or class II mis-matched antigens, and xenoantigens have been demonstrated to prolong donor-specific allograft survival when adoptively transferred into naïve syngeneic mice [6,7,34]. Recently, similar results have been obtained in a semi-allogeneic transplantation model (Torrealba et al., manuscript in preparation). Donor-antigen-activated DN T cells that were cloned from the spleens of both tolerant and naïve mice could prolong donor-specific skin [6], and heart [35] allograft survival when adoptively transferred into naïve syngeneic recipient mice. Interestingly, both primary activated DN T cells and DN T cell clones are able to accumulate in tolerant grafts in an antigen specific manner [28]. In addition, the DN T cells that infiltrate to the graft have been shown to have a more potent ability to suppress anti-donor T cell proliferation than those isolated from the spleens of recipient mice [7]. One hypothesis for this is that the DN T cells comprise a heterogeneous population of cells, wherein only a subset actually possesses immune regulatory function. Hence, DLI may cause activation of the regulatory subset of DN T cells in the peripheral lymphoid organs and subsequent migration of graft-specific DN T cells to the grafted tissue. Another theory is that DN T cells may be able to expand in the graft itself thereby creating a site of highly enriched, functional DN T cells. Finally, one other postulate is that the local graft environment can somehow potentiate the function of DN T cells when compared to the environment found in the peripheral lymphatic system. Further studies are required in order to determine whether any of these models of DN T cell activation and function are accurate.

DN T cell treatment has also been shown to enhance the survival of concordant xenografts. Experiments demonstrated that mice that have been given a short course of anti-CD4 depleting mAb, together with DLI, could permanently accept concordant (ratmouse) cardiac xenografts, whereas mice given either DLI alone, or anti-CD4 mAb alone, or no treatment, all rejected their cardiac xenografts [29]. DN T cells isolated from tolerant xenograft recipients were able to suppress anti-donor T cell proliferation *in vitro*, suggesting that DN T cells play a role in enhanced xenograft survival [29]. The $CD4^+$ T cell population was found to be critical for xenograft rejection in this model as the transfer of $CD4^+$ T cells into $CD4^{-/-}$ recipient mice was required in order for cardiac xenografts to be rejected. Interestingly, adoptive transfer of DN T cells from tolerant mice into xenograft recipient mice could also suppress the *in vivo* proliferation of anti-donor $CD4^+$ T cells and prolong cardiac xenograft survival [7]. Importantly, xenograft recipient-derived DN T cells could only suppress the proliferation of syngeneic anti-donor T cells, but not the proliferation of syngeneic anti-third party T cells. This suggests that DLI can activate DN T cells to enhance xenograft survival in an antigen specific fashion *in vivo*, indicating that DN T cells could provide a novel therapy for specifically tolerizing transplant recipients to donor type antigens.

The function of DN T cells in preventing human transplant rejection has not yet been extensively studied. However, one recent study indicated that the percentage of peripheral blood $CD30^+$DN T cells that could be primed by donor type spleen cells following lung transplantation was increased in recipients with preserved graft function when compared to those that had poor graft function [36]. This suggests that DN T cells may act to enhance graft survival in human patients.

DN T Cells in Autoimmune Diseases

The function of DN T cells within autoimmune diseases has not been extensively studied. However, a few studies indicate that DN T cells may play a role in preventing autoimmune disease development. Priatel et al. demonstrated that the percentage of antigen-specific DN T cells within the T cell population increases in mice that expressed a Tg TCR specific for a self expressed MHC-class-I antigen (Ag-expressing) compared to those that do not express the MHC-class I antigen (Ag-free) [37]. Furthermore, DN T cells from antigen-expressing mice showed an increased ability to lyse syngeneic $CD8^+$ T cells when compared to DN T cells from Ag-free mice. Since Ag-expressing mice do not develop autoimmune disease although they do have autoreactive T cells, this suggests that DN T cells may be involved in inhibiting autoimmune disease development. In addition, experiments using the autoimmune disease-prone lpr mouse model have demonstrated that the DN T cells that accumulate in these mice possess Treg cell activity and can suppress the proliferation of syngeneic $CD4^+$ and $CD8^+$ T cells that express functional Fas receptors [34]. However, since lpr mice have a mutation that prevents functional Fas receptor expression, autoimmune disease may develop because DN T cells require Fas expression on target T cells in order to induce apoptosis. Therefore, DN T cells may increase in lpr mice in a failed attempt to control the systemic autoimmune disease that develops in these mice with age [34,38]. Interestingly, DN T cells also accumulate in humans with mutations in the Fas death pathway, known as autoimmune lymphoproliferative syndrome (ALPS) [39,40]. However, it is not yet known whether the DN T cells from human ALPS patients can also suppress T cells in a Fas-dependent manner, similar to murine DN T cells. Interestingly, human peripheral DN T cells have been shown to express autoreactive TCR $V\beta$ repertoires that cause the deletion of $CD4^+$ and $CD8^+$ T cells [41,42] This suggests that human DN T cells may prevent the development of autoimmune diseases by inducing the deletion of autoreactive T cells.

Recent work has indicated that DN T cell treatment can inhibit autoimmune Type-1 diabetes (T1D) development in mice. P14/RIP-gp mice that express a Tg TCR (P14) specific for gp33 peptides expressed in the pancreas under the control of the rat insulin promoter (RIP-gp33) are tolerant to gp33 expression under nonpathogenic conditions. However, i.v. treatment of P14/RIP-gp mice with gp33 and an anti-CD40 agonist mAb induces T1D which can be prevented by pre-treatment with gp33-activated DN T cells [43]. The specific mechanisms behind DN T cell mediated prevention of T1D in this model remain to be determined; however DN T cells could suppress autologous $CD8^+$ T cell proliferation in response to gp33 and kill gp33-activated autologous $CD8^+$ T cells, suggesting that they may use these mechanisms to prevent T1D development. Taken together these data suggest that DN T cells may be used as a therapy to prevent or treat autoimmune diseases.

DN T Cells in Cancer

Several recent studies have addressed the ability of Treg cells to either promote or prevent the development of cancer. DN T cells have been shown to possess anti-tumor functions. As in the transplantation models, DLI of single MHC class I-mismatched splenocytes was used to activate DN T cells, this time in immunodeficient SCID recipient mice. The recipients were then treated with a lethal dose of syngeneic A20 lymphoma. Mice that received DLI prior to tumor treatment had a significantly lower mortality rate (25%, n = 20) when compared to untreated controls (100%) [44]. Interestingly, although the mice were treated with MHC-mismatched splenocytes, none of the mice that survived the tumor challenge developed GvHD [45]. Furthermore, as in the transplantation models, donor-derived DN T cells increased following the DLI, and these DN T cells were able to suppress the proliferation of both A20 tumor cells, and GvHD causing-anti-host CD8$^+$ T cells *in vitro* [45]. This suggests that DN T cells can suppress both tumor development, and the development of GvHD [44,45]. Importantly, DN T cells have been shown to be responsible for the anti-tumor effect seen in this model as lymphoma development is prevented in mice given a lethal dose of A20 tumor by the transfer of one MHC-class-I mismatched DN T cell clones or primary activated DN T cells [44]. Again, no GvHD was seen in the mice that had been treated with MHC-class I mismatched DN T cells, either alone or in combination with lymphoma treatment, indicating that the anti-tumor effect can occur in the absence of GvHD [44]. Furthermore, Abraham et al. have also identified NK1.1$^-$ DN T cells in the bone marrow of mice that are protected from GvHD by a short course of high dose IL-2 [46]. However, the function of these bone marrow-derived DN T cells has not yet been determined. In addition, preimmunization of bone marrow donor mice with recipient type antigens was shown to decrease GvHD responses in minor antigen mismatched bone marrow reconstituted recipients [47]. The decrease in GvHD was dependent on the presence of the DN T cell population in donor bone marrow as depletion of CD4, CD8 or NK cells did not stimulate GvHD development in recipient mice, whereas depletion of total T cells by anti-CD5 mAb treatment did induce GvHD development [47]. Together, these studies suggest that DN T cells may provide a novel cellular therapy to prevent tumor outgrowth without the induction of GvHD.

However, there is some data to suggest that DN T cell infiltration to tumors may allow for tumor proliferation by preventing the function of other anti-tumor T cells. In one study, cells bearing the DN T cell phenotype were found to infiltrate into tumors and their presence inversely correlated with the survival of tumor treated animals [48]. It is not known whether the tumor itself can play a role in preventing DN T cell function by decreasing MHC or other cell surface marker expression, or by secreting inhibitory cytokines. Clearly, more experiments are needed to determine how to manipulate the function of DN T cells *in vivo* in order to eradicate tumors. Experiments are also currently underway to determine whether human DN T cells, like murine DN T cells can prevent tumor outgrowth without GvHD.

DN T Cells in Tissue Damage and Infectious Diseases

The presence of DN T cells in other disease processes has also been documented. In one study, DN T cells were found to increase in damaged tissues following burn injuries. Interestingly, these DN T cells produced both Th1 and Th2 type cytokines (IFN-γ, IL-2, IL-4 and IL-10), and were shown to have a regulatory effect in mixed lymphocyte cultures [49]. In addition, work done on a model of Listeria monocytogenes infection has shown that DN T cells accumulate in the peritoneal cavity following infection and can contribute to early protection from bacterial infection by secreting cytokines that promote macrophage activation and accumulation [50]. In another study, DN T cells were found to be the dominant T cell population in the uterus and could suppress the proliferation of T cells in response to polyclonal T cell activation [22]. Although experiments did not show that these DN T cells increased in the uterus in response to *Chlamydia trachomatis* infection, or during pregnancy, the authors suggest that the DN T cells may be present in the uterus in order to diminish the CD4 or CD8 T cell responses to these challenges of the immune system [22]. Recent data has also indicated that DN T cells may be involved in controlling intracellular bacterial infections via suppression of macrophage antigen presentation [51]. DN T cells increased following infection with either *Mycobacterium tuberculosis* or *Francisella tularensis* Live Vaccine Strain (LVS) and could inhibit the intracellular growth of these pathogens within macrophages in an antigen-specific fashion *in vitro* [51]. DN T cells have also been shown to increase within human patients during Staphylococcal toxic shock syndrome [52]. However, the function and mechanisms of these DN T cells has not yet been studied.

The presence of DN T cells has also been noted within several models of virus infection. DN T cells increased in the peritoneal cavity, liver and spleen following infection with murine cytomegalovirus (MCMV) [53]. Interestingly, these DN T cells had enhanced production of IFN-γ, TNF-α, Eta-1, and MCP-1, suggesting that DN T cells may have a role in preventing viral infections by producing macrophage-activating cytokines as well as anti-viral cytokines [53]. Interestingly, DN T cells have also been shown to increase in humans during the course of HIV infection [54,55]. Further studies are needed to determine the mechanisms and functions the DN T cells within these infection models. Our data with murine DN T cells suggests that DN T cells can suppress the increased proliferation of autologous T cells in response to LPS and antigen/IL-2 stimulation when compared to antigen/IL-2 stimulation alone (Ford, MS, manuscript in preparation). This suggests that DN T cells can inhibit immune responses in the context of inflammatory stimuli and perhaps they may be involved in preventing the pathologic effects of inflammation.

Mechanisms of DN T Cell Suppression

Cell Contact Mediated Suppression

Experiments have shown that separation of $CD4^+CD25^+$ Treg cells from responder $CD4^+CD25^-$ cells abrogates their suppressive function, indicating that cell-cell contact is required for $CD4^+CD25^+$ Treg cell mediated regulation. Similarly, DN T cell-mediated immune suppression has also been shown to require cell-cell contact since both human and mouse DN T cells that are separated from responder cells using a transwell membrane fail to suppress the proliferation of $CD8^+$ responder cells [6,8]. One of the major pathways that DN T cells have been shown to mediate suppression is via direct cytotoxicity of activated target cells [6,34]. Expression of FasL on murine DN T cells is critical for their cytotoxic function. Inhibiting FasL expression on DN T cells using Fas-Fc fusion protein, or using FasL mutant (gld) DN T cells was shown to significantly inhibit DN T cell mediated killing [6,34]. Furthermore, DN T cells showed an impaired ability to directly kill $CD8^+$ T cells that did not express functional Fas receptors when compared to wild-type

Fas-expressing CD8$^+$ T cells [34]. These data demonstrate that FasL expression on the DN T cell, and Fas expression on the target T cell, is important for DN T cell-mediated regulation of CD8$^+$ T cells.

Several pieces of data suggest that DN T cells may also use other mechanisms to mediate suppression of immune responses. First, the ability of DN T cells to suppress responder T cell proliferation is often much more potent than their ability to cytotoxically lyse T cell targets (unpublished observations). Second, although blocking FasL with Fas-Fc fusion protein significantly reduced DN T cell-mediated killing of CD8$^+$ T cells, blocking of killing is not complete [6,34]. Third, DN T cells have been shown to express Granzyme M [56] and perforin [8], and were able to kill syngeneic T cells and B cells using the perforin/granzyme pathway following activation with xenoantigens [57]. Finally, DN T cells obtained from autoimmune disease-prone lpr mice could inhibit CD4$^+$ and CD8$^+$ T cell proliferation, although to a much lesser extend, even when cell-cell contact was inhibited using a transwell system (Ford, MS, unpublished data). These data suggest that DN T cells may also suppress T cells through as yet undetermined pathways depending on their mode of activation, and the target cell type.

DN T cells have also been shown to suppress the proliferation of CD4$^+$ T cells. Interestingly, some experiments suggest that the mechanism of CD4$^+$ T cell suppression is via a Fas-independent pathway [58]. *In vitro* coculture of DN T cells with CD4$^+$ T cells was shown to inhibit production of IL-2 by the CD4$^+$ T cells in a dose-dependent fashion [58]. Furthermore, DN T cells from Fas-mutant, lpr mice were shown to suppress both syngeneic wild-type Fas expressing, and autologous mutant-Fas expressing CD4$^+$ T cell proliferation to a similar extent [58]. Moreover, in a xenogeneic rat-to-mouse transplantation model, DN T cells were able to suppress cardiac xenograft rejection mediated by CD4$^+$ T cells by inhibiting proliferation and expression of IL-2 compared to recipient mice that did not receive DN T cell treatment and rejected their cardiac xenografts [7]. This data showed for the first time that DN T cells can specifically inhibit the proliferation of CD4$^+$ T cells *in vivo*, and CD4$^+$ T cell IL-2 production *in vitro*.

Neutralization of cytokines, such as IL-4, IL-10 or TGF-β, has been shown to abrogate T cell mediated suppression in some rodent models of autoimmunity, suggesting that some T cells may mediate suppression by releasing these cytokines [59–61]. For instance, the ability of Tr1 cells to suppress the proliferation of naïve CD4$^+$ T cells, antibody production by B cells, and antigen presentation by monocytes and DC *in vitro* can be reversed with neutralizing anti-IL-10 and/or anti-TGF-β antibodies [59,62–64]. The finding that when cell-cell contact is inhibited, wild-type DN T cells cannot suppress T cell proliferation, whereas lpr DN T cells can partially suppress T cell proliferation, suggests that perhaps if DN T cells become chronically activated, such as in the lpr mouse, they may produce some soluble factors that can mediate suppression of T cell responses. As previously outlined, activated murine DN T cells are able to produce TGF-β, TNF-α, and IFN-γ [6,56]. Human DN T cells have also been shown to express high levels of IFN-γ [8]. Some preliminary data has suggested that blocking IFN-γ expression on activated DN T cells can inhibit their ability to kill target cells *in vitro*, although it has not yet been determined if IFN-γ expression affects the overall suppressive function of DN T cells (Ford, MS et al. unpublished observations). Microarray analysis has been used to determine the differences in gene expression between cytotoxic DN T cell clones and those that had lost their cytotoxic function. Several candidate genes were identified that may be involved in the function of DN T cells including FcεRIγ, CXCR5, CD80 and TLR-4 [56]. Interestingly, CXCR5 has been shown to be important for DN T cell mediated homing into allograft tissues [65]. Furthermore, FcεRIγ has recently been demonstrated to be important for DN T cell mediated suppression, and TCR signalling [66]. The potential role that other cytokines and cell surface molecules expressed by DN T cells plays in their ability to mediate immune regulation requires further investigation.

Antigen Specificity

Both mouse and human DN T cells have been shown to suppress immune responses in an antigen-specific manner both *in vitro* and *in vivo*. Experiments suggest that DN T cells require specific interaction between either MHC or TCR on their cell surface, and that of the target cell in order to mediate suppression (Fig. 29.1a and b).

Anti-L^d-TCR expressing DN T cells were shown to kill L^d-expressing tumor cell lines in a dose-dependent fashion, whereas non-L^d-expressing tumor cells were not killed [44]. Similarly, MHC-L^d-expressing target $CD8^+$ T cells, but not third party $CD8^+$ T cells, could be suppressed by DN T cells that had been activated by L^d antigen (Fig. 29.1a) [67]. Interestingly, DN T cells have also been shown to suppress $CD8^+$ T cells through a veto-like mechanism that requires that the TCR on $CD8^+$ T cells interact with MHC expressed on the DN T cell (Fig. 29.1b). In this case, MHC $H-2^{b+}$ DN T cells could kill only $CD8^+$ T cells that expressed anti-$H-2^b$ TCR, and not those that expressed anti-$H-2^s$ TCR [67]. Furthermore, when interaction between TCR and MHC on the DN T and $CD8^+$ T cell is blocked using mAb, DN T cell-mediated killing is inhibited [67]. Likewise, human HLA-A2 restricted DN T cells that were activated by Melan A peptides could only kill $CD8^+$ T cells that were activated by Melan A, but not $CD8^+$ T cells activated by gp-100 [8]. Finally, *in vivo* experiments have suggested that DN T cells require antigen-specific interaction with target cells as DLI preferentially enhances donor-type, but not third party, graft survival [6,28,29]. Furthermore, DN T cells taken from spleen and lymph nodes of DLI treated recipients can enhance donor-type, but not third party graft survival when adoptively transferred into naïve recipient mice [6,7,34]. Taken together, these data suggest that activated DN T cells mediate suppression of immune responses in an antigen-specific fashion.

Acquisition of Alloantigens

Many types of immune cells have the ability to acquire proteins from neighbouring cells, including $CD8^+$ and $CD4^+$ T cells [68–73], B cells [74] and DC [75]. The

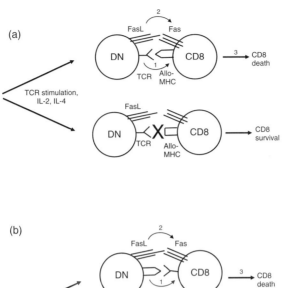

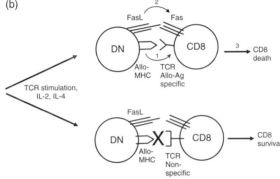

Fig. 29.1 *DN T cell-mediated killing of allogeneic target cells* DN T cells are activated by TCR stimulation, including alloantigen expressed on APC, in the presence of IL-2 and IL-4. (**a**) DN T cells can then interact with allo-MHC expressed on target cells (1). FasL expressed on DN T cells interacts with Fas (2) and the target CD8+ T cell is killed (3). Target cells that do not express MHC that is recognized by the DN T cell-TCR are not killed. (**b**) DN T cells can recognize allogeneic target cells via their MHC interacting with TCR expressed by the CD8+ T cell (1). FasL expressed on DN T cells interacts with Fas (2) and the target CD8+ T cell is killed (3). Target cells that do not express a TCR that can recognize DN T cell expressed MHC are not killed

consequence of acquisition of allo-MHC peptides may be fratricide of CD8+ T cells, linked suppression, or induction of proliferation of naïve T cells [76–80]. Interestingly, T cells that had acquired MHC class II molecules could act as APCs to prime proliferation and IL-2 secretion from resting T cells, whereas interactions with other activated T cells that had acquired alloantigens induced hyporesponsiveness and apoptosis [77]. As outlined previously, DN T cells must be able to interact via either MHC or TCR with the target CD8+ T cell in order for them to suppress or kill (Fig. 29.1). Because of this requirement, it is difficult to imagine a situation whereby two syngeneic MHC expressing DN and CD8+ T cells that carry TCRs reactive against alloantigens would be able to interact. Acquisition of antigens via the TCRs on DN T cells would provide a critical link whereby TCR-specific antigens, either

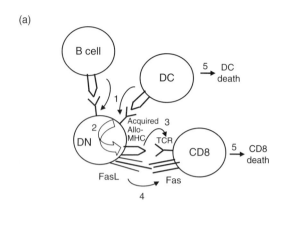

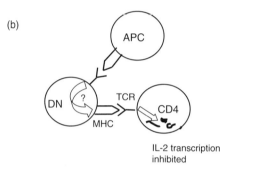

Fig. 29.2 *Mechanisms of syngeneic T cell suppression mediated by DN T cells* (**a**) DN T cells acquire antigen via specific interaction between TCR on the DN T cell and MHC class I/peptide expressed on activated B cells or DC (1). Antigen is then expressed on the surface of DN T cells (2). Antigen reactive $CD8^+$ T cells can then interact specifically with DN T cells via their TCR (3) and FasL on DN T cells interacting with Fas on the activated $CD8^+$ T cell (4). Both target $CD8^+$ T cells and antigen expressing DC are killed upon interacting with DN T cells (5). (**b**) DN T cells may acquire MHC class II expressed antigens from APC and then express them on their cell surface. This may allow for specific interaction between DN T cells and antigen reactive $CD4^+$ T cells via TCR expression on the $CD4^+$ T cell. Proliferation and IL-2 production by $CD4^+$ T cells is prevented in the presence of DN T cells

allogeneic or other peptides, are obtained and expressed on DN T cells, allowing for specific interaction with antigen-reactive T cells via the TCR expressed on the T cell (Fig. 29.2a and b).

Data suggests that acquisition of alloantigens by DN T cells is necessary in order for DN T cell mediated killing of $CD8^+$ T cells to occur. When anti-L^d TCR expressing DN T cells are coincubated with either syngeneic anti-L^d specific Tg $CD8^+$ T cells, or syngeneic $CD8^+$ T cells from anti-male antigen (H-Y) Tg mice in the presence of APC that express both H-Y and L^d, only the anti-L^d specific $CD8^+$ T cells are suppressed [67]. Furthermore, blocking the acquired L^d-antigen on DN T cells using mAb could inhibit the ability of DN T cells to kill target $CD8^+$ T cells [6].

This suggests that acquisition of allo-antigen by DN T cells may be important to allow for suppression of antigen-specific CD8$^+$ T cells.

Subsequently, acquisition of MHC-peptide complexes by DN T cell clones has been shown by flow cytometry [6], and more recently by fluorescence confocal microscopy (Ford, MS and Young, KJ et al., manuscript in preparation). L^d-specific 2C$_{F1}$ DN T cells can acquire and express L^d antigen following coculture with either GFP-L^d transfected Drosophila Fly APC or primary L^d expressing murine APC. Interestingly, DN T cells were able to acquire and present L^d on their cell surface *in vitro* and *in vivo* for a longer period of time than CD8$^+$ T cells [6] and (Ford, MS and Young, KJ, manuscript in preparation). This extended period of antigen expression may give DN T cells an increased window of opportunity to suppress other T cells, by bringing the T cells that are able to recognize the acquired allo-MHC-peptides into cell contact. It is hypothesized that the acquired L^d-molecules on DN T cells may then be directly recognized by the anti-L^d TCR on the responder CD8$^+$ T cells during suppression (Fig. 29.2a).

Data has also suggested that the acquisition of alloantigen by DN T cells is restricted to peptide-MHC complexes that can bind to the TCR expressed on the DN T cell. Alloantigen-activated murine DN T cells were only able to kill target cells that had been activated by the same alloantigen, but not third-party alloantigens [67]. Moreover, acquisition from APC by DN T cells was inhibited by blocking the DN T cell TCR using tetramers, or incubating DN T cells with APC that are not recognized by their TCR (Ford, MS and Young, KJ, manuscript in preparation). Furthermore, A. Mackensen's group has demonstrated using flow cytometry that human HLA-A2 negative DN T cells can acquire allo-HLA-A2-Melan-A peptides from HLA-A2$^+$ DC that have been pulsed *in vitro* with Melan-A peptides [8]. Human DN T cells can then use the acquired Melan-A peptides to trap and kill Melan-A-specific, but not third party gp120 peptide-specific, CD8$^+$ T cells [8]. Further experiments are needed in order to determine the mechanisms of antigen acquisition mediated by murine and human DN T cells. Since the expression of acquired alloantigen on the surface of DN T cells may dictate which cells can come into contact with DN T cells and be suppressed, enhanced knowledge of the molecules and mechanisms that direct this process may allow for DN T cells to be used as an antigen-specific immune therapy.

Migration

Although much has been studied regarding the molecular mechanisms of DN T cell-mediated suppression, the location of DN T cell activation and function is far less characterized. Spleen and lymph node DN T cell numbers and their percentage within the T cell population increases at 7 days following DLI [6,34]. The DN T cell population also increases markedly in donor-type, but not third-party skin grafts that are given to DLI-treated mice [28]. Interestingly, the DN T cell population was shown to infiltrate to donor specific skin grafts in a higher percentage than CD8, CD4, or $\gamma\delta$T cells, or NK cells or monocytes [28]. Moreover, the percentage of DN

T cells relative to other cell populations remained high for over 100 days [28]. This data suggests that DN T cells can migrate to the site of an immune response, and that the infiltration or proliferation of DN T cells occurs in an antigen-specific fashion. Further studies are needed in order to determine whether high numbers of DN T cells actually migrate into graft tissue, or whether small numbers migrate into graft tissue and then proliferate. Recent data suggest that DN T cells, but not CD4 or CD8 T cells, express CXCR5 and can migrate towards CXCL13 expressed on allograft tissue, as blocking CXCR5 using mAb was shown to preferentially inhibit DN T cell migration *in vitro* and *in vivo* [65]. This suggests that DN T cells can migrate towards tissue expressed chemokines, and may use this as a mechanism to increase suppression of immune responses.

Conclusion

DN Treg cells have been shown to be important mediators of immune suppression *in vitro* and *in vivo*. Several studies have demonstrated that DN T cells are a unique population of Treg cells that may provide an innovative treatment to specifically enhance transplanted tissue and organ survival, prevent autoimmune disease development, inhibit tumor cell survival and limit infection pathogenesis. Future studies into the molecular mechanisms leading to DN T cell maturation, activation and suppressive function may allow for the use of DN T cells as a novel therapy for the treatment of immunological disorders in an antigen-specific manner.

Acknowledgments This work was supported by the Canadian Institutes of Health Research and the Canadian Cancer Society.

References

1. van Twuyver, E., Mooijaart, R.J., ten Berge, I.J., van der Horst, A.R., Wilmink, J.M., Kast, W.M., Melief, C.J., and de Waal, L.P., Pretransplantation blood transfusion revisited. *N. Engl. J. Med.* 1991, **325**: 1210–1213.
2. van Twuyver, E., Kast, W.M., Mooijaart, R.J.D., Wilmink, J.M., Melief, C.J.M., and de Waal, L.P., Allograft tolerance induction in adult mice associated with functional deletion of specific CTL precursors. *Transplantation* 1989, **48**: 844–847.
3. van Twuyver, E., Kast, W.M., Mooijaart, R.J.D., Melief, C.J.M., and de Waal, L.P., Induction of transplantation tolerance by intravenous injection of allogeneic lymphocytes across an H-2 class-II mismatch. Different mechanisms operate in tolerization across an H-2 class- I versus H-2 class-II disparity. *Eur. J. Immunol.* 1990, **20**: 441–444.
4. Wood, K.J., Billing, J.S., Binet, I., Bueno, V, and Fry, J., Which donor cells facilitate the induction of specific immunological unresponsiveness to alloantigens in vivo? *Transplantation* 2002, **73**: S16–S18.
5. Yang, L., Du, T.B., Khan, Q., and Zhang, L., Mechanisms of long-term donor-specific allograft survival induced by pretransplant infusion of lymphocytes. *Blood* 1998, **91**: 324–330.

6. Zhang, Z.X., Yang, L., Young, K.J., DuTemple, B., and Zhang, L., Identification of a previously unknown antigen-specific regulatory T cell and its mechanism of suppression. *Nat. Med.* 2000, **6**: 782–789.
7. Chen, W., Zhou, D., Torrealba, J.R., Waddell, T.K., Grant, D., and Zhang, L., Donor lymphocyte infusion induces long-term donor-specific cardiac xenograft survival through activation of recipient double-negative regulatory T cells. *J. Immunol.* 2005, **175**: 3409–3416.
8. Fischer, K., Voelkl, S., Heymann, J., Przybylski, G.K., Mondal, K., Laumer, M., Kunz-Schughart, L., Schmidt, C.A., Andreesen, R., and Mackensen, A., Isolation and characterization of human antigen-specific TCR alpha beta+ CD4(−)CD8− double-negative regulatory T cells. *Blood* 2005, **105**: 2828–2835.
9. Itoh, M., Takahashi, T., Sakaguchi, N., Kuniyasu, Y., Shimizu, J., Otsuka, F., and Sakaguchi, S., Thymus and autoimmunity: production of $CD25^+CD4^+$ naturally anergic and suppressive T cells as a key function of the thymus in maintaining immunologic self-tolerance. *J. Immunol.* 1999, **162**: 5317–5326.
10. Sakaguchi, S., Naturally arising Foxp3-expressing CD25+CD4+ regulatory T cells in immunological tolerance to self and non-self. *Nat. Immunol.* 2005, **6**: 345–352.
11. Fontenot, J.D. and Rudensky, A.Y., A well adapted regulatory contrivance: regulatory T cell development and the forkhead family transcription factor Foxp3. *Nat. Immunol.* 2005, **6**: 331–337.
12. Sakaguchi, S., Takahashi, T., and Nishizuka, Y., Study on cellular events in post-thymectomy autoimmune oophoritis in mice. II. Requirement of Lyt-1 cells in normal female mice for the prevention of oophoritis. *J. Exp. Med.* 1982, **156**: 1577–1586.
13. Watanabe, N., Wang, Y.H., Lee, H.K., Ito, T., Wang, Y.H., Cao, W., and Liu, Y.J., Hassall's corpuscles instruct dendritic cells to induce CD4+CD25+ regulatory T cells in human thymus. *Nature* 2005, **436**: 1181–1185.
14. Cozzo, C., Lerman, M.A., Boesteanu, A., Larkin, J., III, Jordan, M.S., and Caton, A.J., Selection of CD4+CD25+ regulatory T cells by self-peptides. *Curr. Top. Microbiol. Immunol.* 2005, **293**: 3–23.
15. Cozzo, C., Larkin, J., III, and Caton, A.J., Cutting edge: self-peptides drive the peripheral expansion of CD4+CD25+ regulatory T cells. *J. Immunol.* 2003, **171**: 5678–5682.
16. Ribot, J., Romagnoli, P., and van Meerwijk, J.P., Agonist ligands expressed by thymic epithelium enhance positive selection of regulatory T lymphocytes from precursors with a normally diverse TCR repertoire. *J. Immunol.* 2006, **177**: 1101–1107.
17. Cabarrocas, J., Cassan, C., Magnusson, F., Piaggio, E., Mars, L., Derbinski, J., Kyewski, B., Gross, D.A., Salomon, B.L., Khazaie, K., Saoudi, A., and Liblau, R.S., Foxp3+ CD25+ regulatory T cells specific for a neo-self-antigen develop at the double-positive thymic stage. *Proc. Natl. Acad. Sci. U. S. A.* 2006, **103**: 8453–8458.
18. Karim, M., Kingsley, C.I., Bushell, A.R., Sawitzki, B.S., and Wood, K.J., Alloantigen-induced CD25+CD4+ regulatory T cells can develop in vivo from CD25-CD4+ precursors in a thymus-independent process. *J. Immunol.* 2004, **172**: 923–928.
19. Aifantis, I., Bassing, C.H., Garbe, A.I., Sawai, K., Alt, F.W., and von Boehmer, H., The E{delta} enhancer controls the generation of CD4-CD8- {alpha}{beta}TCR-expressing T cells that can give rise to different lineages of {alpha}{beta} T cells. *J. Exp. Med.* 2006, **203**: 1543–1550.
20. Wang, R., Wang-Zhu, Y., and Grey, H., Interactions between double positive thymocytes and high affinity ligands presented by cortical epithelial cells generate double negative thymocytes with T cell regulatory activity. *Proc. Natl. Acad. Sci. U. S. A.* 2002, **99**: 2181–2186.
21. Yamagiwa, S., Sugahara, S., Shimizu, T., Iwanaga, T., Yoshida, Y., Honda, S., Watanabe, H., Suzuki, K., Asakura, H., and Abo, T., The primary site of CD4− 8− B220+ alphabeta T cells in lpr mice: the appendix in normal mice. *J. Immunol.* 1998, **160**: 2665–2674.

22. Johansson, M. and Lycke, N., A unique population of extrathymically derived alpha beta TCR+CD4–CD8– T cells with regulatory functions dominates the mouse female genital tract. *J. Immunol.* 2003, **170**: 1659–1666.
23. Rharbaoui, F., Bruder, D., Vidakovic, M., Ebensen, T., Buer, J., and Guzman, C.A., Characterization of a B220+ lymphoid cell subpopulation with immune modulatory functions in nasal-associated lymphoid tissues. *J. Immunol.* 2005, **174**: 1317–1324.
24. Halder, R.C., Kawamura, T., Bannai, M., Watanabe, H., Kawamura, H., Mannoor, M.K., Morshed, S.R., and Abo, T., Intensive generation of NK1.1- extrathymic T cells in the liver by injection of bone marrow cells isolated from mice with a mutation of polymorphic major histocompatibility complex antigens. *Immunology* 2001, **102**: 450–459.
25. Ohteki, T., Seki, S., Abo, T., and Kumagai, K., Liver is a possible site for the proliferation of abnormal $CD3^+4^-8^-$ double-negative lymphocytes in autoimmune MRL-lpr/lpr mice. *J. Exp. Med.* 1990, **172**: 7–12.
26. Ford, M.S., Zhang, Z.X., Chen, W., and Zhang, L., Double-negative T regulatory cells can develop outside the thymus and do not mature from CD8+ T cell precursors. *J. Immunol.* 2006, **177**: 2803–2809.
27. Erard, F., Wild, M.T., Garcia-Sanz, J.A., and Le Gros, G., Switch of CD8 T cells to non-cytolytic $CD8^-CD4^-$ cells that make TH2 cytokines and help B cells. *Science* 1993, **260**: 1802–1805.
28. Young, K.J., Yang, L.M., Phillips, M.J., and Zhang, L., Donor-lymphocyte infusion induces tolerance by activating systemic and graft-infiltrating double negative T regulatory cells. *Blood* 2002, **100**: 3408–3414.
29. Chen, W.H., Ford, M., Young, K.J., Cybulsky, M., and Zhang, L., The role of DN regulatory T cells in long-term cardiac xenograft survival Induced by pretransplant donor lymphocyte infusion and a short course of depleting anti-CD4 antibody. *J. Immunol.* 2003, **170**: 1846–1853.
30. Harty, J.T. and Badovinac, V.P., Influence of effector molecules on the CD8(+) T cell response to infection. *Curr. Opin. Immunol.* 2002, **14**: 360–365.
31. Khan, Q., Penninger, J.M., Yang, L.M., Marra, L.E.K.I., and Zhang, L., Regulation of apoptosis in mature $\alpha\beta^+$ $CD4^-CD8^-$ antigen-specific suppressor T-cell clones. *J. Immunol.* 1999, **162**: 5860–5867.
32. Chen, W., Ford, M.S., Young, K.J., and Zhang, L., The role and mechanisms of double negative regulatory T cells in the suppression of immune responses. *Cell Mol. Immunol.* 2004, **1**: 328–335.
33. Zhang, Z.X., Stanford, W.L., and Zhang, L., Ly-6A is critical for the function of double negative regulatory T cells. *Eur. J. Immunol.* 2002, **32**: 1584–1592.
34. Ford, M.S., Young, K.J., Zhang, Z.X., Ohashi, P.S., and Zhang, L., The immune regulatory function of lymphoproliferative double negative T cells in vitro and in vivo. *J. Exp. Med.* 2002, **196**: 261–267.
35. Giese, T. and Davidson, W.F., Chronic treatment of C3H-lpr/lpr and C3H-gld/gld mice with anti-CD8 monoclonal antibody prevents the accumulation of double negative T cells but not autoantibody production. *J. Immunol.* 1994, **152**: 2000–2010.
36. Polster, K., Walker, A., Fildes, J., Entwistle, G., Yonan, N., Hutchinson, I.V., and Leonard, C.T., CD4-veCD8-ve CD30+ve T cells are detectable in human lung transplant patients and their proportion of the lymphocyte population after in vitro stimulation with donor spleen cells correlates with preservation of lung physiology. *Transplant. Proc.* 2005, **37**: 2257–2260.
37. Priatel, J.J., Utting, O., and Teh, H.S., TCR/self-antigen interactions drive double-negative T cell peripheral expansion and differentiation into suppressor cells. *J. Immunol.* 2001, **167**: 6188–6194.
38. Benihoud, K., Bonardelle, D., Bobe, P., and Kiger, N., MRL/lpr CD4- CD8- and CD8+ T cells, respectively, mediate Fas-dependent and perforin cytotoxic pathways. *Eur. J. Immunol.* 1997, **27**: 415–420.

39. Canale, V.C. and Smith, C.H., Chronic lymphadenopathy simulating malignant lymphoma. *J. Pediatr.* 1967, **70**: 891–899.
40. Bettinardi, A., Brugnoni, D., Quiros-Roldan, E., Malagoli, A., La Grutta, S., Correra, A., and Notarangelo, L.D., Missense mutations in the Fas gene resulting in autoimmune lymphoproliferative syndrome: a molecular and immunological analysis. *Blood* 1997, **89**: 902–909.
41. Brooks, E.G., Balk, S.P., Aupeix, K., Colonna, M., Strominger, J.L., and Groh-Spies, V., Human T-cell receptor (TCR) alpha/beta + CD4–CD8– T cells express oligoclonal TCRs, share junctional motifs across TCR V beta-gene families, and phenotypically resemble memory T cells. *Proc. Natl. Acad. Sci. U. S. A.* 1993, **90**: 11787–11791.
42. Martinez, C., Marcos, M.A., de Alboran, I.M., Alonso, J.M., de Cid, R., Kroemer, G., and Coutinho, A., Functional double-negative T cells in the periphery express T cell receptor V beta gene products that cause deletion of single-positive T cells. *Eur. J. Immunol.* 1993, **23**: 250–254.
43. Ford, M.S., Chen, W., Wong, S., Li, C., Vanama, R., Elford, A.R., Asa, S.L., Ohashi, P.S., and Zhang, L., Peptide-activated double-negative T cells can prevent autoimmune type-1 diabetes development. *Eur. J. Immunol.* 2007, **37**: 2234–2241.
44. Young, K.J., Kay, L.S., Phillips, M.J., and Zhang, L., Antitumor activity mediated by double-negative T cells. *Cancer Res.* 2003, **63**: 8014–8021.
45. Young, K.J., Du Temple, B., Phillips, M.J., and Zhang, L., Inhibition of graft versus host disease by double negative regulatory T cells. *J. Immunol.* 2003, **171**: 134–141.
46. Abraham, V.S., Sachs, D.H., and Sykes, M., Mechanism of protection from graft-versus-host disease mortality by IL- 2. III. Early reductions in donor T cell subsets and expansion of a $CD3^+CD4^-CD8^-$ cell population. *J. Immunol.* 1992, **148**: 3746–3752.
47. Bruley-Rosset, M., Miconnet, I., Canon, C., and Halle-Pannenko, O., Mlsa generated suppressor cells. I. Suppression is mediated by double- negative (CD3+CD5+CD4-CD8-) alpha/beta T cell receptor-bearing cells. *J. Immunol.* 1990, **145**: 4046–4052.
48. Prins, R.M., Incardona, F., Lau, R., Lee, P., Claus, S., Zhang, W., Black, K.L., and Wheeler, C.J., Characterization of defective CD4–CD8– T cells in murine tumors generated independent of antigen specificity. *J. Immunol.* 2004, **172**: 1602–1611.
49. Matsuo, R., Herndon, D.N., Kobayashi, M., Pollard, R.B., and Suzuki, F., $CD4^- CD8^-$ TCR $\alpha\beta^+$ suppressor T cells demonstrated in mice 1 day after thermal injury. *J. Trauma* 1997, **42**: 635–640.
50. Kadena, T., Matsuzaki, G., Fujise, S., Kishihara, K., Takimoto, H., Sasaki, M., Beppu, M., Nakamura, S., and Nomoto, K., TCR alpha beta+ CD4– CD8– T cells differentiate extrathymically in an lck-independent manner and participate in early response against Listeria monocytogenes infection through interferon-gamma production. *Immunology* 1997, **91**: 511–519.
51. Cowley, S.C., Hamilton, E., Frelinger, J.A., Su, J., Forman, J., and Elkins, K.L., CD4–CD8– T cells control intracellular bacterial infections both in vitro and in vivo. *J. Exp. Med.* 2005, **202**: 309–319.
52. Carulli, G., Lagomarsini, G., Azzara, A., Testi, R., Riccioni, R., and Petrini, M., Expansion of TcRalphabeta+CD3+CD4–CD8– (CD4/CD8 double-negative) T lymphocytes in a case of staphylococcal toxic shock syndrome. *Acta Haematol.* 2004, **111**: 163–167.
53. Hossain, M.S., Takimoto, H., Ninomiya, T., Yoshida, H., Kishihara, K., Matsuzaki, G., Kimura, G., and Nomoto, K., Characterization of CD4– CD8– CD3+ T-cell receptor-alpha/beta+ T cells in murine cytomegalovirus infection. *Immunology* 2000, **101**: 19–29.
54. Mathiot, N.D., Krueger, R., French, M.A., and Price, P., Percentage of CD3+CD4–CD8–gammadeltaTCR- T cells is increased HIV disease. *AIDS Res. Hum. Retroviruses* 2001, **17**: 977–980.
55. Moreau, J.F., Taupin, J.L., Dupon, M., Carron, J.C., Ragnaud, J.M., Marimoutou, C., Bernard, N., Constans, J., Texier-Maugein, J., Barbeau, P., Journot, V., Dabis, F., Bonneville, M., and Pellegrin, J.L., Increases in CD3+CD4–CD8– T lymphocytes in AIDS patients with disseminated Mycobacterium avium-intracellulare complex infection. *J. Infect. Dis.* 1996, **174**: 969–976.

56. Lee, B.P., Mansfield, E., Hsieh, S.C., Hernandez-Boussard, T., Chen, W., Thomson, C.W., Ford, M.S., Bosinger, S.E., Der, S., Zhang, Z.X., Zhang, M., Kelvin, D.J., Sarwal, M.M., and Zhang, L., Expression profiling of murine double-negative regulatory T cells suggest mechanisms for prolonged cardiac allograft survival. *J. Immunol.* 2005, **174**:4535–4544.
57. Zhang, Z.X., Ma, Y., Wang, H., Arp, J., Jiang, J., Huang, X., He, K.M., Garcia, B., Madrenas, J., and Zhong, R., Double-negative T cells, activated by xenoantigen, lyse autologous B and T cells using a perforin/granzyme-dependent, fas-fas ligand-independent pathway. *J. Immunol.* 2006, **177**: 6920–6929.
58. Hamad, A.R., Mohamood, A.S., Trujillo, C.J., Huang, C.T., Yuan, E., and Schneck, J.P., B220+ double-negative T cells suppress polyclonal T cell activation by a Fas-independent mechanism that involves inhibition of IL-2 production. *J. Immunol.* 2003, **171**: 2421–2426.
59. Powrie, F., Carlino, J., Leach, M.W., Mauze, S., and Coffman, R.L., A critical role for transforming growth factor-B but not interleukin 4 in the suppression of T helper type1-mediated colitis by $CD45RB^{low}$ $CD4^{+\,T\,cells}$. *J. Exp. Med.* 1996, **183**: 2669–2674.
60. Asseman, C., Mauze, S., Leach, M.W., Coffman, R.L., and Powrie, F., An essential role for interleukin 10 in the function of regulatory T cells that inhibit intestinal inflammation. *J. Exp. Med.* 1999, **190**: 995–1004.
61. Seddon, B. and Mason, D., Regulatory T cells in the control of autoimmunity: the essential role of transforming growth factor β and interleukin 4 in the prevention of autoimmune thyroiditis in rats by peripheral $CD4^+CD8^-$ Thymocytes. *J. Exp. Med.* 1999, **189**: 279–288.
62. Kitani, A., Chua, K., Nakamura, K., and Strober, W., Activated self-MHC-reactive T cells have the cytokine phenotype of Th3/T regulatory cell 1 T cells. *J. Immunol.* 2000, **165**: 691–702.
63. Groux, H., O'Garra, A., Bigler, M., Rouleau, M., Antonenko, S., de Vries, J.E., and Roncarolo, M.G., A $CD4^+$ T-cell subset inhibits antigen-specific T-cell responses and prevents colitis. *Nature* 1997, **389**: 737–742.
64. Cavani, A., Nasorri, F., Prezzi, C., Sebastiani, S., Albanesi, C., and Girolomoni, G., Human $CD4^+$ T lymphocytes with remarkable regulatory functions on dendritic cells and nickel-specific Th1 immune responses. *J. Invest Dermatol.* 2000, **114**: 295–302.
65. Lee, B.P., Chen, W., Shi, H., Der, S.D., Forster, R., and Zhang, L., CXCR5/CXCL13 interaction is important for double-negative regulatory T cell homing to cardiac allografts. *J. Immunol.* 2006, **176**: 5276–5283.
66. Thomson, C.W., Teft, W.A., Chen, W., Lee, B.P., Madrenas, J., and Zhang, L., FcR{gamma} Presence in TCR Complex of Double-Negative T Cells Is Critical for Their Regulatory Function. *J. Immunol.* 2006, **177**: 2250–2257.
67. Young, K. and Zhang, L., The nature and mechanisms of DN regulatory T-Cell mediated suppression. *Hum. Immunol.* 2002, **63**: 926.
68. Hudrisier, D., Riond, J., Mazarguil, H., Gairin, J.E., and Joly, E., Cutting edge: CTLs rapidly capture membrane fragments from target cells in a TCR signaling-dependent manner. *J. Immunol.* 2001, **166**: 3645–3649.
69. Stinchcombe, J.C., Bossi, G., Booth, S., and Griffiths, G.M., The immunological synapse of CTL contains a secretory domain and membrane bridges. *Immunity* 2001, **15**: 751–761.
70. Hwang, I., Huang, J.F., Kishimoto, H., Brunmark, A., Peterson, P.A., Jackson, M.R., Surh, C.D., Cai, Z., and Sprent, J., T cells can use either T cell receptor or CD28 receptors to absorb and internalize cell surface molecules derived from antigen-presenting cells. *J. Exp. Med.* 2000, **191**: 1137–1148.
71. Sabzevari, H., Kantor, J., Jaigirdar, A., Tagaya, Y., Naramura, M., Hodge, J., Bernon, J., and Schlom, J., Acquisition of CD80 (b7-1) by T cells. *J. Immunol.* 2001, **166**: 2505–2513.
72. Patel, D.M., Arnold, P.Y., White, G.A., Nardella, J.P., and Mannie, M.D., Class II MHC/peptide complexes are released from APC and are acquired by T cell responders during specific antigen recognition. *J. Immunol.* 1999, **163**:5201–5210.

73. Huang, J.F., Yang, Y., Sepulveda, H., Shi, W., Hwang, I., Peterson, P.A., Jackson, M.R., Sprent, J., and Cai, Z., TCR-Mediated internalization of peptide-MHC complexes acquired by T cells. *Science* 1999, **286**: 952–954.
74. Batista, F.D., Iber, D., and Neuberger, M.S., B cells acquire antigen from target cells after synapse formation. *Nature* 2001, **411**: 489–494.
75. Harshyne, L.A., Watkins, S.C., Gambotto, A., and Barratt-Boyes, S.M., Dendritic cells acquire antigens from live cells for cross-presentation to CTL. *J. Immunol.* 2001, **166**: 3717–3723.
76. Hwang, I., Shen, X., and Sprent, J., Direct stimulation of naive T cells by membrane vesicles from antigen-presenting cells: distinct roles for CD54 and B7 molecules. *Proc. Natl. Acad. Sci. U. S. A.* 2003, **100**: 6670–6675.
77. Tsang, J.Y., Chai, J.G., and Lechler, R., Antigen presentation by mouse CD4+ T cells involving acquired MHC class II:peptide complexes: another mechanism to limit clonal expansion? *Blood* 2003, **101**: 2704–2710.
78. Kennedy, R., Undale, A.H., Kieper, W.C., Block, M.S., Pease, L.R., and Celis, E., Direct cross-priming by Th lymphocytes generates memory cytotoxic T cell responses. *J. Immunol.* 2005, **174**: 3967–3977.
79. Zhou, J., Tagaya, Y., Tolouei-Semnani, R., Schlom, J., and Sabzevari, H., Physiological relevance of antigen presentasome (APS), an acquired MHC/costimulatory complex, in the sustained activation of CD4+ T cells in the absence of APCs. *Blood* 2005, **105**: 3238–3246.
80. Game, D.S., Rogers, N.J., and Lechler, R.I., Acquisition of HLA-DR and costimulatory molecules by T cells from allogeneic antigen presenting cells. *Am. J. Transplant.* 2005, **5**: 1614–1625.

Chapter 30
γδ T Cells in Immunoregulation

Long Tang, Ning Kang, and Wei He

Abstract The regulatory function of γδ T cells has not been appreciated until recently. In murine epithelial tissues, dendritic epidermal γδ T cells (DETCs) play a critical regulatory function in skin tissue surveillance and wound repair. It has been shown that this process is mediated by direct lyses of stressed or damaged keratinocytes via a cell-cell contact dependent mechanism. DETC-derived cytokines such as keratinocyte growth factor-1/2 and insulin-like growth factor-1 have also involved in regulating the processes of tissue repair, keratinocyte proliferation and survival, migration and recruitment of inflammatory cells. Moreover, DETCs can directly down-regulate cutaneous inflammation, promote wound healing, and protect against cutaneous malignancy. In humans, Vδ1 cells in tumor-infiltrating lymphocytes (TIL) from human breast cancer have been shown to possess an inhibitory effect on αβ T cells and dendritic cells in a soluble factor-dependent manner. Such immunosuppressive activity could be reversed by human Toll-like receptor 8 ligands both in-vitro and in-vivo, indicating that a unique TLR8 signaling pathway is involved in suppression of regulatory γδ1 TILs. More questions regarding the regulatory functions and the mechanism of suppression of γδ T cells need to be addressed in future studies, and these answers may have important implications in γδ T cell-based adoptive therapy for the treatment of cancers in clinic.

Introduction

γδ T cells, along with αβ T cells and B cells, are the three cell subsets using the somatic DNA rearrangement to assemble the gene segments of variable (V), diversity (D) and joining (J) to encode their defining cell-surface receptors, which are γδ T cell receptor (TCR), αβ TCR and B cell receptor (BCR), respectively [1].

W. He
Department of Immunology, Institute of Basic Medical Sciences, Chinese Academy of Medical Sciences (CAMS) and School of Basic Medicine, Peking Union Medical College (PUMC), 5 Dong Dan San Tiao, Beijing 100005, PR China
e-mail: heweiimu@public.bta.net.cn

Compared to αβ T cells, γδ T cells only represent a small number of T cells in the peripheral blood, but populate mainly as intraepithelial lymphocytes (IEL) in the epithelial tissues [2]. γδ T cells have distinct characteristics in contrast to αβ T cells, such as direct antigen recognition without any requirement for antigen processing, presentation, and the major histocompatibility complex (MHC) molecules [3]. γδ T cells have been long appreciated to act as the first line of immunological defense against microorganism and in immunosurveillance of tumor. Recently, γδ T cells have been shown to exert other functions, such as antigen presentation [4] and immune regulation in the immune response [5]. This review focuses on the regulatory function of γδ T cells, especially their regulatory roles in epithelial tissues and in tumor immunity.

Regulatory Function of γδ T Cells in Epithelial Tissues

As γδ T cells largely reside in the epithelial tissues, it has long been suggested that they could play a role of in the maintenance of tissue integrity and regulation of epithelia homeostasis.

In murine skins, the epithelial resident γδ T cells express tissue-restricted TCRs with limited or invariant diversity. These skin γδ TCR-expressing cells have been coined dendritic epidermal T cells (DETCs) base on their unique dendritic morphology [6, 7]. γδ DETCs express a canonical Vγ5 Vδ1 TCR [8] (according to nomenclature of Heilig and Tonegawa [9]) and are thought to be of thymic origin [10, 11] in contrast to Vγ7-expressed intestinal intraepithelial lymphocytes, which are generated in the absence of functional thymus [12]. The Vγ5 Vδ 1 TCR can recognize antigens expressed by stressed, damaged, or transformed keratinocytes [13, 14]. Thus, DETCs can be directly activated by stressed or damaged keratinocytes and then kill the 'dangerous' keratinocytes to maintain the skin tissue homeostasis. One study has shown that the skin of TCR-δ deficient mice (TCR $\delta^{-/-}$ mice) could still harbor some polyclonal αβ TCR-expressed DETCs, but these TCRαβ$^+$ DETCs failed to respond to damaged keratinocytes, indicating that the Vγ5 Vδ1TCR-expressed DETCs are vital and necessary to the regulatory function of skin tissues [15]. Other studies have shown that TCR-δ deficient mice had profound defects in wound healing and tumor rejection, suggesting that DETCs play a critical regulatory function in skin tissue surveillance and wound repair [16–18]. Apart from direct lysis via a cell-cell contact dependent mechanism, there is another mechanism involving in the regulatory function of DETCs, i.e. through secretion of variety of cytokines, growth factors and chemokines including keratinocyte growth factor-1 (KGF-1) and KGF-2 [19], insulin-like growth factor-1 (IGF-1) [20], interleukin-2 (IL-2) [21], interferon-γ (IFN-γ) [21], and lymphotactin [22], which exert their functions in tissue repair, keratinocyte proliferation and survival, migration and recruitment of inflammatory cells. In-vitro skin organ culture studies have demonstrated that the addition of recombinant KGF could restore normal wound healing in γδ DETC-deficient mice [16], and mice deficient in DETCs had a notable increase in epidermal apoptosis, which could be abrogated by the addition of IGF-1 [20].

Taken together, these data indicate an important role of the chemokines secreted by γδ DETCs. Furthermore, through monitoring the infiltration of leukocytes after full-thickness wounding of mice deficient in γδ DETCs, Jameson JM et al. [23] showed that γδ DETCs-derived KGF was able to induce keratinocytes to produce an extracellular matrix (ECM) molecule – hyaluronan, which played a key role in the inflammatory cell migration associated with wound repair, namely, hyaluronan was capably of recruiting macrophages to the wound sites, thus contributing to the wound healing. This study demonstrated a novel function of skin γδ DETCs in the inflammation through the regulation of the production of ECM. It also indicates that the cross-talk between skin γδ DETCs and keratinocytes is of vital importance in maintaining tissue homeostasis and wound healing. Another study made by Giradi et al. [24], who investigated spontaneous dermatitis occurring in TCRδ$^{-/-}$ mice of different genetic backgrounds also demonstrated a regulatory function of γδ DETCs in cutaneous inflammation. In this study, it has been found that αβ T cell-mediated inflammation is the target for γδ DETCs-mediated downregulation. Taken together, these data suggest that murine γδ DETCs possess immunoregulatory functions in downregulating cutaneous inflammation, in promoting wound healing, and in protecting against cutaneous malignancy.

In human, no phenotypically equivalent to murine γδ T cells have been found in the epidermis, but γδ T cells with limited TCR diversity and distinct subsets from the peripheral blood γδ T cell populations could be found in the dermis [25], suggesting that these cells may perform the similar role as murine γδ DETCs in maintaining the epithelia homeostasis.

Regulatory Function of γδ T Cells in Tumor Immunity

Human γδ T cells are classified as two major subsets, one expressing V γ9 chain paired V δ2 chain [26] (following the nomenclature of Porcelli et al. [27]), which dominate in the peripheral blood accounting 50–95% of all circulating γδ T cells. The other subset expresses Vδ1 chain paired with various Vγ chains, which reside mainly in the epithelial tissues as human IELs [28, 29]. Vγ9/ Vδ2 subsets mainly recognize the small nonpeptide molecules derived from microbes during infection [30–32], while Vδ1 subsets recognize MHC class I-related chain A or B (MICA or MICB), which are induced on stressed or damaged epithelial cells and tumor cells [1, 33]. Thus, Vδ1 T cell subsets are thought to have the immunosurveillance for transform cells and malignancy [34, 35]. However, the functions of γδ T cells are extremely pleiotropic and heterogeneous [36] as evidenced by studies in murine models. One study has shown that the growth of transplanted B16 melanoma could be better controlled in mice deficient in γδ T cells than in wild-type mice [37], suggesting that γδ T cells may possess the suppressive function in tumor immunity.

Recently, one study by Peng et al. [38] has provided a convincing evidence to support the notion. In this study, the authors established a series of T cell clones from tumor-infiltrating lymphocytes (TILs) from human breast cancer using high doses of IL-2 (1,000 U/ml). Surprisingly, a large fraction of these T cell clones exhibited Vδ1 phenotype. Those TIL-derived tumor-specific Vδ1 T cells showed strong inhibitory effect against the activation of naive T cells and effector T cells in-vitro. Cytokine profiling analysis showed that these Vδ1 T cells secreted IFN-γ and GM-CSF, but not IL-2, IL-4, IL-10 or TGF-β, when stimulated with autologous tumor cells or anti-CD3 antibody. Phenotypically, these γδ1 T cells did not express the surface markers that are typically found on the conventional $CD4^+$ regulatory T cells (Tregs), such as Foxp3, CD25 and GITR. Functionally, results from transwell experiments demonstrated that these Vδ1 T cells exerted their suppressive activities through a soluble factor-dependent mechanism yet independent of IL-10 or TGF-β, but not through the cell-cell contact dependent mechanism. These TIL-derived tumor-specific Vδ1 T cells possessed a unique and distinct phenotype and functional features in contrast to the conventional $CD4^+CD25^+$ Tregs. In addition to the suppressive effect on αβ T cells, these Vδ 1 T cells also prevented the maturation of dendritic cells, indicating that they could inhibit tumor immunity at multiple levels. Importantly, their immunosuppressive activity could be reversed by human Toll-like receptor (TLR) 8 ligands both in-vitro and in-vivo, indicating a unique TLR8 signaling pathway links to functional regulation of regulatory Vδ1 TILs.

Taken together, these data reveal a new aspect of biological roles of γδ T cells, especially in epithelial tissues and in tumor immunity. However, there still are more questions than answers. For example, can regulatory γδ T cells be induced from conventional γδ T cells? Is there a cross-talk between γδ T cells with other subsets of Tregs such as naturally occurring $CD4^+CD25^+Foxp3^+$ Tregs and innate NKT cells? More studies are required to address these questions. More importantly, one special concern is that, if γδ T cell adoptive therapy for the treatment of tumor patients could be performed in clinic, as many studies are currently underway to achieve such a goal, the potential regulatory functions of γδ T cells has to be taken into account very seriously to have an effective therapy.

References

1. Tonegawa, S. Somatic generation of antibody diversity. Nature 302:575–581. (1983).
2. Hayday, A. C. γδ cells: A right time and a right place for a conserved third way of protection. Annu Rev Immunol 18:975–1026. (2000).
3. Brenner, M.B., McLean, J., Dialynas, D.P., Strominger, J.L., Smith, J.A., Owen, F.L., Seidman, J.G., Ip, S., Rosen, F., and Krangel, M.S. Identification of a putative second T-cell receptor. Nature 322:145–149. (1986).
4. Brandes M, Willimann K, Moser B. Professional antigen-presentation function by human γδ T cells. Science 309:264–268. (2005).
5. Born WK, Reardon CL, O'Brien RL. The function of gammadelta T cells in innate immunity. Curr Opin Immunol 18(1):31–38. (2006).
6. Bergstresser PR, Tigelaar RE, Dees JH, Streilein JW. Thy-1 antigen-bearing dendritic cells populate murine epidermis. J Invest Dermatol 81:286–288. (1983).
7. Tschachler E, Schuler G, Hutterer J, Leibl H, Wolff K, Stingl G. Expression of Thy-1 antigen by murine epidermal cells. J Invest Dermatol 81:282–285. (1983).
8. Asarnow DM, Kuziel WA, Bonyhadi M, Tigelaar RE, Tucker PW, Allison JP. Limited diversity of gamma/delta antigen receptor genes of Thy-1+ dendritic epidermal cells. Cell 55: 837–847. (1988).

9. Heilig JS, Tonegawa S. Diversity of murine γ genes and expression in fetal and adult lymphocytes. Nature 322:836–840. (1986).
10. Havran WL, Allison JP. Origin of Thy-1+ dendritic epidermal cells of adult mice from fetal thymic precursors. Nature 344:68–70. (1990).
11. Payer E, Elbe A, Stingl G. Circulating CD3+/T cell receptor V γ3+ fetal murine thymocytes home to the skin and give rise to proliferating dendritic epidermal cells. J Immunol 146: 2536–2543. (1991).
12. Bandeira A, Itohara S, Bonneville M, Burlen-Defranoux O, Mota-Santos T, Coutinho A, Tonegawa S. Extrathymic origin of intestinal intraepithelial T cells bearing the γδ T cell receptor. Proc Natl Acad Sci USA 88:43–47. (1991).
13. Havran WL, Chien Y-H, Allison JP. Recognition of self antigens by skin-derived T cells with invariant γδ antigen receptors. Science 252:1430–1432. (1991).
14. Lewis JM, Tigelaar RE. Recognition of an epidermal stress antigen by murine γδ dendritic epidermal T cells (DETC). J Invest Dermatol 96:538A. (1991).
15. Jameson JM, Cauvi G, Witherden DA, Havran WL. A keratinocyte-responsive γδ TCR is necessary for dendritic epidermal T cell activation by damaged keratinocytes and maintenance in the epidermis. J Immunol 172:3573–3579. (2004).
16. Jameson, J. et al. A role for skin γδ T cells in wound repair. Science 296:747–749. (2002).
17. Girardi, M. et al. Regulation of cutaneous malignancy by γδ T cells. Science 294:605–609. (2001).
18. Girardi, M. et al. The distinct contributions of murine T cell receptor (TCR) γδ + and TCR αβ+ T cells to different stages of chemically induced skin cancer. J Exp Med 198:747–755. (2003).
19. Boismenu R, Havran WL. Modulation of epithelial cell growth by intraepithelial γδ T cells. Science 266:1253–1255. (1994).
20. Sharp LL, Jameson JM, Cauvi G, Havran WL. Dendritic epidermal T cells regulate skin homeostasis through local production of insulin-like growth factor 1. Nat Immunol 6:73–79. (2005).
21. Boismenu R, Hobbs MV, Boullier S, Havran WL. Molecular and cellular biology of dendritic epidermal T cells. Semin Immunol 8:323–331. (1996).
22. Boismenu R, Feng L, Xia YY, Chang JCC, Havran WL. Chemokine expression by intraepithelial γδ T cells: Implications for the recruitment of inflammatory cells to damaged epithelia. J Immunol 157:985–992. (1996).
23. Jameson JM, Cauvi G, Sharp LL, Witherden DA, Havran WL. γδ T cell-induced hyaluronan production by epithelial cells regulates inflammation. J Exp Med 201:1269–1279. (2005).
24. Girardi, M. et al. Resident skin-specific γδ T cells provide local, nonredundant regulation of cutaneous inflammation. J Exp Med 195:855–867. (2002).
25. Holtmeier W, Pfander M, Hennemann A, Zollner TM, Kaufmann R, Caspary WF. The TCR-δ repertoire in normal human skin is restricted and distinct from the TCR-δ repertoire in the peripheral blood. J Invest Dermatol 116:275–280. (2001).
26. Hinz T, Wesch D, Halary F, Marx S, Choudhary A, Arden B, Janssen O, Bonneville M, Kabelitz D. Identification of the complete expressed human T-cell receptor Vγ repertoire by flow cytometry. Int Immunol 9:1065–1072. (1997).
27. Porcelli SA, Brenner MB, Band H. Biology of the human γδ T-cell receptor. Immunol Rev 120:137–183. (1991).
28. Deusch K, Luling F, Reich K, Classen M, Wagner H, Pfeffer K. A major fraction of human intraepithelial lymphocytes simultaneously expresses the γ/δ T cell receptor, the CD8 accessory molecule and preferentially uses the Vδ1 gene segment. Eur J Immunol 21:1053–1059. (1991).
29. Holtmeier W. Compartmentalization of γ/δ T cells and their putative role in mucosal immunity. Crit Rev Immunol 23:473–488. (2003).
30. Modlin, R.L., Pirmez, C., Hofman, F.M., Torigian, V., Uyemura, K., Rea, T.H., Bloom, B.R., and Brenner, M.B. Lymphocytes bearing antigen-specific gamma delta T-cell receptors accumulate in human infectious disease lesions. Nature 339:544–548. (1989).

31. Constant, P., Davodeau, F., Peyrat, MA, Poquet, Y., Puzo, G., Bonneville, M., and Fournie, J.J. Stimulation of human gamma delta T cells by nonpeptidic mycobacterial ligands. Science 264:267–270. (1994).
32. Bukowski, J.F., Morita, C.T., and Brenner, M.B. Human gamma delta T cells recognize alkylamines derived from microbes, edible plants, and tea: Implications for innate immunity. Immunity 11:57–65. (1999).
33. Groh, V., Steinle, A., Bauer, S., and Spies, T. Recognition of stress-induced MHC molecules by intestinal epithelial gammadelta T cells. Science 279:1737–1740. (1998).
34. Jameson, J., Witherden, D., and Havran, W.L. T-cell effector mechanisms: Gammadelta and CD1d-restricted subsets. Curr Opin Immunol 15:349–353. (2003).
35. Hayday, A, and Tigelaar, R. Immunoregulation in the tissues by gammadelta T cells. Nat Rev Immunol 3:233–242. (2003).
36. Pennington, D.J., Vermijlen, D., Wise, E.L, Clarke, S.L., Tigelaar, R.E., and Hayday, A.C. The integration of conventional and unconventional T cells that characterizes cell-mediated responses. Adv Immunol 87:27–59. (2005).
37. Ke Y., Kapp, L. M. and Kapp J.A. Inhibition of tumor rejection by $\gamma\delta$ T cells and IL-10. Cell immunol 221:107–114. (2003).
38. Peng G, Wang HY, Peng W, Kiniwa Y, Seo KH, Wang RF. Tumor-infiltrating gammadelta T cells suppress T and dendritic cell function via mechanisms controlled by a unique toll-like receptor signaling pathway. Immunity 27(2):334–48. (2007).

Index

A

Acquired immunological privilege, 327
   See also Transplantation tolerance
Acquired Treg, *see* Adoptive Tregs (aTreg)
Activation induced cell death (AICD), 3
   homeostasis control and, 78
   peripheral, 58–60
   See also Immunological tolerance
Acute allergic response (EAR), 354
   See also Allergic diseases
Acute hepatitis B (AHB), 391
   See also Chronic hepatitis B (CHB); Viral hepatitis
Acute inflammatory responses
   loss of suppression during, 358–360
   See also Allergic diseases
Adaptive Tregs, *see* Adoptive Tregs (aTreg)
Adoptive transfer therapy
   clonal expansion aspects
      antigen-non-specific, 242–243
      antigen-specific, 243–247
   for autoimmune diseases, 236–238
      GVHD, 242–247
   in vivo, 211–216
   issues, 214–216
   transplantation tolerance induction by, 337–338
Adoptive Tregs (aTreg)
   allergic disease and, 360, 363
   autoimmune disease and, 293–294
   hepatitis and, 393–395
   immune tolerance aspects, 156–157
   induction
      copolymer-I (COP-I) role, 297–299
      γ-IFN role, 295–297
      T cell vaccination role, 299–301
      TGF-induced, 114, 120–121
   infections and, 423–424
   mediated immune suppression, 94–95

Th3 cells, 94–95
   therapeutic considerations, 301–302
   Tr1, *see* Tr1 cells
   See also Natural Tregs (nTreg)
Adult T cell leukaemia (ATLL), 438
   See also Cancer
AIDS, 412
   lentiviruses-induced, 413–416
   See also HIV; Viral infections
AIRE
   autoreactive Treg-repertoire aspects, 40
   gene, 10
   regulated antigens, thymic Tregs generation and, 21
Airway hyperreactivity (AHR), 524
   See also Asthma
AKT pathway
   IL-2 signaling and, 80–82
   TGF-β signaling and, 92–93
   Treg cell differentiation and, 140–142
   See also Interleukin-2R (IL-2R) signaling; NF-κB signaling
ALK5, 93
   See also TGF-β signaling
Allergens, 353
   desensitisation immunotherapy (IT), 355
   specific immunotherapy (IT), 364–367
   specific Tregs, iNKT and asthma regulation by, 531–532
   See also Antigen-specific Tregs
Allergic diseases
   acute allergic response (EAR), 354
   allergy, defined, 353–354
   asthmatic, *see* Asthma
   late phase allergic response (LAR), 354
   Th2 responses, 354
   therapies, 354–355
      adjuvants therapy, 366

Allergic diseases (*cont.*)
    allergen desensitisation immunotherapy (IT), 355
    allergen-specific immunotherapy (IT), 364–367
    IL-10, 362–367
    induced Tregs, 360, 363
    naturally Tregs, 360
    non antigen-specific therapies, 366–367
    TGF-β, 360–362, 364
    Tregs in, 353
        allergic responses development prevention, 355–356
        $CD4^+CD25^+$ Tregs animal models, 357–358
        $CD4^+CD25^+Foxp3^+$ Tregs, 356–357
        loss of suppression during acute inflammatory responses, 358–360
        therapeutic application, 360, 363
    XLAAD, 357
    See also Autoimmune diseases; Infections
Alloantigens
    double negative (DN) T cells and, 550–553
    tolerance (Tr1 cells) to, 463
    See also Antigen
αDEC, 33
α-GalCer stimulation, 502–505, 509–512
    See also Invariant NKT (iNKT) cells
α-1,2–mannosidase gene, 314–315
    See also Transplantation
Anergic T cell
    transplantation tolerance, 328, 329
Animal models
    autoimmunity
        immunogenetic determinants of autoimmunity, 177–179
        NOD mouse model of spontaneous diabetes, 174–179
    $CD4^+CD25^+$ Tregs and allergic disease, 357–358
    See also Immune tolerance
Anopheles gambiae, see under Malaria
Anterior chamber associated immune deviation (ACAID), 471–473
    clinical applications for induction of, 480–481
    NKT cells and, 475
    traffic patterns for $CD8^+$ cells induction in eye induced tolerance, 473–475
Anti-CD3 therapy
    for diabetes, 218–219
    See also In vivo approach

Anti-CTLA-4 therapy, 381
    See also Tumor immunotherapy
Antigen
    transplantation tolerance aspects, *See also under* Antigen specificity
        sustained regulation exploits indirect antigen presentation pathway, 326
        tolerising microenvironments creation, 329
        transferable tolerance, 334
    See also Adoptive transfer therapy; Alloantigens; Anterior chamber associated immune deviation (ACAID)
Antigen presenting cells (APC), 205–210, 215–216, 240–241, 326
Antigen specific Tregs
    $CD4^+CD25^+$
        in chronic HBV, 396
        in chronic HCV, 396–397
    for multiple sclerosis (MS), 265, 270
    gene therapy and, 219–220
    in vitro expansion aspects, 217–218
    See also Autoimmune diseases
Antigen specificity, 18, 20, 23–25
    clonal expansion
        antigen-non-specific, 242–243
        antigen-specific, 243–247
    DN T cells suppression mechanism, 550
    effector T cells, 23, 24
    Foxp3–expression, 33
    iNKT and, 500–502
    tolerance
        peripheral, 472
        transplantation, 325, 339–341
    tolerance of Tr1 cells to
        allo-antigens, 463
        non self non harmful antigens, 462–463
        self-antigens, 462
    See also Autoimmune diseases; Immunosuppression
Anti-GITR antibody, 381, 383
    See also Tumor immunotherapy and
Arthritis, *see* Rheumatoid arthritis
Aspergillus Fumigatus
    IL-10 and TGF-β producing Treg cells in, 434–435
    See also Infections
Asthma, 353–354
    AHR feature of, 524
    defined, 523–524
    immune system in, 524
    NKT regulation of, 523

Index

balance between allergen-specific TReg cells, Th2 cells and iNKT cells, 531–532
   in human, 527–528
   in mouse, 527
   iNKT cells and, 527–532
   therapy for, 366
   See also Allergic diseases; Autoimmune diseases
Autoimmune diseases, 170–185
   adaptive $CD4^+CD25^+Foxp3^+$ Tregs and, 293–294
   ALPS, 3
   anti-CD3 therapy, 218–219
   APECED, 10, 40
   DN T cells in, 545–546
   EAE, *see* Experimental autoimmune encephalomyelitis (EAE)
   gene therapy, 219–220
   immune tolerance aspects
      animal models, 174–177
      chemokine directed homing of nTreg cells to sites of inflammation, 183–185
      dysfunctionsal Treg cells in human autoimmunity, 179–181
      regulation of immune responses to microbes, 181–183
      regulatory and autoreactive T cells balance, 170–174
   invariant NKT and, 509–512
   IPEX, 239
   liver diseases (AILD), 399–400, *See also under* Autoimmune liver diseases (AILD)
   lymph node (LN) priming and, 203–208
   multiple sclerosis, 254–256, 265–275
   regulation
      EAE, 255–256
      lupus erythematosus (LE), 259
      rheumatoid arthritis, 258–259
      type 1 diabetes (T1D), 256–258
      Wiskott-Aldrich syndrome (WAS), 259–260
   regulatory and autoreactive T cells, balance between, 170–173
      abnormal central tolerance and peripheral pool of autoreactive T cells, over-riding nTreg cell function and provoking autoimmunity, 173
      can autoreactive T cells be resistant to nTreg cell mediated suppression?, 174
      impaired development or function within nTreg cell compartment, 171–173
   therapies, 232
      adoptive transfer, 236–238, 242–247
      cell-based therapy, 260–261
      gene, 234–236
      generic, 233
      selective, 233–234
      Tregs role in, 210–218, 238–241, 260–261
   Tregs and, 238–241, 399–400
      $CD4^+CD25^+$ Tregs, 254–260
      cell-based therapy, 260
      effect in peripheral tissues, 205
      in vitro therapeutic use, 217–218
      in vivo therapeutic use, 211–216
      natural Tregs, 253
      unifying model, 208–210
   See also Allergic diseases; Autoimmunity; Diabetes; Graft-versus-host disease (GvHD); Infections; Mucosal inflammation
Autoimmune liver diseases (AILD)
   autoimmune hepatitis (AIH), 392, 399–400
   primary biliary cirrhosis (PBC), 399–400
   Tregs and, 399–400
   See also Hepatitis; Hepatocellular carcinoma
Autoimmune lymphoproliferative syndrome (ALPS), 3
Autoimmune polyendocrinopathy, candidiasis, ectodermal dystrophy (APECED), 10, 40
Autoimmune hepatitis (AIH), 399–400
Autoimmunity, 155
   immune tolerance aspects, 174–179
   $CD4^+CD25^+$ regulatory T cells, 10
   CD25– regulatory T cells, 10
   human, 179–181
   IL-2 in, 165
   mouse, 174–179
   Tregs, 3–11
      suppressor T cells (1982–1995) tracking, 5–6
      suppressor T cells (1995–2000) rebirth, 6–7
   See also Autoimmune diseases; Immune tolerance
Autoreactive cells and Tregs
   balance between, 170–174
   See also Immune tolerance

Autospecific TCR repertoire
  peripheral, 35
  thymic, 35
    negative selection of Treg precursors, sensitivity to, 35–37
    positive selection of Treg precursors, sensitivity to, 37–42

B

B lymphocyte-induced maturation protein (Blimp-1), 68–69
  See also Interleukin-2 (IL-2) signaling
Bacterial infections
  Bordetella Pertussis, 433–434
  Helicobacter Pylori, 432–433
  IL-10 and TGF-β producing Tregs in, 431–434
  Lactobacillus, 431
  Mycobacterium Tuberculosis, 431–432
  See also Fungal infections; Parasite infections; Viral infections
Bordetella Pertussis, 433–434
Borrelia burgdorferi, 527
Bystander suppression, 23

C

Cancer, 377–385
  DN T cells in, 546–547
  hepatocellular carcinoma, 391, 398–400
  invariant NKT and, 508–509
  See also Tumor immunotherapy
Candida Albicans, 435–436
CD4 lineage, 43, 45
$CD4^+$ cell
  Foxp3 and immune self-tolerance by immune tolerance aspects
    dominant regulation of peripheral T cell tolerance, 156–157
    induced Treg, 156
    IPEX experiment, 160–161
    IL-2 mediated T cell regulation model, 166–167
    natural Treg, 156–157
    scurfy mice experiment, 160–161
  See also $CD8^+$ Tregs; $CD25^+$ cells; CD28; Double negative (DN) T cells; Invariant natural killer T cells (iNKT)
$CD4^+CD8^{low}CD25^{high}$ cells, 38
$CD4^+CD8^+$ cells, 45
$CD4^+CD25^{hi}$ T cells, 270–273
  See also Multiple sclerosis (MS)

$CD4^+CD25^-$ T cells
  conversion to $CD4^+CD25^+$ Tregs, 111–112
  conversion to $CD4^+CD25^+Foxp3^+$ Tregs (in vivo), 121–123
  IL-6 induced Th17 cell differentiation, 127
  TGF-β induced Th17 cell differentiation, 113–114, 124–127
$CD4^+CD25^-Foxp3^-$ T cells, 33, 117–119
$CD4^+CD25^+$ T cells, 7–8
  adoptive
    TGF-induced, 120–121
    transfer of Tregs for transplantation tolerance in animals, 337–338
  allergy and acute inflammatory responses, 358, 359–360
  autoimmune diseases and, 238–241, 253–254, 399–401
    cell-based therapy, 260
    EAE, 255–256
    lupus erythematosus, 259
    multiple sclerosis, 254–256
    rheumatoid arthritis, 258–259
    type 1 diabetes, 256–258
    Wiskott-Aldrich syndrome (WAS), 259–260
  autoimmunity aspects, 10
  characteristics, 19–20
  ex-vivo expansion of Tregs, 338–339
  generation
    peripheral, 33
    TGF-β mediated, 98–99
    thymic, 31
  graft-versus-host disease (GvHD), 231
  hepatitis and, 392–398
    chronic HBV and, 393–398
    chronic HCV and, 395–398
  hepatocellular carcinoma and, 399–400
  IL-2 signaling and, 77–86
    discrete, 80–82
    in vitro, 81–83
    in vivo, 80, 83–84
    overview, 79–80
    peripheral de novo induction, 84–86
  immune tolerance aspects
    homeostatic functions of IL-2 for nTreg, 162–164
    IL-2 in self-tolerance and autoimmunity, 165–166
  in mice and humans (2000–2003), 7–8
  liver diseases (AILD) and, 399–401
  mucosal inflammation and, 279, 282–288
  multiple sclerosis and, 269–270, 272–273

naïve CD4$^+$CD25$^-$ T cells conversion to, 111–112
natural, 239
positive selection of Treg precursors, sensitivity to, 38
TCR for Tregs induction in human T cells, 124
TGF-induced
  in human T cells, 124
  molecular mechanisms and pathways, 123–124
  phenotype and function, 120–121
  TGF-β mediated Treg generation, 98–99
transplantation tolerance induction by, 333–337
viral infections and, 405–407
  hepatitis viruses, 409–410
  herpesviruses, 407–408
  lentiviruses, 411–413
  lentiviruses-induced AIDS, 413–416
  retroviruses, 410
CD4$^+$CD25$^+$Foxp3$^+$ Tregs
  adaptive
    autoimmune disease and, 293–294
    copolymer-I (COP-I) and, 297–299
    γ-IFN and, 295–297
    induction, 295–301
    mediated immune suppression, 94
    T cell vaccination and, 299–301
    TGF-induced, 114, 120–121
    therapeutic considerations, 301–302
  allergy and, 356–357
  immune tolerance aspects, 164–165
  invariant NKT and, 512–513
  generation
    peripheral, 32
    thymic, 31–32
  naïve CD4$^+$CD25$^-$ T cells conversion to
    in vivo, 121–123
    TCR for (in vivo), 121–122
    TGF-β for (in vivo), 121–123
  naïve CD4$^+$CD25$^-$Foxp3$^-$ T cells conversion to
    CD28 for, 117–119
    IL-2 for, 118
    in vitro, 117–119
    TCR and, 117–119
    TGF-β for, 117–119
  natural, 57, 114–117, 164–170, 239, 253, 282–284, 393
  TCR for Tregs induction in human T cells, 124
TGF-induced
  in human T cells, 124
  molecular mechanisms and pathways, 123–124
  thymic commitment of precursors to Treg lineage commitment, 45
transplantation and
  rejection controlling Tregs characterization, 308–309
  suppression mechanisms used by treg, 311–313
  tolerance aspects, 328
CD4$^+$Foxp3$^+$ Tregs
  generation aspects, 30
    peripheral generation, 34
  immune tolerance aspects, 155
    homeostatic functions of IL-2 for nTreg, 162–164
    IL-2 in self-tolerance and autoimmunity, 165–166
    IL-2 in vivo, expression and cellular sources of, 164–165
  thymic commitment of precursors to Treg lineage commitment, 45
CD8 lineage, 43, 45
CD8αα$^+$ Tregs, 487–489
  CD8αα$^+$ IELs and, 489–490
  priming of, 491–492
  Qa-1 and, 490–491
  suppression mechanisms, 492–493
    cytokines, 493
    negative signaling, 493
    target effectors, killing of, 493–494
CD8$^+$ Tregs
  eye derive tolerance and, 471–481
  See also CD4$^+$ cells; CD25$^+$ cells
CD25^{high}Foxp3$^+$ cells, 37
CD25$^-$Foxp3$^-$ cells, 32–33
CD25$^+$ cells
  characteristics, 19–20
  depletion in patients with cancer, 378–380
  See also CD4$^+$ cells
CD25$^+$Foxp3$^+$ cells
  peripheral generation, 33
  thymic generation, 31
  See also CD4$^+$CD25$^+$Foxp3$^+$ Tregs
CD28
  costimulatory molecules, 144–145
  co-stimulatory signals favoring nTreg cell development and homeostasis, 164
  naïve CD4$^+$CD25$^-$Foxp3$^-$ T cells conversion to CD4$^+$CD25$^+$Foxp3$^+$ Tregs, 117–119

CD28 (cont.)
    signaling for natural $CD4^+CD25^+Foxp3^+$ Tregs generation, 115
    thymic commitment of precursors to Treg lineage commitment, 46–47
    See also $CD4^+$ cells; $CD8^+$ Tregs; $CD25^+$ cells
CD62L expression on $CD4^+CD25^{hi}$ T cells, 272
    See also Multiple sclerosis (MS)
Cellular therapy
    Tr1 cells and, 464–465
    See also Autoimmune diseases; Gene therapy; Immunotherapy
Chronic active EBV (CAEBV), 440
Chronic EAE, 255
Chronic hepatitis B (CHB), 391
    double-edged role of Treg, 397–398
    Tregs in immune suppression in, 393–395
    viral antigen-specific $CD4^+CD25^+$ Treg, 396
    See also Herpesviruses; Retroviruses
Chronic hepatitis C (CHC), 391
    double-edged role of Treg, 397–398
    Tregs in immune suppression in, 395–396
    viral antigen-specific $CD4^+CD25^+$ Treg, 396–397
Cirrhosis
    primary biliary (PBC), 399–400
    See also Liver diseases
Clonal deletion mechanism, 3
    See also Immune tolerance
CNS inflammation, 267–269
    See also Multiple sclerosis (MS)
Copolymer-I (COP-I)
    Tregs induction and, 297–299
    See also Adoptive Tregs (aTreg)
Co-receptor blockade, 324–325
    See also Transplantation tolerance
Costimulatory molecules
    CD-28, 144–145, 164
    signals favoring nTreg cell development and homeostasis, 164
    See also T cell receptor (TCR)
Cutaneous lupus erythematosus (CLE), 259
    See also Systemic lupus erythematosus (SLE)
Cytokines
    $CD8\alpha\alpha^+$ Tregs suppression mechanism and, 493
    natural $CD4^+CD25^+Foxp3^+$ homeostatis and, 57–70

    ploietropic, see TGF-β
    See also Interleukin (IL); Tr1 cells
Cytotoxic T lymphocyte (CTL), 24
Cytotoxic T lymphocyte antigen 4 (CTLA-4), 7–8
    downregulatory signals in nTreg, 167–168
    rejection controlling and, 313
    signaling cascades in Treg, 144–145
    Treg-mediated immune suppression and, 97, 313
    tumor immunotherapy and, 380–381
    viral hepetitis and, 397
    See also Interleukin (IL); TGF-β

D

Day 3 thymectomy experiment, 4–5, 9–10
Day 7 thymectomy experiment, 5
De novo induction
    IL-2 signaling and, 84–86
    See also $CD4^+CD25^+$ Tregs
Deltaretroviruses, 411
    See also Gammaretroviruses
Dendritic cells (DC), 33, 241
    mediated deletion (autoreactive Treg-repertoire aspects), 41
    See also IL-10 signaling; TGF-β
Dendritic epidermal T cells (DETCs), 562, 563
Diabetes
    anti-CD3 therapy for, 218–219
    dysfunctional nTreg cells in human autoimmunity, 179–181
    in vivo Treg mechanisms, 202–203, 211–213
    NOD mouse model of spontaneous, 174–179
    pathogenic T cells dynamics during disease progression, 200–202
    T1D, see Type 1 diabetes (T1D)
    See also Autoimmune diseases; Cancer
Differentiation
    for Tr1 cells generation
        in vitro, 458–459
        in vivo, 460–461
    in vitro, 41, 43, 458–459
    PI3K pathway influence on, 140–142
    Th17 cell, 111, 113–114, 124–127
    thymic vs. peripheral Treg, 34
dnTβRII, 46
Dok-1/2 signaling, 142–143
    See also Phosphatases
Dominant tolerance, 18, 329
Donor transplantation, 345
    See also Transplantation tolerance

Double negative (DN) T cells, 541–554
  activation, 543–544
  development, 542–543
  in autoimmune diseases, 545–546
  in cancer, 546–547
  in GvHD, 546–547
  in infectious diseases, 547–548
  in murine cytomegalovirus (MCMV) infection, 548
  in tissue damage, 547–548
  in transplantation, 544–545
  in type-1 diabetes (T1D), 546
  in vitro culture, 543–544
  in vivo culture, 543–544
  suppression mechanisms
    alloantigens acquisition, 550–553
    antigen specificity, 550
    cell contact mediated suppression, 548–550
    migration, 553–554
  See also Invariant NKT (iNKT) cells; Natural killer T (NKT) cells
Downregulatory signals in natural Treg
  CTLA-4, 167–168
  TGF-β1, 168–170

E

Effector cells
  antigen-specific, 23–24
  TCR-mediated signaling in T, 137–138
  TGF-β regulating, 95–97
Epithelial tissues
  γδ T cells in immunoregulation in, 562–563
Epstein Barr Virus (EBV)
  chronic active EBV (CAEBV), 440
  IL-10 and TGF-β producing Treg cells in, 440–441
  See also Viral infections
Epstein-barr nuclear antigen 1 (EBNA1), 440
Experimental autoimmune encephalomyelitis (EAE)
  chronic, 255
  adaptive tregs induction and
    γ-IFN role in, 295–297
    T cell vaccination, 299–301
  invariant NKT and, 511–512
  MS and, 266, 274
  relapsing-remitting (R-EAE), 255
  Tregs for, 255–256
  See also Autoimmune diseases
Extra-thymic Tregs generation, 21–23
  See also Intra-thymic Tregs generation

Ex-vivo approach
  expansion of Tregs, 338–339
  manipulation of Tregs, 341–342
  See also In vivo approach; In vitro approach; Transplantation tolerance
Eye derive tolerance
  CD8$^+$ Tregs, 471–472
    clinical applications for ACAID induction, 480–481
    NKT cells and ACAID, 475
    peripheral, 478–480
    traffic patterns for CD8$^+$ cells induction, 473–475
  Treg cells generating ocular mechanisms, 475–477
    aqueous humor or its components generating Tregs, 476
    membrane bound molecules suppressing and generating Tregs, 477
  See also anterior chamber associated immune deviation (ACAID); Immune tolerance

F

Fc receptor (FCR), see Anti-CD3 therapy
Foxp3 expression, 8–9
  and NFAT, interactions with, 139
  antigen-specific, 33
  IL-2 signaling and, 82
  immune self-tolerance and, 159–161
    IPEX experiment, 160–161
    scurfy mice experiment, 159–161
    characteristics, 19–20
  mucosal inflammation and, 283, 285–286
  TCR and, 18
  TGF-β mediated, 33–34, 97–99
  thymic commitment of precursors to Treg lineage commitment, 43–45
  Tregs and, 17–20
  Tregs generation aspects
    peripheral, 32
    thymic, 22–23
  tumor immunotherapy and, 383–384
  See also CD4$^+$CD25$^+$Foxp3$^+$ Tregs
Friend virus (FV)
  IL-10 and TGF-β producing Treg cells in, 436, 438
  See also Viral infections
Fungal infections
  Aspergillus Fumigatus, 434–435
  Candida Albicans, 435–436
  IL-10 and TGF-β producing Treg cells in, 434–436

Fungal infections (*cont.*)
   Paracoccodioides Braziliensis, 434
   See also Bacterial infections; Parasite infections; Viral infections

G

γc cytokines, *see* Interleukin (IL)
γδ T cells in immunoregulation, 561–564
   in epithelial tissues, 562–563
   in tumor immunity, 563–564
γ-IFN
   adaptive Tregs induction and, 295–297
   rejection controlling Tregs suppression mechanisms and, 312
   viral infection and
      Epstein-Barr virus (EBV), 441
      friend virus (FV), 436, 438
      hepatitis B virus, 441
      hepatitis C virus, 441–443
      human immunodeficiency virus (HIV), 439–440
      Human T cell lymphotropic virus-1 (HTLV-1), 438
      lymphocytic choriomeningitis virus (LCV), 444
Gammaretroviruses
   $CD4^+CD25^+$ Tregs in, 411
   See also Deltaretroviruses
Gene expression
   alph-1,2–mannosidase, 314–315
   TOAG-1, 314–315
   See also Transplantation tolerance
Gene therapy, 219–220
   See also Cellular therapy; Immunotherapy
GITR, 8
   HSV and, 408
   See also Anti-GITR antibody
Glucocorticoids, 366
   See also Allergic diseases
Grafts tolerance, 327
Graft-versus-host disease (GvHD), 231
   adoptive transfer of Treg, 242–247
      antigen-non-specific clonal expansion, 242–243
      antigen-specific clonal expansion, 243–247
      in vivo, 211, 213–216
   DN T cells in, 546–547
   in vivo adoptive Treg therapy, 211–216
      issue of numbers, 214–216
      issues, 215–216
      phenotype and mode of action, 213–214
   See also Transplantation

Gut inflammation
   TGF-β and, 280–282
   See also mucosal inflammation

H

HAART patients
   HIV infection and, 413
   See also Viral infections
Helicobacter Pylori, 432–433
   See also Bacterial infections
Heligmosomoides Polygyrus, 429–430
   See also Parasite infections
Hemagglutinin (HA), 33
Hematopoietic stem cells (HSC), 464–465
   See also Tr1 cells
Hepatitis
   autoimmune, 391
   $CD4^+CD25^+$ Treg in, 392–393
   CTLA-4 expression and, 397
   HBV, *see* Hepatitis B virus (HBV)
   HCV, *see* Hepatitis C virus (HCV)
   IL-10, 392–393
   TGF-β, 392–393
   Tregs in, 391–392, 393–396
   viral, 392, 396–397
Hepatitis B
   acute (AHB), 391
   chronic (CHB), 391
   See also Hepatitis C virus (HCV)
Hepatitis B virus (HBV), 409–410
   chronic, 391, 393–398
   IL-10 and TGF-β producing Treg cells in, 441
   Tregs in
      as therapeutic target, 400–401
      $CD4^+CD25^+$, 392, 396–398, 409
      double-edged role of $CD4^+CD25^+$, 397–398
      immune suppression, 393–395
      viral antigen-specific $CD4^+CD25^+$ in, 396
Hepatitis C virus (HCV), 409–410
   chronic, 391, 395–398
   IL-10 and TGF-β producing Treg cells in, 441–443
   Tregs in
      as therapeutic target, 400–401
      $CD4^+CD25^+$, 392–393, 396–398, 409–410
      double-edged role of $CD4^+CD25^+$ Treg in, 397–398

Index 577

immune suppression in, 395–396
viral antigen-specific CD4+CD25+
Treg in, 396–397
See also Infections
Hepatitis viruses
HBV, 391–398, 400–401, 409, 441, *See also under* Hepatitis B virus (HBV)
HCV, 391–393, 395–401, 409–410, 441–443, *See also under* Hepatitis C virus (HCV)
See also Herpesviruses; Retroviruses
Hepatocellular carcinoma, 391, 398–400
See also Cancer; Tumor immunotherapy
Herpes simplex virus (HSV-1)
IL-10 and TGF-β producing Treg cells in, 443
invariant NKT and host defense against infection, 506
Herpesviruses, 407–409
CD4+CD25+ Tregs in, 407–408
herpes simplex virus (HSV), 407–408
HSV-1, 407–408
HSV-2, 408
See also Hepatitis viruses; Retroviruses
HIV infections
HAART patients and, 413
IL-10 and TGF-β producing Treg cells in, 439–440
lentiviruses and, 412–416
See also Viral infections
Homeostasis
AICD controls immune, 78
CD4+CD25+Foxp3+, 57–70
co-stimulatory signals favoring nTreg cell development and, 164
peripheral, 62–66
See also IL-2 signaling
Human T cell lymphotropic virus (HTLV-1), 411
associated myelopathy (HAM), 438
IL-10 and TGF-β producing Treg cells in, 438
See also Retroviruses
Human type 1 Tregs, *see* Tr1 cells

I
IDDM, 257–258
IFN-γ, *see* γ-IFN
IL, *see* Interleukin (IL)
Immune dysregulation polyendocrinopathy enteropathy X linked syndrome, *see* IPEX

Immune privilege
defined, 472
See also Anterior chamber associated immune deviation (ACAID)
Immune self-tolerance
by natural CD4+, 157–159
Foxp3 and CD4+
IPEX experiment, 160–161
scurfy mice experiment, 160–161
IL-2 in, 165–166
peripheral, 58–60
See also Immune suppression; Immune tolerance; Transplantation tolerance
Immune suppression
antigen-specific, 23–25, *see also under* Antigen specificity
autoreactive T cells resistance to Treg, 174
by CD8αα+ Tregs, 487–493
by TCRαβ+ Tregs, 487
DN T cells suppression mechanisms
alloantigens acquisition, 550–553
antigen specificity, 550
cell contact mediated suppression, 548–550
migration, 553–554
in chronic HBV, 393–395
in chronic HCV, 395–396
in vitro, 82–83
in vivo, 82–83
linked suppression, 323, 326–329
lymph node (LN) priming and, 205–208
TGF-β and, 92–97
Tr1 cells, 457–458
Tregs and, 93–94
aTreg-mediated, 94–95
IL-2–mediated, 166–167
immune suppression responses, 97
nTreg-mediated, 94–95, 174
See also Immune tolerance
Immune tolerance
abnormal central tolerance and peripheral pool of autoreactive T cells, over-riding nTreg cell function and provoking autoimmunity, 173
AICD of T cells mechanism, 3
animal models of autoimmunity, 174–179
can autoreactive T cells be resistant to nTreg cell mediated suppression?, 174
CD4+ in
dominant regulation of peripheral T cell tolerance, 156–157

Immune tolerance (cont.)
- IL-2 mediated T cell regulation model, 166–167
- induced, 156
- natural, 156–157
- CD4$^+$CD25$^+$ Tregs in, 162–166
  - co-stimulatory signals favoring nTreg cell development and homeostasis, 164
  - homeostatic functions of IL-2 for nTreg, 162–164
  - IL-2 in self-tolerance and autoimmunity, 165–166
- CD4$^+$CD25$^+$Foxp3$^+$ Tregs in
  - IL-2 in vivo, expression and cellular sources of, 164–165
- CD4$^+$Foxp3$^+$ Tregs in, 155
  - homeostatic functions of IL-2 for nTreg, 162–164
  - IL-2 in self-tolerance and autoimmunity, 165–166
  - IL-2 in vivo, expression and cellular sources of, 164–165
- chemokine directed homing of nTreg cells to sites of inflammation, 183–185
- clonal deletion in thymus mechanism, 3
- day 3 neonatal thymectomy experiment, 4–5
- day 7 neonatal thymectomy experiment, 5
- dysfunctional Treg cells in human autoimmunity, 179–181
- IL-2 mediated Treg model
  - homeostatic functions for nTreg, 162–164
  - in vivo, expression and cellular sources of, 164–165
  - self-tolerance and autoimmunity, 165–166
- regulation of immune responses to microbes, 181–183
- regulatory and autoreactive T cells, balance between, 170–174
- suppressor T cells (1982–1995), 5–6
- Tr1 cells tolerance to, 461–464
  - allo-antigens, 463
  - infectious agents and tumors, 463–464
  - non self non harmful antigens, 462–463
  - self antigens, 462
- Tregs, 3–5
  - CD4$^+$, 156–157, 166
  - CD4$^+$CD25$^+$, 162–166
  - CD4$^+$CD25$^+$Foxp3$^+$, 164–165
  - CD4$^+$Foxp3$^+$, 155, 162–166

See also Immune self-tolerance; Immune suppression; Transplantation tolerance

Immunoregulation, 155
- γδ T cells in, 561–564
  - in epithelial tissues, 562–563
  - in tumor immunity, 563–564

See also Immune tolerance

Immunotherapy (IT)
- allergen-specific, 364–367
- Tumor, *see* Tumor immunotherapy

See also Allergic diseases

In vitro approach
- antigen-specific tregs expansion aspects, 217
- CD4$^+$ CD25$^+$ Tregs, 80–83
- differentiation
  - for Tr1 cells generation, 458–459
  - thymic commitment of precursors to Treg lineage commitment and, 41, 43
- generation of Tregs
  - peripheral, 32–33
  - thymic, 32
- IL-2 signaling, 80–83
- naïve CD4$^+$CD25$^-$Foxp3$^-$ T cells conversion to CD4$^+$CD25$^+$Foxp3$^+$ Tregs, 117–119
- transplantation tolerance, 328
- Tregs
  - for autoimmune diseases, 239–241
  - therapy, 218

In vivo approach
- adoptive Treg therapy
  - issue of antigen-specificity, 211–213
  - issue of numbers, 214–216
- differentiation, 460–461
- IL-2 signaling
  - CD4$^+$ CD25$^+$ Tregs, 80–83
  - expression and cellular sources of, 164–165
- inflammation and Treg function, 309
- imaging approach, 238
- naïve CD4$^+$CD25$^-$ T cells conversion to CD4$^+$CD25$^+$Foxp3$^+$ Tregs, 121–123
- suppression aspects, 83–84
- transplantation aspects
  - inflammation and Treg function, 309
  - location of, 309–311
- Tregs
  - diabetes control aspects, 202–203
  - for autoimmune diseases, 239–241

Index

peripheral generation, 33
transplantation and, 309–311
Tregs therapy
adoptive, 211–216
antigen-specific tregs expansion aspects, 218
phenotype and mode of action, 213–214
See also In vitro approach
Inducible Tregs, see Adoptive Tregs (aTregs)
Infections
bacterial
Bordetella Pertussis, 433–434
Helicobacter Pylori, 432–433
Lactobacillus, 431
Mycobacterium Tuberculosis, 431–432
DN T cells in, 547–548
fungal
Aspergillus Fumigatus, 434–435
Candida Albicans, 435–436
Paracoccodioides Braziliensis, 434
invariant NKT and host defense against, 505–507
parasite
Heligmosomoides Polygyrus, 429–430
Leishmaniasis, 426–428
Litomosoides Sigmodontis, 430–431
malaria, 424, 426
Schistosomiasis, 428–429
tolerance
regulation pathway, 328–329
to Tr1 cells, 463–464
transplantation, 343, see also under Transplantation tolerance
Treg
inducible, 423
natural, 421–423
transplantation tolerance, 343
viral, 405
EBV, 440–441
friend virus (FV), 436, 438
HBV, 441
HCV, 441–443
HIV, 439–440
HSV-1, 443
HTLV-1, 438
LCV, 443–444
See also Immune suppression; Immune tolerance
Inflammation
central nervous system (CNS), 267–269, see also Multiple sclerosis (MS)
chemokine directed homing of nTreg cells to sites of, 183–185

mucosal, see Mucosal inflammation
regulation and, 166–167
Treg function in vivo and, 309
See also Autoimmune diseases; Gut inflammation
Inflammatory bowel disease (IBD)
TGF-β disregulation and, 100
See also Mucosal inflammation
Interleukin (IL)
IL-2, see Interleukin-2 (IL-2)
IL-6 induced Th17 cell differentiation, 127
IL-10, see Interleukin-10 (IL-10)
IL-17 production from $CD4^+$ effector, 125
natural $CD4^+CD25^+Foxp3^+$ homeostatis and
IL-7, 69
IL-15, 69
signaling, see Interleukin-2 (IL-2) signaling; Interleukin-2R (IL-2R) signaling
See also Cytotoxic T lymphocyte antigen 4 (CTLA-4); TGF-β
Interleukin-2 (IL-2)
cellular source of, 67–69
for nTreg, homeostatic functions of, 162–164
in self-tolerance and autoimmunity, 165–166
in vivo, expression and cellular sources of, 164–165
mediated T cell regulation model, 166–167
naïve $CD4^+CD25^-Foxp3^-$ T cells conversion to $CD4^+CD25^+Foxp3^+$ Tregs, 118
signaling, see Interleukin-2 (IL-2) signaling
TGF-β suppressing IL-2 production in T cells, 95–97
thymic commitment of precursors to Treg lineage commitment, 46
See also Interleukin-2/Interleukin-2R (IL-2/IL-2R) interaction; Interleukin-10 (IL-10); TGF-β
Interleukin-2 (IL-2) signaling
AICD control of homeostasis and, 78
$CD4^+ CD25^+$ Tregs and, 77–86
peripheral de novo induction, 84–86
discrete signaling in $CD4^+ CD25^+$, 80–82
in vitro signaling, 81
in vitro suppression, 82–83
in vivo signaling, 80
in vivo suppression, 83–84
signaling overview, 79–80

Interleukin-2 (IL-2) signaling (cont.)
  for natural $CD4^+CD25^+Foxp3^+$ Tregs generation, 115–116
  See also Interleukin-2R (IL-2R) signaling
Interleukin-2/Interleukin-2R (IL-2/IL-2R) interaction
  peripheral self-tolerance control aspects, 58–60
  requirement within thymus and peripheral lymphoid tissues, 62–66
  Treg cell function and, 66–67
    IL-2Rα, 66
    IL-2Rβ, 66–67
Interleukin-2R (IL-2R) signaling, 79–80
  for natural $CD4^+CD25^+Foxp3^+$ Tregs generation, 116
  in Treg, 145–146
    $CD4^+CD25^+Foxp3^+$, 61–62, 116
    JAK signaling, 60–62
    MAPK signaling, 60
    PI3K signaling, 60–62
    STAT5 signaling, 60–61
  See also Interleukin-2 (IL-2) signaling; Interleukin-2/Interleukin-2R (IL-2/IL-2R) interaction
Interleukin-10 (IL-10)
  allergic disease and, 362–366
  bacterial infection and, 431–434
  fungal infection and, 434–436
  hepatitis and, 392–393
  parasite infection and, 424–431
  producing T regulatory type 1 cells, see Tr1 cells
  viral infection and, 436–444
  See also Interleukin-2 (IL-2)
Intra-thymic Tregs generation, 20–23
  See also Extra-thymic Tregs generation
Invariant NKT (iNKT) cells, 499, 525–526
  asthma
    forms, 529
    mechanisms, 528
    treatment, 529–530
  asthma regulation, 527–532
    in human, 527, 528
    in mouse, 527
  autoimmune disease and, 509–512
  cancer and, 508–509
  $CD4^+CD25^+FOXP3^+$ Tregs controlling, 512–513
  cell polarization regulation by synthetic ligand analogs, 502–505
  cells and host defense against infection, 505–507
  defined, 500
  natural antigen specificity and, 500–502
  See also Double negative (DN) T cells; Natural killer T (NKT) cells
IPEX, 8–9
  dysfunctional nTreg cells in human autoimmunity, 179–181
  experiment
    Foxp3 and $CD4^+$ immune self-tolerance, 159
    lessons learnt from, 160–161
  Tregs for, 239s
  See also Autoimmune diseases

J

Janus activated kinase (JAK)
  signaling, 60 62
  See also Interleukin-2R (IL-2R) signaling

L

Late phase allergic response (LAR), 354, 366
Leishmania infections
  IL-10 and TGF-β producing Treg cells in, 426–428
  See also Bacterial infections; Parasite infections; Viral infections
Leishmania amazonensis, 427
Leishmania guyanensis, 427–428
Leishmania major, 182–185, 427–428
  invariant NKT and host defense against infection, 507
  See also Autoimmune diseases
Leishmania viannia braziliensis, 426, 428
Lentiviruses
  $CD4^+CD25^+$ Tregs in, 411–416
  HIV infection and, 412–416
  induced AIDS, 413–416
  See also Retroviruses
Lineage commitment, Treg, 41
  CD28 role in, 46–47
  CD4–lineage, 43, 45
  CD8–lineage, 43, 45
  dnTβRII role in, 46
  Foxp3 expression, 43–45
  IL-2 role in, 46
  in vitro T cell differentiation approaches, 43
  instructive model, 44
  stochastic model, 44
  TCR specificity, 45, 47
  TCRαβ, 45
  TCR/ligand interactions, 43–44
  TGF-β role in, 45–46

Index 581

Linked suppression, 323, 329
  transplantation tolerance and, 326
  grafts tolerance and, 327
  See also Immune suppression; Transplantation tolerance
Litomosoides Sigmodontis, 430–431
Liver cirrhosis (LC), 391
Liver diseases, see Autoimmune liver diseases (AILD)
Lupus erythematosus (LE)
  cutaneous (CLE), 259
  systemic (SLE), 181, 211, 259, 512
  Tregs for, 259
  See also Autoimmune diseases
Lymph node (LN)
  priming aspects, 203–208
  See also Diabetes; Immune suppression; Lymphoid tissues
Lymphocytes
  from bone marrow, 4
  from thymus, see Tregs
Lymphocytic choriomeningitis virus (LCV)
  IL-10 and TGF-β producing Treg cells in, 443–444
  See also Viral infections
Lymphoid tissues
  IL-2 requirement within
    peripheral, 62–66
    thymus, 62–66
  See also Lymph node (LN); Lymphocytes

M

Malaria
  Anopheles gambiae, 424
  IL-10 and TGF-β producing Treg cells in, 424, 426
  Plasmodium falciparum, 424, 426
  Plasmodium yoelii, 426
  See also Parasite infections
Memory T cells, 315–316
  See also Transplantation
MHC/peptide specific Treg precursors
  negative selection of, 36–37
  positive selection of, 37–42
Microbes
  regulation of immune responses to, 181–183
  See also Autoimmune diseases; Infections
Mitogen activated protein kinase (MAPK) signaling, 60
  MAPK/Ras signaling pathway, 138–139
  TGF-β signaling and, 92–93
  See also Interleukin-2R (IL-2R) signaling

Molecular signaling
  costimulatory molecules and, 144–145
  IL-2R, 145–146
  NFAT role in FOXP3 interactions, 139
  NF-κB, 143
  PI3K, 140–143
  Ras/MAPK, 138–139
  TCR-mediated
    in T effector cells, 137–138
    mechanistic basis for changes in, 143–144
    signaling downstream, 138
  See also Interleukin-2 (IL-2) signaling; Interleukin-2R (IL-2R) signaling; Tregs
Mouse model, see Animal models
Mucosal inflammation, 280
  FoxP3-expressing Tregs and, 283
  TGF-β
    as Tregs effector molecule, 286–287
    induced Tregs, 285–286
    pathway manipulation as therapy, 287–288
  Tregs for, 279
    natural CD4$^+$CD25$^+$ Tregs, 282, 283, 284
  therapy, 287–288
  See also Allergic diseases; Autoimmune diseases; Gut inflammation; Interleukin-2R (IL-2R) signaling; Multiple sclerosis (MS)
Multiple sclerosis (MS)
  adaptive Tregs induction and
    copolymer-I (COP-I) role in, 297–299
    γ-IFN role in, 297
    T cell vaccination role in, 299, 301
  CD4$^+$CD25 T cells in MS patients, discussion of, 272–273
  CD4$^+$CD25hi T cells
    CD62L expression on, 272
    display impaired function in MS patients, 271–272
    same frequency presence in healthy donors and MS patients, 270–271
  cellular events occurring within and around MS plaques, 267–268
  CNS inflammation regulation, 267–269
  EAE model and, 266, 274
  epitope spreading, 254
  immunopathophysiology, 266–267
  invariant NKT and, 511–512
  molecular mimicry, 254
  Tregs for, 254–255, 265, 269–273

Multiple sclerosis (MS) (Cont.)
  Tregs therapy, 273–275
  See also Allergic diseases; Autoimmune diseases; Gut inflammation; Mucosal inflammation
Murine cytomegalovirus (MCMV)
  double negative (DN) T cells, 548
  invariant NKT and host defense against infection, 506
  See also Viral infections
Mycobacterium tuberculosis
  IL-10 and TGF-β producing Treg cells in, 431–432
  invariant NKT and host defense against infection, 506
  See also Bacterial infections
Myeloid DC, 391

N

Naïve $CD4^+CD25^-$ T cells, see $CD4^+CD25^-$ T cells
Natural killer T (NKT) cells
  ACAID and, 475
  asthma regulation, 523
    asthma forms, 529
    balance between allergen-specific TReg cells, Th2 cells and iNKT cells, 531
    mechanisms, 528
    regulation by Tregs, 530–531
    treatment, 529–530
    in human, 527–528
    in mouse, 527
    suppressive effects of NKT cells, 532
  Borrelia burgdorferi and, 527
  characteristics
    invariant (iNKT), 525–526, See also under Invariant NKT (iNKT) cells
    Type 1, 525
    Type 2, 525
  functions, 526–527
  iNKT, see Invariant NKT (iNKT) cells
  Novosphingobium and, 526–527
  Salmonella and, 527
  Sphingomonas and, 526
  See also Double negative (DN) T cells
Natural Tregs (nTregs)
  allergic disease and, 360
  autoimmune diseases and, 239
  autoimmunity
    animal models, 174–179
    human, 179–181
    regulatory and autoreactive T cells, 173
  $CD4^+$
    dysfunctional Treg cells in human autoimmunity, 179–181
    immune tolerance aspects, 156–159
    regulation of immune responses to microbes, 181–183
  $CD4^+CD25^+$
    autoimmune diseases and, 239, 253
    co-stimulatory signals favoring cell development and homeostasis, 164
    CTLA-4 downregulatory signals in nTreg, 167–168
    hepatitis and, 393
    immune tolerance aspects, 164–166
    mucosal inflammation and, 282–284
    TCR for, 114–115
    TGF-β signaling for, 116–117, 168–170
  $CD4^+CD25^+Foxp3^+$
    CD28 signaling, 115
    homeostatis, 57–70
    IL-2 and, 58–69, 115–116
    IL-2/IL-2R interaction and, 58–60, 116
    peripheral tolerance aspects, 57–60
    TCR for, 114–115
    TGF-β signaling for, 116–117
  CTLA-4 downregulatory signals in, 167–168
  IL-2 mediation and, 166–167
    homeostatic functions, 162–164
    IL-2/IL-2R interaction, 58–60, 116
    in vivo IL-2, expression and cellular sources of, 164–165
    self-tolerance and autoimmunity, 165–166
  immune suppression and, 94
  immune tolerance aspects
    autoimmunity, 173–181
    chemokine directed homing of Treg cells to sites of inflammation, 183–185
    co-stimulatory signals favoring cell development and homeostasis, 164
    impaired development or function within nTreg cell compartment, 171–173
    regulation of immune responses to microbes, 181–183
    regulatory and autoreactive T cells, 173–174
  infections and, 421–423
  TGF-β1 signaling in, 116–117, 168–170
  See also Adoptive Tregs (aTreg)
Negative selection of Treg precursors, 35–37
NFAT

Index

FOXP3 interactions and, 139
  IL-2 signaling and, 82
  Tregs generation, 20–23
Tregs generation and
  NFAT and Foxp3 interaction, 22
  thymic, 22
tumor immunotherapy, 383–384
NF-κB signaling, 143
  See also AKT pathway; Interleukin-2R (IL-2R) signaling
NOD mouse model, 174–179
  See also Animal models
Novosphingobium, 526
  NKT cells and, 527

O

Ontak, 378–380
  See also tumor immunotherapy
Ovarian disease
  in vivo adoptive Treg therapy for, 213–216
  See also Autoimmune diseases

P

Paracoccidioidomycosis (PCM), 434
Paracoccodioides Braziliensis
  IL-10 and TGF-β producing Treg cells in, 434
  See also Fungal infections
Parasite infections
  Heligmosomoides Polygyrus, 429–430
  IL-10 and TGF-β producing Treg cells in, 424–431
  Leishmania, 426–428
  Litomosoides Sigmodontis, 430–431
  malaria, 424, 426
  Schistosoma, 428–429
  See also Bacterial infections; Fungal infections; Viral infections
Peripheral autospecific repertoire, 34–35
Peripheral $CD8^+$ Tregs, 478–480
  See also Eye derive tolerance
Peripheral de novo induction
  $CD4^+$ $CD25^+$ Tregs, 84–86
  See also Interleukin-2 (IL-2) signaling
Peripheral generation of Tregs, 29, 32–33
  $CD4^+CD25^-$Foxp3$^-$ cells, 33
  $CD4^+CD25^+$ cells, 33
  $CD4^+CD25^+$Foxp3$^+$ cells, 32
  $CD4^+$Foxp3$^+$ cells, 34
  $CD25^-$ cells, 32, 33
  $CD25^-$Foxp3$^-$ cells, 33
  $CD25^+$Foxp3$^+$ cells, 33
  cell differentiation aspects, 34
  HA-TCR cells, 33

in vitro, 32–33
in vivo, 33
  See also Thymic generation of Tregs
Peripheral homeostasis
  peripheral lymphoid tissues, 62–66
  See also Interleukin (IL)
Peripheral tissues
  lymphoid tissues, 62–66
  Treg effect in, 205
Peripheral tolerance
  antigen-specific, 472
  dominant regulation of, 156–157
  IL-2/IL-2R interaction controls, 58–60
  natural $CD4^+CD25^+$Foxp3$^+$, 57–60
  See also Anterior chamber associated immune deviation (ACAID); Immune tolerance
Pertussis, see Bordetella Pertussis
Phosphatases
  PTEN, 142
  SHIP, 142
Phosphatidylinositol 3–kinase (PI3K)
  signaling, 60
  See also Interleukin-2 (IL-2) signaling
PI3K pathway
  negative regulators of
    Dok signaling, 142–143
    PTEN, 142
    SHIP, 142
PI3K/AKT pathway
  IL-2 signaling and, 81
  TGF-β signaling and, 92–93
  signaling, 60–62
  Treg cell differentiation and function, influence of, 140–142
  See also NF-κB signaling
Plasmacytoid DC, 391
Plasmodium
  invariant NKT and host defense against infection, 507
  Plasmodium falciparum, 424
  Plasmodium yoelii, 507
  See also Parasite infections
Ploietropic cytokine, see TGF-β
Positive selection of Treg precursors, 37–38, 40–42
Primary biliary cirrhosis (PBC)
  Tregs and, 399–400
  See also Autoimmune liver diseases (AILD)
Privileged microenvironments, 329
  See also Transplantation tolerance

Pseudomonas aeruginosa
    invariant NKT and host defense against infection, 506
    See also Infections
PTEN, 142
    mediated inhibition, IL-2 signaling and, 81
    See also Phosphatases

Q
Qa-1
    CD8$\alpha\alpha^+$ Tregs and, 490–491
    See also Immune suppression

R
Ras/MAPK signaling pathway, 138–139
Recessive tolerance, 18
    See also Immune tolerance
Regulatory T cells, see Tregs
Rejection control
    Tregs characterization, 308–309
    Tregs suppression mechanisms, 311–313
        CTLA-4 role, 313
        IFN-γ role, 312
    See also Transplantation
Relapsing-remitting EAE (R-EAE), 255
Repertoire, see Autospecific TCR repertoire
Resistance, 325
    grafts tolerance and, 327
    transplantation tolerance and, 324, 328
    See also Immune tolerance
Retinoic-acid-related orphan receptor-γt (RORγt), 127
    See also Th17 cell
Retroviruses
    CD4$^+$CD25$^+$ Tregs in, 410–411
    deltaretroviruses, 411
    gammaretroviruses, 411
    See also Hepatitis viruses; Lentiviruses
Rheumatoid arthritis, 258–259
    See also Inflammation

S
Salmonella
    NKT cells and, 527
    See also Infections
Schistosoma
    IL-10 and TGF-β producing Treg cells in, 428–429
    Schistosoma mansoni, 429
Scurfy mice experiment
    Foxp3 and CD4$^+$ immune self-tolerance, 159
    lessons learnt from, 160–161
Self antigens
    tolerance (Tr1 cells) to, 462
    See also Alloantigens
Self-tolerance, see Immune self-tolerance
SHIP, 142
    See also Phosphatases
Signal transducer and activator of transcription 5 (STAT5)
    signaling, 60–62
    STAT5, 61
        IL-2 signaling and, 80–81
        natural CD4$^+$CD25$^+$Foxp3$^+$, 69
    See also Interleukin-2 (IL-2) signaling; Interleukin-2R (IL-2R) signaling
Signaling, see Molecular signaling
Simian immunodeficiency virus (SIV), 439
SMAD family, 282
    See also Gut inflammation
Sphingomonas
    invariant NKT and host defense against infection, 506
    NKT cells and, 526
Suppression, see Immune suppression
Suppressor T cells, 4–7
    rebirth (1995–2000), 6–7
    tracking (1982–1995), 5–6
    See also Tregs
Systemic lupus erythematosus (SLE), 259
    dysfunctional nTreg cells in human autoimmunity, 181
    in vivo adoptive Treg therapy for, 211
    invariant NKT and, 512
    See also Cutaneous lupus erythematosus (CLE)

T
T cell
    DN, see Double negative (DN) T cells
    iNKT, see Invariant NKT (iNKT) cells
    NKT, see Natural killer T (NKT) cells
    regulatory, see Tregs
T cell receptor (TCR)
    for CD4$^+$CD25$^+$Foxp3$^+$ induction in human T cells, 124
    Foxp3 and, 18
    HA-TCR, 33
    mediated signaling
        in T effector cells, 137–138
        mechanistic basis for changes in, 143–144
        signaling downstream of, 138
    naïve CD4$^+$CD25$^-$ T cells
        conversion to CD4$^+$CD25$^+$Foxp3$^+$, 117–119

Index

conversion to in vivo CD4+CD25+Foxp3+, 121–122
repertoire
  peripheral, 34–35
  thymic, 35–42
specificity
  thymic commitment of precursors to Treg lineage commitment, 43–45, 47
  TCRαβ, 20–21
  TCRγδ and, 45
TCRαβ, 20–21, 487
Tregs generation and
  natural CD4+CD25+Foxp3+ Tregs, 114–115
  thymic generation, 21–23
See also Costimulatory molecules
TGF-β
  allergic disease and, 360–364
  dependent signaling, 24
  Foxp3
    expression, 33–34, 97–99
    TGF-induced, 123–124
  gut inflammation and, 280–282
  hepatitis and, 392–393
  IL-2 production in T cells, suppression of, 95–97
  immune suppression aspects, 92–97
  induced Tregs
    in human T cells, 124
    molecular mechanisms and pathways, 123–124
    phenotype and function, 120–121
    TGF requirements for conversion, 121–123
  infection and
    bacterial, 431–434
    fungal, 434–436
    lentiviruses-induced AIDS, 414–416
    parasite, 424–431
    viral, 438–439, 442
  mucosal inflammation and, 279
    TGF-β as Tregs effector molecule, 286–288
    TGF-β induced Tregs, 285–286
  naïve CD4+CD25− T cells conversion to CD4+CD25+Foxp3+ Tregs, 117–119, 121–123
  natural CD4+CD25+Foxp3+
    homeostatis, 70
    Tregs generation, 116, 117
  reciprocal regulation of Tregs and Th17 cells, 111–114

signaling and immune suppression, 92–94
T cell proliferation and effector function, regulation of, 95–97
TGF-β1 downregulatory signals in nTreg, 168–170
Th17 cell differentiation, 124–127
thymic
  commitment of precursors to Treg lineage commitment, 45–46
  Tregs generation and, 22–23
Treg and, 91–92
  and diseases, 99–100
  T cell proliferation and effector function regulation, 95–97
  TGF-β disregulated function aspects, 99–100
  TGF-β mediated Treg generation, 97–99
tumor immunotherapy and, 384–385
See also Cytotoxic T lymphocyte antigen 4 (CTLA-4); Interleukin-2 (IL-2)
TGF-β activating kinase (TAK1), 70
Th1 cells, 124–125
Th2 cells, 124–125
  allergic diseases and, 354
  iNKT and asthma regulation by, 531
Th3 cells, 94–95
See also Adoptive Tregs (aTregs); Tr1 cells
Th17 cell differentiation, 111
  from naïve CD4+CD25− T cells
    IL-6 and TGF-β induced RORrt and, 127
    RORrt for, 127
    TGF-β induced, 113–114, 124–127
  reciprocal regulation of, 111–114
3 neonatal thymectomy experiment, see day 3 neonatal thymectomy experiment, 4
Thymic autospecific repertoire
  negative selection aspects, 35–37
  positive selection aspects, 37–42
Thymic commitment of precursors to Treg lineage commitment, see Lineage commitment, Treg
Thymic generation of Tregs, 29–32
  CD4+CD25+ T cells, 31
  CD4+CD25+Foxp3+ cells, 31–32
  CD25+Foxp3+ cells, 31
  cell differentiation aspects, 34
  extra-, 21–23
  in vitro, 32
  intra-, 20–23
See also Peripheral generation of Tregs

Thymic stromal lymphopoietin (TSLP), 46, 69–70
Thymus lymphoid tissues, 62–66
Thyroiditis, 212
TNF-α
  antibodies, 11
  viral infection and, 438
  See also Autoimmunity
TOAG-1 gene, 314–315
  See also transplantation
Tolerance
  eye derive, *see* Eye derive tolerance
  immune, *see* Immune tolerance
  tranplantation, *see* Tranplantation tolerance
Tolerogenic milieu
  defined, 343–344
  See also Transplantation tolerance
Toll like receptors (TLR), 358
  See also Allergic diseases
Topical spastic paraptesis (TSP), 438
Tr1 cells
  cellular therapy with, 464–465
  distinguishing
    markers other than cytokine production profile, 456–457
    suppressive functions, 457–458
  features, 454–455
  generating
    in vitro differentiation, 458–459
    in vivo differentiation, 460–461
  origins, 455–456
  tolerance to
    allo-antigens, 463
    infectious agents and tumors, 463–464
    non self non harmful antigens, 462–463
    self antigens, 462
  See also Adoptive Tregs (aTregs)
Transforming growth factor, *see* TGF-β
Transplantation
  defined, 307–308
  DN T cells in, 544–545
  memory T cells and, 315–316
  tolerance, *see* Transplantation tolerance
  Tregs in
    alph-1,2–mannosidase gene expression patterns indentification, 314–315
    gene expression patterns indentification, 313–314
    homeostatic proliferation aspects, 315–316
    in vivo Treg location aspects, 309–311
    inflammation and Treg function in vivo, 309
    rejection controlling Tregs characterization, 308–309
    rejection controlling Tregs suppression mechanisms, 311–313
    TOAG-1 gene expression patterns indentification, 314–315
Transplantation tolerance
  adoptive transfer of Tregs in animals, 337–338
  antigen for tolerising microenvironments creation, 329
  antigen specificity for, 339–341
    antigen specific transferable tolerance, 334
    antigen specificity to regulation and, 325
  by co-receptor blockade, 324–325
  $CD4^+CD25^+$ Treg for inducing, 333–337
  grafts tolerance, 327
  infectious tolerance, 325, 328–329
  linked suppression and, 326
  regulation pathway, 327–328
  resistance and, 328
  sustained regulation exploits indirect pathway of antigen presentation, 326
  Treg therapy for transplantation tolerance in humans, 343
    immune monitoring, 345–346
    living related and deceased donor transplantation, 345
    tolerogenic milieu, 344
  Tregs
    and opportunistic infection, 343
    ex-vivo expansion, 338–339
    ex-vivo manipulation, 341–342
    in therapeutic, 323–327
  unifying mechanism to explain tolerance, 326–327
  See also Immune tolerance; Resistance
Tregs
  allergic disease and, 353–366
  antigen-specific effector cells and, 23–24
  asthma and, 530–531
  autoimmune diseases
    adoptive transfer therapy, 236–238, 242–247
    antigen-non-specific clonal expansion aspects, 242–243
    antigen-specific clonal expansion aspects, 243–247
    diabetes control aspects, 199–203
    effect in peripheral tissues, 205

Index

gene therapies, 234–236
GVHD, 231, 242–247
in vitro Tregs, 239–241
in vivo Tregs, 239–241
IPEX, 239
liver diseases (AILD) and, 399–401
lymph node (LN) priming and, 203–208
natural Tregs, 239
selective therapies, 233–234
therapeutic use of Tregs in, 210–218, 400–401
unifying model, 208–210
autoimmunity and, 3–11
autoreactive T cells and Tregs, balance between, 170–174
bystander suppression, 23
characteristics, 19–20
eye derive tolerance and, 471–481
function, 24–25
generation
  extra-thymic, 21–23
  intra-thymic, 20–23
  NFAT role in, 22
  peripheral, 29, 32–33
  TGF-β mediated, 95–100
  thymic, 29–32, 34
hepatitis and, 391–398
hepatocellular carcinoma and, 391, 398–399
human type 1, see Tr1 cells
immune suppression aspects, 24–25, 93–94, 97
iNKT and, 499, 530–531
molecular signaling in, 135–145
mucosal inflammation and, 279, 282–288
multiple sclerosis and, 265, 269–275
TCR specificity and, 20–23
TGF-β and, 91–92, 111–114
transplantation and, 307–316
transplantation tolerance and, 323–346
tumor immunotherapy and, 377–385
types
  acquired/adoptive/induced, see Adoptive Tregs (aTreg)
  natural, see Natural Tregs (nTregs)
See also Immune tolerance; Immune tolerance; Transplantation tolerance
Tuberculosis, see Mycobacterium Tuberculosis
Tumor
  immunity, γδ T cells in immunoregulation in, 563–564
  TGF-β disregulation and, 100
  tolerance (Tr1 cells) to, 463–464

Tumor immunotherapy, 377
  Ontak administration, 378–380
  Tregs and, 377
    anti-CTLA-4 therapy, 381
    anti-GITR antibody expression, 381, 383
    cells depletion in patients with cancer, 378–380
    cells expansion modulation, 384–385
    cells function modification, 380–385
    cells in patients with cancer, 378
    cells trafficking modification, 384
    CTLA-4$^+$ expression, 380–381
    FOXP3 molecular signals modulation, 383–384
  See also Cancer; Hepatoclucllar carcinoma
Type 1 diabetes (T1D)
  control by Tregs, 199–203
    in vivo Treg mechanisms, 202–203
    Tregs and pathogenic T cells dynamics during disease progression, 200–202
  double negative (DN) T cells in, 546
  dysfunctional nTreg cells in human autoimmunity, 179–181
  generic therapy for, 233
  immunogenetic determinants of autoimmunity, 177–179
  in vivo Treg therapy, 211, 213
  invariant NKT and, 511–512
  NOD mouse model of spontaneous, 174–179
  TGF-β disregulation and, 100
  Tregs for, 256–258
  See also Autoimmune diseases

V

Vehicle cells approach
  for autoimmune diseases, 237–238
  See also Viral vectors approach
Viral antigen-specific CD4$^+$CD25$^+$ Treg
  in chronic HBV, 396
  in chronic HCV, 396–397
Viral hepatitis
  CD4$^+$CD25$^+$ Treg in, 392–393
  See also Autoimmune hepatitis; Hepatitis B virus (HBV); Hepatitis C virus (HCV)
Viral infections
  CD4$^+$CD25$^+$ Tregs in, 405–416
    hepatitis viruses, 409, 410
    herpesviruses, 407–408

Viral infections (cont.)
  lentiviruses, 411–413
  lentiviruses-induced AIDS, 413–416
  retroviruses, 410
  viral immunity and immunopathology aspects, 407
  Epstein-Barr virus (EBV), 440–441
  friend virus (FV), 436, 438
  hepatitis B virus, 441
  hepatitis C virus, 441–443
  herpes simplex virus (HSV-1), 443
  human immunodeficiency virus (HIV), 439–440
  human T cell lymphotropic virus-1 (HTLV-1), 438
  IL-10 and TGF-β producing Treg cells in, 436, 438–444
  invariant NKT and host defense against infection, 506
  lymphocytic choriomeningitis virus (LCV), 443–444
  See also Bacterial infections; Fungal infections; Parasite infections
Viral vectors approach
  for autoimmune diseases, 236–237
  See also In vitro approach; In vivo approach

W

Wiskott-Aldrich syndrome (WAS), 259–260
  See also Autoimmune diseases

X

X-linked autoimmune and allergic dysregulation syndrome (XLAAD), 357

Printed in the United States of America